ONE WEEK LOAN

14th

D1312837

BROOKS/COLE
CENGAGE Learning

Australia • Brazil • Japan • Korea • Mexico • Singapore • Spain • United Kingdom • United States

BROOKS/COLE
CENGAGE Learning

Introduction to Probability and Statistics, Fourteenth Edition, International Edition
Mendenhall/Beaver/Beaver

Vice President, Editorial Director:
P.J. Boardman

Publisher: Richard Stratton

Senior Sponsoring Editor: Molly Taylor

Assistant Editor: Shaylin Walsh Hogan

Editorial Assistant: Alexander Gontar

Associate Media Editor: Andrew Coppola

Marketing Director: Mandee Eckersley

Marketing Coordinator: Michael Ledesma

Marketing Communications Manager:
Mary Anne Payumo

Content Project Manager: Jill Quinn

Art Director: Linda May

Manufacturing Planner: Doug Bertke

Rights Acquisition Specialist:
Shalice Shah-Caldwell

Production Service: MPS Limited,
a Macmillan Company

Cover Image: © shutterstock

Compositor: MPS Limited, a Macmillan
Company

International Edition:

ISBN-13: 978-1-133-11150-4

ISBN-10: 1-133-11150-5

Cengage Learning International Offices

Asia
www.cengageasia.com
tel: (65) 6410 1200

Australia/New Zealand
www.cengage.com.au
tel: (61) 3 9685 4111

Brazil
www.cengage.com.br
tel: (55) 11 3665 9900

India
www.cengage.co.in
tel: (91) 11 4364 1111

Latin America
www.cengage.com.mx
tel: (52) 55 1500 6000

UK/Europe/Middle East/Africa
www.cengage.co.uk
tel: (44) 0 1264 332 424

Represented in Canada by Nelson Education, Ltd.
www.nelson.com
tel: (416) 752 9100 / (800) 668 0671

Cengage Learning is a leading provider of customized learning solutions with office locations around the globe, including Singapore, the United Kingdom, Australia, Mexico, Brazil, and Japan. Locate your local office at: **www.cengage.com/global**

For product information and free companion resources:
www.cengage.com/international

Visit your local office: **www.cengage.com/global**

Printed in Canada
1 2 3 4 5 6 7 15 14 13 12 11

Preface

Every time you pick up a newspaper or a magazine, watch TV, or surf the Internet, you encounter statistics. Every time you fill out a questionnaire, register at an online website, or pass your grocery rewards card through an electronic scanner, your personal information becomes part of a database containing your personal statistical information. You cannot avoid the fact that in this information age, data collection and analysis are an integral part of our day-to-day activities. In order to be an educated consumer and citizen, you need to understand how statistics are used and misused in our daily lives.

THE SECRET TO OUR SUCCESS

The first college course in introductory statistics that we ever took used *Introduction to Probability and Statistics* by William Mendenhall. Since that time, this text—currently in the fourteenth edition—has helped several generations of students understand what statistics is all about and how it can be used as a tool in their particular area of application. The secret to the success of *Introduction to Probability and Statistics* is its ability to blend the old with the new. With each revision we try to build on the strong points of previous editions, while always looking for new ways to motivate, encourage, and interest students using new technological tools.

HALLMARK FEATURES OF THE FOURTEENTH EDITION

The fourteenth edition retains the traditional outline for the coverage of descriptive and inferential statistics. This revision maintains the straightforward presentation of the thirteenth edition. In this spirit, we have continued to simplify and clarify the language and to make the language and style more readable and "user friendly"—without sacrificing the statistical integrity of the presentation. Great effort has been taken to explain not only how to apply statistical procedures, but also to explain

- how to meaningfully describe real sets of data
- what the results of statistical tests mean in terms of their practical applications
- how to evaluate the validity of the assumptions behind statistical tests
- what to do when statistical assumptions have been violated

Exercises

In the tradition of all previous editions, the variety and number of real applications in the exercise sets is a major strength of this edition. We have revised the exercise sets to provide new and interesting real-world situations and real data sets, many of which are drawn from current periodicals and journals. The fourteenth edition contains over 1300 problems, many of which are new to this edition. A set of classic exercises compiled from previous editions is available on the website (http://www.cengagebrain.com). Exercises are graduated in level of difficulty; some, involving only basic techniques, can be solved by almost all students, while others, involving practical applications and interpretation of results, will challenge students to use more sophisticated statistical reasoning and understanding.

Organization and Coverage

We believe that Chapters 1 through 10—with the possible exception of Chapter 3— should be covered in the order presented. The remaining chapters can be covered in any order. The analysis of variance chapter precedes the regression chapter, so that the instructor can present the analysis of variance as part of a regression analysis. Thus, the most effective presentation would order these three chapters as well.

Chapters 1–3 present descriptive data analysis for both one and two variables, using both *MINITAB* and Microsoft Excel® graphics. Chapter 4 includes a full presentation of probability and probability distributions. Three optional sections—Counting Rules, the Total Law of Probability, and Bayes' Rule—are placed into the general flow of text, and instructors will have the option of complete or partial coverage. The sections that present event relations, independence, conditional probability, and the Multiplication Rule have been rewritten in an attempt to clarify concepts that often are difficult for students to grasp. As in the thirteenth edition, the chapters on analysis of variance and linear regression include both calculational formulas and computer printouts in the basic text presentation. These chapters can be used with equal ease by instructors who wish to use the "hands-on" computational approach to linear regression and ANOVA and by those who choose to focus on the interpretation of computer-generated statistical printouts.

One important feature in the hypothesis testing chapters involves the emphasis on *p*-values and their use in judging statistical significance. With the advent of computer-generated *p*-values, these probabilities have become essential components in reporting the results of a statistical analysis. As such, the observed value of the test statistic and its *p*-value are presented together at the outset of our discussion of statistical hypothesis testing as equivalent tools for decision-making. Statistical significance is defined in terms of preassigned values of α, and the *p-value approach* is presented as an alternative to the *critical value approach* for testing a statistical hypothesis. Examples are presented using both the *p-value* and *critical value* approaches to hypothesis testing. Discussion of the practical interpretation of statistical results, along with the difference between statistical significance and practical significance, is emphasized in the practical examples in the text.

Special Features of the Fourteenth Edition

- NEED TO KNOW. . .: A special feature of this edition are highlighted sections called "NEED TO KNOW. . ." and identified by this icon. [?] NEED TO KNOW... These sections provide information consisting of definitions, procedures or step-by-step hints on problem solving for specific questions such as "NEED TO KNOW… How to Construct a Relative Frequency Histogram?" or "NEED TO KNOW… How to Decide Which Test to Use?"

- Applets: Easy access to the Internet has made it possible for students to visualize statistical concepts using an interactive webtool called an applet. Applets written by Gary McClelland, author of *Seeing Statistics*™, are found on the CourseMate ⁀CourseMate Website that accompanies the text. Following each applet, appropriate exercises are available that provide visual reinforcement of the concepts presented in the text. Applets allow the user to perform a statistical experiment, to interact with a statistical graph, to change its form, or to access an interactive "statistical table."

- Graphical and numerical data description includes both traditional and EDA methods, using computer graphics generated by *MINITAB 16* for Windows and MS Excel.

FIGURE 2.10

Relative frequency histogram for Example 2.8

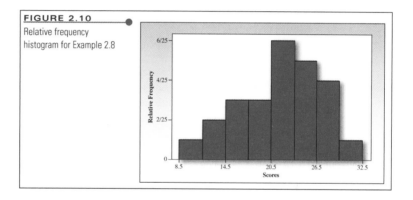

FIGURE 2.18

(b)

	E	F	G	H
	Front Leg Room		*Rear Leg Room*	
Mean		41.333	Mean	28.944
Standard Error		0.344	Standard Error	0.543
Median		41	Median	29.5
Mode		41	Mode	30
Standard Deviation		1.031	Standard Deviation	1.629
Sample Variance		1.063	Sample Variance	2.653
Kurtosis		-0.745	Kurtosis	1.735
Skewness		0.245	Skewness	-1.212
Range		3	Range	5.5
Minimum		40	Minimum	25.5
Maximum		43	Maximum	31
Sum		372	Sum	260.5
Count		9	Count	9

- All examples and exercises in the text contain printouts based on *MINITAB 16* and consistent with earlier versions of *MINITAB* or MS Excel. Printouts are provided for some exercises, while other exercises require the student to obtain solutions without using a computer.

 1.47 Presidential Vetoes Here is a list of the EX0147 44 presidents of the United States along with the number of regular vetoes used by each:[5]

Washington	2	B. Harrison	19
J. Adams	0	Cleveland	42
Jefferson	0	McKinley	6
Madison	5	T. Roosevelt	42
Monroe	1	Taft	30
J. Q. Adams	0	Wilson	33
Jackson	5	Harding	5
Van Buren	0	Coolidge	20
W. H. Harrison	0	Hoover	21
Tyler	6	F. D. Roosevelt	372
Polk	2	Truman	180
Taylor	0	Eisenhower	73
Fillmore	0	Kennedy	12
Pierce	9	L. Johnson	16
Buchanan	4	Nixon	26
Lincoln	2	Ford	48
A. Johnson	21	Carter	13
Grant	45	Reagan	39
Hayes	12	G. H. W. Bush	29
Garfield	0	Clinton	36
Arthur	4	G. W. Bush	11
Cleveland	304	Obama	1

Source: The World Almanac and Book of Facts 2011

Use an appropriate graph to describe the number of vetoes cast by the 44 presidents. Write a summary paragraph describing this set of data.

 **1.48 Windy Cities** Are some cities more EX0148 windy than others? Does Chicago deserve to be

(1950)	121.3	122.3	121.3	122.0	123.0	121.4	123.2	122.1	125.0	122.1
(1960)	122.2	124.0	120.2	121.4	120.0	121.1	122.0	120.3	122.1	121.4
(1970)	123.2	123.1	121.4	119.2[†]	124.0	122.0	121.3	122.1	121.1	122.2
(1980)	122.0	122.0	122.2	122.1	122.2	120.1	122.4	123.2	122.2	125.0
(1990)	122.0	123.0	123.0	122.2	123.3	121.1	121.0	122.4	122.2	123.2
(2000)	121.0	119.97	121.13	121.19	124.06	122.75	121.36	122.17	121.86	122.66
(2010)	124.4									

[†]Record time set by Secretariat in 1973.
Source: www.kentuckyderby.com

a. Do you think there will be a trend in the winning times over the years? Draw a line chart to verify your answer.

b. Describe the distribution of winning times using an appropriate graph. Comment on the shape of the distribution and look for any unusual observations.

1.50 Gulf Oil Spill Cleanup On April 20, EX0150 2010, the United States experienced a major environmental disaster when a Deepwater Horizon drilling rig exploded in the Gulf of Mexico. The number of personnel and equipment used in the Gulf oil spill cleanup, beginning May 2, 2010 (Day 13) through June 9, 2010 (Day 51) is given in the following table.[13]

	Day 13	Day 26	Day 39	Day 51
Number of personnel (1000s)	3.0	17.5	20.0	24.0
Federal Gulf fishing areas closed	3%	8%	25%	32%
Booms laid (miles)	46	315	644	909
Dispersants used (1000 gallons)	156	500	870	1143

TECHNOLOGY TODAY

The Role of Computers in the Fourteenth Edition—TECHNOLOGY TODAY

Computers are now a common tool for college students in all disciplines. Most students are accomplished users of word processors, spreadsheets, and databases, and they have no trouble navigating through software packages in the Windows environment. We believe, however, that advances in computer technology should not turn statistical analyses into a "black box." Rather, we choose to use the computational shortcuts and interactive visual tools that modern technology provides to give us more time to emphasize statistical reasoning as well as the understanding and interpretation of statistical results.

In this edition, students will be able to use computers for both standard statistical analyses and as a tool for reinforcing and visualizing statistical concepts. Both MS Excel and *MINITAB* 16 (consistent with earlier versions of *MINITAB*) are used exclusively as the computer packages for statistical analysis. However, we have chosen to isolate the instructions for generating computer output into individual sections called Technology Today at the end of each chapter. Each discussion uses numerical examples to guide the student through the MS Excel commands and option necessary for the procedures presented in that chapter, and then present the equivalent steps and commands needed to produce the same or similar results using *MINITAB*. We have included screen captures from both MS Excel and *MINITAB* 16, so that the student can actually work through these sections as "mini-labs."

If you do not need "hands-on" knowledge of *MINITAB* or MS Excel, or if you are using another software package, you may choose to skip these sections and simply use the printouts as guides for the basic understanding of computer printouts.

TECHNOLOGY TODAY

Numerical Descriptive Measures in *Excel*

MS Excel provides most of the basic descriptive statistics presented in Chapter a single command on the **Data** tab. Other descriptive statistics can be calculate the **Function** command on the **Formulas** tab.

EXAMPLE 2.15

The following data are the front and rear leg rooms (in inches) for nine differer utility vehicles:[14]

Make & Model	Front Leg Room	Rear Leg Room
Acura MDX	41.0	28.5
Buick Enclave	41.5	30.0
Chevy TrailBlazer	40.0	25.5
Chevy Tahoe Hybrid V8 CVT	41.0	27.5
GMC Terrain 1LT 4-cyl	43.0	31.0

Numerical Descriptive Measures in *MINITAB*

MINITAB provides most of the basic descriptive statistics presented in Chapter 2 using a single command in the drop-down menus.

The following data are the front and rear leg rooms (in inches) for nine different sports utility vehicles:[14]

Make and Model	Front Leg Room	Rear Leg Room
Acura MDX	41.0	28.5
Buick Enclave	41.5	30.0
Chevy TrailBlazer	40.0	25.5
Chevy Tahoe Hybrid V8 CVT	41.0	27.5
GMC Terrain 1LT 4-cyl	43.0	31.0
Honda CR-V	41.0	29.5

Any student who has Internet access can use the applets found on the CourseMate Website to visualize a variety of statistical concepts (access instructions for the CourseMate Website are listed on the Printed Access Card that is an optional bundle with this text). In addition, some of the applets can be used instead of computer software to perform simple statistical analyses. Exercises written specifically for use with these applets also appear on the CourseMate Website. Students can use the applets at home or in a computer lab. They can use them as they read through the text material, once they have finished reading the entire chapter, or as a tool for exam review. Instructors can use the applets as a tool in a lab setting, or use them for visual demonstrations during lectures. We believe that these applets will be a powerful tool that will increase student enthusiasm for, and understanding of, statistical concepts and procedures.

STUDY AIDS

The many and varied exercises in the text provide the best selected tool for students embarking on a first course in statistics. The answers to selected exercises are given in the back of the text. Each application exercise has a title, making it easier for students and instructors to immediately identify both the context of the problem and the area of application.

Use Table 2 to find the following probabilities:

a. $P(x \le 3)$ **b.** $P(x > 3)$

c. $P(x = 3)$ **d.** $P(3 \le x \le 5)$

5.38 Consider a Poisson random variable with $\mu = 0.8$. Use Table 2 to find the following probabilities:

a. $P(x = 0)$ **b.** $P(x \le 2)$

c. $P(x > 2)$ **d.** $P(2 \le x \le 4)$

5.39 Let x be a Poisson random variable with mean $\mu = 2$. Calculate these probabilities:

a. $P(x = 0)$ **b.** $P(x = 1)$

c. $P(x > 1)$ **d.** $P(x = 5)$

APPLICATIONS

5.43 Airport Safety The increased number of small commuter planes in major airports has heightened concern over air safety. An eastern airport has recorded a monthly average of five near misses on landings and takeoffs in the past 5 years.

a. Find the probability that during a given month there are no near misses on landings and takeoffs at the airport.

b. Find the probability that during a given month there are five near misses.

c. Find the probability that there are at least five near-

Students should be encouraged to use the "**NEED TO KNOW. . .**" sections as they occur in the text. The placement of these sections is intended to answer questions as they would normally arise in discussions. In addition, there are numerous hints called "**NEED A TIP?**" that appear in the margins of the text. The tips are short and concise.

 NEED A TIP

Empirical Rule ⇔ mound-shaped data

Tchebysheff ⇔ any shaped data

Is Tchebysheff's Theorem applicable? Yes, because it can be used for any set of data. According to Tchebysheff's Theorem,

- at least 3/4 of the measurements will fall between 10.6 and 32.6.
- at least 8/9 of the measurements will fall between 5.1 and 38.1.

Finally, sections called **Key Concepts and Formulas** appear in each chapter as a review in outline form of the material covered in that chapter.

CHAPTER REVIEW

Key Concepts and Formulas

I. Measures of the Center of a Data Distribution

1. Arithmetic mean (mean) or average
   a. Population: μ
   b. Sample of n measurements: $\bar{x} = \dfrac{\Sigma x_i}{n}$

2. Median; **position** of the median $= .5(n + 1)$

3. Mode

4. The median may be preferred to the mean if the data are highly skewed.

II. Measures of Variability

1. Range: $R = \text{largest} - \text{smallest}$

2. Variance
   a. Population of N measurements:
   $$\sigma^2 = \frac{\Sigma(x_i - \mu)^2}{N}$$
   b. Sample of n measurements:
   $$s^2 = \frac{\Sigma(x_i - \bar{x})^2}{n - 1} = \frac{\Sigma x_i^2 - \dfrac{(\Sigma x_i)^2}{n}}{n - 1}$$

68%, 95%, and 99.7% of the measurements are within one, two, and three standard deviations of the mean, respectively.

IV. Measures of Relative Standing

1. Sample z-score: $z = \dfrac{x - \bar{x}}{s}$

2. pth percentile; $p\%$ of the measurements are smaller, and $(100 - p)\%$ are larger.

3. Lower quartile, Q_1; **position** of $Q_1 = .25(n + 1)$

4. Upper quartile, Q_3; **position** of $Q_3 = .75(n + 1)$

5. Interquartile range: $\text{IQR} = Q_3 - Q_1$

V. The Five-Number Summary and Box Plots

1. The **five-number summary:**

   Min Q_1 Median Q_3 Max

   One-fourth of the measurements in the data set lie between each of the four adjacent pairs of numbers.

2. Box plots are used for detecting outliers and

The CourseMate Website, a password-protected resource that can be accessed with a Printed Access Card (optional bundle item), provides students with an array of study resources, including the complete set of Java applets, the TI Calculator Tech Guide that includes instructions for performing many of the techniques in the text using the popular TI 83/84/89 graphing calculator, an interactive eBook, online Quizzes, flashcards, and more. The data sets (saved in a variety of formats) can be found on the book's website (www.CengageBrain.com) as well as the CourseMate Website. ᶠCourseMate

INSTRUCTOR RESOURCES

The **Instructor's Website** (login.cengage.com), available to adopters of the fourteenth edition, provides a variety of teaching aids, including

- All the material from the CourseMate website including exercises using the Large Data Sets, which is accompanied by three large data sets that can be used throughout the course. A file named "Fortune" contains the revenues (in millions) for the *Fortune* 500 largest U.S. industrial corporations in a recent year; a file named "Batting" contains the batting averages for the National and American baseball league batting champions from 1976 to 2010; and a file named "Blood Pressure" contains the age and diastolic and systolic blood pressures for 965 men and 945 women compiled by the National Institutes of Health.
- Classic exercises with data sets and solutions
- PowerPoint lecture slides
- Applets by Gary McClelland (the complete set of Java applets used for the MyApps exercises on the website)
- TI Calculator Tech Guide, which includes instructions for performing many of the techniques in the text using the Tl-83/84/89 graphing calculators.

Also available for instructors:

Enhanced WebAssign
Enhanced WebAssign is a web-based learning solution that increases student effort and engagement. It helps make statistics relevant and engaging to students by connecting real-world examples to course concepts. When combined with the textual material of *Introduction to Probability and Statistics* (IPS) 14,

- Students receive immediate, detailed explanations for every answer.

- Math and graphing tutorials help students ovecome deficiencies in these crucial areas.

- Enhanced WebAssign allows you to assign, collect, grade, and record homework assignments via the web.

Solution Builder
This online instructor database offers complete worked-out solutions to all exercises in the text, allowing you to create customized, secure solutions printouts (in PDF format) matched exactly to the problems you assign in class. Sign up for access at www.cengage.com/solutionbuilder.

Instructor's Companion Website (login.cengage.com)
Additional instructor resources available online for *Introduction to Probability and Statistics* include the Instructor's Solutions Manual, PowerPoint lectures, ExamView Computerized Testing, Classic Exercises, and TI-83/84/89 calculator Tech Guide which includes instructions for performing many of the techniques in the text using the Tl-83/84/89 graphing calculators.

ACKNOWLEDGMENTS

The authors are grateful to Molly Taylor and the editorial staff of Cengage Learning for their patience, assistance, and cooperation in the preparation of this edition. A special thanks to Gary McClelland for the Java applets used in the text.

Thanks are also due to fourteenth edition reviewers Ronald C. Degges, Bob C. Denton, Dr. Dorothy M. French, Jungwon Mun, Kazuhiko Shinki, Florence P. Shu and thirteenth edition reviewers Bob Denton, Timothy Husband, Rob LaBorde, Craig McBride, Marc Sylvester, Kanapathi Thiru, and Vitaly Voloshin. We wish to thank authors and organizations for allowing us to reprint selected material; acknowledgments are made wherever such material appears in the text.

Robert J. Beaver
Barbara M. Beaver

Brief Contents

Contents

Introduction
What is Statistics?

What is statistics? Have you ever met a statistician? Do you know what a statistician does? Perhaps you are thinking of the person who sits in the broadcast booth at the Rose Bowl, recording the number of pass completions, yards rushing, or interceptions thrown on New Year's Day. Or perhaps the mere mention of the word *statistics* sends a shiver of fear through you. You may think you know nothing about statistics; however, it is almost inevitable that you encounter statistics in one form or another every time you pick up a daily newspaper. Here are some examples concerning the California 2010 elections:

- **Rowdy crowd jeers Whitman.** GOP candidate criticizes unions; earlier stop draws friendlier audience.
  GLENDALE— . . . Whitman, a billionaire, has spent $142 million from her personal fortune in the race so far.
  A Field Poll released Thursday showed her trailing Jerry Brown 49 percent to 39 percent among likely voters.[1]

- **Fiorina calls herself similar to Feinstein, who supports Boxer.**
  MENLO PARK—Republican Carly Fiorina said Friday she would be a like-minded colleague of Democratic Sen. Dianne Feinstein if she unseats Barbara Boxer next week, drawing sharp responses from both Democratic senators.
  . . . Fiorina, the former CEO of Hewlett-Packard Co., disputed a Field Poll released Friday showing Boxer leading her among likely voters, 49 percent to 41 percent.[2]

- **Race for attorney general tight. Field Poll:** Nearly a quarter of those surveyed are undecided. Newsom holds a slim lead over Maldonado for lieutenant governor.

SACRAMENTO—Tuesday's election for attorney general is a tossup, with Democrat Kamala Harris and Republican Steve Cooley virtually tied as Harris gains ground in voter-rich Los Angeles County and among women according to the latest Field Poll.

. . . Today's poll shows Cooley with 39 percent and Harris with 38 percent among likely voters. Almost a quarter of likely voters remain undecided.

. . . Newsom, the mayor of San Francisco, leads Maldonado, who was appointed lieutenant governor this year, 42 percent to 37 percent. A fifth of voters are undecided.

Today's poll was conducted for *The Press-Enterprise* and other California media subscribers. It was conducted October 14 through October 26 and included 1092 voters. It has a margin of error of plus or minus 3.2 percent.[3]

—*The Press-Enterprise,* Riverside, CA

Articles similar to these are commonplace in our newspapers and magazines, and in the period just prior to a presidential or congressional election, a new poll is reported almost every day. The language of these articles are very familiar to us; however, they leave the inquisitive reader with some unanswered questions. How were the people in the poll selected? Will these people give the same response tomorrow? Will they give the same response on election day? Will they even vote? Are these people representative of all those who will vote on election day? It is the job of a statistician to ask these questions and to find answers for them in the language of the poll.

Most Believe "Cover-Up" of JFK Assassination Facts

A majority of the public believes the assassination of President John F. Kennedy was part of a larger conspiracy, not the act of one individual. In addition, most Americans think there was a cover-up of facts about the 1963 shooting. Almost 50 years after JFK's assassination, a FOX news poll shows many Americans disagree with the government's conclusions about the killing. The **Warren Commission** found that **Lee Harvey Oswald** acted alone when he shot Kennedy, but 66 percent of the public today think the assassination was "part of a larger conspiracy" while only 25 percent think it was the "act of one individual."

"For older Americans, the Kennedy assassination was a traumatic experience that began a loss of confidence in government," commented Opinion Dynamics President John Gorman. "Younger people have grown up with movies and documentaries that have pretty much pushed the 'conspiracy' line. Therefore, it isn't surprising there is a fairly solid national consensus that we still don't know the truth."

(The poll asked): "Do you think that we know all the facts about the assassination of President John F. Kennedy or do you think there was a cover-up?"

	We Know All the Facts (%)	There Was a Cover-Up	(Not Sure)
All	14	74	12
Democrats	11	81	8
Republicans	18	69	13
Independents	12	71	17

—www.foxnews.com[4]

When you see an article like this one in a magazine, do you simply read the title and the first paragraph, or do you read further and try to understand the meaning of the numbers? How did the authors get these numbers? Did they really interview every American with each political affiliation? It is the job of the statistician to interpret the language of this study.

Hot News: 98.6 Not Normal

After believing for more than a century that 98.6 was the normal body temperature for humans, researchers now say normal is not normal anymore.

For some people at some hours of the day, 99.9 degrees could be fine. And readings as low as 96 turn out to be highly human.

The 98.6 standard was derived by a German doctor in 1868. Some physicians have always been suspicious of the good doctor's research. His claim: 1 million readings—in an epoch without computers.

So Mackowiak & Co. took temperature readings from 148 healthy people over a three-day period and found that the mean temperature was 98.2 degrees. Only 8 percent of the readings were 98.6.

—The Press-Enterprise[5]

What questions come to your mind when you read this article? How did the researcher select the 148 people, and how can we be sure that the results based on these 148 people are accurate when applied to the general population? How did the researcher arrive at the normal "high" and "low" temperatures given in the article? How did the German doctor record 1 million temperatures in 1868? Again, we encounter a statistical problem with an application to everyday life.

Statistics is a branch of mathematics that has applications in almost every facet of our daily life. It is a new and unfamiliar language for most people, however, and, like any new language, statistics can seem overwhelming at first glance. But once the language of statistics is learned and understood, it provides a powerful tool for data analysis in many different fields of application.

THE POPULATION AND THE SAMPLE

In the language of statistics, one of the most basic concepts is **sampling.** In most statistical problems, a specified number of measurements or data—a **sample**—is drawn from a much larger body of measurements, called the **population.**

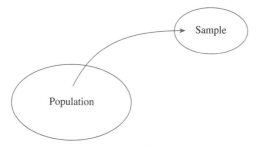

For the body-temperature experiment, the sample is the set of body-temperature measurements for the 148 healthy people chosen by the experimenter. We hope that the sample is representative of a much larger body of measurements—the population— the body temperatures of all healthy people in the world!

Which is of primary interest, the sample or the population? In most cases, we are interested primarily in the population, but the population may be difficult or impossible to enumerate. Imagine trying to record the body temperature of every healthy person on earth or the presidential preference of every registered voter in the United States! Instead, **we try to describe or predict the behavior of the population on the basis of information obtained from a representative sample from that population.**

The words *sample* and *population* have two meanings for most people. For example, you read in the newspapers that a Gallup poll conducted in the United States was based on a sample of 1823 people. Presumably, each person interviewed is asked a particular question, and that person's response represents a single measurement in the sample. Is the sample the set of 1823 people, or is it the 1823 responses that they give?

In statistics, we distinguish between the set of objects on which the measurements are taken and the measurements themselves. To experimenters, the objects on which measurements are taken are called **experimental units.** The sample survey statistician calls them **elements of the sample.**

DESCRIPTIVE AND INFERENTIAL STATISTICS

When first presented with a set of measurements—whether a sample or a population—you need to find a way to organize and summarize it. The branch of statistics that presents techniques for describing sets of measurements is called **descriptive statistics.** You have seen descriptive statistics in many forms: bar charts, pie charts, and line charts presented by a political candidate; numerical tables in the newspaper; or the average rainfall amounts reported by the local television weather forecaster. Computer-generated graphics and numerical summaries are commonplace in our everyday communication.

Definition **Descriptive statistics** consists of procedures used to summarize and describe the important characteristics of a set of measurements.

If the set of measurements is the entire population, you need only to draw conclusions based on the descriptive statistics. However, it might be too expensive or too time consuming to enumerate the entire population. Perhaps enumerating the population would destroy it, as in the case of "time to failure" testing. For these or other reasons, you may have only a sample from the population. By looking at the sample, you want to answer questions about the population as a whole. The branch of statistics that deals with this problem is called **inferential statistics.**

Definition **Inferential statistics** consists of procedures used to make inferences about population characteristics from information contained in a sample drawn from this population.

The **objective of inferential statistics** is to make inferences (that is, draw conclusions, make predictions, make decisions) about the characteristics of a population from information contained in a sample.

ACHIEVING THE OBJECTIVE OF INFERENTIAL STATISTICS: THE NECESSARY STEPS

How can you make inferences about a population using information contained in a sample? The task becomes simpler if you organize the problem into a series of logical steps.

1. **Specify the questions to be answered and identify the population of interest.** In the California election poll, the objective is to determine who will get the most votes on election day. Hence, the population of interest is the set of all votes in the California election. When you select a sample, it is important that

the sample be representative of *this* population, not the population of voter preferences on October 30 or on some other day prior to the election.

2. **Decide how to select the sample.** This is called the *design of the experiment* or the *sampling procedure*. Is the sample representative of the population of interest? For example, if a sample of registered voters is selected from the city of San Francisco, will this sample be representative of all voters in California? Will it be the same as a sample of "likely voters"—those who are likely to actually vote in the election? Is the sample large enough to answer the questions posed in step 1 without wasting time and money on additional information? A good sampling design will answer the questions posed with minimal cost to the experimenter.

3. **Select the sample and analyze the sample information.** No matter how much information the sample contains, you must use an appropriate method of analysis to extract it. Many of these methods, which depend on the sampling procedure in step 2, are explained in the text.

4. **Use the information from step 3 to make an inference about the population.** Many different procedures can be used to make this inference, and some are better than others. For example, 10 different methods might be available to estimate human response to an experimental drug, but one procedure might be more accurate than others. You should use the best inference-making procedure available (many of these are explained in the text).

5. **Determine the reliability of the inference.** Since you are using only a fraction of the population in drawing the conclusions described in step 4, you might be wrong! How can this be? If an agency conducts a statistical survey for you and estimates that your company's product will gain 34% of the market this year, how much confidence can you place in this estimate? Is this estimate accurate to within 1, 5, or 20 percentage points? Is it reliable enough to be used in setting production goals? Every statistical inference should include a measure of reliability that tells you how much confidence you have in the inference.

Now that you have learned a few basic terms and concepts, we again pose the question asked at the beginning of this discussion: Do you know what a statistician does? The statistician's job is to implement all of the preceding steps.

KEYS FOR SUCCESSFUL LEARNING

As you begin to study statistics, you will find that there are many new terms and concepts to be mastered. Since statistics is an applied branch of mathematics, many of these basic concepts are mathematical—developed and based on results from calculus or higher mathematics. However, you do not have to be able to derive results in order to apply them in a logical way. In this text, we use numerical examples and common-sense arguments to explain statistical concepts, rather than more complicated mathematical arguments.

In recent years, computers have become readily available to many students and provide them with an invaluable tool. In the study of statistics, even the beginning student can use packaged programs to perform statistical analyses with a high degree of speed and accuracy. Some of the more common statistical packages available at computer facilities are *MINITAB*™, SAS (Statistical Analysis System), and SPSS

(Statistical Package for the Social Sciences); personal computers will support packages such as *MINITAB*, *MS Excel*, and others. There are even online statistical programs and interactive "applets" on the Internet.

These programs, called **statistical software,** differ in the types of analyses available, the options within the programs, and the forms of printed results (called **output**). However, they are all similar. In this book, we use both *MINITAB* and *Microsoft Excel* as statistical tools. Understanding the basic output of these packages will help you interpret the output from other software systems.

At the end of most chapters, you will find a section called *"Technology Today."* These sections present numerical examples to guide you through the *MINITAB* and *MS Excel* commands and options that are used for the procedures in that chapter. If you are using *MINITAB* or *MS Excel* in a lab or home setting, you may want to work through this section at your own computer so that you become familiar with the hands-on methods in computer analysis. If you do not need hands-on knowledge of *MINITAB* or *MS Excel*, you may choose to skip this section and simply use the computer printouts for analysis as they appear in the text.

Another learning tool called statistical **applets** can be found on the CourseMate Web site. Also found on this Web site are explanatory sections called "Using the Applets," which will help you understand how the applets can be used to visualize many of the chapter concepts. An accompanying section called "Applet APPs" provides some exercises (with solutions) that can be solved using the statistical applets. Whenever there is an applet available for a particular concept or application, you will find an icon in the left margin of the text, together with the name of the appropriate applet.

Most important, using statistics successfully requires common sense and logical thinking. For example, if we want to find the average height of all students at a particular university, would we select our entire sample from the members of the basketball team? In the body-temperature example, the logical thinker would question an 1868 average based on 1 million measurements—when computers had not yet been invented.

As you learn new statistical terms, concepts, and techniques, remember to view every problem with a critical eye and be sure that the rule of common sense applies. Throughout the text, we will remind you of the pitfalls and dangers in the use or misuse of statistics. Benjamin Disraeli once said that there are three kinds of lies: *lies, damn lies,* and *statistics*! Our purpose is to dispel this claim—to show you how to make statistics *work* for you and not *lie* for you!

As you continue through the book, refer back to this introduction periodically. Each chapter will increase your knowledge of statistics and should, in some way, help you achieve one of the steps described here. Each of these steps is essential in attaining the overall objective of inferential statistics: to make inferences about a population using information contained in a sample drawn from that population.

Describing Data with Graphs

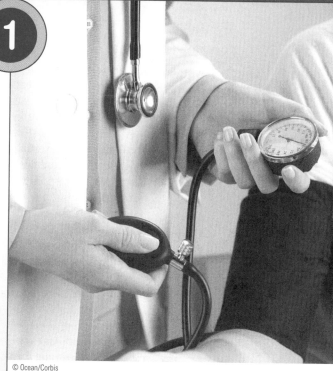

© Ocean/Corbis

GENERAL OBJECTIVES

Many sets of measurements are samples selected from larger populations. Other sets constitute the entire population, as in a national census. In this chapter, you will learn what a *variable* is, how to classify variables into several types, and how measurements or data are generated. You will then learn how to use graphs to describe data sets.

CHAPTER INDEX

NEED TO KNOW...

How to Construct a Stem and Leaf Plot
How to Construct a Relative Frequency Histogram

How Is Your Blood Pressure?

Is your blood pressure normal, or is it too high or too low? The case study at the end of this chapter examines a large set of blood pressure data. You will use graphs to describe these data and compare your blood pressure with that of others of your same age and gender.

7

VARIABLES AND DATA

In Chapters 1 and 2, we will present some basic techniques in *descriptive statistics*—the branch of statistics concerned with describing sets of measurements, both *samples* and *populations*. Once you have collected a set of measurements, how can you display this set in a clear, understandable, and readable form? First, you must be able to define what is meant by measurements or "data" and to categorize the types of data that you are likely to encounter in real life. We begin by introducing some definitions.

Definition A **variable** is a characteristic that changes or varies over time and/or for different individuals or objects under consideration.

For example, body temperature is a variable that changes over time within a single individual; it also varies from person to person. Religious affiliation, ethnic origin, income, height, age, and number of offspring are all variables—characteristics that vary depending on the individual chosen.

In the Introduction, we defined an *experimental unit* or an *element of the sample* as the object on which a measurement is taken. Equivalently, we could define an experimental unit as the object on which a variable is measured. When a variable is actually measured on a set of experimental units, a set of measurements or **data** result.

Definition An **experimental unit** is the individual or object on which a variable is measured. A single **measurement** or data value results when a variable is actually measured on an experimental unit.

If a measurement is generated for every experimental unit in the entire collection, the resulting data set constitutes the *population* of interest. Any smaller subset of measurements is a *sample*.

Definition A **population** is the set of all measurements of interest to the investigator. ·

Definition A **sample** is a subset of measurements selected from the population of interest.

EXAMPLE 1.1

A set of five students is selected from all undergraduates at a large university, and measurements are entered into a spreadsheet as shown in Figure 1.1. Identify the various elements involved in generating this set of measurements.

Solution There are several *variables* in this example. The *experimental unit* on which the variables are measured is a particular undergraduate student on the campus, identified in column A. Five variables are measured for each student: grade point average (GPA), gender, year in college, major, and current number of units enrolled. Each of these characteristics varies from student to student. If we consider the GPAs of all students at this university to be the population of interest, the five GPAs in column B represent a *sample* from this population. If the GPA of each undergraduate student at the university had been measured, we would have generated the entire *population* of measurements for this variable.

FIGURE 1.1

Measurements on five
undergraduate students

	A	B	C	D	E	F
1	Student	GPA	Gender	Year	Major	Number of Units
2	1	2	F	Fr	Psychology	16
3	2	2.3	F	So	Mathematics	15
4	3	2.9	M	So	English	17
5	4	2.7	M	Fr	English	15
6	5	2.6	F	Jr	Business	14

Sheet1 / Sheet2 / Sheet3

The second variable measured on the students is gender, in column C. This variable is somewhat different from GPA, since it can take only one of two values—male (M) or female (F). The population, if it could be enumerated, would consist of a set of Ms and Fs, one for each student at the university. Similarly, the third and fourth variables, year and major, generate nonnumerical data. Year has four categories (Fr, So, Jr, Sr), and major has one category for each undergraduate major on campus. The last variable, current number of units enrolled, is numerically valued, generating a set of numbers rather than a set of qualities or characteristics.

Although we have discussed each variable individually, remember that we have measured each of these five variables on a single experimental unit: the student. Therefore, in this example, a "measurement" really consists of five observations, one for each of the five measured variables. For example, the measurement taken on student 2 produces this observation:

(2.3, F, So, Mathematics, 15)

You can see that there is a difference between a *single* variable measured on a single experimental unit and *multiple* variables measured on a single experimental unit as in Example 1.1.

Definition **Univariate data** result when a single variable is measured on a single experimental unit.

Definition **Bivariate data** result when two variables are measured on a single experimental unit. **Multivariate data** result when more than two variables are measured.

If you measure the body temperatures of 148 people, the resulting data are *univariate*. In Example 1.1, five variables were measured on each student, resulting in *multivariate* data.

TYPES OF VARIABLES

Variables can be classified into one of two types: **qualitative** or **quantitative.**

Definition **Qualitative variables** measure a quality or characteristic on each experimental unit. **Quantitative variables** measure a numerical quantity or amount on each experimental unit.

NEED A TIP?

Qualitative ⇔ "quality" or characteristic

Quantitative ⇔ "quantity" or number

Qualitative variables produce data that can be categorized according to similarities or differences in kind; hence, they are often called **categorical data.** The variables gender, year, and major in Example 1.1 are qualitative variables that produce categorical data. Here are some other examples:

- Political affiliation: Republican, Democrat, Independent
- Taste ranking: excellent, good, fair, poor
- Color of an M&M'S® candy: brown, yellow, red, orange, green, blue

Quantitative variables, often represented by the letter x, produce numerical data, such as those listed here:

- x = Prime interest rate
- x = Number of passengers on a flight from Los Angeles to New York City
- x = Weight of a package ready to be shipped
- x = Volume of orange juice in a glass

Notice that there is a difference in the types of numerical values that these quantitative variables can assume. The number of passengers, for example, can take on only the values $x = 0, 1, 2, \ldots$, whereas the weight of a package can take on any value greater than zero, or $0 < x < \infty$. To describe this difference, we define two types of quantitative variables: **discrete** and **continuous.**

Definition   A **discrete variable** can assume only a finite or countable number of values. A **continuous variable** can assume the infinitely many values corresponding to the points on a line interval.

NEED A TIP?

Discrete ⇔ "listable"

Continuous ⇔ "unlistable"

The name *discrete* relates to the discrete gaps between the possible values that the variable can assume. Variables such as number of family members, number of new car sales, and number of defective tires returned for replacement are all examples of discrete variables. On the other hand, variables such as height, weight, time, distance, and volume are *continuous* because they can assume values at any point along a line interval. For any two values you pick, a third value can always be found between them!

EXAMPLE   **1.2**   Identify each of the following variables as qualitative or quantitative:

1. The most frequent use of your microwave oven (reheating, defrosting, warming, other)
2. The number of consumers who refuse to answer a telephone survey
3. The door chosen by a mouse in a maze experiment (A, B, or C)
4. The winning time for a horse running in the Kentucky Derby
5. The number of children in a fifth-grade class who are reading at or above grade level

NEED A TIP?

Discrete variables often involve the "number of" items in a set.

Solution   Variables 1 and 3 are both *qualitative* because only a quality or characteristic is measured for each individual. The categories for these two variables are shown in parentheses. The other three variables are *quantitative*. Variables 2 and 5 are *discrete* variables that can take on any of the values $x = 0, 1, 2, \ldots$, with a maximum value depending on the number of consumers called or the number of children in the class, respectively. Variable 4, the winning time for a Kentucky Derby horse, is the only *continuous* variable in the list. The winning time, if it could be measured with sufficient accuracy, could be 121 seconds, 121.5 seconds, 121.25 seconds, or any values between any two times we have listed.

Why should you be concerned about different kinds of variables (shown in Figure 1.2) and the data that they generate? The reason is that different types of data require you to use different methods for description, so that the data can be presented clearly and understandably to your audience!

FIGURE 1.2

Types of data ●────

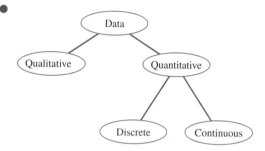

GRAPHS FOR CATEGORICAL DATA

1.3

After the data have been collected, they can be consolidated and summarized to show the following information:

- What values of the variable have been measured
- How often each value has occurred

For this purpose, you can construct a *statistical table* that can be used to display the data graphically as a **data distribution.** The type of graph you choose depends on the type of variable you have measured.

When the variable of interest is *qualitative* or *categorical,* the statistical table is a list of the categories being considered along with a measure of how often each value occurred. You can measure "how often" in three different ways:

- The **frequency,** or number of measurements in each category
- The **relative frequency,** or proportion of measurements in each category
- The **percentage** of measurements in each category

If you let n be the total number of measurements in the set, you can find the relative frequency and percentage using these relationships:

$$\text{Relative frequency} = \frac{\text{Frequency}}{n}$$

$$\text{Percent} = 100 \times \text{Relative frequency}$$

You will find that the sum of the frequencies is always *n,* the sum of the relative frequencies is 1, and the sum of the percentages is 100%.

When the variable is qualitative, the categories should be chosen so that

- a measurement will belong to one and only one category
- each measurement has a category to which it can be assigned

For example, if you categorize meat products according to the type of meat used, you might use these categories: beef, chicken, seafood, pork, turkey, other. To categorize ranks of college faculty, you might use these categories: professor, associate professor, assistant professor, instructor, lecturer, other. The "other" category is included in both cases to allow for the possibility that a measurement cannot be assigned to one of the earlier categories.

Once the measurements have been categorized and summarized in a *statistical table,* you can use either a pie chart or a bar chart to display the distribution of the data. A **pie chart** is the familiar circular graph that shows how the measurements are distributed among the categories. A **bar chart** shows the same distribution of measurements among the categories, with the height of the bar measuring how often a particular category was observed.

EXAMPLE 1.3

In a survey concerning public education, 400 school administrators were asked to rate the quality of education in the United States. Their responses are summarized in Table 1.1. Construct a pie chart and a bar chart for this set of data.

Solution To construct a pie chart, assign one sector of a circle to each category. The angle of each sector should be proportional to the proportion of measurements (or *relative frequency*) in that category. Since a circle contains 360°, you can use this equation to find the angle:

$$\text{Angle} = \text{Relative frequency} \times 360°$$

TABLE 1.1 • **U.S. Education Rating by 400 Educators**

Rating	Frequency
A	35
B	260
C	93
D	12
Total	400

Table 1.2 shows the ratings along with the frequencies, relative frequencies, percentages, and sector angles necessary to construct the pie chart. Figure 1.3 shows the pie chart constructed from the values in the table. While pie charts use percentages to determine the relative sizes of the "pie slices," bar charts usually plot frequency against the categories. A bar chart for these data is shown in Figure 1.4.

TABLE 1.2 ● **Calculations for the Pie Chart in Example 1.3**

Rating	Frequency	Relative Frequency	Percent	Angle
A	35	35/400 = .09	9%	.09 × 360 = 32.4°
B	260	260/400 = .65	65%	234.0°
C	93	93/400 = .23	23%	82.8°
D	12	12/400 = .03	3%	10.8°
Total	400	1.00	100%	360°

The visual impact of these two graphs is somewhat different. The pie chart is used to display the relationship of the parts to the whole; the bar chart is used to emphasize the actual quantity or frequency for each category. Since the categories in this example are ordered "grades" (A, B, C, D), we would not want to rearrange the bars in the chart to change its *shape*. In a pie chart, the order of presentation is irrelevant.

FIGURE 1.3 ●

Pie chart for Example 1.3

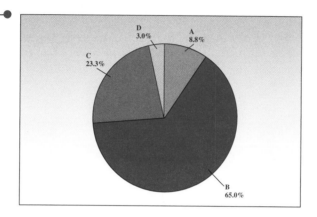

FIGURE 1.4 ●

Bar chart for Example 1.3

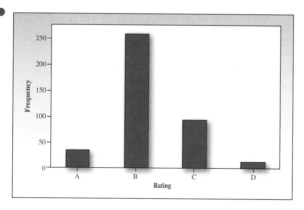

EXAMPLE **1.4**

A snack size bag of peanut M&M'S candies contains 21 candies with the colors listed in Table 1.3. The variable "color" is *qualitative,* so Table 1.4 lists the six categories along with a tally of the number of candies of each color. The last three columns of Table 1.4 show how often each category occurred. Since the categories are colors and have no particular order, you could construct bar charts with many different *shapes* just by reordering the bars. To emphasize that brown is the most frequent color, followed by blue, green, and orange, we order the bars from largest to smallest and create the bar chart in Figure 1.5. A bar chart in which the bars are ordered from largest to smallest is called a **Pareto chart.**

TABLE 1.3 ● **Raw Data: Colors of 21 Candies**

Brown	Green	Brown	Blue
Red	Red	Green	Brown
Yellow	Orange	Green	Blue
Brown	Blue	Blue	Brown
Orange	Blue	Brown	Orange
Yellow			

TABLE 1.4 ● **Statistical Table: M&M'S Data for Example 1.4**

Category	Tally	Frequency	Relative Frequency	Percent
Brown	JHT I	6	6/21	28%
Green	III	3	3/21	14
Orange	III	3	3/21	14
Yellow	II	2	2/21	10
Red	II	2	2/21	10
Blue	JHT	5	5/21	24
Total		21	1	100%

FIGURE 1.5

Pareto chart for Example 1.4

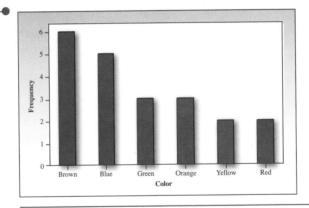

UNDERSTANDING THE CONCEPTS

1.1 Experimental Units Identify the experimental units on which the following variables are measured:

a. Gender of a student

b. Number of errors on a midterm exam

c. Age of a cancer patient

d. Number of flowers on an azalea plant

e. Color of a car entering a parking lot

1.2 Qualitative or Quantitative? Identify each variable as quantitative or qualitative:

a. Amount of time it takes to assemble a simple puzzle

b. Number of students in a first-grade classroom

c. Rating of a newly elected politician (excellent, good, fair, poor)

d. State in which a person lives

1.3 Discrete or Continuous? Identify the following quantitative variables as discrete or continuous:

a. Population in a particular area of the United States

b. Weight of newspapers recovered for recycling on a single day

c. Time to complete a sociology exam

d. Number of consumers in a poll of 1000 who consider nutritional labeling on food products to be important

1.4 Cancer Survival Times A medical researcher wants to estimate the survival time of a patient after the onset of a particular type of cancer and after a particular regimen of radiotherapy.

a. What is the variable of interest to the medical researcher?

b. Is the variable in part a qualitative, quantitative discrete, or quantitative continuous?

c. Identify the population of interest to the medical researcher.

d. Describe how the researcher could select a sample from the population.

e. What problems might arise in sampling from this population?

1.5 Parking on Campus Six vehicles are selected from the vehicles that are issued campus parking permits, and the following data are recorded:

Vehicle	Type	Make	Carpool?	One-way Commute Distance (miles)	Age of Vehicle (years)
1	Car	Honda	No	23.6	6
2	Car	Toyota	No	17.2	3
3	Truck	Toyota	No	10.1	4
4	Van	Dodge	Yes	31.7	2
5	Motor-cycle	Harley-Davidson	No	25.5	1
6	Car	Chevrolet	No	5.4	9

a. What are the experimental units?

b. What are the variables being measured? What types of variables are they?

c. Is this univariate, bivariate, or multivariate data?

1.6 Past U.S. Presidents A data set consists of the ages at death for each of the 38 past presidents of the United States now deceased.

a. Is this set of measurements a population or a sample?

b. What is the variable being measured?

c. Is the variable in part b quantitative or qualitative?

1.7 Voter Attitudes You are a candidate for your state legislature, and you want to survey voter attitudes regarding your chances of winning. Identify the population that is of interest to you and from which you would like to select your sample. How is this population dependent on time?

1.8 Discrete or Continuous? Identify each quantitative variable as discrete or continuous.

a. Number of boating accidents along a 50-mile stretch of the Colorado River

b. Time required to complete a questionnaire

c. Cost of a head of lettuce

d. Number of brothers and sisters you have

e. Yield in kilograms of wheat from a 1-hectare plot in a wheat field

1.9 New Teaching Methods An educational researcher wants to evaluate the effectiveness of a new method for teaching reading to deaf students. Achievement at the end of a period of teaching is measured by a student's score on a reading test.

a. What is the variable to be measured? What type of variable is it?

b. What is the experimental unit?

c. Identify the population of interest to the experimenter.

BASIC TECHNIQUES

1.10 Fifty people are grouped into four categories—A, B, C, and D—and the number of people who fall into each category is shown in the table:

Category	Frequency
A	11
B	14
C	20
D	5

a. What is the experimental unit?

b. What is the variable being measured? Is it qualitative or quantitative?

c. Construct a pie chart to describe the data.

d. Construct a bar chart to describe the data.

e. Does the shape of the bar chart in part d change depending on the order of presentation of the four categories? Is the order of presentation important?

f. What *proportion* of the people are in category B, C, or D?

g. What *percentage* of the people are *not* in category B?

1.11 Jeans A manufacturer of jeans has plants in California, Arizona, and Texas. A group of 25 pairs of jeans is randomly selected from the computerized database, and the state in which each is produced is recorded:

CA	AZ	AZ	TX	CA
CA	CA	TX	TX	TX
AZ	AZ	CA	AZ	TX
CA	AZ	TX	TX	TX
CA	AZ	AZ	CA	CA

a. What is the experimental unit?

b. What is the variable being measured? Is it qualitative or quantitative?

c. Construct a pie chart to describe the data.

d. Construct a bar chart to describe the data.

e. What *proportion* of the jeans are made in Texas?

f. What state produced the most jeans in the group?

g. If you want to find out whether the three plants produced equal numbers of jeans, or whether one produced more jeans than the others, how can you use the charts from parts c and d to help you? What conclusions can *you* draw from these data?

APPLICATIONS

1.12 Election 2012 During the spring of 2010, the news media were already conducting opinion polls that tracked the fortunes of the major candidates hoping to become the president of the United States. One such poll conducted by *CNN/Opinion Research Corporation Poll* showed the following results:[1]

"If Barack Obama were the Democratic Party's candidate and [see below] were the Republican Party's candidate, who would you be more likely to vote for: Obama, the Democrat, or [see below], the Republican?" If unsure: "As of today, who do you lean more toward?"

	Barack Obama (D) %	Mitt Romney (R) %	Neither (vol.) %
4/9–11/10	53	45	1

	Barack Obama (D) %	Mike Huckabee (R) %	Neither (vol.) %
4/9–11/10	54	45	1

	Barack Obama (D) %	Sarah Palin (R) %	Neither (vol.) %
4/9–11/10	55	42	3

	Barack Obama (D) %	Newt Gingrich (R) %	Neither (vol.) %
4/9–11/10	55	43	1

Source: www.pollingreport.com

The results were based on a sample taken April 9–11, 2010, of 907 registered voters nationwide.

a. If the pollsters were planning to use these results to predict the outcome of the 2012 presidential election, describe the population of interest to them.

b. Describe the actual population from which the sample was drawn.

c. Some pollsters prefer to select a sample of "likely" voters. What is the difference between "registered voters" and "likely voters"? Why is this important?

d. Is the sample selected by the pollsters representative of the population described in part a? Explain.

1.13 Want to Be President? Would you want to be the president of the United States? Although many teenagers think that they could grow up to be the president, most don't want the job. In an opinion poll conducted by *ABC News*, nearly 80% of the teens were not interested in the job.[2] When asked "What's the main reason you would not want to be president?" they gave these responses:

Other career plans/no interest	40%
Too much pressure	20%
Too much work	15%
Wouldn't be good at it	14%
Too much arguing	5%

a. Are all of the reasons accounted for in this table? Add another category if necessary.

b. Would you use a pie chart or a bar chart to graphically describe the data? Why?

c. Draw the chart you chose in part b.

d. If you were the person conducting the opinion poll, what other types of questions might you want to investigate?

Data set
EX0114
1.14 Facebook Fanatics The social networking site called *Facebook* has grown quickly since its inception in 2004. In fact, *Facebook's* United States user base grew from 42 million users to 103 million users between 2009 and 2010. The table below shows the age distribution of *Facebook* users (in thousands) as it changed from January 2009 to January 2010.[3]

Age	As of 1/04/2009	As of 1/04/2010
13–17	5675	10,680
18–24	17,192	26,076
25–34	11,255	25,580
35–54	6989	29,918
55+	955	9764
Unknown	23	1068
Total	42,089	103,086

a. Define the variable that has been measured in this table.

b. Is the variable quantitative or qualitative?

c. What do the numbers represent?

d. Construct a pie chart to describe the age distribution of *Facebook* users as of January 4, 2009.

e. Construct a pie chart to describe the age distribution of *Facebook* users as of January 4, 2010.

f. Refer to parts d and e. How would you describe the changes in the age distributions of *Facebook* users during this 1-year period?

1.15 Back to Work How long does it take you to adjust to your normal work routine after coming back from vacation? A bar graph with data from the Snapshots section of *USA Today* is shown below:[4]

a. Are all of the opinions accounted for in the table? Add another category if necessary.

b. Is the bar chart drawn accurately? That is, are the three bars in the correct proportion to each other?

c. Use a pie chart to describe the opinions. Which graph is more interesting to look at?

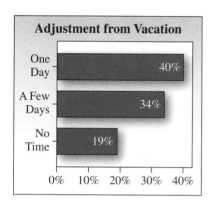

Adjustment from Vacation

GRAPHS FOR QUANTITATIVE DATA

Quantitative variables measure an amount or quantity on each experimental unit. If the variable can take only a finite or countable number of values, it is a *discrete* variable. A variable that can assume an infinite number of values corresponding to points on a line interval is called *continuous*.

Pie Charts and Bar Charts

Sometimes information is collected for a quantitative variable measured on different segments of the population, or for different categories of classification. For example, you might measure the average incomes for people of different age groups, different genders, or living in different geographic areas of the country. In such cases, you can use pie charts or bar charts to describe the data, using the amount measured in each category rather than the frequency of occurrence of each category. The *pie chart* displays how the total quantity is distributed among the categories, and the *bar chart* uses the height of the bar to display the amount in a particular category.

EXAMPLE 1.5

The amount of money expended in fiscal year 2009 by the U.S. Department of Defense in various categories is shown in Table 1.5.[5] Construct both a pie chart and a bar chart to describe the data. Compare the two forms of presentation.

TABLE 1.5 ● **Expenses by Category**

Category	Amount ($ billions)
Military personnel	147.3
Operation and maintenance	259.3
Procurement	129.2
Research and development	79.0
Military construction	17.6
Other	4.3
Total	636.7

Source: The World Almanac and Book of Facts 2011

Solution Two variables are being measured: the category of expenditure (*qualitative*) and the amount of the expenditure (*quantitative*). The bar chart in Figure 1.6 displays the categories on the horizontal axis and the amounts on the vertical axis.

FIGURE 1.6 ●

Bar chart for Example 1.5

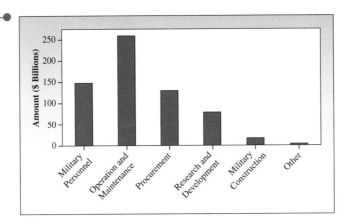

For the pie chart in Figure 1.7, each "pie slice" represents the proportion of the total expenditures ($636.7 billion) corresponding to its particular category. For example, for the research and development category, the angle of the sector is

$$\frac{79.0}{636.7} \times 360° = 44.7°$$

FIGURE 1.7 ●

Pie chart for Example 1.5

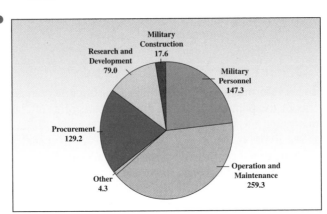

Both graphs show that the largest amounts of money were spent on personnel and operations. Since there is no inherent order to the categories, you are free to rearrange the bars or sectors of the graphs in any way you like. The *shape* of the bar chart has no bearing on its interpretation.

Line Charts

When a quantitative variable is recorded over time at equally spaced intervals (such as daily, weekly, monthly, quarterly, or yearly), the data set forms a **time series.** Time series data are most effectively presented on a **line chart** with time as the horizontal axis. The idea is to try to discern a pattern or **trend** that will likely continue into the future, and then to use that pattern to make accurate predictions for the immediate future.

 EXAMPLE 1.6

In the year 2025, the oldest "baby boomers" (born in 1946) will be 79 years old, and the oldest "Gen Xers" (born in 1965) will be 2 years from Social Security eligibility. How will this affect the consumer trends in the next 25 years? Will there be sufficient funds for "baby boomers" to collect Social Security benefits? The United States *Bureau of the Census* gives projections for the portion of the U.S. population that will be 85 and over in the coming years, as shown in Table 1.6.[5] Construct a line chart to illustrate the data. What is the effect of stretching and shrinking the vertical axis on the line chart?

TABLE 1.6 ● **Population Growth Projections**

Year	2020	2030	2040	2050
85 and over (millions)	6.6	8.7	14.2	19.0

Source: The World Almanac and Book of Facts 2011

 NEED A TIP?

Beware of stretching or shrinking axes when you look at a graph!

Solution The quantitative variable "85 and over" is measured over four time intervals, creating a *time series* that you can graph with a line chart. The time intervals are marked on the horizontal axis and the projections on the vertical axis. The data points are then connected by line segments to form the line charts in Figure 1.8. Notice the marked difference in the vertical scales of the two graphs. *Shrinking* the scale on the vertical axis causes large changes to appear small, and vice versa. To avoid misleading conclusions, you must look carefully at the scales of the vertical and horizontal axes. However, from both graphs you get a clear picture of the steadily increasing number of those 85 and older in the early years of the new millennium.

FIGURE 1.8 ●

Line charts for Example 1.6

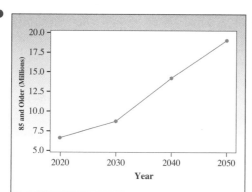 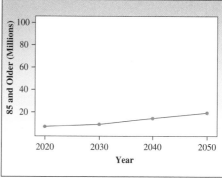

Dotplots

Many sets of quantitative data consist of numbers that cannot easily be separated into categories or intervals of time. You need a different way to graph this type of data!

The simplest graph for quantitative data is the **dotplot.** For a small set of measurements—for example, the set 2, 6, 9, 3, 7, 6—you can simply plot the measurements as points on a horizontal axis, as shown in Figure 1.9(a). For a large data set, however, such as the one in Figure 1.9(b), the dotplot can be uninformative and tedious to interpret.

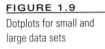

FIGURE 1.9

Dotplots for small and large data sets

(a)

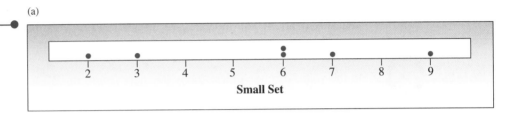

Small Set

MY ONLINE APPLET

Building a Dotplot

(b)

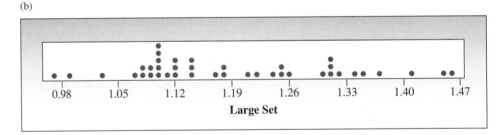

Large Set

Stem and Leaf Plots

Another simple way to display the distribution of a quantitative data set is the **stem and leaf plot.** This plot presents a graphical display of the data using the actual numerical values of each data point.

? NEED TO KNOW...

How to Construct a Stem and Leaf Plot

1. Divide each measurement into two parts: the **stem** and the **leaf.**
2. List the stems in a column, with a vertical line to their right.
3. For each measurement, record the leaf portion in the same row as its corresponding stem.
4. Order the leaves from lowest to highest in each stem.
5. Provide a key to your stem and leaf coding so that the reader can re-create the actual measurements if necessary.

EXAMPLE 1.7

Table 1.7 lists the prices (in dollars) of 19 different brands of walking shoes. Construct a stem and leaf plot to display the distribution of the data.

TABLE 1.7 • **Prices of Walking Shoes**

90	70	70	70	75	70
65	68	60	74	70	95
75	70	68	65	40	65
70					

Solution To create the stem and leaf, you could divide each observation between the ones and the tens place. The number to the left is the stem; the number to the right is the leaf. Thus, for the shoes that cost $65, the stem is 6 and the leaf is 5. The stems, ranging from 4 to 9, are listed in Figure 1.10, along with the leaves for each of the 19 measurements. If you indicate that the leaf unit is 1, the reader will realize that the stem and leaf 6 and 8, for example, represent the number 68, recorded to the nearest dollar.

FIGURE 1.10 •

Stem and leaf plot for the data in Table 1.7

```
4 | 0                  Leaf unit = 1    4 | 0
5 |                                     5 |
6 | 5 8 0 8 5 5        Reordering ──→   6 | 0 5 5 5 8 8
7 | 0 0 0 5 0 4 0 5 0 0                 7 | 0 0 0 0 0 0 0 4 5 5
8 |                                     8 |
9 | 0 5                                 9 | 0 5
```

 NEED A TIP?

Stem | Leaf

Sometimes the available stem choices result in a plot that contains too few stems and a large number of leaves within each stem. In this situation, you can stretch the stems by dividing each one into several lines, depending on the leaf values assigned to them. Stems are usually divided in one of two ways:

- Into two lines, with leaves 0–4 in the first line and leaves 5–9 in the second line
- Into five lines, with leaves 0–1, 2–3, 4–5, 6–7, and 8–9 in the five lines, respectively

EXAMPLE 1.8

The data in Table 1.8 are the weights at birth of 30 full-term babies, born at a metropolitan hospital and recorded to the nearest tenth of a pound.[6] Construct a stem and leaf plot to display the distribution of the data.

TABLE 1.8 • **Birth Weights of 30 Full-Term Newborn Babies**

7.2	7.8	6.8	6.2	8.2
8.0	8.2	5.6	8.6	7.1
8.2	7.7	7.5	7.2	7.7
5.8	6.8	6.8	8.5	7.5
6.1	7.9	9.4	9.0	7.8
8.5	9.0	7.7	6.7	7.7

Solution The data, though recorded to an accuracy of only one decimal place, are measurements of the continuous variable x = weight, which can take on any positive value. By examining Table 1.8, you can quickly see that the highest and lowest weights are 9.4 and 5.6, respectively. But how are the remaining weights distributed?

If you use the decimal point as the dividing line between the stem and the leaf, you have only five stems, which does not produce a very good picture. When you divide each stem into two lines, there are eight stems, since the first line of stem 5 and the second line of stem 9 are empty! This produces a more descriptive plot, as shown in Figure 1.11. For these data, the leaf unit is .1, and the reader can infer that the stem and leaf 8 and 2, for example, represent the measurement $x = 8.2$.

FIGURE 1.11

Stem and leaf plot for the data in Table 1.8

5	8 6
6	1 2
6	8 8 8 7
7	2 2 1
7	8 7 9 5 7 7 5 8 7
8	0 2 2 2
8	5 6 5
9	0 4 0

$\boxed{\text{Reordering}} \rightarrow$

Leaf unit = .1

5	6 8
6	1 2
6	7 8 8 8
7	1 2 2
7	5 5 7 7 7 7 8 8 9
8	0 2 2 2
8	5 5 6
9	0 0 4

If you turn the stem and leaf plot sideways, so that the vertical line is now a horizontal axis, you can see that the data have "piled up" or been "distributed" along the axis in a pattern that can be described as "mound-shaped"—much like a pile of sand on the beach. This plot again shows that the weights of these 30 newborns range between 5.6 and 9.4; many weights are between 7.5 and 8.0 pounds.

Interpreting Graphs with a Critical Eye

Once you have created a graph or graphs for a set of data, what should you look for as you attempt to describe the data?

- First, check the horizontal and vertical **scales,** so that you are clear about what is being measured.
- Examine the **location** of the data distribution. Where on the horizontal axis is the center of the distribution? If you are comparing two distributions, are they both centered in the same place?
- Examine the **shape** of the distribution. Does the distribution have one "peak," a point that is higher than any other? If so, this is the most frequently occurring measurement or category. Is there more than one peak? Are there an approximately equal number of measurements to the left and right of the peak?
- Look for any unusual measurements or **outliers.** That is, are any measurements much bigger or smaller than all of the others? These outliers may not be representative of the other values in the set.

Distributions are often described according to their shapes.

Definition A distribution is **symmetric** if the left and right sides of the distribution, when divided at the middle value, form mirror images.

A distribution is **skewed to the right** if a greater proportion of the measurements lie to the right of the peak value. Distributions that are **skewed right** contain a few unusually large measurements.

A distribution is **skewed to the left** if a greater proportion of the measurements lie to the left of the peak value. Distributions that are **skewed left** contain a few unusually small measurements.

A distribution is **unimodal** if it has one peak; a **bimodal** distribution has two peaks. Bimodal distributions often represent a mixture of two different populations in the data set.

Examine the three dotplots shown in Figure 1.12. Describe these distributions in terms of their locations and shapes.

FIGURE 1.12

Shapes of data distributions for Example 1.9

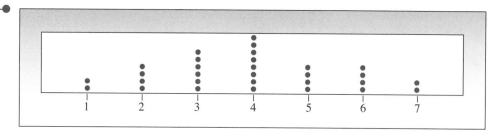

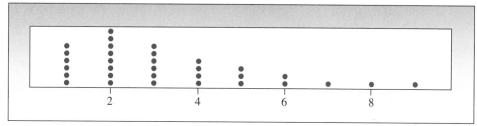

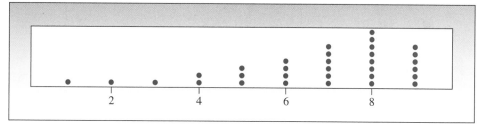

Symmetric ⇔ mirror images
Skewed right ⇔ long right tail
Skewed left ⇔ long left tail

Solution   The first dotplot shows a *relatively symmetric* distribution with a single peak located at $x = 4$. If you were to fold the page at this peak, the left and right halves would *almost* be mirror images. The second dotplot, however, is far from symmetric. It has a long "right tail," meaning that there are a few unusually large observations. If you were to fold the page at the peak, a larger proportion of measurements would be on the right side than on the left. This distribution is *skewed to the right*. Similarly, the third dotplot with the long "left tail" is *skewed to the left*.

An administrative assistant for the athletics department at a local university is monitoring the GPAs for eight members of the women's volleyball team. He enters the GPAs into the database but accidentally misplaces the decimal point in the last entry.

2.8   3.0   3.0   3.3   2.4   3.4   3.0   .21

Use a dotplot to describe the data and uncover the assistant's mistake.

Solution The dotplot of this small data set is shown in Figure 1.13(a). You can clearly see the *outlier* or unusual observation caused by the assistant's data entry error. Once the error has been corrected, as in Figure 1.13(b), you can see the correct distribution of the data set. Since this is a very small set, it is difficult to describe the shape of the distribution, although it seems to have a peak value around 3.0 and it appears to be relatively symmetric.

FIGURE 1.13

Distributions of GPAs for Example 1.10

(a)

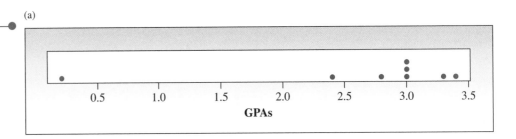

(b)

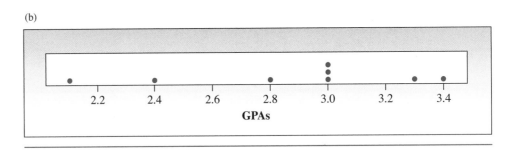

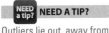 **NEED A TIP?**

Outliers lie out, away from the main body of data.

When comparing graphs created for two data sets, you should compare their *scales of measurement, locations,* and *shapes,* and look for unusual measurements or outliers. Remember that outliers are not always caused by errors or incorrect data entry. Sometimes they provide very valuable information that should not be ignored. You may need additional information to decide whether an outlier is a valid measurement that is simply unusually large or small, or whether there has been some sort of mistake in the data collection. If the scales differ widely, be careful about making comparisons or drawing conclusions that might be inaccurate!

1.5 RELATIVE FREQUENCY HISTOGRAMS

A relative frequency histogram resembles a bar chart, but it is used to graph quantitative rather than qualitative data. The data in Table 1.9 are the birth weights of 30 full-term newborn babies, reproduced from Example 1.8 and shown as a dotplot in Figure 1.14(a). First, divide the interval from the smallest to the largest measurements into subintervals or *classes of equal length.* If you stack up the dots in each subinterval (Figure 1.14(b)), and draw a bar over each stack, you will have created a **frequency histogram** or a **relative frequency histogram,** depending on the scale of the vertical axis.

TABLE 1.9  **Birth Weights of 30 Full-Term Newborn Babies**

7.2	7.8	6.8	6.2	8.2
8.0	8.2	5.6	8.6	7.1
8.2	7.7	7.5	7.2	7.7
5.8	6.8	6.8	8.5	7.5
6.1	7.9	9.4	9.0	7.8
8.5	9.0	7.7	6.7	7.7

FIGURE 1.14

How to construct a
histogram

(a)

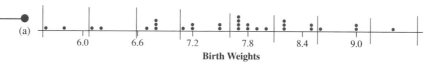

(b)

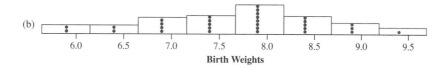

___

Definition   A **relative frequency histogram** for a quantitative data set is a bar graph in which the height of the bar shows "how often" (measured as a proportion or relative frequency) measurements fall in a particular class or subinterval. The classes or subintervals are plotted along the horizontal axis.

___

As a rule of thumb, the number of classes should range from 5 to 12; the more data available, the more classes you need.[†] The classes must be chosen so that each measurement falls into one and only one class. For the birth weights in Table 1.9, we decided to use eight intervals of equal length. Since the total span of the birth weights is

$$9.4 - 5.6 = 3.8$$

the minimum class width necessary to cover the range of the data is $(3.8 \div 8) = .475$. For convenience, we round this approximate width up to .5. Beginning the first interval at the lowest value, 5.6, we form subintervals from 5.6 up to *but not including* 6.1, 6.1 up to *but not including* 6.6, and so on. By using the **method of left inclusion,** and including the left class boundary point but not the right boundary point in the class, we eliminate any confusion about where to place a measurement that happens to fall on a class boundary point.

Table 1.10 shows the eight classes, labeled from 1 to 8 for identification. The boundaries for the eight classes, along with a tally of the number of measurements that fall in each class, are also listed in the table. As with the charts in Section 1.3, you can now measure *how often* each class occurs using *frequency* or *relative frequency.*

___

[†]You can use this table as a guide for selecting an appropriate number of classes. Remember that this is only a guide; you may use more or fewer classes than the table recommends if it makes the graph more descriptive.

Sample Size	25	50	100	200	500
Number of Classes	6	7	8	9	10

To construct the relative frequency histogram, plot the class boundaries along the horizontal axis. Draw a bar over each class interval, with height equal to the relative frequency for that class. The relative frequency histogram for the birth weight data, Figure 1.15, shows at a glance how birth weights are distributed over the interval 5.6 to 9.4.

TABLE 1.10 ● **Relative Frequencies for the Data of Table 1.9**

NEED A TIP?

Relative frequencies add to 1; frequencies add to n.

Class	Class Boundaries	Tally	Class Frequency	Class Relative Frequency
1	5.6 to <6.1	‖	2	2/30
2	6.1 to <6.6	‖	2	2/30
3	6.6 to <7.1	‖‖	4	4/30
4	7.1 to <7.6	‖‖‖	5	5/30
5	7.6 to <8.1	‖‖‖ ‖‖‖	8	8/30
6	8.1 to <8.6	‖‖‖	5	5/30
7	8.6 to <9.1	‖‖	3	3/30
8	9.1 to <9.6	‖	1	1/30

FIGURE 1.15 ●

Relative frequency histogram

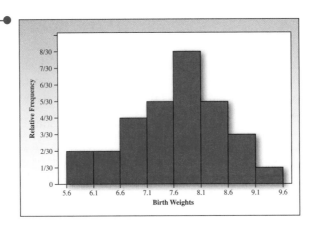

EXAMPLE 1.11 — Twenty-five Starbucks® customers are polled in a marketing survey and asked, "How often do you visit Starbucks in a typical week?" Table 1.11 lists the responses for these 25 customers. Construct a relative frequency histogram to describe the data.

TABLE 1.11 ● **Number of Visits in a Typical Week for 25 Customers**

6	7	1	5	6
4	6	4	6	8
6	5	6	3	4
5	5	5	7	6
3	5	7	5	5

Solution The variable being measured is "number of visits to Starbucks," which is a discrete variable that takes on only integer values. In this case, it is simplest to choose the classes or subintervals as the integer values over the range of observed values: 1, 2, 3, 4, 5, 6, and 7. Table 1.12 shows the classes and their corresponding frequencies and relative frequencies. The relative frequency histogram is shown in Figure 1.16.

TABLE 1.12 • **Frequency Table for Example 1.11**

Number of Visits to Starbucks	Frequency	Relative Frequency
1	1	.04
2	—	—
3	2	.08
4	3	.12
5	8	.32
6	7	.28
7	3	.12
8	1	.04

FIGURE 1.16 •

Relative frequency histogram for Example 1.11

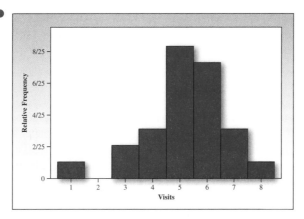

Notice that the distribution is skewed to the left and that there is a gap between 1 and 3.

 NEED TO KNOW...

How to Construct a Relative Frequency Histogram

1. Choose the number of classes, usually between 5 and 12. The more data you have, the more classes you should use.

2. Calculate the approximate class width by dividing the difference between the largest and smallest values by the number of classes.

3. Round the approximate class width up to a convenient number.

4. If the data are discrete, you might assign one class for each integer value taken on by the data. For a large number of integer values, you may need to group them into classes.

5. Locate the class boundaries. The lowest class must include the smallest measurement. Then add the remaining classes using the left inclusion method.

6. Construct a statistical table containing the classes, their frequencies, and their relative frequencies.

7. Construct the histogram like a bar graph, plotting class intervals on the horizontal axis and relative frequencies as the heights of the bars.

A relative frequency histogram can be used to describe the distribution of a set of data in terms of its *location* and *shape,* and to check for *outliers* as you did with other graphs. For example, the birth weight data were relatively symmetric, with no unusual measurements, while the Starbucks data were skewed left. Since the bar constructed above each class represents the *relative frequency* or proportion of the measurements in that class, these heights can be used to give us further information:

- The proportion of the measurements that fall in a particular class or group of classes
- The probability that a measurement drawn at random from the set will fall in a particular class or group of classes

Consider the relative frequency histogram for the birth weight data in Figure 1.15. What proportion of the newborns have birth weights of 7.6 or higher? This involves all classes beyond 7.6 in Table 1.10. Because there are 17 newborns in those classes, the proportion who have birth weights of 7.6 or higher is 17/30, or approximately 57%. This is also the percentage of the total area under the histogram in Figure 1.15 that lies to the right of 7.6.

Suppose you wrote each of the 30 birth weights on a piece of paper, put them in a hat, and drew one at random. What is the chance that this piece of paper contains a birth weight of 7.6 or higher? Since 17 of the 30 pieces of paper fall in this category, you have 17 chances out of 30; that is, the probability is 17/30. The word *probability* is not unfamiliar to you; we will discuss it in more detail in Chapter 4.

Although we are interested in describing the set of $n = 30$ birth weights, we might also be interested in the population from which the sample was drawn, which is the set of birth weights of *all* babies born at this hospital. Or, if we are interested in the weights of newborns in general, we might consider our sample as representative of the population of birth weights for newborns at similar metropolitan hospitals. A sample histogram provides valuable information about the population histogram—the graph that describes the distribution of the entire population. Remember, though, that different samples from the same population will produce *different* histograms, even if you use the same class boundaries. However, you can expect that the sample and population histograms will be similar. As you add more and more data to the sample, the two histograms become more and more alike. If you enlarge the sample to include the entire population, the two histograms will be identical!

1.5 EXERCISES

BASIC TECHNIQUES

1.16 Construct a stem and leaf plot for these
EX0116   50 measurements:

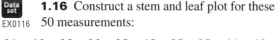

3.1	4.9	2.8	3.6	2.5	4.5	3.5	3.7	4.1	4.9
2.9	2.1	3.5	4.0	3.7	2.7	4.0	4.4	3.7	4.2
3.8	6.2	2.5	2.9	2.8	5.1	1.8	5.6	2.2	3.4
2.5	3.6	5.1	4.8	1.6	3.6	6.1	4.7	3.9	3.9
4.3	5.7	3.7	4.6	4.0	5.6	4.9	4.2	3.1	3.9

a. Describe the shape of the data distribution. Do you see any outliers?

b. Use the stem and leaf plot to find the smallest observation.

c. Find the eighth and ninth largest observations.

1.17 Refer to Exercise 1.16. Construct a relative frequency histogram for the data.

a. Approximately how many class intervals should you use?

b. Suppose you decide to use classes starting at 1.6 with a class width of .5 (i.e., 1.6 to <2.1, 2.1 to <2.6). Construct the relative frequency histogram for the data.

c. What fraction of the measurements are less than 5.1?

d. What fraction of the measurements are larger than 3.6?

e. Compare the relative frequency histogram with the stem and leaf plot in Exercise 1.16. Are the shapes similar?

Data set **1.18** Consider this set of data:

EX0118

4.5	3.2	3.5	3.9	3.5	3.9
4.3	4.8	3.6	3.3	4.3	4.2
3.9	3.7	4.3	4.4	3.4	4.2
4.4	4.0	3.6	3.5	3.9	4.0

a. Construct a stem and leaf plot by using the leading digit as the stem.

b. Construct a stem and leaf plot by using each leading digit twice. Does this technique improve the presentation of the data? Explain.

1.19 A discrete variable can take on only the values 0, 1, or 2. A set of 20 measurements on this variable is shown here:

1	2	1	0	2
2	1	1	0	0
2	2	1	1	0
0	1	2	1	1

a. Construct a relative frequency histogram for the data.

b. What proportion of the measurements are greater than 1?

c. What proportion of the measurements are less than 2?

d. If a measurement is selected at random from the 20 measurements shown, what is the probability that it is a 2?

e. Describe the shape of the distribution. Do you see any outliers?

1.20 Refer to Exercise 1.19.

a. Draw a dotplot to describe the data.

b. How could you define the stem and the leaf for this data set?

c. Draw the stem and leaf plot using your decision from part b.

d. Compare the dotplot, the stem and leaf plot, and the relative frequency histogram (Exercise 1.19). Do they all convey roughly the same information?

1.21 Navigating a Maze An experimental psychologist measured the length of time it took for a rat to successfully navigate a maze on each of 5 days. The results are shown in the table. Create a line chart to describe the data. Do you think that any learning is taking place?

Day	1	2	3	4	5
Time (seconds)	45	43	46	32	25

Data set **1.22 Measuring over Time** The value of a

EX0122 quantitative variable is measured once a year for a 10-year period. Here are the data:

Year	Measurement	Year	Measurement
1	61.5	6	58.2
2	62.3	7	57.5
3	60.7	8	57.5
4	59.8	9	56.1
5	58.0	10	56.0

a. Create a line chart to describe the variable as it changes over time.

b. Describe the measurements using the chart constructed in part a.

1.23 Cheeseburgers Create a dotplot for the number of cheeseburgers consumed in a given week by 10 college students.

4	5	4	2	1
3	3	4	2	7

a. How would you describe the shape of the distribution?

b. What proportion of the students ate more than 4 cheeseburgers that week?

Data set **1.24 Test Scores** The test scores on a

EX0124 100-point test were recorded for 20 students:

61	93	91	86	55	63	86	82	76	57
94	89	67	62	72	87	68	65	75	84

a. Use an appropriate graph to describe the data.

b. Describe the shape and location of the scores.

c. Is the shape of the distribution unusual? Can you think of any reason the distribution of the scores would have such a shape?

APPLICATIONS

Data set **1.25 Survival Times** Altman and Bland

EX0125 report the survival times for patients with active hepatitis, half treated with prednisone and half receiving no treatment.[7] The data that follow are adapted from their data for those treated with prednisone. The survival times are recorded to the nearest month:

8	87	127	147
11	93	133	148
52	97	139	157
57	109	142	162
65	120	144	165

a. Look at the data. Can you guess the approximate shape of the data distribution?

b. Construct a relative frequency histogram for the data. What is the shape of the distribution?

c. Are there any outliers in the set? If so, which survival times are unusually short?

Data set **1.26 A Recurring Illness** The length of time
EX0126 (in months) between the onset of a particular illness and its recurrence was recorded for $n = 50$ patients:

2.1	4.4	2.7	32.3	9.9	9.0	2.0	6.6	3.9	1.6
14.7	9.6	16.7	7.4	8.2	19.2	6.9	4.3	3.3	1.2
4.1	18.4	.2	6.1	13.5	7.4	.2	8.3	.3	1.3
14.1	1.0	2.4	2.4	18.0	8.7	24.0	1.4	8.2	5.8
1.6	3.5	11.4	18.0	26.7	3.7	12.6	23.1	5.6	.4

a. Construct a relative frequency histogram for the data.

b. Would you describe the shape as roughly symmetric, skewed right, or skewed left?

c. Give the fraction of recurrence times less than or equal to 10 months.

Data set **1.27 Education Pays Off!** Education pays
EX0127 off, according to a snapshot provided by the *Bureau of Labor Statistics*.[8] The median weekly earnings for six different levels of education are shown in the table:

Educational Level	Median Weekly Earnings ($)
Less than a high school diploma	454
High school graduate	626
Some college, no degree	699
Associate degree	761
Bachelor's degree	1025
Master's degree	1257
Professional degree	1529
Doctoral degree	1532

Source: Bureau of Labor Statistics, Current Population Survey

a. What graphical methods could you use to describe the data?

b. Select the method from part a that you think best describes the data and create the appropriate graph.

c. How would you summarize the information that you see in the graph regarding educational levels and salary?

Data set **1.28 Preschool** The ages (in months) at
EX0128 which 50 children were first enrolled in a preschool are listed below.

38	40	30	35	39	40	48	36	31	36
47	35	34	43	41	36	41	43	48	40
32	34	41	30	46	35	40	30	46	37
55	39	33	32	32	45	42	41	36	50
42	50	37	39	33	45	38	46	36	31

a. Construct a stem and leaf display for the data.

b. Construct a relative frequency histogram for these data. Start the lower boundary of the first class at 30 and use a class width of 5 months.

c. Compare the graphs in parts a and b. Are there any significant differences that would cause you to choose one as the better method for displaying the data?

d. What proportion of the children were 35 months (2 years, 11 months) or older, but less than 45 months (3 years, 9 months) of age when first enrolled in preschool?

e. If one child were selected at random from this group of children, what is the probability that the child was less than 50 months old (4 years, 2 months) when first enrolled in preschool?

Data set **1.29 Organized Religion** Statistics of the
EX0129 world's religions are only very rough approximations, since many religions do not keep track of their membership numbers. An estimate of these numbers (in millions) is shown in the table.[9]

Religion	Members (millions)	Religion	Members (millions)
Buddhism	376	Judaism	14
Christianity	2100	Sikhism	23
Hinduism	900	Chinese traditional	394
Islam	1500	Other	61
Primal indigenous and African traditional	400		

a. Construct a pie chart to describe the total membership in the world's organized religions.

b. Construct a bar chart to describe the total membership in the world's organized religions.

c. Order the religious groups from the smallest to the largest number of members. Construct a Pareto chart to describe the data. Which of the three displays is most effective?

Data set **1.30 How Long Is the Line?** To decide on
EX0130 the number of service counters needed for
stores to be built in the future, a supermarket chain
wanted to obtain information on the length of time (in
minutes) required to service customers. To find the
distribution of customer service times, a sample of
60 customers' service times was recorded and are
shown here:

3.6	1.9	2.1	.3	.8	.2	1.0	1.4	1.8	1.6
1.1	1.8	.3	1.1	.5	1.2	.6	1.1	.8	1.7
1.4	.2	1.3	3.1	.4	2.3	1.8	4.5	.9	.7
.6	2.8	2.5	1.1	.4	1.2	.4	1.3	.8	1.3
1.1	1.2	.8	1.0	.9	.7	3.1	1.7	1.1	2.2
1.6	1.9	5.2	.5	1.8	.3	1.1	.6	.7	.6

a. Construct a stem and leaf plot for the data.

b. What fraction of the service times are less than or
equal to 1 minute?

c. What is the smallest of the 60 measurements?

1.31 Service Times, continued Refer to Exercise
1.30. Construct a relative frequency histogram for the
supermarket service times.

a. Describe the shape of the distribution. Do you see
any outliers?

b. Assuming that the outliers in this data set are valid
observations, how would you explain them to the
management of the supermarket chain?

c. Compare the relative frequency histogram with the
stem and leaf plot in Exercise 1.30. Do the two
graphs convey the same information?

Data set **1.32 Calcium Content** The calcium (Ca)
EX0132 content of a powdered mineral substance was
analyzed 10 times with the following percent compositions recorded:

.0271	.0282	.0279	.0281	.0268
.0271	.0281	.0269	.0275	.0276

a. Draw a dotplot to describe the data. (HINT: The
scale of the horizontal axis should range from
.0260 to .0290.)

b. Draw a stem and leaf plot for the data. Use the numbers
in the hundredths and thousandths places as the stem.

c. Are any of the measurements inconsistent with the
other measurements, indicating that the technician
may have made an error in the analysis?

Data set **1.33 American Presidents** The following
EX0133 table lists the ages at the time of death for the 38
deceased American presidents from George Washington
to Ronald Reagan:[5]

Washington	67	Arthur	56
J. Adams	90	Cleveland	71
Jefferson	83	B. Harrison	67
Madison	85	McKinley	58
Monroe	73	T. Roosevelt	60
J. Q. Adams	80	Taft	72
Jackson	78	Wilson	67
Van Buren	79	Harding	57
W. H. Harrison	68	Coolidge	60
Tyler	71	Hoover	90
Polk	53	F. D. Roosevelt	63
Taylor	65	Truman	88
Fillmore	74	Eisenhower	78
Pierce	64	Kennedy	46
Buchanan	77	L. Johnson	64
Lincoln	56	Nixon	81
A. Johnson	66	Ford	93
Grant	63	Carter	93
Hayes	70	Reagan	93
Garfield	49		

a. Before you graph the data, try to visualize the
distribution of the ages at death for the presidents.
What shape do you think it will have?

b. Construct a stem and leaf plot for the data.
Describe the shape. Does it surprise you?

c. The five youngest presidents at the time of death
appear in the lower "tail" of the distribution. Three
of the five youngest have one common trait. Identify
the five youngest presidents at death. What common
trait explains these mea-surements?

Data set **1.34 RBC Counts** The red blood cell count of
EX0134 a healthy person was measured on each of
15 days. The number recorded is measured in 10^6 cells
per microliter (μL).

5.4	5.2	5.0	5.2	5.5
5.3	5.4	5.2	5.1	5.3
5.3	4.9	5.4	5.2	5.2

a. Use an appropriate graph to describe the data.

b. Describe the shape and location of the red blood cell
counts.

c. If the person's red blood cell count is measured
today as $5.7 \times 10^6/\mu L$, would you consider
this unusual? What conclusions might you
draw?

Data set **1.35 Batting Champions** The officials of
EX0135 major league baseball have crowned a batting
champion in the National League each year since 1876.
A sample of winning batting averages is listed in the
table:[5]

Year	Name	Average
2000	Todd Helton	.372
1915	Larry Doyle	.320
1917	Edd Roush	.341
1934	Paul Waner	.362
1911	Honus Wagner	.334
1898	Willie Keeler	.379
1924	Roger Hornsby	.424
1963	Tommy Davis	.326
1992	Gary Sheffield	.330
1954	Willie Mays	.345
1975	Bill Madlock	.354
1958	Richie Ashburn	.350
1942	Ernie Lombardi	.330
1948	Stan Musial	.376
1971	Joe Torre	.363
1996	Tony Gwynn	.353
1961	Roberto Clemente	.351
1968	Pete Rose	.335
1885	Roger Connor	.371
2009	Hanley Ramirez	.342

a. Construct a relative frequency histogram to describe the batting averages for these 20 champions.

b. If you were to randomly choose one of the 20 names, what is the chance that you would choose a player whose average was above .400 for his championship year?

Data set **1.36 Top 20 Movies** The table that follows
EX0136 shows the weekend gross ticket sales for the top 20 movies for the weekend of June 25, 2010:[10]

Movie	Weekend Gross ($ millions)
1. Toy Story 3	59.3
2. Grown Ups	40.5
3. Knight and Day	20.1
4. The Karate Kid	15.5
5. The A-Team	6.2
6. Get Him to the Greek	3.1
7. Shrek Forever After	3.1
8. Prince of Persia	2.8
9. Killers	1.9
10. Jonah Hex	1.6
11. Iron Man 2	1.4
12. Sex and the City 2	1.2
13. Marmaduke	1.0
14. Robin Hood	0.6
15. Solitary Man	0.5
16. How to Train Your Dragon	0.5
17. Winter's Bone	0.4
18. Letters to Juliet	0.4
19. Joan Rivers: A Piece of Work	0.4
20. Cyrus	0.3

Source: www.radiofree.com/mov-tops.shtml

a. Draw a stem and leaf plot for the data. Describe the shape of the distribution. Are there any outliers?

b. Construct a dotplot for the data. Which of the two graphs is more informative? Explain.

Data set **1.37 Hazardous Waste** How safe is your
EX0137 neighborhood? Are there any hazardous waste sites nearby? The table shows the number of hazardous waste sites in each of the 50 states and the District of Columbia in the year 2009:[5]

AL	15	HI	3	MA	32	NM	14	SD	2
AK	6	ID	9	MI	69	NY	90	TN	15
AZ	9	IL	48	MN	25	NC	36	TX	50
AR	9	IN	32	MS	6	ND	0	UT	19
CA	96	IA	12	MO	31	OH	41	VT	11
CO	20	KS	12	MT	17	OK	9	VA	31
CT	15	KY	14	NE	13	OR	13	WA	48
DE	15	LA	12	NV	1	PA	97	WV	9
DC	1	ME	12	NH	21	RI	12	WI	39
FL	55	MD	19	NJ	114	SC	26	WY	2
GA	16								

a. What variable is being measured? Is the variable discrete or continuous?

b. Describe the shape of the data distribution using the stem and leaf plot shown here. Identify the unusually large measurements marked "HI" by state.

Stem and Leaf Display: Hazardous Waste

```
Stem-and-leaf of Sites  N = 51
Leaf Unit = 1.0
     6   0   011223
    13   0   6699999
    23   1   1222223344
   (8)   1   55556799
    20   2   01
    18   2   56
    16   3   1122
    12   3   69
    10   4   1
     9   4   88
     7   5   0
     6   5   5

         HI 69, 90, 96, 97, 114
```

c. Can you think of any reason these five states would have a large number of hazardous waste sites? What other variable might you measure to help explain why the data behave as they do?

As you continue to work through the exercises in this chapter, you will become more experienced in recognizing different types of data and in determining the most appropriate graphical method to use. Remember that the type of graphic you use is not as important as the interpretation that accompanies the picture. Look for these important characteristics:

- Location of the center of the data
- Shape of the distribution of data
- Unusual observations in the data set

Using these characteristics as a guide, you can interpret and compare sets of data using graphical methods, which are only the first of many statistical tools that you will soon have at your disposal.

CHAPTER REVIEW

Key Concepts

I. How Data Are Generated

1. Experimental units, variables, measurements
2. Samples and populations
3. Univariate, bivariate, and multivariate data

II. Types of Variables

1. Qualitative or categorical
2. Quantitative
   a. Discrete
   b. Continuous

III. Graphs for Univariate Data Distributions

1. Qualitative or categorical data
   a. Pie charts
   b. Bar charts
2. Quantitative data
   a. Pie and bar charts
   b. Line charts
   c. Dotplots
   d. Stem and leaf plots
   e. Relative frequency histograms
3. Describing data distributions
   a. Shapes—symmetric, skewed left, skewed right, unimodal, bimodal
   b. Proportion of measurements in certain intervals
   c. Outliers

 TECHNOLOGY TODAY

Introduction to *Microsoft Excel*

MS Excel is a spreadsheet program in the Microsoft Office system. It is designed for a variety of analytical applications, including statistical applications. We will assume that you are familiar with Windows, and that you know the basic techniques necessary for executing commands from the tabs, groups, and drop-down menus at the top of the screen. If not, perhaps a lab or teaching assistant can help you to master the basics. The current version of *MS Excel* at the time of this printing is *Excel 2010,* used in the Windows 7 environment. When the program opens, a **spreadsheet** appears (see Figure 1.17), containing rows and columns into which you can enter data. Tabs at the bottom of the screen identify the three spreadsheets available for use; when saved as a collection, these spreadsheets are called a **workbook.**

FIGURE 1.17

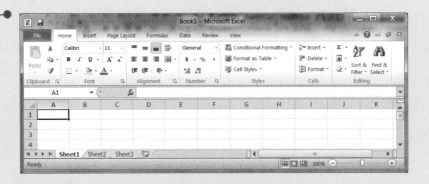

Graphing with *Excel*

Pie charts, bar charts, and **line charts** can all be created in *MS Excel.* Data is entered into an *Excel* spreadsheet, including labels if needed. Highlight the data to be graphed, and then click the chart type that you want on the **Insert** tab in the **Charts** group. Once the chart has been created, it can be edited in a variety of ways to change its appearance.

EXAMPLE 1.12

(Pie and Bar Charts) The class status of 105 students in an introductory statistics class are listed in Table 1.13. The qualitative variable "class status" has been recorded for each student, and the frequencies have already been recorded.

Status of Students in Statistics Class

Status	Freshman	Sophomore	Junior	Senior	Grad Student
Frequency	5	23	32	35	10

1. Enter the *categories* into column A of the first spreadsheet and the *frequencies* into column B. You should have two columns of data, including the labels.

2. Highlight the data, using your left mouse to *click-and-drag* from cell A1 to cell B6 (sometimes written as **A1:B6).** Click the **Insert** tab and select **Pie** in the **Charts** group. In the drop-down list, you will see a variety of styles to choose from. Select the first option to produce the pie chart. Double click on the title "Frequency" and change the title to "Student Status."

3. *Editing the pie chart:* Once the chart has been created, use your mouse to make sure that the chart is selected. You should see a green area above the tabs marked "Chart Tools." Click the **Design** tab, and look at the drop-down lists in the **Chart Layout** and **Chart Styles** groups. These lists allow you to alter the appearance of your chart. In Figure 1.18(a), the pie chart has been changed so that the percentages are shown in the appropriate sectors. By clicking on the legend, we have dragged it so that it is closer to the pie chart.

FIGURE 1.18

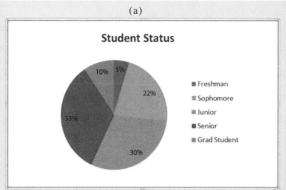

(a)

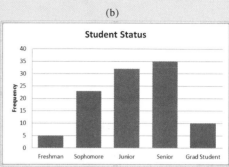

(b)

4. Click on various parts of the pie chart (legend, chart area, sector) and a box with round and/or square handles will appear. Double-click, and a dialog box will appear. You can adjust the appearance of the selected object or region in this box and click **OK.** Click **Cancel** to exit the dialog box without change!

5. Still in the **Design** section, but in the **Type** group, click on **Change Chart Type** and choose the simplest **Column** type. Click **OK** to create a bar chart for the same data set, shown in Figure 1.18(b).

6. *Editing the bar chart:* Again, you can experiment with the various options in the **Chart Layout** and **Chart Styles** groups to change the look of the chart. You can click the entire bar chart ("chart area") or the interior "Plot area" to stretch the chart. You can change colors by double-clicking on the appropriate region. We have chosen a design that allows axis titles and have deleted the "frequency legend entry." We have also chosen to delete the minor gridlines, by clicking the **Layout** tab in the **Chart Tools,** using the **Gridlines** drop-down list, and selecting **Primary Horizontal Gridlines ▶ Major Gridlines.** We have decreased the gaps between the bars by right-clicking on one of the bars, selecting **Format Data Series,** and changing the **Gap Width to 50%.**

EXAMPLE 1.13

(Line Charts) The Dow Jones Industrial Average was monitored at the close of trading for 10 days in a recent year, with the results shown in Table 1.14.

TABLE 1.14 ● **Dow Jones Industrial Average**

Day	1	2	3	4	5	6	7	8	9	10
DJIA	10,636	10,680	10,674	10,653	10,698	10,644	10,378	10,319	10,303	10,302

1. Click the tab at the bottom of the screen marked "Sheet 2." Enter the *Days* into column A of this second spreadsheet and the *DJIA* into column B. You should have two columns of data, including the labels.

2. Highlight the DJIA data in column B, using your left mouse to *click-and-drag* from cell B1 to cell B11 (sometimes written as **B1:B11).** Click the **Insert** tab and select **Line** in the **Charts** group. In the drop-down list, you will see a variety of styles to choose from. Select the first option to produce the line chart.

3. *Editing the line chart:* Again, you can experiment with the various options in the **Chart Layout** and **Chart Styles** groups to change the look of the chart. We have chosen a design that allows titles on both axes, which we have changed to "Day" and "DJIA," and we have deleted the title and the "frequency legend entry." The line chart is shown in Figure 1.19.

FIGURE 1.19

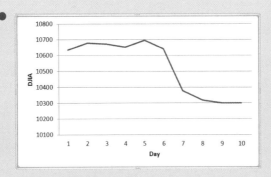

4. *Note:* If your time series involves time periods that are *not equally spaced*, it is better to use a **scatterplot** with points connected to form a line chart. This procedure is described in the *Technology Today* section in Chapter 3 of the text.

EXAMPLE  **(Frequency Histograms)** The top 40 stocks on the over-the-counter (OTC) market, ranked by percentage of outstanding shares traded on a particular day, are listed in Table 1.15.

TABLE 1.15 ● **Percentage of OTC Stocks Traded**

11.88	6.27	5.49	4.81	4.40	3.78	3.44	3.11	2.88	2.68
7.99	6.07	5.26	4.79	4.05	3.69	3.36	3.03	2.74	2.63
7.15	5.98	5.07	4.55	3.94	3.62	3.26	2.99	2.74	2.62
7.13	5.91	4.94	4.43	3.93	3.48	3.20	2.89	2.69	2.61

1. Many of the statistical procedures that we will use in this textbook require the installation of the **Analysis ToolPak** add-in. To load this add-in, click **File ▶ Options ▶ Add-ins.** Select **Analysis ToolPak** and click **OK.**

2. Click the tab at the bottom of the screen marked "Sheet 3." Enter the data into the first column of this spreadsheet and include the label "Stocks" in the first cell.

3. *Excel* refers to the maximum value for each class interval as a **bin.** This means that *Excel* is using a **method of right inclusion,** which is slightly different from the method presented in Section 1.5. For this example, we choose to use the class intervals >2.5–3.5, >3.5–4.5, >4.5–5.5, etc. Enter the *bin values* (3.5, 4.5, 5.5, ..., 12.5) into the second column of the spreadsheet, labeling them as "Percent Traded" in cell **B1.**

4. Select **Data ▶ Data Analysis ▶ Histogram** and click **OK.** The Histogram dialog box will appear, as shown in Figure 1.20.

FIGURE 1.20

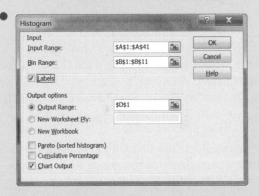

5. Highlight or type in the appropriate Input Range and Bin Range for the data. Notice that you can click the minimize button on the right of the box before you *click-and-drag* to highlight. Click the minimize button again to see the entire dialog box. The Input Range will appear as $A$1:$A$41, with the dollar sign indicating an *absolute cell range*. Make sure to click the "Labels" and "Chart Output" check boxes. Pick a convenient cell location for the output (we picked **D1**) and click **OK.** The frequency table and histogram will appear in the spreadsheet. The histogram (Figure 1.21 (a)) doesn't appear quite like we wanted.

FIGURE 1.21

(b)

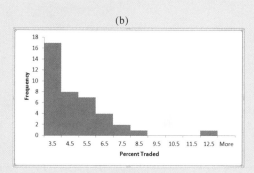

(a)

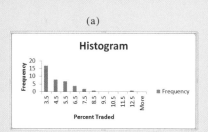

6. ***Editing the histogram:*** Click on the frequency legend entry and press the Delete key. Then select the Data Series by double-clicking on a bar. In the **Series Options** box that appears, change the **Gap Width** to 0% (no gap) and click **Close.** Stretch the graph by dragging the lower right corner, and edit the colors, title, and labels if necessary to finish your histogram, as shown in Figure 1.21 (b). Remember that the numbers shown along the horizontal axis are the **bins,** the upper limit of the class interval, *not the midpoint of the interval.*

7. You can save your *Excel* workbook for use at a later time using **File ▶ Save** or **File ▶ Save As** and naming it "Chapter 1."

TECHNOLOGY TODAY

Introduction to *MINITAB*™

MINITAB computer software is a Windows-based program designed specifically for statistical applications. We will assume that you are familiar with Windows, and that you know the basic techniques necessary for executing commands from the tabs and drop-down menus at the top of the screen. If not, perhaps a lab or teaching assistant can help you to master the basics. The current version of *MINITAB* at the time of this printing is *MINITAB 16,* used in the Windows 7 environment. When the program opens, the main screen (see Figure 1.22) is displayed, containing two windows: the Data window, similar to an *Excel* spreadsheet, and the Session window, in which your results will appear. Just as with *MS Excel, MINITAB* allows you to save worksheets (similar to *Excel* spreadsheets), projects (collections of worksheets), or graphs.

FIGURE 1.22

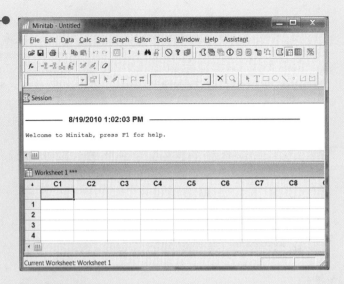

Graphing with *MINITAB*

All of the graphical methods that we have discussed in this chapter can be created in *MINITAB*. Data is entered into a *MINITAB* worksheet, with labels entered in the gray cells just below the column name (C1, C2, etc.) in the Data window.

EXAMPLE 1.15

(Pie and Bar Charts) The class status of 105 students in an introductory statistics class are listed in Table 1.16. The qualitative variable "class status" has been recorded for each student, and the frequencies have already been recorded.

TABLE 1.16

Status of Students in Statistics Class

Status	Freshman	Sophomore	Junior	Senior	Grad Student
Frequency	5	23	32	35	10

1. Enter the *categories* into column C1, with your own descriptive name, perhaps "Status" in the gray cell. Notice that the name **C1** has changed to **C1-T** because you are entering text rather than numbers. Continue by naming column 2 (C2) "Frequency," and enter the five numerical frequencies into C2.

2. To construct a pie chart for these data, click on **Graph ▶ Pie Chart,** and a Dialog box will appear (see Figure 1.23). Click the radio button marked **Chart values from a table.** Then place your cursor in the box marked "Categorical variable." Either (1) highlight C1 in the list at the left and choose **Select,** (2) double-click on C1 in the list at the left, or (3) type C1 in the "Categorical variable" box. Similarly, place the cursor in the box marked "Summary variables" and select C2. Click **Labels** and select the tab marked **Slice Labels.** Check the boxes marked "Category names" and "Percent." When you click **OK** twice, *MINITAB* will create the pie chart in Figure 1.24(a). We have removed the legend by selecting and deleting it.

FIGURE 1.23)

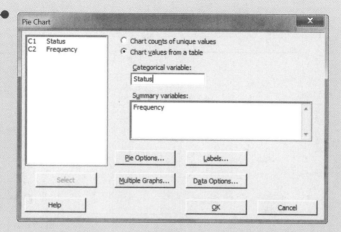

3. As you become more proficient at using the pie chart command, you may want to take advantage of some of the options available. Once the chart is created, *right-click* on the pie chart and select **Edit Pie.** You can change the colors and format of the chart, "explode" important sectors of the pie, and change the order of the categories. If you *right-click* on the pie chart and select **Update Graph Automatically,** the pie chart will automatically update when you change the data in columns C1 and C2 of the *MINITAB* worksheet.

4. If you would rather construct a bar chart, use the command **Graph ▶ Bar Chart.** In the Dialog box that appears, choose **Simple.** Choose an option in the "Bars represent" drop-down list, depending on the way that the data has been entered into the worksheet. For the data in Table 1.13, we choose "Values from a table" and click **OK.** When the Dialog box appears, place your cursor in the "Graph variables" box and select **C2.** Place your cursor in the "Categorical variable" box, and select **C1.** Click **OK** to finish the bar chart, shown in Figure 1.24(b). Once the chart is created, *right-click* on various parts of the bar chart and choose **Edit** to change the look of the chart.

FIGURE 1.24)

(a)

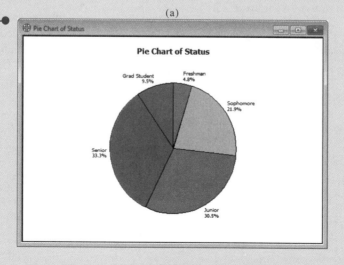

(b)

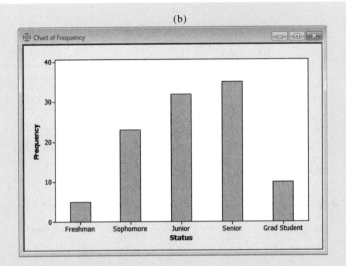

EXAMPLE | 1.16 | **(Line Charts)** The Dow Jones Industrial Average was monitored at the close of trading for 10 days in a recent year with the results shown in Table 1.17.

TABLE 1.17 ● **Dow Jones Industrial Average**

Day	1	2	3	4	5	6	7	8	9	10
DJIA	10,636	10,680	10,674	10,653	10,698	10,644	10,378	10,319	10,303	10,302

1. Although we could simply enter this data into third and fourth columns of the current worksheet, let's create a new worksheet using **File ▶ New ▶ Minitab Worksheet.** Enter the *Days* into column C1 of this second spreadsheet and the *DJIA* into column C2. You should have two columns of data, including the labels.

2. To create the line chart, use **Graph ▶ Time Series Plot ▶ Simple.** In the Dialog box that appears, place your cursor in the "Series" box and select "DJIA" from the list to the left. Under **Time/Scale,** choose "Stamp" and select column **C1** ("Day") in the box labeled "Stamp Columns." Click **OK** twice. You can select the numbered days shown above the line and delete them to obtain the line chart shown in Figure 1.25.

FIGURE 1.25 ●

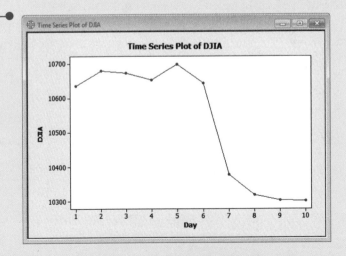

EXAMPLE 1.17

EXAMPLE  **(Dotplots, Stem and Leaf Plots, Histograms)** The top 40 stocks on the over-the-counter (OTC) market, ranked by percentage of outstanding shares traded on a particular day, are listed in Table 1.18. Create a new worksheet (**File ▶ New ▶ Minitab Worksheet**). Enter the data into column C1 and name it "Stocks" in the gray cell just below the C1.

TABLE 1.18 **Percentage of OTC Stocks Traded**

11.88	6.27	5.49	4.81	4.40	3.78	3.44	3.11	2.88	2.68
7.99	6.07	5.26	4.79	4.05	3.69	3.36	3.03	2.74	2.63
7.15	5.98	5.07	4.55	3.94	3.62	3.26	2.99	2.74	2.62
7.13	5.91	4.94	4.43	3.93	3.48	3.20	2.89	2.69	2.61

1. To create a dotplot, use **Graph ▶ Dotplot.** In the Dialog box that appears, choose **One Y ▶ Simple** and click **OK.** To create a stem and leaf plot, use **Graph ▶ Stem-and-Leaf.** For either graph, place your cursor in the "Graph variables" box, and select "Stocks" from the list to the left (see Figure 1.26).

FIGURE 1.26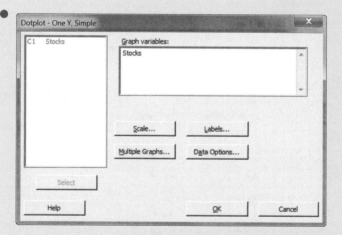

2. You can choose from a variety of formatting options before clicking **OK.** The dotplot appears as a graph, while the stem and leaf plot appears in the Session window. To print either a Graph window or the Session window, click on the window to make it active and use **File ▶ Print Graph** (or **Print Session Window**).

3. To create a histogram, use **Graph ▶ Histogram.** In the Dialog box that appears, choose **Simple** and click **OK,** selecting "Stocks" for the "Graph variables" box. Select **Scale ▶ Y-Scale Type** and click the radio button marked "Frequency." (You can edit the histogram later to show relative frequencies.) Click **OK** twice. Once the histogram has been created, *right-click* on the *Y*-axis and choose **Edit Y-Scale.** Under the tab marked "Scale," you can click the radio button marked "Position of ticks" and type in **0 5 10 15.** Then click the tab marked "Labels," the radio button marked "Specified" and type **0 5/40 10/40 15/40.** Click **OK.** This will reduce the number of ticks on the *y*-axis and change them to relative frequencies. Finally, double-click on the word "Frequency" along the *y*-axis. Change the box marked "Text" to read "Relative frequency" and click **OK.**

4. To adjust the type of boundaries for the histogram, *right-click* on the bars of the histogram and choose **Edit Bars.** Use the tab marked "Binning" to choose either "Cutpoints" or "Midpoints" for the histogram; you can specify the cutpoint or midpoint positions if you want. In this same **Edit** box, you can change the colors, fill type, and font style of the histogram. If you *right-click* on the bars and select **Update Graph Automatically,** the histogram will automatically update when you change the data in the "Stocks" column.

As you become more familiar with *MINITAB* for Windows, you can explore the various options available for each type of graph. It is possible to plot more than one variable at a time, to change the axes, to choose the colors, and to modify graphs in many ways. However, even with the basic default commands, it is clear that the distribution of OTC stocks in Figure 1.27 is highly skewed to the right.

FIGURE 1.27

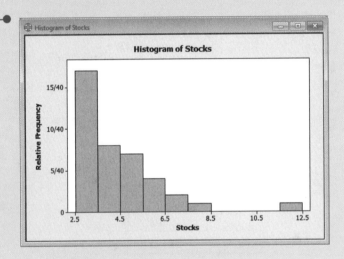

Supplementary Exercises

1.38 Quantitative or Qualitative? Identify each variable as quantitative or qualitative:

a. Ethnic origin of a candidate for public office

b. Score (0–100) on a placement examination

c. Fast-food establishment preferred by a student (McDonald's, Burger King, or Carl's Jr.)

d. Mercury concentration in a sample of tuna

1.39 Symmetric or Skewed? Do you expect the distributions of the following variables to be symmetric or skewed? Explain.

a. Size in dollars of nonsecured loans

b. Size in dollars of secured loans

c. Price of an 8-ounce can of peas

d. Height in inches of freshman women at your university

e. Number of broken taco shells in a package of 100 shells

f. Number of ticks found on each of 50 trapped cottontail rabbits

1.40 Continuous or Discrete, again Identify each variable as continuous or discrete:

a. Number of people in line at a supermarket checkout counter

b. Depth of a snowfall

c. Length of time for a driver to respond when faced with an impending collision

d. Number of aircraft arriving at the Atlanta airport in a given hour

1.41 Continuous or Discrete, again Identify each variable as continuous or discrete:

a. Weight of two dozen shrimp

b. A person's body temperature

c. Number of people waiting for treatment at a hospital emergency room

d. Number of properties for sale by a real estate agency

e. Number of claims received by an insurance company during one day

1.42 Continuous or Discrete? Identify each variable as continuous or discrete:

a. Number of homicides in Detroit during a 1-month period

b. Length of time between arrivals at an outpatient clinic

c. Number of typing errors on a page of manuscript

d. Number of defective lightbulbs in a package containing four bulbs

e. Time required to finish an examination

Data set **1.43 Aqua Running** Aqua running has
EX0143 been suggested as a method of cardiovascular conditioning for injured athletes and others who want a low-impact aerobics program. A study reported in the *Journal of Sports Medicine* investigated the relationship between exercise cadence and heart rate by measuring the heart rates of 20 healthy volunteers at a cadence of 96 steps per minute.[11] The data are listed here:

| 87 | 109 | 79 | 80 | 96 | 95 | 90 | 92 | 96 | 98 |
| 101 | 91 | 78 | 112 | 94 | 98 | 94 | 107 | 81 | 96 |

Construct a stem and leaf plot to describe the data. Discuss the characteristics of the data distribution.

Data set **1.44 Major World Lakes** A lake is a body of
EX0144 water surrounded by land. Hence, some bodies of water named "seas," like the Caspian Sea, are actual salt lakes. In the table that follows, the length in miles is listed for the major natural lakes of the world, excluding the Caspian Sea, which has a length of 760 miles.[5]

Name	Length (mi)	Name	Length (mi)
Superior	350	Titicaca	122
Victoria	250	Nicaragua	102
Huron	206	Athabasca	208
Michigan	307	Reindeer	143
Aral Sea	260	Tonle Sap	70
Tanganyika	420	Turkana	154

Baykal	395	Issyk Kul	115
Great Bear	192	Torrens	130
Nyasa	360	Vänern	91
Great Slave	298	Nettilling	67
Erie	241	Winnipegosis	141
Winnipeg	266	Albert	100
Ontario	193	Nipigon	72
Balkhash	376	Gairdner	90
Ladoga	124	Urmia	90
Maracaibo	133	Manitoba	140
Onega	145	Chad	175
Eyre	90		

Source: The World Almanac and Book of Facts 2011

a. Use a stem and leaf plot to describe the lengths of the world's major lakes.

b. Use a histogram to display these same data. How does this compare to the stem and leaf plot in part a?

c. Are these data symmetric or skewed? If skewed, what is the direction of the skewing?

Data set **1.45 Ages of Pennies** We collected
EX0145 50 pennies and recorded their ages, by calculating AGE = CURRENT YEAR − YEAR ON PENNY.

5	1	9	1	2	20	0	25	0	17
1	4	4	3	0	25	3	3	8	28
5	21	19	9	0	5	0	2	1	0
0	1	19	0	2	0	20	16	22	10
19	36	23	0	1	17	6	0	5	0

a. Before drawing any graphs, try to visualize what the distribution of penny ages will look like. Will it be mound-shaped, symmetric, skewed right, or skewed left?

b. Draw a relative frequency histogram to describe the distribution of penny ages. How would you describe the shape of the distribution?

Data set **1.46 Ages of Pennies, continued** The data
EX0146 below represent the ages of a different set of 50 pennies, again calculated using AGE = CURRENT YEAR − YEAR ON PENNY.

41	9	0	4	3	0	3	8	21	3
2	10	4	0	14	0	25	12	24	19
3	1	14	7	2	4	4	5	1	20
14	9	3	5	3	0	8	17	16	0
0	7	3	5	23	7	28	17	9	2

a. Draw a relative frequency histogram to describe the distribution of penny ages. Is the shape similar to the shape of the relative frequency histogram in Exercise 1.45?

b. Draw a stem and leaf plot to describe the penny ages. Are there any unusually large or small measurements in the set?

Data set **1.47 Presidential Vetoes** Here is a list of the

EX0147 44 presidents of the United States along with the number of regular vetoes used by each:[5]

Washington	2	B. Harrison	19
J. Adams	0	Cleveland	42
Jefferson	0	McKinley	6
Madison	5	T. Roosevelt	42
Monroe	1	Taft	30
J. Q. Adams	0	Wilson	33
Jackson	5	Harding	5
Van Buren	0	Coolidge	20
W. H. Harrison	0	Hoover	21
Tyler	6	F. D. Roosevelt	372
Polk	2	Truman	180
Taylor	0	Eisenhower	73
Fillmore	0	Kennedy	12
Pierce	9	L. Johnson	16
Buchanan	4	Nixon	26
Lincoln	2	Ford	48
A. Johnson	21	Carter	13
Grant	45	Reagan	39
Hayes	12	G. H. W. Bush	29
Garfield	0	Clinton	36
Arthur	4	G. W. Bush	11
Cleveland	304	Obama	1

Source: The World Almanac and Book of Facts 2011

Use an appropriate graph to describe the number of vetoes cast by the 44 presidents. Write a summary paragraph describing this set of data.

Data set **1.48 Windy Cities** Are some cities more

EX0148 windy than others? Does Chicago deserve to be nicknamed "The Windy City"? These data are the average wind speeds (in miles per hour) for 54 selected cities in the United States:[5]

8.9	12.3	10.7	8.4	7.8	11.5	8.2	9.0	8.8
7.1	11.8	10.3	7.7	9.0	10.5	9.1	8.7	8.7
9.1	9.0	10.5	11.2	7.7	8.8	12.2	7.9	8.8
8.7	7.1	8.7	7.6	5.1	35.1	10.5	10.4	11.0
10.2	8.6	10.7	9.6	8.3	8.0	9.5	7.7	9.4
8.7	7.8	10.2	6.9	9.2	10.2	6.2	9.6	12.2

Source: The World Almanac and Book of Facts 2011

a. Construct a relative frequency histogram for the data. (HINT: Choose the class boundaries without including the value $x = 35.1$ in the range of values.)

b. The value $x = 35.1$ was recorded at Mt. Washington, New Hampshire. Does the geography of that city explain the observation?

c. The average wind speed in Chicago is recorded as 10.3 miles per hour. Do you consider this unusually windy?

Data set **1.49 Kentucky Derby** The following data set

EX0149 shows the winning times (in seconds) for the Kentucky Derby races from 1950 to 2010.[12]

(1950)	121.3	122.3	121.3	122.0	123.0	121.4	123.2	122.1	125.0	122.1
(1960)	122.2	124.0	120.2	121.4	120.0	121.1	122.0	120.3	122.1	121.4
(1970)	123.2	123.1	121.4	119.2†	124.0	122.0	121.3	122.1	121.1	122.2
(1980)	122.0	122.0	122.2	122.1	122.2	120.1	122.4	123.2	122.2	125.0
(1990)	122.0	123.0	123.0	122.2	123.3	121.1	121.0	122.4	122.2	123.2
(2000)	121.0	119.97	121.13	121.19	124.06	122.75	121.36	122.17	121.86	122.66
(2010)	124.4									

†Record time set by Secretariat in 1973.
Source: www.kentuckyderby.com

a. Do you think there will be a trend in the winning times over the years? Draw a line chart to verify your answer.

b. Describe the distribution of winning times using an appropriate graph. Comment on the shape of the distribution and look for any unusual observations.

Data set **1.50 Gulf Oil Spill Cleanup** On April 20,

EX0150 2010, the United States experienced a major environmental disaster when a Deepwater Horizon drilling rig exploded in the Gulf of Mexico. The number of personnel and equipment used in the Gulf oil spill cleanup, beginning May 2, 2010 (Day 13) through June 9, 2010 (Day 51) is given in the following table.[13]

	Day 13	Day 26	Day 39	Day 51
Number of personnel (1000s)	3.0	17.5	20.0	24.0
Federal Gulf fishing areas closed	3%	8%	25%	32%
Booms laid (miles)	46	315	644	909
Dispersants used (1000 gallons)	156	500	870	1143
Vessels deployed (100s)	1.0	6.0	14.0	35.0

a. What graphical methods could you use to display these data?

b. Before you draw your graphs, what trends do you see in each of the variables displayed?

c. Use a line chart to display the number of personnel deployed over this 51-day period.

d. Use a bar graph to display the percentage of federal Gulf fishing areas closed.

e. Use a line chart to display the amounts of dispersants used. Is there any underlying straight line relationship over time?

Data set **1.51 Election Results** The 2008 election was

EX0151 a race in which Barack Obama defeated John McCain and other candidates, receiving 53% of the popular vote. The popular vote (in thousands) for Barack Obama in each of the 50 states is listed below:[14]

AL	813	HI	326	MA	1904	NM	472	SD	171
AK	124	ID	236	MI	2873	NY	4805	TN	1087
AZ	1035	IL	3420	MN	1573	NC	2143	TX	3529
AR	422	IN	1374	MS	555	ND	141	UT	328
CA	8274	IA	829	MO	1442	OH	2933	VT	219

CO	1289	KS	515	MT	232	OK	502	VA	1960
CT	998	KY	752	NE	333	OR	1037	WA	1751
DE	255	LA	783	NV	534	PA	3276	WV	304
FL	4282	ME	422	NH	385	RI	297	WI	1677
GA	1844	MD	1629	NJ	2215	SC	862	WY	83

a. By just looking at the table, what shape do you think the data distribution for the popular vote by state will have?

b. Draw a relative frequency histogram to describe the distribution of the popular vote for President Obama in the 50 states.

c. Did the histogram in part b confirm your guess in part a? Are there any outliers? How can you explain them?

1.52 Election Results, continued Refer to EX0152 Exercise 1.51. Listed here is the *percentage* of the popular vote received by President Obama in each of the 50 states:[14]

AL	39	HI	72	MA	62	NM	57	SD	45
AK	38	ID	36	MI	57	NY	63	TN	42
AZ	45	IL	62	MN	54	NC	50	TX	44
AR	39	IN	50	MS	43	ND	45	UT	34
CA	61	IA	54	MO	49	OH	52	VT	68
CO	54	KS	42	MT	47	OK	34	VA	53
CT	61	KY	41	NE	42	OR	57	WA	58
DE	62	LA	40	NV	55	PA	55	WV	43
FL	51	ME	58	NH	54	RI	63	WI	56
GA	47	MD	62	NJ	57	SC	45	WY	33

a. By just looking at the table, what shape do you think the data distribution for the *percentage* of the popular vote by state will have?

b. Draw a relative frequency histogram to describe the distribution. Describe the shape of the distribution and look for outliers. Did the graph confirm your answer to part a?

1.53 Election Results, continued Refer to Exercises 1.51 and 1.52. The accompanying stem and leaf plots were generated using *MINITAB* for the variables named "Popular Vote" and "Percent Vote."

Stem and Leaf Display: Popular Vote

```
Stem-and-leaf of
Popular Vote N = 50
Leaf Unit = 100
17    0 01112222233333444
(10)  0 5555778889
23    1 000234
17    1 5667899
10    2 12
8     2 89
6     3 24
4     3 5

HI  42, 48, 82
```

Stem and Leaf Display: Percent Vote

```
Stem-and-leaf of
Percent Vote N = 50
Leaf Unit = 1.0
3     3 344
7     3 6899
15    4 01222334
22    4 5555779
(9)   5 001234444
19    5 556777788
10    6 11222233
2     6 8
1     7 2
```

a. Describe the shapes of the two distributions. Are there any outliers?

b. Do the stem and leaf plots resemble the relative frequency histograms constructed in Exercises 1.51 and 1.52?

c. Explain why the distribution of the popular vote for President Obama by state is skewed while the percentage of popular votes by state is mound-shaped.

1.54 Student Heights The self-reported EX0154 heights of 105 students in a biostatistics class are described in the relative frequency histogram below.

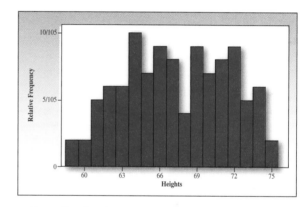

a. Describe the shape of the distribution.

b. Do you see any unusual feature in this histogram?

c. Can you think of an explanation for the two peaks in the histogram? Is there some other factor that is causing the heights to mound up in two separate peaks? What is it?

1.55 Diamonds are Forever! Much of the world's diamond industry is located in Africa, with Russia and Canada also showing large revenues from their diamond mining industry. A visual representation of the various shares of the world's diamond revenues, adapted from *Time Magazine*,[15] is shown below:

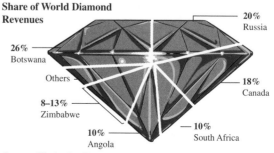

Share of World Diamond Revenues

Source: Kimberley Process

a. Draw a pie chart to describe the various shares of the world's diamond revenues.

b. Draw a bar chart to describe the various shares of the world's diamond revenues.

c. Draw a Pareto chart to describe the various shares of the world's diamond revenues.

d. Which of the charts is the most effective in describing the data?

Data set **1.56 Pulse Rates** A group of 50 biomedical
EX0156 students recorded their pulse rates by counting the number of beats for 30 seconds and multiplying by 2.

```
80  70  88  70  84  66  84   82   66   42
52  72  90  70  96  84  96   86   62   78
60  82  88  54  66  66  80   88   56  104
84  84  60  84  88  58  72   84   68   74
84  72  62  90  72  84  72  110  100   58
```

a. Why are all of the measurements even numbers?

b. Draw a stem and leaf plot to describe the data, splitting each stem into two lines.

c. Construct a relative frequency histogram for the data.

d. Write a short paragraph describing the distribution of the student pulse rates.

Data set **1.57 Starbucks** Students at the University of
EX0157 California, Riverside (UCR), along with many other Californians love their Starbucks! The distances in miles from campus for the 41 Starbucks stores within 10 miles of UCR are shown below:[16]

```
0.4  0.5  4.3  5.2  9.3  4.2  8.3  9.4  7.3  7.1  8.7
6.1  3.7  8.0  4.8  3.9  5.3  9.8  1.2  6.4  9.8
8.0  7.8  9.4  8.4  5.9  9.8  7.3  2.5  9.6  0.7
6.7  7.6  9.7  3.5  9.5  8.8  9.0  7.9  5.2  6.8
```

Construct a relative frequency to describe the distances from the UCR campus, using 10 classes of width 1, starting at 0.0.

a. What is the shape of the histogram? Do you see any unusual features?

b. Can you explain why the histogram looks the way it does?

Data set **1.58 Stressful Times** In the spring of 2010,
EX0158 almost all of the 50 U.S. states plus the District of Columbia were facing drastic financial crises, with many planning across-the-board budget cuts, layoffs, higher education fees, and other strategies to cut next year's expected budget gap. The image that follows shows the expected shortfall next year as a percentage of this year's budget for each of the 50 U.S. states and the District of Columbia.[17]

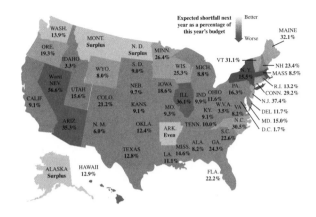

a. Construct a relative frequency histogram to describe the percentages for the 48 states that are expecting to face shortfalls next year.

b. What is the shape of the histogram? Do you see any unusual features in the histogram? If there are any outliers, can you explain them?

c. There are three states, Alaska, Montana, and North Dakota, that are expecting a surplus next year. Can you think of a reason why this might be?

Data set **1.59 An Archeological Find** An article in
EX0159 *Archaeometry* involved an analysis of 26 samples of Romano-British pottery, found at four different kiln sites in the United Kingdom.[18] The samples were analyzed to determine their chemical composition, and the percentage of aluminum oxide in each of the 26 samples is shown in the following table.

Llanederyn		Caldicot	Island Thorns	Ashley Rails
14.4	11.6	11.8	18.3	17.7
13.8	11.1	11.6	15.8	18.3
14.6	13.4		18.0	16.7
11.5	12.4		18.0	14.8
13.8	13.1		20.8	19.1
10.9	12.7			
10.1	12.5			

a. Construct a relative frequency histogram to describe the aluminum oxide content in the 26 pottery samples.

b. What unusual feature do you see in this graph? Can you think of an explanation for this feature?

c. Draw a dotplot for the data, using a letter (L, C, I, or A) to locate the data point on the horizontal scale. Does this help explain the unusual feature in part b?

1.60 The Great Calorie Debate Want to lose weight? You can do it by cutting calories, as long as you get enough nutritional value from the foods that you do eat! Below you will see a visual representation of the number of calories in some of America's favorite foods adapted from an article in *The Press-Enterprise*.[19]

Number of Calories

26	53	140	145	330	800
Hershey's Kiss	Oreo cookie	12-ounce can of Coke	12-ounce bottle of Budweiser beer	Slice of a large Papa John's pepperoni pizza	Burger King Whopper with cheese

a. Comment on the accuracy of the graph shown above. Do the sizes, heights, and volumes of the six items accurately represent the number of calories in the item?

b. Draw an actual bar chart to describe the number of calories in these six food favorites.

1.61 Laptops and Learning An informal experiment was conducted at McNair Academic High School in Jersey City, New Jersey, to investigate the use of laptop computers as a learning tool in the study of algebra.[20] A freshman class of 20 students was given laptops to use at school and at home, while another freshman class of 27 students was not given laptops; however, many of these students were able to use computers at home. The final exam scores for the two classes are shown below.

Laptops		No Laptops		
98	84	63	83	97
97	93	93	52	74
88	57	83	63	88
100	84	86	81	84
100	81	99	91	49
78	83	80	81	89
68	84	78	29	64
47	93	74	72	89
90	57	67	89	70
94	83			

The histograms that follow show the distribution of final exam scores for the two groups.

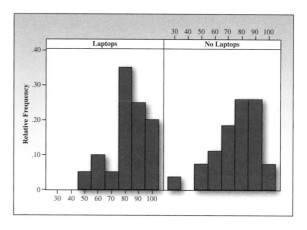

Write a summary paragraph describing and comparing the distribution of final exam scores for the two groups of students.

1.62 Old Faithful The data below are the waiting times between eruptions of the Old Faithful geyser in Yellowstone National Park.[21] Use one of the graphical methods from this chapter to describe the distribution of waiting times. If there are any unusual features in your graph, see if you can think of any practical explanation for them.

56	89	51	79	58	82	52	88	52	78
69	75	77	53	80	54	79	74	65	78
55	87	53	85	61	93	54	76	80	81
59	86	78	71	77	89	45	93	72	71
76	94	75	50	83	82	72	77	75	65
79	72	78	77	79	72	82	74	80	49
75	78	64	80	49	49	88	51	78	85
65	75	77	69	92	91	53	86	49	79
68	87	61	81	55	93	53	84	70	73
93	50	87	77	74	89	87	76	59	80

1.63 Gasoline Tax The following are the 2010 state gasoline tax in cents per gallon for the 50 U.S. states and the District of Columbia.[5]

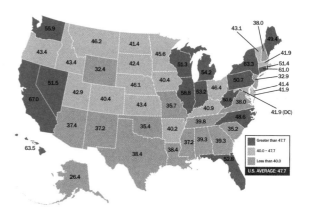

AK	26.4	HI	63.5	MA	41.9	NM	37.2	SD	42.4
AL	39.3	ID	43.4	MI	54.2	NY	63.3	TN	39.8
AR	40.2	IL	58.8	MN	45.6	NC	48.6	TX	38.4
AZ	37.4	IN	53.2	MS	37.2	ND	41.4	UT	42.9
CA	67.0	IA	40.4	MO	35.7	OH	46.4	VT	43.1
CO	40.4	KS	43.4	MT	46.2	OK	35.4	VA	38.0
CT	61.0	KY	40.9	NE	46.1	OR	43.4	WA	55.9
DE	41.4	LA	38.4	NV	51.5	PA	50.7	WV	50.6
DC	41.9	ME	49.4	NH	38.0	RI	51.4	WI	51.3
FL	52.8	MD	41.9	NJ	32.9	SC	35.2	WY	32.4
GA	39.3								

Source: http://www.api.org/statistics/fueltaxes/upload/GASOLINE_ TAX_MAP_ APRIL2010.pdf, July 6, 2010

a. Construct a stem and leaf display for the data.

b. How would you describe the shape of this distribution?

c. Are there states with unusually high or low gasoline taxes? If so, which states are they?

Data set EX0164 **1.64 Hydroelectric Plants** The following data represent the planned rated capacities in megawatts (millions of watts) for the world's 20 largest hydroelectric plants.[5]

18,200	4,500	3,000
12,600	4,200	2,940
10,000	4,200	2,715
8,370	3,840	2,700
6,400	3,444	2,541
6,300	3,300	2,512
6,000	3,100	

Source: The World Almanac and Book of Facts, 2011

a. Construct a stem and leaf display for the data.

b. How would you describe the shape of this distribution?

Data set EX0165 **1.65 Car Colors** The most popular colors for compact and sports cars in a recent year are given in the table.[5]

Color	Percentage	Color	Percentage
Silver	19	White/white pearl	12
Black/black effect	17	Beige/brown	3
Gray	17	Yellow/gold	2
Blue	15	Green	2
Red	12	Other	1

Source: The World Almanac and Book of Facts 2011

Use an appropriate graphical display to describe these data.

Data set EX0166 **1.66 Starbucks** The number of Starbucks coffee shops in cities within 20 miles of the University of California, Riverside is shown in the following table.[16]

City	Starbucks	City	Starbucks
Riverside	18	Ontario	12
Grand Terrace	1	Norco	4
Rialto	6	Fontana	7
Colton	2	Mira Loma	2
San Bernardino	6	Perris	3
Redlands	8	Highland	1
Corona	10	Rancho Cucamonga	10
Yucaipa	3	Lake Elsinore	2
Chino	11	Moreno Valley	5
Upland	2	Montclair	1
Bloomington	1		

a. Draw a dotplot to describe the data.

b. Describe the shape of the distribution.

c. Is there another variable that you could measure that might help to explain why some cities have more Starbucks than others? Explain.

Data set EX0167 **1.67 What's Normal?** The 98.6 degree standard for human body temperature was derived by a German doctor in 1868. In an attempt to verify his claim, Mackowiak, Wasserman, and Levine[22] took temperatures from 148 healthy people over a 3-day period. A data set closely matching the one in Mackowiak's article was derived by Allen Shoemaker, and appears in the *Journal of Statistics Education*.[23] The body temperatures for these 130 individuals are shown in the relative frequency histogram that follows.

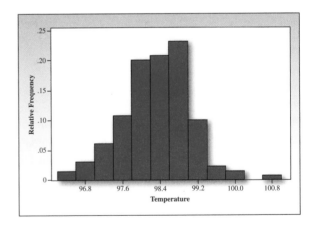

a. Describe the shape of the distribution of temperatures.

b. Are there any unusual observations? Can you think of any explanation for these?

c. Locate the 98.6-degree standard on the horizontal axis of the graph. Does it appear to be near the center of the distribution?

CASE STUDY **How Is Your Blood Pressure?**

 Blood
Pressure

Blood pressure is the pressure that the blood exerts against the walls of the arteries. When physicians or nurses measure your blood pressure, they take two readings. The systolic blood pressure is the pressure when the heart is contracting and therefore pumping. The diastolic blood pressure is the pressure in the arteries when the heart is relaxing. The diastolic blood pressure is always the lower of the two readings. Blood pressure varies from one person to another. It will also vary for a single individual from day to day and even within a given day.

If your blood pressure is too high, it can lead to a stroke or a heart attack. If it is too low, blood will not get to your extremities and you may feel dizzy. Low blood pressure is usually not serious.

So, what should *your* blood pressure be? A systolic blood pressure of 120 would be considered normal. One of 150 would be high. But since blood pressure varies with gender and increases with age, a better gauge of the relative standing of your blood pressure would be obtained by comparing it with the population of blood pressures of all persons of your gender and age in the United States. Of course, we cannot supply you with that data set, but we can show you a very large sample selected from it. The blood pressure data on 1910 persons, 965 men and 945 women between the ages of 15 and 20, are found at the CourseMate Web site. The data are part of a health survey conducted by the National Institutes of Health (NIH). Entries for each person include that person's age and systolic and diastolic blood pressures at the time the blood pressure was recorded.

1. Describe the variables that have been measured in this survey. Are the variables quantitative or qualitative? Discrete or continuous? Are the data univariate, bivariate, or multivariate?

2. What types of graphical methods are available for describing this data set? What types of questions could be answered using various types of graphical techniques?

3. Using the systolic blood pressure data set, construct a relative frequency histogram for the 965 men and another for the 945 women. Use a statistical software package if you have access to one. Compare the two histograms.

4. Consider the 965 men and 945 women as the entire population of interest. Choose a sample of $n = 50$ men and $n = 50$ women, recording their systolic blood pressures and their ages. Draw two relative frequency histograms to graphically display the systolic blood pressures for your two samples. Do the shapes of the histograms resemble the population histograms from part 3?

5. How does your blood pressure compare with that of others of your same gender? Check your systolic blood pressure against the appropriate histogram in part 3 or 4 to determine whether your blood pressure is "normal" or whether it is unusually high or low.

Describing Data
with Numerical
Measures

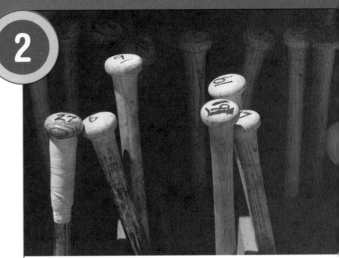

© Joe Sohm-VisionsofAmerica/Photodisc/Getty

 NEED TO KNOW...

How to Calculate Sample Quartiles

The Boys of Summer

Are the baseball champions of today better than those of "yesteryear"? Do players in the National League hit better than players in the American League? The case study at the end of this chapter involves the batting averages of major league batting champions. Numerical descriptive measures can be used to answer these and similar questions.

DESCRIBING A SET OF DATA WITH NUMERICAL MEASURES

2.1

Graphs can help you describe the basic shape of a data distribution; "a picture is worth a thousand words." There are limitations, however, to the use of graphs. Suppose you need to display your data to a group of people and the bulb on the data projector blows out! Or you might need to describe your data over the telephone—no way to display the graphs! You need to find another way to convey a mental picture of the data to your audience.

A second limitation is that graphs are somewhat imprecise for use in statistical inference. For example, suppose you want to use a sample histogram to make inferences about a population histogram. How can you measure the similarities and differences between the two histograms in some concrete way? If they were identical, you could say "They are the same!" But, if they are different, it is difficult to describe the "degree of difference."

One way to overcome these problems is to use **numerical measures,** which can be calculated for either a sample or a population of measurements. You can use the data to calculate a set of *numbers* that will convey a good mental picture of the frequency distribution. These measures are called **parameters** when associated with the population, and they are called **statistics** when calculated from sample measurements.

Definition   Numerical descriptive measures associated with a population of measurements are called **parameters;** those computed from sample measurements are called **statistics.**

MEASURES OF CENTER

2.2

In Chapter 1, we introduced dotplots, stem and leaf plots, and histograms to describe the distribution of a set of measurements on a quantitative variable x. The horizontal axis displays the values of x, and the data are "distributed" along this horizontal line. One of the first important numerical measures is a **measure of center**—a measure along the horizontal axis that locates the center of the distribution.

The birth weight data presented in Table 1.9 ranged from a low of 5.6 to a high of 9.4, with the center of the histogram located in the vicinity of 7.5 (see Figure 2.1). Let's consider some rules for locating the center of a distribution of measurements.

FIGURE 2.1

Center of the birth weight data

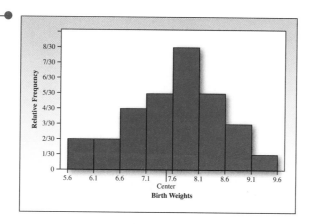

The arithmetic average of a set of measurements is a very common and useful measure of center. This measure is often referred to as the **arithmetic mean,** or simply the **mean,** of a set of measurements. To distinguish between the mean for the sample and the mean for the population, we will use the symbol $\bar{x}$ (x-bar) for a sample mean and the symbol μ (Greek lowercase mu) for the mean of a population.

Definition   The **arithmetic mean** or **average** of a set of n measurements is equal to the sum of the measurements divided by n.

Since statistical formulas often involve adding or "summing" numbers, we use a shorthand symbol to indicate the process of summing. Suppose there are n measurements on the variable x—call them $x_1, x_2, \ldots, x_n$. To add the n measurements together, we use this shorthand notation:

$$\sum_{i=1}^{n} x_i \quad \text{which means } x_1 + x_2 + x_3 + \cdots + x_n$$

The Greek capital sigma (Σ) tells you to add the items that appear to its right, beginning with the number below the sigma ($i = 1$) and ending with the number above ($i = n$). However, since the typical sums in statistical calculations are almost always made on the total set of n measurements, you can use a simpler notation:

Σx_i which means "the sum of all the x measurements"

Using this notation, we write the formula for the sample mean:

NOTATION

Sample mean: $\bar{x} = \dfrac{\Sigma x_i}{n}$

Population mean: μ

EXAMPLE   2.1

Draw a dotplot for the $n = 5$ measurements 2, 9, 11, 5, 6. Find the sample mean and compare its value with what you might consider the "center" of these observations on the dotplot.

Solution   The dotplot in Figure 2.2 seems to be centered between 6 and 8. To find the sample mean, calculate

$$\bar{x} = \frac{\Sigma x_i}{n} = \frac{2 + 9 + 11 + 5 + 6}{5} = 6.6$$

FIGURE 2.2

Dotplot for Example 2.1

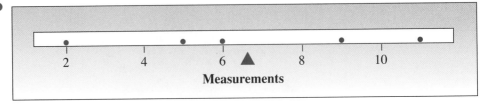

The statistic $\bar{x} = 6.6$ is the balancing point or fulcrum shown on the dotplot. It does seem to mark the center of the data.

NEED A TIP?

mean = balancing point or fulcrum

Remember that samples are measurements drawn from a larger population that is usually unknown. An important use of the sample mean $\bar{x}$ is as an estimator of the unknown population mean μ. The birth weight data in Table 1.9 are a sample from a larger population of birth weights, and the distribution is shown in Figure 2.1. The mean of the 30 birth weights is

$$\bar{x} = \frac{\Sigma x_i}{30} = \frac{227.2}{30} = 7.57$$

shown in Figure 2.1; it marks the balancing point of the distribution. The mean of the entire population of newborn birth weights is unknown, but if you had to guess its value, your best estimate would be 7.57. Although the sample mean $\bar{x}$ changes from sample to sample, the population mean μ stays the same.

A second measure of central tendency is the **median,** which is the value in the middle position in the set of measurements ordered from smallest to largest.

Definition   The **median** m of a set of n measurements is the value of x that falls in the middle position when the measurements are ordered from smallest to largest.

EXAMPLE (2.2)   Find the median for the set of measurements 2, 9, 11, 5, 6.

Solution   Rank the $n = 5$ measurements from smallest to largest:

$$2 \quad 5 \quad 6 \quad 9 \quad 11$$
$$\uparrow$$

The middle observation, marked with an arrow, is in the center of the set, or $m = 6$.

EXAMPLE (2.3)   Find the median for the set of measurements 2, 9, 11, 5, 6, 27.

Solution   Rank the measurements from smallest to largest:

$$2 \quad 5 \quad \boxed{6 \quad 9} \quad 11 \quad 27$$
$$\uparrow$$

NEED A TIP?

Roughly 50% of the measurements are smaller, 50% are larger than the median.

Now there are two "middle" observations, shown in the box. To find the median, choose a value halfway between the two middle observations:

$$m = \frac{6 + 9}{2} = 7.5$$

The value **.5(n + 1)** indicates the **position of the median** in the ordered data set. If the position of the median is a number that ends in the value **.5,** you need to average the two adjacent values.

EXAMPLE (2.4)   For the $n = 5$ ordered measurements from Example 2.2, the position of the median is .5(n + 1) = .5(6) = 3, and the median is the *3rd ordered observation,* or $m = 6$. For the $n = 6$ ordered measurements from Example 2.3, the position of the median is .5(n + 1) = .5(7) = 3.5, and the median is the *average of the 3rd and 4th ordered observations,* or $m = (6 + 9)/2 = 7.5$.

Although both the mean and the median are good measures of the center of a distribution, the median is less sensitive to extreme values or *outliers*. For example, the value $x = 27$ in Example 2.3 is much larger than the other five measurements. The median, $m = 7.5$, is not affected by the outlier, whereas the sample average,

$$\bar{x} = \frac{\Sigma x_i}{n} = \frac{60}{6} = 10$$

is affected; its value is not representative of the remaining five observations.

When a data set has extremely small or extremely large observations, the sample mean is drawn toward the direction of the extreme measurements (see Figure 2.3).

FIGURE 2.3

Relative frequency distributions showing the effect of extreme values on the mean and median

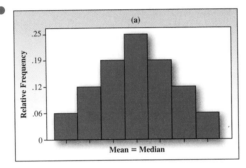
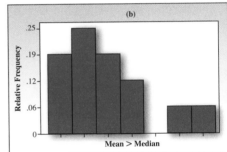

ONLINE APPLET

How Extreme Values Affect the Mean and Median

If a distribution is skewed to the right, the mean shifts to the right; if a distribution is skewed to the left, the mean shifts to the left. The median is not affected by these extreme values because the numerical values of the measurements are not used in its calculation. When a distribution is symmetric, the mean and the median are equal. If a distribution is strongly skewed by one or more extreme values, you should use the median rather than the mean as a measure of center.

Another way to locate the center of a distribution is to look for the value of x that occurs with the highest frequency. This measure of the center is called the **mode.**

Definition   The **mode** is the category that occurs most frequently, or the most frequently occurring value of x. When measurements on a continuous variable have been grouped as a frequency or relative frequency histogram, the class with the highest peak or frequency is called the **modal class,** and the midpoint of that class is taken to be the mode.

The mode is generally used to describe large data sets, whereas the mean and median are used for both large and small data sets. From the data in Example 1.11, reproduced in Table 2.1(a), the mode of the distribution of the number of reported weekly visits to Starbucks for 30 Starbucks customers is 5. The modal class and the value of x occurring with the highest frequency are the same, as shown in Figure 2.4(a).

For the birth weight data in Table 2.1(b), a birth weight of 7.7 occurs four times, and therefore the mode for the distribution of birth weights is 7.7. Using the histogram to find the modal class, you find that the class with the highest peak is the fifth class, from 7.6 to 8.1. Our choice for the mode would be the midpoint of this class, or 7.85. See Figure 2.4(b).

It is possible for a distribution of measurements to have more than one mode. These modes would appear as "local peaks" in the relative frequency distribution. For example, if we were to tabulate the length of fish taken from a lake during one season, we might get a *bimodal distribution,* possibly reflecting a mixture of young and old fish in the population. Sometimes bimodal distributions of sizes or weights reflect a mixture of measurements taken on males and females. In any case, a set or distribution of measurements may have more than one mode.

TABLE 2.1

Starbucks and birth weight data

(a) Starbucks data					(b) Birth weight data				
6	7	1	5	6	7.2	7.8	6.8	6.2	8.2
4	6	4	6	8	8.0	8.2	5.6	8.6	7.1
6	5	6	3	4	8.2	7.7	7.5	7.2	7.7
5	5	5	7	6	5.8	6.8	6.8	8.5	7.5
3	5	7	5	5	6.1	7.9	9.4	9.0	7.8
					8.5	9.0	7.7	6.7	7.7

FIGURE 2.4

Relative frequency histograms for the Starbucks and birth weight data

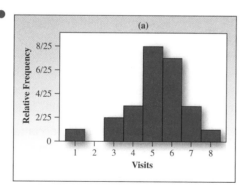
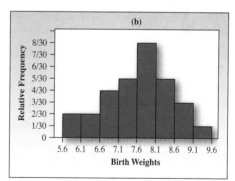

2.2 EXERCISES

BASIC TECHNIQUES

2.1 You are given $n = 5$ measurements: 0, 5, 1, 1, 3.

a. Draw a dotplot for the data. (HINT: If two measurements are the same, place one dot above the other.) Guess the approximate "center."

b. Find the mean, median, and mode.

c. Locate the three measures of center on the dotplot in part a. Based on the relative positions of the mean and median, are the measurements symmetric or skewed?

2.2 You are given $n = 8$ measurements: 3, 2, 5, 6, 4, 4, 3, 5.

a. Find $\bar{x}$.

b. Find m.

c. Based on the results of parts a and b, are the measurements symmetric or skewed? Draw a dotplot to confirm your answer.

2.3 You are given $n = 10$ measurements: 3, 5, 4, 6, 10, 5, 6, 9, 2, 8.

a. Calculate $\bar{x}$.

b. Find m.

c. Find the mode.

APPLICATIONS

2.4 Auto Insurance The cost of automobile insurance has become a sore subject in California because insurance rates are dependent on so many different variables, such as the city in which you live, the number of cars you insure, and the company with which you are insured. The website www.insurance.ca.gov reports the annual 2010 premium for a male, licensed for 6–8 years, who drives a Honda Accord 12,600–15,000 miles per year and has no violations or accidents.[1]

City	GEICO ($)	21st Century ($)
Long Beach	2780	2352
Pomona	2411	2462
San Bernardino	2261	2284
Moreno Valley	2263	2520

Source: www.insurance.ca.gov

a. What is the average premium for GEICO Insurance?

b. What is the average premium for 21st Century Insurance?

c. If you were a consumer, would you be interested in the average premium cost? If not, what would you be interested in?

2.5 DVRs The digital video recorder (DVR) is a common fixture in most American households. In fact, most American households have DVRs, and many have more than one. A sample of 25 households produced the following measurements on x, the number of DVRs in the household:

1	0	2	1	1
1	0	2	1	0
0	1	2	3	2
1	1	1	0	1
3	1	0	1	1

a. Is the distribution of x, the number of DVRs in a household, symmetric or skewed? Explain.

b. Guess the value of the mode, the value of x that occurs most frequently.

c. Calculate the mean, median, and mode for these measurements.

d. Draw a relative frequency histogram for the data set. Locate the mean, median, and mode along the horizontal axis. Are your answers to parts a and b correct?

2.6 Fortune 500 Revenues Ten of the 50 largest businesses in the United States, randomly selected from the *Fortune 500,* are listed below along with their revenues (in millions of dollars):[2]

Company	Revenues ($)	Company	Revenues ($)
General Motors	104,589	Target	65,357
IBM	95,758	Morgan Stanley	31,515
Bank of America	150,450	Johnson & Johnson	61,867
Home Depot	66,176	Apple	36,537
Boeing	68,281	Exxon Mobil	284,650

a. Draw a stem and leaf plot for the data. Are the data skewed?

b. Calculate the mean revenue for these 10 businesses. Calculate the median revenue.

c. Which of the two measures in part b best describes the center of the data? Explain.

2.7 Birth Order and Personality Does birth order have any effect on a person's personality? A report on a study by an MIT researcher indicates that later-born children are more likely to challenge the establishment, more open to new ideas, and more accepting of change.[3] In fact, the number of later-born children is increasing. During the Depression years of the 1930s, families averaged 2.5 children (59% later born), whereas the parents of baby boomers averaged 3 to 4 children (68% later born). What does the author mean by an average of 2.5 children?

2.8 Tuna Fish An article in *Consumer Reports* gives the price—an estimated average for a 6-ounce can or a 7.06-ounce pouch—for 14 different brands of water-packed light tuna, based on prices paid nationally in supermarkets:[4]

.99	1.92	1.23	.85	.65	.53	1.41
1.12	.63	.67	.69	.60	.60	.66

a. Find the average price for the 14 different brands of tuna.

b. Find the median price for the 14 different brands of tuna.

c. Based on your findings in parts a and b, do you think that the distribution of prices is skewed? Explain.

2.9 Sports Salaries As professional sports teams become a more and more lucrative business for their owners, the salaries paid to the players have also increased. In fact, sports superstars are paid astronomical salaries for their talents. If you were asked by a sports management firm to describe the distribution of players' salaries in several different categories of professional sports, what measure of center would you choose? Why?

2.10 Time on Task In a psychological experiment, the time on task was recorded for 10 subjects under a 5-minute time constraint. These measurements are in seconds:

175	190	250	230	240
200	185	190	225	265

a. Find the average time on task.

b. Find the median time on task.

c. If you were writing a report to describe these data, which measure of central tendency would you use? Explain.

2.11 Starbucks The number of Starbucks coffee shops in 21 cities within 20 miles of the University of California, Riverside is shown in the following table.[5]

18	1	6	2	6	8
10	3	11	2	1	12
4	7	2	3	1	10
2	5	1			

a. Find the mean, the median, and the mode.

b. Compare the median and the mean. What can you say about the shape of this distribution?

c. Draw a dotplot for the data. Does this confirm your conclusion about the shape of the distribution from part b?

2.12 Nintendo's Wii The "Wii" is an interactive gaming console popular among many gamers. Its cost can vary dramatically, depending on where it is purchased. The website www.pricegrabber.com lists 14 online sellers with various prices, including shipping and taxes:[6]

Seller	Price ($)	Seller	Price ($)
Buy.com	216.49	Dell	184.86
Sears	222.84	Kmart	222.84
Sam's Club	180.17	EagleDirectUSA	231.04
USA Sales	279.90	Wii4family	262.95
PalaceToys	280.98	QuickShip USA	299.48
Simbaoo7	289.97	BUY-IT-NOW	384.99
jandk425	433.00	SW Evolution	1024.24

a. What is the average price of the Wii for these 14 sellers?

b. What is the median price of the Wii for these 14 sellers?

c. As a consumer, would you be interested in the average price of the Wii? What other variables would be important to you?

2.3 MEASURES OF VARIABILITY

Data sets may have the same center but look different because of the way the numbers *spread out* from the center. Consider the two distributions shown in Figure 2.5. Both distributions are centered at $x = 4$, but there is a big difference in the way the measurements spread out, or *vary*. The measurements in Figure 2.5(a) vary from 3 to 5; in Figure 2.5(b) the measurements vary from 0 to 8.

FIGURE 2.5

Variability or dispersion
of data

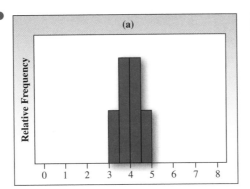

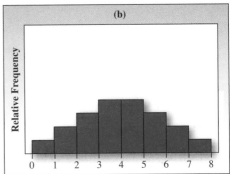

Variability or **dispersion** is a very important characteristic of data. For example, if you were manufacturing bolts, extreme variation in the bolt diameters would cause a high percentage of defective products. On the other hand, if you were trying to discriminate between good and poor accountants, you would have trouble if the examination always produced test grades with little variation, making discrimination very difficult.

Measures of variability can help you create a mental picture of the spread of the data. We will present some of the more important ones. The simplest measure of variation is the **range.**

Definition   The **range, R,** of a set of n measurements is defined as the difference between the largest and smallest measurements.

For example, the measurements 5, 7, 1, 2, 4 vary from 1 to 7. Hence, the range is $7 - 1 = 6$. The range is easy to calculate, easy to interpret, and is an adequate measure of variation for small sets of data. But, for large data sets, the range is not an adequate measure of variability. For example, the two relative frequency distributions in Figure 2.6 have the same range but very different shapes and variability.

FIGURE 2.6

Distributions with equal
range and unequal
variability

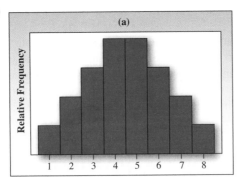

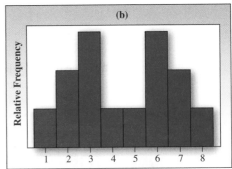

Is there a measure of variability that is more sensitive than the range? Consider again the sample measurements 5, 7, 1, 2, 4, displayed as a dotplot in Figure 2.7. The mean of these five measurements is

$$\bar{x} = \frac{\Sigma x_i}{n} = \frac{19}{5} = 3.8$$

FIGURE 2.7

Dotplot showing the deviations of points from the mean

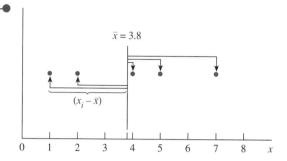

as indicated on the dotplot. The horizontal distances between each dot (measurement) and the mean $\bar{x}$ will help you to measure the variability. If the distances are large, the data are more spread out or more *variable* than if the distances are small. If x_i is a particular dot (measurement), then the **deviation** of that measurement from the mean is $(x_i - \bar{x})$. Measurements to the right of the mean produce positive deviations, and those to the left produce negative deviations. The values of x and the deviations for our example are listed in the first and second columns of Table 2.2.

TABLE 2.2

Computation of $\Sigma(x_i - \bar{x})^2$

x_i	$(x_i - \bar{x})$	$(x_i - \bar{x})^2$
5	1.2	1.44
7	3.2	10.24
1	−2.8	7.84
2	−1.8	3.24
4	.2	.04
19	0.0	22.80

Because the deviations in the second column of the table contain information on variability, one way to combine the five deviations into one numerical measure is to average them. Unfortunately, the average will not work because some of the deviations are positive, some are negative, and the sum is always zero (unless round-off errors have been introduced into the calculations). Note that the deviations in the second column of Table 2.2 sum to zero.

Another possibility might be to disregard the signs of the deviations and calculate the average of their absolute values.[†] This method has been used as a measure of variability in exploratory data analysis and in the analysis of time series data. We prefer, however, to overcome the difficulty caused by the signs of the deviations by working with their sum of squares. From the sum of squared deviations, a single measure called the **variance** is calculated. To distinguish between the variance of a

---

[†]The absolute value of a number is its magnitude, ignoring its sign. For example, the absolute value of −2, represented by the symbol $|-2|$, is 2. The absolute value of 2—that is, $|2|$—is 2.

sample and the variance of a *population,* we use the symbol s^2 for a sample variance and σ^2 (Greek lowercase sigma) for a population variance. *The variance will be relatively large for highly variable data and relatively small for less variable data.*

Definition The **variance of a population** of N measurements is the average of the squares of the deviations of the measurements about their mean μ. The population variance is denoted by σ^2 and is given by the formula

$$\sigma^2 = \frac{\Sigma(x_i - \mu)^2}{N}$$

Most often, you will not have all the population measurements available but will need to calculate the *variance of a sample* of n measurements.

Definition The **variance of a sample** of n measurements is the sum of the squared deviations of the measurements about their mean $\bar{x}$ divided by $(n - 1)$. The sample variance is denoted by s^2 and is given by the formula

$$s^2 = \frac{\Sigma(x_i - \bar{x})^2}{n - 1}$$

For the set of $n = 5$ sample measurements presented in Table 2.2, the square of the deviation of each measurement is recorded in the third column. Adding, we obtain

$$\Sigma(x_i - \bar{x})^2 = 22.80$$

and the sample variance is

$$s^2 = \frac{\Sigma(x_i - \bar{x})^2}{n - 1} = \frac{22.80}{4} = 5.70$$

The variance is measured in terms of the square of the original units of measurement. If the original measurements are in inches, the variance is expressed in square inches. Taking the square root of the variance, we obtain the **standard deviation,** which returns the measure of variability to the original units of measurement.

Definition The **standard deviation** of a set of measurements is equal to the positive square root of the variance.

NEED A TIP?

The variance and the standard deviation *cannot* be negative numbers.

NOTATION

n: number of measurements in the sample

s^2: sample variance

$s = \sqrt{s^2}$: sample standard deviation

N: number of measurements in the population

σ^2: population variance

$\sigma = \sqrt{\sigma^2}$: population standard deviation

NEED A TIP?

If you are using your calculator, make sure to choose the correct key for the sample standard deviation.

For the set of $n = 5$ sample measurements in Table 2.2, the sample variance is $s^2 = 5.70$, so the sample standard deviation is $s = \sqrt{s^2} = \sqrt{5.70} = 2.39$. The more variable the data set is, the larger the value of s.

For the small set of measurements we used, the calculation of the variance is not too difficult. However, for a larger set, the calculations can become very tedious. Most scientific calculators have built-in programs that will calculate $\bar{x}$ and s or μ and σ, so that your computational work will be minimized. The sample or population mean key is usually marked with $\bar{x}$. The sample standard deviation key is usually marked with s, s_x, or σ_{xn-1}, and the population standard deviation key with σ, σ_x, or σ_{xn}. In using any calculator with these built-in function keys, be sure you know which calculation is being carried out by each key!

If you need to calculate s^2 and s by hand, it is much easier to use the alternative computing formula given next. This computational form is sometimes called the **shortcut method for calculating s^2.**

THE COMPUTING FORMULA FOR CALCULATING s^2

$$s^2 = \frac{\Sigma x_i^2 - \dfrac{(\Sigma x_i)^2}{n}}{n - 1}$$

The symbols $(\Sigma x_i)^2$ and Σx_i^2 in the computing formula are shortcut ways to indicate the arithmetic operation you need to perform. You know from the formula for the sample mean that Σx_i is the sum of all the measurements. To find Σx_i^2, you square each individual measurement and then add them together.

$\Sigma x_i^2 = $ Sum of the squares of the individual measurements
$(\Sigma x_i)^2 = $ Square of the sum of the individual measurements

The *sample standard deviation, s,* is the positive square root of s^2.

EXAMPLE (2.5) — Calculate the variance and standard deviation for the five measurements in Table 2.3, which are 5, 7, 1, 2, 4. Use the computing formula for s^2 and compare your results with those obtained using the original definition of s^2.

TABLE 2.3 ● **Table for Simplified Calculation of s^2 and s**

x_i	x_i^2
5	25
7	49
1	1
2	4
4	16
19	95

NEED A TIP?
Don't round off partial
results as you go along!

Solution The entries in Table 2.2 are the individual measurements, x_i, and their squares, x_i^2, together with their sums. Using the computing formula for s^2, you have

$$s^2 = \frac{\Sigma x_i^2 - \frac{(\Sigma x_i)^2}{n}}{n-1} = \frac{95 - \frac{(19)^2}{5}}{4} = \frac{22.80}{4} = 5.70$$

and $s = \sqrt{s^2} = \sqrt{5.70} = 2.39$, as before.

Why Divide by n-1

You may wonder why you need to divide by $(n-1)$ rather than n when computing the sample variance. Just as we used the sample mean $\bar{x}$ to estimate the population mean μ, you may want to use the sample variance s^2 to estimate the population variance σ^2. It turns out that the sample variance s^2 with $(n-1)$ in the denominator provides better estimates of σ^2 than would an estimator calculated with n in the denominator. **For this reason, we always divide by $(n-1)$ when computing the sample variance s^2 and the sample standard deviation s.**

Now that you have learned how to compute the variance and standard deviation, remember these points:

- The value of s is always greater than or equal to zero.
- The larger the value of s^2 or s, the greater the variability of the data set.
- If s^2 or s is equal to zero, all the measurements must have the same value.
- In order to measure the variability in the same units as the original observations, we compute the standard deviation $s = \sqrt{s^2}$.

This information allows you to compare several sets of data with respect to their locations and their variability. How can you use these measures to say something more specific about a single set of data? The theorem and rule presented in the next section will help answer this question.

2.3 EXERCISES

BASIC TECHNIQUES

2.13 You are given $n = 5$ measurements: 2, 1, 1, 3, 5.

a. Calculate the sample mean, $\bar{x}$.

b. Calculate the sample variance, s^2, using the formula given by the definition.

c. Find the sample standard deviation, s.

d. Find s^2 and s using the computing formula. Compare the results with those found in parts b and c.

2.14 Refer to Exercise 2.13.

a. Use the data entry method in your scientific calculator to enter the five measurements. Recall the

proper memories to find the sample mean and standard deviation.

b. Verify that the calculator provides the same values for $\bar{x}$ and s as in Exercise 2.13, parts a and c.

2.15 You are given $n = 8$ measurements: 4, 1, 3, 1, 3, 1, 2, 2.

a. Find the range.

b. Calculate $\bar{x}$.

c. Calculate s^2 and s using the computing formula.

d. Use the data entry method in your calculator to find $\bar{x}$, s, and s^2. Verify that your answers are the same as those in parts b and c.

2.16 You are given $n = 8$ measurements: 3, 1, 5, 6, 4, 4, 3, 5.

a. Calculate the range.

b. Calculate the sample mean.

c. Calculate the sample variance and standard deviation.

d. Compare the range and the standard deviation. The range is approximately how many standard deviations?

APPLICATIONS

2.17 An Archeological Find, again An article in *Archaeometry* involved an analysis of 26 samples of Romano-British pottery found at four different kiln sites in the United Kingdom.[7] The samples were analyzed to determine their chemical composition. The percentage of iron oxide in each of five samples collected at the Island Thorns site was:

    1.28,   2.39,   1.50,   1.88,   1.51

a. Calculate the range.

b. Calculate the sample variance and the standard deviation using the computing formula.

c. Compare the range and the standard deviation. The range is approximately how many standard deviations?

Data set **2.18 Utility Bills in Southern California**
EX0218 The monthly utility bills for a household in Riverside, California, were recorded for 12 consecutive months starting in January 2010:

Month	Amount ($)	Month	Amount ($)
January	288.02	July	311.20
February	230.60	August	370.23
March	216.85	September	368.57
April	243.74	October	301.79
May	236.96	November	271.99
June	288.57	December	298.12

a. Calculate the range of the utility bills for the year 2010.

b. Calculate the average monthly utility bill for the year 2010.

c. Calculate the standard deviation for the 2010 utility bills.

2.4 ON THE PRACTICAL SIGNIFICANCE OF THE STANDARD DEVIATION

We now introduce a useful theorem developed by the Russian mathematician Tchebysheff. Proof of the theorem is not difficult, but we are more interested in its application than its proof.

Tchebysheff's Theorem Given a number k greater than or equal to 1 and a set of n measurements, at least $[1 - (1/k^2)]$ of the measurements will lie within k standard deviations of their mean.

Tchebysheff's Theorem applies to *any set of measurements* and can be used to describe either a sample or a population. We will use the notation appropriate for populations, but you should realize that we could just as easily use the mean and the standard deviation for the sample.

    The idea involved in Tchebysheff's Theorem is illustrated in Figure 2.8. An interval is constructed by measuring a distance $k\sigma$ on either side of the mean μ. The number k can be any number as long as it is greater than or equal to 1. Then Tchebysheff's Theorem states that at least $1 - (1/k^2)$ of the total number n measurements lies in the constructed interval.

FIGURE 2.8 ●

Illustrating Tchebysheff's
Theorem

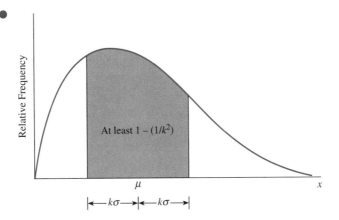

In Table 2.4, we choose a few numerical values for k and compute $[1 - (1/k^2)]$.

TABLE 2.4 ●   **Illustrative Values of $[1 - (1/k^2)]$**

k	$1 - (1/k^2)$
1	$1 - 1 = 0$
2	$1 - 1/4 = 3/4$
3	$1 - 1/9 = 8/9$

From the calculations in Table 2.4, the theorem states:

- At least none of the measurements lie in the interval $\mu - \sigma$ to $\mu + \sigma$.
- At least 3/4 of the measurements lie in the interval $\mu - 2\sigma$ to $\mu + 2\sigma$.
- At least 8/9 of the measurements lie in the interval $\mu - 3\sigma$ to $\mu + 3\sigma$.

Although the first statement is not at all helpful, the other two values of k provide valuable information about the proportion of measurements that fall in certain intervals. The values $k = 2$ and $k = 3$ are not the only values of k you can use; for example, the proportion of measurements that fall within $k = 2.5$ standard deviations of the mean is at least $1 - [1/(2.5)^2] = .84$.

EXAMPLE 2.6 The mean and variance of a sample of $n = 25$ measurements are 75 and 100, respectively. Use Tchebysheff's Theorem to describe the distribution of measurements.

Solution   You are given $\bar{x} = 75$ and $s^2 = 100$. The standard deviation is $s = \sqrt{100} = 10$. The distribution of measurements is centered about $\bar{x} = 75$, and Tchebysheff's Theorem states:

- *At least* 3/4 of the 25 measurements lie in the interval $\bar{x} \pm 2s = 75 \pm 2(10)$ —that is, 55 to 95.
- *At least* 8/9 of the measurements lie in the interval $\bar{x} \pm 3s = 75 \pm 3(10)$—that is, 45 to 105.

Since Tchebysheff's Theorem applies to *any* distribution, it is very conservative. This is why we emphasize "at least $1 - (1/k^2)$" in this theorem.

Another rule for describing the variability of a data set does not work for *all* data sets, but it does work very well for data that "pile up" in the familiar mound shape shown in Figure 2.9. The closer your data distribution is to the mound-shaped curve in Figure 2.9, the more accurate the rule will be. Since mound-shaped data distributions occur quite frequently in nature, the rule can often be used in practical applications. For this reason, we call it the **Empirical Rule.**

FIGURE 2.9 ●

Mound-shaped distribution

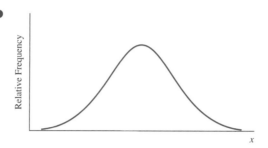

Empirical Rule   Given a distribution of measurements that is approximately mound-shaped:

The interval $(\mu \pm \sigma)$ contains approximately 68% of the measurements.

The interval $(\mu \pm 2\sigma)$ contains approximately 95% of the measurements.

The interval $(\mu \pm 3\sigma)$ contains approximately 99.7% of the measurements.

NEED A TIP?

Remember these three numbers:

68—95—99.7

The mound-shaped distribution shown in Figure 2.9 is commonly known as the **normal distribution** and will be discussed in detail in Chapter 6.

EXAMPLE 2.7

In a time study conducted at a manufacturing plant, the length of time to complete a specified operation is measured for each of $n = 40$ workers. The mean and standard deviation are found to be 12.8 and 1.7, respectively. Describe the sample data using the Empirical Rule.

Solution   To describe the data, calculate these intervals:

$(\bar{x} \pm s) = 12.8 \pm 1.7$ or   11.1 to 14.5

$(\bar{x} \pm 2s) = 12.8 \pm 2(1.7)$ or   9.4 to 16.2

$(\bar{x} \pm 3s) = 12.8 \pm 3(1.7)$ or   7.7 to 17.9

According to the Empirical Rule, you expect approximately 68% of the measurements to fall into the interval from 11.1 to 14.5, approximately 95% to fall into the interval from 9.4 to 16.2, and approximately 99.7% to fall into the interval from 7.7 to 17.9.

If you doubt that the distribution of measurements is mound-shaped, or if you wish for some other reason to be conservative, you can apply Tchebysheff's Theorem and be absolutely certain of your statements. Tchebysheff's Theorem tells you that at least 3/4 of the measurements fall into the interval from 9.4 to 16.2 and at least 8/9 into the interval from 7.7 to 17.9.

EXAMPLE 2.8

Student teachers are trained to develop lesson plans, on the assumption that the written plan will help them to perform successfully in the classroom. In a study to assess the relationship between written lesson plans and their implementation in the classroom, 25 lesson plans were scored on a scale of 0 to 34 according to a Lesson Plan Assessment Checklist. The 25 scores are shown in Table 2.5. Use Tchebysheff's Theorem and the Empirical Rule (if applicable) to describe the distribution of these assessment scores.

TABLE 2.5 ● **Lesson Plan Assessment Scores**

26.1	26.0	14.5	29.3	19.7
22.1	21.2	26.6	31.9	25.0
15.9	20.8	20.2	17.8	13.3
25.6	26.5	15.7	22.1	13.8
29.0	21.3	23.5	22.1	10.2

Solution Use your calculator or the computing formulas to verify that $\bar{x} = 21.6$ and $s = 5.5$. The appropriate intervals are calculated and listed in Table 2.6. We have also referred back to the original 25 measurements and counted the actual number of measurements that fall into each of these intervals. These frequencies and relative frequencies are shown in Table 2.6.

TABLE 2.6 ● **Intervals $\bar{x} \pm ks$ for the Data of Table 2.5**

k	Interval $\bar{x} \pm ks$	Frequency in Interval	Relative Frequency
1	16.1–27.1	16	.64
2	10.6–32.6	24	.96
3	5.1–38.1	25	1.00

 NEED A TIP?

Empirical Rule ⇔ mound-shaped data

Tchebysheff ⇔ any shaped data

Is Tchebysheff's Theorem applicable? Yes, because it can be used for any set of data. According to Tchebysheff's Theorem,

- at least 3/4 of the measurements will fall between 10.6 and 32.6.

- at least 8/9 of the measurements will fall between 5.1 and 38.1.

You can see in Table 2.6 that Tchebysheff's Theorem is true for these data. In fact, the proportions of measurements that fall into the specified intervals exceed the lower bound given by this theorem.

Is the Empirical Rule applicable? You can check for yourself by drawing a graph—either a stem and leaf plot or a histogram. The relative frequency histogram in Figure 2.10 shows that the distribution is *relatively* mound-shaped, so the Empirical Rule should work *relatively well.* That is,

- approximately 68% of the measurements will fall between 16.1 and 27.1.

- approximately 95% of the measurements will fall between 10.6 and 32.6.

- approximately 99.7% of the measurements will fall between 5.1 and 38.1.

The relative frequencies in Table 2.6 closely approximate those specified by the Empirical Rule.

FIGURE 2.10

Relative frequency
histogram for Example 2.8

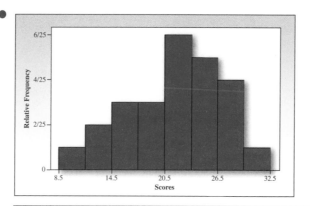

USING TCHEBYSHEFF'S THEOREM AND THE EMPIRICAL RULE

Tchebysheff's Theorem can be proven mathematically. It applies to any set of measurements—sample or population, large or small, mound-shaped or skewed.

Tchebysheff's Theorem gives a *lower bound* to the fraction of measurements to be found in an interval constructed as $\bar{x} \pm ks$. *At least* $1 - (1/k^2)$ of the measurements will fall into this interval, and probably more!

The Empirical Rule is a "rule of thumb" that can be used as a descriptive tool only when the data tend to be roughly mound-shaped (the data tend to pile up near the center of the distribution).

When you use these two tools for describing a set of measurements, Tchebysheff's Theorem will always be satisfied, but it is a very conservative estimate of the fraction of measurements that fall into a particular interval. If it is appropriate to use the Empirical Rule (mound-shaped data), this rule will give you a more accurate estimate of the fraction of measurements that fall into the interval.

2.5   A CHECK ON THE CALCULATION OF *s*

Tchebysheff's Theorem and the Empirical Rule can be used to detect gross errors in the calculation of *s*. Roughly speaking, these two tools tell you that *most of the time,* measurements lie within *two* standard deviations of their mean. This interval is marked off in Figure 2.11, and it implies that the total range of the measurements, from smallest to largest, should be somewhere around four standard deviations. This is, of course, a very rough approximation, but it can be very useful in checking for large errors in your calculation of *s*. If the range, *R,* is about four standard deviations, or 4*s,* you can write

$$R \approx 4s \quad \text{or} \quad s \approx \frac{R}{4}$$

The computed value of *s* using the shortcut formula should be of roughly the same order as the approximation.

FIGURE 2.11

Range approximation to *s*

EXAMPLE 2.9 — Use the range approximation to check the calculation of s for Table 2.3.

Solution The range of the five measurements—5, 7, 1, 2, 4—is

$$R = 7 - 1 = 6$$

Then

$$s \approx \frac{R}{4} = \frac{6}{4} = 1.5$$

This is the same order as the calculated value $s = 2.39$.

 NEED A TIP?

$s \approx R/4$ gives only an **approximate** value for s.

The range approximation is *not* intended to provide an accurate value for s. Rather, its purpose is to detect gross errors in calculating, such as the failure to divide the sum of squares of deviations by $(n - 1)$ or the failure to take the square root of s^2. If you make one of these mistakes, your answer will be many times larger than the range approximation of s.

EXAMPLE 2.10 — Use the range approximation to determine an approximate value for the standard deviation for the data in Table 2.5.

Solution The range $R = 31.9 - 10.2 = 21.7$. Then

$$s \approx \frac{R}{4} = \frac{21.7}{4} = 5.4$$

Since the exact value of s is 5.5 for the data in Table 2.5, the approximation is very close.

The range for a sample of n measurements will depend on the sample size, n. For larger values of n, a larger range of the x values is expected. The range for large samples (say, $n = 50$ or more observations) may be as large as $6s$, whereas the range for small samples (say, $n = 5$ or less) may be as small as or smaller than $2.5s$.

The range approximation for s can be improved if it is known that the sample is drawn from a mound-shaped distribution of data. Thus, the calculated s should not differ substantially from the range divided by the appropriate ratio given in Table 2.7.

TABLE 2.7 ● **Divisor for the Range Approximation of s**

Number of Measurements	Expected Ratio of Range to s
5	2.5
10	3
25	4

2.5 EXERCISES

BASIC TECHNIQUES

2.19 A set of $n = 10$ measurements consists of the values 5, 2, 3, 6, 1, 2, 4, 5, 1, 3.

a. Use the range approximation to estimate the value of s for this set. (HINT: Use the table at the end of Section 2.5.)

b. Use your calculator to find the actual value of s. Is the actual value close to your estimate in part a?

c. Draw a dotplot of this data set. Are the data mound-shaped?

d. Can you use Tchebysheff's Theorem to describe this data set? Why or why not?

e. Can you use the Empirical Rule to describe this data set? Why or why not?

2.20 A set of data has a mean of 75 and a standard deviation of 5. You know nothing else about the size of the data set or the shape of the data distribution.

a. What can you say about the proportion of measurements that fall between 60 and 90?

b. What can you say about the proportion of measurements that fall between 65 and 85?

c. What can you say about the proportion of measurements that are less than 65?

2.21 A distribution of measurements is relatively mound-shaped with mean 50 and standard deviation 10.

a. What proportion of the measurements will fall between 40 and 60?

b. What proportion of the measurements will fall between 30 and 70?

c. What proportion of the measurements will fall between 30 and 60?

d. If a measurement is chosen at random from this distribution, what is the probability that it will be greater than 60?

2.22 Suppose you want to create a mental picture of the relative frequency histogram for a large data set consisting of 1000 observations, and you know that the mean and standard deviation of the data set are 36 and 3, respectively.

a. If you are fairly certain that the relative frequency distribution of the data is mound-shaped, how might you picture the relative frequency distribution? (HINT: Use the Empirical Rule.)

b. If you have no prior information concerning the shape of the relative frequency distribution, what can you say about the relative frequency histogram? (HINT: Construct intervals $\bar{x} \pm ks$ for several choices of k.)

APPLICATIONS

2.23 Driving Emergencies The length of time required for an automobile driver to respond to a particular emergency situation was recorded for $n = 10$ drivers. The times (in seconds) were .5, .8, 1.1, .7, .6, .9, .7, .8, .7, .8.

a. Scan the data and use the procedure in Section 2.5 to find an approximate value for s. Use this value to check your calculations in part b.

b. Calculate the sample mean $\bar{x}$ and the standard deviation s. Compare with part a.

Data set **2.24 Packaging Hamburger Meat** The data
EX0224 listed here are the weights (in pounds) of 27 packages of ground beef in a supermarket meat display:

1.08	.99	.97	1.18	1.41	1.28	.83
1.06	1.14	1.38	.75	.96	1.08	.87
.89	.89	.96	1.12	1.12	.93	1.24
.89	.98	1.14	.92	1.18	1.17	

a. Construct a stem and leaf plot or a relative frequency histogram to display the distribution of weights. Is the distribution relatively mound-shaped?

b. Find the mean and standard deviation of the data set.

c. Find the percentage of measurements in the intervals $\bar{x} \pm s$, $\bar{x} \pm 2s$, and $\bar{x} \pm 3s$.

d. How do the percentages obtained in part c compare with those given by the Empirical Rule? Explain.

e. How many of the packages weigh exactly 1 pound? Can you think of any explanation for this?

2.25 Breathing Rates Is your breathing rate normal? Actually, there is no standard breathing rate for humans. It can vary from as low as 4 breaths per minute to as high as 70 or 75 for a person engaged in strenuous exercise. Suppose that the resting breathing rates for college-age students have a relative frequency distribution that is mound-shaped, with a mean equal to 12 and a standard deviation of 2.3 breaths per minute.

What fraction of all students would have breathing rates in the following intervals?

a. 9.7 to 14.3 breaths per minute

b. 7.4 to 16.6 breaths per minute

c. More than 18.9 or less than 5.1 breaths per minute

Data set **2.26 Ore Samples** A geologist collected 20 EX0226 different ore samples, all the same weight, and randomly divided them into two groups. She measured the titanium (Ti) content of the samples using two different methods.

Method 1					Method 2				
.011	.013	.013	.015	.014	.011	.016	.013	.012	.015
.013	.010	.013	.011	.012	.012	.017	.013	.014	.015

a. Construct stem and leaf plots for the two data sets. Visually compare their centers and their ranges.

b. Calculate the sample means and standard deviations for the two sets. Do the calculated values confirm your visual conclusions from part a?

Data set **2.27 Social Security Numbers** A group of EX0227 70 students were asked to record the last digit of their social security number.

1	6	9	1	5	9	0	2	8	4
0	7	3	4	2	3	5	8	4	2
3	2	0	0	2	1	2	7	7	4
0	0	9	9	5	3	8	4	7	4
6	6	9	0	2	6	2	9	5	8
5	1	7	7	7	8	7	5	1	8
3	4	1	9	3	8	6	6	6	6

a. Draw a relative frequency histogram using the values 0 through 9 as the class midpoints. What is the shape of the distribution? Based on the shape, what would be your best estimate for the mean of the data set?

b. Use the range approximation to guess the value of s for this set.

c. Use your calculator to find the actual values of $\bar{x}$ and s. Compare with your estimates in parts a and b.

2.28 Tuna Fish, again Refer to Exercise 2.8 and data set EX0208. The prices of a 6-ounce can or a 7.06-ounce pouch for 14 different brands of water-packed light tuna, based on prices paid nationally in supermarkets are reproduced here.[4]

.99	1.92	1.23	.85	.65	.53	1.41
1.12	.63	.67	.69	.60	.60	.66

a. Use the range approximation to find an estimate of s.

b. How does it compare to the computed value of s?

2.29 Survival Times A group of laboratory animals is infected with a particular form of bacteria, and their survival time is found to average 32 days, with a standard deviation of 36 days.

a. Visualize the distribution of survival times. Do you think that the distribution is relatively mound-shaped, skewed right, or skewed left? Explain.

b. Within what limits would you expect at least 3/4 of the measurements to lie?

2.30 Survival Times, continued Refer to Exercise 2.29. You can use the Empirical Rule to see why the distribution of survival times could not be mound-shaped.

a. Find the value of x that is exactly one standard deviation below the mean.

b. If the distribution is in fact mound-shaped, approximately what percentage of the measurements should be less than the value of x found in part a?

c. Since the variable being measured is time, is it possible to find any measurements that are more than one standard deviation below the mean?

d. Use your answers to parts b and c to explain why the data distribution cannot be mound-shaped.

Data set **2.31 Timber Tracts** To estimate the amount EX0231 of lumber in a tract of timber, an owner decided to count the number of trees with diameters exceeding 12 inches in randomly selected 50-by-50-foot squares. Seventy 50-by-50-foot squares were chosen, and the selected trees were counted in each tract. The data are listed here:

7	8	7	10	4	8	6	8	9	10
9	6	4	9	10	9	8	8	7	9
3	9	5	9	9	8	7	5	8	8
10	2	7	4	8	5	10	7	7	7
9	6	8	8	8	7	8	9	6	8
6	11	9	11	7	7	11	7	9	13
10	8	8	5	9	9	8	5	9	8

a. Construct a relative frequency histogram to describe the data.

b. Calculate the sample mean $\bar{x}$ as an estimate of μ, the mean number of timber trees for all 50-by-50-foot squares in the tract.

c. Calculate s for the data. Construct the intervals $\bar{x} \pm s$, $\bar{x} \pm 2s$, and $\bar{x} \pm 3s$. Calculate the percentage of squares falling into each of the three intervals, and compare with the corresponding percentages given by the Empirical Rule and Tchebysheff's Theorem.

2.32 Social Security Numbers, continued Refer to the data set in Exercise 2.27.

a. Find the percentage of measurements in the intervals $\bar{x} \pm s$, $\bar{x} \pm 2s$, and $\bar{x} \pm 3s$.

b. How do the percentages obtained in part a compare with those given by the Empirical Rule? Should they be approximately the same? Explain.

Data set **2.33 Old Faithful** The data below are 30 EX0233 waiting times between eruptions of the Old Faithful geyser in Yellowstone National Park.[8]

| 56 | 89 | 51 | 79 | 58 | 82 | 52 | 88 | 52 | 78 | 69 | 75 | 77 | 72 | 71 |
| 55 | 87 | 53 | 85 | 61 | 93 | 54 | 76 | 80 | 81 | 59 | 86 | 78 | 71 | 77 |

a. Calculate the range.

b. Use the range approximation to approximate the standard deviation of these 30 measurements.

c. Calculate the sample standard deviation s.

d. What proportion of the measurements lie within two standard deviations of the mean? Within three standard deviations of the mean? Do these proportions agree with the proportions given in Tchebysheff's Theorem?

Data set **2.34 The President's Kids** The table EX0234 below shows the names of the 43 presidents of the United States along with the number of their children.[9]

Washington	0	Van Buren	4	Buchanan	0
Adams	5	W.H. Harrison	10	Lincoln	4
Jefferson	6	Tyler*	15	A. Johnson	5
Madison	0	Polk	0	Grant	4
Monroe	2	Taylor	6	Hayes	8
J.Q. Adams	4	Fillmore*	2	Garfield	7
Jackson	0	Pierce	3	Arthur	3
Cleveland	5	Coolidge	2	Nixon	2
B. Harrison*	3	Hoover	2	Ford	4
McKinley	2	F.D. Roosevelt	6	Carter	4
T. Roosevelt*	6	Truman	1	Reagan*	4
Taft	3	Eisenhower	2	G.H.W. Bush	6
Wilson*	3	Kennedy	3	Clinton	1
Harding	0	L.B. Johnson	2	G.W. Bush	2
				Obama	2

*Married twice   Source: The World Almanac and Book of Facts 2011

a. Construct a relative frequency histogram to describe the data. How would you describe the shape of this distribution?

b. Calculate the mean and the standard deviation for the data set.

c. Construct the intervals $\bar{x} \pm s$, $\bar{x} \pm 2s$, and $\bar{x} \pm 3s$. Find the percentage of measurements falling into these three intervals and compare with the corresponding percentages given by Tchebysheff's Theorem and the Empirical Rule.

2.35 An Archeological Find, again Refer to Exercise 2.17. The percentage of iron oxide in each of five pottery samples collected at the Island Thorns site was:

1.28   2.39   1.50   1.88   1.51

a. Use the range approximation to find an estimate of s, using an appropriate divisor from Table 2.7.

b. Calculate the standard deviation s. How close did your estimate come to the actual value of s?

Data set **2.36 Aaron Rodgers** The number of passes EX0236 completed by Aaron Rodgers, quarterback for the Minnesota Vikings, was recorded for each of the 15 regular season games that he played in the fall of 2010 (www.ESPN.com):[10]

| 19 | 19 | 34 | 12 | 27 | 18 | 21 | 15 |
| 27 | 22 | 26 | 21 | 7 | 25 | 19 | |

a. Draw a stem and leaf plot to describe the data.

b. Calculate the mean and standard deviation for Aaron Rodgers' per game pass completions.

c. What proportion of the measurements lie within two standard deviations of the mean?

CALCULATING THE MEAN AND STANDARD DEVIATION FOR GROUPED DATA (OPTIONAL)

2.37 Suppose that some measurements occur more than once and that the data $x_1, x_2, \ldots, x_k$ are arranged in a frequency table as shown here:

Observations	Frequency f_i
x_1	f_1
x_2	f_2
.	.
.	.
.	.
x_k	f_k

The formulas for the mean and variance for grouped data are

$$\bar{x} = \frac{\Sigma x_i f_i}{n}, \qquad \text{where } n = \Sigma f_i$$

and

$$s^2 = \frac{\Sigma x_i^2 f_i - \dfrac{(\Sigma x_i f_i)^2}{n}}{n-1}$$

Notice that if each value occurs once, these formulas reduce to those given in the text. Although these formulas for grouped data are primarily of value when you have a large number of measurements, demonstrate their use for the sample 1, 0, 0, 1, 3, 1, 3, 2, 3, 0, 0, 1, 1, 3, 2.

a. Calculate $\bar{x}$ and s^2 directly, using the formulas for ungrouped data.

b. The frequency table for the $n = 15$ measurements is as follows:

x	f
0	4
1	5
2	2
3	4

Calculate $\bar{x}$ and s^2 using the formulas for grouped data. Compare with your answers to part a.

2.38 International Baccalaureate High school students in an International Baccalaureate (IB) program are placed in accelerated or advanced courses and must take IB examinations in each of six subject areas at the end of their junior or senior year. Students are scored on a scale of 1–7, with 1–2 being poor, 3 mediocre, 4 average, and 5–7 excellent. During its first year of operation at John W. North High School in Riverside, California, 17 juniors attempted the IB economics exam, with these results:

Exam Grade	Number of Students
7	1
6	4
5	4
4	4
3	4

Calculate the mean and standard deviation for these scores.

2.39 A Skewed Distribution To illustrate the utility of the Empirical Rule, consider a distribution that is heavily skewed to the right, as shown in the accompanying figure.

a. Calculate $\bar{x}$ and s for the data shown. (NOTE: There are 10 zeros, 5 ones, and so on.)

b. Construct the intervals $\bar{x} \pm s$, $\bar{x} \pm 2s$, and $\bar{x} \pm 3s$ and locate them on the frequency distribution.

c. Calculate the proportion of the $n = 25$ measurements that fall into each of the three intervals. Compare with Tchebysheff's Theorem and the Empirical Rule. Note that, although the proportion that falls into the interval $\bar{x} \pm s$ does not agree closely with the Empirical Rule, the proportions that fall into the intervals $\bar{x} \pm 2s$ and $\bar{x} \pm 3s$ agree very well. Many times this is true, even for non-mound-shaped distributions of data.

Distribution for Exercise 2.39

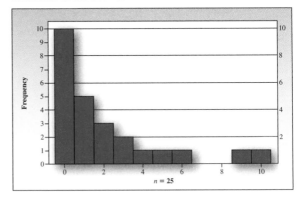

$n = 25$

MEASURES OF RELATIVE STANDING

2.6

Sometimes you need to know the position of one observation relative to others in a set of data. For example, if you took an examination with a total of 35 points, you might want to know how your score of 30 compared to the scores of the other students in the class. The mean and standard deviation of the scores can be used to calculate a **z-score,** which measures the relative standing of a measurement in a data set.

Definition   The **sample z-score** is a measure of relative standing defined by

$$z\text{-score} = \frac{x - \bar{x}}{s}$$

 NEED A TIP?

Positive *z*-score ⇔ *x* is above the mean.

Negative *z*-score ⇔ *x* is below the mean.

A **z-score measures the distance between an observation and the mean, measured in units of standard deviation.** For example, suppose that the mean and standard deviation of the test scores (based on a total of 35 points) are 25 and 4, respectively. The *z*-score for your score of 30 is calculated as follows:

$$z\text{-score} = \frac{x - \bar{x}}{s} = \frac{30 - 25}{4} = 1.25$$

Your score of 30 lies 1.25 standard deviations above the mean ($30 = \bar{x} + 1.25s$).

The *z*-score is a valuable tool for determining whether a particular observation is likely to occur quite frequently or whether it is unlikely and might be considered an **outlier.**

According to Tchebysheff's Theorem and the Empirical Rule,

- at least 75% and more likely 95% of the observations lie within two standard deviations of their mean: their *z*-scores are between -2 and $+2$. *Observations with z-scores exceeding 2 in absolute value happen about 5% of the time for mound-shaped data and are considered somewhat unlikely.*

- at least 89% and more likely 99.7% of the observations lie within three standard deviations of their mean: their *z*-scores are between -3 and $+3$. *Observations with z-scores exceeding 3 in absolute value happen less than 1% of the time for mound-shaped data and are considered very unlikely.*

NEED A TIP?

z-Scores above 3 in absolute value are very unusual.

You should look carefully at any observation that has a *z*-score exceeding 3 in absolute value. Perhaps the measurement was recorded incorrectly or does not belong to the population being sampled. Perhaps it is just a highly unlikely observation, but a valid one nonetheless!

EXAMPLE   **2.11**

Consider this sample of $n = 10$ measurements:

$$1, 1, 0, 15, 2, 3, 4, 0, 1, 3$$

The measurement $x = 15$ appears to be unusually large. Calculate the *z*-score for this observation and state your conclusions.

Solution   Calculate $\bar{x} = 3.0$ and $s = 4.42$ for the $n = 10$ measurements. Then the *z*-score for the suspected outlier, $x = 15$, is calculated as

$$z\text{-score} = \frac{x - \bar{x}}{s} = \frac{15 - 3}{4.42} = 2.71$$

Hence, the measurement $x = 15$ lies 2.71 standard deviations above the sample mean, $\bar{x} = 3.0$. Although the *z*-score does not exceed 3, it is close enough so that you might suspect that $x = 15$ is an outlier. You should examine the sampling procedure to see whether $x = 15$ is a faulty observation.

A **percentile** is another measure of relative standing and is most often used for large data sets. (Percentiles are not very useful for small data sets.)

Definition   A set of n measurements on the variable x has been arranged in order of magnitude. The **pth percentile** is the value of x that is greater than $p\%$ of the measurements and is less than the remaining $(100 - p)\%$.

EXAMPLE 2.12

Suppose you have been notified that your score of 610 on the Verbal Graduate Record Examination placed you at the 60th percentile in the distribution of scores. Where does your score of 610 stand in relation to the scores of others who took the examination?

Solution   Scoring at the 60th percentile means that 60% of all the examination scores were lower than your score and 40% were higher.

For any data distribution, regardless of its shape, the 60th percentile for the variable x is a point on the *horizontal axis* of the data distribution that is greater than 60% of the measurements and less than the others. That is, 60% of the measurements are less than the 60th percentile and 40% are greater (see Figure 2.12). Since the total area under the distribution is 100%, 60% of the area is to the left and 40% of the area is to the right of the 60th percentile. Remember that the median, m, of a set of data is the middle measurement; that is, 50% of the measurements are smaller and 50% are larger than the median. Thus, the *median is the same as the 50th percentile!*

FIGURE 2.12

The 60th percentile shown on the relative frequency histogram for a data set

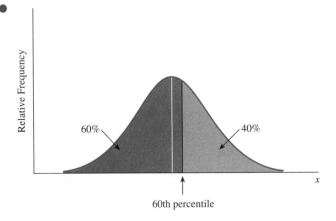

The 25th and 75th percentiles, called the **lower** and **upper quartiles**, along with the median (the 50th percentile), locate points that divide the data into four sets, each containing an equal number of measurements. Twenty-five percent of the measurements will be less than the lower (first) quartile, 50% will be less than the median (the second quartile), and 75% will be less than the upper (third) quartile. Thus, the median and the lower and upper quartiles are located at points on the x-axis so that the area under the relative frequency histogram for the data is partitioned into four equal areas, as shown in Figure 2.13.

FIGURE 2.13

Location of quartiles

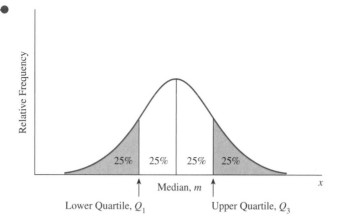

Definition   A set of n measurements on the variable x has been arranged in order of magnitude. The **lower quartile (first quartile), Q_1,** is the value of x that is greater than one-fourth of the measurements and is less than the remaining three-fourths. The **second quartile** is the median. The **upper quartile (third quartile), Q_3,** is the value of x that is greater than three-fourths of the measurements and is less than the remaining one-fourth.

For small data sets, it is often impossible to divide the set into four groups, each of which contains exactly 25% of the measurements. For example, when $n = 10$, you would need to have $2\frac{1}{2}$ measurements in each group! Even when you can perform this task (for example, if $n = 12$), there are many numbers that would satisfy the preceding definition, and could therefore be considered "quartiles." To avoid this ambiguity, we use the following rule to locate sample quartiles.

CALCULATING SAMPLE QUARTILES

- When the measurements are arranged in order of magnitude, the **lower quartile, Q_1,** is the value of x in position $.25(n + 1)$, and the **upper quartile, Q_3,** is the value of x in position $.75(n + 1)$.

- When $.25(n + 1)$ and $.75(n + 1)$ are not integers, the quartiles are found by interpolation, using the values in the two adjacent positions.[†]

EXAMPLE 2.13   Find the lower and upper quartiles for this set of measurements:

16, 25, 4, 18, 11, 13, 20, 8, 11, 9

Solution   Rank the $n = 10$ measurements from smallest to largest:

4, 8, 9, 11, 11, 13, 16, 18, 20, 25

[†]This definition of quartiles is consistent with the one used in *MINITAB 16* and *MS Excel 2010*. Some textbooks use ordinary rounding when finding quartile positions, whereas others compute sample quartiles as the medians of the upper and lower halves of the data set.

Calculate

$$\text{Position of } Q_1 = .25(n + 1) = .25(10 + 1) = 2.75$$
$$\text{Position of } Q_3 = .75(n + 1) = .75(10 + 1) = 8.25$$

Since these positions are not integers, the lower quartile is taken to be the value 3/4 of the distance between the second and third ordered measurements, and the upper quartile is taken to be the value 1/4 of the distance between the eighth and ninth ordered measurements. Therefore,

$$Q_1 = 8 + .75(9 - 8) = 8 + .75 = 8.75$$

and

$$Q_3 = 18 + .25(20 - 18) = 18 + .5 = 18.5$$

Because the median and the quartiles divide the data distribution into four parts, each containing approximately 25% of the measurements, Q_1 and Q_3 are the upper and lower boundaries for the middle 50% of the distribution. We can measure the range of this "middle 50%" of the distribution using a numerical measure called the **interquartile range.**

Definition The **interquartile range (IQR)** for a set of measurements is the difference between the upper and lower quartiles; that is, IQR $= Q_3 - Q_1$.

For the data in Example 2.13, IQR $= Q_3 - Q_1 = 18.50 - 8.75 = 9.75$. We will use the IQR along with the quartiles and the median in the next section to construct another graph for describing data sets.

 NEED TO KNOW...

How to Calculate Sample Quartiles

1. Arrange the data set in order of magnitude from smallest to largest.
2. Calculate the quartile positions:
   - Position of Q_1: $.25(n + 1)$
   - Position of Q_3: $.75(n + 1)$
3. If the positions are integers, then Q_1 and Q_3 are the values in the ordered data set found in those positions.
4. If the positions in step 2 are not integers, find the two measurements in positions just above and just below the calculated position. Calculate the quartile by finding a value either one-fourth, one-half, or three-fourths of the way between these two measurements.

Many of the numerical measures that you have learned are easily found using computer programs or even graphics calculators. The *MINITAB* command **Stat ▶ Basic Statistics ▶ Display Descriptive Statistics** or the *Excel* command **Data ▶ Data Analysis ▶ Descriptive Statistics** (see the "Technology Today" section at the end of this chapter) produce output containing the mean, the standard deviation, the median, the maximum and minimum, as well as the values of other statistics that we have not discussed. The data from Example 2.13 produced the *MINITAB* output shown in Figure 2.14. Notice that the quartiles are identical to the hand-calculated values in that example.

FIGURE 2.14 ●

MINITAB output for the data in Example 2.13

Descriptive Statistics: x

Variable	N	N*	Mean	SE Mean	StDev	Minimum	Q1	Median	Q3	Maximum
X	10	0	13.50	1.98	6.28	4.00	8.75	12.00	18.50	25.00

THE FIVE-NUMBER SUMMARY AND THE BOX PLOT

2.7

The median and the upper and lower quartiles shown in Figure 2.13 divide the data into four sets, each containing an equal number of measurements. If we add the largest number (Max) and the smallest number (Min) in the data set to this group, we will have a set of numbers that provide a quick and rough summary of the data distribution.

The **five-number summary** consists of the smallest number, the lower quartile, the median, the upper quartile, and the largest number, presented in order from smallest to largest:

   Min Q_1 **Median** Q_3 **Max**

By definition, one-fourth of the measurements in the data set lie between each of the four adjacent pairs of numbers.

The five-number summary can be used to create a simple graph called a **box plot** to visually describe the data distribution. From the box plot, you can quickly detect any skewness in the shape of the distribution and see whether there are any outliers in the data set. An outlier may result from transposing digits when recording a measurement, from incorrectly reading an instrument dial, from a malfunctioning piece of equipment, or from other problems.

Even when there are no recording or observational errors, a data set may contain one or more valid measurements that, for one reason or another, differ markedly from the others in the set. These outliers can cause a marked distortion in commonly used numerical measures such as $\bar{x}$ and s. In fact, outliers may themselves contain important information not shared with the other measurements in the set. Therefore, isolating outliers, if they are present, is an important step in any preliminary analysis of a data set. The box plot is designed expressly for this purpose.

TO CONSTRUCT A BOX PLOT

• Calculate the median, the upper and lower quartiles, and the IQR for the data set.

• Draw a horizontal line representing the scale of measurement. Form a box just above the horizontal line with the right and left ends at Q_1 and Q_3. Draw a vertical line through the box at the location of the median.

A box plot is shown in Figure 2.15.

FIGURE 2.15

Box plot

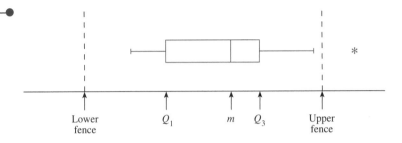

In Section 2.6, the z-score provided boundaries for finding unusually large or small measurements. You looked for z-scores greater than 2 or 3 in absolute value. The box plot uses the IQR to create imaginary "fences" to separate outliers from the rest of the data set:

DETECTING OUTLIERS—OBSERVATIONS THAT ARE BEYOND:

• Lower fence: $Q_1 - 1.5(\text{IQR})$
• Upper fence: $Q_3 + 1.5(\text{IQR})$

The upper and lower fences are shown with broken lines in Figure 2.15, but they are not usually drawn on the box plot. Any measurement beyond the upper or lower fence is an **outlier;** the rest of the measurements, inside the fences, are not unusual. Finally, the box plot marks the range of the data set using "whiskers" to connect the smallest and largest measurements (*excluding outliers*) to the box.

TO FINISH THE BOX PLOT

• Mark any **outliers** with an asterisk (*) on the graph.
• Extend horizontal lines called "whiskers" from the ends of the box to the smallest and largest observations that are *not* outliers.

EXAMPLE 2.14

As American consumers become more careful about the foods they eat, food processors try to stay competitive by avoiding excessive amounts of fat, cholesterol, and sodium in the foods they sell. The following data are the amounts of sodium per slice (in milligrams) for each of eight brands of regular American cheese. Construct a box plot for the data and look for outliers.

340, 300, 520, 340, 320, 290, 260, 330

Solution   The $n = 8$ measurements are first ranked from smallest to largest:

260, 290, 300, 320, 330, 340, 340, 520

The positions of the median, Q_1, and Q_3 are

$$.5(n + 1) = .5(9) = 4.5$$
$$.25(n + 1) = .25(9) = 2.25$$
$$.75(n + 1) = .75(9) = 6.75$$

so that $m = (320 + 330)/2 = 325$, $Q_1 = 290 + .25(10) = 292.5$, and $Q_3 = 340$. The interquartile range is calculated as

$$\text{IQR} = Q_3 - Q_1 = 340 - 292.5 = 47.5$$

Calculate the upper and lower fences:

$$\text{Lower fence: } 292.5 - 1.5(47.5) = 221.25$$
$$\text{Upper fence: } 340 + 1.5(47.5) = 411.25$$

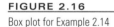

Building a Box Plot

The value $x = 520$, a brand of cheese containing 520 milligrams of sodium, is the only *outlier,* lying beyond the upper fence.

The box plot for the data is shown in Figure 2.16. The outlier is marked with an asterisk (*). Once the outlier is excluded, we find (from the ranked data set) that the smallest and largest measurements are $x = 260$ and $x = 340$. These are the two values that form the whiskers. Since the value $x = 340$ is the same as Q_3, there is no whisker on the right side of the box.

FIGURE 2.16

Box plot for Example 2.14

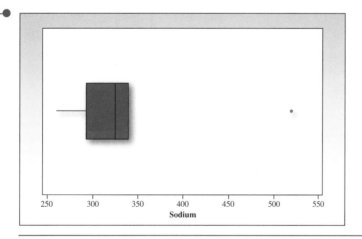

You can use the box plot to describe the shape of a data distribution by looking at the position of the median line compared to Q_1 and Q_3, the left and right ends of the box. If the median is close to the middle of the box, the distribution is fairly symmetric, providing equal-sized intervals to contain the two middle quarters of the data. If the median line is to the left of center, the distribution is skewed to the right; if the median is to the right of center, the distribution is skewed to the left. Also, for most skewed distributions, the whisker on the skewed side of the box tends to be longer than the whisker on the other side.

Figure 2.17 shows two box plots, one for the sodium contents of the eight brands of cheese in Example 2.14, and another for five brands of fat-free cheese with these sodium contents:

$$300, \quad 300, \quad 320, \quad 290, \quad 180$$

Look at the long whisker on the left side of both box plots and the position of the median lines. Both distributions are skewed to the left; that is, there are a few unusually small measurements. The regular cheese data, however, also show one brand ($x = 520$) with an unusually large amount of sodium. In general, it appears that the sodium content of the fat-free brands is lower than that of the regular brands, but the variability of the sodium content for regular cheese (excluding the outlier) is less than that of the fat-free brands.

FIGURE 2.17

Box plots for regular and fat-free cheese

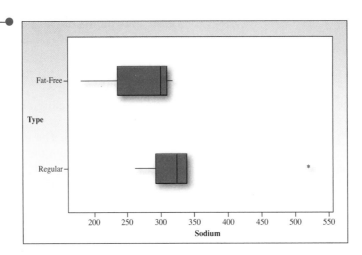

2.7 EXERCISES

BASIC TECHNIQUES

2.40 Given the following data set: 2.3, 1.0, 2.1, 6.5, 2.8, 8.8, 1.7, 2.9, 4.4, 5.1, 2.0

a. Find the positions of the lower and upper quartiles.

b. Sort the data from smallest to largest and find the lower and upper quartiles.

c. Calculate the IQR.

2.41 Find the five-number summary and the IQR for these data:

19, 12, 16, 0, 14, 9, 6, 1, 12, 13, 10, 19, 7, 5, 8

2.42 Given the following data set: 8, 7, 1, 4, 6, 6, 4, 5, 7, 6, 3, 0

a. Find the five-number summary and the IQR.

b. Calculate $\bar{x}$ and s.

c. Calculate the z-score for the smallest and largest observations. Is either of these observations unusually large or unusually small?

2.43 Given the following data set: .23, .30, .35, .41, .56, .58, .76, .80

a. Find the lower and upper quartiles.

b. Calculate the IQR.

c. Calculate the lower and upper fences. Are there any outliers?

2.44 Construct a box plot for these data and identify any outliers:

25, 22, 26, 23, 27, 26, 28, 18, 25, 24, 12

2.45 Construct a box plot for these data and identify any outliers:

3, 9, 10, 2, 6, 7, 5, 8, 6, 6, 4, 9, 22

APPLICATIONS

2.46 If you scored at the 69th percentile on a placement test, how does your score compare with others?

2.47 Mercury Concentration in Dolphins Environmental scientists are increasingly concerned with the accumulation of toxic elements in marine mammals and the transfer of such elements to the animals' offspring. The striped dolphin (*Stenella coeruleoalba*), considered to be a top predator in the marine food chain, was the subject of one such study. The mercury concentrations (micrograms/gram) in the livers of 28 male striped dolphins were as follows:

1.70	183.00	221.00	286.00
1.72	168.00	406.00	315.00
8.80	218.00	252.00	241.00
5.90	180.00	329.00	397.00
101.00	264.00	316.00	209.00
85.40	481.00	445.00	314.00
118.00	485.00	278.00	318.00

a. Calculate the five-number summary for the data.

b. Construct a box plot for the data.

c. Are there any outliers?

d. If you knew that the first four dolphins were all less than 3 years old, while all the others were more than 8 years old, would this information help explain the difference in the magnitude of those four observations? Explain.

2.48 Hamburger Meat The weights (in pounds) of the 27 packages of ground beef from Exercise 2.24 (see data set EX0224) are listed here in order from smallest to largest:

.75	.83	.87	.89	.89	.89	.92
.93	.96	.96	.97	.98	.99	1.06
1.08	1.08	1.12	1.12	1.14	1.14	1.17
1.18	1.18	1.24	1.28	1.38	1.41	

a. Confirm the values of the mean and standard deviation, calculated in Exercise 2.24 as $\bar{x} = 1.05$ and $s = .17$.

b. The two largest packages of meat weigh 1.38 and 1.41 pounds. Are these two packages unusually heavy? Explain.

c. Construct a box plot for the package weights. What does the position of the median line and the length of the whiskers tell you about the shape of the distribution?

2.49 Comparing NFL Quarterbacks How EX0249 does Aaron Rodgers, quarterback for the 2011 Super Bowl winners, the Green Bay Packers, compare to Drew Brees, quarterback for the 2010 Super Bowl winners, the New Orleans Saints? The table below shows the number of completed passes for each athlete during the 2010 NFL football season:[11]

Aaron Rodgers			Drew Brees		
19	21	7	27	37	25
19	15	25	28	34	29
34	27	19	30	27	35
12	22		33	29	22
27	26		24	23	
18	21		21	24	

a. Calculate five-number summaries for the number of passes completed by both Aaron Rodgers and Drew Brees.

b. Construct box plots for the two sets of data. Are there any outliers? What do the box plots tell you about the shapes of the two distributions?

c. Write a short paragraph comparing the number of pass completions for the two quarterbacks.

2.50 Presidential Vetoes The set of presidential vetoes in Exercise 1.47 and data set EX0147 is listed here, along with a box plot generated by *MINITAB*. Use the box plot to describe the shape of the distribution and identify any outliers.

Washington	2	B. Harrison	19
J. Adams	0	Cleveland	42
Jefferson	0	McKinley	6
Madison	5	T. Roosevelt	42
Monroe	1	Taft	30
J. Q. Adams	0	Wilson	33
Jackson	5	Harding	5
Van Buren	0	Coolidge	20
W. H. Harrison	0	Hoover	21
Tyler	6	F. D. Roosevelt	372
Polk	2	Truman	180
Taylor	0	Eisenhower	73
Fillmore	0	Kennedy	12
Pierce	9	L. Johnson	16
Buchanan	4	Nixon	26
Lincoln	2	Ford	48
A. Johnson	21	Carter	13
Grant	45	Reagan	39
Hayes	12	G. H. W. Bush	29
Garfield	0	Clinton	36
Arthur	4	G. W. Bush	11
Cleveland	304	Obama	1

Source: The World Almanac and Book of Facts 2011

Box plot for Exercise 2.50

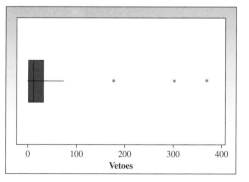

2.51 Survival Times Altman and Bland report the survival times for patients with active hepatitis, half treated with prednisone and half receiving no treatment.[12] The survival times (in months) (Exercise 1.25 and EX0125) are adapted from their data for those treated with prednisone.

8	87	127	147
11	93	133	148
52	97	139	157
57	109	142	162
65	120	144	165

a. Can you tell by looking at the data whether it is roughly symmetric? Or is it skewed?

b. Calculate the mean and the median. Use these measures to decide whether or not the data are symmetric or skewed.

c. Draw a box plot to describe the data. Explain why the box plot confirms your conclusions in part b.

Data set **2.52 Utility Bills in Southern California,**
EX0252 **again** The monthly utility bills for a household in Riverside, California, were recorded for 12 consecutive months starting in January 2010:

Month	Amount ($)	Month	Amount ($)
January	288.02	July	311.20
February	230.60	August	370.23
March	216.85	September	368.57
April	243.74	October	301.79
May	236.96	November	271.99
June	288.57	December	298.12

a. Construct a box plot for the monthly utility costs.

b. What does the box plot tell you about the distribution of utility costs for this household?

2.53 What's Normal? again Refer to Exercise 1.67 and data set EX0167. In addition to the normal body temperature in degrees Fahrenheit for the 130 individuals, the data record the gender of the individuals. Box plots for the two groups, male and female, are shown below:[13]

Box plots for Exercise 2.53

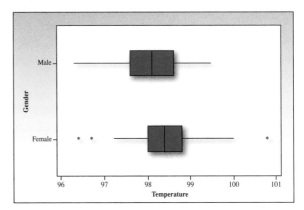

How would you describe the similarities and differences between male and female temperatures in this data set?

CHAPTER REVIEW

Key Concepts and Formulas

I. Measures of the Center of a Data Distribution

1. Arithmetic mean (mean) or average

   a. Population: μ

   b. Sample of n measurements: $\bar{x} = \dfrac{\Sigma x_i}{n}$

2. Median; **position** of the median $= .5(n + 1)$

3. Mode

4. The median may be preferred to the mean if the data are highly skewed.

II. Measures of Variability

1. Range: $R = $ largest $-$ smallest

2. Variance

   a. Population of N measurements:

   $$\sigma^2 = \frac{\Sigma(x_i - \mu)^2}{N}$$

   b. Sample of n measurements:

   $$s^2 = \frac{\Sigma(x_i - \bar{x})^2}{n - 1} = \frac{\Sigma x_i^2 - \dfrac{(\Sigma x_i)^2}{n}}{n - 1}$$

3. Standard deviation

   a. Population: $\sigma = \sqrt{\sigma^2}$

   b. Sample: $s = \sqrt{s^2}$

4. A rough approximation for s can be calculated as $s \approx R/4$. The divisor can be adjusted depending on the sample size.

III. Tchebysheff's Theorem and the Empirical Rule

1. Use Tchebysheff's Theorem for any data set, regardless of its shape or size.

   a. At least $1 - (1/k^2)$ of the measurements lie within k standard deviations of the mean.

   b. This is only a lower bound; there may be more measurements in the interval.

2. The Empirical Rule can be used only for relatively mound-shaped data sets. Approximately 68%, 95%, and 99.7% of the measurements are within one, two, and three standard deviations of the mean, respectively.

IV. Measures of Relative Standing

1. Sample z-score: $z = \dfrac{x - \bar{x}}{s}$

2. pth percentile; $p\%$ of the measurements are smaller, and $(100 - p)\%$ are larger.

3. Lower quartile, Q_1; **position** of $Q_1 = .25\,(n + 1)$

4. Upper quartile, Q_3; **position** of $Q_3 = .75\,(n + 1)$

5. Interquartile range: IQR $= Q_3 - Q_1$

V. The Five-Number Summary and Box Plots

1. The **five-number summary:**

   | Min | Q_1 | Median | Q_3 | Max |

   One-fourth of the measurements in the data set lie between each of the four adjacent pairs of numbers.

2. Box plots are used for detecting outliers and shapes of distributions.

3. Q_1 and Q_3 form the ends of the box. The median line is in the interior of the box.

4. Upper and lower fences are used to find outliers, observations that lie outside these fences.

   a. **Lower fence:** $Q_1 - 1.5(\text{IQR})$

   b. **Upper fence:** $Q_3 + 1.5(\text{IQR})$

5. **Outliers** are marked on the box plot with an asterisk (*).

6. **Whiskers** are connected to the box from the smallest and largest observations that are *not* outliers.

7. Skewed distributions usually have a long whisker *in the direction of the skewness,* and the median line is drawn *away from the direction of the skewness.*

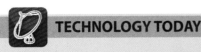

Numerical Descriptive Measures in *Excel*

MS Excel provides most of the basic descriptive statistics presented in Chapter 2 using a single command on the **Data** tab. Other descriptive statistics can be calculated using the **Function** command on the **Formulas** tab.

EXAMPLE 2.15

The following data are the front and rear leg rooms (in inches) for nine different sports utility vehicles:[14]

Make & Model	Front Leg Room	Rear Leg Room
Acura MDX	41.0	28.5
Buick Enclave	41.5	30.0
Chevy TrailBlazer	40.0	25.5
Chevy Tahoe Hybrid V8 CVT	41.0	27.5
GMC Terrain 1LT 4-cyl	43.0	31.0
Honda CR-V	41.0	29.5
Hyundai Tucson	42.5	29.5
Kia Sportage	40.0	29.0
Lexus GX	42.0	30.0

1. Since the data involve two variables and a third labeling variable, enter the data into the first three columns of an *Excel* spreadsheet, using the labels in the table. Select **Data ▶ Data Analysis ▶ Descriptive Statistics**, and highlight or type the **Input range** (the data in the second and third columns) into the Descriptive Statistics Dialog box (Figure 2.18(a)). Type an Output location, make sure the check boxes for "Labels in First Row" and "Summary Statistics" are both checked, and click **OK**. The summary statistics (Figure 2.18(b)) will appear in the selected location in your spreadsheet.

FIGURE 2.18

(a)

(b)

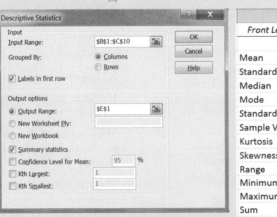

	E	F		G	H
	Front Leg Room			*Rear Leg Room*	
Mean		41.333	Mean		28.944
Standard Error		0.344	Standard Error		0.543
Median		41	Median		29.5
Mode		41	Mode		30
Standard Deviation		1.031	Standard Deviation		1.629
Sample Variance		1.063	Sample Variance		2.653
Kurtosis		-0.745	Kurtosis		1.735
Skewness		0.245	Skewness		-1.212
Range		3	Range		5.5
Minimum		40	Minimum		25.5
Maximum		43	Maximum		31
Sum		372	Sum		260.5
Count		9	Count		9

2. You may notice that some of the cells in the spreadsheet are overlapping. To adjust this, highlight the affected columns and click the **Home** tab. In the **Cells** group, choose **Format ▶ AutoFit Column Width**. You may want to modify the

appearance of the output by decreasing the decimal accuracy in certain cells. Highlight the appropriate cells and click the **Decrease Decimal** icon $\overset{\bullet0}{.\overset{}{00}}$ (**Home** tab, **Number** group) to modify the output. We have displayed the accuracy to three decimal places.

3. Notice that the sample quartiles, Q_1 and Q_3, are not given in the *Excel* output in Figure 2.18(b). You can calculate the quartiles using the function command. Place your cursor into an empty cell and select **Formulas ▶ More Functions ▶ Statistical ▶ QUARTILE.EXC**. Highlight the appropriate cells in the box marked "Array" and type an integer (0 = min, 1 = first quartile, 2 = median, 3 = third quartile, or 4 = max) in the box marked "Quart." The quartile (calculated using this textbook's method) will appear in the cell you have chosen. An alternative method for calculating the quartiles will be used if you select **Formulas ▶ More Functions ▶ Statistical ▶ QUARTILE.INC**. (NOTE: This function is called **QUARTILE** in *Excel 2007* and earlier versions.) Using the two quartiles, you can calculate the IQR and construct a box plot by hand.

Numerical Descriptive Measures in *MINITAB*

MINITAB provides most of the basic descriptive statistics presented in Chapter 2 using a single command in the drop-down menus.

The following data are the front and rear leg rooms (in inches) for nine different sports utility vehicles:[14]

Make and Model	Front Leg Room	Rear Leg Room
Acura MDX	41.0	28.5
Buick Enclave	41.5	30.0
Chevy TrailBlazer	40.0	25.5
Chevy Tahoe Hybrid V8 CVT	41.0	27.5
GMC Terrain 1LT 4-cyl	43.0	31.0
Honda CR-V	41.0	29.5
Hyundai Tucson	42.5	29.5
Kia Sportage	40.0	29.0
Lexus GX	42.0	30.0

1. Since the data involve two variables and a third labeling variable, enter the data into the first three columns of a *MINITAB* worksheet, using the labels in the table. Using the drop-down menus, select **Stat ▶ Basic Statistics ▶ Display Descriptive Statistics.** The Dialog box is shown in Figure 2.19(a).

FIGURE 2.19

(a)

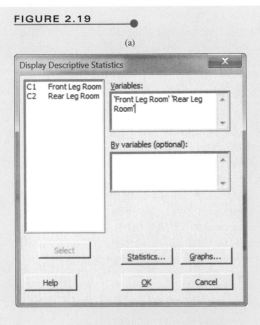

(b)

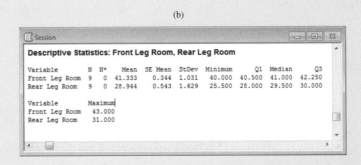

2. Now click on the Variables box and **select** both columns from the list on the left. (You can click on the **Graphs** option and choose one of several graphs if you like. You may also click on the **Statistics** option to select the statistics you would like to see displayed.) Click **OK.** A display of descriptive statistics for both columns will appear in the Session window (see Figure 2.19(b)). You may print this output using **File ▶ Print Session Window** if you choose.

3. To examine the distribution of the two variables and look for outliers, you can create box plots using the command **Graph ▶ Boxplot ▶ One Y ▶ Simple.** Click **OK.** Select the appropriate column of measurements in the Dialog box (see Figure 2.20(a)). You can change the appearance of the box plot in several ways. **Scale ▶ Axes and Ticks** will allow you to transpose the axes and orient the box plot horizontally, when you check the box marked "Transpose value and category scales." **Multiple Graphs** provides printing options for multiple box plots. **Labels** will let you annotate the graph with titles and footnotes. If you have entered data into the worksheet as a frequency distribution (values in one column, frequencies in another), the **Data Options** will allow the data to be read in that format. The box plot for the rear leg rooms is shown in Figure 2.20(b).

4. Save this worksheet in a file called "Leg Room" before exiting *MINITAB.* We will use it again in Chapter 3.

FIGURE 2.20 ●

(a)

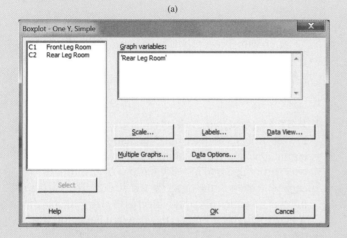

(b)

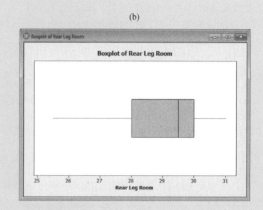

Supplementary Exercises

Data set EX0254 **2.54 Raisins** The number of raisins in each of 14 miniboxes (1/2-ounce size) was counted for a generic brand and for Sunmaid brand raisins. The two data sets are shown here:

Generic Brand				Sunmaid			
25	26	25	28	25	29	24	24
26	28	28	27	28	24	28	22
26	27	24	25	25	28	30	27
26	26			28	24		

a. What are the mean and standard deviation for the generic brand?

b. What are the mean and standard deviation for the Sunmaid brand?

c. Compare the centers and variabilities of the two brands using the results of parts a and b.

2.55 Raisins, continued Refer to Exercise 2.54.

a. Find the median, the upper and lower quartiles, and the IQR for each of the two data sets.

b. Construct two box plots on the same horizontal scale to compare the two sets of data.

c. Draw two stem and leaf plots to depict the shapes of the two data sets. Do the box plots in part b verify these results?

d. If we can assume that none of the boxes of raisins are being underfilled (that is, they all weigh approximately 1/2 ounce), what do your results say about the average number of raisins for the two brands?

Data set EX0256 **2.56 TV Viewers** The number of television viewing hours per household and the prime viewing times are two factors that affect television advertising income. A random sample of 25 households in a particular viewing area produced the following estimates of viewing hours per household:

3.0	6.0	7.5	15.0	12.0
6.5	8.0	4.0	5.5	6.0
5.0	12.0	1.0	3.5	3.0
7.5	5.0	10.0	8.0	3.5
9.0	2.0	6.5	1.0	5.0

a. Scan the data and use the range to find an approximate value for s. Use this value to check your calculations in part b.

b. Calculate the sample mean $\bar{x}$ and the sample standard deviation s. Compare s with the approximate value obtained in part a.

c. Find the percentage of the viewing hours per household that falls into the interval $\bar{x} \pm 2s$. Compare with the corresponding percentage given by the Empirical Rule.

2.57 A Recurring Illness Refer to Exercise 1.26 and data set EX0126. The lengths of time (in months) between the onset of a particular illness and its recurrence were recorded:

2.1	4.4	2.7	32.3	9.9
9.0	2.0	6.6	3.9	1.6
14.7	9.6	16.7	7.4	8.2
19.2	6.9	4.3	3.3	1.2
4.1	18.4	.2	6.1	13.5
7.4	.2	8.3	.3	1.3
14.1	1.0	2.4	2.4	18.0
8.7	24.0	1.4	8.2	5.8
1.6	3.5	11.4	18.0	26.7
3.7	12.6	23.1	5.6	.4

a. Find the range.

b. Use the range approximation to find an approximate value for s.

c. Compute s for the data and compare it with your approximation from part b.

2.58 Tuna Fish, again Refer to Exercise 2.8. The prices of a 6-ounce can or a 7.06-ounce pouch for 14 different brands of water-packed light tuna, based on prices paid nationally in supermarkets, are reproduced here.[4]

.99	1.92	1.23	.85	.65	.53	1.41
1.12	.63	.67	.69	.60	.60	.66

a. Calculate the five-number summary.

b. Construct a box plot for the data. Are there any outliers?

c. The value $x = 1.92$ looks large in comparison to the other prices. Use a z-score to decide whether this is an unusually expensive brand of tuna.

2.59 A Recurring Illness, again Find the median and the lower and upper quartiles for the data on times until recurrence of an illness in Exercise 2.57. Use these descriptive measures to construct a box plot for the data. Use the box plot to describe the data distribution.

2.60 A Recurring Illness, continued Refer to Exercise 2.57.

a. Examine the data and count the number of observations that fall into the intervals $\bar{x} \pm s$, $\bar{x} \pm 2s$, and $\bar{x} \pm 3s$.

b. Do the percentages that fall into these intervals agree with Tchebysheff's Theorem? With the Empirical Rule?

c. Why might the Empirical Rule be unsuitable for describing these data?

2.61 Electrolysis An analytical chemist wanted to use electrolysis to determine the number of moles of cupric ions in a given volume of solution. The solution was partitioned into $n = 30$ portions of .2 milliliter each, and each of the portions was tested. The average number of moles of cupric ions for the $n = 30$ portions was found to be .17 mole; the standard deviation was .01 mole.

a. Describe the distribution of the measurements for the $n = 30$ portions of the solution using Tchebysheff's Theorem.

b. Describe the distribution of the measurements for the $n = 30$ portions of the solution using the Empirical Rule. (Do you expect the Empirical Rule to be suitable for describing these data?)

c. Suppose the chemist had used only $n = 4$ portions of the solution for the experiment and obtained the readings .15, .19, .17, and .15. Would the Empirical Rule be suitable for describing the $n = 4$ measurements? Why?

2.62 Chloroform According to the EPA, chloroform, which in its gaseous form is suspected of being a cancer-causing agent, is present in small quantities in all of the country's 240,000 public water sources. If the mean and standard deviation of the amounts of chloroform present in the water sources are 34 and 53 micrograms per liter, respectively, describe the distribution for the population of all public water sources.

2.63 Achievement Tests Mathematics achievement test scores for 400 students were found to have a mean and a variance equal to 600 and 4900, respectively. If the distribution of test scores was mound-shaped, approximately how many of the scores would fall into the interval 530 to 670? Approximately how many scores would be expected to fall into the interval 460 to 740?

2.64 Sleep and the College Student How much sleep do you get on a typical school night? A group of 10 college students were asked to report the number of hours that they slept on the previous night with the following results:

$$7, \quad 6, \quad 7.25, \quad 7, \quad 8.5, \quad 5, \quad 8, \quad 7, \quad 6.75, \quad 6$$

a. Find the mean and the standard deviation of the number of hours of sleep for these 10 students.

b. Calculate the z-score for the largest value ($x = 8.5$). Is this an unusually sleepy college student?

c. What is the most frequently reported measurement? What is the name for this measure of center?

d. Construct a box plot for the data. Does the box plot confirm your results in part b? [HINT: Since the z-score and the box plot are two unrelated methods for detecting outliers, and use different types of statistics, they do not necessarily have to (but usually do) produce the same results.]

Data set **2.65 Gas Mileage** The miles per gallon (mpg) for each of 20 medium-sized cars selected from a production line during the month of March follow.

23.1	21.3	23.6	23.7
20.2	24.4	25.3	27.0
24.7	22.7	26.2	23.2
25.9	24.7	24.4	24.2
24.9	22.2	22.9	24.6

a. What are the maximum and minimum miles per gallon? What is the range?

b. Construct a relative frequency histogram for these data. How would you describe the shape of the distribution?

c. Find the mean and the standard deviation.

d. Arrange the data from smallest to largest. Find the z-scores for the largest and smallest observations. Would you consider them to be outliers? Why or why not?

e. What is the median?

f. Find the lower and upper quartiles.

2.66 Gas Mileage, continued Refer to Exercise 2.65. Construct a box plot for the data. Are there any outliers? Does this conclusion agree with your results in Exercise 2.65?

2.67 Polluted Seawater Petroleum pollution in seas and oceans stimulates the growth of some types of bacteria. A count of petroleumlytic micro-organisms (bacteria per 100 milliliters) in 10 portions of seawater gave these readings:

49, 70, 54, 67, 59, 40, 61, 69, 71, 52

a. Guess the value for s using the range approximation.

b. Calculate $\bar{x}$ and s and compare with the range approximation of part a.

c. Construct a box plot for the data and use it to describe the data distribution.

2.68 Basketball Attendances at a high school's basketball games were recorded and found to have a sample mean and variance of 420 and 25, respectively. Calculate $\bar{x} \pm s$, $\bar{x} \pm 2s$, and $\bar{x} \pm 3s$ and then state the approximate fractions of measurements you would expect to fall into these intervals according to the Empirical Rule.

2.69 SAT Tests The College Board's verbal and mathematics scholastic aptitude tests are scored on a scale of 200 to 800. It seems reasonable to assume that a distribution of all test scores, either verbal or math, is mound-shaped. If σ is the standard deviation of one of these distributions, what is the largest value (approximately) that σ might assume? Explain.

2.70 Summer Camping A favorite summer pastime for many Americans is camping. In fact, camping has become so popular at the California beaches that reservations must sometimes be made months in advance! Data from a *USA Today* Snapshot is shown below.[15]

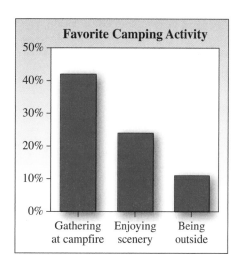

The Snapshot also reports that men go camping 2.9 times a year, women go 1.7 times a year; and men are more likely than women to want to camp more often. What does the magazine mean when they talk about 2.9 or 1.7 times a year?

2.71 Long-Stemmed Roses A strain of long-stemmed roses has an approximate normal distribution with a mean stem length of 15 inches and standard deviation of 2.5 inches.

a. If one accepts as "long-stemmed roses" only those roses with a stem length greater than 12.5 inches, what percentage of such roses would be unacceptable?

b. What percentage of these roses would have a stem length between 12.5 and 20 inches?

2.72 Drugs for Hypertension A pharmaceutical company wishes to know whether an experimental drug being tested in its laboratories has any effect on systolic blood pressure. Fifteen randomly selected subjects were given the drug, and their systolic blood pressures (in millimeters) are recorded.

172	148	123
140	108	152
123	129	133
130	137	128
115	161	142

a. Guess the value of s using the range approximation.

b. Calculate $\bar{x}$ and s for the 15 blood pressures.

c. Find two values, a and b, such that at least 75% of the measurements fall between a and b.

2.73 Lumber Rights A company interested in lumbering rights for a certain tract of slash pine trees is told that the mean diameter of these trees is 14 inches with a standard deviation of 2.8 inches. Assume the distribution of diameters is roughly mound-shaped.

a. What fraction of the trees will have diameters between 8.4 and 22.4 inches?

b. What fraction of the trees will have diameters greater than 16.8 inches?

2.74 Social Ambivalence The following data represent the social ambivalence scores for 15 people as measured by a psychological test. (The higher the score, the stronger the ambivalence.)

9	13	12
14	15	11
10	4	10
8	19	13
11	17	9

a. Guess the value of s using the range approximation.

b. Calculate $\bar{x}$ and s for the 15 social ambivalence scores.

c. What fraction of the scores actually lie in the interval $\bar{x} \pm 2s$?

2.75 TV Commercials The mean duration of television commercials on a given network is 75 seconds, with a standard deviation of 20 seconds. Assume that durations are approximately normally distributed.

a. What is the approximate probability that a commercial will last less than 35 seconds?

b. What is the approximate probability that a commercial will last longer than 55 seconds?

2.76 Parasites in Foxes A random sample of 100 foxes was examined by a team of veterinarians to determine the prevalence of a particular type of parasite. Counting the number of parasites per fox, the veterinarians found that 69 foxes had no parasites, 17 had one parasite, and so on. A frequency tabulation of the data is given here:

Number of Parasites, x	0	1	2	3	4	5	6	7	8
Number of Foxes, f	69	17	6	3	1	2	1	0	1

a. Construct a relative frequency histogram for x, the number of parasites per fox.

b. Calculate $\bar{x}$ and s for the sample.

c. What fraction of the parasite counts fall within two standard deviations of the mean? Within three standard deviations? Do these results agree with Tchebysheff's Theorem? With the Empirical Rule?

2.77 College Teachers Consider a population consisting of the number of teachers per college at small 2-year colleges. Suppose that the number of teachers per college has an average $\mu = 175$ and a standard deviation $\sigma = 15$.

a. Use Tchebysheff's Theorem to make a statement about the percentage of colleges that have between 145 and 205 teachers.

b. Assume that the population is normally distributed. What fraction of colleges have more than 190 teachers?

2.78 Is It Accurate? From the following data, a student calculated s to be .263. On what grounds might we doubt his accuracy? What is the correct value (to the nearest hundredth)?

17.2 17.1 17.0 17.1 16.9 17.0 17.1 17.0 17.3 17.2
17.1 17.0 17.1 16.9 17.0 17.1 17.3 17.2 17.4 17.1

2.79 Homerun Kings In the summer of 2001, Barry Bonds began his quest to break Mark McGwire's record of 70 home runs hit in a single season. At the end of the 2003 major league baseball season, the number of home runs hit per season by each of four major league superstars over each player's career were recorded, and are shown in the box plots below:[16]

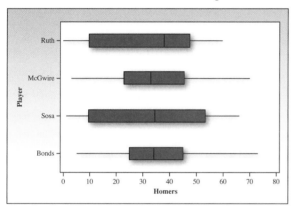

Write a short paragraph comparing the home run hitting patterns of these four players.

2.80 Barry Bonds In the seasons that followed his 2001 record-breaking season, Barry Bonds hit 46, 45, 45, 5, 26, and 28 homers, respectively, until he retired from major league baseball in 2007 (www.ESPN.com).[16] Two box plots, one of Bond's homers through 2001, and a second including the years 2002–2007 follow.

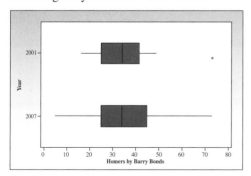

The statistics used to construct these box plots are given in the table.

Years	Min	Q_1	Median	Q_3	IQR	Max	n
2001	16	25.00	34.00	41.50	16.5	73	16
2007	5	25.00	34.00	45.00	20.0	73	22

a. Calculate the upper fences for both of these box plots.

b. Can you explain why the record number of homers is an outlier in the 2001 box plot, but not in the 2007 box plot?

2.81 Ages of Pennies Here are the ages of 50 pennies from Exercise 1.45 and data set EX0145. The data have been sorted from smallest to largest.

0	0	0	0	0	0	0	0	0	0
0	0	1	1	1	1	1	1	2	2
2	3	3	3	4	4	5	5	5	5
6	8	9	9	10	16	17	17	19	19
19	20	20	21	22	23	25	25	28	36

a. What is the average age of the pennies?

b. What is the median age of the pennies?

c. Based on the results of parts a and b, how would you describe the age distribution of these 50 pennies?

d. Construct a box plot for the data set. Are there any outliers? Does the box plot confirm your description of the distribution's shape?

2.82 Snapshots Here are a few facts reported as Snapshots in *USA Today.*

- About 12% of America's volunteers spend more than 5 hours per week volunteering.[17]

- Fifty-eight percent of all cars in operation are at least 8 years old.[18]

- Twenty-two percent of all fans are willing to pay $75 or more for a ticket to one of the top 100 concert tours.[19]

Identify the variable x being measured, and any percentiles you can determine from this information.

2.83 Breathing Patterns Research psychologists are interested in finding out whether a person's breathing patterns are affected by a particular experimental treatment. To determine the general respiratory patterns of the $n = 30$ people in the study, the researchers collected some baseline measurements—the total ventilation in liters of air per minute adjusted for body size—for each person before the treatment. The data are shown here, along with some descriptive tools generated by *MINITAB* and *MS Excel.*

5.23	4.79	5.83	5.37	4.35	5.54	6.04	5.48	6.58	4.82
5.92	5.38	6.34	5.12	5.14	4.72	5.17	4.99	4.51	5.70
4.67	5.77	5.84	6.19	5.58	5.72	5.16	5.32	4.96	5.63

Descriptive Statistics: Liters

```
Variable    N    N*     Mean   SE Mean    StDev
Liters     30    0    5.3953   0.0997    0.5462

Minimum     Q1  Median      Q3  Variable  Maximum
 4.3500 4.9825  5.3750  5.7850  Liters     6.5800
```

Stem and Leaf Display: Liters

```
Stem-and-leaf of Liters   N = 30
Leaf Unit = 0.10

 1    4  3
 2    4  5
 5    4  677
 8    4  899
12    5  1111
(4)   5  2333
14    5  455
11    5  6777
 7    5  889
 4    6  01
 2    6  3
 1    6  5
```

MS Excel Descriptive Statistics

Liters	
Mean	5.3953
Standard Error	0.0997
Median	5.3750
Mode	#N/A
Standard Deviation	0.5462
Sample Variance	0.2983
Kurtosis	20.4069
Skewness	0.1301
Range	2.23
Minimum	4.35
Maximum	6.58
Sum	161.86
Count	30

a. Summarize the characteristics of the data distribution using the computer output.

b. Does the Empirical Rule provide a good description of the proportion of measurements that fall within two or three standard deviations of the mean? Explain.

c. How large or small does a ventilation measurement have to be before it is considered unusual?

Data set **2.84 Arranging Objects** The following data
EX0284   are the response times in seconds for $n = 25$ first graders to arrange three objects by size.

```
5.2  3.8  5.7  3.9  3.7
4.2  4.1  4.3  4.7  4.3
3.1  2.5  3.0  4.4  4.8
3.6  3.9  4.8  5.3  4.2
4.7  3.3  4.2  3.8  5.4
```

a. Find the mean and the standard deviation for these 25 response times.

b. Order the data from smallest to largest.

c. Find the z-scores for the smallest and largest response times. Is there any reason to believe that these times are unusually large or small? Explain.

2.85 Arranging Objects, continued Refer to Exercise 2.84.

a. Find the five-number summary for this data set.

b. Construct a box plot for the data.

c. Are there any unusually large or small response times identified by the box plot?

d. Construct a stem and leaf display for the response times. How would you describe the shape of the distribution? Does the shape of the box plot confirm this result?

CASE STUDY

 Batting

The Boys of Summer

Which baseball league has had the best hitters? Many of us have heard of baseball greats like Stan Musial, Hank Aaron, Roberto Clemente, and Pete Rose of the National League and Ty Cobb, Babe Ruth, Ted Williams, Rod Carew, and Wade Boggs of the American League. But have you ever heard of Willie Keeler, who batted .432 for the Baltimore Orioles, or Nap Lajoie, who batted .422 for the Philadelphia A's? The batting averages for the batting champions of the National and American Leagues are given on the CourseMate Web site.

The batting averages for the National League begin in 1876 with Roscoe Barnes, whose batting average was .403 when he played with the Chicago Cubs. The last entry for the National League is for the year 2010, when Carlos Gonzalez of the Colorado Rockies averaged .336. The American League records begin in 1901 with Nap Lajoie of the Philadelphia A's, who batted .422, and end in 2010 with Josh Hamilton of the Texas Rangers, who batted .359.[9] How can we summarize the information in this data set?

1. Use *MS Excel, MINITAB,* or another statistical software package to describe the batting averages for the American and National League batting champions. Generate any graphics that may help you in interpreting these data sets.

2. Does one league appear to have a higher percentage of hits than the other? Do the batting averages of one league appear to be more variable than the other?

3. Are there any outliers in either league?

4. Summarize your comparison of the two baseball leagues.

Describing Bivariate Data

© Janis Christie/Photodisc/Getty Images

 NEED TO KNOW...

How to Calculate the Correlation Coefficient
How to Calculate the Regression Line

Do You *Think* Your Dishes Are *Really* Clean?

Does the price of an appliance, such as a dishwasher, convey something about its quality? In the case study at the end of this chapter, we rank 48 different brands of dishwashers according to their prices, and then we rate them on various characteristics, such as how the dishwasher performs, how much noise it makes, its cost for either gas or electricity, its cycle time, and its water use. The techniques presented in this chapter will help to answer our question.

3.1 BIVARIATE DATA

Very often researchers are interested in more than just one variable that can be measured during their investigation. For example, an auto insurance company might be interested in the number of vehicles owned by a policyholder as well as the number of drivers in the household. An economist might need to measure the amount spent per week on groceries in a household and also the number of people in that household. A real estate agent might measure the selling price of a residential property and the square footage of the living area.

When two variables are measured on a single experimental unit, the resulting data are called **bivariate data.** How should you display these data? Not only are both variables important when studied separately, but you also may want to explore the *relationship between the two variables.* Methods for graphing bivariate data, whether the variables are qualitative or quantitative, allow you to study the two variables together. As with *univariate data,* you use different graphs depending on the type of variables you are measuring.

 NEED A TIP?

"**Bi**" means "two."
Bivariate data generate pairs of measurements.

3.2 GRAPHS FOR CATEGORICAL VARIABLES

When at least one of the two variables is *qualitative* or *categorical*, you can use either simple or more intricate pie charts, line charts, and bar charts to display and describe the data. Sometimes you will have one qualitative and one quantitative variable that have been measured in two different populations or groups. In this case, you can use two **side-by-side pie charts** or a bar chart in which the bars for the two populations are placed side by side. Another option is to use a **stacked bar chart,** in which the bars for each category are stacked on top of each other.

EXAMPLE   3.1

Are professors in private colleges paid more than professors at public colleges? The data in Table 3.1 were collected from a sample of 400 college professors whose rank, type of college, and salary were recorded.[1] The number in each cell is the average salary (in thousands of dollars) for all professors who fell into that category. Use a graph to answer the question posed for this sample.

TABLE 3.1 ● **Salaries of Professors by Rank and Type of College**

	Full Professor	Associate Professor	Assistant Professor
Public	98.1	72.7	61.5
Private	110.5	74.9	61.6

Source: Digest of Educational Statistics

Solution   To display the average salaries of these 400 professors, you can use a side-by-side bar chart, as shown in Figure 3.1. The height of the bars is the average salary, with each pair of bars along the horizontal axis representing a different professorial rank. Salaries are substantially higher for full professors in private colleges, however, there are less striking differences at the lower two ranks.

FIGURE 3.1 ●

Comparative bar charts for
Example 3.1

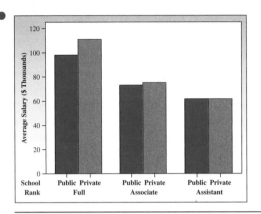

EXAMPLE **3.2** — Along with the salaries for the 400 college professors in Example 3.1, the researcher recorded two qualitative variables for each professor: rank and type of college. Table 3.2 shows the number of professors in each of the $2 \times 3 = 6$ categories. Use comparative charts to describe the data. Do the private colleges employ as many high-ranking professors as the public colleges do?

TABLE 3.2 ● **Number of Professors by Rank and Type of College**

	Full Professor	Associate Professor	Assistant Professor	Total
Public	24	57	69	150
Private	60	78	112	250

Solution The numbers in the table are not quantitative measurements on a single experimental unit (the professor). They are *frequencies,* or counts, of the number of professors who fall into each category. To compare the numbers of professors at public and private colleges, you might draw two pie charts and display them side by side, as in Figure 3.2.

FIGURE 3.2 ●

Comparative pie charts for
Example 3.2

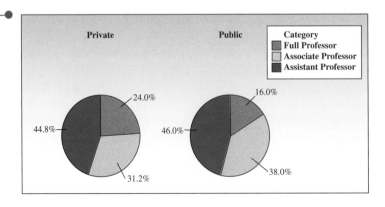

Alternatively, you could draw either a stacked or a side-by-side bar chart. The stacked bar chart is shown in Figure 3.3.

FIGURE 3.3

Stacked bar chart for
Example 3.2

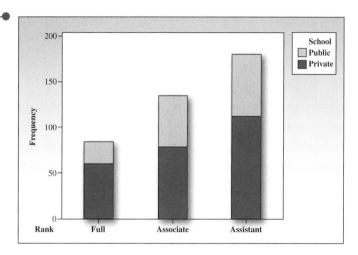

Although the graphs are not strikingly different, you can see that public colleges have fewer full professors and more associate professors than private colleges. The reason for these differences is not clear, but you might speculate that private colleges, with their higher salaries, are able to attract more full professors. Or perhaps public colleges are not as willing to promote professors to the higher-paying ranks. In any case, the graphs provide a means for comparing the two sets of data.

You can also compare the distributions for public versus private colleges by creating *conditional data distributions*. These conditional distributions are shown in Table 3.3. One distribution shows the proportion of professors in each of the three ranks under the *condition* that the college is public, and the other shows the proportions under the *condition* that the college is private. These *relative frequencies* are easier to compare than the *actual frequencies* and lead to the same conclusions:

- The proportion of assistant professors is roughly the same for both public and private colleges.
- Public colleges have a smaller proportion of full professors and a larger proportion of associate professors.

TABLE 3.3 ● **Proportions of Professors by Rank for Public and Private Colleges**

	Full Professor	Associate Professor	Assistant Professor	Total
Public	$\frac{24}{150} = .16$	$\frac{57}{150} = .38$	$\frac{69}{150} = .46$	1.00
Private	$\frac{60}{250} = .24$	$\frac{78}{250} = .31$	$\frac{112}{250} = .45$	1.00

3.2 **EXERCISES**

BASIC TECHNIQUES

3.1 Gender Differences Male and female respondents to a questionnaire about gender differences are categorized into three groups according to their answers to the first question:

	Group 1	Group 2	Group 3
Men	37	49	72
Women	7	50	31

a. Create side-by-side pie charts to describe these data.
b. Create a side-by-side bar chart to describe these data.
c. Draw a stacked bar chart to describe these data.
d. Which of the three charts best depicts the difference or similarity of the responses of men and women?

3.2 State-by-State A group of items are categorized according to a certain attribute—X, Y, Z—and according to the state in which they are produced:

	X	Y	Z
New York	20	5	5
California	10	10	5

a. Create a comparative (side-by-side) bar chart to compare the numbers of items of each type made in California and New York.
b. Create a stacked bar chart to compare the numbers of items of each type made in the two states.
c. Which of the two types of presentation in parts a and b is more easily understood? Explain.
d. What other graphical methods could you use to describe the data?

3.3 Consumer Spending The table below shows the average amounts spent per week by men and women in each of four spending categories:

	A	B	C	D
Men	$54	$27	$105	$22
Women	21	85	100	75

a. What possible graphical methods could you use to compare the spending patterns of women and men?
b. Choose two different methods of graphing and display the data in graphical form.
c. What can you say about the similarities or differences in the spending patterns for men and women?
d. Which of the two methods used in part b provides a better descriptive graph?

APPLICATIONS

3.4 M&M'S The color distributions for two snack-size bags of M&M'S® candies, one plain and one peanut, are displayed in the table. Choose an appropriate graphical method and compare the distributions.

	Brown	Yellow	Red	Orange	Green	Blue
Plain	15	14	12	4	5	6
Peanut	6	2	2	3	3	5

3.5 How Much Free Time? When you were growing up, did you feel that you did not have enough free time? Parents and children have differing opinions on this subject. A research group surveyed 198 parents and 200 children and recorded their responses to the question, "How much free time does your child have?" or "How much free time do you have?" The responses are shown in the table below:[2]

	Just the Right Amount	Not Enough	Too Much	Don't Know
Parents	138	14	40	6
Children	130	48	16	6

a. Define the sample and the population of interest to the researchers.
b. Describe the variables that have been measured in this survey. Are the variables qualitative or quantitative? Are the data univariate or bivariate?
c. What do the entries in the cells represent?
d. Use comparative pie charts to compare the responses for parents and children.
e. What other graphical techniques could be used to describe the data? Would any of these techniques be more informative than the pie charts constructed in part d?

3.6 Consumer Price Index The price of living in the United States has increased dramatically in the past decade, as demonstrated by the consumer price indexes (CPIs) for housing and transportation. These CPIs are listed in the table for the years 1996 through the first half of 2010.[3]

Year	1996	1997	1998	1999	2000	2001	2002	2003
Housing	152.8	156.8	160.4	163.9	169.6	176.4	180.3	184.8
Transportation	143.0	144.3	141.6	144.4	153.3	154.3	152.9	157.6

Year	2004	2005	2006	2007	2008	2009	2010
Housing	189.5	195.7	201.6	209.6	216.3	217.1	215.9
Transportation	163.1	173.9	181.4	184.7	195.5	179.3	192.2

a. Create side-by-side comparative bar charts to describe the CPIs over time.

b. Draw two line charts on the same set of axes to describe the CPIs over time.

c. What conclusions can you draw using the two graphs in parts a and b? Which is the most effective?

Data set **3.7 How Big Is the Household?** A local
EX0307 chamber of commerce surveyed 126 households within its city and recorded the type of residence and the number of family members in each of the households. The data are shown in the table.

	Type of Residence		
Family Members	Apartment	Duplex	Single Residence
1	8	10	2
2	15	4	14
3	9	5	24
4 or more	6	1	28

a. Use a side-by-side bar chart to compare the number of family members living in each of the three types of residences.

b. Use a stacked bar chart to compare the number of family members living in each of the three types of residences.

c. What conclusions can you draw using the graphs in parts a and b?

Data set **3.8 Facebook, again** Not only is the
EX0308 *Facebook* social networking site growing rapidly in the United States, but the composition of *Facebook* members depends on both age and gender. During a 1-month period in early 2010, *Facebook* reported its growth by both age and gender, as shown in the table.[4]

	Growth (number of users)		
Age category	Female	Male	Total
13–17	270,900	121,280	392,180
18–25	445,920	653,060	1,098,980
26–34	570,920	154,600	725,520
35–44	365,740	305,260	671,000
45–54	90,240	36,680	126,920
55–65	64,960	83,480	148,440
Total	1,808,680	1,354,360	3,163,040

a. Construct a stacked bar chart to display the *Facebook* growth given in the table.

b. Construct two comparative pie charts to display the *Facebook* growth given in the table.

c. Write a short paragraph summarizing the information that can be gained by looking at these graphs. Which of the two types of comparative graphs is more effective?

SCATTERPLOTS FOR TWO QUANTITATIVE VARIABLES

3.3

When both variables to be displayed on a graph are *quantitative,* one variable is plotted along the horizontal axis and the second along the vertical axis. The first variable is often called *x* and the second is called *y,* so that the graph takes the form of a plot on the (*x, y*) axes, which is familiar to most of you. Each pair of data values is plotted as a point on this two-dimensional graph, called a **scatterplot.** It is the two-dimensional extension of the dotplot we used to graph one quantitative variable in Section 1.4.

You can describe the relationship between two variables, *x* and *y,* using the patterns shown in the scatterplot.

ONLINE APPLET
Building a Scatterplot

- **What type of pattern do you see?** Is there a constant upward or downward trend that follows a straight-line pattern? Is there a curved pattern? Is there no pattern at all, but just a random scattering of points?

- **How strong is the pattern?** Do all of the points follow the pattern exactly, or is the relationship only weakly visible?

- **Are there any unusual observations?** An outlier is a point that is far from the cluster of the remaining points. Do the points cluster into groups? If so, is there an explanation for the observed groupings?

EXAMPLE **3.3**

The number of household members, x, and the amount spent on groceries per week, y, are measured for six households in a local area. Draw a scatterplot of these six data points.

x	2	2	3	4	1	5
y	$95.75	$110.19	$118.33	$150.92	$85.86	$180.62

Solution   Label the horizontal axis x, the vertical axis y, and plot the points using the coordinates (x, y) for each of the six pairs. The scatterplot in Figure 3.4 shows the six pairs marked as dots. You can see a pattern even with only six data pairs. The cost of weekly groceries increases with the number of household members in an apparent straight-line relationship.

Suppose you found that a seventh household with two members spent $165 on groceries. This observation is shown as an X in Figure 3.4. It does not fit the linear pattern of the other six observations and is classified as an outlier. Possibly these two people were having a party the week of the survey!

FIGURE 3.4

Scatterplot for
Example 3.3

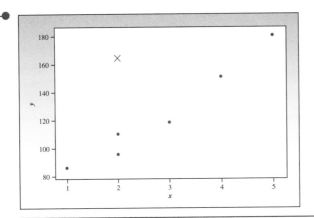

EXAMPLE **3.4**

A distributor of table wines conducted a study of the relationship between price and demand using a type of wine that ordinarily sells for $10.00 per bottle. He sold this wine in 10 different marketing areas over a 12-month period, using five different price levels—from $10 to $14. The data are given in Table 3.4. Construct a scatterplot for the data, and use the graph to describe the relationship between price and demand.

TABLE 3.4 ● **Cases of Wine Sold at Five Price Levels**

Cases Sold per 10,000 Population	Price per Bottle
23, 21	$10
19, 18	11
15, 17	12
19, 20	13
25, 24	14

Solution The 10 data points are plotted in Figure 3.5. As the price increases from $10 to $12, the demand decreases. However, as the price continues to increase, from $12 to $14, the demand begins to *increase*. The data show a curved pattern, with the relationship changing as the price changes. How do you explain this relationship? Possibly, the increased price is a signal of increased quality for the consumer, which causes the increase in demand once the cost exceeds $12. You might be able to think of other reasons, or perhaps some other variable, such as the income of people in the marketing areas, that may be causing the change.

FIGURE 3.5

Scatterplot for
Example 3.4

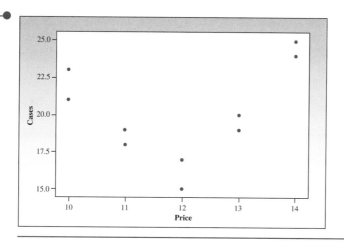

NUMERICAL MEASURES FOR QUANTITATIVE BIVARIATE DATA

3.4

A constant rate of increase or decrease is perhaps the most common pattern found in bivariate scatterplots. The scatterplot in Figure 3.4 exhibits this *linear* pattern—that is, a straight line with the data points lying both above and below the line and within a fixed distance from the line. When this is the case, we say that the two variables exhibit a *linear relationship*.

EXAMPLE **3.5**

The data in Table 3.5 are the size of the living area (in square feet), x, and the selling price, y, of 12 residential properties. The scatterplot in Figure 3.6 shows a linear pattern in the data.

TABLE 3.5 **Living Area and Selling Price of 12 Properties**

Residence	x (sq. ft.)	y (in thousands)
1	1360	$278.5
2	1940	375.7
3	1750	339.5
4	1550	329.8
5	1790	295.6
6	1750	310.3
7	2230	460.5
8	1600	305.2
9	1450	288.6
10	1870	365.7
11	2210	425.3
12	1480	268.8

FIGURE 3.6

Scatterplot of x versus y
for Example 3.5

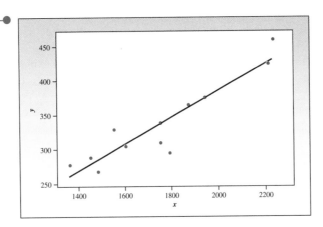

For the data in Example 3.5, you could describe each variable, x and y, individually using descriptive measures such as the means ($\bar{x}$ and $\bar{y}$) or the standard deviations (s_x and s_y). However, these measures do not describe the relationship between x and y for a particular residence—that is, how the size of the living space affects the selling price of the home. A simple measure that serves this purpose is called the **correlation coefficient,** denoted by r, and is defined as

$$r = \frac{s_{xy}}{s_x s_y}$$

The quantities s_x and s_y are the standard deviations for the variables x and y, respectively, which can be found by using the statistics function on your calculator or the computing formula in Section 2.3. The new quantity s_{xy} is called the **covariance** between x and y and is defined as

Exploring Correlation

$$s_{xy} = \frac{\Sigma(x_i - \bar{x})(y_i - \bar{y})}{n - 1}$$

There is also a computing formula for the covariance:

$$s_{xy} = \frac{\Sigma x_i y_i - \dfrac{(\Sigma x_i)(\Sigma y_i)}{n}}{n - 1}$$

where $\Sigma x_i y_i$ is the sum of the products $x_i y_i$ for each of the n pairs of measurements. How does this quantity detect and measure a linear pattern in the data?

Look at the signs of the cross products $(x_i - \bar{x})(y_i - \bar{y})$ in the numerator of r, or s_{xy}. When a data point (x, y) is in either area I or III in the scatterplot shown in Figure 3.7,

FIGURE 3.7

The signs of the cross
products $(x_i - \bar{x})(y_i - \bar{y})$
in the covariance formula

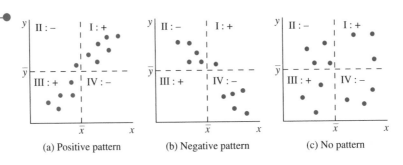

(a) Positive pattern   (b) Negative pattern   (c) No pattern

the cross product will be positive; when a data point is in area II or IV, the cross product will be negative. We can draw these conclusions:

- If most of the points are in areas I and III (forming a positive pattern), s_{xy} and r will be positive.

- If most of the points are in areas II and IV (forming a negative pattern), s_{xy} and r will be negative.

- If the points are scattered across all four areas (forming *no* pattern), s_{xy} and r will be close to 0.

NEED A TIP?

$r > 0 \Leftrightarrow$ positive *linear* relationship

$r < 0 \Leftrightarrow$ negative *linear* relationship

$r \approx 0 \Leftrightarrow$ no *linear* relationship

Most scientific and graphics calculators can compute the correlation coefficient, r, when the data are entered in the proper way. Check your calculator manual for the proper sequence of entry commands. Computer programs are also programmed to perform these calculations. The output in Figure 3.8 shows the covariance and correlation coefficient for x and y in Example 3.5. In the covariance table, you will find these values:

$$s_{xy} = 15,545.20 \qquad s_x^2 = 79,233.33 \qquad s_y^2 = 3571.16$$

and in the correlation output, you find $r = .924$.

However you decide to calculate the correlation coefficient, it can be shown that the value of r always lies between -1 and 1. When r is positive, x increases when y increases, and vice versa. When r is negative, x decreases when y increases, or x increases when y decreases. When r takes the value 1 or -1, all the points lie exactly on a straight line. If $r = 0$, then there is no apparent linear relationship between the two variables. The closer the value of r is to 1 or -1, the stronger the linear relationship between the two variables.

FIGURE 3.8

MINITAB output of covariance and *EXCEL* correlation output for Example 3.5

```
Covariances: x, y

           x         y
x   79233.33
y   15545.20   3571.16
```

```
Correlations: x, y

           x         y
x           1
y     0.92414         1
```

EXAMPLE 3.6

Find the correlation coefficient for the number of square feet of living area and the selling price of a home for the data in Example 3.5.

Solution Three quantities are needed to calculate the correlation coefficient. The standard deviations of the x and y variables are found using a calculator with a statistical function. You can verify that $s_x = 281.4842$ and $s_y = 59.7592$. Finally,

$$s_{xy} = \frac{\Sigma x_i y_i - \dfrac{(\Sigma x_i)(\Sigma y_i)}{n}}{n - 1}$$

$$= \frac{7,240,383 - \dfrac{(20,980)(4043.5)}{12}}{11} = 15,545.19697$$

This agrees with the value given in the printout in Figure 3.8(a). Then

$$r = \frac{s_{xy}}{s_x s_y} = \frac{15{,}545.19697}{(281.4842)(59.7592)} = .9241$$

which also agrees with the value of the correlation coefficient given in Figure 3.8(b). (You may wish to verify the value of r using your calculator.) This value of r is fairly close to 1, which indicates that the linear relationship between these two variables is very strong. Additional information about the correlation coefficient and its role in analyzing linear relationships, along with alternative calculation formulas, can be found in Chapter 12.

NEED A TIP?

x "explains" y or y "depends on" x.

x is the **explan**atory or independent variable.

y is the response or **depend**ent variable.

Sometimes the two variables, x and y, are related in a particular way. It may be that the value of y depends on the value of x; that is, the value of x in some way explains the value of y. For example, the cost of a home (y) may *depend* on its amount of floor space (x); a student's grade point average (x) may *explain* her score on an achievement test (y). In these situations, we call y the **dependent variable,** while x is called the **independent variable.**

If one of the two variables can be classified as the dependent variable y and the other as x, and if the data exhibit a straight-line pattern, it is possible to describe the relationship relating y to x using a straight line given by the equation

$$y = a + bx$$

as shown in Figure 3.9.

FIGURE 3.9

The graph of a straight line

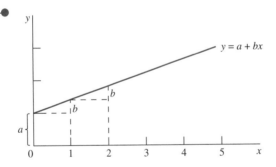

ONLINE APPLET

How a Line Works

As you can see, a is where the line crosses or intersects the y-axis: a is called the *y-intercept*. You can also see that for every one-unit increase in x, y increases by an amount b. The quantity b determines whether the line is increasing ($b > 0$), decreasing ($b < 0$), or horizontal ($b = 0$) and is appropriately called the **slope** of the line.

When plotting the (x, y) points for two variables x and y, the points generally do not fall exactly on a straight line, but they may show a trend that could be described as a linear pattern. We can describe this trend by fitting a line as best we can through the points. This best-fitting line relating y to x, often called the **regression** or **least-squares line,** is found by minimizing the sum of the squared differences between the data points and the line itself, as shown in Figure 3.10. The formulas for computing b and a, which are derived mathematically, follow.

COMPUTING FORMULAS FOR THE LEAST-SQUARES REGRESSION LINE

$$b = r\left(\frac{s_y}{s_x}\right) \quad \text{and} \quad a = \bar{y} - b\bar{x}$$

and the least-squares regression line is: $y = a + bx$

FIGURE 3.10

The best-fitting line

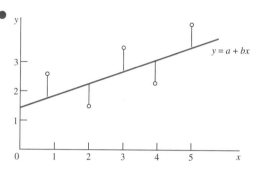

Since s_x and s_y are both positive, b and r have the same sign, so that:

NEED A TIP?

Remember that r and b have the same sign!

- When r is positive, so is b, and the line is increasing with x.
- When r is negative, so is b, and the line is decreasing with x.
- When r is close to 0, then b is close to 0.

EXAMPLE **3.7**

Find the best-fitting line relating $y =$ starting hourly wage to $x =$ number of years of work experience for the following data. Plot the line and the data points on the same graph.

x	2	3	4	5	6	7
y	$6.00	7.50	8.00	12.00	13.00	15.50

Solution   Use the data entry method for your calculator to find these descriptive statistics for the bivariate data set:

$$\bar{x} = 4.5 \quad \bar{y} = 10.333 \quad s_x = 1.871 \quad s_y = 3.710 \quad r = .980$$

NEED A TIP?

Use the regression line to predict y for a given value of x.

Then

$$b = r\left(\frac{s_y}{s_x}\right) = .980\left(\frac{3.710}{1.871}\right) = 1.9432389 \approx 1.943$$

and

$$a = \bar{y} - b\bar{x} = 10.333 - 1.943(4.5) = 1.590$$

Therefore, the best-fitting line is $y = 1.590 + 1.943x$. The plot of the regression line and the actual data points are shown in Figure 3.11.

The best-fitting line can be used to estimate or predict the value of the variable y when the value of x is known. For example, if a person applying for a job has 3 years of work experience (x), what would you predict his starting hourly wage (y) to be? From the best-fitting line in Figure 3.11, the best estimate would be

$$y = a + bx = 1.590 + 1.943(3) = 7.419$$

FIGURE 3.11

Fitted line and data points for Example 3.7

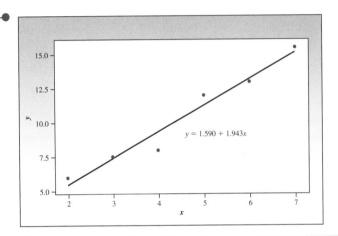

? NEED TO KNOW...

How to Calculate the Correlation Coefficient

1. First, create a table or use your calculator to find Σx, Σy, and Σxy.
2. Calculate the covariance, s_{xy}.
3. Use your calculator or the computing formula from Chapter 2 to calculate s_x and s_y.
4. Calculate $r = \dfrac{s_{xy}}{s_x s_y}$.

How to Calculate the Regression Line

1. First, calculate $\bar{y}$ and $\bar{x}$. Then, calculate $r = \dfrac{s_{xy}}{s_x s_y}$.

2. Find the slope, $b = r\left(\dfrac{s_y}{s_x}\right)$ and the y-intercept, $a = \bar{y} - b\bar{x}$.

3. Write the regression line by substituting the values for a and b into the equation: $y = a + bx$.

When should you describe the linear relationship between x and y using the correlation coefficient r, and when should you use the regression line $y = a + bx$? The regression approach is used when the values of x are set in advance and then the corresponding value of y is measured. The correlation approach is used when an experimental unit is selected at random and then measurements are made on both variables x and y. This technical point will be taken up in Chapter 12, which addresses regression analysis.

Most data analysts begin any data-based investigation by examining plots of the variables involved. If the relationship between two variables is of interest, data analysts can also explore bivariate plots in conjunction with numerical measures of location, dispersion, and correlation. Graphs and numerical descriptive measures are only the first of many statistical tools you will soon have at your disposal.

3.4 **EXERCISES**

BASIC TECHNIQUES

3.9 Suppose that the relationship between two variables x and y can be described by the regression line $y = 2.0 + 0.5x$.

a. What is the change in y for a one-unit change in x?

b. Do the values of y increase or decrease as x increases?

c. At what point does the line cross the y-axis? What is the name given to this value?

d. If $x = 2.5$, use the least squares equation to predict the value of y. What value would you predict if $x = 4.0$?

3.10 Consider this set of bivariate data: (1, 6), (3, 2), and (2, 4).

a. Calculate the covariance s_{xy}.

b. Calculate the correlation coefficient r.

c. Calculate the equation of the regression line using the computing formulas.

d. Plot the three points and the straight line on a scatterplot. Does the line pass through the middle of the three points?

 3.11 A set of bivariate data consists of these
EX0311 measurements on two variables, x and y:

$(3, 6)$ $(5, 8)$ $(2, 6)$ $(1, 4)$ $(4, 7)$ $(4, 6)$

a. Draw a scatterplot to describe the data.

b. Does there appear to be a relationship between x and y? If so, how do you describe it?

c. Calculate the correlation coefficient, r, using the computing formula given in this section.

d. Find the best-fitting line using the computing formulas. Graph the line on the scatterplot from part a. Does the line pass through the middle of the points?

3.12 Refer to Exercise 3.11.

a. Use the data entry method in your scientific calculator to enter the six pairs of measurements. Recall the proper memories to find the correlation coefficient, r, the y-intercept, a, and the slope, b, of the line.

b. Verify that the calculator provides the same values for r, a, and b as in Exercise 3.11.

 3.13 Consider this set of bivariate data:
EX0313

x	1	2	3	4	5	6
y	5.6	4.6	4.5	3.7	3.2	2.7

a. Draw a scatterplot to describe the data.

b. Does there appear to be a relationship between x and y? If so, how do you describe it?

c. Calculate the correlation coefficient, r. Does the value of r confirm your conclusions in part b? Explain.

 3.14 The value of a quantitative variable is
EX0314 measured once a year for a 10-year period:

Year	Measurement	Year	Measurement
1	61.5	6	58.2
2	62.3	7	57.5
3	60.7	8	57.5
4	59.8	9	56.1
5	58.0	10	56.0

a. Draw a scatterplot to describe the variable as it changes over time.

b. Describe the measurements using the graph constructed in part a.

c. Use this *MINITAB* output to calculate the correlation coefficient, r:

MINITAB output for Exercise 3.14

Covariances

	x	y
x	9.16667	
y	−6.42222	4.84933

d. Find the best-fitting line using the results of part c. Verify your answer using the data entry method in your calculator.

e. Plot the best-fitting line on your scatterplot from part a. Describe the fit of the line.

APPLICATIONS

Data set **3.15 Grocery Costs** These data relating the
EX0315 amount spent on groceries per week and the
number of household members are from Example 3.3:

x	2	2	3	4	1	5
y	$95.75	$110.19	$118.33	$150.92	$85.86	$180.62

a. Find the best-fitting line for these data.

b. Plot the points and the best-fitting line on the same
graph. Does the line summarize the information in
the data points?

c. What would you estimate a household of six to
spend on groceries per week? Should you use the fitted line to estimate this amount? Why or why not?

Data set **3.16 Real Estate Prices** The data relating
EX0316 the square feet of living space and the selling
price of 12 residential properties given in Example 3.5
are reproduced here. First, find the best-fitting line that
describes these data, and then plot the line and the data
points on the same graph. Comment on the goodness
of the fitted line in describing the selling price of a residential property as a linear function of the square feet
of living area.

Residence	x (sq. ft.)	y (in thousands)
1	1360	$278.5
2	1940	375.7
3	1750	339.5
4	1550	329.8
5	1790	295.6
6	1750	310.3
7	2230	460.5
8	1600	305.2
9	1450	288.6
10	1870	365.7
11	2210	425.3
12	1480	268.8

Data set **3.17 Disabled Students** A social skills
EX0317 training program, reported in *Psychology in the
Schools,* was implemented for seven students with
mild handicaps in a study to determine whether the
program caused improvement in pre/post measures and
behavior ratings.[5] For one such test, these are the
pretest and posttest scores for the seven students:

Student	Pretest	Posttest
Earl	101	113
Ned	89	89
Jasper	112	121
Charlie	105	99
Tom	90	104
Susie	91	94
Lori	89	99

a. Draw a scatterplot relating the posttest score to the
pretest score.

b. Describe the relationship between pretest and
posttest scores using the graph in part a. Do you see
any trend?

c. Calculate the correlation coefficient and interpret its
value. Does it reinforce any relationship that was
apparent from the scatterplot? Explain.

Data set **3.18 Chirping Crickets** Male crickets chirp
EX0318 by rubbing their front wings together, and their
chirping is temperature dependent. Crickets chirp
faster with increasing temperature and slower with
decreasing temperatures. The table below shows the
number of chirps per second for a cricket, recorded at
10 different temperatures.

Chirps per second	20	16	19	18	18	16	14	17	15	16
Temperature	88	73	91	85	82	75	69	82	69	83

a. Which of the two variables (temperature and number
of chirps) is the independent variable, and which is
the dependent variable?

b. Plot the data using a scatterplot. How would you
describe the relationship between temperature and
number of chirps?

c. Find the least-squares line relating the number of
chirps to the temperature.

d. If a cricket is monitored at a temperature of 80
degrees, what would you predict his number of
chirps would be?

Data set **3.19 How to Choose a TV** As technology
EX0319 improves, your choice of televisions becomes
more complicated. Should you choose an LCD TV, a
plasma TV, and with rear or front projection? In the
table below, *Consumer Reports*[6] gives the prices and
screen sizes for the top 10 LCD TVs in the 46-inch
and higher categories. Does the price of an LCD TV
depend on the size of the screen?

Brand	Price ($)	Size	Brand	Price ($)	Size
Sony Bravia KDL-52NX800	2340	52	Sony Bravia KDL-46XBR10	2500	46
Samsung LN55C650	1600	55	Samsung UN46C8000	2200	46
Vizio VF550M	1330	55	Vizio SV472XVT	1400	47
Sony Bravia KDL-60EX700	2700	60	Samsung UN46C7000	2100	46
Sharp Aquos LED LC-52LE700UN	1620	52	LG 47LD450	900	47

a. Which of the two variables (price and size) is the
independent variable, and which is the dependent
variable?

b. Construct a scatterplot for the data. Does the
relationship appear to be linear?

3.20 LCD TVs, continued Refer to Exercise 3.19. Suppose we assume that the relationship between x and y is linear.

a. Find the correlation coefficient, r. What does this value tell you about the strength and direction of the relationship between size and price?

b. Refer to part a. Would it be reasonable to construct a regression line used to predict the price of an LCD TV based on the size of the screen?

CHAPTER REVIEW

Key Concepts

I. Bivariate Data

1. Both qualitative and quantitative variables
2. Describing each variable separately
3. Describing the relationship between the two variables

II. Describing Two Qualitative Variables

1. Side-by-side pie charts
2. Comparative line charts
3. Comparative bar charts
   a. Side-by-side
   b. Stacked
4. Relative frequencies to describe the relationship between the two variables

III. Describing Two Quantitative Variables

1. Scatterplots
   a. Linear or nonlinear pattern

b. Strength of relationship
c. Unusual observations: clusters and outliers

2. Covariance and correlation:

Covariance: $s_{xy} = \dfrac{\Sigma xy - \dfrac{(\Sigma x)(\Sigma y)}{n}}{n-1}$

Correlation: $r = \dfrac{s_{xy}}{s_x s_y}$

3. The best-fitting regression line
   a. Calculating the slope and y-intercept

   $b = r\left(\dfrac{s_y}{s_x}\right)$ and $a = \bar{y} - b\bar{x}$

   b. Graphing the line
   c. Using the line for prediction

 TECHNOLOGY TODAY

Describing Bivariate Data in *Excel*

MS Excel provides different graphical techniques for *qualitative* and *quantitative* bivariate data, as well as commands for obtaining bivariate descriptive measures when the data are quantitative.

 **EXAMPLE** 3.8   **(Comparative Line and Bar Charts)** Suppose that the 105 students whose status was tabulated in Example 1.12 were from the University of California, Riverside, and that another 100 students from an introductory statistics class at UC Berkeley were also interviewed. Table 3.6 shows the status distribution for both sets of students.

TABLE 3.6   Status of Students in a Statistics Class at UCR and UCB

	Freshman	Sophomore	Junior	Senior	Grad Student
Frequency (UCR)	5	23	32	35	10
Frequency (UCB)	10	35	24	25	6

1. Enter the data into an *Excel* spreadsheet *just as it appears in the table, including the labels.* Highlight the data in the spreadsheet, click the **Insert** tab and select **Line** in the **Charts** group. In the drop-down list, you will see a variety of styles to choose from. Select the first option to produce the line chart.

2. *Editing the line chart:* Again, you can experiment with the various options in the **Chart Layout** and **Chart Styles** groups to change the look of the chart. We have chosen a design that allows a title on the vertical axis; we have added the title and have changed the "line style" of the UCR students to a "dashed" style, by double-clicking on that line. The line chart is shown in Figure 3.12(a).

FIGURE 3.12 ──●

(a)

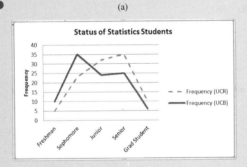

(b)

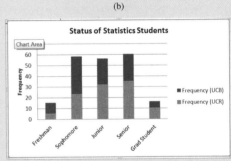

3. Once the line chart has been created, right-click on the chart area and select **Change Chart Type.** Then choose either **Stacked Column** or **Clustered Column.** The comparative bar chart (a stacked bar chart), with the same editing that you chose for the line chart, will appear as shown in Figure 3.12(b).

EXAMPLE ⬤ 3.9 ── **(Scatterplots, Correlation, and the Regression Line)** The data from Example 2.15 give the front and rear leg rooms (in inches) for nine different sports utility vehicles:[7]

Make and Model	Front Leg Room	Rear Leg Room
Acura MDX	41.0	28.5
Buick Enclave	41.5	30.0
Chevy TrailBlazer	40.0	25.5
Chevy Tahoe Hybrid V8 CVT	41.0	27.5
GMC Terrain 1LT 4-cyl	43.0	31.0
Honda CR-V	41.0	29.5
Hyundai Tucson	42.5	29.5
Kia Sportage	40.0	29.0
Lexus GX	42.0	30.0

1. If you did not save the *Excel* spreadsheet from Chapter 2, enter the data into the first three columns of another *Excel* spreadsheet, using the labels in the table. Highlight the front and rear leg room data (columns B and C), click the **Insert** tab and select **Scatter** in the **Charts** group, and select the first option in the drop-down list. The scatterplot appears as in Figure 3.13(a), and will need to be edited!

FIGURE 3.13

(a)

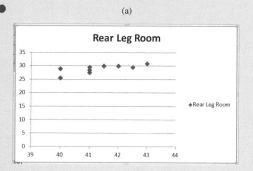

(b)

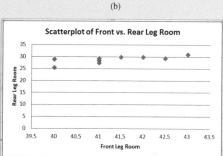

2. ***Editing the scatterplot:*** With the scatterplot selected, look in the drop-down list in the **Chart Layouts** group. Find a layout that allows titles on both axes (we chose layout 1) and select it. Label the axes, remove the "legend entry" and retitle the chart as "Scatterplot of Front vs. Rear Leg Room." The scatterplot now appears in Figure 3.13(b). The plot is still not optimal, since *Excel* chooses to use zero as the lower limit of the vertical scale, causing the points to cluster at the top of the plot. To adjust this, double-click on the vertical axis. In the **Format Axis** Dialog Box, change the **Minimum** to **Fixed,** type **25** in the box, and click **Close**. (You can make a similar adjustment to the horizontal axis if needed.)

3. To plot the best-fitting line, simply right-click on one of the data points and select **Add Trendline**. In the Dialog box that opens, make sure that the radio button marked "Linear" is selected, and check the boxes marked "Display Equation on Chart" and "Display R-squared value on Chart". The final scatterplot is shown in Figure 3.14.

FIGURE 3.14

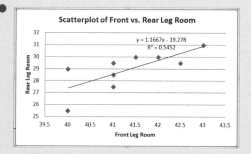

4. To find the sample correlation coefficient, r, you can use the command **Data ▶ Data Analysis ▶ Correlation,** selecting the two appropriate columns for the Input Range, clicking "Labels in First Row," and selecting an appropriate Output

Range. When you click **OK,** the correlation matrix will appear in the spreadsheet.

5. *(ALTERNATE PROCEDURE)* You can also place your cursor in the cell in which you want the correlation coefficient to appear. Select **Formulas ▶ More Functions ▶ Statistical ▶ CORREL** or click the "Insert Function" icon f_x at the top of the spreadsheet, choosing **CORREL** fiom the **Statistical** category. Highlight or type the cell ranges for the two variables in the boxes marked "Array 1" and "Array 2" and click **OK.** For our example, the value is $r = .738$.

Describing Bivariate Data in *MINITAB*

MINITAB provides different graphical techniques for *qualitative* and *quantitative* bivariate data, as well as commands for obtaining bivariate descriptive measures when the data are quantitative.

EXAMPLE 3.10 — (**Comparative Line and Bar Charts**) Suppose that the 105 students whose status was tabulated in Example 1.12 were from the University of California, Riverside, and that another 100 students from an introductory statistics class at UC Berkeley were also interviewed. Table 3.7 shows the status distribution for both sets of students.

TABLE 3.7   **Status of Students in a Statistics Class at UCR and UCB**

	Freshman	Sophomore	Junior	Senior	Grad Student
Frequency (UCR)	5	23	32	35	10
Frequency (UCB)	10	35	24	25	6

1. Enter the data into a *MINITAB* worksheet as you did in Chapter 1, using your Chapter 1 project as a base if you have saved it. Column C1 will contain the 10 "Frequencies" and column C2 will contain the student "Status" corresponding to each frequency. Create a third column C3 called "College," and enter either **UCR** or **UCB** as appropriate. You can use the familiar Windows cut-and-paste commands if you like.

2. To graphically describe the UCR/UCB student data, you can use comparative pie charts—one for each school (see Chapter 1). Alternatively, you can use either stacked or side-by-side bar charts. Use **Graph ▶ Bar Chart.**

3. In the "Bar Charts" Dialog box (Figure 3.15(a)), select **Values from a Table** in the drop-down list and click either **Stack** or **Cluster** in the row marked "One Column of Values." Click **OK.** In the next Dialog box (Figure 3.15(b)), select "Frequency" for the **Graph variables** box and "Status" and "College" for the **Categorical variable for grouping** box. Click **OK.**

4. Once the bar chart is displayed (Figure 3.16), you can *right-click* on various items in the bar chart to edit. If you *right-click* on the bars and select **Update Graph Automatically,** the bar chart will automatically update when you change the data in the *MINITAB* worksheet.

FIGURE 3.15

(a)

(b)

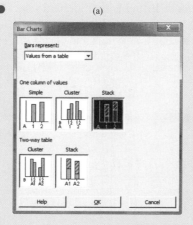
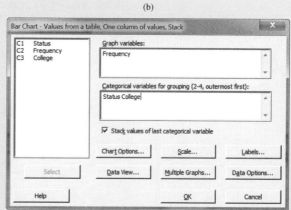

FIGURE 3.16

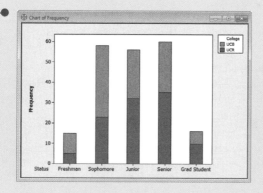

EXAMPLE 3.11

(**Scatterplots, Correlation, and the Regression Line**) The data from Example 2.15 give the front and rear leg rooms (in inches) for nine different sports utility vehicles:[7]

Make and Model	Front Leg Room	Rear Leg Room
Acura MDX	41.0	28.5
Buick Enclave	41.5	30.0
Chevy TrailBlazer	40.0	25.5
Chevy Tahoe Hybrid V8 CVT	41.0	27.5
GMC Terrain 1LT 4-cyl	43.0	31.0
Honda CR-V	41.0	29.5
Hyundai Tucson	42.5	29.5
Kia Sportage	40.0	29.0
Lexus GX	42.0	30.0

1. If you did not save the *MINITAB* worksheet from Chapter 2, enter the data into the first three columns of another *MINITAB* worksheet, using the labels in the table. To examine the relationship between the front and rear leg rooms, you can plot the data and numerically describe the relationship with the correlation coefficient and the best-fitting line.

2. Select **Stat ▶ Regression ▶ Fitted Line Plot,** and select "Front Leg Room" and "Rear Leg Room" for **Y** and **X,** respectively (see Figure 3.17(a)). Make sure that the radio button next to **Linear** is selected, and click **OK.** The plot of the nine data points and the best-fitting line will be generated as in Figure 3.17(b).

FIGURE 3.17

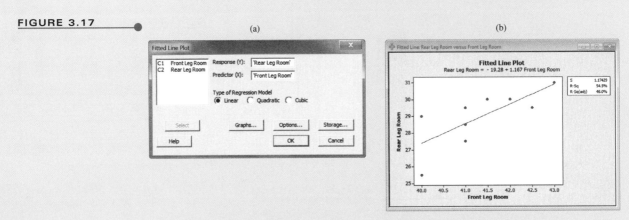

3. To calculate the correlation coefficient, use **Stat ▶ Basic Statistics ▶ Correlation,** selecting "Front Leg Room" and "Rear Leg Room" for the Variables box. To select both variables at once, hold the **Shift** key down as you highlight the variables and then click **Select.** Click **OK,** and the correlation coefficient will appear in the Session window (see Figure 3.18). Notice the relatively strong positive correlation and the positive slope of the regression line, indicating that a sports utility vehicle with a large front leg room will also tend to have a large rear leg room.

FIGURE 3.18

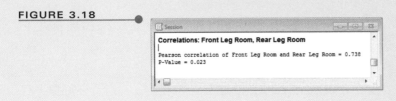

Supplementary Exercises

Data set
EX0321 **3.21 Professor Asimov** Professor Isaac Asimov was one of the most prolific writers of all time. He wrote nearly 500 books during a 40-year career prior to his death in 1992. In fact, as his career progressed, he became even more productive in terms of the number of books written within a given period of time.[8] These data are the times (in months) required to write his books, in increments of 100:

Number of Books	100	200	300	400	490
Time (in months)	237	350	419	465	507

a. Plot the accumulated number of books as a function of time using a scatterplot.

b. Describe the productivity of Professor Asimov in light of the data set graphed in part a. Does the relationship between the two variables seem to be linear?

Data set
EX0322 **3.22 Cheese, Please!** Health-conscious Americans often consult the nutritional information on food packages in an attempt to avoid foods with large amounts of fat, sodium, or cholesterol. The following information was taken from eight different brands of American cheese slices:

Brand	Fat (g)	Satu-rated Fat (g)	Choles-terol (mg)	Sodium (mg)	Calories
Kraft Deluxe American	7	4.5	20	340	80
Kraft Velveeta Slices	5	3.5	15	300	70
Private Selection	8	5.0	25	520	100
Ralphs Singles	4	2.5	15	340	60
Kraft 2% Milk Singles	3	2.0	10	320	50
Kraft Singles American	5	3.5	15	290	70
Borden Singles	5	3.0	15	260	60
Lake to Lake American	5	3.5	15	330	70

a. Which pairs of variables do you expect to be strongly related?

b. Draw a scatterplot for fat and saturated fat. Describe the relationship.

c. Draw a scatterplot for fat and calories. Compare the pattern to that found in part b.

d. Draw a scatterplot for fat versus sodium and another for cholesterol versus sodium. Compare the patterns. Are there any clusters or outliers?

e. For the pairs of variables that appear to be linearly related, calculate the correlation coefficients.

f. Write a paragraph to summarize the relationships you can see in these data. Use the correlations and the patterns in the four scatterplots to verify your conclusions.

3.23 Facebook Stats In Exercise 1.14, we looked at the age distribution of users of the social networking site *Facebook* (in thousands) as it changed from January 2009 to January 2010. The online article also provided insight into other demographics of *Facebook* users during this time period.[9] Two bar charts, constructed from the data, are shown below.

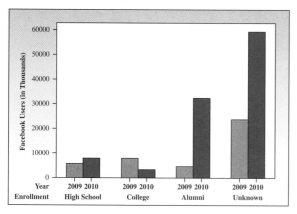

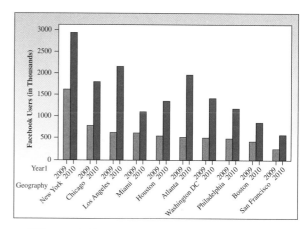

a. What variables have been measured in this study? Are the variables qualitative or quantitative?

b. Describe the populations of interest. Do these data represent populations or samples drawn from a population?

c. What type of graphical presentation has been used? What other type could have been used?

d. How would you describe the changes in the geographical and educational distributions of *Facebook* users during this 1-year period?

3.24 Cheese, again! The demand for healthy foods that are low in fats and calories has resulted in a large number of "low-fat" and "fat-free" products at the supermarket. The table shows the numbers of calories and the amounts of sodium (in milligrams) per slice for five different brands of fat-free American cheese.

Brand	Sodium (mg)	Calories
Kraft Fat Free Singles	300	30
Ralphs Fat Free Singles	300	30
Borden Fat Free	320	30
Healthy Choice Fat Free	290	30
Smart Beat American	180	25

a. Draw a scatterplot to describe the relationship between the amount of sodium and the number of calories.

b. Describe the plot in part a. Do you see any outliers? Do the rest of the points seem to form a pattern?

c. Based *only* on the relationship between sodium and calories, can you make a clear decision about which of the five brands to buy? Is it reasonable to base your choice on only these two variables? What other variables should you consider?

3.25 Peak Current Using a chemical procedure called *differential pulse polarography*, a

EX0325

chemist measured the peak current generated (in microamperes) when a solution containing a given amount of nickel (in parts per billion) is added to a buffer. The data are shown here:

x = Ni (ppb)	y = Peak Current (μA)
19.1	.095
38.2	.174
57.3	.256
76.2	.348
95	.429
114	.500
131	.580
150	.651
170	.722

Use a graph to describe the relationship between x and y. Add any numerical descriptive measures that are appropriate. Write a paragraph summarizing your results.

3.26 Movie Money How much money do movies make on a single weekend? Does this amount in any way predict the movie's success or failure, or is the movie's total monetary success more dependent on the number of weeks that the movie remains in the movie theaters? In a recent week, the following data was collected for the top 16 movies in theaters that weekend.[10]

TW	Title	Studio	Weekend Gross ($ millions)	Theater Count	Average ($)	Total Gross ($ millions)	Production Budget ($ millions)	Week
1	The Expendables	LGF	17.0	3270	5189	65.4	80	2
2	Vampires Suck	Fox	12.2	3233	3774	18.6	20	1
3	Eat Pray Love	Sony	12.1	3082	3930	47.2	60	2
4	Lottery Ticket	WB	10.7	1973	5399	10.7	17	1
5	The Other Guys	Sony	10.2	3472	2927	88.3	100	3
6	Piranha 3D	W/Dim.	10.1	2470	4092	10.1	24	1
7	The Switch	Mira.	8.4	2012	4193	8.4	—	1
8	Nanny McPhee Returns	Uni.	8.4	2784	3020	8.4	35	1
9	Inception	WB	7.8	2401	3265	262.0	160	6
10	Scott Pilgrim vs. the World	Uni.	5.2	2820	1845	20.9	60	2
11	Despicable Me	Uni.	4.7	2236	2085	231.1	69	7
12	Dinner for Schmucks	P/DW	3.5	2149	1638	65.8	69	4
13	Salt	Sony	3.4	1794	1901	109.9	110	5
14	Step Up 3-D	BV	3.2	1592	1979	36.9	30	3
15	Cats & Dogs: The Revenge of Kitty Galore	WB	1.7	1580	1077	39.7	85	4
16	Toy Story 3	BV	1.5	730	2086	403.8	200	10

a. Which pairs of variables in the table do you think will have a positive correlation? Which pairs will have a negative correlation? Explain.

b. Draw a scatterplot relating the number of weeks in release to the gross to date. How would you describe the relationship between these two variables?

c. Draw a scatterplot relating the weekend gross to the number of theaters in which the movie is being shown. How would you describe the relationship between these two variables?

d. Draw a scatterplot relating the number of theaters in which the movie is being shown to the per theater average. How would you describe the relationship between these two variables?

3.27 Movie Money, continued The data from Exercise 3.26 were entered into a *MINITAB* worksheet, and the following output was obtained.

Covariances:

	Weekend Gross	Theaters	Average	Total Gross	Week
Weekend Gross	19.6				
Theaters	2550.7	550521.6			
Average	5093.3	421558.5	1700161.7		
Total Gross	-194.1	-38787.2	-43562.8	12839.1	
Week	-7.1	-1152.8	-1852.7	278.0	6.8

a. Use the *MINITAB* output or the original data to find the correlation between the number of weeks in release and the total gross to date.

b. For the pair of variables described in part a, which of the variables would you classify as the independent variable? The dependent variable?

c. Use the *MINITAB* output or the original data to find the correlation between the weekend gross and the number of theaters in which the movie is being shown. Find the correlation between the number of theaters in which the movie is being shown and the per theater average.

d. Do the correlations found in part c confirm your answer to Exercise 3.26a? What might be the practical reasons for the direction and strength of the correlations in part c?

3.28 Heights and Gender Refer to Exercise 1.54 and data set EX0154. When the heights of these 105 students were recorded, their gender was also recorded.

a. What variables have been measured in this experiment? Are they qualitative or quantitative?

b. Look at the histogram from Exercise 1.54 along with the comparative box plots shown below. Do the box plots help to explain the two local peaks in the histogram? Explain.

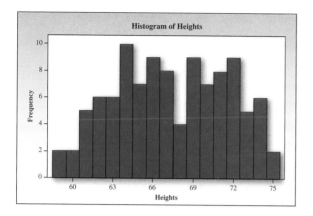

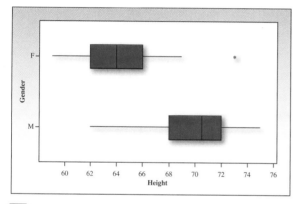

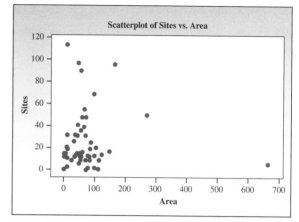

MINITAB printout for Exercise 3.29

Covariances: Sites, Area

	Sites	Area
Sites	702.776	
Area	-72.176	9346.603

a. Is there any clear pattern in the scatterplot? Describe the relationship between number of waste sites and the size of the state.

b. Use the *MINITAB* output to calculate the correlation coefficient. Does this confirm your answer to part a?

c. Are there any outliers or clusters in the data? If so, can you explain them?

d. What other variables could you consider in trying to understand the distribution of hazardous waste sites in the United States?

3.29 Hazardous Waste The data in Exercise 1.37 gave the number of hazardous waste sites in each of the 50 states and the District of Columbia in 2009.[11] Suspecting that there might be a relationship between the number of waste sites and the size of the state (in thousands of square miles), researchers recorded both variables and generated a scatterplot.

EX0329

State	Sites	Area	State	Sites	Area	State	Sites	Area
AL	15	52	KY	14	40	ND	0	71
AK	6	663	LA	12	52	OH	41	45
AZ	9	114	ME	12	35	OK	9	70
AR	9	53	MD	19	12	OR	13	98
CA	96	164	MA	32	11	PA	97	46
CO	20	104	MI	69	97	RI	12	2
CT	15	6	MN	25	87	SC	26	32
DE	15	2	MS	6	48	SD	2	77
DC	1	0	MO	31	70	TN	15	42
FL	55	66	MT	17	147	TX	50	269
GA	16	59	NE	13	77	UT	19	85
HI	3	11	NV	1	111	VT	11	10
ID	9	84	NH	21	9	VA	31	43
IL	48	58	NJ	114	9	WA	48	71
IN	32	36	NM	14	122	WV	9	24
IA	12	56	NY	90	55	WI	39	65
KS	12	82	NC	36	54	WY	2	98

3.30 Aaron Rodgers, again The number of passes completed and the total number of passing yards were recorded for Aaron Rodgers for each of the 15 regular season games that he played in the fall of 2010:[12]

EX0330

Week	Completions	Total Yards	Week	Completions	Total Yards
1	19	188	9	27	289
2	19	255	11	22	301
3	34	316	12	26	344
4	12	181	13	21	298
5	27	293	14	7	46
6	18	313	16	25	404
7	21	295	17	19	229
8	15	170			

Source: www.ESPN.com

a. Draw a scatterplot to describe the relationship between number of completions and total passing yards for Aaron Rodgers.

b. Describe the plot in part a. Do you see any outliers? Do the rest of the points seem to form a pattern?

c. Calculate the correlation coefficient, r, between the number of completions and total passing yards.

d. What is the regression line for predicting total number of passing yards y based on the total number of completions x?

e. If Aaron Rodgers had 20 pass completions in his next game, what would you predict his total number of passing yards to be?

3.31 Pottery, continued In Exercise 1.59, we analyzed the percentage of aluminum oxide in 26 samples of Romano-British pottery found at four different kiln sites in the United Kingdom.[13] Since one of the sites only provided two measurements, that site is eliminated, and comparative box plots of aluminum oxide at the other three sites are shown.

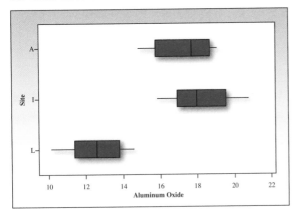

a. What two variables have been measured in this experiment? Are they qualitative or quantitative?

b. How would you compare the amount of aluminum oxide in the samples at the three sites?

3.32 Pottery, continued Here is the percentage of aluminum oxide, the percentage of iron oxide, and the percentage of magnesium oxide in five samples collected at Ashley Rails in the United Kingdom.
EX0332

Sample	Al	Fe	Mg
1	17.7	1.12	0.56
2	18.3	1.14	0.67
3	16.7	0.92	0.53
4	14.8	2.74	0.67
5	19.1	1.64	0.60

a. Find the correlation coefficients describing the relationships between aluminum and iron oxide content, between iron oxide and magnesium oxide, and between aluminum oxide and magnesium oxide.

b. Write a sentence describing the relationships between these three chemicals in the pottery samples.

3.33 Gestation Times and Longevity The table below shows the gestation time in days and the average longevity in years for a variety of
EX0333

mammals in captivity; the potential life span of animals is rarely attained for animals in the wild.[11]

Animal	Gestation (days)	Avg Longevity (yrs)	Animal	Gestation (days)	Avg Longevity (yrs)
Ass	365	12	Hippopotamus	238	41
Baboon	187	20	Horse	330	20
Bear (black)	219	18	Kangaroo (gray)	36	7
Bear (grizzly)	225	25	Leopard	98	12
Bear (polar)	240	20	Lion	100	15
Beaver	105	5	Monkey (rhesus)	166	15
Bison	285	15	Moose	240	12
Camel	406	12	Mouse (meadow)	21	3
Cat (domestic)	63	12	Mouse (dom.white)	19	3
Chimpanzee	230	20	Opossum (American)	13	1
Chipmunk	31	6	Pig (domestic)	112	10
Cow	284	15	Puma	90	12
Deer (whitetailed)	201	8	Rabbit (domestic)	31	5
Dog (domestic)	61	12	Rhinoceros (black)	450	15
Elephant (African)	660	35	Rhinoceros (white)	480	20
Elephant (Asian)	645	40	Sea lion (California)	350	12
Elk	250	15	Sheep (domestic)	154	12
Fox (red)	52	7	Squirrel (gray)	44	10
Giraffe	457	10	Tiger	105	16
Goat (domestic)	151	8	Wolf (maned)	63	5
Gorilla	258	20	Zebra (Grant's)	365	15
Guinea pig	68	4			

Source: The World Almanac and Book of Facts 2011

a. Draw a scatterplot for the data.

b. Describe the form, direction, and strength for the pattern in the scatterplot.

c. Are there any outliers or other unusual data points in the set? If so, to which animals do these data points correspond?

d. Remove the outliers or unusual data points from the set, and reconstruct the scatterplot. Does it appear that a straight line is appropriate for describing the data?

3.34 Housing Blues As the United States fell further and further into a recession in the years 2007–2009, the number of families who had to default on their mortgages and those who actually lost their homes grew. However, as we recover from the recession, *Moody's Economy* predicts that these numbers will fall.[14] The table below, adapted from their report, shows their approximate estimates and predictions for the years 2005–2012.
EX0334

Year	Defaults (thousands)	Lost Homes (thousands)
2005	750	350
2006	800	400
2007	1350	800
2008	2600	1650
2009	3700	2000
2010	3400	2400
2011	2200	1350
2012	1000	600

a. What variables have been measured in this report? Are they qualitative or quantitative?

b. Draw side-by-side comparative bar charts to describe the number of defaults and lost homes, categorized by year.

c. Draw two line charts on the same set of axes to describe the same number over the period 2005–2012.

d. What conclusions can you draw using the two graphs in parts b and c? Which is more effective?

3.35 Armspan and Height Leonardo da Vinci (1452–1519) drew a sketch of a man, indicating that a person's armspan (measuring across the back with arms outstretched to make a "T") is roughly equal to the person's height. To test this claim, we measured eight people with the following results:

Person	1	2	3	4
Armspan (inches)	68	62.25	65	69.5
Height (inches)	69	62	65	70

Person	5	6	7	8
Armspan (inches)	68	69	62	60.25
Height (inches)	67	67	63	62

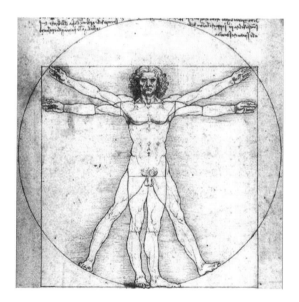

a. Draw a scatterplot for armspan and height. Use the same scale on both the horizontal and vertical axes. Describe the relationship between the two variables.

b. Calculate the correlation coefficient relating armspan and height.

c. If you were to calculate the regression line for predicting height based on a person's armspan, how would you estimate the slope of this line?

d. Find the regression line relating armspan to a person's height.

e. If a person has an armspan of 62 inches, what would you predict the person's height to be?

3.36 Midterm Scores When a student performs poorly on a midterm exam, the student sometimes is convinced that their score is an anomaly and that they will do much better on the second midterm. The data below show the midterm scores (out of 100 points) for eight students in an introductory statistics class.

Student	Midterm 1	Midterm 2
1	70	88
2	58	52
3	85	84
4	82	74
5	70	80
6	40	36
7	85	48
8	85	96

a. Construct a scatterplot for the data.

b. Describe the pattern that you see in the scatterplot. Are there any clusters or outliers? If so, how would you explain them?

3.37 Midterm Scores, continued Refer to Exercise 3.36.

a. Calculate r, the correlation coefficient between the two midterm scores. How would you describe the relationship between scores on the first and second midterms?

b. Calculate the regression line for predicting a student's score on the second midterm exam based on the student's score on the first midterm.

c. Using the regression line from part b, predict a student's score on the second midterm if his score on the first midterm was 85.

3.38 Test Interviews Of two personnel evaluation techniques available, the first requires a 2-hour test-interview while the second can be completed in less than an hour. The scores for each of the eight individuals who took both tests are given in the next table.

Applicant	Test 1 (x)	Test 2 (y)
1	75	38
2	89	56
3	60	35
4	71	45
5	92	59
6	105	70
7	55	31
8	87	52

a. Construct a scatterplot for the data.

b. Describe the form, direction, and strength of the pattern in the scatterplot.

3.39 Test Interviews, continued Refer to Exercise 3.38.

a. Find the correlation coefficient, r, to describe the relationship between the two tests.

b. Would you be willing to use the second and quicker test rather than the longer test-interview to evaluate personnel? Explain.

3.40 Rain and Snow Is there a correlation
EX0340 between the amount of rain and the amount of snow that falls in a particular location? The table below shows the average annual rainfall (inches) and the average annual snowfall (inches) for 10 cities in the United States.[15]

City	Rainfall (inches)	Snowfall (inches)
Billings, MT	14.77	56.9
Casper, WY	13.03	77.8
Concord, NH	37.60	64.5
Fargo, ND	21.19	40.8
Kansas City, MO	37.98	19.9
Juneau, AK	58.33	97.0
Memphis, TN	54.65	5.1
New York, NY	49.69	28.6
Portland, OR	37.07	6.5
Springfield, IL	35.56	23.2

Source: Time Almanac 2007

a. Construct a scatterplot for the data.

b. Calculate the correlation coefficient r between rainfall and snowfall. Describe the form, direction, and strength of the relationship between rainfall and snowfall.

c. Are there any outliers in the scatterplot? If so, which city does this outlier represent?

d. Remove the outlier that you found in part c from the data set and recalculate the correlation coefficient r for the remaining nine cities. Does the correlation between rainfall and snowfall change, and, if so, in what way?

3.41 Smartphones The table below shows
EX0341 the prices of nine *Verizon* smartphones along with their overall score (on a scale of 0–100) in a consumer rating survey presented by *Consumer Reports*.[16]

Brand and Model	Price($)	Overall Score
Motorola Droid X	200	75
Motorola Droid	150	73
HTC Droid	200	73
LG Ally	50	72
Samsung Omnia II	50	71
HTC Imagio	100	70
Motorola Devour	80	70
Blackberry Storm2 9550	150	70
Palm Pre Plus	50	66

a. Plot the nine data points using a scatterplot. Describe the form, direction, and strength of the relationship between price and overall score.

b. Calculate r, the correlation coefficient between price and overall score.

c. Find the regression line for predicting the overall score of a smartphone based on its price.

CASE STUDY

Dish-
washers

Are Your Dishes *Really* Clean?

Does the price of an appliance convey something about its quality? Forty-eight different dishwashers were ranked on characteristics ranging from an overall satisfaction score, washing (x_1), energy use (x_2), noise (x_3), ease of use (x_4), and cycle time (in minutes).[17] The Bosch (SHE55M1[2]UC) had the highest performance score of 82 while the GE (GLD4408R[WW]) had the lowest at 53. Ratings pictograms were converted to numerical values for $x_1, \ldots, x_4$, where 5 = Excellent, 4 = Very good, 3 = Good, 2 = Fair, and 1 = Poor. Use a statistical computer package to explore the relationships between various pairs of variables in the table.

Excellent Very good Good Fair Poor

Brand and Model	Price $	Overall Score	Washing (x_1)	Energy Use (x_2)	Noise (x_3)	Ease of Use (x_4)	Cycle Time (min.)	x_1	x_2	x_3	x_4
Amana ADB1600AW[W]	350	61					130	4	4	2	3
Asko D3531	1600	80					145	4	5	5	4
Asko D5233XXL[HS]	1500	56	NA				180	*	4	4	4
Asko D5253XXL	1300	77					180	5	4	4	4
Bosch SHE55M1[2]UC	850	82					120	5	5	4	4
Bosch SHE6AP0[2]UC	600	75					135	5	4	3	4
Bosch SHX43P1[2]UC	800	77					115	5	5	3	4
Bosch SHX45P0[5]UC	900	79					115	5	5	4	4
Bosch SHX65P0[5]UC	1150	75					120	4	4	4	4
Bosch SHX6AP0[2]UC	700	77					110	5	5	4	4
Bosch SHX98M0[9]UC	1550	82					115	5	4	5	4
Dacor Epicure ED24[S]	1550	69					110	4	4	4	3
Electrolux Wave-Touch EWDW6505G[W]	1200	65					135	4	4	4	5
Frigidaire Gallery FGBD2431K[W]	350	70					155	5	3	3	3
Frigidaire Gallery FGBD2432K[W]	380	68					145	4	4	3	4
Frigidaire Gallery FGHD2433K[F]	500	66					135	4	4	3	4
GE GDWF100R[WW]	600	70					120	5	4	3	4
GE GLD4408R[WW]	400	53					135	3	4	2	3
GE GLD7400R[WW]	600	62					110	4	4	3	5
GE Profile PDWT500R[WW]	1300	77					110	5	4	4	5
Hotpoint HDA3600R[WW]	300	53					115	3	5	1	2
Jenn-Air JDB3200AW[W]	1100	64					125	4	5	4	3
Kenmore 1318[2]	840	79					145	5	5	4	4
Kenmore 1324[2]	410	58					125	3	4	3	3
Kenmore 1344[2]	300	60					110	4	3	2	3
Kenmore 1348[2]	500	77					120	5	4	3	3
Kenmore 1374[2]	650	80					125	5	4	3	4
Kenmore 1389[2]	500	78					135	5	4	3	4
Kenmore Elite UltraWash HE 1312[2]	780	79					140	5	5	4	4
Kenmore Elite UltraWash HE 1315[2]	1100	81					145	5	5	4	4
Kenmore Pro 1317[3]	1280	79					145	5	5	5	4
KitchenAid KUDE50CV[SS]	1200	76					125	5	5	4	4
KitchenAid KUDE60FV[WH]	1340	78					135	5	5	4	4
KitchenAid KUDE70CV[SS]	1300	81					140	5	5	5	4
KitchenAid KUDS30IV[WH]	675	77					120	5	4	3	4
KitchenAid KUDS40CV[WH]	990	79					115	5	4	4	4
LG LDF6920[WW]	700	79					125	5	4	4	4
LG Steam LDF7932[ST]	1000	81					130	5	4	4	5
Maytag MDB7609AW[W]	450	75					120	5	4	3	2
Maytag MDB8959AW[W]	750	71					120	4	4	3	5
Maytag MTB4709AW[W]	400	69					125	5	4	2	3
Miele Inspira G2142SC[WH]	1150	76					145	5	5	4	3

		Excellent ◉	Very good ◒	Good ○	Fair ◓	Poor ●

Brand and Model	Price $	Overall Score	Washing (x_1)	Energy Use (x_2)	Noise (x_3)	Ease of Use (x_4)	Cycle Time (min.)	x_1	x_2	x_3	x_4
Whirlpool DU1030XTX[Q]	350	68	◉	◒	◓	○	130	5	4	2	3
Whirlpool DU1055XTV[Q]	400	76	◉	◒	○	○	125	5	4	3	3
Whirlpool DU1300XTV[Q]	420	72	◉	○	○	○	140	5	3	3	3
Whirlpool Gold GU2300XTV[Q]	550	78	◉	◒	○	◒	135	5	4	3	4
Whirlpool Gold GU2800XTV[Q]	700	77	◉	◉	○	◒	155	5	5	3	4
Whirlpool Gold GU3600XTV[Q]	800	77	◒	◉	◒	◒	145	4	5	4	4

Source: © 2007 by Consumers Union of U.S., Inc., Yonkers, NY 10703-1057, a nonprofit organization. Reprinted with permission from the September 2007 issue of *Consumer Reports*® for educational purposes only. No commercial use or reproduction permitted. www.ConsumerReports.org®.

1. Look at the variables Price, Score, and Cycle Time individually. What can you say about symmetry? About outliers?

2. Look at all the variables in pairs. Which pairs are positively correlated? Negatively correlated? Are there any pairs that exhibit little or no correlation? Are some of these results counterintuitive?

3. Does the price of an appliance, specifically a dishwasher, convey something about its quality? Which variables did you use in arriving at your answer?

Probability and Probability Distributions

© Tammie Arroyo/Getty Images

GENERAL OBJECTIVES

Now that you have learned to describe a data set, how can you use sample data to draw conclusions about the sampled populations? The technique involves a statistical tool called *probability*. To use this tool correctly, you must first understand how it works. The first part of this chapter will present the basic concepts with simple examples.

The variables that we measured in Chapters 1 and 2 can now be redefined as random variables, whose values depend on the chance selection of the elements in the sample. Using probability as a tool, you can create probability distributions that serve as models for discrete random variables, and you can describe these random variables using a mean and standard deviation similar to those in Chapter 2.

? NEED TO KNOW...

**How to Calculate the Probability of an Event
The Difference between Mutually Exclusive and Independent Events**

Probability and Decision Making in the Congo

In his exciting novel *Congo*, author Michael Crichton describes an expedition racing to find boron-coated blue diamonds in the rain forests of eastern Zaire. Can probability help the heroine Karen Ross in her search for the Lost City of Zinj? The case study at the end of this chapter involves Ross's use of probability in decision-making situations.

THE ROLE OF PROBABILITY IN STATISTICS

4.1

Probability and statistics are related in an important way. Probability is used as a *tool;* it allows you to evaluate the reliability of your conclusions about the population when you have only sample information. Consider these situations:

- When you toss a single coin, you will see either a head (H) or a tail (T). If you toss the coin repeatedly, you will generate an infinitely large number of Hs and Ts—the entire population. What does this population look like? If the coin is fair, then the population should contain 50% Hs and 50% Ts. Now toss the coin one more time. What is the chance of getting a head? Most people would say that the "probability" or chance is 1/2.

- Now suppose you are not sure whether the coin is fair; that is, you are not sure whether the makeup of the population is 50–50. You decide to perform a simple experiment. You toss the coin $n = 10$ times and observe 10 heads in a row. Can you conclude that the coin is fair? Probably not, because if the coin were fair, observing 10 heads in a row would be very *unlikely;* that is, the "probability" would be very small. It is more *likely* that the coin is biased.

As in the coin-tossing example, statisticians use probability in two ways. When the population is *known,* probability is used to describe the likelihood of observing a particular sample outcome. When the population is *unknown* and only a sample from that population is available, probability is used in making statements about the makeup of the population—that is, in making statistical inferences.

In Chapters 4–7, you will learn many different ways to calculate probabilities. You will assume that the population is *known* and calculate the probability of observing various sample outcomes. Once you begin to use probability for statistical inference in Chapter 8, the population will be *unknown* and you will use your knowledge of probability to make reliable inferences from sample information. We begin with some simple examples to help you grasp the basic concepts of probability.

EVENTS AND THE SAMPLE SPACE

4.2

Data are obtained by observing either uncontrolled events in nature or by observing events in controlled situations. We use the term **experiment** to describe either method of data collection.

Definition An **experiment** is the process by which an observation (or measurement) is obtained.

The observation or measurement generated by an experiment may or may not produce a numerical value. Here are some examples of experiments:

- Recording a test grade
- Measuring daily rainfall
- Interviewing a householder to obtain his or her opinion on a greenbelt zoning ordinance

- Testing a printed circuit board to determine whether it is a defective product or an acceptable product
- Tossing a coin and observing the face that appears

When an experiment is performed, what we observe is an outcome called a **simple event,** often denoted by the capital E with a subscript.

Definition   A **simple event** is the outcome that is observed on a single repetition of the experiment.

Experiment: Toss a die and observe the number that appears on the upper face. List the simple events in the experiment.

Solution   When the die is tossed once, there are six possible outcomes. These are the simple events, listed below.

Event E_1: Observe a 1	Event E_4: Observe a 4
Event E_2: Observe a 2	Event E_5: Observe a 5
Event E_3: Observe a 3	Event E_6: Observe a 6

We can now define an **event** as a collection of simple events, often denoted by a capital letter.

Definition   An **event** is a collection of simple events.

EXAMPLE
continued   4.1

We can define the events A and B for the die-tossing experiment:

A: Observe an odd number

B: Observe a number less than 4

Since event A occurs if the upper face is 1, 3, or 5, it is a collection of three simple events and we write $A = \{E_1, E_3, E_5\}$. Similarly, the event B occurs if the upper face is 1, 2, or 3 and is defined as a collection or set of these three simple events: $B = \{E_1, E_2, E_3\}$.

Sometimes when one event occurs, it means that another event cannot.

Definition   Two events are **mutually exclusive** if, when one event occurs, the other cannot, and vice versa.

In the die-tossing experiment, events A and B are *not* mutually exclusive, because they have two outcomes in common—if the number on the upper face of the die is a 1 or a 3. Both events A and B will occur if either E_1 or E_3 is observed when the experiment is performed. In contrast, the six simple events $E_1, E_2, \ldots, E_6$ form a set of all mutually exclusive outcomes of the experiment. When the experiment is performed once, one and only one of these simple events can occur.

Definition   The set of all simple events is called the sample space, S.

Sometimes it helps to visualize an experiment using a picture called a **Venn diagram,** shown in Figure 4.1. The white box represents the *sample space,* which contains all of the *simple events,* represented by labeled points. Since an event is a collection of one or more simple events, the appropriate points are circled and labeled with the event letter. For the die-tossing experiment, the sample space is $S = \{E_1, E_2, E_3, E_4, E_5, E_6\}$ or, more simply, $S = \{1, 2, 3, 4, 5, 6\}$. The events $A = \{1, 3, 5\}$ and $B = \{1, 2, 3\}$ are circled in the Venn diagram.

FIGURE 4.1

Venn diagram for die tossing

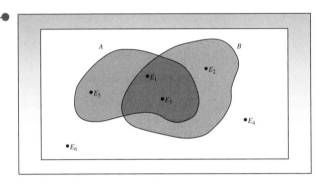

EXAMPLE 4.2

Experiment: Toss a single coin and observe the result. These are the simple events:

E_1: Observe a head (H)

E_2: Observe a tail (T)

The sample space is $S = \{E_1, E_2\}$, or, more simply, $S = \{H, T\}$.

EXAMPLE 4.3

Experiment: Record a person's blood type. The four mutually exclusive possible outcomes are these simple events:

E_1: Blood type A

E_2: Blood type B

E_3: Blood type AB

E_4: Blood type O

The sample space is $S = \{E_1, E_2, E_3, E_4\}$, or $S = \{A, B, AB, O\}$.

Some experiments can be generated in stages, and the sample space can be displayed in a **tree diagram.** Each successive level of branching on the tree corresponds to a step required to generate the final outcome.

EXAMPLE 4.4

A medical technician records a person's blood type and Rh factor. List the simple events in the experiment.

Solution   For each person, a two-stage procedure is needed to record the two variables of interest. The tree diagram is shown in Figure 4.2. The eight simple events in the tree diagram form the sample space, $S = \{A+, A-, B+, B-, AB+, AB-, O+, O-\}$.

FIGURE 4.2

Tree diagram for
Example 4.4

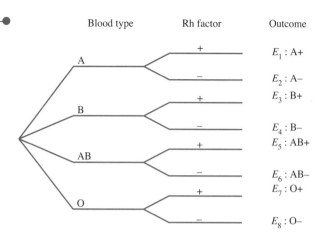

An alternative way to display the simple events is to use a **probability table,** as shown in Table 4.1. The columns and rows show the possible outcomes at the first and second stages, respectively, and the simple events are shown in the cells of the table.

TABLE 4.1

Probability Table for Example 4.4

Rh Factor	Blood Type			
	A	B	AB	O
Negative	A−	B−	AB−	O−
Positive	A+	B+	AB+	O+

CALCULATING PROBABILITIES USING SIMPLE EVENTS

4.3

The probability of an event A is a measure of our belief that the event A will occur. One practical way to interpret this measure is with the concept of *relative frequency.* Recall from Chapter 1 that if an experiment is performed n times, then the relative frequency of a particular occurrence—say, A—is

$$\text{Relative frequency} = \frac{\text{Frequency}}{n}$$

where the frequency is the number of times the event A occurred. If you let n, the number of repetitions of the experiment, become larger and larger ($n \to \infty$), you will eventually generate the entire population. In this population, the relative frequency of the event A is defined as the **probability of event A;** that is,

$$P(A) = \lim_{n \to \infty} \frac{\text{Frequency}}{n}$$

Since $P(A)$ behaves like a relative frequency, $P(A)$ must be a proportion lying between 0 and 1; $P(A) = 0$ if the event A never occurs, and $P(A) = 1$ if the event A always occurs. The closer $P(A)$ is to 1, the more likely it is that A will occur.

For example, if you tossed a balanced, six-sided die an infinite number of times, you would expect the relative frequency for any of the six values, $x = 1, 2, 3, 4, 5, 6$, to be 1/6. Needless to say, it would be very time-consuming, if not impossible, to repeat an experiment an infinite number of times. For this reason, there are alternative methods for calculating probabilities that make use of the relative frequency concept.

An important consequence of the relative frequency definition of probability involves the simple events. Since the simple events are mutually exclusive, their probabilities must satisfy two conditions.

REQUIREMENTS FOR SIMPLE-EVENT PROBABILITIES

- Each probability must lie between 0 and 1.
- The sum of the probabilities for all simple events in S equals 1.

When it is possible to write down the simple events associated with an experiment and to determine their respective probabilities, we can find the probability of an event A as follows:

Definition The **probability of an event** A is equal to the sum of the probabilities of the simple events contained in A.

EXAMPLE 4.5

Toss two fair coins and record the outcome. Find the probability of observing exactly one head in the two tosses.

Solution To list the simple events in the sample space, you can use a tree diagram as shown in Figure 4.3. The letters H and T mean that you observed a head or a tail, respectively, on a particular toss. To assign probabilities to each of the four simple events, you need to remember that the coins are fair. Therefore, any of the four simple events is as likely as any other. Since the sum of the four simple events must be 1, each must have probability $P(E_i) = 1/4$. The simple events in the sample space are shown in Table 4.2, along with their *equally likely probabilities*. To find $P(A) = P(\text{observe exactly one head})$, you need to find all the simple events that result in event A—namely, E_2 and E_3:

$$P(A) = P(E_2) + P(E_3)$$

$$= \frac{1}{4} + \frac{1}{4} = \frac{1}{2}$$

 **NEED A TIP?**

Probabilities must lie between 0 and 1.

FIGURE 4.3

Tree diagram for Example 4.5

 NEED A TIP?

The probabilities of all the simple events must add to 1.

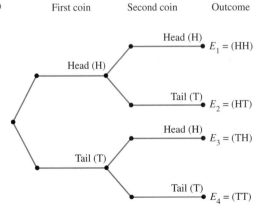

TABLE 4.2 ● **Simple Events and Their Probabilities**

Event	First Coin	Second Coin	$P(E_i)$
E_1	H	H	1/4
E_2	H	T	1/4
E_3	T	H	1/4
E_4	T	T	1/4

EXAMPLE 4.6

The proportions of blood phenotypes A, B, AB, and O in the population of all Caucasians in the United States are reported as .40, .11, .04, and .45, respectively.[1] If a single Caucasian is chosen randomly from the population, what is the probability that he or she will have either type A or type AB blood?

Solution The four simple events, A, B, AB, and O, do *not* have equally likely probabilities. Their probabilities are found using the relative frequency concept as

$$P(A) = .40 \qquad P(B) = .11 \qquad P(AB) = .04 \qquad P(O) = .45$$

The event of interest consists of two simple events, so

$$P(\text{person is either type A or type AB}) = P(A) + P(AB)$$

$$= .40 + .04 = .44$$

EXAMPLE 4.7

A candy dish contains one yellow and two red candies. You close your eyes, choose two candies one at a time from the dish, and record their colors. What is the probability that both candies are red?

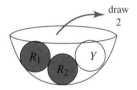
draw 2

Solution Since no probabilities are given, you must list the simple events in the sample space. The two-stage selection of the candies suggests a tree diagram, shown in Figure 4.4. There are two red candies in the dish, so you can use the letters R_1, R_2, and Y to indicate that you have selected the first red, the second red, or the yellow candy, respectively. Since you closed your eyes when you chose the candies, all six choices should be *equally likely* and are assigned probability 1/6. If A is the event that both candies are red, then

$$A = \{R_1R_2, R_2R_1\}$$

Thus,

$$P(A) = P(R_1R_2) + P(R_2R_1)$$

$$= \frac{1}{6} + \frac{1}{6} = \frac{1}{3}$$

FIGURE 4.4 ●

Tree diagram for Example 4.7

 NEED A TIP?

A tree diagram helps to find simple events.

Branch = step toward outcome

Following branches ⇒ list of simple events

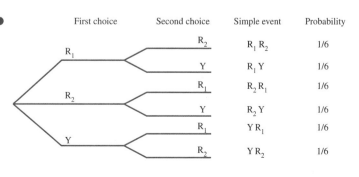

First choice	Second choice	Simple event	Probability
R_1	R_2	$R_1 R_2$	1/6
	Y	$R_1 Y$	1/6
R_2	R_1	$R_2 R_1$	1/6
	Y	$R_2 Y$	1/6
Y	R_1	$Y R_1$	1/6
	R_2	$Y R_2$	1/6

NEED TO KNOW...

How to Calculate the Probability of an Event

1. List all the simple events in the sample space.
2. Assign an appropriate probability to each simple event.
3. Determine which simple events result in the event of interest.
4. Sum the probabilities of the simple events that result in the event of interest.

In your calculation, you must always be careful that you satisfy these two conditions:

- Include all simple events in the sample space.
- Assign realistic probabilities to the simple events.

When the sample space is large, it is easy to unintentionally omit some of the simple events. If this happens, or if your assigned probabilities are wrong, your answers will not be useful in practice.

One way to determine the required number of simple events is to use the counting rules presented in the next optional section. These rules can be used to solve more complex problems, which generally involve a large number of simple events. If you need to master only the basic concepts of probability, you may choose to skip the next section.

 EXERCISES

BASIC TECHNIQUES

4.1 Tossing a Die An experiment involves tossing a single die. These are some events:

  A: Observe a 2

  B: Observe an even number

  C: Observe a number greater than 2

  D: Observe both A and B

  E: Observe A or B or both

  F: Observe both A and C

a. List the simple events in the sample space.
b. List the simple events in each of the events A through F.
c. What probabilities should you assign to the simple events?
d. Calculate the probabilities of the six events A through F by adding the appropriate simple-event probabilities.

4.2 A sample space S consists of five simple events with these probabilities:

$$P(E_1) = P(E_2) = .15 \qquad P(E_3) = .4$$
$$P(E_4) = 2P(E_5)$$

a. Find the probabilities for simple events E_4 and E_5.
b. Find the probabilities for these two events:

$$A = \{E_1, E_3, E_4\}$$
$$B = \{E_2, E_3\}$$

c. List the simple events that are either in event A or event B or both.
d. List the simple events that are in both event A and event B.

4.3 A sample space contains 10 simple events: E_1, $E_2, \ldots, E_{10}$. If $P(E_1) = 3P(E_2) = .45$ and the remaining simple events are equiprobable, find the probabilities of these remaining simple events.

4.4 Free Throws A particular basketball player hits 70% of her free throws. When she tosses a pair of free throws, the four possible simple events and three of their associated probabilities are as given in the table:

Simple Event	Outcome of First Free Throw	Outcome of Second Free Throw	Probability
1	Hit	Hit	.49
2	Hit	Miss	?
3	Miss	Hit	.21
4	Miss	Miss	.09

a. Find the probability that the player will hit on the first throw and miss on the second.

b. Find the probability that the player will hit on at least one of the two free throws.

4.5 Four Coins A jar contains four coins: a nickel, a dime, a quarter, and a half-dollar. Three coins are randomly selected from the jar.

a. List the simple events in *S*.

b. What is the probability that the selection will contain the half-dollar?

c. What is the probability that the total amount drawn will equal 60¢ or more?

4.6 Preschool or Not? On the first day of kindergarten, the teacher randomly selects 1 of his 25 students and records the student's gender, as well as whether or not that student had gone to preschool.

a. How would you describe the experiment?

b. Construct a tree diagram for this experiment. How many simple events are there?

c. The table below shows the distribution of the 25 students according to gender and preschool experience. Use the table to assign probabilities to the simple events in part b.

	Male	Female
Preschool	8	9
No Preschool	6	2

d. What is the probability that the randomly selected student is male? What is the probability that the student is a female and did not go to preschool?

4.7 The Urn Problem A bowl contains three red and two yellow balls. Two balls are randomly selected and their colors recorded. Use a tree diagram to list the 20 simple events in the experiment, keeping in mind the order in which the balls are drawn.

4.8 The Urn Problem, continued Refer to Exercise 4.7. A ball is randomly selected from the bowl containing three red and two yellow balls. Its color is noted, and the ball is returned to the bowl before a second ball is selected. List the additional five simple events that must be added to the sample space in Exercise 4.7.

APPLICATIONS

4.9 Need Eyeglasses? A survey classified a large number of adults according to whether they were judged to need eyeglasses to correct their reading vision and whether they used eyeglasses when reading. The proportions falling into the four categories are shown in the table. (Note that a small proportion, .02, of adults used eyeglasses when in fact they were judged not to need them.)

Judged to Need Eyeglasses	Used Eyeglasses for Reading	
	Yes	No
Yes	.44	.14
No	.02	.40

If a single adult is selected from this large group, find the probability of each event:

a. The adult is judged to need eyeglasses.

b. The adult needs eyeglasses for reading but does not use them.

c. The adult uses eyeglasses for reading whether he or she needs them or not.

4.10 Roulette The game of roulette uses a wheel containing 38 pockets. Thirty-six pockets are numbered 1, 2, . . . , 36, and the remaining two are marked 0 and 00. The wheel is spun, and a pocket is identified as the "winner." Assume that the observance of any one pocket is just as likely as any other.

a. Identify the simple events in a single spin of the roulette wheel.

b. Assign probabilities to the simple events.

c. Let *A* be the event that you observe either a 0 or a 00. List the simple events in the event *A* and find *P(A)*.

d. Suppose you placed bets on the numbers 1 through 18. What is the probability that one of your numbers is the winner?

4.11 Jury Duty Three people are randomly selected to report for jury duty. The gender of each person is noted by the county clerk.

a. Define the experiment.

b. List the simple events in S.

c. If each person is just as likely to be a man as a woman, what probability do you assign to each simple event?

d. What is the probability that only one of the three is a man?

e. What is the probability that all three are women?

4.12 Jury Duty II Refer to Exercise 4.11. Suppose that there are six prospective jurors, four men and two women, who might be impaneled to sit on the jury in a criminal case. Two jurors are randomly selected from these six to fill the two remaining jury seats.

a. List the simple events in the experiment (HINT: There are 15 simple events if you ignore the order of selection of the two jurors.)

b. What is the probability that both impaneled jurors are women?

4.13 Tea Tasters A food company plans to conduct an experiment to compare its brand of tea with that of two competitors. A single person is hired to taste and rank each of three brands of tea, which are unmarked except for identifying symbols A, B, and C.

a. Define the experiment.

b. List the simple events in S.

c. If the taster has no ability to distinguish a difference in taste among teas, what is the probability that the taster will rank tea type A as the most desirable? As the least desirable?

4.14 Creation Which of the following comes closest to your views on the origin and development of human beings? Do you believe that human beings have developed over millions of years from less advanced forms of life, but that God has guided the process? Do you think that human beings have developed over millions of years, and that God had no part in the process? Or do you believe that God created humans in their present form within the last 10,000 years or so? When asked these questions, the proportions of Americans with varying opinions are approximately as shown in the table.[2]

Opinion	Proportion
Guided by God	.36
God had no part	.13
God created in present form	.46
No opinion	.05

Source: Adapted from www.pollingreport.com

Suppose that one person is randomly selected and his or her opinion on this question is recorded.

a. What are the simple events in the experiment?

b. Are the simple events in part a equally likely? If not, what are the probabilities?

c. What is the probability that the person feels that God had some part in the creation of humans?

d. What is the probability that the person feels that God had no part in the process?

4.15 Fruit Flies In a genetics experiment, the researcher mated two *Drosophila* fruit flies and observed the traits of 300 offspring. The results are shown in the table.

	Wing Size	
Eye Color	Normal	Miniature
Normal	140	6
Vermillion	3	151

One of these offspring is randomly selected and observed for the two genetic traits.

a. What is the probability that the fly has normal eye color and normal wing size?

b. What is the probability that the fly has vermillion eyes?

c. What is the probability that the fly has either vermillion eyes or miniature wings, or both?

4.16 100-Meter Run Four equally qualified runners, John, Bill, Ed, and Dave, run a 100-meter sprint, and the order of finish is recorded.

a. How many simple events are in the sample space?

b. If the runners are equally qualified, what probability should you assign to each simple event?

c. What is the probability that Dave wins the race?

d. What is the probability that Dave wins and John places second?

e. What is the probability that Ed finishes last?

USEFUL COUNTING RULES (OPTIONAL)

4.4

Suppose that an experiment involves a large number N of simple events and you know that all the simple events are *equally likely*. Then each simple event has probability $1/N$, and the probability of an event A can be calculated as

$$P(A) = \frac{n_A}{N}$$

where n_A is the number of simple events that result in the event A. In this section, we present three simple rules that can be used to count either N, the number of simple events in the sample space, or n_A, the number of simple events in event A. Once you have obtained these counts, you can find $P(A)$ without actually listing all the simple events.

THE *mn* RULE

Consider an experiment that is performed in two stages. If the first stage can be accomplished in m ways and for each of these ways, the second stage can be accomplished in n ways, then there are mn ways to accomplish the experiment.

For example, suppose that you can order a car in one of three styles and in one of four paint colors. To find out how many options are available, you can think of first picking one of the $m = 3$ styles and then selecting one of the $n = 4$ paint colors. Using the *mn* Rule, as shown in Figure 4.5, you have $mn = (3)(4) = 12$ possible options.

FIGURE 4.5

Style–color combinations

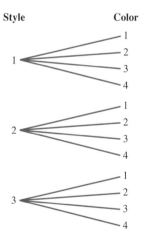

EXAMPLE 4.8

Two dice are tossed. How many simple events are in the sample space S?

Tossing Dice

Solution The first die can fall in one of $m = 6$ ways, and the second die can fall in one of $n = 6$ ways. Since the experiment involves two stages, forming the pairs of numbers shown on the two faces, the total number of simple events in S is

$$mn = (6)(6) = 36$$

EXAMPLE 4.9

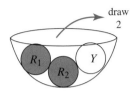

A candy dish contains one yellow and two red candies. Two candies are selected one at a time from the dish, and their colors are recorded. How many simple events are in the sample space S?

Solution The first candy can be chosen in $m = 3$ ways. Since one candy is now gone, the second candy can be chosen in $n = 2$ ways. The total number of simple events is

$$mn = (3)(2) = 6$$

These six simple events were listed in Example 4.7.

We can extend the *mn* Rule for an experiment that is performed in more than two stages.

THE EXTENDED *mn* RULE

If an experiment is performed in k stages, with n_1 ways to accomplish the first stage, n_2 ways to accomplish the second stage, . . . , and n_k ways to accomplish the kth stage, then the number of ways to accomplish the experiment is

$$n_1 n_2 n_3 \cdots n_k$$

EXAMPLE 4.10

How many simple events are in the sample space when three coins are tossed?

Solution Each coin can land in one of two ways. Hence, the number of simple events is

$$(2)(2)(2) = 8$$

EXAMPLE 4.11

A truck driver can take three routes from city A to city B, four from city B to city C, and three from city C to city D. If, when traveling from A to D, the driver must drive from A to B to C to D, how many possible A-to-D routes are available?

Solution Let

$$n_1 = \text{Number of routes from } A \text{ to } B = 3$$
$$n_2 = \text{Number of routes from } B \text{ to } C = 4$$
$$n_3 = \text{Number of routes from } C \text{ to } D = 3$$

Then the total number of ways to construct a complete route, taking one subroute from each of the three groups, (A to B), (B to C), and (C to D), is

$$n_1 n_2 n_3 = (3)(4)(3) = 36$$

A second useful counting rule follows from the *mn* Rule and involves **orderings** or **permutations.** For example, suppose you have three books, A, B, and C, but you have room for only two on your bookshelf. In how many ways can you select and arrange the two books? There are three choices for the two books—A and B, A and C, or B and C—but each of the pairs can be arranged in two ways on the shelf. All the permutations of the two books, chosen from three, are listed in Table 4.3. The *mn*

Rule implies that there are 6 ways, because the first book can be chosen in $m = 3$ ways and the second in $n = 2$ ways, so the result is $mn = 6$.

TABLE 4.3 ● **Permutations of Two Books Chosen from Three**

Combinations of Two	Reordering of Combinations
AB	BA
AC	CA
BC	CB

In how many ways can you arrange all three books on your bookshelf? These are the six permutations:

ABC   ACB   BAC
BCA   CAB   CBA

Since the first book can be chosen in $n_1 = 3$ ways, the second in $n_2 = 2$ ways, and the third in $n_3 = 1$ way, the total number of orderings is $n_1 n_2 n_3 = (3)(2)(1) = 6$.

Rather than applying the *mn* Rule each time, you can find the number of orderings using a general formula involving *factorial notation*.

A COUNTING RULE FOR PERMUTATIONS

The number of ways we can arrange n distinct objects, taking them r at a time, is

$$P_r^n = \frac{n!}{(n - r)!}$$

where $n! = n(n - 1)(n - 2) \cdots (3)(2)(1)$ and $0! \doteq 1$.

Since r objects are chosen, this is an *r-stage* experiment. The first object can be chosen in n ways, the second in $(n - 1)$ ways, the third in $(n - 2)$ ways, and the rth in $(n - r + 1)$ ways. We can simplify this awkward notation using the counting rule for permutations because

$$\frac{n!}{(n - r)!} = \frac{n(n - 1)(n - 2) \cdots (n - r + 1)(n - r) \cdots (2)(1)}{(n - r) \cdots (2)(1)}$$

$$= n(n - 1) \cdots (n - r + 1)$$

A SPECIAL CASE: ARRANGING n ITEMS

The number of ways to arrange an entire set of n distinct items is $P_n^n = n!$

EXAMPLE 4.12 Three lottery tickets are drawn from a total of 50. If the tickets will be distributed to each of three employees in the order in which they are drawn, the order will be important. How many simple events are associated with the experiment?

Solution The total number of simple events is

$$P_3^{50} = \frac{50!}{47!} = 50(49)(48) = 117,600$$

EXAMPLE (4.13)

A piece of equipment is composed of five parts that can be assembled in any order. A test is to be conducted to determine the time necessary for each order of assembly. If each order is to be tested once, how many tests must be conducted?

Solution   The total number of tests is

$$P_5^5 = \frac{5!}{0!} = 5(4)(3)(2)(1) = 120$$

When we counted the number of permutations of the two books chosen for your bookshelf, we used a systematic approach:

- First we counted the number of *combinations* or pairs of books to be chosen.
- Then we counted the number of ways to arrange the two chosen books on the shelf.

Sometimes the ordering or arrangement of the objects is not important, but only the objects that are chosen. In this case, you can use a counting rule for **combinations.** For example, you may not care in what order the books are placed on the shelf, but only which books you are able to shelve. When a five-person committee is chosen from a group of 12 students, the order of choice is unimportant because all five students will be equal members of the committee.

A COUNTING RULE FOR COMBINATIONS

The number of distinct combinations of n distinct objects that can be formed, taking them r at a time, is

$$C_r^n = \frac{n!}{r!(n-r)!}$$

The number of *combinations* and the number of *permutations* are related:

$$C_r^n = \frac{P_r^n}{r!}$$

You can see that C_r^n results when you divide the number of permutations by $r!$, the number of ways of rearranging each distinct group of r objects chosen from the total n.

EXAMPLE (4.14)

A printed circuit board may be purchased from five suppliers. In how many ways can three suppliers be chosen from the five?

Solution   Since it is important to know only which three have been chosen, not the order of selection, the number of ways is

$$C_3^5 = \frac{5!}{3!2!} = \frac{(5)(4)}{2} = 10$$

The next example illustrates the use of counting rules to solve a probability problem.

EXAMPLE (4.15)

Five manufacturers produce a certain electronic device, whose quality varies from manufacturer to manufacturer. If you were to select three manufacturers at random, what is the chance that the selection would contain exactly two of the best three?

Solution The simple events in this experiment consist of all possible combinations of three manufacturers, chosen from a group of five. Of these five, three have been designated as "best" and two as "not best." You can think of a candy dish containing three red and two yellow candies, from which you will select three, as illustrated in Figure 4.6. The total number of simple events N can be counted as the number of ways to choose three of the five manufacturers, or

$$N = C_3^5 = \frac{5!}{3!2!} = 10$$

FIGURE 4.6

Illustration for
Example 4.15

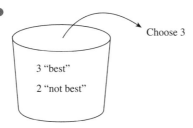

Choose 3

3 "best"

2 "not best"

Since the manufacturers are selected at random, any of these 10 simple events will be *equally likely,* with probability 1/10. But how many of these simple events result in the event?

A: Exactly two of the "best" three

You can count n_A, the number of events in A, in two steps because event A will occur when you select two of the "best" three and one of the two "not best." There are

$$C_2^3 = \frac{3!}{2!1!} = 3$$

ways to accomplish the first stage and

$$C_1^2 = \frac{2!}{1!1!} = 2$$

ways to accomplish the second stage. Applying the *mn* Rule, we find there are $n_A = (3)(2) = 6$ of the 10 simple events in event A and $P(A) = n_A/N = 6/10$.

Many other counting rules are available in addition to the three presented in this section. If you are interested in this topic, you should consult one of the many textbooks on combinatorial mathematics.

4.4 EXERCISES

BASIC TECHNIQUES

4.17 You have *two* groups of distinctly different items, 10 in the first group and 8 in the second. If you select one item from each group, how many different pairs can you form?

4.18 You have *three* groups of distinctly different items, four in the first group, seven in the second, and three in the third. If you select one item from each group, how many different triplets can you form?

4.19 Permutations Evaluate the following *permutations.* (HINT: Your scientific calculator may have a function that allows you to calculate permutations and combinations quite easily.)

a. P_3^5 **b.** P_9^{10} **c.** P_6^6 **d.** P_1^{20}

4.20 Combinations Evaluate these *combinations:*

a. C_3^5 **b.** C_9^{10} **c.** C_6^6 **d.** C_1^{20}

4.21 Choosing People In how many ways can you select five people from a group of eight if the order of selection is important?

4.22 Choosing People, again In how many ways can you select two people from a group of 20 if the order of selection is not important?

4.23 Dice Three dice are tossed. How many simple events are in the sample space?

4.24 Coins Four coins are tossed. How many simple events are in the sample space?

4.25 The Urn Problem, again Three balls are selected from a box containing 10 balls. The order of selection is not important. How many simple events are in the sample space?

APPLICATIONS

4.26 Dinner at Gerard's A French restaurant in Riverside, California, offers a special summer menu in which, for a fixed dinner cost, you can choose from one of two salads, one of two entrees, and one of two desserts. How many different dinners are available?

4.27 Itineraries A businessman in New York is preparing an itinerary for a visit to six major cities. The distance traveled, and hence the cost of the trip, will depend on the order in which he plans his route. How many different itineraries (and trip costs) are possible?

4.28 What to Wear? You own 4 pairs of jeans, 12 clean T-shirts, and 4 wearable pairs of sneakers. How many outfits (jeans, T-shirt, and sneakers) can you create?

4.29 A Card Game Three students are playing a card game. They decide to choose the first person to play by each selecting a card from the 52-card deck and looking for the highest card in value and suit. They rank the suits from lowest to highest: clubs, diamonds, hearts, and spades.

a. If the card is replaced in the deck after each student chooses, how many possible configurations of the three choices are possible?

b. How many configurations are there in which each student picks a different card?

c. What is the probability that all three students pick exactly the same card?

d. What is the probability that all three students pick different cards?

4.30 Vacation Plans Your family vacation involves a cross-country air flight, a rental car, and a hotel stay in Boston. If you can choose from four major air carriers, five car rental agencies, and three major hotel chains, how many options are available for your vacation accommodations?

4.31 Playing Poker Five cards are selected from a 52-card deck for a poker hand.

a. How many simple events are in the sample space?

b. A *royal flush* is a hand that contains the A, K, Q, J, and 10, all in the same suit. How many ways are there to get a royal flush?

c. What is the probability of being dealt a royal flush?

4.32 Traffic Problems Two city council members are to be selected from a total of five to form a subcommittee to study the city's traffic problems.

a. How many different subcommittees are possible?

b. If all possible council members have an equal chance of being selected, what is the probability that members Smith and Jones are both selected?

4.33 A Hospital Survey A study is to be conducted in a hospital to determine the attitudes of nurses toward various administrative procedures. If a sample of 10 nurses is to be selected from a total of 90, how many different samples can be selected? (HINT: Is order important in determining the makeup of the sample to be selected for the survey?)

4.34 Poker II Refer to Exercise 4.31. You have a poker hand containing four of a kind.

a. How many possible poker hands can be dealt?

b. In how many ways can you receive four cards of the same face value *and* one card from the other 48 available cards?

c. What is the probability of being dealt four of a kind?

4.35 The WNBA Professional basketball is now a reality for women basketball players in the United States. There are two conferences in the WNBA, each with six teams, as shown in the table below.[3]

Western Conference	Eastern Conference
Minnesota Lynx	Atlanta Dream
Phoenix Mercury	Indiana Fever
Tulsa Shock	New York Liberty
Los Angeles Sparks	Washington Mystics
Seattle Storm	Connecticut Sun
San Antonio Silver Stars	Chicago Sky

Source: www.espn.com

Two teams, one from each conference, are randomly selected to play an exhibition game.

a. How many pairs of teams can be chosen?

b. What is the probability that the two teams are Los Angeles and New York?

c. What is the probability that the Western Conference team is not from California?

4.36 100-Meter Run, again Refer to Exercise 4.14, in which a 100-meter sprint is run by John, Bill, Ed, and Dave. Assume that all of the runners are equally qualified, so that any order of finish is equally likely. Use the *mn* Rule or permutations to answer these questions:

a. How many orders of finish are possible?

b. What is the probability that Dave wins the sprint?

c. What is the probability that Dave wins and John places second?

d. What is the probability that Ed finishes last?

4.37 Gender Bias? A woman brought a complaint of gender discrimination to an eight-member human relations advisory board. The board, composed of five women and three men, voted 5–3 in favor of the plaintiff, the five women voting for the plaintiff and the three men against. Has the board been affected by gender bias? That is, if the vote in favor of the plaintiff was 5–3 and the board members were not biased by gender, what is the probability that the vote would split along gender lines (five women for, three men against)?

4.38 Cramming A student prepares for an exam by studying a list of 10 problems. She can solve 6 of them. For the exam, the instructor selects 5 questions at random from the list of 10. What is the probability that the student can solve all 5 problems on the exam?

4.39 Monkey Business A monkey is given 12 blocks: 3 shaped like squares, 3 like rectangles, 3 like triangles, and 3 like circles. If it draws three of each kind in order—say, 3 triangles, then 3 squares, and so on—would you suspect that the monkey associates identically shaped figures? Calculate the probability of this event.

EVENT RELATIONS AND PROBABILITY RULES

4.5

Sometimes the event of interest can be formed as a combination of several other events. Let *A* and *B* be two events defined on the sample space *S*. Here are three important relationships between events.

Definition The **union** of events *A* and *B*, denoted by $A \cup B$, is the event that either *A* or *B* or both occur.

Definition The **intersection** of events *A* and *B*, denoted by $A \cap B$, is the event that both *A* and *B* occur.[†]

Definition The **complement** of an event *A*, denoted by A^c, is the event that *A* *does not* occur.

Figures 4.7, 4.8, and 4.9 show Venn diagram representations of $A \cup B$, $A \cap B$, and A^c, respectively. Any simple event in the shaded area is a possible outcome resulting in the appropriate event. One way to find the probabilities of the union, the intersection, or the complement is to sum the probabilities of all the associated simple events.

[†]Some authors use the notation *AB*.

FIGURE 4.7

Venn diagram of $A \cup B$

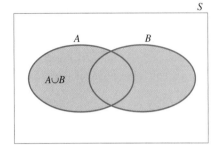

FIGURE 4.8

Venn diagram $A \cap B$

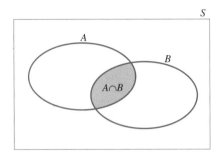

  NEED A TIP?

Intersection ⇔ "both . . . and" or just "and"

Union ⇔ "either . . . or . . . or both" or just "or"

FIGURE 4.9

The complement of an event

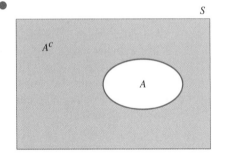

EXAMPLE 4.16 — Two fair coins are tossed, and the outcome is recorded. These are the events of interest:

  A: Observe at least one head

  B: Observe at least one tail

Define the events A, B, $A \cap B$, $A \cup B$, and A^c as collections of simple events, and find their probabilities.

Solution Recall from Example 4.5 that the simple events for this experiment are

  E_1: HH (head on first coin, head on second)
  E_2: HT
  E_3: TH
  E_4: TT

and that each simple event has probability 1/4. Event A, at least one head, occurs if E_1, E_2, or E_3 occurs, so that

$$A = \{E_1, E_2, E_3\} \qquad P(A) = \frac{3}{4}$$

and

$$A^c = \{E_4\} \qquad P(A^c) = \frac{1}{4}$$

Similarly,

$$B = \{E_2, E_3, E_4\} \qquad\qquad P(B) = \frac{3}{4}$$

$$A \cap B = \{E_2, E_3\} \qquad\qquad P(A \cap B) = \frac{1}{2}$$

$$A \cup B = \{E_1, E_2, E_3, E_4\} \qquad P(A \cup B) = \frac{4}{4} = 1$$

Note that $(A \cup B) = S$, the sample space, and is thus certain to occur.

The concept of unions and intersections can be extended to more than two events. For example, the union of three events A, B, and C, which is written as $A \cup B \cup C$, is the set of simple events that are in A or B or C or in any combination of those events. Similarly, the intersection of three events A, B, and C, which is written as $A \cap B \cap C$, is the collection of simple events that are common to the three events A, B, and C.

Calculating Probabilities for Unions and Complements

When we can write the event of interest in the form of a union, a complement, or an intersection, there are special probability rules that can simplify our calculations. The first rule deals with *unions* of events.

THE ADDITION RULE

Given two events, A and B, the probability of their union, $A \cup B$, is equal to

$$P(A \cup B) = P(A) + P(B) - P(A \cap B)$$

Notice in the Venn diagram in Figure 4.10 that the sum $P(A) + P(B)$ double counts the simple events that are common to both A and B. Subtracting $P(A \cap B)$ gives the correct result.

FIGURE 4.10

The Addition Rule

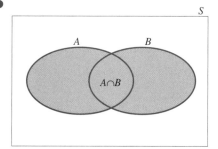

When two events A and B are **mutually exclusive** or **disjoint,** it means that when A occurs, B cannot, and vice versa. This means that the probability that they both occur, $P(A \cap B)$, must be zero. Figure 4.11 is a Venn diagram representation of two such events with no simple events in common.

FIGURE 4.11

Two disjoint or mutually exclusive events

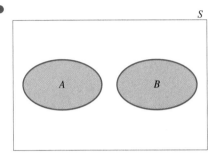

When two events A and B are **mutually exclusive,** then $P(A \cap B) = 0$ and the Addition Rule simplifies to

$$P(A \cup B) = P(A) + P(B)$$

 NEED A TIP?

Remember, mutually exclusive $\Leftrightarrow P(A \cap B) = 0$.

The second rule deals with *complements* of events. You can see from the Venn diagram in Figure 4.9 that A and A^c are mutually exclusive and that $A \cup A^c = S$, the entire sample space. It follows that

$$P(A) + P(A^c) = 1 \text{ and } P(A^c) = 1 - P(A)$$

RULE FOR COMPLEMENTS

$$P(A^c) = 1 - P(A)$$

EXAMPLE  **4.17**

An oil-prospecting firm plans to drill two exploratory wells. Past evidence is used to assess the possible outcomes listed in Table 4.4.

TABLE 4.4 ● **Outcomes for Oil-Drilling Experiment**

Event	Description	Probability
A	Neither well produces oil or gas	.80
B	Exactly one well produces oil or gas	.18
C	Both wells produce oil or gas	.02

Find $P(A \cup B)$ and $P(B \cup C)$.

Solution By their definition, events A, B, and C are jointly mutually exclusive because the occurrence of one event precludes the occurrence of either of the other two. Therefore,

$$P(A \cup B) = P(A) + P(B) = .80 + .18 = .98$$

and

$$P(B \cup C) = P(B) + P(C) = .18 + .02 = .20$$

The event $A \cup B$ can be described as the event that *at most* one well produces oil or gas, and $B \cup C$ describes the event that *at least* one well produces gas or oil.

EXAMPLE 4.18

In a telephone survey of 1000 adults, respondents were asked their opinion about the cost of a college education. The respondents were classified according to whether they currently had a child in college and whether they thought the loan burden for most college students is too high, the right amount, or too little. The proportions responding in each category are shown in the **probability table** in Table 4.5. Suppose one respondent is chosen at random from this group.

TABLE 4.5 ● **Probability Table**

	Too High (A)	Right Amount (B)	Too Little (C)
Child in College (D)	.35	.08	.01
No Child in College (E)	.25	.20	.11

1. What is the probability that the respondent has a child in college?
2. What is the probability that the respondent does not have a child in college?
3. What is the probability that the respondent has a child in college or thinks that the loan burden is too high or both?

Solution Table 4.5 gives the probabilities for the six simple events in the cells of the table. For example, the entry in the top left corner of the table is the probability that a respondent has a child in college *and* thinks the loan burden is too high $(A \cap D)$.

1. The event that a respondent has a child in college will occur regardless of his or her response to the question about loan burden. That is, event D consists of the simple events in the first row:

$$P(D) = .35 + .08 + .01 = .44$$

In general, the probabilities of *marginal* events such as D and A are found by summing the probabilities in the appropriate row or column.

2. The event that the respondent does not have a child in college is the complement of the event D denoted by D^c. The probability of D^c is found as

$$P(D^c) = 1 - P(D)$$

Using the result of part 1, we have

$$P(D^c) = 1 - .44 = .56$$

3. The event of interest is $P(A \cup D)$. Using the Addition Rule

$$P(A \cup D) = P(A) + P(D) - P(A \cap D)$$
$$= .60 + .44 - .35$$
$$= .69$$

INDEPENDENCE, CONDITIONAL PROBABILITY, AND THE MULTIPLICATION RULE

4.6

In Example 4.18, we were able to use the Addition Rule to calculate $P(A \cup D)$ because we could find $P(A \cap D)$ directly from the probability table. Sometimes, however, the intersection probability is unknown. In this situation, there is a probability rule that can be used to calculate the probability of the intersection of several events. This rule depends on the important statistical concept of **independent** or **dependent events.**

Definition   Two events, *A* and *B*, are said to be **independent** if and only if the probability of event *B* is not influenced or changed by the occurrence of event *A*, or vice versa.

Colorblindness   Suppose a researcher notes a person's gender and whether or not the person is colorblind to red and green. Does the probability that a person is colorblind change, depending on whether the person is male or not? Define two events:

   A: Person is a male

   B: Person is colorblind

In this case, since colorblindness is a male sex-linked characteristic, the probability that a man is colorblind will be greater than the probability that a person chosen from the general population will be colorblind. The probability of event *B*, that a person is colorblind, depends on whether or not event *A*, that the person is a male, has occurred. We say that *A* and *B* are *dependent events.*

Tossing Dice   On the other hand, consider tossing a single die two times, and define two events:

   A: Observe a 2 on the first toss

   B: Observe a 2 on the second toss

If the die is fair, the probability of event *A* is $P(A) = 1/6$. Consider the probability of event *B*. Regardless of whether event *A* has or has not occurred, the probability of observing a 2 on the second toss is still 1/6. We could write:

   $P(B$ given that A occurred$) = 1/6$

   $P(B$ given that A did not occur$) = 1/6$

Since the probability of event *B* is not changed by the occurrence of event *A*, we say that *A* and *B* are *independent events.*

   The probability of an event *A*, given that the event *B* has occurred, is called the **conditional probability of *A*, given that *B* has occurred,** denoted by $P(A|B)$. The vertical bar is read "given" and the events appearing to the right of the bar are those that you know have occurred. We will use these probabilities to calculate the probability that *both A and B* occur when the experiment is performed.

THE GENERAL MULTIPLICATION RULE

The probability that *both A and B* occur when the experiment is performed is

   $P(A \cap B) = P(A)P(B|A)$

or

   $P(A \cap B) = P(B)P(A|B)$

EXAMPLE 4.19

In a color preference experiment, eight toys are placed in a container. The toys are identical except for color—two are red, and six are green. A child is asked to choose two toys *at random*. What is the probability that the child chooses the two red toys?

Solution You can visualize the experiment using a tree diagram as shown in Figure 4.12. Define the following events:

R: Red toy is chosen

G: Green toy is chosen

FIGURE 4.12

Tree diagram for Example 4.19

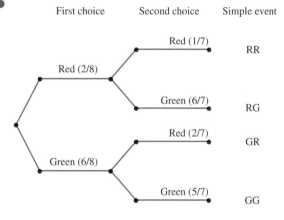

The event *A* (both toys are red) can be constructed as the intersection of two events:

A = (R on first choice) ∩ (R on second choice)

Since there are only two red toys in the container, the probability of choosing red on the first choice is 2/8. However, once this red toy has been chosen, the probability of red on the second choice is *dependent* on the outcome of the first choice (see Figure 4.12). If the first choice was red, the probability of choosing a second red toy is only 1/7 because there is only one red toy among the seven remaining. If the first choice was green, the probability of choosing red on the second choice is 2/7 because there are two red toys among the seven remaining. Using this information and the Multiplication Rule, you can find the probability of event *A*.

$$P(A) = P(\text{R on first choice} \cap \text{R on second choice})$$
$$= P(\text{R on first choice})\, P(\text{R on second choice}|\text{R on first})$$
$$= \left(\frac{2}{8}\right)\left(\frac{1}{7}\right) = \frac{2}{56} = \frac{1}{28}$$

The solution in Example 4.19 was possible only because we knew *P*(R on second choice|R on first choice). If you don't know the conditional probability, *P*(*A*|*B*), you may be able to calculate it by using the Multiplication Rule in a slightly different form. Just *rearrange the terms* in the Multiplication Rule.

CONDITIONAL PROBABILITIES

The conditional probability of event A, given that event B has occurred is

$$P(A|B) = \frac{P(A \cap B)}{P(B)} \quad \text{if} \quad P(B) \neq 0$$

The conditional probability of event B, given that event A has occurred is

$$P(B|A) = \frac{P(A \cap B)}{P(A)} \quad \text{if} \quad P(A) \neq 0$$

Notice that, in this form, you need to know $P(A \cap B)$!

Colorblindness, continued   Suppose that in the general population, there are 51% men and 49% women, and that the proportions of colorblind men and women are shown in the probability table below:

	Men(B)	Women (B^C)	Total
Colorblind (A)	.04	.002	.042
Not Colorblind (A^C)	.47	.488	.958
Total	.51	.49	1.00

If a person is drawn at random from this population and is found to be a man (event B), what is the probability that the man is colorblind (event A)? If we know that the event B has occurred, we must restrict our focus to only the 51% of the population that is male. The probability of being colorblind, given that the person is male, is 4% of the 51%, or

$$P(A|B) = \frac{P(A \cap B)}{P(B)} = \frac{.04}{.51} = .078$$

What is the probability of being colorblind, given that the person is female? Now we are restricted to only the 49% of the population that is female, and

$$P(A|B^C) = \frac{P(A \cap B^C)}{P(B^C)} = \frac{.002}{.49} = .004$$

Notice that the probability of event A changed, depending on whether event B occured. This indicates that these two events are *dependent*.

When two events are **independent**—that is, if the probability of event B is the same, whether or not event A has occurred, then event A does not affect event B and

$$P(B|A) = P(B)$$

The Multiplication Rule can now be simplified.

THE MULTIPLICATION RULE FOR INDEPENDENT EVENTS

If two events A and B are independent, the probability that *both A and B* occur is

$$P(A \cap B) = P(A)P(B)$$

Similarly, if A, B, and C are mutually independent events (all pairs of events are independent), then the probability that A, B, and C all occur is

$$P(A \cap B \cap C) = P(A)P(B)P(C)$$

Coin Tosses at Football Games   A football team is involved in two over-time periods during a given game, so that there are three coin tosses. If the coin is fair, what is the probability that they lose all three tosses?

Solution   If the coin is fair, the event can be described in three steps:

A: lose the first toss

B: lose the second toss

C: lose the third toss

Since the tosses are independent, and since $P(\text{win}) = P(\text{lose}) = .5$ for any of the three tosses,

$$P(A \cap B \cap C) = P(A)P(B)P(C) = (.5)(.5)(.5) = .125$$

How can you check to see if two events are independent or dependent? The easiest solution is to redefine the concept of **independence** in a more formal way.

CHECKING FOR INDEPENDENCE

Two events A and B are said to be **independent** if and only if either

$P(A \cap B) = P(A)P(B)$

or

$P(B|A) = P(B)$ or equivalently, $P(A|B) = P(A)$

Otherwise, the events are said to be **dependent.**

EXAMPLE 4.20

Toss two coins and observe the outcome. Define these events:

A: Head on the first coin

B: Tail on the second coin

Are events A and B independent?

Solution   From previous examples, you know that $S = \{HH, HT, TH, TT\}$. Use these four simple events to find

NEED A TIP?

Remember,
independence ⇔
$P(A \cap B) = P(A)P(B)$.

$$P(A) = \frac{1}{2}, P(B) = \frac{1}{2}, \text{ and } P(A \cap B) = \frac{1}{4}.$$

Since $P(A)P(B) = \left(\frac{1}{2}\right)\left(\frac{1}{2}\right) = \frac{1}{4}$ and $P(A \cap B) = \frac{1}{4}$, we have $P(A)P(B) = P(A \cap B)$

and the two events must be independent.

EXAMPLE 4.21

Refer to the probability table in Example 4.18, which is reproduced below.

	Too High (A)	Right Amount (B)	Too Little (C)
Child in College (D)	.35	.08	.01
No Child in College (E)	.25	.20	.11

Are events D and A independent? Explain.

Solution

1. Use the probability table to find $P(A \cap D) = .35$, $P(A) = .60$, and $P(D) = .44$. Then

$$P(A)P(D) = (.60)(.44) = .264 \text{ and } P(A \cap D) = .35$$

Since these two probabilities are not the same, events A and D are *dependent*.

2. Alternately, calculate

$$P(A|D) = \frac{P(A \cap D)}{P(D)} = \frac{.35}{.44} = .80$$

Since $P(A|D) = .80$ and $P(A) = .60$, we are again led to the conclusion that events A and D are *dependent*.

3. Equivalently,

$$P(D|A) = \frac{P(A \cap D)}{P(A)} = \frac{.35}{.60} = .58$$

while $P(D) = .44$. Again we see that A and D are dependent events.

 NEED TO KNOW...

The Difference between Mutually Exclusive and Independent Events

Many students find it hard to tell the difference between *mutually exclusive* and *independent* events.

- When two events are *mutually exclusive* or *disjoint,* they cannot both happen together when the experiment is performed. Once the event B has occurred, event A cannot occur, so that $P(A|B) = 0$, or vice versa. The occurrence of event B certainly affects the probability that event A can occur.

- Therefore, mutually exclusive events must be *dependent*.

- When two events are *mutually exclusive* or *disjoint,* $P(A \cap B) = 0$ and $P(A \cup B) = P(A) + P(B)$.

- When two events are *independent,* $P(A \cap B) = P(A)P(B)$, and $P(A \cup B) = P(A) + P(B) - P(A)P(B)$.

Using probability rules to calculate the probability of an event requires some experience and ingenuity. You need to express the event of interest as a union or intersection (or the combination of both) of two or more events whose probabilities are known or easily calculated. Often you can do this in different ways; the key is to find the right combination.

EXAMPLE **4.22**   Two cards are drawn from a deck of 52 cards. Calculate the probability that the draw includes an ace and a ten.

Solution   Consider the event of interest:

$$A: \text{Draw an ace and a ten}$$

Then $A = B \cup C$, where

B: Draw the ace on the first draw and the ten on the second
C: Draw the ten on the first draw and the ace on the second

Events B and C were chosen to be mutually exclusive and also to be intersections of events with known probabilities; that is,

$$B = B_1 \cap B_2 \text{ and } C = C_1 \cap C_2$$

where

B_1: Draw an ace on the first draw

B_2: Draw a ten on the second draw

C_1: Draw a ten on the first draw

C_2: Draw an ace on the second draw

Applying the Multiplication Rule, you get

$$P(B_1 \cap B_2) = P(B_1)P(B_2|B_1)$$

$$= \left(\frac{4}{52}\right)\left(\frac{4}{51}\right)$$

and

$$P(C_1 \cap C_2) = \left(\frac{4}{52}\right)\left(\frac{4}{51}\right)$$

Then, applying the Addition Rule,

$$P(A) = P(B) + P(C)$$

$$= \left(\frac{4}{52}\right)\left(\frac{4}{51}\right) + \left(\frac{4}{52}\right)\left(\frac{4}{51}\right) = \frac{8}{663}$$

Check each composition carefully to be certain that it is actually equal to the event of interest.

4.6   EXERCISES

BASIC TECHNIQUES

4.40 An experiment can result in one of five equally likely simple events, $E_1, E_2, \ldots, E_5$. Events A, B, and C are defined as follows:

$A: E_1, E_3$ $P(A) = .4$

$B: E_1, E_2, E_4, E_5$ $P(B) = .8$

$C: E_3, E_4$ $P(C) = .4$

Find the probabilities associated with the following events by listing the simple events in each.

a. A^c b. $A \cap B$ c. $B \cap C$

d. $A \cup B$ e. $B|C$ f. $A|B$

g. $A \cup B \cup C$ h. $(A \cap B)^c$

4.41 Refer to Exercise 4.40. Use the definition of a complementary event to find these probabilities:

a. $P(A^c)$ b. $P((A \cap B)^c)$

Do the results agree with those obtained in Exercise 4.40?

4.42 Refer to Exercise 4.40. Use the definition of conditional probability to find these probabilities:

a. $P(A|B)$ **b.** $P(B|C)$

Do the results agree with those obtained in Exercise 4.40?

4.43 Refer to Exercise 4.40. Use the Addition and Multiplication Rules to find these probabilities:

a. $P(A \cup B)$ **b.** $P(A \cap B)$ **c.** $P(B \cap C)$

Do the results agree with those obtained in Exercise 4.40?

4.44 Refer to Exercise 4.40.

a. Are events A and B independent?

b. Are events A and B mutually exclusive?

4.45 Suppose $P(A) = .1$ and $P(B) = .5$.

a. If $P(A|B) = .1$, what is $P(A \cap B)$?

b. If $P(A|B) = .1$, are A and B independent?

c. If $P(A \cap B) = 0$, are A and B independent?

d. If $P(A \cup B) = .65$, are A and B mutually exclusive?

4.46 Dice An experiment consists of tossing a single die and observing the number of dots that show on the upper face. Events A, B, and C are defined as follows:

   A: Observe a number less than 4

   B: Observe a number less than or equal to 2

   C: Observe a number greater than 3

Find the probabilities associated with the events below using either the simple event approach or the rules and definitions from this section.

a. S **b.** $A|B$ **c.** B

d. $A \cap B \cap C$ **e.** $A \cap B$ **f.** $A \cap C$

g. $B \cap C$ **h.** $A \cup C$ **i.** $B \cup C$

4.47 Refer to Exercise 4.46.

a. Are events A and B independent? Mutually exclusive?

b. Are events A and C independent? Mutually exclusive?

4.48 Two fair dice are tossed.

a. What is the probability that the sum of the number of dots shown on the upper faces is equal to 7? To 11?

b. What is the probability that you roll "doubles"—that is, both dice have the same number on the upper face?

c. What is the probability that both dice show an odd number?

4.49 Suppose that $P(A) = .4$ and $P(B) = .2$. If events A and B are independent, find these probabilities:

a. $P(A \cap B)$ **b.** $P(A \cup B)$

4.50 Suppose that $P(A) = .3$ and $P(B) = .5$. If events A and B are mutually exclusive, find these probabilities:

a. $P(A \cap B)$ **b.** $P(A \cup B)$

4.51 Suppose that $P(A) = .4$ and $P(A \cap B) = .12$.

a. Find $P(B|A)$.

b. Are events A and B mutually exclusive?

c. If $P(B) = .3$, are events A and B independent?

4.52 An experiment can result in one or both of events A and B with the probabilities shown in this probability table:

	A	A^c
B	.34	.46
B^c	.15	.05

Find the following probabilities:

a. $P(A)$ **b.** $P(B)$ **c.** $P(A \cap B)$

d. $P(A \cup B)$ **e.** $P(A|B)$ **f.** $P(B|A)$

4.53 Refer to Exercise 4.52.

a. Are events A and B mutually exclusive? Explain.

b. Are events A and B independent? Explain.

APPLICATIONS

4.54 Drug Testing Many companies are now testing prospective employees for drug use. However, opponents claim that this procedure is unfair because the tests themselves are not 100% reliable. Suppose a company uses a test that is 98% accurate—that is, it correctly identifies a person as a drug user or nonuser with probability .98—and to reduce the chance of error, each job applicant is required to take two tests. If the outcomes of the two tests on the same person are independent events, what are the probabilities of these events?

a. A nonuser fails both tests.

b. A drug user is detected (i.e., he or she fails at least one test).

c. A drug user passes both tests.

4.55 Grant Funding Suppose a group of research proposals was evaluated by a panel of experts to decide whether or not they were worthy of funding. When these same proposals were submitted to a second independent panel of experts, the decision to fund was reversed in 30% of the cases. If the probability that a proposal is judged worthy of funding by the first panel is .2, what are the probabilities of these events?

a. A worthy proposal is approved by both panels.

b. A worthy proposal is disapproved by both panels.

c. A worthy proposal is approved by one panel.

4.56 Drug Offenders A study of drug offenders who have been treated for drug abuse suggests that the likelihood of conviction within a 2-year period after treatment may depend on the offender's education. The proportions of the total number of cases that fall into four education/conviction categories are shown in the table below.

| | Status Within 2 Years After Treatment | | |
Education	Convicted	Not Convicted	Totals
10 Years or More	.10	.30	.40
9 Years or Less	.27	.33	.60
Totals	.37	.63	1.00

Suppose a single offender is selected from the treatment program. Here are the events of interest:

  A: The offender has 10 or more years of education

  B: The offender is convicted within 2 years after completion of treatment

Find the appropriate probabilities for these events:

a. *A*   **b.** *B*   **c.** *A ∩ B*

d. *A ∪ B*   **e.** *A^c*   **f.** *(A ∪ B)^c*

g. *(A ∩ B)^c* **h.** *A* given that *B* has occurred

i. *B* given that *A* has occurred

4.57 Use the probabilities of Exercise 4.56 to show that these equalities are true:

a. $P(A \cap B) = P(A)P(B|A)$

b. $P(A \cap B) = P(B)P(A|B)$

c. $P(A \cup B) = P(A) + P(B) - P(A \cap B)$

4.58 The Birthday Problem Two people enter a room and their birthdays (ignoring years) are recorded.

a. Identify the nature of the simple events in *S*.

b. What is the probability that the two people have a specific pair of birthdates?

c. Identify the simple events in event *A:* Both people have the same birthday.

d. Find *P(A)*.   **e.** Find *P(A^c)*.

4.59 The Birthday Problem, continued If *n* people enter a room, find these probabilities:

  A: None of the people have the same birthday

  B: At least two of the people have the same birthday

Solve for

a. *n* = 3   **b.** *n* = 4

[NOTE: Surprisingly, *P(B)* increases rapidly as *n* increases. For example, for *n* = 20, *P(B)* = .411; for *n* = 40, *P(B)* = .891.]

4.60 Starbucks or Peet's®? A college student frequents one of two coffee houses on campus, choosing Starbucks 70% of the time and Peet's 30% of the time. Regardless of where she goes, she buys a cafe mocha on 60% of her visits.

a. The next time she goes into a coffee house on campus, what is the probability that she goes to Starbucks and orders a cafe mocha?

b. Are the two events in part a independent? Explain.

c. If she goes into a coffee house and orders a cafe mocha, what is the probability that she is at Peet's?

d. What is the probability that she goes to Starbucks or orders a cafe mocha or both?

4.61 Inspection Lines A certain manufactured item is visually inspected by two different inspectors. When a defective item comes through the line, the probability that it gets by the first inspector is .1. Of those that get past the first inspector, the second inspector will "miss" 5 out of 10. What fraction of the defective items get by both inspectors?

4.62 Smoking and Cancer A survey of people in a given region showed that 20% were smokers. The probability of death due to lung cancer, given that a person smoked, was roughly 10 times the probability of death due to lung cancer, given that a person did not smoke. If the probability of death due to lung cancer in the region is .006, what is the probability of death due to lung cancer given that a person is a smoker?

4.63 Smoke Detectors A smoke-detector system uses two devices, A and B. If smoke is present, the probability that it will be detected by device A is .95; by device B, .98; and by both devices, .94.

a. If smoke is present, find the probability that the smoke will be detected by device A or device B or both devices.

b. Find the probability that the smoke will not be detected.

4.64 Plant Genetics In 1865, Gregor Mendel suggested a theory of inheritance based on the science of genetics. He identified heterozygous individuals for flower color that had two alleles (one r = recessive white color allele and one R = dominant red color allele). When these individuals were mated, 3/4 of the offspring were observed to have red flowers and 1/4 had white flowers. The table summarizes this mating; each parent gives one of its alleles to form the gene of the offspring.

	Parent 2	
Parent 1	r	R
r	rr	rR
R	Rr	RR

We assume that each parent is equally likely to give either of the two alleles and that, if either one or two of the alleles in a pair is dominant (R), the offspring will have red flowers.

a. What is the probability that an offspring in this mating has at least one dominant allele?

b. What is the probability that an offspring has at least one recessive allele?

c. What is the probability that an offspring has one recessive allele, given that the offspring has red flowers?

4.65 Soccer Injuries During the inaugural season of Major League Soccer in the United States, the medical teams documented 256 injuries that caused a loss of participation time to the player. The results of this investigation, reported in *The American Journal of Sports Medicine,* are shown in the table.[4]

Severity	Practice (P)	Game (G)	Total
Minor (A)	66	88	154
Moderate (B)	23	44	67
Major (C)	12	23	35
Total	101	155	256

If one individual is drawn at random from this group of 256 soccer players, find the following probabilities:

a. $P(A)$ b. $P(G)$ c. $P(A \cap G)$

d. $P(G|A)$ e. $P(G|B)$ f. $P(G|C)$

g. $P(C|P)$ h. $P(B^c)$

4.66 Choosing a Mate Men and women often disagree on how they think about selecting a mate. Suppose that a poll of 1000 individuals in their twenties gave the following responses to the question of whether it is more important for their future mate to be able to communicate their feelings (F) than it is for that person to make a good living (G).

	Feelings (F)	Good Living (G)	Totals
Men (M)	.35	.20	.55
Women (W)	.36	.09	.45
Totals	.71	.29	1.00

If an individual is selected at random from this group of 1000 individuals, calculate the following probabilities:

a. $P(F)$ b. $P(G)$ c. $P(F|M)$

d. $P(F|W)$ e. $P(M|F)$ f. $P(W|G)$

4.67 Kobe and Lamar Two stars of the *LA Lakers* are very different when it comes to making free throws. ESPN.com reports that Kobe Bryant makes 85% of his free throw shots while Lamar Odum makes 62% of his free throws.[5] Assume that the free throws are independent and that each player shoots two free throws during a team practice.

a. What is the probability that Kobe makes both of his free throws?

b. What is the probability that Lamar makes exactly one of his two free throws?

c. What is the probability that Kobe makes both of his free throws and Lamar makes neither of his?

4.68 Golfing Player A has entered a golf tournament but it is not certain whether player B will enter. Player A has probability 1/6 of winning the tournament if player B enters and probability 3/4 of winning if player B does not enter the tournament. If the probability that player B enters is 1/3, find the probability that player A wins the tournament.

4.7 # BAYES' RULE (OPTIONAL)

Colorblindness Let us reconsider the experiment involving colorblindness from Section 4.6. Notice that the two events

     B: the person selected is a man

     B^C: the person selected is a woman

taken together make up the sample space *S*, consisting of both men and women. Since colorblind people can be either male or female, the event *A*, which is that a person is colorblind, consists of both those simple events that are in *A* **and** *B* and those simple events that are in *A* **and** B^C. Since these two *intersections* are *mutually exclusive*, you can write the event *A* as

$$A = (A \cap B) \cup (A \cap B^C)$$

and

$$P(A) = P(A \cap B) + P(A \cap B^C)$$
$$= .04 + .002 = .042$$

Suppose now that the sample space can be partitioned into *k* subpopulations, $S_1, S_2, S_3, \ldots, S_k$, that, as in the colorblindness example, are **mutually exclusive and exhaustive;** that is, taken together they make up the entire sample space. In a similar way, you can express an event *A* as

$$A = (A \cap S_1) \cup (A \cap S_2) \cup (A \cap S_3) \cup \cdots \cup (A \cap S_k)$$

Then

$$P(A) = P(A \cap S_1) + P(A \cap S_2) + P(A \cap S_3) + \cdots + P(A \cap S_k)$$

This is illustrated for *k* = 3 in Figure 4.13.

FIGURE 4.13

Decomposition of event *A*

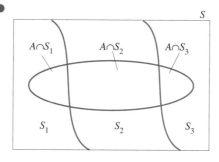

You can go one step further and use the Multiplication Rule to write $P(A \cap S_i)$ as $P(S_i)P(A|S_i)$, for $i = 1, 2, \ldots, k$. The result is known as the **Law of Total Probability.**

LAW OF TOTAL PROBABILITY

Given a set of events $S_1, S_2, S_3, \ldots, S_k$ that are mutually exclusive and exhaustive and an event *A*, the probability of the event *A* can be expressed as

$$P(A) = P(S_1)P(A|S_1) + P(S_2)P(A|S_2) + P(S_3)P(A|S_3) + \cdots + P(S_k)P(A|S_k)$$

EXAMPLE   4.23   Sneakers are no longer just for the young. In fact, most adults own multiple pairs of sneakers. Table 4.6 gives the fraction of U.S. adults 20 years of age and older who own five or more pairs of wearable sneakers, along with the fraction of the U.S. adult population 20 years or older in each of five age groups.[6] Use the Law of Total Probability to determine the unconditional probability of an adult 20 years and older owning five or more pairs of wearable sneakers.

TABLE 4.6 ● **Probability Table**

	Groups and Ages				
	G_1 20–24	G_2 25–34	G_3 35–49	G_4 50–64	G_5 $\geq$ 65
Fraction with $\geq$ 5 Pairs	.26	.20	.13	.18	.14
Fraction of U.S. Adults 20 and Older	.09	.18	.30	.25	.18

Solution Let A be the event that a person chosen at random from the U.S. adult population 20 years of age and older owns five or more pairs of wearable sneakers. Let $G_1, G_2, \ldots, G_5$ represent the event that the person selected belongs to each of the five age groups, respectively. Since the five groups are *exhaustive,* you can write the event A as

$$A = (A \cap G_1) \cup (A \cap G_2) \cup (A \cap G_3) \cup (A \cap G_4) \cup (A \cap G_5)$$

Using the Law of Total Probability, you can find the probability of A as

$$\begin{aligned} P(A) &= P(A \cap G_1) + P(A \cap G_2) + P(A \cap G_3) + P(A \cap G_4) + P(A \cap G_5) \\ &= P(G_1)P(A|G_1) + P(G_2)P(A|G_2) + P(G_3)P(A|G_3) \\ &\quad + P(G_4)P(A|G_4) + P(G_5)P(A|G_5) \end{aligned}$$

From the probabilities in Table 4.6,

$$\begin{aligned} P(A) &= (.09)(.26) + (.18)(.20) + (.30)(.13) + (.25)(.18) + (.18)(.14) \\ &= .0234 + .0360 + .0390 + .0450 + .0252 = .1686 \end{aligned}$$

The *unconditional probability* that a person selected at random from the population of U.S. adults 20 years of age and older owns at least five pairs of wearable sneakers is about .17. Notice that the Law of Total Probability is a weighted average of the probabilities within each group, with weights .09, .18, .30, .25, and .18, which reflect the relative sizes of the groups.

Often you need to find the conditional probability of an event B, given that an event A has occurred. One such situation occurs in screening tests, which used to be associated primarily with medical diagnostic tests but are now finding applications in a variety of fields. Automatic test equipment is routinely used to inspect parts in high-volume production processes. Steroid testing of athletes, home pregnancy tests, and AIDS testing are some other applications. Screening tests are evaluated on the probability of a false negative or a false positive, and both of these are *conditional probabilities.*

A **false positive** is the event that the test is positive for a given condition, given that the person does not have the condition. A **false negative** is the event that the test is negative for a given condition, given that the person has the condition. You can evaluate these conditional probabilities using a formula derived by the probabilist Thomas Bayes.

The experiment involves selecting a sample from one of k subpopulations that are mutually exclusive and exhaustive. Each of these subpopulations, denoted by $S_1, S_2, \ldots, S_k$, has a selection probability $P(S_1), P(S_2), P(S_3), \ldots, P(S_k)$, called *prior probabilities.* An event A is observed in the selection. What is the probability that the sample came from subpopulation S_i, given that A has occurred?

You know from Section 4.6 that $P(S_i|A) = [P(A \cap S_i)]/P(A)$, which can be rewritten as $P(S_i|A) = [P(S_i)P(A|S_i)]/P(A)$. Using the Law of Total Probability to rewrite $P(A)$, you have

$$P(S_i|A) = \frac{P(S_i)P(A|S_i)}{P(S_1)P(A|S_1) + P(S_2)P(A|S_2) + P(S_3)P(A|S_3) + \cdots + P(S_k)P(A|S_k)}$$

These new probabilities are often referred to as *posterior probabilities*—that is, probabilities of the subpopulations (also called *states of nature*) that have been updated after observing the sample information contained in the event A. Bayes suggested that if the prior probabilities are unknown, they can be taken to be $1/k$, which implies that each of the events S_1 through S_k is equally likely.

BAYES' RULE

Let $S_1, S_2, \ldots, S_k$ represent k mutually exclusive and exhaustive subpopulations with prior probabilities $P(S_1), P(S_2), \ldots, P(S_k)$. If an event A occurs, the posterior probability of S_i given A is the conditional probability

$$P(S_i|A) = \frac{P(S_i)P(A|S_i)}{\sum_{j=1}^{k} P(S_j)P(A|S_j)}$$

for $i = 1, 2, \ldots, k$.

EXAMPLE 4.24 Refer to Example 4.23. Find the probability that the person selected was 65 years of age or older, given that the person owned at least five pairs of wearable sneakers.

Solution You need to find the conditional probability given by

$$P(G_5|A) = \frac{P(A \cap G_5)}{P(A)}$$

You have already calculated $P(A) = .1686$ using the Law of Total Probability. Therefore,

$$P(G_5|A) =$$

$$\frac{P(G_5)P(A|G_5)}{P(G_1)P(A|G_1) + P(G_2)P(A|G_2) + P(G_3)P(A|G_3) + P(G_4P(A|G_4) + P(G_5)P(A|G_5)}$$

$$= \frac{(.18)(.14)}{(.09)(.26) + (.18)(.20) + (.30)(.13) + (.25)(.18) + (.18)(.14)}$$

$$= \frac{.0252}{.1686} = .1495$$

In this case, the posterior probability of .15 is somewhat larger than the prior probability of .13 (from Table 4.6). This group *a priori* was the second smallest, and only a small proportion of this segment had five or more pairs of wearable sneakers.

What is the posterior probability for those aged 35 to 49? For this group of adults, we have

$$P(G_3|A) = \frac{(.30)(.13)}{(.09)(.26) + (.18)(.20) + (.30)(.13) + (.25)(.18) + (.18)(.14)}$$

$$= \frac{.0390}{.1686} = .2313$$

This posterior probability of .23 is substantially less than the prior probability of .30. In effect, this group was *a priori* the largest segment of the population sampled, but at the same time, the proportion of individuals in this group who had at least five pairs of wearable sneakers was the smallest of any of the groups. These two facts taken together cause a downward adjustment of almost one-third in the *a priori* probability of .30.

 4.7 **EXERCISES**

BASIC TECHNIQUES

4.69 Bayes' Rule A sample is selected from one of two populations, S_1 and S_2, with probabilities $P(S_1) = .7$ and $P(S_2) = .3$. If the sample has been selected from S_1, the probability of observing an event A is $P(A|S_1) = .2$. Similarly, if the sample has been selected from S_2, the probability of observing A is $P(A|S_2) = .3$.

a. If a sample is randomly selected from one of the two populations, what is the probability that event A occurs?

b. If the sample is randomly selected and event A is observed, what is the probability that the sample was selected from population S_1? From population S_2?

4.70 Bayes' Rule II If an experiment is conducted, one and only one of three mutually exclusive events S_1, S_2, and S_3 can occur, with these probabilities:

$$P(S_1) = .2 \qquad P(S_2) = .5 \qquad P(S_3) = .3$$

The probabilities of a fourth event A occurring, given that event S_1, S_2, or S_3 occurs, are

$$P(A|S_1) = .2 \quad P(A|S_2) = .1 \quad P(A|S_3) = .3$$

If event A is observed, find $P(S_1|A)$, $P(S_2|A)$, and $P(S_3|A)$.

4.71 Law of Total Probability A population can be divided into two subgroups that occur with probabilities 60% and 40%, respectively. An event A occurs 30% of the time in the first subgroup and 50% of the time in the second subgroup. What is the unconditional probability of the event A, regardless of which subgroup it comes from?

APPLICATIONS

4.72 Violent Crime City crime records show that 20% of all crimes are violent and 80% are nonviolent, involving theft, forgery, and so on. Ninety percent of violent crimes are reported versus 70% of nonviolent crimes.

a. What is the overall reporting rate for crimes in the city?

b. If a crime in progress is reported to the police, what is the probability that the crime is violent? What is the probability that it is nonviolent?

c. Refer to part b. If a crime in progress is reported to the police, why is it more likely that it is a nonviolent crime? Wouldn't violent crimes be more likely to be reported? Can you explain these results?

4.73 Worker Error A worker-operated machine produces a defective item with probability .01 if the worker follows the machine's operating instructions exactly, and with probability .03 if he does not. If the worker follows the instructions 90% of the time, what proportion of all items produced by the machine will be defective?

4.74 Airport Security Suppose that, in a particular city, airport A handles 50% of all airline traffic, and airports B and C handle 30% and 20%, respectively.

The detection rates for weapons at the three airports are .9, .8, and .85, respectively. If a passenger at one of the airports is found to be carrying a weapon through the boarding gate, what is the probability that the passenger is using airport A? Airport C?

4.75 Football Strategies A particular football team is known to run 30% of its plays to the left and 70% to the right. A linebacker on an opposing team notes that the right guard shifts his stance most of the time (80%) when plays go to the right and that he uses a balanced stance the remainder of the time. When plays go to the left, the guard takes a balanced stance 90% of the time and the shift stance the remaining 10%. On a particular play, the linebacker notes that the guard takes a balanced stance.

a. What is the probability that the play will go to the left?

b. What is the probability that the play will go to the right?

c. If you were the linebacker, which direction would you prepare to defend if you saw the balanced stance?

4.76 Cheating on Your Taxes? Suppose 5% of all people filing the long income tax form seek deductions that they know are illegal, and an additional 2% incorrectly list deductions because they are unfamiliar with income tax regulations. Of the 5% who are guilty of cheating, 80% will deny knowledge of the error if confronted by an investigator. If the filer of the long form is confronted with an unwarranted deduction and he or she denies the knowledge of the error, what is the probability that he or she is guilty?

4.77 Medical Diagnostics Medical case histories indicate that different illnesses may produce identical symptoms. Suppose a particular set of symptoms, which we will denote as event H, occurs only when any one of three illnesses—A, B, or C—occurs. (For the sake of simplicity, we will assume that illnesses A, B, and C are mutually exclusive.) Studies show these probabilities of getting the three illnesses:

$$P(A) = .01$$
$$P(B) = .005$$
$$P(C) = .02$$

The probabilities of developing the symptoms H, given a specific illness, are

$$P(H|A) = .90$$
$$P(H|B) = .95$$
$$P(H|C) = .75$$

Assuming that an ill person shows the symptoms H, what is the probability that the person has illness A?

4.78 No Pass, No Play Under the "no pass, no play" rule for athletes, an athlete who fails a course is disqualified from participating in sports activities during the next grading period. Suppose the probability that an athlete who has not previously been disqualified will be disqualified is .15 and the probability that an athlete who has been disqualified will be disqualified again in the next time period is .5. If 30% of the athletes have been disqualified before, what is the unconditional probability that an athlete will be disqualified during the next grading period?

4.79 Screening Tests Suppose that a certain disease is present in 10% of the population, and that there is a screening test designed to detect this disease if present. The test does not always work perfectly. Sometimes the test is negative when the disease is present, and sometimes it is positive when the disease is absent. The table below shows the proportion of times that the test produces various results.

	Test Is Positive (P)	Test Is Negative (N)
Disease Present (D)	.08	.22
Disease Absent (D^c)	.05	.85

a. Find the following probabilities from the table: $P(D)$, $P(D^c)$, $P(N|D^c)$, $P(N|D)$.

b. Use Bayes' Rule and the results of part a to find $P(D|N)$.

c. Use the definition of conditional probability to find $P(D|N)$. (Your answer should be the same as the answer to part b.)

d. Find the probability of a false positive, that the test is positive, given that the person is disease-free.

e. Find the probability of a false negative, that the test is negative, given that the person has the disease.

f. Are either of the probabilities in parts d or e large enough that you would be concerned about the reliability of this screening method? Explain.

DISCRETE RANDOM VARIABLES AND THEIR PROBABILITY DISTRIBUTIONS

In Chapter 1, *variables* were defined as characteristics that change or vary over time and/or for different individuals or objects under consideration. *Quantitative variables* generate numerical data, whereas *qualitative variables* generate categorical data. However, even qualitative variables can generate numerical data if the categories are numerically coded to form a scale. For example, if you toss a single coin, the qualitative outcome could be recorded as "0" if a head and "1" if a tail.

Random Variables

A numerically valued variable x will vary or change depending on the particular outcome of the experiment being measured. For example, suppose you toss a die and measure x, the number observed on the upper face. The variable x can take on any of six values—1, 2, 3, 4, 5, 6—depending on the *random* outcome of the experiment. For this reason, we refer to the variable x as a **random variable.**

Definition   A variable x is a **random variable** if the value that it assumes, corresponding to the outcome of an experiment, is a chance or random event.

You can think of many examples of random variables:

- x = Number of defects on a *randomly selected* piece of furniture
- x = SAT score for a *randomly selected* college applicant
- x = Number of telephone calls received by a crisis intervention hotline during a *randomly selected* time period

As in Chapter 1, quantitative random variables are classified as either *discrete* or *continuous,* according to the values that x can assume. It is important to distinguish between discrete and continuous random variables because different techniques are used to describe their distributions. We focus on discrete random variables in the remainder of this chapter; continuous random variables are the subject of Chapter 6.

Probability Distributions

In Chapters 1 and 2, you learned how to construct the *relative frequency distribution* for a set of numerical measurements on a variable x. The distribution gave this information about x:

- What values of x occurred
- How often each value of x occurred

You also learned how to use the mean and standard deviation to measure the center and variability of this data set.

In this chapter, we defined *probability* as the limiting value of the relative frequency as the experiment is repeated over and over again. Now we define the **probability distribution** for a random variable x as the *relative frequency distribution* constructed for the entire population of measurements.

Definition   The **probability distribution** for a discrete random variable is a formula, table, or graph that gives the possible values of x, and the probability $p(x)$ associated with each value of x.

The values of x represent mutually exclusive numerical events. Summing $p(x)$ over all values of x is equivalent to adding the probabilities of all simple events and therefore equals 1.

REQUIREMENTS FOR A DISCRETE PROBABILITY DISTRIBUTION

- $0 \leq p(x) \leq 1$
- $\Sigma\, p(x) = 1$

EXAMPLE 4.25

Toss two fair coins and let x equal the number of heads observed. Find the probability distribution for x.

Solution The simple events for this experiment with their respective probabilities are listed in Table 4.7. Since $E_1 = HH$ results in two heads, this simple event results in the value $x = 2$. Similarly, the value $x = 1$ is assigned to E_2, and so on.

TABLE 4.7 ● **Simple Events and Probabilities in Tossing Two Coins**

Simple Event	Coin 1	Coin 2	$P(E_i)$	x
E_1	H	H	1/4	2
E_2	H	T	1/4	1
E_3	T	H	1/4	1
E_4	T	T	1/4	0

For each value of x, you can calculate $p(x)$ by adding the probabilities of the simple events in that event. For example, when $x = 0$, simple event E_4 occurs, so that

$$p(0) = P(E_4) = \frac{1}{4}$$

and when $x = 1$,

$$p(1) = P(E_2) + P(E_3) = \frac{1}{2}$$

The values of x and their respective probabilities, $p(x)$, are listed in Table 4.8. Notice that the probabilities add to 1.

TABLE 4.8 ● **Probability Distribution for x (x = Number of Heads)**

x	Simple Events in x	$p(x)$
0	E_4	1/4
1	E_2, E_3	1/2
2	E_1	1/4
	$\Sigma\, p(x) = 1$	

ONLINE APPLET

Flipping Coins

The probability distribution in Table 4.8 can be graphed using the methods of Section 1.5 to form the **probability histogram** in Figure 4.14.[†] The three values of the random variable x are located on the horizontal axis, and the probabilities $p(x)$ are

---

[†]The probability distribution in Table 4.8 can also be presented using a formula, which is given in Section 5.2.

located on the vertical axis (replacing the relative frequencies used in Chapter 1). Since the width of each bar is 1, the area under the bar is the probability of observing the particular value of x and the total area equals 1.

FIGURE 4.14

Probability histogram for Example 4.25

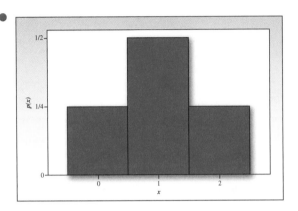

The Mean and Standard Deviation for a Discrete Random Variable

The probability distribution for a discrete random variable looks very similar to the relative frequency distribution discussed in Chapter 1. The difference is that the relative frequency distribution describes a *sample* of n measurements, whereas the probability distribution is constructed as a model for the *entire population* of measurements. Just as the mean $\bar{x}$ and the standard deviation s measured the center and spread of the sample data, you can calculate similar measures to describe the center and spread of the population.

The population mean, which measures the average value of x in the population, is also called the **expected value** of the random variable x and is written as $E(x)$. It is the value that you would *expect* to observe on *average* if the experiment is repeated over and over again. The formula for calculating the population mean is easier to understand by example. Toss those two fair coins again, and let x be the number of heads observed. We constructed this probability distribution for x:

x	0	1	2
$p(x)$	1/4	1/2	1/4

Suppose the experiment is repeated a large number of times—say, $n = 4{,}000{,}000$ times. Intuitively, you would expect to observe approximately 1 million zeros, 2 million ones, and 1 million twos. Then the average value of x would equal

$$\frac{\text{Sum of measurements}}{n} = \frac{1{,}000{,}000(0) + 2{,}000{,}000(1) + 1{,}000{,}000(2)}{4{,}000{,}000}$$

$$= \left(\frac{1}{4}\right)(0) + \left(\frac{1}{2}\right)(1) + \left(\frac{1}{4}\right)(2)$$

Note that the first term in this sum is $(0)p(0)$, the second is equal to $(1)p(1)$, and the third is $(2)p(2)$. The average value of x, then, is

$$\Sigma x p(x) = 0 + \frac{1}{2} + \frac{2}{4} = 1$$

This result provides some intuitive justification for the definition of the expected value of a discrete random variable x.

Definition Let x be a discrete random variable with probability distribution $p(x)$. The mean or **expected value of x** is given as

$$\mu = E(x) = \Sigma x p(x)$$

where the elements are summed over all values of the random variable x.

We could use a similar argument to justify the formulas for the **population variance** σ^2 and the **population standard deviation** σ. These numerical measures describe the spread or variability of the random variable using the "average" or "expected value" of $(x - \mu)^2$, the squared deviations of the x-values from their mean μ.

Definition Let x be a discrete random variable with probability distribution $p(x)$ and mean μ. The **variance of x** is

$$\sigma^2 = E[(x - \mu)^2] = \Sigma(x - \mu)^2 p(x)$$

where the summation is over all values of the random variable x.[†]

Definition The **standard deviation σ of a random variable x** is equal to the positive square root of its variance.

EXAMPLE 4.26

An electronics store sells a particular model of a laptop computer. There are only four laptops in stock, and the manager wonders what today's demand for this particular model will be. She learns from the marketing department that the probability distribution for x, the daily demand for the laptop, is as shown in the table. Find the mean, variance, and standard deviation of x. Is it likely that five or more customers will want to buy the laptop today?

x	0	1	2	3	4	5
$p(x)$	.10	.40	.20	.15	.10	.05

Solution Table 4.9 shows the values of x and $p(x)$, along with the individual terms used in the formulas for μ and σ^2. The sum of the values in the third column is

$$\mu = \Sigma x p(x) = (0)(.10) + (1)(.40) + \cdots + (5)(.05) = 1.90$$

while the sum of the values in the fifth column is

$$\sigma^2 = \Sigma(x - \mu)^2 p(x)$$
$$= (0 - 1.9)^2(.10) + (1 - 1.9)^2(.40) + \cdots + (5 - 1.9)^2(.05) = 1.79$$

and

$$\sigma = \sqrt{\sigma^2} = \sqrt{1.79} = 1.34$$

[†]It can be shown (proof omitted) that $\sigma^2 = \Sigma(x - \mu)^2 p(x) = \Sigma x^2 p(x) - \mu^2$. This result is analogous to the computing formula for the sum of squares of deviations given in Chapter 2.

TABLE 4.9 ● **Calculations for Example 4.26**

x	p(x)	xp(x)	(x − μ)²	(x − μ)² p(x)
0	.10	.00	3.61	.361
1	.40	.40	.81	.324
2	.20	.40	.01	.002
3	.15	.45	1.21	.1815
4	.10	.40	4.41	.441
5	.05	.25	9.61	.4805
Totals	1.00	μ = 1.90		σ² = 1.79

The graph of the probability distribution is shown in Figure 4.15. Since the distribution is approximately mound-shaped, approximately 95% of all measurements should lie within *two* standard deviations of the mean—that is,

$$\mu \pm 2\sigma \Rightarrow 1.90 \pm 2(1.34) \qquad \text{or} \ -.78 \text{ to } 4.58$$

Since $x = 5$ lies outside this interval, you can say it is unlikely that five or more customers will want to buy the laptop today. In fact, $P(x \geq 5)$ is exactly .05, or 1 time in 20.

FIGURE 4.15 ●

Probability distribution for Example 4.26

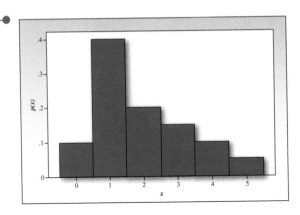

EXAMPLE 4.27 — In a lottery conducted to benefit a local charity, 8000 tickets are to be sold at $10 each. The prize is a $24,000 compact car. If you purchase two tickets, what is your expected gain?

Solution Your gain x may take one of two values. You will either lose $20 (i.e., your "gain" will be −$20) or win $23,980, with probabilities 7998/8000 and 2/8000, respectively. The probability distribution for the gain x is shown in the table:

x	p(x)
−$20	7998/8000
$23,980	2/8000

The expected gain will be

$$\mu = \Sigma \, xp(x)$$

$$= (-\$20)\left(\frac{7998}{8000}\right) + (\$23,980)\left(\frac{2}{8000}\right) = -\$14$$

Recall that the expected value of x is the average of the theoretical population that would result if the lottery were repeated an infinitely large number of times. If this were done, your average or expected gain per lottery ticket would be a loss of $14.

EXAMPLE 4.28

Determine the yearly premium for a $10,000 insurance policy covering an event that, over a long period of time, has occurred at the rate of 2 times in 100. Let x equal the yearly financial gain to the insurance company resulting from the sale of the policy, and let C equal the unknown yearly premium. Calculate the value of C such that the expected gain $E(x)$ will equal zero. Then C is the premium required to break even. To this, the company would add administrative costs and profit.

Solution The first step in the solution is to determine the values that the gain x may take and then to determine $p(x)$. If the event does not occur during the year, the insurance company will gain the premium of $x = C$ dollars. If the event does occur, the gain will be negative; that is, the company will lose $10,000 less the premium of C dollars already collected. Then $x = -(10,000 - C)$ dollars. The probabilities associated with these two values of x are 98/100 and 2/100, respectively. The probability distribution for the gain is shown in the table:

x = Gain	$p(x)$
C	98/100
$-(10,000 - C)$	2/100

Since the company wants the insurance premium C such that, in the long run (for many similar policies), the mean gain will equal zero, you can set the expected value of x equal to zero and solve for C. Then

$$\mu = E(x) = \Sigma x p(x)$$

$$= C\left(\frac{98}{100}\right) + [-10,000 + C]\left(\frac{2}{100}\right) = 0$$

or

$$\frac{98}{100}C + \frac{2}{100}C - 200 = 0$$

Solving this equation for C, you obtain $C = \$200$. Therefore, if the insurance company charged a yearly premium of $200, the average gain calculated for a large number of similar policies would equal zero. The actual premium would equal $200 plus administrative costs and profit.

The method for calculating the expected value of x for a continuous random variable is similar to what you have done, but in practice it involves the use of calculus. Nevertheless, the basic results concerning expectations are the same for continuous and discrete random variables. For example, regardless of whether x is continuous or discrete, $\mu = E(x)$ and $\sigma^2 = E[(x - \mu)^2]$.

4.8 EXERCISES

BASIC TECHNIQUES

4.80 Discrete or Continuous? Identify the following as discrete or continuous random variables:

a. Total number of points scored in a football game

b. Shelf life of a particular drug

c. Height of the ocean's tide at a given location

d. Length of a 2-year-old black bass

e. Number of aircraft near-collisions in a year

4.81 Discrete or Continuous? II Identify the following as discrete or continuous random variables:

a. Increase in length of life attained by a cancer patient as a result of surgery

b. Tensile breaking strength (in pounds per square inch) of 1-inch-diameter steel cable

c. Number of deer killed per year in a state wildlife preserve

d. Number of overdue accounts in a department store at a particular time

e. Your blood pressure

4.82 Dice Let x equal the number observed on the throw of a single balanced die.

a. Find and graph the probability distribution for x.

b. What is the average or expected value of x?

c. What is the standard deviation of x?

d. Locate the interval $\mu \pm 2\sigma$ on the x-axis of the graph in part a. What proportion of all the measurements would fall into this range?

4.83 Probability Distribution II A random variable x can assume five values: 0, 1, 2, 3, 4. A portion of the probability distribution is shown here:

x	0	1	2	3	4
$p(x)$	.1	.3	.3	?	.1

a. Find $p(3)$.

b. Construct a probability histogram for $p(x)$.

c. Calculate the population mean, variance, and standard deviation.

d. What is the probability that x is greater than 2?

e. What is the probability that x is 3 or less?

4.84 Probability Distribution I A random variable x has this probability distribution:

x	0	1	2	3	4	5
$p(x)$	.1	.3	.4	.1	?	.05

a. Find $p(4)$.

b. Construct a probability histogram to describe $p(x)$.

c. Find μ, σ^2, and σ.

d. Locate the interval $\mu \pm 2\sigma$ on the x-axis of the histogram. What is the probability that x will fall into this interval?

e. If you were to select a very large number of values of x from the population, would most fall into the interval $\mu \pm 2\sigma$? Explain.

4.85 Grocery Visits Let x represent the number of times a customer visits a grocery store in a 1-week period. Assume this is the probability distribution of x:

x	0	1	2	3
$p(x)$	.1	.4	.4	.1

Find the expected value of x, the average number of times a customer visits the store.

4.86 If you toss a pair of dice, the sum T of the numbers appearing on the upper faces of the dice can assume the value of an integer in the interval $2 \le T \le 12$.

a. Find the probability distribution for T. Display this probability distribution in a table.

b. Construct a probability histogram for $P(T)$. How would you describe the shape of this distribution?

APPLICATIONS

4.87 RU Texting? The proportion of adults (18 years or more) who admit to texting while driving is 47%.[7] Suppose you randomly select three adult drivers and ask if they text while driving.

a. Find the probability distribution for x, the number of drivers in the sample who admit to texting while driving.

b. Construct a probability histogram for $p(x)$.

c. What is the probability that exactly one of the three drivers texts while driving?

d. What are the population mean and standard deviation for the random variable x?

4.88 Which Key Fits? A key ring contains four office keys that are identical in appearance, but only one will open your office door. Suppose you randomly select one key and try it. If it does not fit, you randomly select one of the three remaining keys. If it does not fit, you randomly select one of the last two. Each different sequence that could occur in selecting the keys represents one of a set of equiprobable simple events.

a. List the simple events in S and assign probabilities to the simple events.

b. Let x equal the number of keys that you try before you find the one that opens the door ($x = 1, 2, 3, 4$). Then assign the appropriate value of x to each simple event.

c. Calculate the values of $p(x)$ and display them in a table.

d. Construct a probability histogram for $p(x)$.

4.89 Gender Bias? A company has five applicants for two positions: two women and three men. Suppose that the five applicants are equally qualified and that no preference is given for choosing either gender. Let x equal the number of women chosen to fill the two positions.

a. Find $p(x)$.

b. Construct a probability histogram for x.

4.90 Tennis, Anyone? Two tennis professionals, A and B, are scheduled to play a match; the winner is the first player to win three sets in a total that cannot exceed five sets. The event that A wins any one set is independent of the event that A wins any other, and the probability that A wins any one set is equal to .6. Let x equal the total number of sets in the match; that is, $x = 3$, 4, or 5. Find $p(x)$.

4.91 Drilling Oil Wells Past experience has shown that, on the average, only 1 in 10 wells drilled hits oil. Let x be the number of drillings until the first success (oil is struck). Assume that the drillings represent independent events.

a. Find $p(1)$, $p(2)$, and $p(3)$.

b. Give a formula for $p(x)$.

c. Graph $p(x)$.

4.92 Defective Equipment A piece of electronic equipment contains six computer chips, two of which are defective. Three chips are selected at random, removed from the piece of equipment, and inspected. Let x equal the number of defectives observed, where $x = 0$, 1, or 2. Find the probability distribution for x. Express the results graphically as a probability histogram.

4.93 Tennis, again In Exercise 4.92 you found the probability distribution for x, the number of sets required to play a best-of-five-sets match, given that the probability that A wins any one set—call this $P(A)$—is .6.

a. Find the expected number of sets required to complete the match for $P(A) = .6$.

b. Find the expected number of sets required to complete the match when the players are of equal ability—that is, $P(A) = .5$.

c. Find the expected number of sets required to complete the match when the players differ greatly in ability—that is, say, $P(A) = .9$.

d. What is the relationship between $P(A)$ and $E(x)$, the expected number of sets required to complete the match?

4.94 The PGA One professional golfer plays best on short-distance holes. Experience has shown that the numbers x of shots required for 3-, 4-, and 5-par holes have the probability distributions shown in the table:

Par-3 Holes		Par-4 Holes		Par-5 Holes	
x	$p(x)$	x	$p(x)$	x	$p(x)$
2	.12	3	.14	4	.04
3	.80	4	.80	5	.80
4	.06	5	.04	6	.12
5	.02	6	.02	7	.04

What is the golfer's expected score on these holes?

a. A par-3 hole

b. A par-4 hole

c. A par-5 hole

4.95 Insuring Your Diamonds You can insure a $50,000 diamond for its total value by paying a premium of D dollars. If the probability of loss in a given year is estimated to be .01, what premium should the insurance company charge if it wants the expected gain to equal $1000?

4.96 FDA Testing The maximum patent life for a new drug is 17 years. Subtracting the length of time required by the FDA for testing and approval of the drug provides the actual patent life of the drug—that is, the length of time that a company has to recover research and development costs and make a profit. Suppose the distribution of the lengths of patent life for new drugs is as shown here:

Years, x	3	4	5	6	7	8
$p(x)$	.03	.05	.07	.10	.14	.20

Years, x	9	10	11	12	13
$p(x)$	.18	.12	.07	.03	.01

a. Find the expected number of years of patent life for a new drug.

b. Find the standard deviation of x.

c. Find the probability that x falls into the interval $\mu \pm 2\sigma$.

4.97 Coffee Breaks Most coffee drinkers take a little time each day for their favorite beverage, and many take more than one coffee break every day. The table below, adapted from a Snapshot in *USA Today*, shows the probability distribution for x, the number of daily coffee breaks taken per day by coffee drinkers.[8]

x	0	1	2	3	4	5
$p(x)$	.28	.37	.17	.12	.05	.01

a. What is the probability that a randomly selected coffee drinker would take no coffee breaks during the day?

b. What is the probability that a randomly selected coffee drinker would take more than two coffee breaks during the day?

c. Calculate the mean and standard deviation for the random variable x.

d. Find the probability that x falls into the interval $\mu \pm 2\sigma$.

4.98 Shipping Charges From experience, a shipping company knows that the cost of delivering a small package within 24 hours is $14.80. The company charges $15.50 for shipment but guarantees to refund the charge if delivery is not made within 24 hours. If the company fails to deliver only 2% of its packages within the 24-hour period, what is the expected gain per package?

4.99 Actuaries A CEO is considering buying an insurance policy to cover possible losses incurred by marketing a new product. If the product is a complete failure, a loss of $800,000 would be incurred; if it is only moderately successful, a loss of $250,000 would be incurred. Insurance actuaries have determined that the probabilities that the product will be a failure or only moderately successful are .01 and .05, respectively. Assuming that the CEO is willing to ignore all other possible losses, what premium should the insurance company charge for a policy in order to break even?

CHAPTER REVIEW

Key Concepts and Formulas

I. Experiments and the Sample Space

1. Experiments, events, mutually exclusive events, simple events
2. The sample space
3. Venn diagrams, tree diagrams, probability tables

II. Probabilities

1. Relative frequency definition of probability
2. Properties of probabilities
   a. Each probability lies between 0 and 1.
   b. Sum of all simple-event probabilities equals 1.
3. $P(A)$, the sum of the probabilities for all simple events in A

III. Counting Rules

1. mn Rule; extended mn Rule
2. Permutations: $P_r^n = \dfrac{n!}{(n-r)!}$
3. Combinations: $C_r^n = \dfrac{n!}{r!(n-r)!}$

IV. Event Relations

1. Unions and intersections
2. Events

   a. Disjoint or mutually exclusive:
   $P(A \cap B) = 0$
   b. Complementary: $P(A) = 1 - P(A^c)$
3. Conditional probability: $P(A|B) = \dfrac{P(A \cap B)}{P(B)}$
4. Independent and dependent events
5. Addition Rule:
   $P(A \cup B) = P(A) + P(B) - P(A \cap B)$
6. Multiplication Rule: $P(A \cap B) = P(A)P(B|A)$
7. Law of Total Probability
8. Bayes' Rule

V. Discrete Random Variables and Probability Distributions

1. Random variables, discrete and continuous
2. Properties of probability distributions
   a. $0 \le p(x) \le 1$
   b. $\Sigma p(x) = 1$
3. Mean or expected value of a discrete random variable: $\mu = \Sigma x p(x)$
4. Variance and standard deviation of a discrete random variable: $\sigma^2 = \Sigma(x - \mu)^2 p(x)$ and $\sigma = \sqrt{\sigma^2}$

TECHNOLOGY TODAY

Discrete Probability Distributions in *MS Excel*

Although *MS Excel* cannot help you solve the types of general probability problems presented in this chapter, it is useful for calculating the mean, variance, and standard deviation of the random variable x. In Chapters 5 and 6, we will use *Excel* to calculate exact probabilities for three special cases: the binomial, the Poisson, and the normal random variables.

EXAMPLE 4.29

Suppose you have this general discrete probability distribution:

x	0	1	3	5
$p(x)$	.25	.35	.25	.15

1. Enter the values of x and $p(x)$ into columns A and B of a new *Excel* spreadsheet. Then create two columns—column C (named "$x*p(x)$") and column D (named "$xsq*p(x)$"). You can now use the **Function** f_x command to fill in columns C and D. In *Excel*, an "equals" sign indicates that you are going to type an equation (or insert a function). Hence, in cell C2, we type:

   =A2*B2

   Then, to copy this formula to the remaining three cells in column C, simply click on cell C2, grab the square in the lower right corner of the cell with your mouse, and drag to cell C5 to copy.

2. To fill in column D, type the following equation into cell D2: = A2*A2*B2 and then copy this formula to the remaining three cells in column D as explained above.

3. Finally, use the first three cells in column F to type the names "Mean," "Variance," and "Std Dev." Again, use the equation (or insert function) commands. In cell G1 (Mean), type: =SUM(C2:C5); in cell G2 (Variance), type: =SUM(D2:D5)-(G1*G1); and in cell G3 (Standard Deviation), type: =SQRT(G2). The resulting spreadsheet is shown in Figure 4.16.

FIGURE 4.16

	A	B	C	D	E	F	G
1	x	p(x)	x*p(x)	xsq*p(x)		Mean	1.85
2	0	0.25	0	0		Variance	2.9275
3	1	0.35	0.35	0.35		Std Dev	1.710994
4	3	0.25	0.75	2.25			
5	5	0.15	0.75	3.75			

Discrete Probability Distributions in *MINITAB*

Although *MINITAB* cannot help you solve the types of general probability problems presented in this chapter, it is useful for graphing the probability distribution $p(x)$ for a general discrete random variable x when the probabilities are known, and for calculating the mean, variance, and standard deviation of the random variable x. In Chapters 5 and 6, we will use *MINITAB* to calculate exact probabilities for three special cases: the binomial, the Poisson, and the normal random variables.

EXAMPLE 4.30   Suppose you have this general probability distribution:

x	0	1	3	5
$p(x)$	.25	.35	.25	.15

1. Enter the values of x and $p(x)$ into columns C1 and C2 of a new *MINITAB* worksheet. In the gray boxes just below C3, C4, and C5, respectively, type the names "Mean," "Variance," and "Std Dev." You can now use the **Calc ▶ Calculator** command to calculate μ, σ^2, and σ and to store the results in columns C3–C5 of the worksheet.

2. Use the same approach for all three parameters. In the Calculator dialog box, select "Mean" as the column in which to store μ. In the Expression box, use the Functions list, the calculator keys, and the variables list on the left to highlight, select, and create the expression for the mean (see Figure 4.17(a)):

SUM('x'*'p(x)')

MINITAB will multiply each row element in C1 times the corresponding row element in C2, sum the resulting products, and store the result in C3! You can check the result by hand if you like.

3. The formulas for the variance and standard deviation are selected in a similar way:

$$\text{Variance:} \quad \text{SUM(('x' - 'Mean')**2*'p(x)')}$$
$$\text{Std Dev: SQRT('Variance')}$$

4. To see the tabular form of the probability distribution and the three parameters, use **Data ▶ Display Data** and select all five columns. Click **OK** and the results will be displayed in the Session window, as shown in Figure 4.17(b).

FIGURE 4.17

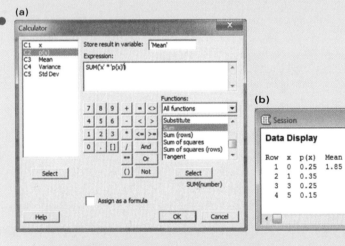

The probability histogram can be plotted using the *MINITAB* command **Graph ▶ Scatterplot ▶ Simple ▶ OK.** In the Scatterplot dialog box, select 'p(x)' for **Y variables** and 'x' for **X variables.** To display the discrete probability bars, click on **Data View,** uncheck the box marked "Symbols," and check the box marked "Project Lines." Click **OK** twice to see the plot. You will see a single straight line projected at each of the four values of x. If you want the plot to look more like the discrete probability histograms in Section 4.8, position your cursor on one of the lines, right-click the mouse and choose "Edit Project Lines." Under the "Attributes" tab, select **Custom** and change the line size to **75**. Click **OK**. If the bar width is not satisfactory, you can readjust the line size. Finally, right-click on the X-axis, choose "Edit X Scale" and select **−.5** and **5.5** for the minimum and maximum **Scale Ranges**. Click **OK**. The probability histogram is shown in Figure 4.18.

Locate the mean on the graph. Is it at the center of the distribution? If you mark off two standard deviations on either side of the mean, do most of the possible values of x fall into this interval?

FIGURE 4.18 ●

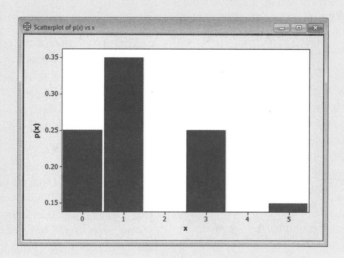

Supplementary Exercises

Starred (*) exercises are optional.

4.100 DVRs A retailer sells two styles of high-priced digital video recorders (DVR) that experience indicates are in equal demand. (Fifty percent of all potential customers prefer style 1, and 50% favor style 2.) If the retailer stocks four of each, what is the probability that the first four customers seeking a DVR all purchase the same style?

4.101 Whistle Blowers Although there is legal protection for "whistle blowers"—employees who report illegal or unethical activities in the workplace—it has been reported that approximately 23% of those who reported fraud suffered reprisals such as demotion or poor performance ratings. Suppose the probability that an employee will fail to report a case of fraud is .69. Find the probability that an employee who observes a case of fraud will report it and will subsequently suffer some form of reprisal.

4.102 Aspirin Two cold tablets are unintentionally placed in a box containing two aspirin tablets. The

four tablets are identical in appearance. One tablet is selected at random from the box and is swallowed by the first patient. A tablet is then selected at random from the three remaining tablets and is swallowed by the second patient. Define the following events as specific collections of simple events:

a. The sample space S

b. The event A that the first patient obtained a cold tablet

c. The event B that exactly one of the two patients obtained a cold tablet

d. The event C that neither patient obtained a cold tablet

4.103 Refer to Exercise 4.102. By summing the probabilities of simple events, find $P(A)$, $P(B)$, $P(A \cap B)$, $P(A \cup B)$, $P(C)$, $P(A \cap C)$, and $P(A \cup C)$.

4.104 Heavy Equipment A heavy-equipment salesman can contact either one or two customers per day with probabilities 1/3 and 2/3, respectively. Each contact will result in either no sale or a $50,000 sale with probabilities 9/10 and 1/10, respectively. What is the expected value of his daily sales?

4.105 Interstate Commerce A shipping container contains seven complex electronic systems. Unknown to the purchaser, three are defective. Two of the seven are selected for thorough testing and are then classified as defective or nondefective. What is the probability that no defectives are found?

4.106 Playing the Slots A slot machine has three slots; each will show a cherry, a lemon, a star, or a bar when spun. The player wins if all three slots show the same three items. If each of the four items is equally likely to appear on a given spin, what is your probability of winning?

4.107 Fire Insurance A county containing a large number of rural homes is thought to have 60% of those homes insured against fire. Four rural home-owners are chosen at random from the entire population, and x are found to be insured against fire. Find the probability distribution for x. What is the probability that at least three of the four will be insured?

4.108 Racial Bias? Four union men, two from a minority group, are assigned to four distinctly different one-man jobs, which can be ranked in order of desirability.

a. Define the experiment.

b. List the simple events in S.

c. If the assignment to the jobs is unbiased—that is, if any one ordering of assignments is as probable as any other—what is the probability that the two men from the minority group are assigned to the least desirable jobs?

4.109 Roulette Exercise 4.10 described the game of roulette. Suppose you bet $5 on a single number—say, the number 18. The payoff on this type of bet is usually 35 to 1. What is your expected gain?

4.110 Plant Genetics Refer to the experiment conducted by Gregor Mendel in Exercise 4.64. Suppose you are interested in following two independent traits in snap peas—seed texture (S = smooth, s = wrinkled) and seed color (Y = yellow, y = green)—in a second-generation cross of heterozygous parents. Remember that the capital letter represents the dominant trait. Complete the table with the gene pairs for both traits. All possible pairings are equally likely.

		Seed Color		
Seed Texture	yy	yY	Yy	YY
ss	(ss yy)	(ss yY)		
sS				
Ss				
SS				

a. What proportion of the offspring from this cross will have smooth yellow peas?

b. What proportion of the offspring will have smooth green peas?

c. What proportion of the offspring will have wrinkled yellow peas?

d. What proportion of the offspring will have wrinkled green peas?

e. Given that an offspring has smooth yellow peas, what is the probability that this offspring carries one s allele? One s allele *and* one y allele?

4.111 Profitable Stocks An investor has the option of investing in three of five recommended stocks. Unknown to her, only two will show a substantial profit within the next 5 years. If she selects the three stocks at random (giving every combination of three stocks an equal chance of selection), what is the probability that she selects the two profitable stocks? What is the probability that she selects only one of the two profitable stocks?

4.112 Fire Alarms A fire-detection device uses three temperature-sensitive cells acting independently of one another in such a manner that any one or more can activate the alarm. Each cell has a probability $p = .8$ of activating the alarm when the temperature reaches 135°F or higher. Let x equal the number of cells activating the alarm when the temperature reaches 135°F.

a. Find the probability distribution of x.

b. Find the probability that the alarm will function when the temperature reaches 135°F.

c. Find the expected value and the variance for the random variable x.

4.113 A Reticent Salesman A salesperson figures that the probability of her making a sale during the first contact with a client is .4 but improves to .55 on the second contact if the client did not buy during the first contact. Suppose this salesperson makes one and only one callback to any client. If she contacts a client, calculate the probabilities for these events:

a. The client will buy.

b. The client will not buy.

4.114 MRIs An article in *The American Journal of Sports Medicine* compared the results of magnetic resonance imaging (MRI) evaluation with arthroscopic surgical evaluation of cartilage tears at two sites in the knees of 35 patients. The $2 \times 35 = 70$ examinations produced the classifications shown in the table.[10] Actual tears were confirmed by arthroscopic surgical examination.

	Tears	No Tears	Total
MRI Positive	27	0	27
MRI Negative	4	39	43
Total	31	39	70

a. What is the probability that a site selected at random has a tear and has been identified as a tear by MRI?

b. What is the probability that a site selected at random has no tear and has been identified as having a tear?

c. What is the probability that a site selected at random has a tear and has not been identified by MRI?

d. What is the probability of a positive MRI, given that there is a tear?

e. What is the probability of a false negative—that is, a negative MRI, given that there is a tear?

4.115 Guided Missiles The failure rate for a guided missile control system is 1 in 1000. Suppose that a duplicate, but completely independent, control system is installed in each missile so that, if the first fails, the second can take over. The reliability of a missile is the probability that it does not fail. What is the reliability of the modified missile?

4.116 Rental Trucks A rental truck agency services its vehicles on a regular basis, routinely checking for mechanical problems. Suppose that the agency has six moving vans, two of which need to have new brakes. During a routine check, the vans are tested one at a time.

a. What is the probability that the last van with brake problems is the fourth van tested?

b. What is the probability that no more than four vans need to be tested before both brake problems are detected?

c. Given that one van with bad brakes is detected in the first two tests, what is the probability that the remaining van is found on the third or fourth test?

4.117 Pennsylvania Lottery Probability played a role in the rigging of the April 24, 1980, Pennsylvania state lottery. To determine each digit of the three-digit winning number, each of the numbers 0, 1, 2, . . . , 9 is written on a Ping-Pong ball, the 10 balls are blown into a compartment, and the number selected for the digit is the one on the ball that floats to the top of the machine. To alter the odds, the conspirators injected a liquid into all balls used in the game except those numbered 4 and 6, making it almost certain that the lighter balls would be selected and determine the digits in the winning number. They then proceeded to buy lottery tickets bearing the potential winning numbers. How many potential winning numbers were there (666 was the eventual winner)?

*4.118 Lottery, continued Refer to Exercise 4.117. Hours after the rigging of the Pennsylvania state lottery was announced on September 19, 1980, Connecticut state lottery officials were stunned to learn that *their* winning number for the day was 666.

a. All evidence indicates that the Connecticut selection of 666 was pure chance. What is the probability that a 666 would be drawn in Connecticut, given that a 666 had been selected in the April 24, 1980, Pennsylvania lottery?

b. What is the probability of drawing a 666 in the April 24, 1980, Pennsylvania lottery (remember, this drawing was rigged) *and* a 666 on the September 19, 1980, Connecticut lottery?

*4.119 ACL/MCL Tears *The American Journal of Sports Medicine* published a study of 810 women collegiate rugby players with two common knee injuries: medial cruciate ligament (MCL) sprains and anterior cruciate ligament (ACL) tears.[9] For backfield players, it was found that 39% had MCL sprains and 61% had ACL tears. For forwards, it was found that 33% had MCL sprains and 67% had ACL tears. Since a rugby team consists of eight forwards and seven backs, you can assume that 47% of the players with knee injuries are backs and 53% are forwards.

a. Find the unconditional probability that a rugby player selected at random from this group of players has experienced an MCL sprain.

b. Given that you have selected a player who has an MCL sprain, what is the probability that the player is a forward?

c. Given that you have selected a player who has an ACL tear, what is the probability that the player is a back?

4.120 Bus or Subway A man takes either a bus or the subway to work with probabilities .3 and .7, respectively. When he takes the bus, he is late 30% of the days. When he takes the subway, he is late 20% of the days. If the man is late for work on a particular day, what is the probability that he took the bus?

4.121 The Match Game Two men each toss a coin. They obtain a "match" if either both coins are heads or both are tails. Suppose the tossing is repeated three times.

a. What is the probability of three matches?

b. What is the probability that all six tosses (three for each man) result in tails?

c. Coin tossing provides a model for many practical experiments. Suppose that the coin tosses represent the answers given by two students for three specific true–false questions on an examination. If the two students gave three matches for answers, would the low probability found in part a suggest collusion?

4.122 Blood Pressure Meds A research physician compared the effectiveness of two blood pressure drugs A and B by administering the two drugs to each of four pairs of identical twins. Drug A was given to one member of a pair; drug B to the other. If, in fact, there is no difference in the effects of the drugs, what is the probability that the drop in the blood pressure reading for drug A exceeds the corresponding drop in the reading for drug B for all four pairs of twins? Suppose drug B created a greater drop in blood pressure than drug A for each of the four pairs of twins. Do you think this provides sufficient evidence to indicate that drug B is more effective in lowering blood pressure than drug A?

4.123 Work Tenure Suppose the probability of remaining with a particular company 10 years or longer is 1/6. A man and a woman start work at the company on the same day.

a. What is the probability that the man will work there less than 10 years?

b. What is the probability that both the man and the woman will work there less than 10 years? (Assume they are unrelated and their lengths of service are independent of each other.)

c. What is the probability that one or the other or both will work 10 years or longer?

4.124 Contract Negotiations Experience has shown that, 50% of the time, a particular union–management contract negotiation led to a contract

settlement within a 2-week period, 60% of the time the union strike fund was adequate to support a strike, and 30% of the time both conditions were satisfied. What is the probability of a contract settlement given that the union strike fund is adequate to support a strike? Is settlement of a contract within a 2-week period dependent on whether the union strike fund is adequate to support a strike?

4.125 Waiting Times Suppose that at a particular supermarket the probability of waiting 5 minutes or longer for checkout at the cashier's counter is .2. On a given day, a man and his wife decide to shop individually at the market, each checking out at different cashier counters. They both reach cashier counters at the same time.

a. What is the probability that the man will wait less than 5 minutes for checkout?

b. What is probability that both the man and his wife will be checked out in less than 5 minutes? (Assume that the checkout times for the two are independent events.)

c. What is the probability that one or the other or both will wait 5 minutes or longer?

4.126 Accident Insurance Accident records collected by an automobile insurance company give the following information: The probability that an insured driver has an automobile accident is .15; if an accident has occurred, the damage to the vehicle amounts to 20% of its market value with probability .80, 60% of its market value with probability .12, and a total loss with probability .08. What premium should the company charge on a $22,000 car so that the expected gain by the company is zero?

4.127 Mass Transit Only 40% of all people in a community favor the development of a mass transit system. If four citizens are selected at random from the community, what is the probability that all four favor the mass transit system? That none favors the mass transit system?

4.128 Quality Control A quality-control plan calls for accepting a large lot of crankshaft bearings if a sample of seven is drawn and none are defective. What is the probability of accepting the lot if none in the lot are defective? If 1/10 are defective? If 1/2 are defective?

4.129 Blood Tests To reduce the cost of detecting a disease, blood tests are conducted on a pooled sample of blood collected from a group of n people. If no indication of the disease is present in the pooled blood sample, none have the disease. If analysis of the pooled blood sample indicates that the disease is present, the blood of each individual must be tested. The individual tests are conducted in sequence. If, among a group of five people, one person has the disease, what is the probability that six blood tests (including the pooled test) are required to detect the single diseased person? If two people have the disease, what is the probability that six tests are required to locate both diseased people?

4.130 Tossing a Coin How many times should a coin be tossed to obtain a probability equal to or greater than .9 of observing at least one head?

4.131 Flextime A survey to determine the availability of flextime schedules in the California workplace provided the following information for 220 firms located in two California cities.

City	Flextime Schedule		
	Available	Not Available	Total
A	39	75	114
B	25	81	106
Totals	64	156	220

A company is selected at random from this pool of 220 companies.

a. What is the probability that the company is located in city A?

b. What is the probability that the company is located in city B and offers flextime work schedules?

c. What is the probability that the company does not have flextime schedules?

d. What is the probability that the company is located in city B, given that the company has flextime schedules available?

4.132 A Color Recognition Experiment An experiment is run as follows—the colors red, yellow, and blue are each flashed on a screen for a short period of time. A subject views the colors and is asked to choose the one he feels was flashed for the longest time. The experiment is repeated three times with the same subject.

a. If all the colors were flashed for the same length of time, find the probability distribution for x, the number of times that the subject chose the color red. Assume that his three choices are independent.

b. Construct the probability histogram for the random variable x.

4.133 Pepsi™ or Coke™? A taste-testing experiment is conducted at a local supermarket, where passing shoppers are asked to taste two soft-drink samples—one Pepsi and one Coke—and state their preference. Suppose that four shoppers are chosen at random and asked to participate in the experiment, and that there is actually no difference in the taste of the two brands.

a. What is the probability that all four shoppers choose Pepsi?

b. What is the probability that exactly one of the four shoppers chooses Pepsi?

4.134 Viruses A certain virus afflicted the families in three adjacent houses in a row of 12 houses. If houses were randomly chosen from a row of 12 houses, what is the probability that the three houses would be adjacent? Is there reason to believe that this virus is contagious?

4.135 Orchestra Politics The board of directors of a major symphony orchestra has voted to create a committee for the purpose of handling employee complaints. The committee will consist of the president and vice president of the symphony board and two orchestra representatives. The two orchestra representatives will be randomly selected from a list of six volunteers, consisting of four men and two women.

a. Find the probability distribution for x, the number of women chosen to be orchestra representatives.

b. What is the probability that both orchestra representatives will be women?

c. Find the mean and variance for the random variable x.

4.136 Independence and Mutually Exclusive Suppose that $P(A) = .3$ and $P(B) = .4$.

a. If $P(A \cap B) = .12$ are A and B independent? Justify your answer.

b. If $P(A \cup B) = .7$ what is $P(A \cap B)$? Justify your answer.

c. If A and B are independent, what is $P(A|B)$?

d. If A and B are mutually exclusive, what is $P(A|B)$?

4.137 Bringing Home the Bacon The following information reflects the results of a survey reported by Mya Frazier in an *Ad Age Insights* white paper.[11] Working spouses were asked "Who is the household breadwinner?" Suppose that one person is selected at random from these 200 individuals.

	You	Spouse or Significant Other	About Equal	Totals
Men	64	16	20	100
Women	32	45	23	100
Totals	96	61	43	200

a. What is the probability that this person will identify his/herself as the household breadwinner?

b. What is the probability that the person selected will be a man who indicates that he and his spouse/significant other are equal breadwinners?

c. If the person selected indicates that the spouse or significant other is the breadwinner, what is the probability that the person is a man?

CASE STUDY ## Probability and Decision Making in the Congo

In his exciting novel *Congo,* Michael Crichton describes a search by Earth Resources Technology Service (ERTS), a geological survey company, for deposits of boron-coated blue diamonds, diamonds that ERTS believes to be the key to a new generation of optical computers.[12] In the novel, ERTS is racing against an international consortium to find the Lost City of Zinj, a city that thrived on diamond mining and existed several thousand years ago (according to African fable), deep in the rain forests of eastern Zaire.

After the mysterious destruction of its first expedition, ERTS launches a second expedition under the leadership of Karen Ross, a 24-year-old computer genius who is accompanied by Professor Peter Elliot, an anthropologist; Amy, a talking gorilla; and the famed mercenary and expedition leader, "Captain" Charles Munro. Ross's efforts to find the city are blocked by the consortium's offensive actions, by the deadly rain forest, and by hordes of "talking" killer gorillas whose perceived mission is to defend the diamond mines. Ross overcomes these obstacles by using space-age computers to evaluate the probabilities of success for all possible circumstances and all possible actions that the expedition might take. At each stage of the expedition, she is able to quickly evaluate the chances of success.

At one stage in the expedition, Ross is informed by her Houston headquarters that their computers estimate that she is 18 hours and 20 minutes behind the competing Euro-Japanese team, instead of 40 hours ahead. She changes plans and decides to have the 12 members of her team—Ross, Elliot, Munro, Amy, and eight native porters—parachute into a volcanic region near the estimated location of Zinj. As Crichton relates, "Ross had double-checked outcome probabilities from the Houston computer, and the results were unequivocal. The probability of a successful jump was .7980, meaning that there was approximately one chance in five that someone would be badly hurt. However, given a successful jump, the probability of expedition success was .9943, making it virtually certain that they would beat the consortium to the site."

Keeping in mind that this is an excerpt from a novel, let us examine the probability, .7980, of a successful jump. If you were one of the 12-member team, what is the probability that you would successfully complete your jump? In other words, if the probability of a successful jump by all 12 team members is .7980, what is the probability that a single member could successfully complete the jump?

Several Useful Discrete Distributions

© Kim Steele/Photodisc/Getty Images

A Mystery: Cancers Near a Reactor

Is the Pilgrim I nuclear reactor responsible for an increase in cancer cases in the surrounding area? A political controversy was set off when the Massachusetts Department of Public Health found an unusually large number of cases in a 4-mile-wide coastal strip just north of the nuclear reactor in Plymouth, Massachusetts. The case study at the end of this chapter examines how this question can be answered using one of the discrete probability distributions presented here.

INTRODUCTION

Examples of *discrete random variables* can be found in a variety of everyday situations and across most academic disciplines. However, there are three discrete probability distributions that serve as *models* for a large number of these applications. In this chapter we study the binomial, the Poisson, and the hypergeometric probability distributions and discuss their usefulness in different physical situations.

THE BINOMIAL PROBABILITY DISTRIBUTION

A coin-tossing experiment is a simple example of an important discrete random variable called the **binomial random variable.** Many practical experiments result in data similar to the head or tail outcomes of the coin toss. For example, consider the political polls used to predict voter preferences in elections. Each sampled voter can be compared to a coin because the voter may be in favor of our candidate—a "head"—or not—a "tail." In most cases, the proportion of voters who favor our candidate does not equal 1/2; that is, the coin is not fair. In fact, the proportion of voters who favor our candidate is exactly what the poll is designed to measure!

Here are some other situations that are similar to the coin-tossing experiment:

- A sociologist is interested in the proportion of elementary school teachers who are men.
- A soft drink marketer is interested in the proportion of cola drinkers who prefer her brand.
- A geneticist is interested in the proportion of the population who possess a gene linked to Alzheimer's disease.

Each sampled person is analogous to tossing a coin, but the probability of a "head" is not necessarily equal to 1/2. Although these situations have different practical objectives, they all exhibit the common characteristics of the **binomial experiment.**

Definition A **binomial experiment** is one that has these five characteristics:

1. The experiment consists of n identical trials.
2. Each trial results in one of two outcomes. For lack of a better name, the one outcome is called a success, S, and the other a failure, F.
3. The probability of success on a single trial is equal to p and remains the same from trial to trial. The probability of failure is equal to $(1 - p) = q$.
4. The trials are independent.
5. We are interested in x, the number of successes observed during the n trials, for $x = 0, 1, 2, \ldots, n$.

EXAMPLE 5.1

Suppose there are approximately 1,000,000 adults in a county and an unknown proportion p favors term limits for politicians. A sample of 1000 adults will be chosen in such a way that every one of the 1,000,000 adults has an equal chance of being selected, and each adult is asked whether he or she favors term limits. (The ultimate objective of this survey is to estimate the unknown proportion p, a problem that we will discuss in Chapter 8.) Is this a binomial experiment?

Solution Does the experiment have the five binomial characteristics?

1. A "trial" is the choice of a single adult from the 1,000,000 adults in the county. This sample consists of $n = 1000$ identical trials.

2. Since each adult will either favor or not favor term limits, there are two outcomes that represent the "successes" and "failures" in the binomial experiment.[†]

3. The probability of success, p, is the probability that an adult favors term limits. Does this probability remain the same for each adult in the sample? For all practical purposes, the answer is *yes*. For example, if 500,000 adults in the population favor term limits, then the probability of a "success" when the first adult is chosen is 500,000/1,000,000 = 1/2. When the second adult is chosen, the probability p changes slightly, depending on the first choice. That is, there will be either 499,999 or 500,000 successes left among the 999,999 adults. In either case, p is still approximately equal to 1/2.

4. The independence of the trials is guaranteed because of the large group of adults from which the sample is chosen. The probability of an adult favoring term limits does not change depending on the responses of previously chosen people.

5. The random variable x is the number of adults in the sample who favor term limits.

Because the survey satisfies the five characteristics reasonably well, for all practical purposes it can be viewed as a binomial experiment.

EXAMPLE (5.2) A patient fills a prescription for a 10-day regimen of 2 pills daily. Unknown to the pharmacist and the patient, the 20 tablets consist of 18 pills of the prescribed medication and 2 pills that are the generic equivalent of the prescribed medication. The patient selects two pills at random for the first day's dosage. If we check the selection and record the number of pills that are generic, is this a binomial experiment?

Solution Again, check the sampling procedure for the characteristics of a binomial experiment.

1. A "trial" is the selection of a pill from the 20 in the prescription. This experiment consists of $n = 2$ trials.

2. Each trial results in one of two outcomes. Either the pill is generic (call this a "success") or not (a "failure").

3. Since the pills in a prescription bottle can be considered randomly "mixed," the unconditional probability of drawing a generic pill on a given trial would be 2/20.

4. The condition of independence between trials is *not* satisfied, because the probability of drawing a generic pill on the second trial is dependent on the first trial. For example, if the first pill drawn is generic, then there is only 1 generic pill in the remaining 19. Therefore,

$$P(\text{generic on trial } 2 \mid \text{generic on trial } 1) = 1/19$$

[†]Although it is traditional to call the two possible outcomes of a trial "success" and "failure," they could have been called "head" and "tail," "red" and "white," or any other pair of words. Consequently, the outcome called a "success" does not need to be viewed as a success in the ordinary use of the word.

If the first selection *does not* result in a generic pill, then there are still 2 generic pills in the remaining 19, and the probability of a "success" (a generic pill) changes to

$$P(\text{generic on trial 2}|\text{no generic on trial 1}) = 2/19$$

Therefore, the trials are dependent and the sampling does not represent a binomial experiment.

Think about the difference between these two examples. When the sample (the n identical trials) came from a large population, the probability of success p stayed about the same from trial to trial. When the population size N was small, the probability of success p changed quite dramatically from trial to trial, and the experiment *was not* binomial.

RULE OF THUMB

If the sample size is large relative to the population size—in particular, if $n/N \geq .05$—then the resulting experiment is not binomial.

In Chapter 4, we tossed two fair coins and constructed the probability distribution for x, the number of heads—a binomial experiment with $n = 2$ and $p = .5$. The general binomial probability distribution is constructed in the same way, but the procedure gets complicated as n gets large. Fortunately, the probabilities $p(x)$ follow a general pattern. This allows us to use a single formula to find $p(x)$ for any given value of x.

THE BINOMIAL PROBABILITY DISTRIBUTION

A binomial experiment consists of n identical trials with probability of success p on each trial. The probability of k successes in n trials is

$$P(x = k) = C_k^n p^k q^{n-k} = \frac{n!}{k!(n-k)!} p^k q^{n-k}$$

for values of $k = 0, 1, 2, \ldots, n$. The symbol C_k^n equals

$$\frac{n!}{k!(n-k)!}$$

where $n! = n(n-1)(n-2) \cdots (2)(1)$ and $0! \equiv 1$.

The general formulas for μ, σ^2, and σ given in Chapter 4 can be used to derive the following simpler formulas for the binomial mean and standard deviation.

MEAN AND STANDARD DEVIATION
FOR THE BINOMIAL RANDOM VARIABLE

The random variable x, the number of successes in n trials, has a probability distribution with this center and spread:

$$\text{Mean:} \quad \mu = np$$
$$\text{Variance:} \quad \sigma^2 = npq$$
$$\text{Standard deviation:} \quad \sigma = \sqrt{npq}$$

EXAMPLE 5.3

Find $P(x = 2)$ for a binomial random variable with $n = 10$ and $p = .1$.

Solution $P(x = 2)$ is the probability of observing 2 successes and 8 failures in a sequence of 10 trials. You might observe the 2 successes first, followed by 8 consecutive failures:

S, S, F, F, F, F, F, F, F, F

NEED A TIP?

$n! =$
$n(n - 1)(n - 2) \ldots (2)(1)$
For example,
$5! = 5(4)(3)(2)(1) = 120$
and $0! \equiv 1$.

Since p is the probability of success and q is the probability of failure, this particular sequence has probability

$$ppqqqqqqqq = p^2 q^8$$

However, many *other* sequences also result in $x = 2$ successes. The binomial formula uses C_2^{10} to count the number of sequences and gives the exact probability when you use the binomial formula with $k = 2$:

$$P(x = 2) = C_2^{10}(.1)^2(.9)^{10-2}$$

$$= \frac{10!}{2!(10 - 2)!}(.1)^2(.9)^8 = \frac{10(9)}{2(1)}(.01)(.430467) = .1937$$

You could repeat the procedure in Example 5.3 for each value of x—0, 1, 2, . . . , 10—and find all the values of $p(x)$ necessary to construct a probability histogram for x. This would be a long and tedious job, but the resulting graph would look like Figure 5.1(a). You can check the height of the bar for $x = 2$ and find $p(2) = P(x = 2) = .1937$. The graph is skewed right; that is, most of the time you will observe small values of x. The mean or "balancing point" is around $x = 1$; in fact, you can use the formula to find the exact mean:

$$\mu = np = 10(.1) = 1$$

Figures 5.1(b) and 5.1(c) show two other binomial distributions with $n = 10$ but with different values of p. Look at the shapes of these distributions. When $p = .5$, the distribution is exactly symmetric about the mean, $\mu = np = 10(.5) = 5$. When $p = .9$, the distribution is the "mirror image" of the distribution for $p = .1$ and is skewed to the left.

FIGURE 5.1

Binomial probability distributions

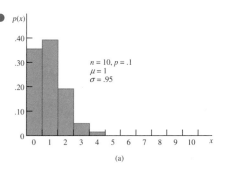

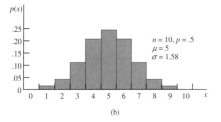

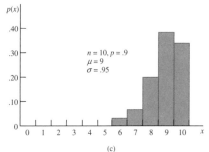

EXAMPLE (5.4) Over a long period of time, it has been observed that a professional basketball player can make a free throw on a given trial with probability equal to .8. Suppose he shoots four free throws.

1. What is the probability that he will make exactly two free throws?
2. What is the probability that he will make at least one free throw?

Solution A "trial" is a single free throw, and you can define a "success" as a basket and a "failure" as a miss, so that $n = 4$ and $p = .8$. If you assume that the player's chance of making the free throw does not change from shot to shot, then the number x of times that he makes the free throw is a *binomial random variable*.

1. $P(x = 2) = C_2^4(.8)^2(.2)^2$

$$= \frac{4!}{2!2!}(.64)(.04) = \frac{4(3)(2)(1)}{2(1)(2)(1)}(.64)(.04) = .1536$$

The probability is .1536 that he will make exactly two free throws.

2. $P(\text{at least one}) = P(x \geq 1) = p(1) + p(2) + p(3) + p(4)$
$$= 1 - p(0)$$
$$= 1 - C_0^4(.8)^0(.2)^4$$
$$= 1 - .0016 = .9984.$$

Although you could calculate $P(x = 1)$, $P(x = 2)$, $P(x = 3)$, and $P(x = 4)$ to find this probability, using the complement of the event makes your job easier; that is,

$$P(x \geq 1) = 1 - P(x < 1) = 1 - P(x = 0).$$

Can you think of any reason your assumption of independent trials might be wrong? If the player learns from his previous attempt (i.e., he adjusts his shooting according to his last attempt), then his probability p of making the free throw may change, possibly increase, from shot to shot. The trials would *not* be independent and the experiment would *not* be binomial.

Calculating binomial probabilities can become tedious even for relatively small values of n. As n gets larger, it becomes almost impossible without the help of a calculator or computer. Fortunately, both of these tools are available to us. Computer-generated tables of **cumulative binomial probabilities** are given in Table 1 of Appendix I for values of n ranging from 2 to 25 and for selected values of p. These probabilities can also be generated using *MINITAB, MS Excel,* or the Java applets on the CourseMate Web site.

NEED A TIP?
Use Table 1 in Appendix I rather than the binomial formula whenever possible. *This is an easier way!*

Cumulative binomial probabilities differ from the *individual* binomial probabilities that you calculated with the binomial formula. Once you find the column of probabilities for the correct values of n and p in Table 1, the row marked k gives the sum of all the binomial probabilities from $x = 0$ to $x = k$. Table 5.1 shows part of Table 1 for $n = 5$ and $p = .6$. If you look in the row marked $k = 3$, you will find

$$P(x \leq 3) = p(0) + p(1) + p(2) + p(3) = .663$$

TABLE 5.1 ● **Portion of Table 1 in Appendix I for $n = 5$**

							p							
k	.01	.05	.10	.20	.30	.40	.50	.60	.70	.80	.90	.95	.99	k
0	—	—	—	—	—	—	—	.010	—	—	—	—	—	0
1	—	—	—	—	—	—	—	.087	—	—	—	—	—	1
2	—	—	—	—	—	—	—	.317	—	—	—	—	—	2
3	—	—	—	—	—	—	—	.663	—	—	—	—	—	3
4	—	—	—	—	—	—	—	.922	—	—	—	—	—	4
5	—	—	—	—	—	—	—	1.000	—	—	—	—	—	5

If the probability you need to calculate is not in this form, you will need to think of a way to rewrite your probability to make use of the tables!

EXAMPLE **5.5** Use the cumulative binomial table for $n = 5$ and $p = .6$ to find the probabilities of these events:

1. Exactly three successes
2. Three or more successes

Solution

1. When $k = 3$ in Table 5.1, the tabled value is

$$P(x \le 3) = p(0) + p(1) + p(2) + p(3)$$

Since you want only $P(x = 3) = p(3)$, you must subtract out the unwanted probability:

$$P(x \le 2) = p(0) + p(1) + p(2)$$

which is found in Table 5.1 with $k = 2$. Then

$$P(x = 3) = P(x \le 3) - P(x \le 2)$$
$$= .663 - .317 = .346$$

2. To find P(three or more successes) $= P(x \ge 3)$ using Table 5.1, you must use the complement of the event of interest. Write

$$P(x \ge 3) = 1 - P(x < 3) = 1 - P(x \le 2)$$

You can find $P(x \le 2)$ in Table 5.1 with $k = 2$. Then

$$P(x \ge 3) = 1 - P(x \le 2)$$
$$= 1 - .317 = .683$$

EXAMPLE **5.6**

Refer to Example 5.5 and the binomial random variable x with $n = 5$ and $p = .6$. Use the cumulative binomial table to find the remaining binomial probabilities, $p(0)$, $p(1)$, $p(2)$, $p(4)$, and $p(5)$. Construct the probability histogram for the random variable x and describe its shape and location.

Solution

1. You can find $P(x = 0)$ directly from Table 5.1 with $k = 0$. That is, $p(0) = .010$.
2. The other probabilities can be found by subtracting successive entries in Table 5.1. Then

$$P(x = 1) = P(x \le 1) - P(x = 0) = .087 - .010 = .077$$
$$P(x = 2) = P(x \le 2) - P(x \le 1) = .317 - .087 = .230$$
$$P(x = 4) = P(x \le 4) - P(x \le 3) = .922 - .663 = .259$$
$$P(x = 5) = P(x \le 5) - P(x \le 4) = 1.000 - .922 = .078$$

The probability histogram is shown in Figure 5.2. The distribution is relatively mound-shaped, with a center around 3.

FIGURE 5.2

Binomial probability distribution for Example 5.6.

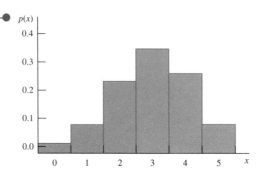

❓ NEED TO KNOW...

How to Use Table 1 to Calculate Binomial Probabilities

1. Find the necessary values of n and p. Isolate the appropriate column in Table 1.
2. Table 1 gives $P(x \le k)$ in the row marked k. Rewrite the probability you need so that it is in this form.

   - List the values of x in your event.
   - From the list, write the event as either the difference of two probabilities:
     $$P(x \le a) - P(x \le b) \quad \text{for } a > b$$
     or the complement of the event:
     $$1 - P(x \le a)$$
     or just the event itself:
     $$P(x \le a) \text{ or } P(x < a) = P(x \le a - 1)$$

EXAMPLE 5.7

MY ONLINE APPLET

Calculating Binomial
Probabilities

A regimen consisting of a daily dose of vitamin C was tested to determine its effectiveness in preventing the common cold. Ten people who were following the prescribed regimen were observed for a period of 1 year. Eight survived the winter without a cold. Suppose the probability of surviving the winter without a cold is .5 when the vitamin C regimen is not followed. What is the probability of observing eight or more survivors, given that the regimen is ineffective in increasing resistance to colds?

Solution If you assume that the vitamin C regimen is ineffective, then the probability p of surviving the winter without a cold is .5. The probability distribution for x, the number of survivors, is

$$p(x) = C_x^{10}(.5)^x(.5)^{10-x}$$

You have learned several ways to find $P(8 \text{ or more survivors}) = P(x \geq 8)$. You will get the same results with any of these methods; choose the most convenient method for your particular problem.

1. *The binomial formula:*

$$
\begin{aligned}
P(8 \text{ or more}) &= p(8) + p(9) + p(10) \\
&= C_8^{10}(.5)^{10} + C_9^{10}(.5)^{10} + C_{10}^{10}(.5)^{10} \\
&= .055
\end{aligned}
$$

2. *The cumulative binomial tables:* Find the column corresponding to $p = .5$ in the table for $n = 10$:

$$
\begin{aligned}
P(8 \text{ or more}) = P(x \geq 8) &= 1 - P(x \leq 7) \\
&= 1 - .945 = .055
\end{aligned}
$$

3. *Output from MINITAB or MS Excel:* The outputs shown in Figures 5.3(a) and 5.3(b) give the **cumulative distribution function,** which are the same probabilities you found in the cumulative binomial tables. The **probability density function** gives the individual binomial probabilities, which you found using the binomial formula.

FIGURE 5.3(a) ●
MINITAB output for
Example 5.7

Cumulative Distribution Function

Binomial with n = 10 and p = 0.5

x	P(X <= x)
0	0.00098
1	0.01074
2	0.05469
3	0.17187
4	0.37695
5	0.62305
6	0.82813
7	0.94531
8	0.98926
9	0.99902
10	1.00000

Probability Density Function

Binomial with n = 10 and p = 0.5

x	P(X = x)
0	0.000977
1	0.009766
2	0.043945
3	0.117188
4	0.205078
5	0.246094
6	0.205078
7	0.117188
8	0.043945
9	0.009766
10	0.000977

FIGURE 5.3(b)

Excel output for
Example 5.7

x	P(X<= x)	P(X = x)
0	0.000977	0.000977
1	0.010742	0.009766
2	0.054688	0.043945
3	0.171875	0.117188
4	0.376953	0.205078
5	0.623047	0.246094
6	0.828125	0.205078
7	0.945313	0.117188
8	0.989258	0.043945
9	0.999023	0.009766
10	1	0.000977

Using the cumulative distribution function, calculate

$$P(x \geq 8) = 1 - P(x \leq 7)$$
$$= 1 - .94531 = .05469$$

Or, using the probability density function, calculate

$$P(x \geq 8) = p(8) + p(9) + p(10)$$
$$= .043945 + .009766 + .000977 = .05469$$

EXAMPLE **5.8**

Would you rather take a multiple-choice or a full recall test? If you have absolutely no knowledge of the material, you will score zero on a full recall test. However, if you are given five choices for each question, you have at least one chance in five of guessing correctly! If a multiple-choice exam contains 100 questions, each with five possible answers, what is the expected score for a student who is guessing on each question? Within what limits will the "no-knowledge" scores fall?

Solution If x is the number of correct answers on the 100-question exam, the probability of a correct answer, p, is one in five, so that $p = .2$. Since the student is randomly selecting answers, the $n = 100$ answers are independent, and the expected score for this binomial random variable is

$$\mu = np = 100(.2) = 20 \quad \text{correct answers}$$

To evaluate the spread or variability of the scores, you can calculate

$$\sigma = \sqrt{npq} = \sqrt{100(.2)(.8)} = 4$$

Then, using your knowledge of variation from Tchebysheff's Theorem and the Empirical Rule, you can make these statements:

- A large proportion of the scores will lie within two standard deviations of the mean, or from $20 - 8 = 12$ to $20 + 8 = 28$.

- Almost all the scores will lie within three standard deviations of the mean, or from $20 - 12 = 8$ to $20 + 12 = 32$.

The "guessing" option gives the student a better score than the zero score on the full recall test, but the student still will not pass the exam. What other options does the student have?

5.2 EXERCISES

BASIC TECHNIQUES

5.1 Consider a binomial random variable with $n = 8$ and $p = .7$. Let x be the number of successes in the sample.

a. Find the probability that x is 3 or less.
b. Find the probability that x is 3 or more.
c. Find $P(x < 3)$.
d. Find $P(x = 3)$.
e. Find $P(3 \le x \le 5)$.

5.2 Consider a binomial random variable with $n = 9$ and $p = .3$. Let x be the number of successes in the sample.

a. Find the probability that x is exactly 2.
b. Find the probability that x is less than 2.
c. Find $P(x > 2)$.
d. Find $P(2 \le x \le 4)$.

5.3 Evaluate these binomial probabilities:

a. $C_2^8(.3)^2(.7)^6$ **b.** $C_0^4(.05)^0(.95)^4$
c. $C_3^{10}(.5)^3(.5)^7$ **d.** $C_1^7(.2)^1(.8)^6$

5.4 Evaluate these binomial probabilities:

a. $C_0^8(.2)^0(.8)^8$ **b.** $C_1^8(.2)^1(.8)^7$ **c.** $C_2^8(.2)^2(.8)^6$
d. $P(x \le 1)$ when $n = 8$, $p = .2$
e. P(two or fewer successes)

5.5 Let x be a binomial random variable with $n = 7$, $p = .3$. Find these values:

a. $P(x = 4)$ **b.** $P(x \le 1)$ **c.** $P(x > 1)$
d. $\mu = np$ **e.** $\sigma = \sqrt{npq}$

5.6 Use the formula for the binomial probability distribution to calculate the values of $p(x)$ and construct the probability histogram for x when $n = 6$ and $p = .2$. [HINT: Calculate $P(x = k)$ for seven different values of k.]

5.7 Refer to Exercise 5.6. Construct the probability histogram for a binomial random variable x with $n = 6$ and $p = .8$. Use the results of Exercise 5.6; do not recalculate all the probabilities.

5.8 If x has a binomial distribution with $p = .5$, will the shape of the probability distribution be symmetric, skewed to the left, or skewed to the right?

5.9 Let x be a binomial random variable with $n = 10$ and $p = .4$. Find these values:

a. $P(x = 4)$ **b.** $P(x \ge 4)$ **c.** $P(x > 4)$
d. $P(x \le 4)$ **e.** $\mu = np$ **f.** $\sigma = \sqrt{npq}$

5.10 Use Table 1 in Appendix I to find the sum of the binomial probabilities from $x = 0$ to $x = k$ for these cases:

a. $n = 10$, $p = .1$, $k = 3$
b. $n = 15$, $p = .6$, $k = 7$
c. $n = 25$, $p = .5$, $k = 14$

5.11 Use Table 1 in Appendix I to evaluate the following probabilities for $n = 6$ and $p = .8$:

a. $P(x \ge 4)$ **b.** $P(x = 2)$
c. $P(x < 2)$ **d.** $P(x > 1)$

Verify these answers using the values of $p(x)$ calculated in Exercise 5.7.

5.12 Find $P(x \le k)$ for each of the following cases:

a. $n = 20$, $p = .05$, $k = 2$
b. $n = 15$, $p = .7$, $k = 8$
c. $n = 10$, $p = .9$, $k = 9$

5.13 Use Table 1 in Appendix I to find the following:

a. $P(x < 12)$ for $n = 20$, $p = .5$
b. $P(x \le 6)$ for $n = 15$, $p = .4$
c. $P(x > 4)$ for $n = 10$, $p = .4$
d. $P(x \ge 6)$ for $n = 15$, $p = .6$
e. $P(3 < x < 7)$ for $n = 10$, $p = .5$

5.14 Find the mean and standard deviation for a binomial distribution with these values:

a. $n = 1000$, $p = .3$ **b.** $n = 400$, $p = .01$
c. $n = 500$, $p = .5$ **d.** $n = 1600$, $p = .8$

5.15 Find the mean and standard deviation for a binomial distribution with $n = 100$ and these values of p:

a. $p = .01$ **b.** $p = .9$ **c.** $p = .3$
d. $p = .7$ **e.** $p = .5$

5.16 In Exercise 5.15, the mean and standard deviation for a binomial random variable were calculated for a fixed sample size, $n = 100$, and for different values of p. Graph the values of the standard deviation for the five values of p given in Exercise 5.15. For what value of p does the standard deviation seem to be a maximum?

5.17 Let x be a binomial random variable with $n = 20$ and $p = .1$.

a. Calculate $P(x \leq 4)$ using the binomial formula.

b. Calculate $P(x \leq 4)$ using Table 1 in Appendix I.

c. Use the *Excel* output below to calculate $P(x \leq 4)$. Compare the results of parts a, b, and c.

d. Calculate the mean and standard deviation of the random variable x.

e. Use the results of part d to calculate the intervals $\mu \pm \sigma$, $\mu \pm 2\sigma$, and $\mu \pm 3\sigma$. Find the probability that an observation will fall into each of these intervals.

f. Are the results of part e consistent with Tchebysheff's Theorem? With the Empirical Rule? Why or why not?

Excel output for Exercise 5.17: Binomial with $n = 20$ and $p = .1$

x	p(x)	x	p(x)
0	0.1216	11	7E-07
1	0.2702	12	5E-08
2	0.2852	13	4E-09
3	0.1901	14	2E-10
4	0.0898	15	9E-12
5	0.0319	16	3E-13
6	0.0089	17	8E-15
7	0.0020	18	2E-16
8	0.0004	19	2E-18
9	0.0001	20	1E-20
10	0.0000		

APPLICATIONS

5.18 The Urn Problem A jar contains five balls: three red and two white. Two balls are randomly selected without replacement from the jar, and the number x of red balls is recorded. Explain why x is or is not a binomial random variable. (HINT: Compare the characteristics of this experiment with the characteristics of a binomial experiment given in this section.) If the experiment is binomial, give the values of n and p.

5.19 The Urn Problem, continued Refer to Exercise 5.18. Assume that the sampling was conducted with replacement. That is, assume that the first ball was selected from the jar, observed, and then replaced, and that the balls were then mixed before the second ball was selected. Explain why x, the number of red balls observed, is or is not a binomial random variable.

If the experiment is binomial, give the values of n and p.

5.20 Chicago Weather A meteorologist in Chicago recorded the number of days of rain during a 30-day period. If the random variable x is defined as the number of days of rain, does x have a binomial distribution? If not, why not? If so, are both values of n and p known?

5.21 Telemarketers A market research firm hires operators to conduct telephone surveys. The computer randomly dials a telephone number, and the operator asks the respondent whether or not he has time to answer some questions. Let x be the number of telephone calls made until the first respondent is willing to answer the operator's questions. Is this a binomial experiment? Explain.

5.22 SAT Scores In 2010, the average overall SAT score (Critical Reading, Math, and Writing) for college-bound students in the United States was 1509 out of 2400. Suppose that 45% of all high school graduates took this test, and that 100 high school graduates are randomly selected from throughout the United States.[1] Which of the following random variables has an approximate binomial distribution? If possible, give the values for n and p.

a. The number of students who took the SAT.

b. The scores of the 100 students on the SAT.

c. The number of students who scored above average on the SAT.

d. The amount of time it took the students to complete the SAT.

5.23 Security Systems A home security system is designed to have a 99% reliability rate. Suppose that nine homes equipped with this system experience an attempted burglary. Find the probabilities of these events:

a. At least one of the alarms is triggered.

b. More than seven of the alarms are triggered.

c. Eight or fewer alarms are triggered.

5.24 Blood Types In a certain population, 85% of the people have Rh-positive blood. Suppose that two people from this population get married. What is the probability that they are both Rh-negative, thus making it inevitable that their children will be Rh-negative?

5.25 Car Colors Car color preferences change over the years and according to the particular model that the customer selects. In a recent year, suppose that 10% of all luxury cars sold were black. If 25 cars of that year and type are randomly selected, find the following probabilities:

a. At least five cars are black.

b. At most six cars are black.

c. More than four cars are black.

d. Exactly four cars are black.

e. Between three and five cars (inclusive) are black.

f. More than 20 cars are not black.

5.26 O Canada! The National Hockey League (NHL) has 80% of its players born outside the United States, and of those born outside the United States, 50% are born in Canada.[2] Suppose that $n = 12$ NHL players were selected at random. Let x be the number of players in the sample who were born outside of the United States so that $p = .8$. Find the following probabilities:

Just over 50% of the foreign players in the NHL are from Canada

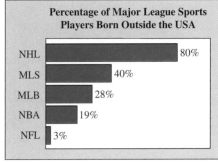

Percentage of Major League Sports Players Born Outside the USA

a. At least five or more of the sampled players were born outside the United States.

b. Exactly seven of the players were born outside the United States.

c. Fewer than six were born outside the United States.

5.27 Medical Bills Records show that 30% of all patients admitted to a medical clinic fail to pay their bills and that eventually the bills are forgiven. Suppose $n = 4$ new patients represent a random selection from the large set of prospective patients served by the clinic. Find these probabilities:

a. All the patients' bills will eventually have to be forgiven.

b. One will have to be forgiven.

c. None will have to be forgiven.

5.28 Medical Bills II Refer to Exercise 5.27 where 30% of all admitted patients fail to pay their bills and the debts are eventually forgiven. Suppose that the clinic treats 2000 different patients over a period of 1 year, and let x be the number of forgiven debts.

a. What is the mean (expected) number of debts that have to be forgiven?

b. Find the variance and standard deviation of x.

c. What can you say about the probability that x will exceed 700? (HINT: Use the values of μ and σ, along with Tchebysheff's Theorem.)

5.29 Whitefly Infestation Suppose that 10% of the fields in a given agricultural area are infested with the sweet potato whitefly. One hundred fields in this area are randomly selected and checked for whitefly.

a. What is the average number of fields sampled that are infested with whitefly?

b. Within what limits would you expect to find the number of infested fields, with probability approximately 95%?

c. What might you conclude if you found that $x = 25$ fields were infested? Is it possible that one of the characteristics of a binomial experiment is not satisfied in this experiment? Explain.

5.30 Color Preferences in Mice In a psychology experiment, the researcher designs a maze in which a mouse must choose one of two paths, colored either red or blue, at each of 10 intersections. At the end of the maze, the mouse is given a food reward. The researcher counts the number of times the mouse chooses the red path. If you were the researcher, how would you use this count to decide whether the mouse has any preference for color?

5.31 Back Pain Six in 10 adults say lower back pain substantially limits their athletic activities.[3] A random sample of $n = 8$ adults were asked if lower back pain was a limiting factor in their athletic activities. The printout below shows the cumulative and individual probabilities for a binomial random variable with $n = 8$ and $p = .6$.

MINITAB Output for Exercise 5.31

Cumulative Distribution Function		Probability Density Function	
Binomial with $n = 8$ and $p = 0.6$		Binomial with $n = 8$ and $p = 0.6$	
x	P(X <= x)	x	P(X = x)
0	0.00066	0	0.000655
1	0.00852	1	0.007864
2	0.04981	2	0.041288
3	0.17367	3	0.123863
4	0.40591	4	0.232243
5	0.68461	5	0.278692
6	0.89362	6	0.209019
7	0.98320	7	0.089580
8	1.00000	8	0.016796

a. Use the binomial formula to find the probability that all eight indicate that lower back pain was a limiting factor in their athletic activities.

b. Confirm the results of part a using the printout.

c. What is the probability that at most seven individuals give lower back pain as a limiting factor in their athletic activities?

5.32 Fast Food and Gas Stations Forty percent of all Americans who travel by car look for gas stations and food outlets that are close to or visible from the highway. Suppose a random sample of $n = 25$ Americans who travel by car are asked how they determine where to stop for food and gas. Let x be the number in the sample who respond that they look for gas stations and food outlets that are close to or visible from the highway.

a. What are the mean and variance of x?

b. Calculate the interval $\mu \pm 2\sigma$. What values of the binomial random variable x fall into this interval?

c. Find $P(6 \le x \le 14)$. How does this compare with the fraction in the interval $\mu \pm 2\sigma$ for any distribution? For mound-shaped distributions?

5.33 Taste Test for PTC The taste test for PTC (phenylthiocarbamide) is a favorite exercise for every human genetics class. It has been established that a single gene determines the characteristic, and that 70% of Americans are "tasters," while 30% are "nontasters." Suppose that 20 Americans are randomly chosen and are tested for PTC.

a. What is the probability that 17 or more are "tasters"?

b. What is the probability that 15 or fewer are "tasters"?

5.34 Man's Best Friend According to the Humane Society of the United States, there are approximately 77.5 million owned dogs in the United States, and approximately 40% of all U.S. households own at least one dog.[4] Suppose that the 40% figure is correct and that 15 households are randomly selected for a pet ownership survey.

a. What is the probability that exactly eight of the households have at least one dog?

b. What is the probability that at most four of the households have at least one dog?

c. What is the probability that more than 10 households have at least one dog?

THE POISSON PROBABILITY DISTRIBUTION

5.3

Another discrete random variable that has numerous practical applications is the **Poisson random variable.** Its probability distribution provides a good model for data that represent the number of occurrences of a specified event in a given unit of time or space. Here are some examples of experiments for which the random variable x can be modeled by the Poisson random variable:

- The number of calls received by a technical support specialist during a given period of time
- The number of bacteria per small volume of fluid
- The number of customer arrivals at a checkout counter during a given minute
- The number of machine breakdowns during a given day
- The number of traffic accidents on a section of freeway during a given time period

In each example, x **represents the number of events that occur in a period of time or space during which an average of μ such events can be expected to occur.** The only assumptions needed when one uses the Poisson distribution to model experiments such as these are that the counts or events occur **randomly and independently** of one another. The formula for the Poisson probability distribution, as well as its mean and variance, are given next.

THE POISSON PROBABILITY DISTRIBUTION

Let μ be the average number of times that an event occurs in a certain period of time or space. The probability of k occurrences of this event is

$$P(x = k) = \frac{\mu^k e^{-\mu}}{k!}$$

for values of $k = 0, 1, 2, 3, \ldots$. The mean and standard deviation of the Poisson random variable x are

Mean: μ Standard deviation: $\sigma = \sqrt{\mu}$

NEED A TIP?

Use either the Poisson formula or Table 2 to calculate Poisson probabilities.

The symbol $e = 2.71828\ldots$ is evaluated using your scientific calculator, which should have a function such as e^x. For each value of k, you can obtain the individual probabilities for the Poisson random variable, just as you did for the binomial random variable.

EXAMPLE **5.9**

The average number of traffic accidents on a certain section of highway is two per week. Assume that the number of accidents follows a Poisson distribution with $\mu = 2$.

1. Find the probability of no accidents on this section of highway during a 1-week period.
2. Find the probability of at most three accidents on this section of highway during a 2-week period.

Solution

1. The average number of accidents per week is $\mu = 2$. Therefore, the probability of no accidents on this section of highway during a given week is

$$P(x = 0) = p(0) = \frac{2^0 e^{-2}}{0!} = e^{-2} = .135335$$

2. During a 2-week period, the average number of accidents on this section of highway is $2(2) = 4$. The probability of at most three accidents during a 2-week period is

$$P(x \le 3) = p(0) + p(1) + p(2) + p(3)$$

where

$$p(0) = \frac{4^0 e^{-4}}{0!} = .018316 \qquad p(2) = \frac{4^2 e^{-4}}{2!} = .146525$$

$$p(1) = \frac{4^1 e^{-4}}{1!} = .073263 \qquad p(3) = \frac{4^3 e^{-4}}{3!} = .195367$$

Therefore,

$$P(x \le 3) = .018316 + .073263 + .146525 + .195367 = .433471$$

Once the values for $p(x)$ have been calculated, you can use them to construct a probability histogram for the random variable x. Graphs of the Poisson probability distribution for $\mu = .5, 2,$ and 4 are shown in Figure 5.4.

FIGURE 5.4

Poisson probability distributions for μ = .5, 2, and 4

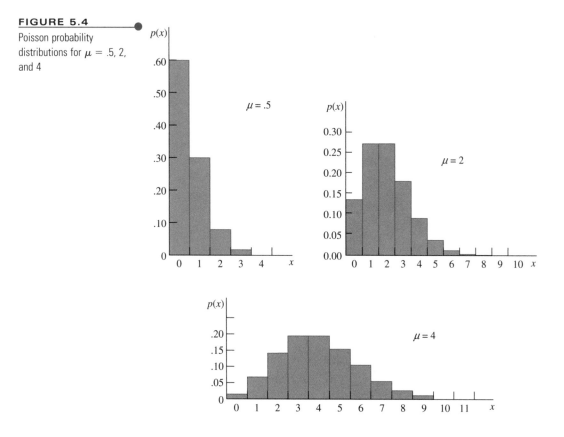

Alternatively, you can use **cumulative Poisson tables** (Table 2 in Appendix I) or the cumulative or individual probabilities generated by *MINITAB* or *MS Excel*. All of these options are usually more convenient than hand calculation. The procedures are similar to those used for the binomial random variable.

 NEED TO KNOW...

How to Use Table 2 to Calculate Poisson Probabilities

1. Find the necessary value of μ. Isolate the appropriate column in Table 2.
2. Table 2 gives $P(x \leq k)$ in the row marked k. Rewrite the probability you need so that it is in this form.

   - List the values of x in your event.
   - From the list, write the event as either the difference of two probabilities:

$$P(x \leq a) - P(x \leq b) \text{ for } a > b$$

or the complement of the event:

$$1 - P(x \leq a)$$

or just the event itself:

$$P(x \leq a) \text{ or } P(x < a - 1)$$

EXAMPLE  5.10

Refer to Example 5.9, where we calculated probabilities for a Poisson distribution with $\mu = 2$ and $\mu = 4$. Use the cumulative Poisson table to find the probabilities of these events:

1. No accidents during a 1-week period.
2. At most three accidents during a 2-week period.

Solution
A portion of Table 2 in Appendix I is shown in Figure 5.5.

FIGURE 5.5

Portion of Table 2 in Appendix I

k	μ				
	2.0	2.5	3.0	3.5	4.0
0	.135	.082	.055	.033	.018
1	.406	.287	.199	.136	.092
2	.677	.544	.423	.321	.238
3	.857	.758	.647	.537	.433
4	.947	.891	.815	.725	.629
5	.983	.958	.916	.858	.785
6	.995	.986	.966	.935	.889
7	.999	.996	.988	.973	.949
8	1.000	.999	.996	.990	.979
9		1.000	.999	.997	.992
10			1.000	.999	.997
11				1.000	.999
12					1.000

1. From Example 5.9, the average number of accidents in a 1-week period is $\mu = 2.0$. Therefore, the probability of no accidents in a 1-week period can be read directly from Table 2 in the column marked "2.0" as $P(x = 0) = p(0) = .135$.

2. The average number of accidents in a 2-week period is $2(2) = 4$. Therefore, the probability of at most three accidents in a 2-week period is found in Table 2, indexing $\mu = 4.0$ and $k = 3$ as $P(x \leq 3) = .433$.

Both of these probabilities match the calculations done in Example 5.9, correct to three decimal places.

In Section 5.2, we used the cumulative binomial tables to simplify the calculation of binomial probabilities. Unfortunately, in practical situations, n is often large and no tables are available.

THE POISSON APPROXIMATION
TO THE BINOMIAL DISTRIBUTION

The Poisson probability distribution provides a simple, easy-to-compute, and accurate approximation to binomial probabilities when n is large and $\mu = np$ is small, preferably with $np < 7$. An approximation suitable for larger values of $\mu = np$ will be given in Chapter 6.

EXAMPLE **5.11**

Suppose a life insurance company insures the lives of 5000 men aged 42. If actuarial studies show the probability that any 42-year-old man will die in a given year to be .001, find the exact probability that the company will have to pay $x = 4$ claims during a given year.

Solution The exact probability is given by the binomial distribution as

$$P(x = 4) = p(4) = \frac{5000!}{4!4996!}(.001)^4(.999)^{4996}$$

for which binomial tables are not available. To compute $P(x = 4)$ without the aid of a scientific calculator or a computer would be very time-consuming, but the Poisson distribution can be used to provide a good approximation to $P(x = 4)$. Computing $\mu = np = (5000)(.001) = 5$ and substituting into the formula for the Poisson probability distribution, we have

$$p(4) \approx \frac{\mu^4 e^{-\mu}}{4!} = \frac{5^4 e^{-5}}{4!} = \frac{(625)(.006738)}{24} = .175$$

The value of $p(4)$ could also be obtained using Table 2 in Appendix I with $\mu = 5$ as

$$p(4) = P(x \le 4) - P(x \le 3) = .440 - .265 = .175$$

EXAMPLE **5.12**

A manufacturer of power lawn mowers buys 1-horsepower, two-cycle engines in lots of 1000 from a supplier. She then equips each of the mowers produced by her plant with one of the engines. History shows that the probability of any one engine from that supplier proving unsatisfactory is .001. In a shipment of 1000 engines, what is the probability that none is defective? Three are? Four are?

Solution This is a binomial experiment with $n = 1000$ and $p = .001$. The expected number of defectives in a shipment of $n = 1000$ engines is $\mu = np = (1000)(.001) = 1$. Since this is a binomial experiment with $np < 7$, the probability of x defective engines in the shipment may be approximated by

$$P(x = k) = p(k) = \frac{\mu^k e^{-\mu}}{k!} = \frac{1^k e^{-1}}{k!} = \frac{e^{-1}}{k!}$$

Therefore,

$$p(0) \approx \frac{e^{-1}}{0!} = \frac{.368}{1} = .368$$

$$p(3) \approx \frac{e^{-1}}{3!} = \frac{.368}{6} = .061$$

$$p(4) \approx \frac{e^{-1}}{4!} = \frac{.368}{24} = .015$$

The individual Poisson probabilities for $\mu = 1$ along with the individual binomial probabilities for $n = 1000$ and $p = .001$ and $x = 0, 1, \ldots, 10$ were generated by *MS Excel* and are shown in Figure 5.6. The individual probabilities, even though they are computed with totally different formulas, are almost the same. The exact binomial probabilities are in the left section of Figure 5.6, and the Poisson approximations are on the right.

FIGURE 5.6 ●

Excel output of binomial
and Poisson probabilities

x	Binomial p(x)	x	Poisson p(x)
0	0.3677	0	0.3679
1	0.3681	1	0.3679
2	0.1840	2	0.1839
3	0.0613	3	0.0613
4	0.0153	4	0.0153
5	0.0030	5	0.0031
6	0.0005	6	0.0005
7	0.0001	7	0.0001
8	0.0000	8	0.0000
9	0.0000	9	0.0000
10	0.0000	10	0.0000

5.3 EXERCISES

BASIC TECHNIQUES

5.35 Consider a Poisson random variable with $\mu = 2.5$. Use the Poisson formula to calculate the following probabilities:

a. $P(x = 0)$ **b.** $P(x = 1)$

c. $P(x = 2)$ **d.** $P(x \leq 2)$

5.36 Consider a Poisson random variable with $\mu = 3$. Use the Poisson formula to calculate the following probabilities:

a. $P(x = 0)$ **b.** $P(x = 1)$

c. $P(x > 1)$

5.37 Consider a Poisson random variable with $\mu = 3$. Use Table 2 to find the following probabilities:

a. $P(x \leq 3)$ **b.** $P(x > 3)$

c. $P(x = 3)$ **d.** $P(3 \leq x \leq 5)$

5.38 Consider a Poisson random variable with $\mu = 0.8$. Use Table 2 to find the following probabilities:

a. $P(x = 0)$ **b.** $P(x \leq 2)$

c. $P(x > 2)$ **d.** $P(2 \leq x \leq 4)$

5.39 Let x be a Poisson random variable with mean $\mu = 2$. Calculate these probabilities:

a. $P(x = 0)$ **b.** $P(x = 1)$

c. $P(x > 1)$ **d.** $P(x = 5)$

5.40 Let x be a Poisson random variable with mean $\mu = 2.5$. Use Table 2 in Appendix I to calculate these probabilities:

a. $P(x \geq 5)$ **b.** $P(x < 6)$

c. $P(x = 2)$ **d.** $P(1 \leq x \leq 4)$

5.41 Poisson vs. Binomial Let x be a binomial random variable with $n = 20$ and $p = .1$.

a. Calculate $P(x \leq 2)$ using Table 1 in Appendix I to obtain the exact binomial probability.

b. Use the Poisson approximation to calculate $P(x \leq 2)$.

c. Compare the results of parts a and b. Is the approximation accurate?

5.42 Poisson vs. Binomial II To illustrate how well the Poisson probability distribution approximates the binomial probability distribution, calculate the Poisson approximate values for $p(0)$ and $p(1)$ for a binomial probability distribution with $n = 25$ and $p = .05$. Compare the answers with the exact values obtained from Table 1 in Appendix I.

APPLICATIONS

5.43 Airport Safety The increased number of small commuter planes in major airports has heightened concern over air safety. An eastern airport has recorded a monthly average of five near misses on landings and takeoffs in the past 5 years.

a. Find the probability that during a given month there are no near misses on landings and takeoffs at the airport.

b. Find the probability that during a given month there are five near misses.

c. Find the probability that there are at least five near misses during a particular month.

5.44 Intensive Care The number x of people entering the intensive care unit at a particular hospital on any one day has a Poisson probability distribution with mean equal to five persons per day.

a. What is the probability that the number of people entering the intensive care unit on a particular day is two? Less than or equal to two?

b. Is it likely that x will exceed 10? Explain.

5.45 Accident Prone According to a study conducted by the Department of Pediatrics at the University of California, San Francisco, children who are injured two or more times tend to sustain these injuries during a relatively limited time, usually 1 year or less. If the average number of injuries per year for school-age children is two, what are the probabilities of these events?

a. A school-age child will sustain two injuries during the year.

b. A school-age child will sustain two or more injuries during the year.

c. A school-age child will sustain at most one injury during the year.

5.46 Accident Prone, continued Refer to Exercise 5.45.

a. Calculate the mean and standard deviation for x, the number of injuries per year sustained by a school-age child.

b. Within what limits would you expect the number of injuries per year to fall?

5.47 Bacteria in Water Samples If a drop of water is placed on a slide and examined under a microscope, the number x of a specific type of bacteria present has been found to have a Poisson probability distribution. Suppose the maximum permissible count per water specimen for this type of bacteria is five. If the mean count for your water supply is two and you test a single specimen, is it likely that the count will exceed the maximum permissible count? Explain.

5.48 E. coli Outbreaks An outbreak of *E. coli* infections in August of 2010 occurred in three Washington state day care centers. There were eight confirmed and six suspected cases of *E. coli*, with over 70 children awaiting test results.[5] Outbreaks of *E. coli* infections for 2009 are reported to be less than 1 per 100,000, down from 2.5 per 100,000 reported earlier.[6] Using the rate of 1 per 100,000, find the following probabilities.

a. What is the probability that at most two outbreaks per 100,000 are reported across the United States this year?

b. What is the probability that more than three outbreaks per 100,000 are reported across the United States this year?

THE HYPERGEOMETRIC PROBABILITY DISTRIBUTION

5.4

Suppose you are selecting a sample of elements from a population and you record whether or not each element possesses a certain characteristic. You are recording the typical "success" or "failure" data found in the binomial experiment. The sample survey of Example 5.1 and the sampling for defectives of Example 5.2 are practical illustrations of these sampling situations.

If the number of elements in the population is large relative to the number in the sample (as in Example 5.1), the probability of selecting a success on a single trial is equal to the proportion p of successes in the population. Because the population is large in relation to the sample size, this probability will remain constant (for all practical purposes) from trial to trial, and the number x of successes in the sample will follow a binomial probability distribution. However, if the number of elements in the population is small in relation to the sample size ($n/N \geq .05$), the probability of a success for a given trial is dependent on the outcomes of preceding trials. Then the number x of successes follows what is known as a **hypergeometric probability distribution.**

It is easy to visualize the **hypergeometric random variable x** by thinking of a bowl containing M red balls and $N - M$ white balls, for a *total of N* balls in the bowl. You select n balls from the bowl and record x, the number of red balls that you see. If you now define a "success" to be a red ball, you have an example of the hypergeometric random variable x.

The formula for calculating the probability of exactly k successes in n trials is given next.

THE HYPERGEOMETRIC PROBABILITY DISTRIBUTION

A population contains M successes and $N - M$ failures. The probability of exactly k successes in a random sample of size n is

$$P(x = k) = \frac{C_k^M C_{n-k}^{N-M}}{C_n^N}$$

for values of k that depend on N, M, and n with

$$C_n^N = \frac{N!}{n!(N - n)!}$$

The mean and variance of a hypergeometric random variable are very similar to those of a binomial random variable with a correction for the finite population size:

$$\mu = n\left(\frac{M}{N}\right)$$

$$\sigma^2 = n\left(\frac{M}{N}\right)\left(\frac{N - M}{N}\right)\left(\frac{N - n}{N - 1}\right)$$

EXAMPLE 5.13

A case of wine has 12 bottles, 3 of which contain spoiled wine. A sample of 4 bottles is randomly selected from the case.

1. Find the probability distribution for x, the number of bottles of spoiled wine in the sample.
2. What are the mean and variance of x?

Solution For this example, $N = 12$, $n = 4$, $M = 3$, and $(N - M) = 9$. Then

$$p(x) = \frac{C_x^3 C_{4-x}^9}{C_4^{12}}$$

NEED A TIP?

Draw 4

3S
9G

1. The possible values for x are 0, 1, 2, and 3, with probabilities

$$p(0) = \frac{C_0^3 C_4^9}{C_4^{12}} = \frac{1(126)}{495} = .25$$

$$p(1) = \frac{C_1^3 C_3^9}{C_4^{12}} = \frac{3(84)}{495} = .51$$

$$p(2) = \frac{C_2^3 C_2^9}{C_4^{12}} = \frac{3(36)}{495} = .22$$

$$p(3) = \frac{C_3^3 C_1^9}{C_4^{12}} = \frac{1(9)}{495} = .02$$

2. The mean is given by

$$\mu = 4\left(\frac{3}{12}\right) = 1$$

and the variance is

$$\sigma^2 = 4\left(\frac{3}{12}\right)\left(\frac{9}{12}\right)\left(\frac{12-4}{11}\right) = .5455$$

EXAMPLE **5.14**

A particular industrial product is shipped in lots of 20. Testing to determine whether an item is defective is costly; hence, the manufacturer samples production rather than using a 100% inspection plan. A sampling plan constructed to minimize the number of defectives shipped to customers calls for sampling five items from each lot and rejecting the lot if more than one defective is observed. (If the lot is rejected, each item in the lot is then tested.) If a lot contains four defectives, what is the probability that it will be accepted?

Solution Let x be the number of defectives in the sample. Then $N = 20$, $M = 4$, $(N - M) = 16$, and $n = 5$. The lot will be rejected if $x = 2$, 3, or 4. Then

$$P(\text{accept the lot}) = P(x \leq 1) = p(0) + p(1) = \frac{C_0^4 C_5^{16}}{C_5^{20}} + \frac{C_1^4 C_4^{16}}{C_5^{20}}$$

$$= \frac{\left(\frac{4!}{0!4!}\right)\left(\frac{16!}{5!11!}\right)}{\frac{20!}{5!15!}} + \frac{\left(\frac{4!}{1!3!}\right)\left(\frac{16!}{4!12!}\right)}{\frac{20!}{5!15!}}$$

$$= \frac{91}{323} + \frac{455}{969} = .2817 + .4696 = .7513$$

5.4 EXERCISES

BASIC TECHNIQUES

5.49 Evaluate these probabilities:

a. $\dfrac{C_1^2 C_1^1}{C_2^3}$ b. $\dfrac{C_0^4 C_2^2}{C_2^6}$ c. $\dfrac{C_2^2 C_1^2}{C_3^4}$

5.50 Let x be the number of successes observed in a sample of $n = 4$ items selected from a population of $N = 8$. Suppose that of the $N = 8$ items, 5 are considered "successes."

a. Find the probability of observing all successes.

b. Find the probability of observing one success.

c. Find the probability of observing at most two successes.

5.51 Evaluate these probabilities:

a. $\dfrac{C_1^3 C_1^2}{C_2^5}$ b. $\dfrac{C_2^4 C_1^3}{C_3^7}$ c. $\dfrac{C_4^5 C_0^3}{C_4^8}$

5.52 Let x be the number of successes observed in a sample of $n = 5$ items selected from $N = 10$. Suppose that, of the $N = 10$ items, 6 are considered "successes."

a. Find the probability of observing no successes.

b. Find the probability of observing at least two successes.

c. Find the probability of observing exactly two successes.

5.53 Let x be a hypergeometric random variable with $N = 15$, $n = 3$, and $M = 4$.

a. Calculate $p(0)$, $p(1)$, $p(2)$, and $p(3)$.

b. Construct the probability histogram for x.

c. Use the formulas given in Section 5.4 to calculate $\mu = E(x)$ and σ^2.

d. What proportion of the population of measurements fall into the interval $(\mu \pm 2\sigma)$? Into the interval

$(\mu \pm 3\sigma)$? Do these results agree with those given by Tchebysheff's Theorem?

5.54 Candy Choices A candy dish contains five blue and three red candies. A child reaches up and selects three candies without looking.

a. What is the probability that there are two blue and one red candies in the selection?

b. What is the probability that the candies are all red?

c. What is the probability that the candies are all blue?

APPLICATIONS

5.55 Defective Computer Chips A piece of electronic equipment contains six computer chips, two of which are defective. Three computer chips are randomly chosen for inspection, and the number of defective chips is recorded. Find the probability distribution for x, the number of defective computer chips. Compare your results with the answers obtained in Exercise 4.90.

5.56 Seed Treatments Seeds are often treated with a fungicide for protection in poor-draining, wet environments. In a small-scale trial prior to a large-scale experiment to determine what dilution of the fungicide to apply, five treated seeds and five untreated seeds were planted in clay soil and the number of plants emerging from the treated and untreated seeds were recorded. Suppose the dilution was not effective and only four plants emerged. Let x represent the number of plants that emerged from treated seeds.

a. Find the probability that $x = 4$.

b. Find $P(x \le 3)$.

c. Find $P(2 \le x \le 3)$.

5.57 Teaching Credentials In southern California, a growing number of persons pursuing a teaching credential are choosing paid internships over traditional student teaching programs. A group of eight candidates for three teaching positions consisted of five paid interns and three traditional student teachers. Let us assume that all eight candidates are equally qualified for the positions. Let x represent the number of paid interns who are hired for these three positions.

a. Does x have a binomial distribution or a hypergeometric distribution? Support your answer.

b. Find the probability that three paid interns are hired for these positions.

c. What is the probability that none of the three hired was a paid intern?

d. Find $P(x \le 1)$

5.58 Gender Bias? A company has five applicants for two positions: two women and three men. Suppose that the five applicants are equally qualified and that no preference is given for choosing either gender. Let x equal the number of women chosen to fill the two positions.

a. Write the formula for $p(x)$, the probability distribution of x.

b. What are the mean and variance of this distribution?

c. Construct a probability histogram for x.

CHAPTER REVIEW

Key Concepts and Formulas

I. The Binomial Random Variable

1. **Five characteristics:** n identical independent trials, each resulting in either *success* (S) or *failure* (F); probability of success is p and remains constant from trial to trial; and x is the number of successes in n trials

2. **Calculating binomial probabilities**

   a. Formula: $P(x = k) = C_k^n p^k q^{n-k}$

   b. Cumulative binomial tables

   c. Individual and cumulative probabilities using *MINITAB* and *MS Excel*

3. Mean of the binomial random variable:
   $\mu = np$

4. Variance and standard deviation: $\sigma^2 = npq$
   and $\sigma = \sqrt{npq}$

II. The Poisson Random Variable

1. The number of events that occur in a period of time or space, during which an average of μ such events are expected to occur

2. **Calculating Poisson probabilities**

   a. Formula: $P(x = k) = \dfrac{\mu^k e^{-\mu}}{k!}$

b. Cumulative Poisson tables

c. Individual and cumulative probabilities using *MINITAB* and *MS Excel*

3. Mean of the Poisson random variable:
$E(x) = \mu$

4. Variance and standard deviation: $\sigma^2 = \mu$
and $\sigma = \sqrt{\mu}$

5. Binomial probabilities can be approximated with Poisson probabilities when $np < 7$, using $\mu = np$.

III. The Hypergeometric Random Variable

1. The number of successes in a sample of size n from a finite population containing M successes and $N - M$ failures

2. Formula for the probability of k successes in n trials:
$$P(x = k) = \frac{C_k^M C_{n-k}^{N-M}}{C_n^N}$$

3. Mean of the hypergeometric random variable:
$$\mu = n\left(\frac{M}{N}\right)$$

4. Variance and standard deviation:
$$\sigma^2 = n\left(\frac{M}{N}\right)\left(\frac{N-M}{N}\right)\left(\frac{N-n}{N-1}\right) \text{ and } \sigma = \sqrt{\sigma^2}$$

 TECHNOLOGY TODAY

Binomial and Poisson Probabilities in *Microsoft Excel*

For a random variable that has either a binomial or a Poisson probability distribution, *MS Excel* has been programmed to calculate either exact probabilities—$P(x = k)$—for a given value of k or cumulative probabilities—$P(x \leq k)$—for a given value of k. You must specify which distribution you are using and the necessary parameters: n and p for the binomial distribution and μ for the Poisson distribution.

Binomial Probabilities

1. Consider a binomial distribution with $n = 10$ and $p = .25$. The *value* of p does not appear in Table 1 of Appendix I, but you can use *Excel* to generate the entire probability distribution as well as the cumulative probabilities by entering the numbers 0–10 in column A.

One way to quickly enter a set of consecutive integers in a column is to do the following:

- Name columns A, B, and C as "x," "P(x = k)", and P(x<=k), respectively.
- Enter the first two values of x—0 and 1—to create a pattern in column A.
- Use your mouse to highlight the first two integers. Then grab the square handle in the lower right corner of the highlighted area. Drag the handle down to continue the pattern.
- As you drag, you will see an integer appear in a small rectangle. Release the mouse when you have the desired number of integers—in this case, 10.

2. Once the necessary values of x have been entered, place your cursor in the cell corresponding to $p(0)$, cell B2 in the spreadsheet. Select the **Insert Function** icon ⨍. In the drop-down list, select the **Statistical** category, select the

BINOM.DIST function and click **OK**. (NOTE: This function is called BINOMDIST in *Excel 2007* and earlier versions.) The Dialog box shown in Figure 5.7 will appear.

FIGURE 5.7

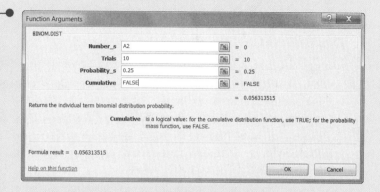

3. You must type in or select numbers or cell locations for each of the four boxes. When you place your cursor in the box, you will see a description of the necessary input for that box. Enter the address of the cell corresponding to $x = 0$ (cell A2) in the first box, the value of n in the second box, the value of p in the third box, and the word FALSE in the fourth box to calculate $P(x = k)$.

4. The resulting probability is marked as "Formula result = .056313515" at the bottom of the box, and when you click **OK**, the probability $P(x = 0)$ will appear in cell B2. To obtain the other probabilities, simply place your cursor in cell B2, grab the square handle in the lower right corner of the cell and drag the handle down to copy the formula into the other nine cells. *MS Excel* will automatically adjust the cell location in the formula as you copy.

5. If you want to generate the cumulative probabilities, $P(x \leq k)$, place your cursor in the cell corresponding to $P(x \leq 0)$, cell C2 in the spreadsheet. Then select **Insert Function ▶ Statistical ▶ BINOM.DIST,** and click **OK**. Continue as in steps 3 and 4, but type TRUE in the fourth line of the Dialog box to calculate $P(x \leq k)$. The resulting output is shown in Figure 5.8.

FIGURE 5.8

◢	A	B	C
1	x	P(x = k)	P(x <= k)
2	0	0.0563	0.0563
3	1	0.1877	0.2440
4	2	0.2816	0.5256
5	3	0.2503	0.7759
6	4	0.1460	0.9219
7	5	0.0584	0.9803
8	6	0.0162	0.9965
9	7	0.0031	0.9996
10	8	0.0004	1.0000
11	9	0.0000	1.0000
12	10	0.0000	1.0000

6. What value k is such that only 5% of the values of x exceed this value (and 95% are less than or equal to k)? Place your cursor in an empty cell, select **Insert Function ▶ Statistical ▶ BINOM.INV**, and click **OK**. (NOTE: This function is new to *Excel 2010*.) The resulting Dialog box will calculate what is sometimes called the **inverse cumulative probability.** Type **10** in the first box, **.25** in the second box, and **.95** in the third box. When you click **OK**, the number **5** will

appear in the empty cell. This is the smallest value of k for which $P(x \leq k)$ is greater than or equal to .95. Refer to Figure 5.8 and notice that $P(x \leq 5) = .9803$ so that $P(x > 5) = 1 - .9803 = .0197$. Hence, if you observed a value of $x = 5$, this would be an unusual observation.

Poisson Probabilities

1. The procedures for calculating individual or cumulative probabilities and probability distributions for the Poisson random variable are similar to those used for the binomial distribution.

2. To find Poisson probabilities $P(x = k)$ or $P(x \leq k)$ select **Insert Function** ▶ **Statistical** ▶ **POISSON.DIST,** and click **OK.** (NOTE: This function is called **POISSON** in *Excel 2007* and earlier versions.) Enter the values for k, μ, and **FALSE/TRUE** before clicking **OK.**

3. There is no **inverse cumulative probability** as there was for the binomial distribution.

Binomial and Poisson Probabilities in *MINITAB*

For a random variable that has either a binomial or a Poisson probability distribution, *MINITAB* has been programmed to calculate either exact probabilities—$P(X = x)$—for a given value of x or the cumulative probabilities—$P(X \leq x)$—for a given value of x. (NOTE: *MINITAB* uses the notation "X" for the random variable and "x" for a particular value of the random variable.) You must specify which distribution you are using and the necessary parameters: n and p for the binomial distribution and μ for the Poisson distribution.

Binomial Probabilities

1. Consider a binomial distribution with $n = 10$ and $p = .25$. The *value* of p does not appear in Table 1 of Appendix I, but you can use *MINITAB* to generate the entire probability distribution as well as the cumulative probabilities by entering the numbers 0 to 10 in column A.

2. One way to quickly enter a set of consecutive integers in a column is to do the following:

   • Name columns C1, C2, and C3 as "x," "P(X = x)," and P(X<=x), respectively.

   • Enter the first two values of x—0 and 1—to create a pattern in column C1.

   • Use your mouse to highlight the first two integers. Then grab the square handle in the lower right corner of the highlighted area. Drag the handle down to continue the pattern.

   • As you drag, you will see an integer appear in a small rectangle. Release the mouse when you have the desired number of integers—in this case, $\boxed{10}$.

3. Once the necessary values of x have been entered, use **Calc** ▶ **Probability Distributions** ▶ **Binomial** to generate the Dialog box shown in Figure 5.9.

FIGURE 5.9

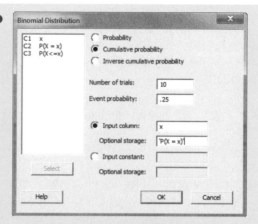

4. Type the number of trials and the value of p (Event probability) in the appropriate boxes, select C1 ('x') for the input column, and select C2 ('P(X = x)') for the Optional storage. Make sure that the radio button marked "Probability" is selected. The probability distribution for x will appear in column C2 when you click **OK.** (NOTE: If you do not select a column for the Optional storage, the results will be displayed in the Session window.)
5. If you want to generate the cumulative probabilities, $P(x \le k)$, again use **Calc ▶ Probability Distributions ▶ Binomial** to generate the Dialog box. This time, select the radio button marked "Cumulative probability" and select C3 (P(X<=x)) for the Optional storage in the Dialog box (Figure 5.9). The cumulative probability distribution will appear in column C3 when you click **OK.** You can display both distributions in the Session window using **Data ▶ Display Data,** selecting columns C1–C3 and clicking **OK.** The results are shown in Figure 5.10(a).

(a)

FIGURE 5.10

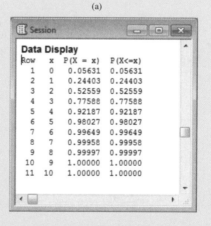

(b)

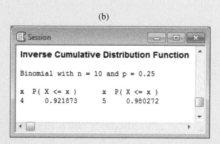

6. What value x is such that only 5% of the values of the random variable X exceed this value (and 95% are less than or equal to x)? Again, use **Calc ▶ Probability Distributions ▶ Binomial** to generate the Dialog box. This time, select the radio button marked "Inverse cumulative probability" and enter the probability **.95** into the "Input constant" box (Figure 5.9). When you click **OK,** the values of x on either side of the ".95 mark" will appear in the Session window as shown in Figure 5.10(b). Hence, if you observed a value of $x = 5$, this would be an unusual observation, because $P(x > 5) = 1 - .980272 = .019728$.

Poisson Probabilities

1. The procedures for calculating individual or cumulative probabilities and probability distributions for the Poisson random variable are similar to those used for the binomial distribution.

2. To find Poisson probabilities $P(X = x)$ or $P(X \leq x)$, use **Calc ▶ Probability Distributions ▶ Poisson** to generate the Dialog box. Enter the value for the mean μ, choose the appropriate radio button, and input either a column or a constant to indicate the value(s) of X for which you want to calculate a probability before clicking **OK.**

3. The **inverse cumulative probability** calculates the values of x such that $P(X \leq x) = C$, where C is a constant probability, between 0 and 1. Follow the steps described for the binomial random variable in step 6 above.

Supplementary Exercises

5.59 List the five identifying characteristics of the binomial experiment.

5.60 Under what conditions can the Poisson random variable be used to approximate the probabilities associated with the binomial random variable? What application does the Poisson distribution have other than to estimate certain binomial probabilities?

5.61 Under what conditions would you use the hypergeometric probability distribution to evaluate the probability of x successes in n trials?

5.62 Tossing a Coin A balanced coin is tossed three times. Let x equal the number of heads observed.

a. Use the formula for the binomial probability distribution to calculate the probabilities associated with $x = 0, 1, 2,$ and 3.

b. Construct the probability distribution.

c. Find the mean and standard deviation of x, using these formulas:

$$\mu = np$$
$$\sigma = \sqrt{npq}$$

d. Using the probability distribution in part b, find the fraction of the population measurements lying within one standard deviation of the mean. Repeat for two standard deviations. How do your results agree with Tchebysheff's Theorem and the Empirical Rule?

5.63 Coins, continued Refer to Exercise 5.62. Suppose the coin is definitely unbalanced and the probability of a head is equal to $p = .1$. Follow the instructions in parts a, b, c, and d. Note that the probability

distribution loses its symmetry and becomes skewed when p is not equal to 1/2.

5.64 Cancer Survivor Rates The 10-year survival rate for bladder cancer is approximately 50%. If 20 people who have bladder cancer are properly treated for the disease, what is the probability that:

a. At least 1 will survive for 10 years?

b. At least 10 will survive for 10 years?

c. At least 15 will survive for 10 years?

5.65 Garbage Collection A city commissioner claims that 80% of all people in the city favor private garbage collection in contrast to collection by city employees. To check the 80% claim, you randomly sample 25 people and find that x, the number of people who support the commissioner's claim, is 22.

a. What is the probability of observing at least 22 who support the commissioner's claim if, in fact, $p = .8$?

b. What is the probability that x is exactly equal to 22?

c. Based on the results of part a, what would you conclude about the claim that 80% of all people in the city favor private collection? Explain.

5.66 Integers If a person is given the choice of an integer from 0 to 9, is it more likely that he or she will choose an integer near the middle of the sequence than one at either end?

a. If the integers are equally likely to be chosen, find the probability distribution for x, the number chosen.

b. What is the probability that a person will choose a 4, 5, or 6?

c. What is the probability that a person will not choose a 4, 5, or 6?

5.67 Integers II Refer to Exercise 5.66. Twenty people are asked to select a number from 0 to 9. Eight of them choose a 4, 5, or 6.

a. If the choice of any one number is as likely as any other, what is the probability of observing eight or more choices of the numbers 4, 5, or 6?

b. What conclusions would you draw from the results of part a?

5.68 Checking In Fewer Americans are really getting away while on vacation. In fact, among small business owners, more than half (51%) say they check in with the office at least once a day while on vacation; only 27% say they cut the cord completely.[7] If 20 small business owners are randomly selected, and we assume that exactly half check in with the office at least once a day, then $n = 20$ and $p = .5$. Find the following probabilities.

a. Exactly 16 say that they check in with the office at least once a day while on vacation.

b. Between 15 and 18 (inclusive) say they check in with the office at least once a day while on vacation.

c. Five or fewer say that they check in with the office at least once a day while on vacation. Would this be an unlikely occurrence?

5.69 Psychosomatic Problems A psychiatrist believes that 80% of all people who visit doctors have problems of a psychosomatic nature. She decides to select 25 patients at random to test her theory.

a. Assuming that the psychiatrist's theory is true, what is the expected value of x, the number of the 25 patients who have psychosomatic problems?

b. What is the variance of x, assuming that the theory is true?

c. Find $P(x \leq 14)$. (Use tables and assume that the theory is true.)

d. Based on the probability in part c, if only 14 of the 25 sampled had psychosomatic problems, what conclusions would you make about the psychiatrist's theory? Explain.

5.70 Student Fees A student government states that 80% of all students favor an increase in student fees to subsidize a new recreational area. A random sample of $n = 25$ students produced 15 in favor of increased fees. What is the probability that 15 or fewer in the sample would favor the issue if student government is correct? Do the data support the student government's assertion, or does it appear that the percentage favoring an increase in fees is less than 80%?

5.71 Gray Hair on Campus College campuses are graying! According to a recent article, one in four college students is aged 30 or older. Assume that the 25% figure is accurate, that your college is representative of colleges at large, and that you sample $n = 200$ students, recording x, the number of students age 30 or older.

a. What are the mean and standard deviation of x?

b. If there are 35 students in your sample who are age 30 or older, would you be willing to assume that the 25% figure is representative of your campus? Explain.

5.72 Probability of Rain To check the accuracy of a particular weather forecaster, records were checked only for those days when the forecaster predicted rain "with 30% probability." A check of 25 of those days indicated that it rained on 10 of the 25.

a. If the forecaster is accurate, what is the appropriate value of p, the probability of rain on one of the 25 days?

b. What are the mean and standard deviation of x, the number of days on which it rained, assuming that the forecaster is accurate?

c. Calculate the z-score for the observed value, $x = 10$. [HINT: Recall from Section 2.6 that $z\text{-score} = (x - \mu)/\sigma$.]

d. Do these data disagree with the forecast of a "30% probability of rain"? Explain.

5.73 What's for Breakfast? A packaging experiment is conducted by placing two different package designs for a breakfast food side by side on a supermarket shelf. On a given day, 25 customers purchased a package of the breakfast food from the supermarket. Let x equal the number of buyers who choose the second package design.

a. If there is no preference for either of the two designs, what is the value of p, the probability that a buyer chooses the second package design?

b. If there is no preference, use the results of part a to calculate the mean and standard deviation of x.

c. If 5 of the 25 customers choose the first package design and 20 choose the second design, what do you conclude about the customers' preference for the second package design?

5.74 Diabetes in Children Insulin-dependent diabetes (IDD) among children occurs most frequently in persons of northern European descent. The incidence ranges from a low of 1–2 cases per 100,000 per year to a high of more than 40 per 100,000 in parts of Finland.[9] Let us assume that an area in Europe has an incidence of 5 cases per 100,000 per year.

a. Can the distribution of the number of cases of IDD in this area be approximated by a Poisson distribution? If so, what is the mean?

b. What is the probability that the number of cases of IDD in this area is less than or equal to 3 per 100,000?

c. What is the probability that the number of cases is greater than or equal to 3 but less than or equal to 7 per 100,000?

d. Would you expect to observe 10 or more cases of IDD per 100,000 in this area in a given year? Why or why not?

5.75 Plant Genetics A peony plant with red petals was crossed with another plant having streaky petals. The probability that an offspring from this cross has red flowers is .75. Let x be the number of plants with red petals resulting from ten seeds from this cross that were collected and germinated.

a. Does the random variable x have a binomial distribution? If not, why not? If so, what are the values of n and p?

b. Find $P(x \geq 9)$.

c. Find $P(x \leq 1)$.

d. Would it be unusual to observe one plant with red petals and the remaining nine plants with streaky petals? If these experimental results actually occurred, what conclusions could you draw?

5.76 Dominant Traits The alleles for black (B) and white (b) feather color in chickens show incomplete dominance; individuals with the gene pair Bb have "blue" feathers. When one individual that is homozygous dominant (BB) for this trait is mated with an individual that is homozygous recessive (bb) for this trait, 1/2 will carry the gene pair Bb. Let x be the number of chicks with "blue" feathers in a sample of $n = 20$ chicks resulting from this type of cross.

a. Does the random variable x have a binomial distribution? If not, why not? If so, what are the values of n and p?

b. What is the mean number of chicks with "blue" feathers in the sample?

c. What is the probability of observing fewer than five chicks with "blue" feathers?

d. What is the probability that the number of chicks with "blue" feathers is greater than or equal to 10 but less than or equal to 12?

5.77 Football Coin Tosses During the 1992 football season, the Los Angeles Rams (now the St. Louis Rams) had a bizarre streak of coin-toss losses. In fact, they lost the call 11 weeks in a row.[8]

a. The Rams' computer system manager said that the odds against losing 11 straight tosses are 2047 to 1. Is he correct?

b. After these results were published, the Rams lost the call for the next two games, for a total of 13 straight losses. What is the probability of this happening if, in fact, the coin was fair?

5.78 Plant Density One model for plant competition assumes that there is a zone of resource depletion around each plant seedling. Depending on the size of the zones and the density of the plants, the zones of resource depletion may overlap with those of other seedlings in the vicinity. When the seeds are randomly dispersed over a wide area, the number of neighbors that a seedling may have usually follows a Poisson distribution with a mean equal to the density of seedlings per unit area. Suppose that the density of seedlings is four per square meter (m²).

a. What is the probability that a given seedling has no neighbors within 1 m²?

b. What is the probability that a seedling has at most three neighbors per m²?

c. What is the probability that a seedling has five or more neighbors per m²?

d. Use the fact that the mean and variance of a Poisson random variable are equal to find the proportion of neighbors that would fall into the interval $\mu \pm 2\sigma$. Comment on this result.

5.79 Problems with Your New Smartphone? A new study by Square Trade indicates that smartphones are 50% more likely to malfunction than simple phones over a 3-year period.[10] Of smartphone failures, 30% are related to internal components not working, and overall, there is a 31% chance of having your smartphone fail over 3 years. Suppose that smartphones are shipped in cartons of $N = 50$ phones. Before shipment $n = 10$ phones are selected from each carton and the carton is shipped if none of the selected phones are defective. If one or more are found to be defective, the whole carton is tested.

a. What is the probability distribution of x, the number of defective phones related to internal components not working in the sample of $n = 10$ phones?

b. What is the probability that the carton will be shipped if two of the $N = 50$ smartphones in the carton have defective internal components?

c. What is the probability that the carton will be shipped if it contains four defectives? Six defectives?

5.80 Conservative Spenders A Snapshot in *USA Today* shows that 60% of consumers say they have become more conservative spenders.[12] When asked "What would you do first if you won $1 million tomorrow?" the answers had to do with somewhat conservative measures like "hire a financial advisor," or "pay off my credit card," or "pay off my mortgage." Suppose a random sample of $n = 15$ consumers is selected and the number x of those who say they have become conservative spenders recorded.

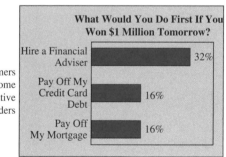

60% of consumers say they have become more conservative spenders

What Would You Do First If You Won $1 Million Tomorrow?

Hire a Financial Adviser — 32%
Pay Off My Credit Card Debt — 16%
Pay Off My Mortgage — 16%

a. What is the probability that more than six consumers say they have become conservative spenders?

b. What is the probability that fewer than five of those sampled have become conservative spenders?

c. What is the probability that exactly nine of those sampled are now conservative spenders.

5.81 Tay–Sachs Disease Tay–Sachs disease is a genetic disorder that is usually fatal in young children. If both parents are carriers of the disease, the probability that their offspring will develop the disease is

approximately .25. Suppose a husband and wife are both carriers of the disease and the wife is pregnant on three different occasions. If the occurrence of Tay–Sachs in any one offspring is independent of the occurrence in any other, what are the probabilities of these events?

a. All three children will develop Tay–Sachs disease.

b. Only one child will develop Tay–Sachs disease.

c. The third child will develop Tay–Sachs disease, given that the first two did not.

5.82 Dark Chocolate Despite reports that dark chocolate is beneficial to the heart, 47% of adults still prefer milk chocolate to dark chocolate.[11] Suppose a random sample of $n = 5$ adults is selected and asked whether they prefer milk chocolate to dark chocolate.

a. What is the probability that all five adults say that they prefer milk chocolate to dark chocolate?

b. What is the probability that exactly three of the five adults say they prefer milk chocolate to dark chocolate?

c. What is the probability that at least one adult prefers milk chocolate to dark chocolate?

5.83 The Triangle Test A procedure often used to control the quality of name-brand food products utilizes a panel of five "tasters." Each member of the panel tastes three samples, two of which are from batches of the product known to have the desired taste and the other from the latest batch. Each taster selects the sample that is different from the other two. Assume that the latest batch does have the desired taste, and that there is no communication between the tasters.

a. If the latest batch tastes the same as the other two batches, what is the probability that the taster picks it as the one that is different?

b. What is the probability that exactly one of the tasters picks the latest batch as different?

c. What is the probability that at least one of the tasters picks the latest batch as different?

5.84 Do You Return Your Questionnaires? A public opinion research firm claims that approximately 70% of those sent questionnaires respond by returning the questionnaire. Twenty such questionnaires are sent out, and assume that the president's claim is correct.

a. What is the probability that exactly 10 of the questionnaires are filled out and returned?

b. What is the probability that at least 12 of the questionnaires are filled out and returned?

c. What is the probability that at most 10 of the questionnaires are filled out and returned?

5.85 Questionnaires, continued Refer to Exercise 5.84. If $n = 20$ questionnaires are sent out,

a. What is the average number of questionnaires that will be returned?

b. What is the standard deviation of the number of questionnaires that will be returned?

c. If $x = 10$ of the 20 questionnaires are returned to the company, would you consider this to be an unusual response? Explain.

5.86 Poultry Problems A preliminary investigation reported that approximately 30% of locally grown poultry were infected with an intestinal parasite that, though not harmful to those consuming the poultry, decreased the usual weight growth rates in the birds. A diet supplement believed to be effective against this parasite was added to the bird's food. Twenty-five birds were examined after having the supplement for at least two weeks, and three birds were still found to be infested with the parasite.

a. If the diet supplement is ineffective, what is the probability of observing three or fewer birds infected with the intestinal parasite?

b. If in fact the diet supplement was effective and reduced the infection rate to 10%, what is the probability observing three or fewer infected birds?

5.87 Machine Breakdowns In a food processing and packaging plant, there are, on the average, two packaging machine breakdowns per week. Assume the weekly machine breakdowns follow a Poisson distribution.

a. What is the probability that there are no machine breakdowns in a given week?

b. Calculate the probability that there are no more than two machine breakdowns in a given week.

5.88 Bad Wiring Improperly wired control panels were mistakenly installed on two of eight large automated machine tools. It is uncertain which of the machine tools have the defective panels, and a sample of four tools is randomly chosen for inspection. What is the probability that the sample will include no defective panels? Both defective panels?

5.89 Stressed Out A subject is taught to do a task in two different ways. Studies have shown that when subjected to mental strain and asked to perform the task, the subject most often reverts to the method first learned, regardless of whether it was easier or more difficult. If the probability that a subject returns to the first method learned is .8 and six subjects are tested, what is the probability that at least five of the subjects revert to their first learned method when asked to perform their task under stress?

5.90 Safe Drivers? Evidence shows that the probability that a driver will be involved in a serious automobile accident during a given year is .01. A particular corporation employs 100 full-time traveling sales reps. Based on this evidence, use the Poisson approximation to the binomial distribution to find the probability that exactly two of the sales reps will be involved in a serious automobile accident during the coming year.

5.91 Earthquakes! Suppose that one out of every 10 homeowners in the state of California has invested in earthquake insurance. If 15 homeowners are randomly chosen to be interviewed,

a. What is the probability that at least one had earthquake insurance?

b. What is the probability that four or more have earthquake insurance?

c. Within what limits would you expect the number of homeowners insured against earthquakes to fall?

5.92 Enrolling in College A West Coast university has found that about 90% of its accepted applicants for enrollment in the freshman class will actually enroll. In 2012, 1360 applicants were accepted to the university. Within what limits would you expect to find the size of the freshman class at this university in the fall of 2012?

5.93 Eating on the Run How do you survive when there's no time to eat—fast food, no food, a protein bar, candy? A Snapshot in *USA Today* indicates that 36% of women aged 25–55 say that, when they are too busy to eat, they get fast food from a drive-thru.[13] A random sample of 100 women aged 25–55 is selected.

a. What is the average number of women who say they eat fast food when they're too busy to eat?

b. What is the standard deviation for the number of women who say they eat fast food when they're too busy to eat?

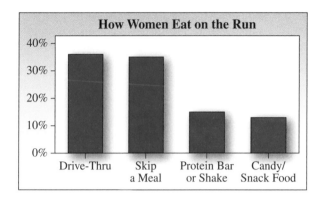

c. If 49 of the women in the sample said they eat fast food when they're too busy to eat, would this be an unusual occurrence? Explain.

5.94 Drivable Streets According to a *USA Today* Snapshot, drivers say fixing or repaving streets is the best way to make their communities drivable—better than building new roads or adding lanes.[14] Suppose that $n = 15$ drivers are randomly selected and x is the number who say that improved road conditions would make their communities more drivable. Let $p = .4$ when finding probabilities associated with any following outcomes:

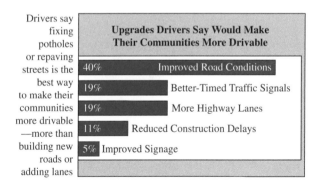

a. What is the probability distribution for x?

b. What is $P(x \leq 4)$?

c. Find the probability that x exceeds 5.

d. What is the largest value of c for which $P(x \leq c) \leq .5$?

5.95 Credit Problems? The recession has caused many people to use their credit cards far less. In fact, in the United States, 60% of consumers say they are committed to living with fewer credit cards.[15] A sample of $n = 400$ consumers with credit cards are randomly selected.

a. What is the average number of consumers in the sample who said they are committed to living with fewer credit cards?

b. What is the standard deviation of the number in the sample who said they are committed to living with fewer credit cards?

c. Within what range would you expect to find the number in the sample who said they are committed to living with fewer credit cards?

d. If only 200 of the sample of consumers said they were committed to living with fewer credit cards, would you consider this unusual? Explain. What conclusion might you draw from this sample information?

5.96 Successful Surgeries A new surgical procedure is said to be successful 80% of the time. Suppose the operation is performed five times and the results are assumed to be independent of one another. What are the probabilities of these events?

a. All five operations are successful.

b. Exactly four are successful.

c. Less than two are successful.

5.97 Surgery, continued Refer to Exercise 5.96. If less than two operations were successful, how would you feel about the performance of the surgical team?

5.98 Engine Failure Suppose the four engines of a commercial aircraft are arranged to operate independently and that the probability of in-flight failure of a single engine is .01. What is the probability of the following events on a given flight?

a. No failures are observed.

b. No more than one failure is observed.

5.99 McDonald's or Burger King? Suppose that 50% of all young adults prefer McDonald's to Burger King when asked to state a preference. A group of 10 young adults were randomly selected and their preferences recorded.

a. What is the probability that more than 6 preferred McDonald's?

b. What is the probability that between 4 and 6 (inclusive) preferred McDonald's?

c. What is the probability that between 4 and 6 (inclusive) preferred Burger King?

5.100 Vacation Destinations High gas prices may keep some American vacationers closer to home. However, when given a choice of getaway spots, 66% of U.S. leisure travelers indicated that they would like to visit national parks.[16] A random sample of $n = 100$ leisure travelers is selected.

a. What is the average of x, the number of travelers in the sample who indicate they would like to visit national parks? What is the standard deviation of x?

b. Would it be unlikely to find only 50 or fewer of those sampled who indicated they would like to visit national parks?

CASE STUDY **A Mystery: Cancers Near a Reactor**

How safe is it to live near a nuclear reactor? Men who lived in a coastal strip that extends 20 miles north from a nuclear reactor in Plymouth, Massachusetts, developed some forms of cancer at a rate 50% higher than the statewide rate, according to a study endorsed by the Massachusetts Department of Public Health and reported in the May 21, 1987, edition of the *New York Times.*[17]

The cause of the cancers is a mystery, but it was suggested that the cancer was linked to the Pilgrim I reactor, which had been shut down for 13 months because of management problems. Boston Edison, the owner of the reactor, acknowledged radiation releases in the mid-1970s that were just above permissible levels. If the reactor was in fact responsible for the excessive cancer rate, then the currently acknowledged level of radiation required to cause cancer would have to change. However, confounding the mystery was the fact that women in this same area were seemingly unaffected.

In his report, Dr. Sidney Cobb, an epidemiologist, noted the connection between the radiation releases at the Pilgrim I reactor and 52 cases of hematopoietic cancers. The report indicated that this unexpectedly large number might be attributable to airborne radioactive effluents from Pilgrim I, concentrated along the coast by wind patterns and not dissipated, as assumed by government regulators. How unusual was this number of cancer cases? That is, statistically speaking, is 52 a highly improbable number of cases? If the answer is yes, then either some external factor (possibly radiation) caused this unusually large number, or we have observed a very rare event!

The Poisson probability distribution provides a good approximation to the distributions of variables such as the number of deaths in a region due to a rare disease, the number of accidents in a manufacturing plant per month, or the number of airline crashes per month. Therefore, it is reasonable to assume that the Poisson distribution provides an appropriate model for the number of cancer cases in this instance.

1. If the 52 reported cases represented a rate 50% higher than the statewide rate, what is a reasonable estimate of μ, the average number of such cancer cases statewide?

2. Based on your estimate of μ, what is the estimated standard deviation of the number of cancer cases statewide?

3. What is the z-score for the $x = 52$ observed cases of cancer? How do you interpret this z-score in light of the concern about an elevated rate of hematopoietic cancers in this area?

6

The Normal Probability Distribution

Laurence Gough/Shutterstock.com

GENERAL OBJECTIVES

In Chapters 4 and 5, you learned about discrete random variables and their probability distributions. In this chapter, you will learn about continuous random variables and their probability distributions and about one very important continuous random variable—the normal. You will learn how to calculate normal probabilities and, under certain conditions, how to use the normal probability distribution to approximate the binomial probability distribution. Then, in Chapter 7 and in the chapters that follow, you will see how the normal probability distribution plays a central role in statistical inference.

CHAPTER INDEX

NEED TO KNOW...

How to Use Table 3 to Calculate Probabilities under the Standard Normal Curve
How to Calculate Binomial Probabilities Using the Normal Approximation

"Are You Going to Curve the Grades?"

"Curving the grades" doesn't necessarily mean that you will receive a higher grade on a test, although many students would like to think so! Curving actually refers to a method of assigning the letter grades A, B, C, D, or F using fixed proportions of the grades corresponding to each of the letter grades. One such curving technique assumes that the distribution of the grades is approximately normal and uses these proportions.

Letter Grade	A	B	C	D	F
Proportion of grades	10%	20%	40%	20%	10%

In the case study at the end of this chapter, we will examine this and other assigned proportions for curving grades.

PROBABILITY DISTRIBUTIONS FOR CONTINUOUS RANDOM VARIABLES

6.1

When a random variable x is discrete, you can assign a positive probability to each value that x can take and get the probability distribution for x. The sum of all the probabilities associated with the different values of x is 1. However, not all experiments result in random variables that are discrete.

Continuous random variables, such as heights and weights, length of life of a particular product, or experimental laboratory error, can assume the infinitely many values corresponding to points on a line interval. If you try to assign a positive probability to each of these uncountable values, the probabilities will no longer sum to 1, as with discrete random variables. Therefore, you must use a different approach to generate the probability distribution for a continuous random variable.

Suppose you have a set of measurements on a continuous random variable, and you create a relative frequency histogram to describe their distribution. For a small number of measurements, you could use a small number of classes; then as more and more measurements are collected, you can use more classes and reduce the class width. The outline of the histogram will change slightly, for the most part becoming less and less irregular, as shown in Figure 6.1. As the number of measurements becomes very large and the class widths become very narrow, the relative frequency histogram appears more and more like the smooth curve shown in Figure 6.1(d). This smooth curve describes the **probability distribution of the continuous random variable.**

FIGURE 6.1

Relative frequency histograms for increasingly large sample sizes

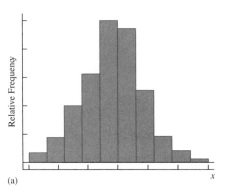

(a)

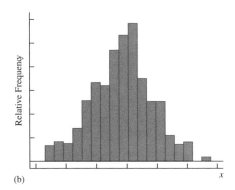

(b)

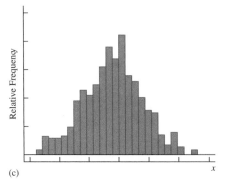

(c)

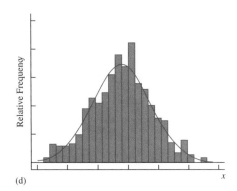

(d)

How can you create a model for this probability distribution? A continuous random variable can take on any of an infinite number of values on the real line, much like the infinite number of grains of sand on a beach. The probability distribution is created by distributing one unit of probability along the line, much as you might distribute a handful of sand. The probability—grains of sand or measurements—will pile up in certain places, and the result is the probability distribution shown in Figure 6.2. The depth or **density** of the probability, which varies with x, may be described by a mathematical formula $f(x)$, called the **probability distribution** or **probability density function** for the random variable x.

FIGURE 6.2

The probability distribution $f(x)$; $P(a < x < b)$ is equal to the shaded area under the curve

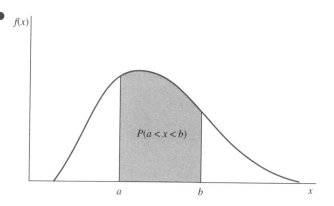

Remember that, for discrete random variables, (1) the sum of all the probabilities $p(x)$ equals 1 and (2) the probability that x falls into a certain interval is the sum of all the probabilities in that interval. Continuous random variables have some parallel characteristics listed next.

NEED A TIP?

For continuous random variables,

area = probability.

- The area under a continuous probability distribution is equal to 1.
- The probability that x will fall into a particular interval—say, from a to b—is equal to the area under the curve between the two points a and b. This is the shaded area in Figure 6.2.

There is also one important difference between discrete and continuous random variables. Consider the probability that x equals some particular value—say, a. Since there is no area above a single point—say, $x = a$—in the probability distribution for a continuous random variable, our definition implies that the probability is 0.

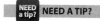

NEED A TIP?

Area under the curve equals 1.

- $P(x = a) = 0$ for continuous random variables.
- This implies that $P(x \geq a) = P(x > a)$ and $P(x \leq a) = P(x < a)$.
- This is *not* true in general for discrete random variables.

How do you choose the model—that is, the probability distribution $f(x)$—appropriate for a given experiment? Many types of continuous curves are available for modeling. Some are mound-shaped, like the one in Figure 6.1(d), but others are not. In general, try to pick a model that meets these criteria:

- It fits the accumulated body of data.
- It allows you to make the best possible inferences using the data.

EXAMPLE **6.1**

The *uniform random variable** is used to model the behavior of a continuous random variable whose values are uniformly or evenly distributed over a given interval. For example, the error x introduced by rounding an observation to the nearest inch would probably have a uniform distribution over the interval from $-.5$ to $.5$. The probability density function $f(x)$ would be "flat" as shown in Figure 6.3. The height of the rectangle is set at 1, so that the total area under the probability distribution is 1.

FIGURE 6.3 ●

A uniform probability distribution

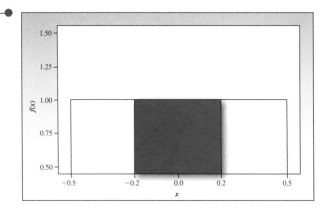

What is the probability that the rounding error is less than $.2$ in magnitude?

Solution   This probability corresponds to the area under the distribution between $x = -.2$ and $x = .2$. Since the height of the rectangle is 1,

$$P(-.2 < x < .2) = [.2 - (-.2)] \times 1 = .4$$

EXAMPLE **6.2**

The *exponential random variable** is used to model continuous random variables such as waiting times or lifetimes associated with electronic components. For example, the waiting time at a supermarket checkout counter has an exponential distribution with an average waiting time of 5 minutes. The probability density function $f(x) = .2e^{-.2x}$ is shown in Figure 6.4. To find areas under this curve, you can use the fact that $P(x > a) = e^{-.2a}$ for $a > 0$. What is the probability that you have to wait 10 minutes or more at the checkout counter?

FIGURE 6.4 ●

An exponential probability distribution

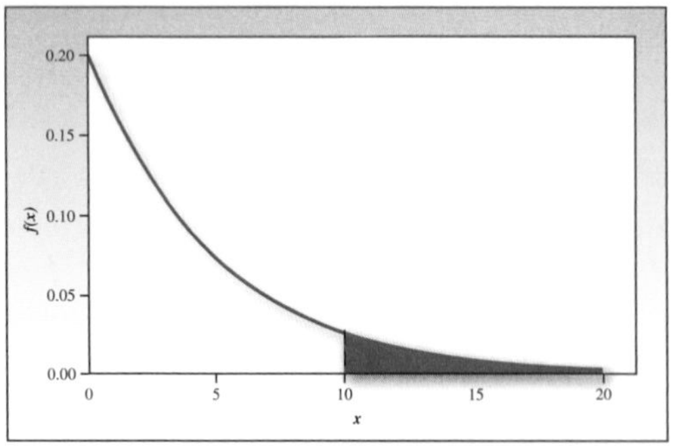

*The probability density function (*pdf*) for the *general uniform random variable x* is $f(x) = 1/(b - a)$ for $a \leq x \leq b$ and with mean $\mu = (a + b)/2$ and $\sigma^2 = (b - a)^2/12$. For the *exponential random variable x*, the *pdf* is $f(x) = \lambda e^{-\lambda x}$ for $x \geq 0$, $\lambda > 0$, with $\mu = 1/\lambda$ and $\sigma^2 = 1/\lambda^2$.

Solution   The probability to be calculated is the area shaded in Figure 6.4. Use the general formula for $P(x > a)$ to find

$$P(x > 10) = e^{-.2(10)} = .135$$

Your model may not always fit the experimental situation perfectly, but you should try to choose a model that *best fits* the population relative frequency histogram. The better the model approximates reality, the better your inferences will be. Fortunately, many continuous random variables have mound-shaped frequency distributions, such as the data in Figure 6.1(d). The **normal probability distribution** provides a good model for describing this type of data.

THE NORMAL PROBABILITY DISTRIBUTION

6.2

Continuous probability distributions can assume a variety of shapes. However, a large number of random variables observed in nature possess a frequency distribution that is approximately mound-shaped or, as the statistician would say, is approximately a normal probability distribution. The formula or probability density function (pdf) that generates this distribution is shown next.

NORMAL PROBABILITY DISTRIBUTION

$$f(x) = \frac{1}{\sigma\sqrt{2\pi}} e^{-(x-\mu)^2/(2\sigma^2)} \qquad -\infty < x < \infty$$

The symbols e and π are mathematical constants given approximately by 2.7183 and 3.1416, respectively; μ and σ ($\sigma > 0$) are parameters that represent the population mean and standard deviation, respectively.

The graph of a normal probability distribution with mean μ and standard deviation σ is shown in Figure 6.5. The mean μ locates the *center* of the distribution, and the distribution is *symmetric* about its mean μ. Since the total area under the normal probability distribution is equal to 1, this symmetry implies that the area to the right of μ is .5 and the area to the left of μ is also .5.

The *shape* of the distribution is determined by σ, the population standard deviation. As you can see in Figure 6.6, large values of σ reduce the height of the curve and increase the spread; small values of σ increase the height of the curve

FIGURE 6.5

Normal probability
distribution

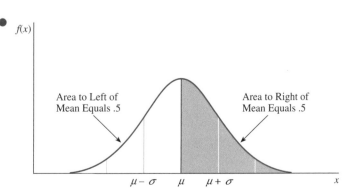

FIGURE 6.6 ●

Normal probability distributions with differing values of μ and σ

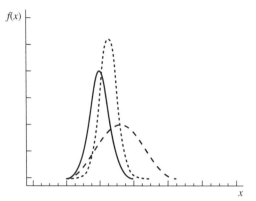

ONLINE APPLET

Visualizing Normal Curves

and reduce the spread. Figure 6.6 shows three normal probability distributions with different means and standard deviations. Notice the differences in shape and location.

You rarely find a variable with values that are infinitely small ($-\infty$) or infinitely large ($+\infty$). Even so, many *positive* random variables (such as heights, weights, and times) have distributions that are well approximated by a normal distribution. According to the Empirical Rule, almost all values of a normal random variable lie in the interval $\mu \pm 3\sigma$. As long as the values within three standard deviations of the mean are *positive,* the normal distribution provides a good model to describe the data.

6.3 TABULATED AREAS OF THE NORMAL PROBABILITY DISTRIBUTION

To find the probability that a normal random variable x lies in the interval from a to b, we need to find the area under the normal curve between the points a and b (see Figure 6.2). However (see Figure 6.6), there are an infinitely large number of normal distributions—one for each different mean and standard deviation. A separate table of areas for each of these curves is obviously impractical. Instead, we use a standardization procedure that allows us to use the same table for all normal distributions.

The Standard Normal Random Variable

A normal random variable x is **standardized** by expressing its value as the number of standard deviations (σ) it lies to the left or right of its mean μ. This is really just a change in the units of measure that we use, as if we were measuring in inches rather than in centimeters! The standardized normal random variable, z, is defined as

$$z = \frac{x - \mu}{\sigma}$$

NEED A TIP?

Area under the *z*-curve equals 1.

or equivalently,

$$x = \mu + z\sigma$$

From the formula for z, we can draw these conclusions:

- When x is less than the mean μ, the value of z is negative.
- When x is greater than the mean μ, the value of z is positive.
- When $x = \mu$, the value of $z = 0$.

The probability distribution for z, shown in Figure 6.7, is called the **standardized normal distribution** because its mean is 0 and its standard deviation is 1. Values of z on the left side of the curve are negative, while values on the right side are positive. The area under the standard normal curve to the left of a specified value of z—say, z_0—is the probability $P(z \leq z_0)$. This **cumulative area** is recorded in Table 3 of Appendix I and is shown as the shaded area in Figure 6.7. An abbreviated version of Table 3 is given in Table 6.1. Notice that the table contains both positive and negative values of z. The left-hand column of the table gives the value of z correct to the tenths place; the second decimal place for z, corresponding to hundredths, is given across the top row.

FIGURE 6.7

Standardized normal distribution

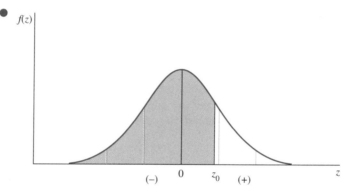

Abbreviated Version of Table 3 in Appendix I

TABLE 6.1

Table 3. Areas Under the Normal Curve

z	.00	.01	.02	.03	...	.09
−3.4	.0003	.0003	.0003	.0003		
−3.3	.0005	.0005	.0005	.0004		
−3.2	.0007	.0007	.0006	.0006		
−3.1	.0010	.0009	.0009	.0009		
−3.0	.0013	.0013	.0013	.0012	...	.0010
−2.9	.0019	.	.	.		
−2.8	.0026	.	.	.		
−2.7	.0035	.	.	.		
−2.6	.0047					
−2.5	.0062					
.	.					
.	.					
.	.					
−2.0	.0228					
.	.					
.	.					
.	.					

Table 3. Areas Under the Normal Curve (*continued*)

z	.00	.01	.02	.03	.04	. . .	.09
0.0	.5000	.5040	.5080	.5120	.5160		
0.1	.5398	.5438	.5478	.5517	.5557		
0.2	.5793	.5832	.5871	.5910	.5948		
0.3	.6179	.6217	.6255	.6293	.6331		
0.4	.6554	.6591	.6628	.6664	.6700	. . .	.6879
0.5	.6915	.	.	.			
0.6	.7257	.	.	.			
0.7	.7580	.	.	.			
0.8	.7881						
0.9	.8159						
.	.						
.	.						
2.0	.9772						

EXAMPLE 6.3 Find $P(z \leq 1.63)$. This probability corresponds to the area to the left of a point $z = 1.63$ standard deviations to the right of the mean (see Figure 6.8).

NEED A TIP?

$P(z \leq 1.63) = P(z < 1.63)$

Solution The area is shaded in Figure 6.8. Since Table 3 in Appendix I gives areas under the normal curve to the left of a specified value of z, you simply need to find the tabled value for $z = 1.63$. Proceed down the left-hand column of the table to $z = 1.6$ and across the top of the table to the column marked .03. The intersection of this row and column combination gives the area .9484, which is $P(z \leq 1.63)$.

FIGURE 6.8

Area under the standard normal curve for Example 6.3

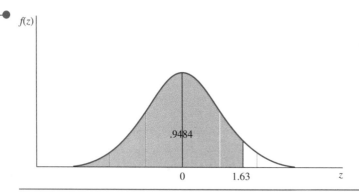

Areas to the left of $z = 0$ are found using negative values of z.

EXAMPLE 6.4 Find $P(z \geq -.5)$. This probability corresponds to the area to the *right* of a point $z = -.5$ standard deviation to the left of the mean (see Figure 6.9).

FIGURE 6.9

Area under the standard normal curve for Example 6.4

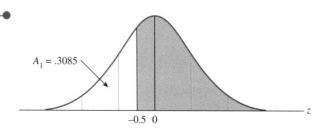

Solution   The area given in Table 3 in Appendix I is the area to the left of a specified value of z. Indexing $z = -.5$ in Table 3, we can find the area A_1 to the *left* of $-.5$ to be .3085.

Since the area under the curve is 1, we find

$$P(z \geq -.5) = 1 - A_1 = 1 - .3085 = .6915.$$

EXAMPLE 6.5

Find $P(-.5 \leq z \leq 1.0)$. This probability is the area between $z = -.5$ and $z = 1.0$, as shown in Figure 6.10.

FIGURE 6.10

Area under the standard normal curve for Example 6.5

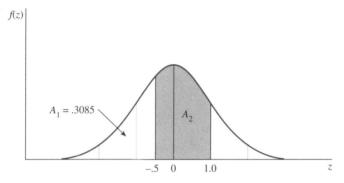

Solution   The area required is the shaded area A_2 in Figure 6.10. From Table 3 in Appendix I, you can find the area to the left of $z = -.5$ ($A_1 = .3085$) and the area to the left of $z = 1.0$ ($A_1 + A_2 = .8413$). To find the area marked A_2, we subtract the two entries:

$$A_2 = (A_1 + A_2) - A_1 = .8413 - .3085 = .5328$$

That is, $P(-.5 \leq z \leq 1.0) = .5328$.

 NEED TO KNOW...

How to Use Table 3 to Calculate Probabilities under the Standard Normal Curve

- To calculate the area to the left of a z-value, find the area directly from Table 3.
- To calculate the area to the right of a z-value, find the area in Table 3, and subtract from 1.
- To calculate the area between two values of z, find the two areas in Table 3, and subtract one area from the other.

EXAMPLE 6.6

Find the probability that a normally distributed random variable will fall within these ranges:

1. One standard deviation of its mean
2. Two standard deviations of its mean

Solution

1. Since the standard normal random variable z measures the distance from the mean in units of standard deviations, you need to find

$$P(-1 \le z \le 1) = .8413 - .1587 = .6826$$

Remember that you calculate the area between two z-values by subtracting the tabled entries for the two values.

2. As in part 1, $P(-2 \le z \le 2) = .9772 - .0228 = .9544$.

These probabilities agree with the approximate values of 68% and 95% in the Empirical Rule from Chapter 2.

EXAMPLE 6.7

Find the value of z—say z_0—such that .95 of the area is within $\pm z_0$ standard deviations of the mean.

 NEED A TIP?

We know the area. Work from the inside of the table out.

Solution The shaded area in Figure 6.11 is the area within $\pm z_0$ standard deviations of the mean, which needs to be equal to .95. The "tail areas" under the curve are not shaded, and have a combined area of $1 - .95 = .05$. Because of the symmetry of the normal curve, these two tail areas have the same area, so that $A_1 = .05/2 = .025$ in Figure 6.11. Thus, the entire *cumulative area* to the left of $-z_0$ equals $A_1 = .025$. This area is found in the interior of Table 3 in Appendix I in the row corresponding to $z = -1.9$ and the .06 column. Hence, $-z_0 = -1.96$ or $z_0 = 1.96$. Note that this result is very close to the approximate value, $z = 2$, used in the Empirical Rule.

FIGURE 6.11

Area under the standard normal curve for Example 6.7

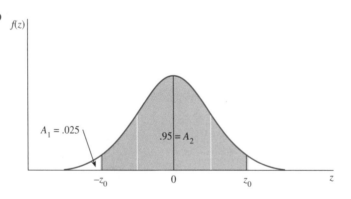

Calculating Probabilities for a General Normal Random Variable

Most of the time, the probabilities you are interested in will involve x, a normal random variable with mean μ and standard deviation σ. You must then *standardize* the interval of interest, writing it as the equivalent interval in terms of z, the standard normal random variable. Once this is done, the probability of interest is the area that you find using the *standard normal probability distribution*.

EXAMPLE **6.8**

Let x be a normally distributed random variable with a mean of 10 and a standard deviation of 2. Find the probability that x lies between 11 and 13.6.

Solution   The interval from $x = 11$ to $x = 13.6$ must be standardized using the formula for z. When $x = 11$,

NEED A TIP?

Always draw a picture—it helps!

$$z = \frac{x - \mu}{\sigma} = \frac{11 - 10}{2} = .5$$

and when $x = 13.6$,

$$z = \frac{x - \mu}{\sigma} = \frac{13.6 - 10}{2} = 1.8$$

The desired probability is therefore $P(.5 \leq z \leq 1.8)$, the area lying between $z = .5$ and $z = 1.8$, as shown in Figure 6.12. From Table 3 in Appendix I, you find that the area to the left of $z = .5$ is .6915, and the area to the left of $z = 1.8$ is .9641. The desired probability is the difference between these two probabilities, or

$$P(.5 \leq z \leq 1.8) = .9641 - .6915 = .2726$$

FIGURE 6.12

Area under the standard normal curve for Example 6.8

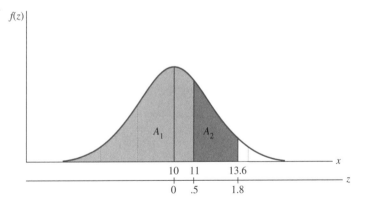

ONLINE APPLET

Normal Probability Distributions & Normal Probabilities and z-Scores?

EXAMPLE **6.9**

Studies show that gasoline use for compact cars sold in the United States is normally distributed, with a mean of 35.5 miles per gallon (mpg) and a standard deviation of 4.5 mpg. What percentage of compacts get 40 mpg or more?

Solution   The proportion of compacts that get 40 mpg or more is given by the shaded area in Figure 6.13. To solve this problem, you must first find the z-value corresponding to $x = 40$ by calculating.

$$z = \frac{x - \mu}{\sigma} = \frac{40 - 35.5}{4.5} = 1.0$$

From Table 3 in Appendix I, the area A_1 to the left of $z = 1.0$ is .8413. Then the proportion of compacts that get 40 mpg or more is equal to

$$P(x \geq 40) = 1 - P(z < 1) = 1 - .8413 = .1587$$

The percentage exceeding 40 mpg is

$$100(.1587) = 15.87\%$$

FIGURE 6.13

Area under the standard normal curve for Example 6.9

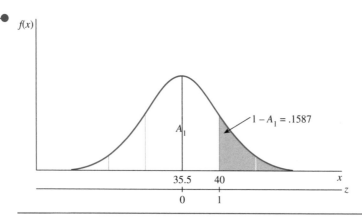

EXAMPLE 6.10 Refer to Example 6.9. In times of scarce energy resources, a competitive advantage is given to an automobile manufacturer who can produce a car that has substantially better fuel economy than the competitors' cars. If a manufacturer wishes to develop a compact car that outperforms 95% of the current compacts in fuel economy, what must the gasoline use rate for the new car be?

Solution The gasoline use rate x has a normal distribution with a mean of 35.5 mpg and a standard deviation of 4.5 mpg. You need to find a particular value—say, x_0—such that

$$P(x \leq x_0) = .95$$

This is the 95th percentile of the distribution of gasoline use rate x. Since the only information you have about normal probabilities is in terms of the standard normal random variable z, start by standardizing the value of x_0:

$$z_0 = \frac{x_0 - 35.5}{4.5}$$

Since the value of z_0 corresponds to x_0, it must *also* have area .95 to its left, as shown in Figure 6.14. If you look in the interior of Table 3 in Appendix I, you will find that the area .9500 is exactly halfway between the areas for $z = 1.64$ and $z = 1.65$. Thus, we take z_0 to be halfway between 1.64 and 1.65, or

$$z_0 = \frac{x_0 - 35.5}{4.5} = 1.645$$

Solving for x_0, you obtain

$$x_0 = \mu + z_0\sigma = 35.5 + (1.645)(4.5) = 42.9$$

FIGURE 6.14

Area under the standard normal curve for Example 6.10

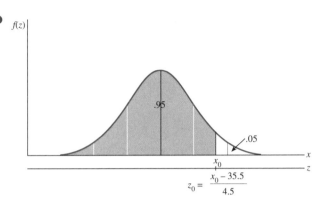

The manufacturer's new compact car must therefore get 42.9 mpg to outperform 95% of the compact cars currently available on the U.S. market.

6.3 EXERCISES

BASIC TECHNIQUES

6.1 Consider a standard normal random variable with $\mu = 0$ and standard deviation $\sigma = 1$. Use Table 3 to find the following probabilities:

a. $P(z < 2)$ **b.** $P(z > 1.16)$
c. $P(-2.33 < z < 2.33)$ **d.** $P(z < 1.88)$

6.2 Find these probabilities associated with the standard normal random variable z:

a. $P(z > 5)$ **b.** $P(-3 < z < 3)$
c. $P(z < 2.81)$ **d.** $P(z > 2.81)$

6.3 Calculate the area under the standard normal curve to the left of these values:

a. $z = 1.6$ **b.** $z = 1.83$
c. $z = .90$ **d.** $z = 4.18$

6.4 Find these probabilities for the standard normal random variable z:

a. $P(z < 2.33)$ **b.** $P(z < 1.645)$
c. $P(z > 1.96)$ **d.** $P(-2.58 < z < 2.58)$

6.5 Find the following probabilities for the standard normal random variable z:

a. $P(-1.43 < z < .68)$ **b.** $P(.58 < z < 1.74)$
c. $P(-1.55 < z < -.44)$ **d.** $P(z > 1.34)$
e. $P(z < -4.32)$

6.6 Calculate the area under the standard normal curve between these values:

a. $z = -1.4$ and $z = 1.4$ **b.** $z = -3.0$ and $z = 3.0$

6.7 a. Find a z_0 such that $P(z > z_0) = .025$.
b. Find a z_0 such that $P(z < z_0) = .9251$.

6.8 Find a z_0 such that $P(-z_0 < z < z_0) = .8262$.

6.9 a. Find a z_0 that has area .9505 to its left.
b. Find a z_0 that has area .05 to its left.

6.10 a. Find a z_0 such that $P(-z_0 < z < z_0) = .90$.
b. Find a z_0 such that $P(-z_0 < z < z_0) = .99$.

6.11 Find the following *percentiles* for the standard normal random variable z:

a. 90th percentile **b.** 95th percentile
c. 98th percentile **d.** 99th percentile

6.12 A normal random variable x has mean $\mu = 10$ and standard deviation $\sigma = 2$. Find the probability associated with each of the following intervals.

a. $x > 13.5$ **b.** $x < 8.2$ **c.** $9.4 < x < 10.6$

6.13 A normal random variable x has mean $\mu = 1.2$ and standard deviation $\sigma = .15$. Find the probability associated with each of the following intervals.

a. $1.00 < x < 1.10$ **b.** $x > 1.38$
c. $1.35 < x < 1.50$

6.14 A normal random variable x has an unknown mean μ and standard deviation $\sigma = 2$. If the probability that x exceeds 7.5 is .8023, find μ.

6.15 A normal random variable x has mean 35 and standard deviation 10. Find a value of x that has area

.01 to its right. This is the *99th percentile* of this normal distribution.

6.16 A normal random variable x has mean 50 and standard deviation 15. Would it be unusual to observe the value $x = 0$? Explain your answer.

6.17 A normal random variable x has an unknown mean and standard deviation. The probability that x exceeds 4 is .9772, and the probability that x exceeds 5 is .9332. Find μ and σ.

APPLICATIONS

6.18 Hamburger Meat The meat department at a local supermarket specifically prepares its "1-pound" packages of ground beef so that there will be a variety of weights, some slightly more and some slightly less than 1 pound. Suppose that the weights of these "1-pound" packages are normally distributed with a mean of 1.00 pound and a standard deviation of .15 pound.

a. What proportion of the packages will weigh more than 1 pound?

b. What proportion of the packages will weigh between .95 and 1.05 pounds?

c. What is the probability that a randomly selected package of ground beef will weigh less than .80 pound?

d. Would it be unusual to find a package of ground beef that weighs 1.45 pounds? How would you explain such a large package?

6.19 Human Heights Human heights are one of many biological random variables that can be modeled by the normal distribution. Assume that the heights of American men have a mean of 69.5 inches and a standard deviation of 3.5 inches.

a. What proportion of all men will be taller than 6'0"? (HINT: Convert the measurements to inches.)

b. What is the probability that a randomly selected man will be between 5'8" and 6'1" tall?

c. President Barack Obama is 6'1". Is this an unusual height?

d. Of the 43 presidents elected from 1789 through 2008, 18 were 6'0" or taller.[1] Would you consider this to be unusual, given the proportion found in part a?

6.20 Christmas Trees The diameters of Douglas firs grown at a Christmas tree farm are normally distributed with a mean of 4 inches and a standard deviation of 1.5 inches.

a. What proportion of the trees will have diameters between 3 and 5 inches?

b. What proportion of the trees will have diameters less than 3 inches?

c. Your Christmas tree stand will expand to a diameter of 6 inches. What proportion of the trees will not fit in your Christmas tree stand?

6.21 Cerebral Blood Flow Cerebral blood flow (CBF) in the brains of healthy people is normally distributed with a mean of 74 and a standard deviation of 16.

a. What proportion of healthy people will have CBF readings between 60 and 80?

b. What proportion of healthy people will have CBF readings above 100?

c. If a person has a CBF reading below 40, he is classified as at risk for a stroke. What proportion of healthy people will mistakenly be diagnosed as "at risk"?

6.22 Braking Distances For a car traveling 30 miles per hour (mph), the distance required to brake to a stop is normally distributed with a mean of 50 feet and a standard deviation of 8 feet. Suppose you are traveling 30 mph in a residential area and a car moves abruptly into your path at a distance of 60 feet.

a. If you apply your brakes, what is the probability that you will brake to a stop within 40 feet or less? Within 50 feet or less?

b. If the only way to avoid a collision is to brake to a stop, what is the probability that you will avoid the collision?

6.23 Elevator Capacities Suppose that you must establish regulations concerning the maximum number of people who can occupy an elevator. A study indicates that if eight people occupy the elevator, the probability distribution of the total weight of the eight people is approximately normally distributed with a mean equal to 1200 pounds and a standard deviation of 99 pounds. What is the probability that the total weight of eight people exceeds 1300 pounds? 1500 pounds?

6.24 A Phosphate Mine The discharge of suspended solids from a phosphate mine is normally distributed, with a mean daily discharge of 27 milligrams per liter (mg/l) and a standard deviation of 14 mg/l. On what proportion of days will the daily discharge exceed 50 mg/l?

6.25 Sunflowers An experimenter publishing in the *Annals of Botany* investigated whether the stem diameters of the dicot sunflower would change depending on whether the plant was left to sway freely in the wind or was artificially supported.[2] Suppose that the unsupported stem diameters at the base of a particular species of sunflower plant have a normal distribution with an average diameter of 35 millimeters (mm) and a standard deviation of 3 mm.

a. What is the probability that a sunflower plant will have a basal diameter of more than 40 mm?

b. If two sunflower plants are randomly selected, what is the probability that both plants will have a basal diameter of more than 40 mm?

c. Within what limits would you expect the basal diameters to lie, with probability .95?

d. What diameter represents the 90th percentile of the distribution of diameters?

6.26 Loading Grain A grain loader can be set to discharge grain in amounts that are normally distributed, with mean μ bushels and standard deviation equal to 25.7 bushels. If a company wishes to use the loader to fill containers that hold 2000 bushels of grain and wants to overfill only one container in 100, at what value of μ should the company set the loader?

6.27 Economic Forecasts One method of arriving at economic forecasts is to use a consensus approach. A forecast is obtained from each of a large number of analysts, and the average of these individual forecasts is the consensus forecast. Suppose the individual 2013 January prime interest rate forecasts of economic analysts are approximately normally distributed with the mean equal to 4.75% and a standard deviation equal to 0.2%. If a single analyst is randomly selected from among this group, what is the probability that the analyst's forecast of the prime rate will take on these values?

a. Exceed 4.25%

b. Be less than 4.375%

6.28 Tax Audits How does the IRS decide on the percentage of income tax returns to audit for each state? Suppose they do it by randomly selecting 50 values from a normal distribution with a mean equal to 1.55% and a standard deviation equal to .45%. (Computer programs are available for this type of sampling.)

a. What is the probability that a particular state will have more than 2.5% of its income tax returns audited?

b. What is the probability that a state will have less than 1% of its income tax returns audited?

6.29 Bacteria in Drinking Water Suppose the numbers of a particular type of bacteria in samples of 1 milliliter (ml) of drinking water tend to be approximately normally distributed, with a mean of 85 and a standard deviation of 9. What is the probability that a given 1-ml sample will contain more than 100 bacteria?

6.30 Tennis Anyone? A stringer of tennis rackets has found that the actual string tension achieved for any individual racket will vary as much as 6 pounds per square inch from the desired tension set on the stringing machine. If the stringer wishes to string at a tension lower than that specified by a customer only 5% of the time, how much above or below the customer's specified tension should the stringer set the stringing machine? (NOTE: Assume that the distribution of string tensions produced by the stringing machine is normally distributed, with a mean equal to the tension set on the machine and a standard deviation equal to 2 pounds per square inch.)

6.31 How Many Words? A publisher has discovered that the number of words contained in new manuscripts is normally distributed, with a mean equal to 20,000 words in excess of that specified in the author's contract and a standard deviation of 10,000 words. If the publisher wants to be almost certain (say, with a probability of .95) that the manuscript will have less than 100,000 words, what number of words should the publisher specify in the contract?

6.32 Breathing Rates The number of times x an adult human breathes per minute when at rest has a probability distribution that is approximately normal, with the mean equal to 16 and the standard deviation equal to 4. If a person is selected at random and the number x of breaths per minute while at rest is recorded, what is the probability that x will exceed 22?

6.33 Mall Rats An article in *American Demographics* claims that more than twice as many shoppers are out shopping on the weekends than during the week.[3] Not only that, such shoppers also spend more money on their purchases on Saturdays and Sundays! Suppose that the amount of money spent at shopping centers between 4 P.M. and 6 P.M. on Sundays has a normal distribution with mean $85 and with a standard deviation of $20. A shopper is randomly selected on a Sunday between 4 P.M. and 6 P.M. and asked about his spending patterns.

a. What is the probability that he has spent more than $95 at the mall?

b. What is the probability that he has spent between $95 and $115 at the mall?

c. If two shoppers are randomly selected, what is the probability that both shoppers have spent more than $115 at the mall?

6.34 Pulse Rates What's a *normal* pulse rate? That depends on a variety of factors. Pulse rates between 60 and 100 beats per minute are considered normal for children over 10 and adults.[4] Suppose that these pulse

rates are approximately normally distributed with a mean of 78 and a standard deviation of 12.

a. What proportion of adults will have pulse rates between 60 and 100?

b. What is the 95th percentile for the pulse rates of adults?

c. Would a pulse rate of 110 be considered unusual? Explain.

THE NORMAL APPROXIMATION TO THE BINOMIAL PROBABILITY DISTRIBUTION (OPTIONAL)

In Chapter 5, you learned three ways to calculate probabilities for the binomial random variable x:

- Using the binomial formula, $P(x = k) = C_k^n p^k q^{n-k}$
- Using the cumulative binomial tables
- Using *MS Excel* and *MINITAB*

The binomial formula produces lengthy calculations, and the tables are available for only certain values of n and p. There is another option available when $np < 7$; the Poisson probabilities can be used to approximate $P(x = k)$. When this approximation *does not work* and n is large, the normal probability distribution provides another approximation for binomial probabilities.

> ## THE NORMAL APPROXIMATION TO THE BINOMIAL PROBABILITY DISTRIBUTION
>
> Let x be a binomial random variable with n trials and probability p of success. The probability distribution of x is approximated using a normal curve with
>
> $$\mu = np \quad \text{and} \quad \sigma = \sqrt{npq}$$
>
> This approximation is adequate as long as n is large and p is not too close to 0 or 1.

Since the normal distribution is continuous, the area under the curve at any single point is equal to 0. Keep in mind that this result applies only to continuous random variables. Because the binomial random variable x is a discrete random variable, the probability that x takes some specific value—say, $x = 11$—will not necessarily equal 0.

Figures 6.15 and 6.16 show the binomial probability histograms for $n = 25$ with $p = .5$ and $p = .1$, respectively. The distribution in Figure 6.15 is exactly symmetric. If you superimpose a normal curve with the same mean, $\mu = np$, and the same standard deviation, $\sigma = \sqrt{npq}$, over the top of the bars, it "fits" quite well; that is, the

FIGURE 6.15

The binomial probability distribution for $n = 25$ and $p = .5$ and the approximating normal distribution with $\mu = 12.5$ and $\sigma = 2.5$

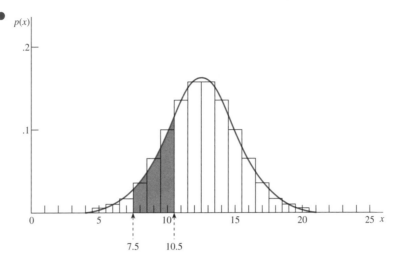

FIGURE 6.16

The binomial probability distribution and the approximating normal distribution for $n = 25$ and $p = .1$

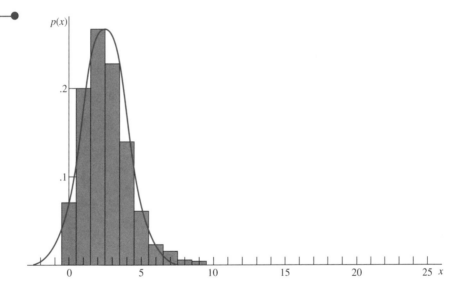

areas under the curve are almost the same as the areas under the bars. However, when the probability of success, p, gets small and the distribution is skewed, as in Figure 6.16, the symmetric normal curve no longer fits very well. If you try to use the normal curve areas to approximate the area under the bars, your approximation will not be very good.

EXAMPLE   6.11   Use the normal curve to approximate the probability that $x = 8$, 9, or 10 for a binomial random variable with $n = 25$ and $p = .5$. Compare this approximation to the exact binomial probability.

Solution   You can find the exact binomial probability for this example because there are cumulative binomial tables for $n = 25$. From Table 1 in Appendix I,

$$P(x = 8, 9, \text{ or } 10) = P(x \leq 10) - P(x \leq 7) = .212 - .022 = .190$$

To use the normal approximation, first find the appropriate mean and standard deviation for the normal curve:

$$\mu = np = 25(.5) = 12.5$$

$$\sigma = \sqrt{npq} = \sqrt{25(.5)(.5)} = 2.5$$

NEED A TIP?

Only use the continuity correction if *x* has a **binomial** distribution!

The probability that you need corresponds to the area of the three rectangles lying over $x = 8$, 9, and 10. The equivalent area under the normal curve lies between $x = 7.5$ (the lower edge of the rectangle for $x = 8$) and $x = 10.5$ (the upper edge of the rectangle for $x = 10$). This area is shaded in Figure 6.15.

To find the normal probability, follow the procedures of Section 6.3. First you standardize each interval endpoint:

$$z = \frac{x - \mu}{\sigma} = \frac{7.5 - 12.5}{2.5} = -2.0$$

$$z = \frac{x - \mu}{\sigma} = \frac{10.5 - 12.5}{2.5} = -.8$$

Then the approximate probability (shaded in Figure 6.17) is found from Table 3 in Appendix I:

$$P(-2.0 < z < -.8) = .2119 - .0228 = .1891$$

You can compare the approximation, .1891, to the actual probability, .190. They are quite close!

FIGURE 6.17

Area under the normal curve for Example 6.11

ONLINE APPLET

Normal Approximation to Binomial Probabilities

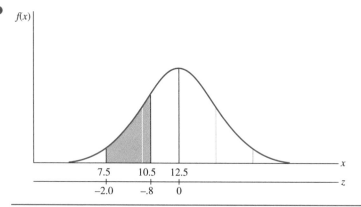

You must be careful not to exclude half of the two extreme probability rectangles when you use the normal approximation to the binomial probability distribution. This adjustment, called the **continuity correction,** helps account for the fact that you are approximating a *discrete random variable* with a *continuous* one. If you forget the correction, your approximation will not be very good! Use this correction only for *binomial probabilities;* do not try to use it when the random variable is already continuous, such as a height or weight.

How can you tell when it is appropriate to use the normal approximation to binomial probabilities? The normal approximation works well when the binomial histogram is roughly symmetric. This happens when the binomial distribution is not "bunched up" near 0 or n—that is, when it can spread out at least two standard deviations from its mean without exceeding its limits, 0 and n. Using this criterion, you can derive this simple rule of thumb:

RULE OF THUMB

The normal approximation to the binomial probabilities will be adequate if both

$np > 5$ and $nq > 5$

NEED TO KNOW...

How to Calculate Binomial Probabilities Using the Normal Approximation

- Find the necessary values of n and p. Calculate $\mu = np$ and $\sigma = \sqrt{npq}$.
- Write the probability you need in terms of x and locate the appropriate area on the curve.
- Correct the value of x by $\pm.5$ to include the entire block of probability for that value. This is the *continuity correction.*
- Convert the necessary x-values to z-values using

$$z = \frac{x \pm .5 - np}{\sqrt{npq}}$$

- Use Table 3 in Appendix I to calculate the approximate probability.

EXAMPLE **6.12**

The reliability of an electrical fuse is the probability that a fuse, chosen at random from production, will function under its designed conditions. A random sample of 1000 fuses was tested and $x = 27$ defectives were observed. Calculate the approximate probability of observing 27 or more defectives, assuming that the fuse reliability is .98.

Solution   The probability of observing a defective when a single fuse is tested is $p = .02$, given that the fuse reliability is .98. Then

$$\mu = np = 1000(.02) = 20$$
$$\sigma = \sqrt{npq} = \sqrt{1000(.02)(.98)} = 4.43$$

The probability of 27 or more defective fuses, given $n = 1000$, is

$$P(x \geq 27) = p(27) + p(28) + p(29) + \cdots + p(999) + p(1000)$$

It is appropriate to use the normal approximation to the binomial probability because

$$np = 1000(.02) = 20 \quad \text{and} \quad nq = 1000(.98) = 980$$

 NEED A TIP?

If np and nq are both greater than 5, you can use the normal approximation.

are both greater than 5. The normal area used to approximate $P(x \geq 27)$ is the area under the normal curve to the right of 26.5, so that the entire rectangle for $x = 27$ is included. Then, the z-value corresponding to $x = 26.5$ is

$$z = \frac{x - \mu}{\sigma} = \frac{26.5 - 20}{4.43} = \frac{6.5}{4.43} = 1.47$$

and the area to the left of $z = 1.47$ is equal to .9292, as shown in Figure 6.18. Since the total area under the curve is 1, you have

$$P(x \geq 27) \approx P(z \geq 1.47) = 1 - .9292 = .0708$$

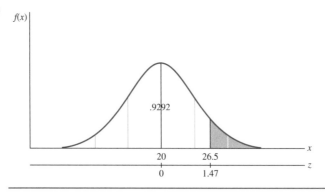

EXAMPLE 6.13

A producer of soft drinks was fairly certain that her brand had a 10% share of the soft drink market. In a market survey involving 2500 consumers of soft drinks, $x = 211$ expressed a preference for her brand. If the 10% figure is correct, find the probability of observing 211 or fewer consumers who prefer her brand of soft drink.

Solution If the producer is correct, then the probability that a consumer prefers her brand of soft drink is $p = .10$. Calculate

$$\mu = np = 2500(.10) = 250$$
$$\sigma = \sqrt{npq} = \sqrt{2500(.10)(.90)} = 15$$

The probability of observing 211 or fewer who prefer her brand is

$$P(x \leq 211) = p(0) + p(1) + \cdots + p(210) + p(211)$$

The normal approximation to this probability is the area to the left of 211.5 under a normal curve with a mean of 250 and a standard deviation of 15. First calculate

$$z = \frac{x - \mu}{\sigma} = \frac{211.5 - 250}{15} = -2.57$$

Then

$$P(x \leq 211) \approx P(z < -2.57) = .0051$$

The probability of observing a sample value of 211 or less when $p = .10$ is so small that you can conclude that one of two things has occurred: Either you have observed an unusual sample even though really $p = .10$, *or* the sample reflects that the actual value of p is less than .10 and perhaps closer to the observed sample proportion, $211/2500 = .08$.

6.4 EXERCISES

BASIC TECHNIQUES

6.35 Consider a binomial random variable x with $n = 25$ and $p = .6$.

a. Can the normal approximation be used to approximate probabilities in this case? Why or why not?

b. What are the mean and standard deviation of x?

c. Using the correction for continuity, approximate $P(x > 9)$.

6.36 Consider a binomial random variable x with $n = 45$ and $p = .05$.

a. Are np and nq both larger than 5?

b. Based on your answer to part a, can we use the normal approximation to approximate the binomial probabilities associated with x? If not, is there another possible approximation we could use?

6.37 Let x be a binomial random variable with $n = 25$ and $p = .3$.

a. Is the normal approximation appropriate for this binomial random variable?

b. Find the mean and standard deviation for x.

c. Use the normal approximation to find $P(6 \leq x \leq 9)$.

d. Use Table 1 in Appendix I to find the exact probability $P(6 \leq x \leq 9)$. Compare the results of parts c and d. How close was your approximation?

6.38 Let x be a binomial random variable with $n = 15$ and $p = .5$.

a. Is the normal approximation appropriate?

b. Find $P(x \geq 6)$ using the normal approximation.

c. Find $P(x > 6)$ using the normal approximation.

d. Find the exact probabilities for parts b and c, and compare these with your approximations.

6.39 Let x be a binomial random variable with $n = 100$ and $p = .2$. Find approximations to these probabilities:

a. $P(x > 22)$ **b.** $P(x \geq 22)$

c. $P(20 < x < 25)$ **d.** $P(x \leq 25)$

6.40 Let x be a binomial random variable with $n = 25$, $p = .2$.

a. Use Table 1 in Appendix I to calculate $P(4 \leq x \leq 6)$.

b. Find μ and σ for the binomial probability distribution, and use the normal distribution to approximate the probability $P(4 \leq x \leq 6)$. Note that this value

is a good approximation to the exact value of $P(4 \leq x \leq 6)$ even though $np = 5$.

6.41 Suppose the random variable x has a binomial distribution corresponding to $n = 20$ and $p = .30$. Use Table 1 of Appendix I to calculate these probabilities:

a. $P(x = 5)$ **b.** $P(x \geq 7)$

6.42 Refer to Exercise 6.41. Use the normal approximation to calculate $P(x = 5)$ and $P(x \geq 7)$. Compare with the exact values obtained from Table 1 in Appendix I.

6.43 Consider a binomial experiment with $n = 20$ and $p = .4$. Calculate $P(x \geq 10)$ using each of these methods:

a. Table 1 in Appendix I

b. The normal approximation to the binomial probability distribution

6.44 Find the normal approximation to $P(355 \leq x \leq 360)$ for a binomial probability distribution with $n = 400$ and $p = .9$.

APPLICATIONS

6.45 How Can I Help? Are you helping to save the environment? A *USA Today* Snapshot found that about 78% of Americans believe that recycling trash makes the biggest difference in protecting the environment.[5]

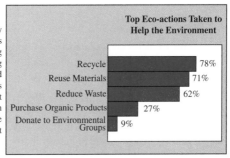

Suppose a random sample of $n = 50$ adults are polled and asked if they believed that recycling made the biggest difference in protecting our environment. Let us assume that the 78% figure is, in fact, correct. What are the probabilities for the following events?

a. Fewer than 30 individuals believe that recycling makes the biggest difference?

b. More than 40 individuals believe that recycling makes the biggest difference?

c. More than 10 individuals believe that recycling *does not* make the biggest difference?

6.46 Genetic Defects Data collected over a long period of time show that a particular genetic defect occurs in 1 of every 1000 children. The records of a medical clinic show $x = 60$ children with the defect in a total of 50,000 examined. If the 50,000 children were a random sample from the population of children represented by past records, what is the probability of observing a value of x equal to 60 or more? Would you say that the observation of $x = 60$ children with genetic defects represents a rare event?

6.47 No Shows Airlines and hotels often grant reservations in excess of capacity to minimize losses due to no-shows. Suppose the records of a hotel show that, on the average, 10% of their prospective guests will not claim their reservation. If the hotel accepts 215 reservations and there are only 200 rooms in the hotel, what is the probability that all guests who arrive to claim a room will receive one?

6.48 Lung Cancer Compilation of large masses of data on lung cancer shows that approximately 1 of every 40 adults acquires the disease. Workers in a certain occupation are known to work in an air-polluted environment that may cause an increased rate of lung cancer. A random sample of $n = 400$ workers shows 19 with identifiable cases of lung cancer. Do the data provide sufficient evidence to indicate a higher rate of lung cancer for these workers than for the national average?

6.49 Tall or Short? Do Americans tend to vote for the taller of the two major candidates in a presidential election? In 49 of our presidential elections for which the heights of all the major-party candidates are known, 26 of the winners were taller than their opponents.[1] Assume that Americans are not biased by a candidate's height and that the winner is just as likely to be taller or shorter than his opponent.

a. Is the observed number of taller winners in the U.S. presidential election unusual? Find the approximate probability of finding 26 or more of the 49 pairs in which the taller candidate wins.

b. Based on your answer to part a, can you conclude that Americans might consider a candidate's height when casting their ballot?

6.50 The Rh Factor In a certain population, 15% of the people have Rh-negative blood. A blood bank serving this population receives 92 blood donors on a particular day.

a. What is the probability that 10 or fewer are Rh-negative?

b. What is the probability that 15 to 20 (inclusive) of the donors are Rh-negative?

c. What is the probability that more than 80 of the donors are Rh-positive?

6.51 Pepsi's Market Share Two of the biggest soft drink rivals, Pepsi and Coke, are very concerned about their market shares. The pie chart that follows claims that PepsiCo's share of the beverage market is 25%.[6] Assume that this proportion will be close to the probability that a person selected at random indicates a preference for a Pepsi product when choosing a soft drink.

U.S. Refreshment Beverage Market Share

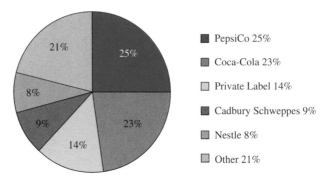

A group of $n = 500$ consumers is selected and the number preferring a Pepsi product is recorded. Use the normal curve to approximate the following binomial probabilities.

a. Exactly 150 consumers prefer a Pepsi product.

b. Between 120 and 150 consumers (inclusive) prefer a Pepsi product.

c. Fewer than 150 consumers prefer a Pepsi product.

d. Would it be unusual to find that 232 of the 500 consumers preferred a Pepsi product? If this were to occur, what conclusions would you draw?

6.52 Ready, Set, Relax! In a study conducted for the Center for a New American Dream, *Time* magazine reports that 60% of Americans felt pressure to work too much, and 80% wished for more family time.[7] Assume that these percentages are correct for all

Americans, and that a random sample of 25 Americans is selected.

a. Use Table 1 in Appendix I to find the probability that more than 20 felt pressure to work too much.

b. Use the normal approximation to the binomial distribution to aproximate the probability in part a. Compare your answer with the exact value from part a.

c. Use Table 1 in Appendix I to find the probability that between 15 and 20 (inclusive) wished for more family time.

d. Use the normal approximation to the binomial distribution to approximate the probability in part c. Compare your answer with the exact value from part c.

6.53 We said, "Relax!" The article in *Time* magazine[7] (Exercise 6.52) also reported that 80% of men and 62% of women put in more than 40 hours a week on the job. Assume that these percentages are correct for all Americans, and that a random sample of 50 working women is selected.

a. What is the average number of women who put in more than 40 hours a week on the job?

b. What is the standard deviation for the number of women who put in more than 40 hours a week on the job?

c. Suppose that in our sample of 50 working women, there are 25 who work more than 40 hours a week. Would you consider this to be an unusual occurrence? Explain.

CHAPTER REVIEW

Key Concepts and Formulas

I. Continuous Probability Distributions

   1. Continuous random variables

   2. Probability distributions or probability density functions

     a. Curves are smooth.

     b. Area under the curve equals 1.

     c. The area under the curve between a and b represents the probability that x falls between a and b.

     d. $P(x = a) = 0$ for continuous random variables.

II. The Normal Probability Distribution

   1. Symmetric about its mean μ

   2. Shape determined by its standard deviation σ

III. The Standard Normal Distribution

   1. The standard normal random variable z has mean 0 and standard deviation 1.

   2. Any normal random variable x can be transformed to a standard normal random variable using

$$z = \frac{x - \mu}{\sigma}$$

   3. Convert necessary values of x to z.

   4. Use Table 3 in Appendix I to compute standard normal probabilities.

   5. Several important z-values have right-tail areas as follows:

Right-Tail Area	.005	.01	.025	.05	.10
z-Value	2.58	2.33	1.96	1.645	1.28

TECHNOLOGY TODAY

Normal Probabilities in *Microsoft Excel*

When the random variable of interest has a normal probability distribution, you can generate the following probabilities using the following functions:

- **NORM.DIST** and **NORM.S.DIST:** Generate cumulative probabilities— $P(x \le x_0)$ for a general normal random variable or $P(z \le z_0)$ for a standard normal random variable. (NOTE: These functions are called **NORMDIST** and **NORMSDIST** in *Excel 2007* and earlier versions.)

- **NORM.INV** and **NORM.S.INV:** Generate inverse cumulative probabilities— the value x_0 such that the area to its left under the general normal probability distribution is equal to a, or the value z_0 such that the area to its left under the standard normal probability distribution is equal to a. (NOTE: These functions are called **NORMINV** and **NORMSINV** in *Excel 2007* and earlier versions.)

You must specify which normal distribution you are using and, if it is a general normal random variable, the values for the mean μ and the standard deviation σ. As in Chapter 5, you must also specify the values for x_0, z_0, or a, depending on the function you are using.

EXAMPLE 6.14

For a standard normal random variable z, find $P(1.2 < z < 1.96)$. Find the value z_0 with area .025 to its right.

1. Name columns A and B of an *Excel* spreadsheet as "z0", and "P(z<=z0)", respectively. Then enter the two values for z_0 (1.2 and 1.96) in cells A2 and A3. To generate cumulative probabilities for these two values, first place your cursor in cell B2. Select **Insert Function** 𝑓ₓ ▶ **Statistical** ▶ **NORM.S.DIST** and click **OK.** The Dialog box shown in Figure 6.19 will appear.

FIGURE 6.19

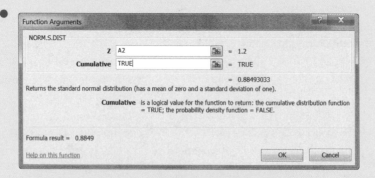

2. Enter the location of first value of z_0 (cell A2) into the first box and the word TRUE into the second box. The resulting probability is marked as "Formula result = .8849" at the bottom of the box, and when you click **OK,** the probability $P(z \le 1.2)$ will appear in cell B2. To obtain the other probability, simply place your cursor in cell B2, grab the square handle in the lower right corner of the cell and drag the handle down to copy the formula into the other cell and obtain $P(z \le 1.96) = .9750$. *MS Excel* has automatically adjusted the cell location in the formula as you copied.

3. To find $P(1.2 < z < 1.96)$, remember that the cumulative probability is the area to the left of the given value of z. Hence,

$$P(1.2 < z < 1.96) = P(z < 1.96) - P(z < 1.2) = .9750 - .8849 = .0901.$$

You can check this calculation using Table 3 in Appendix I if you wish!

4. To calculate inverse cumulative probabilities, place your cursor in an empty cell, select **Insert Function ▶ Statistical ▶ NORM.S.INV** and click **OK.** We need a value z_0 with area .025 to its right, or area .975 to its left. Enter **.975** in the box marked "Probability" and notice the "Formula Result = 1.959963985," which will appear in the empty cell when you click **OK.** This value, when rounded to two decimal places, is the familiar $z_0 = 1.96$ used in Example 6.7.

EXAMPLE 6.15

Suppose that the average birth weights of babies born at hospitals owned by a major health maintenance organization (HMO) are approximately normal with mean 6.75 pounds and standard deviation 0.54 pound. What proportion of babies born at these hospitals weigh between 6 and 7 pounds? Find the 95th percentile of these birth weights.

1. Name columns A and B of an *Excel* spreadsheet as "x0", and "P(x<=x0)", respectively. Then enter the two values for x_0 (6 and 7) in cells A2 and A3. Proceed as in Example 6.14, this time selecting **Insert Function ▶ Statistical ▶ NORM.DIST** and clicking **OK.** The Dialog box shown in Figure 6.20 will appear.

FIGURE 6.20

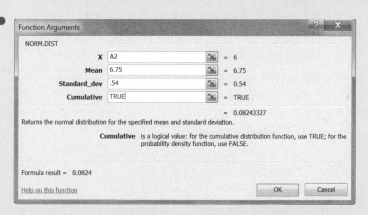

2. Enter the location of first value of x_0 (cell A2) into the first box, the appropriate mean and standard deviation in the second and third boxes, and the word TRUE into the fourth box. The resulting probability is marked as "Formula result = .0824" at the bottom of the box, and when you click **OK,** the probability $P(x \le 6)$ will appear in cell B2. To obtain the other probability, simply place your cursor in cell B2, grab the square handle in the lower right corner of the cell and drag the handle down to copy the formula into the other cell and obtain $P(x \le 7) = .6783$.

3. Finally, use the values calculated by *Excel* to calculate

$$P(6 < x < 7) = P(x < 7) - P(x < 6) = .6783 - .0824 = .5959.$$

4. To calculate the 95th percentile, place your cursor in an empty cell, select **Insert Function ▶ Statistical ▶ NORM.INV** and click **OK.** We need a value x_0 with area .95 to its left. Enter **.95** in the box marked "Probability," the appropriate mean and standard deviation (see Figure 6.21), and notice the "Formula Result = 7.638220959," which will appear in the empty cell when you click **OK.**

FIGURE 6.21

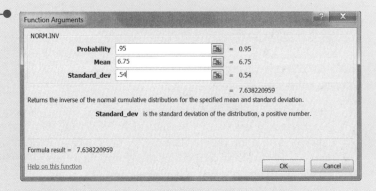

That is, 95% of all babies born at these hospitals weigh 7.638 pounds or less. Would you consider a baby who weighs 9 pounds to be unusually large?

Normal Probabilities in *MINITAB*

When the random variable of interest has a normal probability distribution, you can generate the following probabilities:

- Cumulative probabilities—$P(X \le x)$ for a given value of x. (NOTE: *MINITAB* uses the notation "X" for the random variable and "x" for a particular value of the random variable.)

- Inverse cumulative probabilities—the value x such that the area to its left under the normal probability distribution is equal to a.

You must specify which normal distribution you are using and the values for the mean μ and the standard deviation σ. As in Chapter 5, you have the option of specifying only one single value of x (or a) or several values of x (or a), which should be stored in a column (say, C1) of the *MINITAB* worksheet.

EXAMPLE 6.16

For a standard normal random variable z, find $P(1.2 < z < 1.96)$. Find the value z_0 with area .025 to its right.

1. Name columns C1 and C2 of a *MINITAB* worksheet as "x", and "P(X<=x)", respectively. Then enter the two values for x (1.2 and 1.96) in the first two cells of column C1. To generate cumulative probabilities for these two values, select **Calc ▶ Probability Distributions ▶ Normal** and the Dialog box shown in Figure 6.22 will appear.

FIGURE 6.22

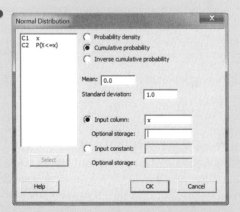

2. By default, *MINITAB* chooses $\mu = 0$ and $\sigma = 1$ as the mean and standard deviation of the standard normal z distribution, so you need only to enter the Input column (C1) and make sure that the radio button marked "Cumulative probability" is selected. If you do not specify a column for "Optional storage," *MINITAB* will display the results in the Session window, shown in Figure 6.23(a).

FIGURE 6.23

(a)

(b)

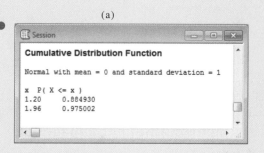

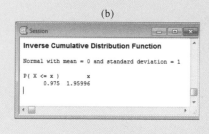

3. To find $P(1.2 < z < 1.96)$, remember that the cumulative probability is the area to the left of the given value of z. Hence,

$$P(1.2 < z < 1.96) = P(z < 1.96) - P(z < 1.2) = .975002 - .884930 = .090072.$$

You can check this calculation using Table 3 in Appendix I if you wish!

4. To calculate inverse cumulative probabilities, select **Calc ▶ Probability Distributions ▶ Normal**, and click the radio button marked "Inverse cumulative probability," shown in Figure 6.22. We need a value z_0 with area .025 to its right, or area .975 to its left. Enter **.975** in the box marked "Input constant" and click **OK.** The value of z_0 will appear in the Session window, shown in Figure 6.23(b). This value, when rounded to two decimal places, is the familiar $z_0 = 1.96$ used in Example 6.7.

EXAMPLE 6.17

Suppose that the average birth weights of babies born at hospitals owned by a major health maintenance organization (HMO) are approximately normal with mean 6.75 pounds and standard deviation 0.54 pound. What proportion of babies born at these hospitals weigh between 6 and 7 pounds? Find the 95th percentile of these birth weights.

1. Enter the two values for x (6 and 7) in the first two cells of column C1. Proceed as in Example 6.16, again selecting **Calc ▶ Probability Distributions ▶ Normal.** This time, enter the values for the mean ($\mu = 6.75$) and standard deviation ($\sigma = .54$) in the appropriate boxes, and select column C1 ("x") for the Input column. Make sure that the radio button marked "Cumulative probability" is selected and click **OK.** In the Session window, you will see that $P(x \le 7) = .678305$ and $P(x \le 6) = .082433$.

2. Finally, use the values calculated by *MINITAB* to calculate

$$P(6 < x < 7) = P(x < 7) - P(x < 6) = .678305 - .082433 = .595872.$$

3. To calculate the 95th percentile, selecting **Calc ▶ Probability Distributions ▶ Normal,** enter the values for the mean ($\mu = 6.75$) and standard deviation ($\sigma = .54$) in the appropriate boxes, and make sure that the radio button marked "Inverse

cumulative probability" is selected. We need a value x_0 with area .95 to its left. Enter **.95** in the box marked "Input constant" and click **OK.** In the Session window, you will see the 95th percentile, as shown in Figure 6.24.

FIGURE 6.24

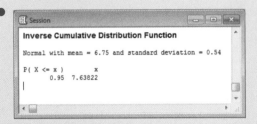

That is, 95% of all babies born at these hospitals weigh 7.63822 pounds or less. Would you consider a baby who weighs 9 pounds to be unusually large?

Supplementary Exercises

6.54 Calculate the area under the standard normal curve to the left of these values:
a. $z = -.90$ **b.** $z = 2.34$
c. $z = 5.4$

6.55 Calculate the area under the standard normal curve between these values:
a. $z = -2.0$ and $z = 2.0$ **b.** $z = -2.3$ and $z = -1.5$

6.56 Find the following probabilities for the standard normal random variable z:
a. $P(-1.96 \leq z \leq 1.96)$ **b.** $P(z > 1.96)$
c. $P(z < -1.96)$

6.57 a. Find a z_0 such that $P(z > z_0) = .9750$.
b. Find a z_0 such that $P(z > z_0) = .3594$.

6.58 a. Find a z_0 such that $P(-z_0 \leq z \leq z_0) = .95$.
b. Find a z_0 such that $P(-z_0 \leq z \leq z_0) = .98$.
c. Find a z_0 such that $P(-z_0 \leq z \leq z_0) = .90$.
d. Find a z_0 such that $P(-z_0 \leq z \leq z_0) = .99$.

6.59 A normal random variable x has mean $\mu = 5$ and standard deviation $\sigma = 2$. Find the probabilities associated with the following intervals:
a. $1.2 < x < 10$ **b.** $x > 7.5$ **c.** $x \leq 0$

6.60 Let x be a binomial random variable with $n = 36$ and $p = .54$. Use the normal approximation to find:
a. $P(x \leq 25)$ **b.** $P(15 \leq x \leq 20)$ **c.** $P(x > 30)$

6.61 Using Table 3 in Appendix I, calculate the area under the standard normal curve to the left of the following:
a. $z = 1.2$ **b.** $z = -.9$
c. $z = 1.46$ **d.** $z = -.42$

6.62 Find the following probabilities for the standard normal random variable:
a. $P(.3 < z < 1.56)$ **b.** $P(-.2 < z < .2)$

6.63 a. Find the probability that z is greater than $-.75$.
b. Find the probability that z is less than 1.35.

6.64 Find z_0 such that $P(z > z_0) = .5$.

6.65 Find the probability that z lies between $z = -1.48$ and $z = 1.48$.

6.66 Find z_0 such that $P(-z_0 < z < z_0) = .5$. What percentiles do $-z_0$ and z_0 represent?

6.67 Drill Bits It is estimated that the mean life span of oil-drilling bits is 75 hours. Suppose an oil exploration company purchases drill bits that have a life span that is approximately normally distributed with a mean equal to 75 hours and a standard deviation equal to 12 hours.
a. What proportion of the company's drill bits will fail before 60 hours of use?
b. What proportion will last at least 60 hours?
c. What proportion will have to be replaced after more than 90 hours of use?

6.68 Faculty Ages The influx of new ideas into a college or university, introduced primarily by new young faculty, is becoming a matter of concern because of the increasing ages of faculty members. That is, the distribution of faculty ages is shifting upward; there is a shortage of vacant positions and an oversupply of PhDs. If the retirement age at most universities is 65, would you expect the distribution of faculty ages to be normal? Explain.

6.69 Bearing Diameters A machine operation produces bearings whose diameters are normally distributed, with mean and standard deviation equal to .498 and .002, respectively. If specifications require that the bearing diameter equal .500 inch $\pm$ .004 inch, what fraction of the production will be unacceptable?

6.70 Used Cars A used-car dealership has found that the length of time before a major repair is required on the cars it sells is normally distributed with a mean equal to 10 months and a standard deviation of 3 months. If the dealer wants only 5% of the cars to fail before the end of the guarantee period, for how many months should the cars be guaranteed?

6.71 Restaurant Sales The daily sales total (excepting Saturday) at a small restaurant has a probability distribution that is approximately normal, with a mean μ equal to $1230 per day and a standard deviation σ equal to $120.

a. What is the probability that the sales will exceed $1400 for a given day?

b. The restaurant must have at least $1000 in sales per day to break even. What is the probability that on a given day the restaurant will not break even?

6.72 Washers The life span of a type of automatic washer is approximately normally distributed with mean and standard deviation equal to 10.5 and 3.0 years, respectively. If this type of washer is guaranteed for a period of 5 years, what fraction will need to be repaired and/or replaced?

6.73 Garage Door Openers Most users of automatic garage door openers activate their openers at distances that are normally distributed with a mean of 30 feet and a standard deviation of 11 feet. To minimize interference with other remote-controlled devices, the manufacturer is required to limit the operating distance to 50 feet. What percentage of the time will users attempt to operate the opener outside its operating limit?

6.74 How Long Is the Test? The average length of time required to complete a college achievement test was found to equal 70 minutes with a standard deviation of 12 minutes. When should the test be terminated if you wish to allow sufficient time for 90% of the students to complete the test? (Assume that the time required to complete the test is normally distributed.)

6.75 Servicing Automobiles The length of time required to run a 5000-mile check and to service an automobile has a mean equal to 1.4 hours and a standard deviation of .7 hour. Suppose that the service department plans to service 50 automobiles per 8-hour day and that, in order to do so, it must spend no more than an average of 1.6 hours per automobile. What proportion of all days will the service department have to work overtime?

6.76 TV Viewers An advertising agency has stated that 20% of all television viewers watch a given program. In a random sample of 1000 viewers, $x = 184$ viewers were watching the program. Do these data present sufficient evidence to contradict the advertiser's claim?

6.77 Forecasting Earnings A researcher notes that senior corporation executives are not very accurate forecasters of their own annual earnings. He states that his studies of a large number of company executive forecasts "showed that the average estimate missed the mark by 15%."

a. Suppose the distribution of these forecast errors has a mean of 15% and a standard deviation of 10%. Is it likely that the distribution of forecast errors is approximately normal?

b. Suppose the probability is .5 that a corporate executive's forecast error exceeds 15%. If you were to sample the forecasts of 100 corporate executives, what is the probability that more than 60 would be in error by more than 15%?

6.78 Filling Soda Cups A soft drink machine can be regulated to discharge an average of μ ounces per cup. If the ounces of fill are normally distributed, with standard deviation equal to .3 ounce, give the setting for μ so that 8-ounce cups will overflow only 1% of the time.

6.79 Light Bulbs A manufacturing plant uses light bulbs whose life spans are normally distributed, with mean and standard deviation equal to 500 and 50 hours, respectively. In order to minimize the number of bulbs that burn out during operating hours, all the bulbs are replaced after a given period of operation. How often should the bulbs be replaced if we wish no more than

1% of the bulbs to burn out between replacement periods?

6.80 The Freshman Class The admissions office of a small college is asked to accept deposits from a number of qualified prospective freshmen so that, with probability about .95, the size of the freshman class will be less than or equal to 120. Suppose the applicants constitute a random sample from a population of applicants, 80% of whom would actually enter the freshman class if accepted.

a. How many deposits should the admissions counselor accept?

b. If applicants in the number determined in part a are accepted, what is the probability that the freshman class size will be less than 105?

6.81 No Shows An airline finds that 5% of the persons making reservations on a certain flight will not show up for the flight. If the airline sells 160 tickets for a flight that has only 155 seats, what is the probability that a seat will be available for every person holding a reservation and planning to fly?

6.82 Long Distance It is known that 30% of all calls coming into a telephone exchange are long-distance calls. If 200 calls come into the exchange, what is the probability that at least 50 will be long-distance calls?

6.83 Plant Genetics In Exercise 5.75, a cross between two peony plants—one with red petals and one with streaky petals—produced offspring plants with red petals 75% of the time. Suppose that 100 seeds from this cross were collected and germinated, and x, the number of plants with red petals, was recorded.

a. What is the exact probability distribution for x?

b. Is it appropriate to approximate the distribution in part a using the normal distribution? Explain.

c. Use an appropriate method to find the approximate probability that between 70 and 80 (inclusive) offspring plants have red flowers.

d. What is the probability that 53 or fewer offspring plants had red flowers? Is this an unusual occurrence?

e. If you actually observed 53 of 100 offspring plants with red flowers, and if you were certain that the genetic ratio 3:1 was correct, what other explanation could you give for this unusual occurrence?

6.84 Suppliers A or B? A purchaser of electric relays buys from two suppliers, A and B. Supplier A supplies two of every three relays used by the company. If 75 relays are selected at random from those in use by the company, find the probability that at most

48 of these relays come from supplier A. Assume that the company uses a large number of relays.

6.85 Snacking and TV Psychologists believe that excessive eating may be associated with emotional states (being upset or bored) and environmental cues (watching television, reading, and so on). To test this theory, suppose you randomly selected 60 persons and matched them by weight and gender in pairs. For a period of 2 weeks, one of each pair is required to spend evenings reading novels of interest to him or her, while the other spends each evening watching television. The calorie count for all snack and drink intake for the evenings is recorded for each person, and you record $x = 19$, the number of pairs for which the television watchers' calorie intake exceeded the intake of the readers. If there is no difference in the effects of television and reading on calorie intake, the probability p that the calorie intake of one member of a pair exceeds that of the other member is .5. Do these data provide sufficient evidence to indicate a difference between the effects of television watching and reading on calorie intake? (HINT: Calculate the z-score for the observed value, $x = 19$.)

6.86 Gestation Times *The Biology Data Book* reports that the gestation time for human babies averages 278 days with a standard deviation of 12 days.[8] Suppose that these gestation times are normally distributed.

a. Find the upper and lower quartiles for the gestation times.

b. Would it be unusual to deliver a baby after only 6 months of gestation? Explain.

6.87 Tax Audits In Exercise 6.28, we suggested that the IRS assigns auditing rates per state by randomly selecting 50 auditing percentages from a normal distribution with a mean equal to 1.55% and a standard deviation of .45%.

a. What is the probability that a particular state would have more than 2% of its tax returns audited?

b. What is the expected value of x, the number of states that will have more than 2% of their income tax returns audited?

c. Is it likely that as many as 15 of the 50 states will have more than 2% of their income tax returns audited?

6.88 Your Favorite Sport Among the 10 most popular sports, men include competition-type sports—pool and billiards, basketball, and softball—whereas women include aerobics, running, hiking, and calisthenics. However, the top recreational activity for men was still the relaxing sport of fishing, with 41% of those surveyed indicating that they had fished during the year. Suppose 180 randomly selected men are asked whether they had fished in the past year.

a. What is the probability that fewer than 50 had fished?

b. What is the probability that between 50 and 75 (inclusive) had fished?

c. If the 180 men selected for the interview were selected by the marketing department of a sporting-goods company based on information obtained from their mailing lists, what would you conclude about the reliability of their survey results?

6.89 Introvert or Extrovert? A psychological introvert–extrovert test produced scores that had a normal distribution with a mean and standard deviation of 75 and 12, respectively. If we wish to designate the *highest* 15% as extroverts, what would be the proper score to choose as the cutoff point?

6.90 Normal Distribution? The chest measurements for 5738 Scottish militiamen in the early 19th century are given below.[9] Chest sizes are measured in inches, and each observation reports the number of soldiers with that chest size.

Count	Chest	Count	Chest
3	33	934	41
18	34	658	42
81	35	370	43
185	36	92	44
420	37	50	45
749	38	21	46
1073	39	4	47
1079	40	1	48

Notice the approximate normality of the histogram of the 5738 chest measurements.

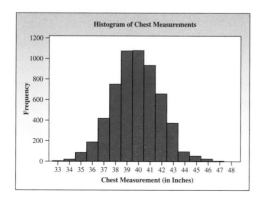

a. The mean of this distribution is $\bar{x} = 39.83$ and the standard deviation is $s = 2.05$. What is the 95th percentile of this distribution based on a normal curve with $\mu = 39.83$ and $\sigma = 2.05$?

b. Find the empirical estimate of the 95th percentile and compare with your answer in part a. (HINT: The 95th percentile will be in position $.95(n + 1) = .95 \times 5739 = 5452.05$ from the left tail of the distribution or in position $5738 - 5452.05 = 285.95$ from the right tail of the distribution.)

c. Find the 90th percentile of this distribution based on a normal curve with $\mu = 39.83$ and $\sigma = 2.05$. What is the value of the empirical 90th percentile? How does it compare with the value assuming normality?

6.91 Normal Distribution? continued Assume that the chest measurements in Exercise 6.90 are normally distributed with a mean of $\mu = 39.83$ and standard deviation of $\sigma = 2.05$.

a. What proportion of the observations would lie between 36.5 and 43.5 inches?

b. Between what two measurements would 95% of the observations lie?

c. What are the actual proportions for parts a and b using the data directly? Comment on the accuracy of the proportions found using assumed normality of the chest measurements.

6.92 Normal Temperatures In Exercise 1.67, Allen Shoemaker derived a distribution of human body temperatures, which has a distinct mound-shape.[10] Suppose we assume that the temperatures of healthy humans is approximately normal with a mean of 98.6 degrees and a standard deviation of 0.8 degrees.

a. If a healthy person is selected at random, what is the probability that the person has a temperature above 99.0 degrees?

b. What is the 95th percentile for the body temperatures of healthy humans?

6.93 Cellphone Etiquette A Snapshot in *USA Today* indicates that 51% of Americans say the average person is not very considerate of others when talking on a cellphone.[11] Suppose that 100 Americans are randomly selected. Find the approximate probability that 60 or more Americans would indicate that the average person is not very considerate of others when talking on a cellphone.

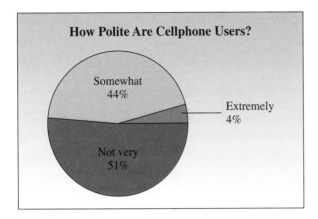

How Polite Are Cellphone Users?

Somewhat 44%

Extremely 4%

Not very 51%

6.94 Stamps Philatelists (stamp collectors) often buy stamps at or near retail prices, but, when they sell, the price is considerably lower. For example, depending on the mix of a collection, condition, demand, economic conditions, etc., a collection may sell at $x\%$ of the retail price, where x is normally distributed with a mean equal to 45% and a standard deviation of 4.5%. If a philatelist has a collection to sell that has a retail value of $30,000, what is the probability that the philatelist receives these amounts for the collection?
a. More than $15,000   **b.** Less than $15,000
c. Less than $12,000

6.95 Test Scores The scores on a national achievement test were approximately normally distributed, with a mean of 540 and a standard deviation of 110.

a. If you achieved a score of 680, how far, in standard deviations, did your score depart from the mean?

b. What percentage of those who took the examination scored higher than you?

6.96 Faculty Salaries Although faculty salaries at colleges and universities in the United States continue to rise, they do not always keep pace with the cost of living nor with salaries in the private sector. In 2010, the National Center for Educational Statistics indicated that the average salary for Assistant Professors at public 4-year colleges was $63,441.[12] Suppose that these salaries are normally distributed with a standard deviation of $4000.

a. What proportion of assistant professors at public 4-year colleges will have salaries less than $55,000?

b. What proportion of these professors will have salaries between $55,000 and $65,000?

6.97 Transplanting Cells Briggs and King developed the technique of nuclear transplantation, in which the nucleus of a cell from one of the later stages of the development of an embryo is transplanted into a zygote (a single-cell fertilized egg) to see whether the nucleus can support normal development. If the probability that a single transplant from the early gastrula stage will be successful is .65, what is the probability that more than 70 transplants out of 100 will be successful?

CASE STUDY *"Are You Going to Curve the Grades?"*

Very often, at the end of an exam that seemed particularly difficult, students will ask the professor, "Are you going to curve the grades?" Unfortunately, "curving the grades" doesn't necessarily mean that you will receive a higher grade on a test, although you might like to think so! Curving grades is actually a technique whereby a fixed proportion of the highest grades receive As (even if the highest grade is a failing grade on a percentage basis), and a fixed proportion of the lowest grades receive Fs (even if the lowest score is a passing grade on a percentage basis). The B, C, and D grades are also assigned according to fixed proportion. One such allocation uses the following proportions.

Letter Grade	A	B	C	D	F
Proportion of grades	10%	20%	40%	20%	10%

1. If the average C grade is centered on the average grade for all students, and if we assume that the grades are normally distributed, how many standard deviations on each side of the mean will designate the C grades?

2. How many standard deviations on either side of the mean will be the cutoff points for the B and D grades?

A histogram of the grades for an introductory Statistics class together with summary statistics follows.

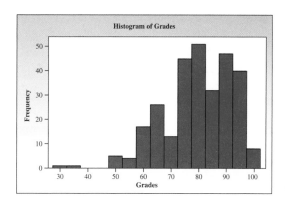

Descriptive Statistics: Grades

Variable	N	N*	Mean	StDev	Minimum	Q1	Median	Q3	Maximum
Grades	290	0	79.972	12.271	31.000	73.000	82.000	88.000	100.000

For ease of calculation, round the number of standard deviations for C grades to ±.5 standard deviations and for B and D grades to ±1.5 standard deviations.

3. Find the cutoff points for A, B, C, D, and F grades corresponding to these rounded values.

4. If you had a score of 92 on the exam and you had the choice of curving the grades or using the absolute standard of 90–100 for an A, 80–89 for a B, 70–79 for a C, and so on, what would be your choice? Explain your reasoning. Is the skewness of the distribution of grades a problem?

© PictureNet/CORBIS

7

Sampling Distributions

GENERAL OBJECTIVES

In the past several chapters, we studied *populations* and the *parameters* that describe them. These populations were either discrete or continuous, and we used *probability* as a tool for determining how likely certain sample outcomes might be. In this chapter, our focus changes as we begin to study *samples* and the *statistics* that describe them. These sample statistics are used to make inferences about the corresponding population parameters. This chapter involves sampling and sampling distributions, which describe the behavior of sample statistics in repeated sampling.

CHAPTER INDEX

 NEED TO KNOW...

When the Sample Size is Large Enough to Use the Central Limit Theorem
How to Calculate Probabilities for the Sample Mean $\bar{x}$
How to Calculate Probabilities for the Sample Proportion $\hat{p}$

Sampling the Roulette at Monte Carlo

How would you like to try your hand at gambling without the risk of losing? You could do it by simulating the gambling process, making imaginary bets, and observing the results. This technique, called a Monte Carlo procedure, is the topic of the case study at the end of this chapter.

INTRODUCTION

 **NEED A TIP?**

Parameter ⇔ Population
Statistic ⇔ Sample

In the previous three chapters, you have learned a lot about probability distributions, such as the binomial and normal distributions. The shape of the normal distribution is determined by its mean μ and its standard deviation σ, whereas the shape of the binomial distribution is determined by p. These numerical descriptive measures—called **parameters**—are needed to calculate the probability of observing sample results.

In practical situations, you may be able to decide which *type* of probability distribution to use as a model, but the values of the *parameters* that specify its *exact form* are unknown. Here are two examples:

- A pollster is sure that the responses to his "agree/disagree" questions will follow a binomial distribution, but p, the proportion of those who "agree" in the population, is unknown.

- An agronomist believes that the yield per acre of a variety of wheat is approximately normally distributed, but the mean μ and standard deviation σ of the yields are unknown.

In these cases, you must rely on the *sample* to learn about these parameters. The proportion of those who "agree" in the pollster's sample provides information about the actual value of p. The mean and standard deviation of the agronomist's sample approximate the actual values of μ and σ. If you want the sample to provide *reliable information* about the population, however, you must select your sample in a certain way!

SAMPLING PLANS AND EXPERIMENTAL DESIGNS

The way a sample is selected is called the **sampling plan** or **experimental design** and determines the quantity of information in the sample. Knowing the sampling plan used in a particular situation will often allow you to measure the reliability or goodness of your inference.

Simple random sampling is a commonly used sampling plan in which every sample of size n has the same chance of being selected. For example, suppose you want to select a sample of size $n = 2$ from a population containing $N = 4$ objects. If the four objects are identified by the symbols x_1, x_2, x_3, and x_4, there are six distinct pairs that could be selected, as listed in Table 7.1. If the sample of $n = 2$ observations is selected so that each of these six samples has the same chance of selection, given by 1/6, then the resulting sample is called a **simple random sample**, or just a **random sample**.

TABLE 7.1 ● **Ways of Selecting a Sample of Size 2 from 4 Objects**

Sample	Observations in Sample
1	x_1, x_2
2	x_1, x_3
3	x_1, x_4
4	x_2, x_3
5	x_2, x_4
6	x_3, x_4

Definition   If a sample of n elements is selected from a population of N elements using a sampling plan in which each of the possible samples has the same chance of selection, then the sampling is said to be **random** and the resulting sample is a **simple random sample.**

Perfect random sampling is difficult to achieve in practice. If the size of the population N is small, you might write each of N numbers on a poker chip, mix the chips, and select a sample of n chips. The numbers that you select correspond to the n measurements that appear in the sample. Since this method is not always very practical, a simpler and more reliable method uses **random numbers**—digits generated so that the values 0 to 9 occur randomly and with equal frequency. These numbers can be generated by computer or may even be available on your scientific calculator. Alternatively, Table 10 in Appendix I is a table of random numbers that you can use to select a *random sample.*

EXAMPLE 7.1

A computer database at a downtown law firm contains files for $N = 1000$ clients. The firm wants to select $n = 5$ files for review. Select a simple random sample of five files from this database.

Solution   You must first label each file with a number from 1 to 1000. Perhaps the files are stored alphabetically, and the computer has already assigned a number to each. Then generate a sequence of 10 three-digit random numbers. If you are using Table 10 of Appendix I, select a random starting point and use a portion of the table similar to the one shown in Table 7.2. The random starting point ensures that you will not use the same sequence over and over again.

The first three digits of Table 7.2 indicate the number of the first file to be reviewed. The random number 001 corresponds to file #1, and the last file, #1000, corresponds to the random number 000. Using Table 7.2, you would choose the five files numbered 155, 450, 32, 882, and 350 for review. Alternately, you might choose to read across the lines, and choose files 155, 350, 989, 450, and 369 for review.

TABLE 7.2   ● **Portion of a Table of Random Numbers**

15574	35026	98924
45045	36933	28630
03225	78812	50856
88292	26053	21121

The situation described in Example 7.1 is called an **observational study** because the data already existed before you decided to *observe* or describe their characteristics. Most sample surveys, in which information is gathered with a questionnaire, fall into this category. Computer databases make it possible to assign identification numbers to each element even when the population is large and to select a simple random sample. However, when conducting a sample survey, you must be careful to watch for these frequently occurring problems:

- **Nonresponse:** You have carefully selected your random sample and sent out your questionnaires, but only 50% of those surveyed return their questionnaires. Are the responses you received still representative of the entire population, or are they **biased** because only those people who were particularly opinionated about the subject chose to respond?

- **Undercoverage:** You have selected your random sample using land-line telephone records as a database. Does the database you used systematically exclude certain segments of the population—perhaps those who use only cell phones or have unlisted numbers?
- **Wording bias:** Your questionnaire may have questions that are too complicated or tend to confuse the reader. Possibly the questions are sensitive in nature—for example, "Have you ever used drugs?" or "Have you ever cheated on your income tax?"—and the respondents will not answer truthfully.

Methods have been devised to solve some of these problems, but only if you know that they exist. If your survey is *biased* by any of these problems, then your conclusions will not be very reliable, even though you did select a random sample!

Some research involves **experimentation,** in which an experimental condition or *treatment* is imposed on the *experimental units.* Selecting a simple random sample is more difficult in this situation.

EXAMPLE 7.2

A research chemist is testing a new method for measuring the amount of titanium (Ti) in ore samples. She chooses 10 ore samples of the same weight for her experiment. Five of the samples will be measured using a standard method, and the other five using the new method. Use random numbers to assign the 10 ore samples to the new and standard groups. Do these data represent a simple random sample from the population?

Solution   There are really two populations in this experiment. They consist of titanium measurements, using either the new or standard method, for *all possible* ore samples of this weight. These populations do not exist in fact; they are **hypothetical populations,** envisioned in the mind of the researcher. Thus, it is impossible to select a simple random sample using the methods of Example 7.1. Instead, the researcher selects what she believes are 10 *representative* ore samples and hopes that these samples will *behave as if* they had been randomly selected from the two populations.

The researcher can, however, randomly select the five samples to be measured with each method. Number the samples from 1 to 10. The five samples selected for the new method may correspond to 5 one-digit random numbers. Use this sequence of random digits generated on a scientific calculator:

948247817184610

Since you cannot select the same ore sample twice, you must skip any digit that has already been chosen. Ore samples 9, 4, 8, 2, and 7 will be measured using the new method. The other samples—1, 3, 5, 6, and 10—will be measured using the standard method.

In addition to *simple random sampling,* there are other sampling plans that involve randomization and therefore provide a probabilistic basis for inference making. Three such plans are based on *stratified, cluster,* and *systematic sampling.*

When the population consists of two or more subpopulations, called **strata,** a sampling plan that ensures that each subpopulation is represented in the sample is called a **stratified random sample.**

Definition   **Stratified random sampling** involves selecting a simple random sample from each of a given number of subpopulations, or **strata.**

Citizens' opinions about the construction of a performing arts center could be collected using a stratified random sample with city voting wards as strata. National polls usually involve some form of stratified random sampling with states as strata.

Another form of random sampling is used when the available sampling units are groups of elements, called **clusters.** For example, a household is a *cluster* of individuals living together. A city block or a neighborhood might be a convenient sampling unit and might be considered a *cluster* for a given sampling plan.

Definition A **cluster sample** is a simple random sample of clusters from the available clusters in the population.

When a particular cluster is included in the sample, a census of every element in the cluster is taken.

Sometimes the population to be sampled is ordered, such as an alphabetized list of people with driver's licenses, a list of utility users arranged by service addresses, or a list of customers by account numbers. In these and other situations, one element is chosen at random from the first k elements, and then every kth element thereafter is included in the sample.

Definition A **1-in-k systematic random sample** involves the random selection of one of the first k elements in an ordered population, and then the systematic selection of every kth element thereafter.

Not all sampling plans, however, involve random selection. You have probably heard of the nonrandom telephone polls in which those people who wish to express support for a question call one "900 number" and those opposed call a second "900 number." Each person must pay for his or her call. It is obvious that those people who call do not represent the population at large. This type of sampling plan is one form of a **convenience sample**—a sample that can be easily and simply obtained without random selection. Advertising for subjects who will be paid a fee for participating in an experiment produces a convenience sample. **Judgment sampling** allows the sampler to decide who will or will not be included in the sample. **Quota sampling,** in which the makeup of the sample must reflect the makeup of the population on some preselected characteristic, often has a nonrandom component in the selection process. **Remember that nonrandom samples can be described but cannot be used for making inferences!**

NEED A TIP?

All sampling plans used for making inferences must involve *randomization*!

7.2 EXERCISES

BASIC TECHNIQUES

7.1 A population consists of $N = 500$ experimental units. Use a random number table to select a random sample of $n = 20$ experimental units. (HINT: Since you need to use three-digit numbers, you can assign 2 three-digit numbers to each of the sampling units in the manner shown in the table.) What is the probability that each experimental unit is selected for inclusion in the sample?

Experimental Units	Random Numbers
1	001, 501
2	002, 502
3	003, 503
4	004, 504
.	.
.	.
.	.
499	499, 999
500	500, 000

7.2 A political analyst wishes to select a sample of $n = 20$ people from a population of 2000. Use the random number table to identify the people to be included in the sample.

7.3 A population contains 50,000 voters. Use the random number table to identify the voters to be included in a random sample of $n = 15$.

7.4 A small city contains 20,000 voters. Use the random number table to identify the voters to be included in a random sample of $n = 15$.

7.5 Every 10th Person A random sample of public opinion in a small town was obtained by selecting every 10th person who passed by the busiest corner in the downtown area. Will this sample have the characteristics of a random sample selected from the town's citizens? Explain.

7.6 Parks and Recreation A questionnaire was mailed to 1000 registered municipal voters selected at random. Only 500 questionnaires were returned, and of the 500 returned, 360 respondents were strongly opposed to a surcharge proposed to support the city Parks and Recreation Department. Are you willing to accept the 72% figure as a valid estimate of the percentage in the city who are opposed to the surcharge? Why or why not?

7.7 DMV Lists In many states, lists of possible jurors are assembled from voter registration lists and Department of Motor Vehicles records of licensed drivers and car owners. In what ways might this list not cover certain sectors of the population adequately?

7.8 Sex and Violence One question on a survey questionnaire is phrased as follows: "Don't you agree that there is too much sex and violence during prime TV viewing hours?" Comment on possible problems with the responses to this question. Suggest a better way to pose the question.

APPLICATIONS

7.9 Omega-3 Fats Contrary to current thought about omega-3 fatty acids, new research shows that the beneficial fats may not help reduce second heart attacks in heart attack survivors. The study included 4837 men and women being treated for heart disease. The experimental group received an additional 400 mg of the fats daily.[1] Suppose that this experiment was repeated with 50 individuals in the control group and 50 individuals in the experimental group. Determine

a randomization scheme to assign the 100 individuals to the two groups.

7.10 Racial Bias? Does the race of an interviewer matter? This question was investigated by Chris Gilberg and colleagues and reported in an issue of *Chance* magazine.[2] The interviewer asked, "Do you feel that affirmative action should be used as an occupation selection criteria?" with possible answers of yes or no.

a. What problems might you expect with responses to this question when asked by interviewers of different ethnic origins?

b. When people were interviewed by an African-American, the response was about 70% in favor of affirmative action, approximately 35% when interviewed by an Asian, and approximately 25% when interviewed by a Caucasian. Do these results support your answer in part a?

7.11 Native American Youth The *American Journal of Human Biology* reported on a study of a dietary assessment tool for use in the population of urban Native American youth.[3] The subjects were Native American youth attending an after-school program in Minneapolis, MN. All 61 children between the ages of 9 and 13 who satisfied the requirements of the study objectives were included in the experiment.

a. Describe the sampling plan used to select study participants.

b. What chance mechanism was used to select this sample of 61 Native American 9- to 13-year-old individuals?

c. Can valid inferences be made using the results of this study? Why or why not?

d. If you had to devise an alternative sampling plan, what would you change?

7.12 Tai Chi and Fibromyalgia A small new study shows that tai chi, an ancient Chinese practice of exercise and meditation, may relieve symptoms of chronic painful fibromyalgia. The study assigned 66 fibromyalgia patients to take either a 12-week tai chi class, or attend a wellness education class.[4]

a. Provide a randomization scheme to assign 66 subjects to the two groups.

b. Will your randomization scheme result in equal-sized groups? Explain.

7.13 Going to the Moon Two different Gallup Polls were conducted for *CNN/USA Today,* both of which involved people's feelings about the U.S.

space program.[5] Here is a question from each poll, along with the responses of the sampled Americans:

Space Exploration

CNN/USA Today/Gallup Poll. Nationwide:

"Would you favor or oppose a new U.S. space program that would send astronauts to the moon?" Form A (N = 510, MoE ± 5)

	Favor %	Oppose %	No Opinion %
12/03	53	45	2

"Would you favor or oppose the U.S. government spending billions of dollars to send astronauts to the moon?" Form B (N = 494, MoE ± 5)

	Favor %	Oppose %	No Opinion %
12/03	31	67	2

---

a. Read the two poll questions. Which of the two wordings is more unbiased? Explain.

b. Look at the responses for the two different polls. How would you explain the large differences in the percentages either favoring or opposing the new program?

7.14 Ask America A 2003 nationwide policy survey titled "Ask America" was sent by the National Republican Congressional Committee to voters in the Forty-fourth Congressional District, asking for opinions on a variety of political issues.[6] Here are some questions from the survey:

- In recent years has the federal government grown more or less intrusive in your personal and business affairs?

- Is President Bush right in trying to rein in the size and scope of the federal government against the wishes of the big government Democrats?

- Do you believe the death penalty is a deterrent to crime?

- Do you agree that the obstructionist Democrats should not be allowed to gain control of the U.S. Congress in the upcoming elections?

Comment on the effect of wording bias on the responses gathered using this survey.

STATISTICS AND SAMPLING DISTRIBUTIONS

7.3

When you select a random sample from a population, the numerical descriptive measures you calculate from the sample are called **statistics.** These statistics vary or change for each different random sample you select; that is, they are *random variables.* The probability distributions for statistics are called **sampling distributions** because, in repeated sampling, they provide this information:

- What values of the statistic can occur.

- How often each value occurs.

Definition   The **sampling distribution of a statistic** is the probability distribution for the possible values of the statistic that results when random samples of size *n* are repeatedly drawn from the population.

There are three ways to find the sampling distribution of a statistic:

1. Derive the distribution *mathematically* using the laws of probability.

2. Use a *simulation* to approximate the distribution. That is, draw a large number of samples of size *n,* calculating the value of the statistic for each sample, and tabulate the results in a relative frequency histogram. When the number of

samples is large, the histogram will be very close to the theoretical sampling distribution.

3. Use *statistical theorems* to derive exact or approximate sampling distributions.

The next example demonstrates how to derive the sampling distributions of two statistics for a very small population.

EXAMPLE 7.3

A population consists of $N = 5$ numbers: 3, 6, 9, 12, 15. If a random sample of size $n = 3$ is selected without replacement, find the sampling distributions for the sample mean $\bar{x}$ and the sample median m.

Solution   You are sampling from the population shown in Figure 7.1. It contains five distinct numbers and each is equally likely, with probability $p(x) = 1/5$. You can easily find the population mean and median as

$$\mu = \frac{3 + 6 + 9 + 12 + 15}{5} = 9 \quad \text{and} \quad M = 9$$

FIGURE 7.1

Probability histogram for the $N = 5$ population values in Example 7.3

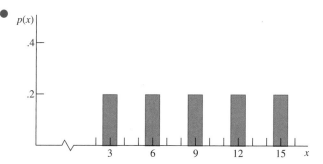

NEED A TIP?

Sampling distributions can be either discrete or continuous.

There are 10 possible random samples of size $n = 3$ and each is equally likely, with probability 1/10. These samples, along with the calculated values of $\bar{x}$ and m for each, are listed in Table 7.3. You will notice that some values of $\bar{x}$ are more likely than others because they occur in more than one sample. For example,

$$P(\bar{x} = 8) = \frac{2}{10} = .2 \quad \text{and} \quad P(m = 6) = \frac{3}{10} = .3$$

The values in Table 7.3 are tabulated, and the sampling distributions for $\bar{x}$ and m are shown in Table 7.4 and Figure 7.2.

Since the population of $N = 5$ values is symmetric about the value $x = 9$, both the *population mean* and the *median* equal 9. It would seem reasonable, therefore, to consider using either $\bar{x}$ or m as a possible estimator of $M = \mu = 9$. Which estimator would you choose? From Table 7.3, you see that, in using m as an estimator, you would be in error by $9 - 6 = 3$ with probability .3 or by $9 - 12 = -3$ with probability .3. That is, the error in estimation using m would be 3 with probability .6. In using $\bar{x}$, however, an error of 3 would occur with probability only .2. On these grounds alone, you may wish to use $\bar{x}$ as an estimator in preference to m.

TABLE 7.3 • **Values of $\bar{x}$ and m for Simple Random Sampling when $n = 3$ and $N = 5$**

Sample	Sample Values	$\bar{x}$	m
1	3, 6, 9	6	6
2	3, 6, 12	7	6
3	3, 6, 15	8	6
4	3, 9, 12	8	9
5	3, 9, 15	9	9
6	3, 12, 15	10	12
7	6, 9, 12	9	9
8	6, 9, 15	10	9
9	6, 12, 15	11	12
10	9, 12, 15	12	12

TABLE 7.4 • **Sampling Distributions for (a) the Sample Mean and (b) the Sample Median**

(a) $\bar{x}$	$p(\bar{x})$	(b) m	$p(m)$
6	.1	6	.3
7	.1	9	.4
8	.2	12	.3
9	.2		
10	.2		
11	.1		
12	.1		

FIGURE 7.2 •

Probability histograms for the sampling distributions of the sample mean, $\bar{x}$, and the sample median, m, in Example 7.3

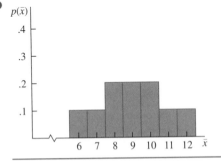
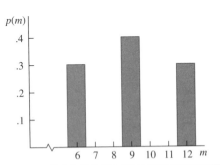

It was not too difficult to derive these sampling distributions in Example 7.3 because the number of elements in the population was very small. When this is not the case, you may need to use one of these methods:

- Use a simulation to approximate the sampling distribution empirically.
- Rely on statistical theorems and theoretical results.

One important statistical theorem that describes the sampling distribution of statistics that are sums or averages is presented in the next section.

THE CENTRAL LIMIT THEOREM

The **Central Limit Theorem** states that, under rather general conditions, sums and means of random samples of measurements drawn from a population tend to have an approximately normal distribution. Suppose you toss a balanced die $n = 1$ time. The random variable x is the number observed on the upper face. This familiar random variable can take six values, each with probability 1/6, and its probability distribution is shown in Figure 7.3. The shape of the distribution is *flat* or *uniform* and symmetric about the mean $\mu = 3.5$, with a standard deviation $\sigma = 1.71$. (See Section 4.8 and Exercise 4.84.)

FIGURE 7.3 ●

Probability distribution for x, the number appearing on a single toss of a die

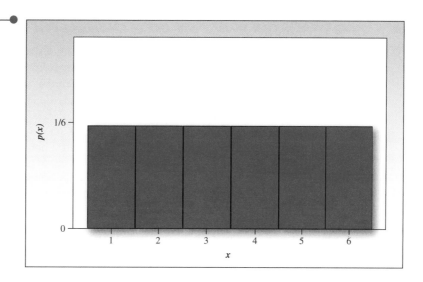

Now, take a sample of size $n = 2$ from this population; that is, toss two dice and record the sum of the numbers on the two upper faces, $\Sigma x_i = x_1 + x_2$. Table 7.5 shows the 36 possible outcomes, each with probability 1/36. The sums are tabulated, and each of the possible sums is divided by $n = 2$ to obtain an average. The result is the **sampling distribution** of $\bar{x} = \Sigma x_i/n$, shown in Figure 7.4. You should notice the dramatic difference in the shape of the sampling distribution. It is now roughly mound-shaped but still symmetric about the mean $\mu = 3.5$.

TABLE 7.5 ● **Sums of the Upper Faces of Two Dice**

Second Die	First Die					
	1	2	3	4	5	6
1	2	3	4	5	6	7
2	3	4	5	6	7	8
3	4	5	6	7	8	9
4	5	6	7	8	9	10
5	6	7	8	9	10	11
6	7	8	9	10	11	12

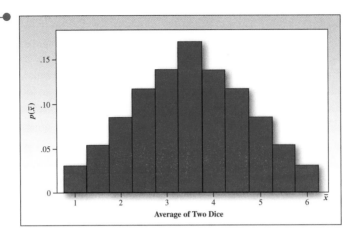

Using a similar procedure, we generated the sampling distributions of $\bar{x}$ when $n = 3$ and $n = 4$. For $n = 3$, the sampling distribution in Figure 7.5 clearly shows the mound shape of the normal probability distribution, still centered at $\mu = 3.5$. Notice also that the spread of the distribution is slowly *decreasing* as the sample size *n increases.* Figure 7.6 dramatically shows that the distribution of $\bar{x}$ is approximately normally distributed based on a sample as small as $n = 4$. This phenomenon is the result of an important statistical theorem called the **Central Limit Theorem (CLT).**

FIGURE 7.5

Sampling distribution of $\bar{x}$
for $n = 3$ dice

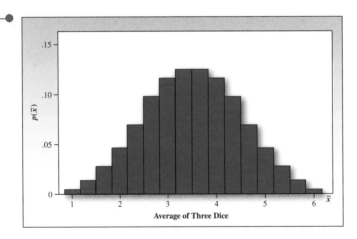

FIGURE 7.6

Sampling distribution of $\bar{x}$
for $n = 4$ dice

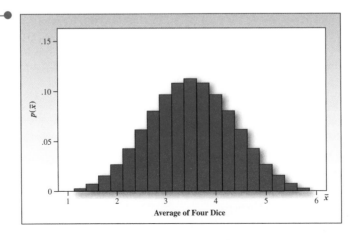

Central Limit Theorem	If random samples of n observations are drawn from a nonnormal population with finite mean μ and standard deviation σ, then, when n is large, the sampling distribution of the sample mean $\bar{x}$ is approximately normally distributed, with mean μ and standard deviation

$$\frac{\sigma}{\sqrt{n}}$$

The approximation becomes more accurate as n becomes large.

Regardless of its shape, the sampling distribution of $\bar{x}$ always has a mean identical to the mean of the sampled population and a standard deviation equal to the population standard deviation σ divided by $\sqrt{n}$. Consequently, *the spread of the distribution of sample means is considerably less than the spread of the sampled population.*

The Central Limit Theorem can be restated to apply to the **sum of the sample measurements** Σx_i, which, as n becomes large, also has an approximately normal distribution with mean $n\mu$ and standard deviation $\sigma\sqrt{n}$.

The important contribution of the Central Limit Theorem is in statistical inference. Many estimators that are used to make inferences about population parameters are sums or averages of the sample measurements. When the sample size is sufficiently large, you can expect these estimators to have sampling distributions that are approximately normal. You can then use the normal distribution to describe the behavior of these estimators in repeated sampling and evaluate the probability of observing certain sample results. As in Chapter 6, these probabilities are calculated using the standard normal random variable

$$z = \frac{\text{Estimator} - \text{Mean}}{\text{Standard deviation}}$$

As you reread the Central Limit Theorem, you may notice that the approximation is valid as long as the sample size n is "large"—but how large is "large"? Unfortunately, there is no clear answer to this question. The appropriate value of n depends on the shape of the population from which you sample as well as on how you want to use the approximation. However, these guidelines will help:

NEED A TIP?

The sampling distribution of $\bar{x}$ always has a mean μ and standard deviation $\sigma/\sqrt{n}$. The CLT helps describe its **shape.**

❓ NEED TO KNOW...

When the Sample Size is Large Enough to Use the Central Limit Theorem

- If the sampled population is **normal,** then the sampling distribution of $\bar{x}$ will also be normal, no matter what sample size you choose. This result can be proven theoretically, but it should not be too difficult for you to accept without proof.
- When the sampled population is approximately **symmetric,** the sampling distribution of $\bar{x}$ becomes approximately normal for relatively small values of n. Remember how rapidly the "flat" distribution in the dice example became mound-shaped ($n = 3$).
- When the sampled population is **skewed,** the sample size n must be larger, with n at least 30 before the sampling distribution of $\bar{x}$ becomes approximately normal.

These guidelines suggest that, for many populations, the sampling distribution of $\bar{x}$ will be approximately normal for moderate sample sizes; an exception to this rule occurs in sampling a binomial population when either p or $q = (1 - p)$ is very small. As specific applications of the Central Limit Theorem arise, we will give you the appropriate sample size n.

THE SAMPLING DISTRIBUTION OF THE SAMPLE MEAN

7.5

If the population mean μ is unknown, you might choose several *statistics* as an estimator; the sample mean $\bar{x}$ and the sample median m are two that readily come to mind. Which should you use? Consider these criteria in choosing the estimator for μ:

- Is it easy or hard to calculate?
- Does it produce estimates that are generally too high or too low?
- Is it more or less variable than other possible estimators?

The sampling distributions for $\bar{x}$ and m with $n = 3$ for the small population in Example 7.3 showed that, in terms of these criteria, the sample mean performed better than the sample median as an estimator of μ. In many situations, the sample mean $\bar{x}$ has desirable properties as an estimator that are not shared by other competing estimators; therefore, it is more widely used.

THE SAMPLING DISTRIBUTION OF THE SAMPLE MEAN, $\bar{x}$

- If a random sample of n measurements is selected from a population with mean μ and standard deviation σ, the sampling distribution of the sample mean $\bar{x}$ will have mean μ and standard deviation[*]

$$\frac{\sigma}{\sqrt{n}}$$

- If the population has a *normal* distribution, the sampling distribution of $\bar{x}$ will be *exactly* normally distributed, *regardless of the sample size, n.*
- If the population distribution is *nonnormal,* the sampling distribution of $\bar{x}$ will be *approximately* normally distributed for large samples (by the Central Limit Theorem). Conservatively, we require $n \geq 30$.

---

[*]When repeated samples of size n are randomly selected from a *finite* population with N elements whose mean is μ and whose variance is σ^2, the standard deviation of $\bar{x}$ is

$$\frac{\sigma}{\sqrt{n}} \sqrt{\frac{N - n}{N - 1}}$$

where σ^2 is the population variance. When N is large relative to the sample size n, $\sqrt{(N - n)(N - 1)}$ is approximately equal to 1, and the standard deviation of $\bar{x}$ is

$$\frac{\sigma}{\sqrt{n}}$$

Standard Error

Definition   The standard deviation of a statistic used as an estimator of a population parameter is also called the **standard error of the estimator** (abbreviated **SE**) because it refers to the precision of the estimator. Therefore, the standard deviation of $\bar{x}$—given by $\sigma/\sqrt{n}$—is referred to as the **standard error of the mean** (abbreviated as $SE(\bar{x})$, SEM, or sometimes just SE).

 NEED TO KNOW...

How to Calculate Probabilities for the Sample Mean $\bar{x}$

If you know that the sampling distribution of $\bar{x}$ is *normal* or *approximately normal,* you can describe the behavior of the sample mean $\bar{x}$ by calculating the probability of observing certain values of $\bar{x}$ in repeated sampling.

1. Find μ and calculate $SE(\bar{x}) = \sigma/\sqrt{n}$.
2. Write down the event of interest in terms of $\bar{x}$, and locate the appropriate area on the normal curve.
3. Convert the necessary values of $\bar{x}$ to z-values using

$$z = \frac{\bar{x} - \mu}{\sigma/\sqrt{n}}$$

4. Use Table 3 in Appendix I to calculate the probability.

EXAMPLE 7.4

The duration of Alzheimer's disease from the onset of symptoms until death ranges from 3 to 20 years; the average is 8 years with a standard deviation of 4 years. The administrator of a large medical center randomly selects the medical records of 30 deceased Alzheimer's patients from the medical center's database, and records the average duration. Find the approximate probabilities for these events:

1. The average duration is less than 7 years.
2. The average duration exceeds 7 years.
3. The average duration lies within 1 year of the population mean $\mu = 8$.

Solution   Since the administrator has selected a random sample from the database at this medical center, he can draw conclusions about only past, present, or future patients with Alzheimer's disease at this medical center. If, on the other hand, this medical center can be considered representative of other medical centers in the country, it may be possible to draw more far-reaching conclusions.

What can you say about the shape of the sampled population? It is not symmetric, because the mean $\mu = 8$ does not lie halfway between the maximum and minimum values. Since the mean is closer to the minimum value, the distribution is skewed to the right, with a few patients living a long time after the onset of the disease. Regardless of the shape of the population distribution, however, the sampling distribution of $\bar{x}$ has a mean $\mu = 8$ and standard deviation $\sigma/\sqrt{n} = 4/\sqrt{30} = .73$. In addition,

NEED A TIP?

If x is normal, $\bar{x}$ is normal for any *n*.

If x is not normal, $\bar{x}$ is approximately normal for large *n*.

Normal Probabilities for
Means

because the sample size is $n = 30$, the Central Limit Theorem ensures the *approximate normality* of the sampling distribution of $\bar{x}$.

1. The probability that $\bar{x}$ is less than 7 is given by the shaded area in Figure 7.7. To find this area, you need to calculate the value of z corresponding to $\bar{x} = 7$:

$$z = \frac{\bar{x} - \mu}{\sigma/\sqrt{n}} = \frac{7 - 8}{.73} = -1.37$$

From Table 3 in Appendix I, you can find the cumulative area corresponding to $z = -1.37$ and

$$P(\bar{x} < 7) = P(z < -1.37) = .0853$$

FIGURE 7.7

The probability that $\bar{x}$
is less than 7 for
Example 7.4

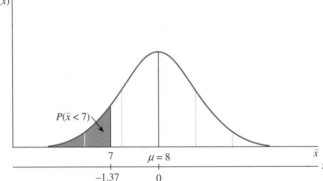

(NOTE: You must use $\sigma/\sqrt{n}$ (not σ) in the formula for z because you are finding an area under the sampling distribution for $\bar{x}$, not under the probability distribution for x.)

2. The event that $\bar{x}$ exceeds 7 is the complement of the event that $\bar{x}$ is less than 7. Thus, the probability that $\bar{x}$ exceeds 7 is

$$P(\bar{x} > 7) = 1 - P(\bar{x} \leq 7)$$
$$= 1 - .0853 = .9147$$

NEED A TIP?

Remember that for
continuous random
variables, there is no
probability
assigned to a single
point. Therefore,
$P(\bar{x} \leq 7) = P(\bar{x} < 7)$.

3. The probability that $\bar{x}$ lies within 1 year of $\mu = 8$ is the shaded area in Figure 7.8. The z-value corresponding to $\bar{x} = 7$ is $z = -1.37$, from part 1, and the z-value for $\bar{x} = 9$ is

$$z = \frac{\bar{x} - \mu}{\sigma/\sqrt{n}} = \frac{9 - 8}{.73} = 1.37$$

The probability of interest is

$$P(7 < \bar{x} < 9) = P(-1.37 < z < 1.37)$$
$$= .9147 - .0853 = .8294$$

The probability that $\bar{x}$ lies within 1 year of $\mu = 8$ for Example 7.4

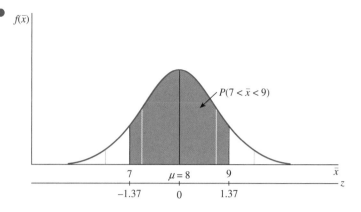

EXAMPLE 7.5

To avoid difficulties with the Federal Trade Commission or state and local consumer protection agencies, a beverage bottler must make reasonably certain that 12-ounce bottles actually contain 12 ounces of beverage. To determine whether a bottling machine is working satisfactorily, one bottler randomly samples 10 bottles per hour and measures the amount of beverage in each bottle. The mean $\bar{x}$ of the 10 fill measurements is used to decide whether to readjust the amount of beverage delivered per bottle by the filling machine.

If records show that the amount of fill per bottle is normally distributed, with a standard deviation of .2 ounce, and if the bottling machine is set to produce a mean fill per bottle of 12.1 ounces, what is the approximate probability that the sample mean $\bar{x}$ of the 10 test bottles is less than 12 ounces?

Solution The mean of the sampling distribution of the sample mean $\bar{x}$ is identical to the mean of the population of bottle fills—namely, $\mu = 12.1$ ounces—and the standard error of $\bar{x}$ is

$$ \text{SE} = \frac{\sigma}{\sqrt{n}} = \frac{.2}{\sqrt{10}} = .063 $$

(NOTE: σ is the standard deviation of the population of bottle fills, and n is the number of bottles in the sample.) Since the amount of fill is normally distributed, $\bar{x}$ is also normally distributed, as shown in Figure 7.9.

To find the probability that $\bar{x}$ is less than 12 ounces, express the value $\bar{x} = 12$ in units of standard deviations:

$$ z = \frac{\bar{x} - \mu}{\sigma/\sqrt{n}} = \frac{12 - 12.1}{.063} = -1.59 $$

Then

$$ P(\bar{x} < 12) = P(z < -1.59) = .0559 \approx .056 $$

Thus, if the machine is set to deliver an average fill of 12.1 ounces, the mean fill $\bar{x}$ of a sample of 10 bottles will be less than 12 ounces with a probability equal to .056.

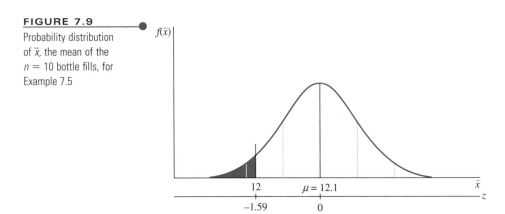

FIGURE 7.9

Probability distribution of $\bar{x}$, the mean of the $n = 10$ bottle fills, for Example 7.5

When this danger signal occurs ($\bar{x}$ is less than 12), the bottler takes a larger sample to recheck the setting of the filling machine.

7.5 EXERCISES

BASIC TECHNIQUES

7.15 Random samples of size n were selected from populations with the means and variances given here. Find the mean and standard deviation of the sampling distribution of the sample mean in each case:

a. $n = 36$, $\mu = 10$, $\sigma^2 = 9$

b. $n = 100$, $\mu = 5$, $\sigma^2 = 4$

c. $n = 8$, $\mu = 120$, $\sigma^2 = 1$

7.16 Refer to Exercise 7.15.

a. If the sampled populations are normal, what is the sampling distribution of $\bar{x}$ for parts a, b, and c?

b. According to the Central Limit Theorem, if the sampled populations are *not* normal, what can be said about the sampling distribution of $\bar{x}$ for parts a, b, and c?

Data set **7.17** A population consists of $N = 5$ numbers: EX0717 1, 3, 5, 6, and 7. It can be shown that the mean and standard deviation for this population are $\mu = 4.4$ and $\sigma = 2.15$, respectively.

a. Construct a probability histogram for this population.

b. Use the random number table, Table 10 in Appendix I, to select a random sample of size $n = 10$ with replacement from the population. Calculate the sample mean, $\bar{x}$. Repeat this procedure, calculating the sample mean $\bar{x}$ for your second sample.

(HINT: Assign the random digits 0 and 1 to the measurement $x = 1$; assign digits 2 and 3 to the measurement $x = 3$, and so on.)

c. To simulate the sampling distribution of $\bar{x}$, we have selected 50 more samples of size $n = 10$ with replacement, and have calculated the corresponding sample means. Construct a relative frequency histogram for these 50 values of $\bar{x}$. What is the shape of this distribution?

4.8	4.2	4.2	4.5	4.3	4.3	5.0	4.0	3.3	4.7
3.0	5.9	5.7	4.2	4.4	4.8	5.0	5.1	4.8	4.2
4.6	4.1	3.4	4.9	4.1	4.0	3.7	4.3	4.3	4.5
5.0	4.6	4.1	5.1	3.4	5.9	5.0	4.3	4.5	3.9
4.4	4.2	4.2	5.2	5.4	4.8	3.6	5.0	4.5	4.9

7.18 Refer to Exercise 7.17.

a. Use the data entry method in your calculator to find the mean and standard deviation of the 50 values of $\bar{x}$ given in Exercise 7.17, part c.

b. Compare the values calculated in part a to the theoretical mean μ and the theoretical standard deviation $\sigma/\sqrt{n}$ for the sampling distribution of $\bar{x}$. How close do the values calculated from the 50 measurements come to the theoretical values?

7.19 A random sample of n observations is selected from a population with standard deviation $\sigma = 1$. Calculate the standard error of the mean (SE) for these values of n:

a. $n = 1$ **b.** $n = 2$ **c.** $n = 4$

d. $n = 9$ **e.** $n = 16$ **f.** $n = 25$
g. $n = 100$

7.20 Refer to Exercise 7.19. Plot the standard error of the mean (SE) versus the sample size n and connect the points with a smooth curve. What is the effect of increasing the sample size on the standard error?

7.21 A random sample of size $n = 49$ is selected from a population with mean $\mu = 53$ and standard deviation $\sigma = 21$.

a. What will be the approximate shape of the sampling distribution of $\bar{x}$?

b. What will be the mean and standard deviation of the sampling distribution of $\bar{x}$?

7.22 Refer to Exercise 7.21. Find the probability that the sample mean is greater than 55.

7.23 A random sample of size $n = 40$ is selected from a population with mean $\mu = 100$ and standard deviation $\sigma = 20$.

a. What will be the approximate shape of the sampling distribution of $\bar{x}$?

b. What will be the mean and standard deviation of the sampling distribution of $\bar{x}$?

7.24 Refer to Exercise 7.23. Find the probability that the sample mean is between 105 and 110.

7.25 Suppose a random sample of $n = 25$ observations is selected from a population that is normally distributed with mean equal to 106 and standard deviation equal to 12.

a. Give the mean and the standard deviation of the sampling distribution of the sample mean $\bar{x}$.

b. Find the probability that $\bar{x}$ exceeds 110.

c. Find the probability that the sample mean deviates from the population mean $\mu = 106$ by no more than 4.

APPLICATIONS

7.26 Faculty Salaries Suppose that college faculty with the rank of professor at public 2-year institutions earn an average of $71,802 per year[7] with a standard deviation of $4000. In an attempt to verify this salary level, a random sample of 60 professors was selected from a personnel database for all 2-year institutions in the United States.

a. Describe the sampling distribution of the sample mean $\bar{x}$.

b. Within what limits would you expect the sample average to lie, with probability .95?

c. Calculate the probability that the sample mean $\bar{x}$ is greater than $73,000?

d. If your random sample actually produced a sample mean of $73,000, would you consider this unusual? What conclusion might you draw?

7.27 Measurement Error When research chemists perform experiments, they may obtain slightly different results on different replications, even when the experiment is performed identically each time. These differences are due to a phenomenon called "measurement error."

a. List some variables in a chemical experiment that might cause some small changes in the final response measurement.

b. If you want to make sure that your measurement error is small, you can replicate the experiment and take the sample average of all the measurements. To decrease the amount of variability in your average measurement, should you use a large or a small number of replications? Explain.

7.28 Tomatoes Explain why the weight of a package of one dozen tomatoes should be approximately normally distributed if the dozen tomatoes represent a random sample.

7.29 Bacteria in Water Use the Central Limit Theorem to explain why a Poisson random variable— say, the number of a particular type of bacteria in a cubic foot of water—has a distribution that can be approximated by a normal distribution when the mean μ is large. (HINT: One cubic foot of water contains 1728 cubic inches of water.)

7.30 Paper Strength A paper manufacturer requires a minimum strength of 20 pounds per square inch. To check on the quality of the paper, a random sample of 10 pieces of paper is selected each hour from the previous hour's production and a strength measurement is recorded for each. Assume that the strength measurements are normally distributed with a standard deviation $\sigma = 2$ pounds per square inch.

a. What is the approximate sampling distribution of the sample mean of $n = 10$ test pieces of paper?

b. If the mean of the population of strength measurements is 21 pounds per square inch, what is the approximate probability that, for a random sample of $n = 10$ test pieces of paper, $\bar{x} < 20$?

c. What value would you select for the mean paper strength μ in order that $P(\bar{x} < 20)$ be equal to .001?

7.31 Potassium Levels The normal daily human potassium requirement is in the range of 2000 to 6000 milligrams (mg), with larger amounts required during hot summer weather. The amount of potassium in food varies, but bananas are often associated with high potassium, with approximately 422 mg in a medium-sized banana[8]. Suppose the distribution of potassium in a banana is normally distributed, with mean equal to 422 mg and standard deviation equal to 13 mg per banana. You eat $n = 3$ bananas per day, and T is the total number of milligrams of potassium you receive from them.

a. Find the mean and standard deviation of T.

b. Find the probability that your total daily intake of potassium from the three bananas will exceed 1300 mg. (HINT: Note that T is the sum of three random variables, x_1, x_2, and x_3, where x_1 is the amount of potassium in banana number 1, etc.)

7.32 Deli Sales The total daily sales, x, in the deli section of a local market is the sum of the sales generated by a fixed number of customers who make purchases on a given day.

a. What kind of probability distribution do you expect the total daily sales to have? Explain.

b. For this particular market, the average sale per customer in the deli section is $8.50 with $\sigma = 2.50. If 30 customers make deli purchases on a given day, give the mean and standard deviation of the probability distribution of the total daily sales, x.

7.33 Normal Temperatures In Exercise 1.67, Allen Shoemaker derived a distribution of human body temperatures with a distinct mound shape.[9] Suppose we assume that the temperatures of healthy humans are approximately normal with a mean of 98.6° and a standard deviation of 0.8°.

a. If 130 healthy people are selected at random, what is the probability that the average temperature for these people is 98.25° or lower?

b. Would you consider an average temperature of 98.25° to be an unlikely occurrence, given that the true average temperature of healthy people is 98.6°? Explain.

7.34 Sports and Achilles Tendon Injuries Sports that involve a significant amount of running, jumping, or hopping put participants at risk for Achilles tendinopathy (AT), an inflammation and thickening of the Achilles tendon. A study in *The American Journal of Sports Medicine* looked at the diameter (in mm) of the affected and nonaffected tendons for patients who participated in these types of sports activities.[10] Suppose that the Achilles tendon diameters in the general population have a mean of 5.97 millimeters (mm) with a standard deviation of 1.95 mm.

a. What is the probability that a randomly selected sample of 31 patients would produce an average diameter of 6.5 mm or less for the nonaffected tendon?

b. When the diameters of the affected tendon were measured for a sample of 31 patients, the average diameter was 9.80. If the average tendon diameter in the population of patients with AT is no different than the average diameter of the non-affected tendons (5.97 mm), what is the probability of observing an average diameter of 9.80 or higher?

c. What conclusions might you draw from the results of part b?

THE SAMPLING DISTRIBUTION OF THE SAMPLE PROPORTION

7.6

There are many practical examples of the binomial random variable x. One common application involves consumer preference or opinion polls, in which we use a random sample of n people to estimate the proportion p of people in the population who have a specified characteristic. If x of the sampled people have this characteristic, then the sample proportion

$$\hat{p} = \frac{x}{n}$$

NEED A TIP?

Q: How do you know if it's binomial or not?
A: Look to see if the measurement taken on a single experimental unit in the sample is a "success/failure" type. If so, it's probably binomial.

can be used to estimate the population proportion p.[†]

The binomial random variable x has a probability distribution $p(x)$, described in Chapter 5, with mean np and standard deviation $\sqrt{npq}$. Since $\hat{p}$ is simply the value of x, expressed as a proportion $\left(\hat{p} = \dfrac{x}{n}\right)$, the sampling distribution of $\hat{p}$ is identical to the probability distribution of x, except that it has a new scale along the horizontal axis (Figure 7.10).

FIGURE 7.10

Sampling distribution of the binomial random variable x and the sample proportion $\hat{p}$

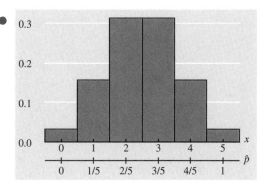

Because of this change of scale, the mean and standard deviation of $\hat{p}$ are also rescaled, so that the mean of the sampling distribution of $\hat{p}$ is p, and its standard error is

$$\text{SE}(\hat{p}) = \sqrt{\frac{pq}{n}} \qquad \text{where } q = 1 - p$$

Finally, just as we can approximate the probability distribution of x with a normal distribution when the sample size n is large, we can do the same with the sampling distribution of $\hat{p}$.

PROPERTIES OF THE SAMPLING DISTRIBUTION OF THE SAMPLE PROPORTION, $\hat{p}$

- If a random sample of n observations is selected from a binomial population with parameter p, then the sampling distribution of the sample proportion

$$\hat{p} = \frac{x}{n}$$

will have a mean

$$p$$

and a standard deviation

$$\text{SE}(\hat{p}) = \sqrt{\frac{pq}{n}} \qquad \text{where } q = 1 - p$$

- When the sample size n is large, the sampling distribution of $\hat{p}$ can be approximated by a normal distribution. The approximation will be adequate if $np > 5$ and $nq > 5$.

[†]A "hat" placed over the symbol of a population parameter denotes a statistic used to estimate the population parameter. For example, the symbol $\hat{p}$ denotes the sample proportion.

EXAMPLE 7.6

In a survey, 500 mothers and fathers were asked about the importance of sports for boys and girls. Of the parents interviewed, 60% agreed that the genders are equal and should have equal opportunities to participate in sports. Describe the sampling distribution of the sample proportion $\hat{p}$ of parents who agree that the genders are equal and should have equal opportunities.

Solution You can assume that the 500 parents represent a random sample of the parents of all boys and girls in the United States and that the true proportion in the population is equal to some unknown value that you can call p. The sampling distribution of $\hat{p}$ can be approximated by a normal distribution,[††] with mean equal to p (see Figure 7.11) and standard error

$$SE(\hat{p}) = \sqrt{\frac{pq}{n}}$$

FIGURE 7.11

The sampling distribution for $\hat{p}$ based on a sample of $n = 500$ parents for Example 7.6

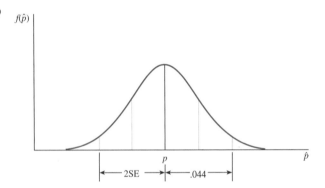

You can see from Figure 7.11 that the sampling distribution of $\hat{p}$ is centered over its mean p. Even though you do not know the exact value of p (the sample proportion $\hat{p} = .60$ may be larger or smaller than p), an approximate value for the standard deviation of the sampling distribution can be found using the sample proportion $\hat{p} = .60$ to approximate the unknown value of p. Thus,

$$SE = \sqrt{\frac{pq}{n}} \approx \sqrt{\frac{\hat{p}\hat{q}}{n}}$$

$$= \sqrt{\frac{(.60)(.40)}{500}} = .022$$

Therefore, approximately 95% of the time, $\hat{p}$ will fall within $2SE \approx .044$ of the (unknown) value of p.

[††]Checking the conditions that allow the normal approximation to the distribution of $\hat{p}$, you can see that $n = 500$ is adequate for values of p near .60 because $n\hat{p} = 300$ and $n\hat{q} = 200$ are both greater than 5.

NEED TO KNOW...

How to Calculate Probabilities for the Sample Proportion $\hat{p}$

1. Find the necessary values of n and p.
2. Check whether the normal approximation to the binomial distribution is appropriate ($np > 5$ and $nq > 5$).
3. Write down the event of interest in terms of $\hat{p}$, and locate the appropriate area on the normal curve.
4. Convert the necessary values of $\hat{p}$ to z-values using

$$z = \frac{\hat{p} - p}{\sqrt{\dfrac{pq}{n}}}$$

5. Use Table 3 in Appendix I to calculate the probability.

EXAMPLE 7.7

Refer to Example 7.6. Suppose the proportion p of parents in the population is actually equal to .55. What is the probability of observing a sample proportion as large as or larger than the observed value $\hat{p} = .60$?

Solution Figure 7.12 shows the sampling distribution of $\hat{p}$ when $p = .55$, with the observed value $\hat{p} = .60$ located on the horizontal axis. The probability of observing a sample proportion $\hat{p}$ equal to or larger than .60 is approximated by the shaded area in the upper tail of this normal distribution with

$$p = .55$$

and

$$\text{SE} = \sqrt{\frac{pq}{n}} = \sqrt{\frac{(.55)(.45)}{500}} = .0222$$

FIGURE 7.12

The sampling distribution of $\hat{p}$ for $n = 500$ and $p = .55$ for Example 7.7

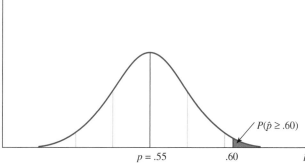

To find this shaded area, first calculate the z-value corresponding to $\hat{p} = .60$:

$$z = \frac{\hat{p} - p}{\sqrt{pq/n}} = \frac{.60 - .55}{.0222} = 2.25$$

Using Table 3 in Appendix I, you find

$$P(\hat{p} > .60) \approx P(z > 2.25) = 1 - .9878 = .0122$$

That is, if you were to select a random sample of $n = 500$ observations from a population with proportion p equal to .55, the probability that the sample proportion $\hat{p}$ would be as large as or larger than .60 is only .0122.

When the normal distribution was used in Chapter 6 to approximate the binomial probabilities associated with x, a correction of $\pm.5$ was applied to improve the approximation. The equivalent correction here is $\pm(.5/n)$. For example, for $\hat{p} = .60$ the value of z with the correction is

$$z_1 = \frac{(.60 - .001) - .55}{\sqrt{\frac{(.55)(.45)}{500}}} = 2.20$$

with $P(\hat{p} > .60) \approx .0139$. To two-decimal-place accuracy, this value agrees with the earlier result. When n is large, the effect of using the correction is generally negligible. You should solve problems in this and the remaining chapters *without* the correction factor unless you are specifically instructed to use it.

7.6 EXERCISES

BASIC TECHNIQUES

7.35 Random samples of size n were selected from binomial populations with population parameters p given here. Find the mean and the standard deviation of the sampling distribution of the sample proportion $\hat{p}$ in each case:

a. $n = 100, p = .3$

b. $n = 400, p = .1$

c. $n = 250, p = .6$

7.36 Is it appropriate to use the normal distribution to approximate the sampling distribution of $\hat{p}$ in the following circumstances?

a. $n = 50, p = .05$

b. $n = 75, p = .1$

c. $n = 250, p = .99$

7.37 Random samples of size $n = 75$ were selected from a binomial population with $p = .4$. Use the normal distribution to approximate the following probabilities:

a. $P(\hat{p} \le .43)$

b. $P(.35 \le \hat{p} \le .43)$

7.38 Random samples of size $n = 500$ were selected from a binomial population with $p = .1$.

a. Is it appropriate to use the normal distribution to approximate the sampling distribution of $\hat{p}$? Check to make sure the necessary conditions are met.

Using the results of part a, find these probabilities:

b. $\hat{p} > .12$

c. $\hat{p} < .10$

d. $\hat{p}$ lies within .02 of p

7.39 Calculate SE($\hat{p}$) for $n = 100$ and these values of p:

a. $p = .01$ **b.** $p = .10$ **c.** $p = .30$

d. $p = .50$ **e.** $p = .70$ **f.** $p = .90$

g. $p = .99$

h. Plot SE($\hat{p}$) versus p on graph paper and sketch a smooth curve through the points. For what value of p is the standard deviation of the sampling distribution of $\hat{p}$ a maximum? What happens to the standard error when p is near 0 or near 1.0?

7.40 A random sample of size $n = 50$ is selected from a binomial distribution with population proportion $p = .7$.

a. What will be the approximate shape of the sampling distribution of $\hat{p}$?

b. What will be the mean and standard deviation (or standard error) of the sampling distribution of $\hat{p}$?

c. Find the probability that the sample proportion $\hat{p}$ is less than .8.

7.41 A random sample of size $n = 80$ is selected from a binomial distribution with population proportion $p = .25$.

a. What will be the approximate shape of the sampling distribution of $\hat{p}$?

b. What will be the mean and standard deviation (or standard error) of the sampling distribution of $\hat{p}$?

c. Find the probability that the sample proportion $\hat{p}$ is between .18 and .44.

7.42 a. Is the normal approximation to the sampling distribution of $\hat{p}$ appropriate when $n = 400$ and $p = .8$?

b. Use the results of part a to find the probability that $\hat{p}$ is greater than .83.

c. Use the results of part a to find the probability that $\hat{p}$ lies between .76 and .84.

APPLICATIONS

7.43 Eco-Friendly Recycling trash, reducing waste, and reusing materials are eco-actions that will help the environment. According to a *USA Today* snapshot (Exercise 6.45), 78% of respondents list recycling as the leading way to help our environment.[11] Suppose that a random sample of $n = 100$ adults is selected and that the 78% figure is correct.

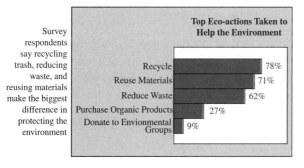

a. Does the distribution of $\hat{p}$, the sample proportion of adults who list recycling as the leading way to help the environment have an approximate normal distribution? If so, what is its mean and standard deviation?

b. What is the probability that the sample proportion $\hat{p}$ is less than 75%?

c. What is the probability that $\hat{p}$ lies in the interval .7 to .75?

d. What might you conclude about p if the sample proportion were less than .65?

7.44 Road Trip! Parents with children list a GPS system (28%) and a DVD player (28%) as "must have" accessories for a road trip.[12] Suppose a sample of $n = 1000$ parents are randomly selected and asked what devices they would like to have for a family road trip. Let $\hat{p}$ be the proportion of parents in the sample who choose either a GPS system or a DVD player.

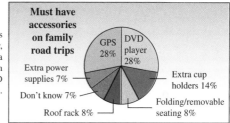

a. If $p = .28 + .28 = .56$, what is the exact distribution of $\hat{p}$? How can you approximate the distribution of $\hat{p}$?

b. What is the probability that $\hat{p}$ exceeds .6?

c. What is the probability that $\hat{p}$ lies between .5 and .6?

d. Would a sample percentage of $\hat{p} = .7$ contradict the reported value of .56?

7.45 M&M'S An advertiser claims that the average percentage of brown M&M'S candies in a package of milk chocolate M&M'S is 13%. Suppose you randomly select a package of milk chocolate M&M'S that contains 55 candies and determine the proportion of brown candies in the package.

a. What is the approximate distribution of the sample proportion of brown candies in a package that contains 55 candies?

b. What is the probability that the sample percentage of brown candies is less than 20%?

c. What is the probability that the sample percentage exceeds 35%?

d. Within what range would you expect the sample proportion to lie about 95% of the time?

7.46 Fido's in the Car It seems that driving with a pet in the car is the third worst driving distraction, behind talking on the phone and texting.

Source: USA Today, 19 August 2010, p. 8A

According to an American Automobile Association study, 80% of drivers admit to driving with a pet in the car, and of those, 20% allow their dogs to sit on their laps.[13] Suppose that you randomly select a sample of 100 drivers who have admitted to driving with a pet in their car.

a. What is the probability that 25% or more of the drivers allow their dogs to sit on their laps?

b. What is the probability that 10% or fewer of the drivers allow their dogs to sit on their laps?

c. Would it be unusual to find that 35% of the drivers allow their dogs to sit on their laps?

7.47 Oh, Nuts! Are you a chocolate "purist," or do you like other ingredients in your chocolate? *American Demographics* reports that almost 75% of consumers like traditional ingredients such as nuts or caramel in their chocolate. They are less enthusiastic about the taste of mint or coffee that provide more distinctive flavors.[14] A random sample of 200 consumers is selected and the number who like nuts or caramel in their chocolate is recorded.

a. What is the approximate sampling distribution for the sample proportion $\hat{p}$? What are the mean and standard deviation for this distribution?

b. What is the probability that the sample percentage is greater than 80%?

c. Within what limits would you expect the sample proportion to lie about 95% of the time?

A SAMPLING APPLICATION: STATISTICAL PROCESS CONTROL (OPTIONAL)

Statistical process control (SPC) methodology was developed to monitor, control, and improve products and services. For example, steel bearings must conform to size and hardness specifications, industrial chemicals must have a low prespecified level of impurities, and accounting firms must minimize and ultimately eliminate incorrect bookkeeping entries. It is often said that statistical process control consists of 10% statistics, 90% engineering and common sense. We can statistically monitor a process mean and tell when the mean falls outside preassigned limits, but we cannot tell *why* it is out of control. Answering this last question requires knowledge of the process and problem-solving ability—the other 90%!

Product quality is usually monitored using statistical control charts. Measurements on a process variable change or vary over time. The cause of this change is said to be *assignable* if it can be found and corrected. Other variation—small haphazard changes due to alteration in the production environment—that is not controllable is regarded as *random variation*. If the variation in a process variable is solely random, the process is said to be *in control*.

The first objective in statistical process control is to eliminate assignable causes of variation in the process variable and then get the process in control. The next step is to reduce variation and get the measurements on the process variable within *specification limits,* the limits within which the measurements on usable items or services must fall.

Once a process is in control and is producing a satisfactory product, the process variables are monitored with **control charts.** Samples of n items are drawn from the process at specified intervals of time, and a sample statistic is computed. These statistics are plotted on the control chart, so that the process can be checked for shifts in the process variable that might indicate control problems.

A Control Chart for the Process Mean: The x̄ Chart

Assume that n items are randomly selected from a production process at equal intervals and that measurements are recorded on the process variable. If the process is in control, the sample means should vary about the population mean μ in a random manner. Moreover, according to the Central Limit Theorem, the sampling distribution of $\bar{x}$ should be approximately normal, so that almost all of the values of $\bar{x}$ fall into the interval $(\mu \pm 3 \text{ SE}) = \mu \pm 3(\sigma/\sqrt{n})$. Although the exact values of μ and σ are unknown, you can obtain accurate estimates by using the sample measurements.

Every control chart has a *centerline* and *control limits.* The centerline for the **x̄ chart** is the estimate of μ, the grand average of all the sample statistics calculated from the measurements on the process variable. The upper and lower *control limits* are placed three standard deviations above and below the centerline. If you monitor the process mean based on k samples of size n taken at regular intervals, the centerline is $\bar{\bar{x}}$, the average of the sample means, and the control limits are at $\bar{\bar{x}} \pm 3(\sigma/\sqrt{n})$, with σ estimated by s, the standard deviation of the nk measurements.

EXAMPLE 7.8

A statistical process control monitoring system samples the inside diameters of $n = 4$ bearings each hour. Table 7.6 provides the data for $k = 25$ hourly samples. Construct an $\bar{x}$ chart for monitoring the process mean.

Solution The sample mean was calculated for each of the $k = 25$ samples. For example, the mean for sample 1 is

$$\bar{x} = \frac{.992 + 1.007 + 1.016 + .991}{4} = 1.0015$$

The sample means are shown in the last column of Table 7.6. The centerline is located at the average of the sample means, or

$$\bar{\bar{x}} = \frac{24.9675}{25} = .9987$$

The calculated value of s, the sample standard deviation of all $nk = 4(25) = 100$ observations, is $s = .011458$, and the estimated standard error of the mean of $n = 4$ observations is

$$\frac{s}{\sqrt{n}} = \frac{.011458}{\sqrt{4}} = .005729$$

TABLE 7.6 ● **25 Hourly Samples of Bearing Diameters, n = 4 Bearings per Sample**

Sample	Sample Measurements				Sample Mean, $\bar{x}$
1	.992	1.007	1.016	.991	1.00150
2	1.015	.984	.976	1.000	.99375
3	.988	.993	1.011	.981	.99325
4	.996	1.020	1.004	.999	1.00475
5	1.015	1.006	1.002	1.001	1.00600
6	1.000	.982	1.005	.989	.99400
7	.989	1.009	1.019	.994	1.00275
8	.994	1.010	1.009	.990	1.00075
9	1.018	1.016	.990	1.011	1.00875
10	.997	1.005	.989	1.001	.99800
11	1.020	.986	1.002	.989	.99925
12	1.007	.986	.981	.995	.99225
13	1.016	1.002	1.010	.999	1.00675
14	.982	.995	1.011	.987	.99375
15	1.001	1.000	.983	1.002	.99650
16	.992	1.008	1.001	.996	.99925
17	1.020	.988	1.015	.986	1.00225
18	.993	.987	1.006	1.001	.99675
19	.978	1.006	1.002	.982	.99200
20	.984	1.009	.983	.986	.99050
21	.990	1.012	1.010	1.007	1.00475
22	1.015	.983	1.003	.989	.99750
23	.983	.990	.997	1.002	.99300
24	1.011	1.012	.991	1.008	1.00550
25	.987	.987	1.007	.995	.99400

The upper and lower control limits are found as

$$UCL = \bar{\bar{x}} + 3\frac{s}{\sqrt{n}} = .9987 + 3(.005729) = 1.015887$$

and

$$LCL = \bar{\bar{x}} - 3\frac{s}{\sqrt{n}} = .9987 - 3(.005729) = .981513$$

Figure 7.13 shows the $\bar{x}$ chart constructed from the data. If you assume that the samples used to construct the $\bar{x}$ chart were collected when the process was in con-

FIGURE 7.13 ●

$\bar{x}$ chart for
Example 7.8

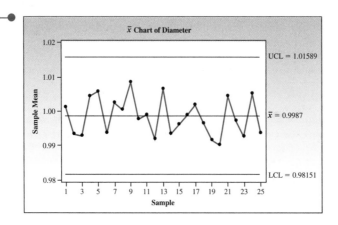

trol, the chart can now be used to detect changes in the process mean. Sample means are plotted periodically, and if a sample mean falls outside the control limits, the process should be checked to locate the cause of the unusually large or small mean.

A Control Chart for the Proportion Defective: The p Chart

Sometimes the observation made on an item is simply whether or not it meets specifications; thus, it is judged to be defective or nondefective. If the fraction defective produced by the process is p, then x, the number of defectives in a sample of n items, has a binomial distribution.

To monitor a process for defective items, samples of size n are selected at periodic intervals and the sample proportion $\hat{p}$ is calculated. When the process is in control, $\hat{p}$ should fall into the interval $p \pm 3\text{SE}$, where p is the proportion of defectives in the population (or the process fraction defective) with standard error

$$\text{SE} = \sqrt{\frac{pq}{n}} = \sqrt{\frac{p(1-p)}{n}}$$

The process fraction defective is unknown but can be estimated by the average of the k sample proportions:

$$\bar{p} = \frac{\Sigma \hat{p}_i}{k}$$

and the standard error is estimated by

$$\text{SE} = \sqrt{\frac{\bar{p}(1-\bar{p})}{n}}$$

The centerline for the **p chart** is located at $\bar{p}$, and the upper and lower control limits are

$$\text{UCL} = \bar{p} + 3\sqrt{\frac{\bar{p}(1-\bar{p})}{n}}$$

and

$$\text{LCL} = \bar{p} - 3\sqrt{\frac{\bar{p}(1-\bar{p})}{n}}$$

EXAMPLE 7.9

A manufacturer of ballpoint pens randomly samples 400 pens per day and tests each to see whether the ink flow is acceptable. The proportions of pens judged defective each day over a 40-day period are listed in Table 7.7. Construct a control chart for the proportion $\hat{p}$ defective in samples of $n = 400$ pens selected from the process.

Solution The estimate of the process proportion defective is the average of the $k = 40$ sample proportions in Table 7.7. Therefore, the centerline of the control chart is located at

$$\bar{p} = \frac{\Sigma \hat{p}_i}{k} = \frac{.0200 + .0125 + \cdots + .0225}{40} = \frac{.7600}{40} = .019$$

TABLE 7.7 ● **Proportions of Defectives in Samples of *n* = 400 Pens**

Day	Proportion	Day	Proportion	Day	Proportion	Day	Proportion
1	.0200	11	.0100	21	.0300	31	.0225
2	.0125	12	.0175	22	.0200	32	.0175
3	.0225	13	.0250	23	.0125	33	.0225
4	.0100	14	.0175	24	.0175	34	.0100
5	.0150	15	.0275	25	.0225	35	.0125
6	.0200	16	.0200	26	.0150	36	.0300
7	.0275	17	.0225	27	.0200	37	.0200
8	.0175	18	.0100	28	.0250	38	.0150
9	.0200	19	.0175	29	.0150	39	.0150
10	.0250	20	.0200	30	.0175	40	.0225

An estimate of SE, the standard error of the sample proportions, is

$$\sqrt{\frac{\bar{p}(1 - \bar{p})}{n}} = \sqrt{\frac{(.019)(.981)}{400}} = .00683$$

and 3SE = (3)(.00683) = .0205. Therefore, the upper and lower control limits for the *p* chart are located at

$$\text{UCL} = \bar{p} + 3\text{SE} = .0190 + .0205 = .0395$$

and

$$\text{LCL} = \bar{p} - 3\text{SE} = .0190 - .0205 = -.0015$$

Or, since *p* cannot be negative, LCL = 0.

The *p* control chart is shown in Figure 7.14. Note that all 40 sample proportions fall within the control limits. If a sample proportion collected at some time in the future falls outside the control limits, the manufacturer should be concerned about an increase in the defective rate. He should take steps to look for the possible causes of this increase.

FIGURE 7.14 ●

p chart for
Example 7.9

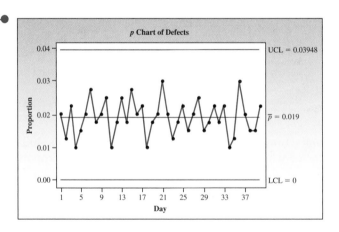

Other commonly used control charts are the *R chart,* which is used to monitor variation in the process variable by using the sample range, and the *c chart,* which is used to monitor the number of defects per item.

7.7 EXERCISES

BASIC TECHNIQUES

7.48 The sample means were calculated for 30 samples of size $n = 10$ for a process that was judged to be in control. The means of the 30 $\bar{x}$-values and the standard deviation of the combined 300 measurements were $\bar{\bar{x}} = 20.74$ and $s = .87$, respectively.

a. Use the data to determine the upper and lower control limits for an $\bar{x}$ chart.

b. What is the purpose of an $\bar{x}$ chart?

c. Construct an $\bar{x}$ chart for the process and explain how it can be used.

7.49 The sample means were calculated for 40 samples of size $n = 5$ for a process that was judged to be in control. The means of the 40 values and the standard deviation of the combined 200 measurements were $\bar{\bar{x}} = 155.9$ and $s = 4.3$, respectively.

a. Use the data to determine the upper and lower control limits for an $\bar{x}$ chart.

b. Construct an $\bar{x}$ chart for the process and explain how it can be used.

7.50 Explain the difference between an $\bar{x}$ chart and a p chart.

7.51 Samples of $n = 100$ items were selected hourly over a 100-hour period, and the sample proportion of defectives was calculated each hour. The mean of the 100 sample proportions was .035.

a. Use the data to find the upper and lower control limits for a p chart.

b. Construct a p chart for the process and explain how it can be used.

7.52 Samples of $n = 200$ items were selected hourly over a 100-hour period, and the sample proportion of defectives was calculated each hour. The mean of the 100 sample proportions was .041.

a. Use the data to find the upper and lower control limits for a p chart.

b. Construct a p chart for the process and explain how it can be used.

APPLICATIONS

7.53 Black Jack A gambling casino records and plots the mean daily gain or loss from five blackjack tables on an $\bar{x}$ chart. The overall mean of the sample means and the standard deviation of the combined data over 40 weeks were $\bar{\bar{x}} = \$10,752$ and $s = \$1605$, respectively.

a. Construct an $\bar{x}$ chart for the mean daily gain per blackjack table.

b. How can this $\bar{x}$ chart be of value to the manager of the casino?

7.54 Coal-Burning Power Plant A coal-burning power plant tests and measures three specimens of coal each day to monitor the percentage of ash in the coal. The overall mean of 30 daily sample means and the combined standard deviation of all the data were $\bar{\bar{x}} = 7.24$ and $s = .07$, respectively. Construct an $\bar{x}$ chart for the process and explain how it can be of value to the manager of the power plant.

 EX0755 **7.55 Lumber Specs** The manager of a building-supplies company randomly samples incoming lumber to see whether it meets quality specifications. From each shipment, 100 pieces of 2×4 lumber are inspected and judged according to whether they are first (acceptable) or second (defective) grade. The proportions of second-grade 2×4s recorded for 30 shipments were as follows:

.14	.21	.19	.18	.23	.20	.25	.19	.22	.17
.21	.15	.23	.12	.19	.22	.15	.26	.22	.21
.14	.20	.18	.22	.21	.13	.20	.23	.19	.26

Construct a control chart for the proportion of second-grade 2×4s in samples of 100 pieces of lumber. Explain how the control chart can be of use to the manager of the building-supplies company.

7.56 Brass Rivets A producer of brass rivets randomly samples 400 rivets each hour and calculates the proportion of defectives in the sample. The mean sample proportion calculated from 200 samples was equal to .021. Construct a control chart for the proportion of defectives in samples of 400 rivets. Explain how the control chart can be of value to a manager.

EX0757 **7.57 Nuclear Power Plant** The data in the table are measures of the radiation in air particulates at a nuclear power plant. Four measurements were recorded at weekly intervals over a 26-week period. Use the data to construct an $\bar{x}$ chart and plot the 26 values of $\bar{x}$. Explain how the chart can be used.

Week	Radiation			
1	.031	.032	.030	.031
2	.025	.026	.025	.025
3	.029	.029	.031	.030
4	.035	.037	.034	.035
5	.022	.024	.022	.023
6	.030	.029	.030	.030
7	.019	.019	.018	.019
8	.027	.028	.028	.028
9	.034	.032	.033	.033
10	.017	.016	.018	.018
11	.022	.020	.020	.021
12	.016	.018	.017	.017
13	.015	.017	.018	.017
14	.029	.028	.029	.029
15	.031	.029	.030	.031
16	.014	.016	.016	.017
17	.019	.019	.021	.020
18	.024	.024	.024	.025
19	.029	.027	.028	.028
20	.032	.030	.031	.030
21	.041	.042	.038	.039
22	.034	.036	.036	.035
23	.021	.022	.024	.022
24	.029	.029	.030	.029
25	.016	.017	.017	.016
26	.020	.021	.020	.022

7.58 Baseball Bats A hardwoods manufacturing plant has a production line designed to produce baseball bats weighing 32 ounces. During a period of time when the production process was known to be in statistical control, the average bat weight was found to be 31.7 ounces. The observed data were gathered from 50 samples, each consisting of 5 measurements. The standard deviation of all samples was found to be $s = .2064$ ounces. Construct an $\bar{x}$-chart to monitor the 32-ounce bat production process.

7.59 More Baseball Bats Refer to Exercise 7.58 and suppose that during a day when the state of the 32-ounce bat production process was unknown, the following measurements were obtained at hourly intervals.

Hour	$\bar{x}$	Hour	$\bar{x}$
1	31.6	4	33.1
2	32.5	5	31.6
3	33.4	6	31.8

Each measurement represents a statistic computed from a sample of five bat weights selected from the production process during a certain hour. Use the control chart constructed in Exercise 7.58 to monitor the process.

CHAPTER REVIEW

Key Concepts and Formulas

I. Sampling Plans and Experimental Designs

1. Simple random sampling
   a. Each possible sample of size n is equally likely to occur.
   b. Use a computer or a table of random numbers.
   c. Problems are nonresponse, undercoverage, and wording bias.

2. Other sampling plans involving randomization
   a. Stratified random sampling
   b. Cluster sampling
   c. Systematic 1-in-k sampling

3. Nonrandom sampling
   a. Convenience sampling
   b. Judgment sampling
   c. Quota sampling

II. Statistics and Sampling Distributions

1. Sampling distributions describe the possible values of a statistic and how often they occur in repeated sampling.

2. Sampling distributions can be derived mathematically, approximated empirically, or found using statistical theorems.

3. The Central Limit Theorem states that sums and averages of measurements from a nonnormal population with finite mean μ and standard deviation σ have approximately normal distributions for large samples of size n.

III. Sampling Distribution of the Sample Mean

1. When samples of size n are randomly drawn from a normal population with mean μ and variance σ^2, the sample mean $\bar{x}$ has a normal distribution with mean μ and standard deviation $\sigma/\sqrt{n}$.

2. When samples of size n are randomly drawn from a nonnormal population with mean μ and variance σ^2, the Central Limit Theorem ensures that the sample mean $\bar{x}$ will have an approximately normal distribution with mean μ and standard deviation $\sigma/\sqrt{n}$ when n is large ($n \geq 30$).

3. Probabilities involving the sample mean can be calculated by standardizing the value of $\bar{x}$ using z:

$$z = \frac{\bar{x} - \mu}{\sigma/\sqrt{n}}$$

IV. Sampling Distribution of the Sample Proportion

1. When samples of size n are drawn from a binomial population with parameter p, the sample proportion $\hat{p}$ will have an approximately normal distribution with mean p and standard deviation $\sqrt{pq/n}$ as long as $np > 5$ and $nq > 5$.

2. Probabilities involving the sample proportion can be calculated by standardizing the value $\hat{p}$ using z:

$$z = \frac{\hat{p} - p}{\sqrt{\dfrac{pq}{n}}}$$

V. Statistical Process Control

1. To monitor a quantitative process, use an $\bar{x}$ chart. Select k samples of size n and calculate the overall mean $\bar{\bar{x}}$ and the standard deviation s of all nk measurements. Create upper and lower control limits as

$$\bar{\bar{x}} \pm 3\frac{s}{\sqrt{n}}$$

If a sample mean exceeds these limits, the process is out of control.

2. To monitor a *binomial* process, use a p chart. Select k samples of size n and calculate the average of the sample proportions as

$$\bar{p} = \frac{\Sigma \hat{p}_i}{k}$$

Create upper and lower control limits as

$$\bar{p} \pm 3\sqrt{\frac{\bar{p}(1 - \bar{p})}{n}}$$

If a sample proportion exceeds these limits, the process is out of control.

 TECHNOLOGY TODAY

The Central Limit Theorem at Work—*Microsoft Excel*

Microsoft Excel can be used to explore the way the Central Limit Theorem works in practice. Remember that, according to the Central Limit Theorem, if random samples of size n are drawn from a nonnormal population with mean μ and standard deviation σ, then when n is large, the sampling distribution of the sample mean $\bar{x}$ will be approximately normal with the same mean μ and with standard error $\sigma/\sqrt{n}$. Let's try sampling from a nonnormal population using *Excel*.

In a new spreadsheet, generate 100 samples of size $n = 30$ from a continuous uniform distribution (Example 6.1) over the interval (0, 10). Label column A as "Sample" and enter the numbers 1 to 100 in that column. Then select **Data ▶ Data Analysis ▶ Random Number Generation,** to obtain the Dialog box in Figure 7.15(a). Type **30** for the number of variables and 100 for the number of random numbers. In the drop-down "Distribution" list, choose uniform, with parameters between **0** and **10**. We will leave the first row of our spreadsheet empty, starting the "Output Range" at cell B2. Press **OK** to see the 100 random samples of size $n = 30$. You can look at the distribution of the entire set of data using **Data ▶ Data Analysis ▶ Histogram,** choosing bins 1, 2, ..., 9, 10 and using the procedures described in the "Technology Today" section in Chapter 2. For our data, the distribution, shown in Figure 7.15(b) is not mound-shaped, but is fairly flat, as expected for the uniform distribution.

For the uniform distribution that we have used, the mean and standard deviation are $\mu = 5$ and $\sigma = 2.89$, respectively. Check the descriptive statistics for the 30×100

FIGURE 7.15

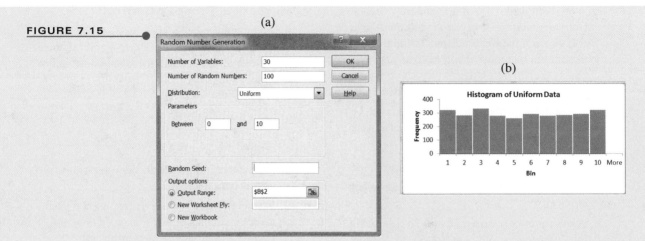

(a)

(b)

$= 3000$ measurements (use the functions **=AVERAGE(B2:AE101)** and **=STDEV (B2:AE101)**), and you will find that the 100 observations have a sample mean and standard deviation *close to* but not exactly equal to $\mu = 5$ and $\sigma = 2.89$, respectively.

Now, generate 100 values of $\bar{x}$ based on samples of size $n = 30$ by creating a column of means for the 100 rows. First, label column AF as "x-bar" and place your cursor in cell AF2. Use **Insert Function ▶ Statistical ▶ Average** (or type **=AVERAGE (B2:AE2)**) to obtain the first average. Then copy the formula into the other 99 cells in column AF. You can now look at the distribution of these 100 sample means using **Data ▶ Data Analysis ▶ Histogram** and choosing bins 1, 1.5, 2, 2.5, ..., 9, 9.5, 10. The distribution for our 100 sample means is shown in Figure 7.16.

FIGURE 7.16

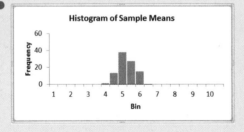

Notice the distinct mound shape of the distribution in Figure 7.16 compared to the original distribution in Figure 7.15(b). Also, if you check the mean and standard deviation for the 100 sample means in column AE, you will find that they are not too different from the theoretical values, $\mu = 5$ and $\sigma/\sqrt{n} = 2.89/\sqrt{30} = .53$. (For our data, the sample mean is 4.98 and the standard deviation is .49.) Since we had only 100 samples, our results are not *exactly* equal to the theoretical values. If we had generated an *infinite* number of samples, we would have gotten an exact match. This is the Central Limit Theorem at work!

The Central Limit Theorem at Work—*MINITAB*

MINITAB provides a perfect tool for exploring the way the Central Limit Theorem works in practice. Remember that, according to the Central Limit Theorem, if random samples of size n are drawn from a nonnormal population with mean μ and standard deviation σ, then when n is large, the sampling distribution of the sample mean $\bar{x}$ will be approximately normal with the same mean μ and with standard error $\sigma/\sqrt{n}$. Let's try sampling from a nonnormal population with the help of *MINITAB*.

In a new *MINITAB* worksheet, generate 100 samples of size $n = 30$ from a non-normal distribution called the exponential distribution. Use **Calc ▶ Random Data ▶ Exponential.** Type **100** for the number of rows of data, and store the results in C1–C30 (see Figure 7.17(a)). Leave the mean at the default of 1.0, the threshold at 0.0, and click **OK.** The data are generated and stored in the worksheet. Use **Graph ▶ Histogram ▶ Simple** to look at the distribution of some of the data—say, C1 (as in Figure 7.17(b)). Notice that the distribution is not mound-shaped; it is highly skewed to the right.

(a) (b)

FIGURE 7.17

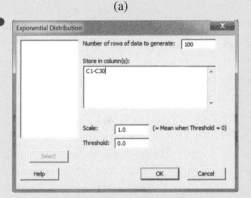
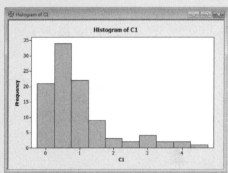

For the exponential distribution that we have used, the mean and standard deviation are $\mu = 1$ and $\sigma = 1$, respectively. Check the descriptive statistics for one of the columns (use **Stat ▶ Basic Statistics ▶ Display Descriptive Statistics**), and you will find that the 100 observations have a sample mean and standard deviation that are both *close to* but not exactly equal to 1. Now, generate 100 values of $\bar{x}$ based on samples of size $n = 30$ by creating a column of means for the 100 rows. Use **Calc ▶ Row Statistics,** and select **Mean.** To average the entries in all 30 columns, select or type **C1–C30** in the Input variables box, and store the results in **C31** (see Figure 7.18(a)). You can now look at the distribution of the sample means using **Graph ▶ Histogram ▶ Simple,** selecting **C31** and clicking **OK.** The distribution of the 100 sample means generated for our example is shown in Figure 7.18(b).

(a) (b)

FIGURE 7.18

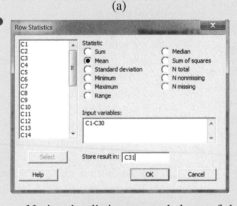
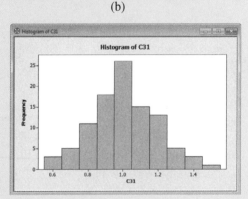

Notice the distinct mound shape of the distribution in Figure 7.18(b) compared to the original distribution in Figure 7.17(b). Also, if you check the descriptive statistics for C31, you will find that the mean and standard deviation of our 100 sample means are not too different from the theoretical values, $\mu = 1$ and $\sigma/\sqrt{n} = 1/\sqrt{30} = .18$. (For our data, the sample mean is 1.0024 and the standard deviation is .1813.) Since we had only 100 samples, our results are not *exactly* equal to the theoretical values. If we had generated an *infinite* number of samples, we would have gotten an exact match. This is the Central Limit Theorem at work!

Supplementary Exercises

7.60 A finite population consists of four elements: 6, 1, 3, 2.

a. How many different samples of size $n = 2$ can be selected from this population if you sample without replacement? (Sampling is said to be *without replacement* if an element cannot be selected twice for the same sample.)

b. List the possible samples of size $n = 2$.

c. Compute the sample mean for each of the samples given in part b.

d. Find the sampling distribution of $\bar{x}$. Use a probability histogram to graph the sampling distribution of $\bar{x}$.

e. If all four population values are equally likely, calculate the value of the population mean μ. Do any of the samples listed in part b produce a value of $\bar{x}$ exactly equal to μ?

7.61 Refer to Exercise 7.60. Find the sampling distribution for $\bar{x}$ if random samples of size $n = 3$ are selected *without replacement*. Graph the sampling distribution of $\bar{x}$.

7.62 Suppose a random sample of $n = 5$ observations is selected from a population that is normally distributed, with mean equal to 1 and standard deviation equal to .36.

a. Give the mean and standard deviation of the sampling distribution of $\bar{x}$.

b. Find the probability $\bar{x}$ that exceeds 1.3.

c. Find the probability that the sample mean $\bar{x}$ is less than .5.

d. Find the probability that the sample mean deviates from the population mean $\mu = 1$ by more than .4.

7.63 Batteries A certain type of automobile battery is known to last an average of 1110 days with a standard deviation of 80 days. If 400 of these batteries are selected, find the following probabilities for the average length of life of the selected batteries:

a. The average is between 1100 and 1110.

b. The average is greater than 1120.

c. The average is less than 900.

7.64 Lead Pipes Studies indicate that drinking water supplied by some old lead-lined city piping systems may contain harmful levels of lead. An important study of the Boston water supply system showed that the distribution of lead content readings for individual water specimens had a mean and standard deviation of approximately .033 milligrams per liter (mg/l) and .10 mg/l, respectively.[15]

a. Explain why you believe this distribution is or is not normally distributed.

b. Because the researchers were concerned about the shape of the distribution in part a, they calculated the average daily lead levels at 40 different locations on each of 23 randomly selected days. What can you say about the shape of the distribution of the average daily lead levels from which the sample of 23 days was taken?

c. What are the mean and standard deviation of the distribution of average lead levels in part b?

7.65 Biomass Studies[16] indicate that the earth's vegetative mass, or biomass for tropical woodlands, thought to be about 35 kilograms per square meter (kg/m^2), may in fact be too high and that tropical biomass values vary regionally—from about 5 to 55 kg/m^2. Suppose you measure the tropical biomass in 400 randomly selected square-meter plots.

a. Approximate σ, the standard deviation of the biomass measurements.

b. What is the probability that your sample average is within two units of the true average tropical biomass?

c. If your sample average is $\bar{x} = 31.75$, what would you conclude about the overestimation that concerns the scientists?

7.66 Hard Hats The safety requirements for hard hats worn by construction workers and others, established by the American National Standards Institute (ANSI), specify that each of three hats pass the following test. A hat is mounted on an aluminum head form. An 8-pound steel ball is dropped on the hat from a height of 5 feet, and the resulting force is measured at the bottom of the head form. The force exerted on the head form by each of the three hats must be less than 1000 pounds, and the average of the three must be less than 850 pounds. (The relationship between this test and actual human head damage is unknown.) Suppose the exerted force is normally distributed, and hence a sample mean of three force measurements is normally distributed. If a random sample of three hats is selected from a shipment with a mean equal to 900 and $\sigma = 100$, what is the probability that the sample mean will satisfy the ANSI standard?

7.67 Imagery and Memory A research psychologist is planning an experiment to determine whether the use of imagery—picturing a word in your mind—

affects people's ability to memorize. He wants to use two groups of subjects: a group that memorizes a set of 20 words using the imagery technique, and a control group that does not use imagery.

a. Use a randomization technique to divide a group of 20 subjects into two groups of equal size.

b. How can the researcher randomly select the group of 20 subjects?

c. Suppose the researcher offers to pay subjects $50 each to participate in the experiment and uses the first 20 students who apply. Would this group behave as if it were a simple random sample of size $n = 20$?

7.68 Same-Sex Marriage The results of a *CBS News Poll* concerning views on same-sex marriage and gay rights given in the table that follows show that there is no consensus on this issue among Americans.[17]

CBS News Poll. August 20–24, 2010. N = 1,082 adults nationwide. MoE ± 3.

"Which comes closest to your view? Gay couples should be allowed to legally marry. OR, Gay couples should be allowed to form civil unions but not legally marry. OR, There should be no legal recognition of a gay couple's relationship."

| | Legal | | No Legal | |
	Marriage (%)	Civil Unions (%)	Recognition (%)	Unsure (%)
All adults	40	30	25	5
Republicans	25	34	37	4
Democrats	46	27	20	7
Independents	44	29	21	6

---

a. Is this an observational study or a planned experiment?

b. Is there the possibility of problems in responses arising because of the somewhat sensitive nature of the subject? What kinds of biases might occur?

7.69 Sprouting Radishes A biology experiment was designed to determine whether sprouting radish seeds inhibit the germination of lettuce seeds.[18] Three 10-centimeter petri dishes were used. The first contained 26 lettuce seeds, the second contained 26 radish seeds, and the third contained 13 lettuce seeds and 13 radish seeds.

a. Assume that the experimenter had a package of 50 radish seeds and another of 50 lettuce seeds.

Devise a plan for randomly assigning the radish and lettuce seeds to the three treatment groups.

b. What assumptions must the experimenter make about the packages of 50 seeds in order to assure randomness in the experiment?

7.70 9/11 A study of about $n = 1000$ individuals in the United States during September 21–22, 2001, revealed that 43% of the respondents indicated that they were less willing to fly following the events of September 11, 2001.[19]

a. Is this an observational study or a designed experiment?

b. What problems might or could have occurred because of the sensitive nature of the subject? What kinds of biases might have occurred?

7.71 Telephone Service Suppose a telephone company executive wishes to select a random sample of $n = 20$ out of 7000 customers for a survey of customer attitudes concerning service. If the customers are numbered for identification purposes, indicate the customers whom you will include in your sample. Use the random number table and explain how you selected your sample.

7.72 Rh Positive The proportion of individuals with an Rh-positive blood type is 85%. You have a random sample of $n = 500$ individuals.

a. What are the mean and standard deviation of $\hat{p}$, the sample proportion with Rh-positive blood type?

b. Is the distribution of $\hat{p}$ approximately normal? Justify your answer.

c. What is the probability that the sample proportion $\hat{p}$ exceeds 82%?

d. What is the probability that the sample proportion lies between 83% and 88%?

e. 99% of the time, the sample proportion would lie between what two limits?

7.73 What survey design is used in each of these situations?

a. A random sample of $n = 50$ city blocks is selected, and a census is done for each single-family dwelling on each block.

b. The highway patrol stops every 10th vehicle on a given city artery between 9:00 A.M. and 3:00 P.M. to perform a routine traffic safety check.

c. One hundred households in each of four city wards are surveyed concerning a pending city tax relief referendum.

d. Every 10th tree in a managed slash pine plantation is checked for pine needle borer infestation.

e. A random sample of $n = 1000$ taxpayers from the city of San Bernardino is selected by the Internal Revenue Service and their tax returns are audited.

7.74 Elevator Loads The maximum load (with a generous safety factor) for the elevator in an office building is 2000 pounds. The relative frequency distribution of the weights of all men and women using the elevator is mound-shaped (slightly skewed to the heavy weights), with mean μ equal to 150 pounds and standard deviation σ equal to 35 pounds. What is the largest number of people you can allow on the elevator if you want their total weight to exceed the maximum weight with a small probability (say, near .01)? (HINT: Use the alternative statement of the Central Limit Theorem for Σx_i given in Section 7.4.)

7.75 Wiring Packages The number of wiring packages that can be assembled by a company's employees has a normal distribution, with a mean equal to 16.4 per hour and a standard deviation of 1.3 per hour.

a. What are the mean and standard deviation of the number x of packages produced per worker in an 8-hour day?

b. Do you expect the probability distribution for x to be mound-shaped and approximately normal? Explain.

c. What is the probability that a worker will produce at least 135 packages per 8-hour day?

7.76 Wiring Packages, continued Refer to Exercise 7.75. Suppose the company employs 10 assemblers of wiring packages.

a. Find the mean and standard deviation of the company's daily (8-hour day) production of wiring packages.

b. What is the probability that the company's daily production is less than 1280 wiring packages per day?

Data set
EX0777 **7.77 Defective Lightbulbs** The table lists the number of defective 60-watt lightbulbs found in samples of 100 bulbs selected over 25 days from a manufacturing process. Assume that during this time the manufacturing process was not producing an excessively large fraction of defectives.

Day	1	2	3	4	5	6	7	8	9	10
Defectives	4	2	5	8	3	4	4	5	6	1

Day	11	12	13	14	15	16	17	18	19	20
Defectives	2	4	3	4	0	2	3	1	4	0

Day	21	22	23	24	25
Defectives	2	2	3	5	3

a. Construct a p chart to monitor the manufacturing process, and plot the data.

b. How large must the fraction of defective items be in a sample selected from the manufacturing process before the process is assumed to be out of control?

c. During a given day, a sample of 100 items is selected from the manufacturing process and 15 defective bulbs are found. If a decision is made to shut down the manufacturing process in an attempt to locate the source of the implied controllable variation, explain how this decision might lead to erroneous conclusions.

7.78 Lightbulbs, continued A hardware store chain purchases large shipments of lightbulbs from the manufacturer described in Exercise 7.77 and specifies that each shipment must contain no more than 4% defectives. When the manufacturing process is in control, what is the probability that the hardware store's specifications are met?

7.79 Lightbulbs, again Refer to Exercise 7.77. During a given week the number of defective bulbs in each of five samples of 100 were found to be 2, 4, 9, 7, and 11. Is there reason to believe that the production process has been producing an excessive proportion of defectives at any time during the week?

Data set
EX0780 **7.80 Canned Tomatoes** During long production runs of canned tomatoes, the average weights (in ounces) of samples of five cans of standard-grade tomatoes in pureed form were taken at 30 control points during an 11-day period. These results are shown in the table.[20] When the machine is performing normally, the average weight per can is 21 ounces with a standard deviation of 1.20 ounces.

a. Compute the upper and lower control limits and the centerline for the $\bar{x}$ chart.

b. Plot the sample data on the $\bar{x}$ chart and determine whether the performance of the machine is in control.

Sample Number	Average Weight	Sample Number	Average Weight
1	23.1	16	21.4
2	21.3	17	20.4
3	22.0	18	22.8
4	21.4	19	21.1
5	21.8	20	20.7
6	20.6	21	21.6
7	20.1	22	22.4
8	21.4	23	21.3
9	21.5	24	21.1
10	20.2	25	20.1
11	20.3	26	21.2
12	20.1	27	19.9
13	21.7	28	21.1
14	21.0	29	21.6
15	21.6	30	21.3

Source: Adapted from J. Hackl, *Journal of Quality Technology,* April 1991. Used with permission.

7.81 Pepsi or Coke? The battle for consumer preference continues between Pepsi and Coke. How can you make your preferences known? There is a web page where you can vote for one of these colas if you click on the link that says PAY CASH for your opinion. Explain why the respondents do not represent a random sample of the opinions of purchasers or drinkers of these drinks. Explain the types of distortions that could creep into an Internet opinion poll.

7.82 Total Packing Weight Packages of food whose average weight is 16 ounces with a standard deviation of 0.6 ounces are shipped in boxes of 24 packages. If the package weights are approximately normally distributed, what is the probability that a box of 24 packages will weigh more than 392 ounces (24.5 pounds)?

7.83 Filling Soda Cans A bottler of soft drinks packages cans in six-packs. Suppose that the fill per can has an approximate normal distribution with a mean of 12 fluid ounces and a standard deviation of 0.2 fluid ounces.

a. What is the distribution of the total fill for a case of 24 cans?

b. What is the probability that the total fill for a case is less than 286 fluid ounces?

c. If a six-pack of soda can be considered a random sample of size $n = 6$ from the population, what is the probability that the average fill per can for a six-pack of soda is less than 11.8 fluid ounces?

7.84 Strawberries An experimenter wants to find an appropriate temperature at which to store fresh strawberries to minimize the loss of ascorbic acid. There are 20 storage containers, each with controllable temperature, in which strawberries can be stored. If two storage temperatures are to be used, how would the experimenter assign the 20 containers to one of the two storage temperatures?

7.85 Electronic Components A manufacturing process is designed to produce an electronic component for use in small portable television sets. The components are all of standard size and need not conform to any measurable characteristic, but are sometimes inoperable when emerging from the manufacturing process. Fifteen samples were selected from the process at times when the process was known to be in statistical control. Fifty components were observed within each sample, and the number of inoperable components was recorded.

$$6, 7, 3, 5, 6, 8, 4, 5, 7, 3, 1, 6, 5, 4, 5$$

Construct a *p* chart to monitor the manufacturing process.

CASE STUDY **Sampling the Roulette at Monte Carlo**

The technique of simulating a process that contains random elements and repeating the process over and over to see how it behaves is called a **Monte Carlo procedure.** It is widely used in business and other fields to investigate the properties of an operation that is subject to random effects, such as weather, human behavior, and so on. For example, you could model the behavior of a manufacturing company's inventory by creating, on paper, daily arrivals, and departures of manufactured products from the company's warehouse. Each day a random number of items produced by the company would be received into inventory. Similarly, each day a random number of orders of varying random sizes would be shipped. Based on the input and output of items, you could calculate the inventory—that is, the number of items on hand at the end of each day. The values of the random variables, the number of items produced, the number of orders, and the number of items per order needed for each day's simulation would be obtained from theoretical distributions of observations that closely model the corresponding distributions of the variables that have been observed over time in the manufacturing operation. By repeating the simulation of the supply, the shipping, and the calculation of daily inventory for a large number of days (a sampling of what might really happen), you can observe the behavior of the plant's daily inventory. The Monte Carlo procedure is particularly valuable because it enables the manufacturer to see how the daily inventory would behave when certain changes are made in the supply pattern or in some other aspect of the operation that could be controlled.

In an article entitled "The Road to Monte Carlo," Daniel Seligman comments on the Monte Carlo method, noting that although the technique is widely used in business schools to study capital budgeting, inventory planning, and cash flow management, no one seems to have used the procedure to study how well we might do if we were to gamble at Monte Carlo.[21]

To follow up on this thought, Seligman programmed his personal computer to simulate the game of roulette. Roulette involves a wheel with its rim divided into 38 pockets. Thirty-six of the pockets are numbered 1 to 36 and are alternately colored red and black. The two remaining pockets are colored green and are marked 0 and 00. To play the game, you bet a certain amount of money on one or more pockets. The wheel is spun and turns until it stops. A ball falls into a slot on the wheel to indicate the winning number. If you have money on that number, you win a specified amount.

For example, if you were to play the number 20, the payoff is 35 to 1. If the wheel does not stop at that number, you lose your bet. Seligman decided to see how his nightly gains (or losses) would fare if he were to bet $5 on each turn of the wheel and repeat the process 200 times each night. He did this 365 times, thereby simulating the outcomes of 365 nights at the casino. Not surprisingly, the mean "gain" per $1000 evening for the 365 nights was a *loss* of $55, the average of the winnings retained by the gambling house. The surprise, according to Seligman, was the extreme variability of the nightly "winnings." Seven times out of the 365 evenings, the fictitious gambler lost the $1000 stake, and only once did he win a maximum of $1160. On 141 nights, the loss exceeded $250.

1. To evaluate the results of Seligman's Monte Carlo experiment, first find the probability distribution of the gain x on a single $5 bet.

2. Find the expected value and variance of the gain x from part 1.

3. Find the expected value and variance for the evening's gain, the sum of the gains or losses for the 200 bets of $5 each.

4. Use the results of part 2 to evaluate the probability of 7 out of 365 evenings resulting in a loss of the total $1000 stake.

5. Use the results of part 3 to evaluate the probability that the largest evening's winnings were as great as $1160.

Large-Sample Estimation

© Associated Press

GENERAL OBJECTIVE

In previous chapters, you learned about the probability distributions of random variables and the sampling distributions of several statistics that, for large sample sizes, can be approximated by a normal distribution according to the Central Limit Theorem. This chapter presents a method for estimating population parameters and illustrates the concept with practical examples. The Central Limit Theorem and the sampling distributions presented in Chapter 7 play a key role in evaluating the reliability of the estimates.

CHAPTER INDEX

❓ NEED TO KNOW...

How to Estimate a Population Mean or Proportion
How to Choose the Sample Size

How Reliable Is That Poll?

Do the national polls conducted by the Gallup and Harris organizations, the news media, and others provide accurate estimates of the percentages of people in the United States who have a variety of eating habits? The case study at the end of this chapter examines the reliability of a poll conducted by *CBS News* using the theory of large-sample estimation.

WHERE WE'VE BEEN

The first seven chapters of this book have given you the building blocks you will need to understand statistical inference and how it can be applied in practical situations. The first three chapters were concerned with using descriptive statistics, both graphical and numerical, to describe and interpret sets of measurements. In the next three chapters, you learned about probability and probability distributions—the basic tools used to describe *populations* of measurements. The binomial and the normal distributions were emphasized as important for practical applications.

The seventh chapter provided the link between probability and statistical inference. Many statistics are either sums or averages calculated from sample measurements. The Central Limit Theorem states that, even if the sampled populations are not normal, the sampling distributions of these *statistics* will be approximately normal when the sample size *n* is large. These statistics are the tools you use for *inferential statistics*—making inferences about a population using information contained in a sample.

WHERE WE'RE GOING— STATISTICAL INFERENCE

Inference—specifically, decision making and prediction—is centuries old and plays a very important role in most peoples' lives. Here are some applications:

- The government needs to predict short- and long-term interest rates.
- A broker wants to forecast the behavior of the stock market.
- A metallurgist wants to decide whether a new type of steel is more resistant to high temperatures than the current type.
- A consumer wants to estimate the selling price of her house before putting it on the market.

Parameter ⇔ Population
Statistic ⇔ Sample

There are many ways to make these decisions or predictions, some subjective and some more objective in nature. How good will your predictions or decisions be? Although you may feel that your own built-in decision-making ability is quite good, experience suggests that this may not be the case. It is the job of the mathematical statistician to provide methods of statistical inference making that are better and more reliable than just subjective guesses.

Statistical inference is concerned with making decisions or predictions about **parameters**—the numerical descriptive measures that characterize a population. Three parameters you encountered in earlier chapters are the population mean μ, the population standard deviation σ, and the binomial proportion p. In statistical inference, a practical problem is restated in the framework of a population with a specific parameter of interest. For example, the metallurgist could measure the *average* coefficients of expansion for both types of steel and then compare their values.

Methods for making inferences about population parameters fall into one of two categories:

- **Estimation:** Estimating or predicting the value of the parameter
- **Hypothesis testing:** Making a decision about the value of a parameter based on some preconceived idea about what its value might be

EXAMPLE **8.1**   The circuits in computers and other electronics equipment consist of one or more printed circuit boards (PCB), and computers are often repaired by simply replacing one or more defective PCBs. In an attempt to find the proper setting of a plating process applied to one side of a PCB, a production supervisor might *estimate* the average thickness of copper plating on PCBs using samples from several days of operation. Since he has no knowledge of the average thickness μ before observing the production process, his is an *estimation* problem.

EXAMPLE **8.2**   The supervisor in Example 8.1 is told by the plant owner that the thickness of the copper plating must not be less than .001 inch in order for the process to be in control. To decide whether or not the process is in control, the supervisor might formulate a test. He could *hypothesize* that the process is in control—that is, assume that the average thickness of the copper plating is .001 or greater—and use samples from several days of operation to decide whether or not his hypothesis is correct. The supervisor's decision-making approach is called a *test of hypothesis*.

Which method of inference should be used? That is, should the parameter be estimated, or should you test a hypothesis concerning its value? The answer is dictated by the practical question posed and is often determined by personal preference. Since both estimation and tests of hypotheses are used frequently in scientific literature, we include both methods in this and the next chapter.

A statistical problem, which involves planning, analysis, and inference making, is incomplete without a measure of the **goodness of the inference.** That is, how accurate or reliable is the method you have used? If a stockbroker predicts that the price of a stock will be $80 next Monday, will you be willing to take action to buy or sell your stock without knowing how reliable her prediction is? Will the prediction be within $1, $2, or $10 of the actual price next Monday? Statistical procedures are important because they provide two types of information:

- Methods for making the inference
- A numerical measure of the goodness or reliability of the inference

8.3 TYPES OF ESTIMATORS

To estimate the value of a population parameter, you can use information from the sample in the form of an **estimator.** Estimators are calculated using information from the sample observations, and hence, by definition they are also *statistics.*

Definition   An **estimator** is a rule, usually expressed as a formula, that tells us how to calculate an estimate based on information in the sample.

Estimators are used in two different ways:

- **Point estimation:** Based on sample data, a single number is calculated to estimate the population parameter. The rule or formula that describes this calculation is called the **point estimator,** and the resulting number is called a **point estimate.**

- **Interval estimation:** Based on sample data, two numbers are calculated to form an interval within which the parameter is expected to lie. The rule or formula that describes this calculation is called the **interval estimator,** and the resulting pair of numbers is called an **interval estimate** or **confidence interval.**

EXAMPLE 8.3

A veterinarian wants to estimate the average weight gain per month of 4-month-old golden retriever pups that have been placed on a lamb and rice diet. The *population* consists of the weight gains per month of all 4-month-old golden retriever pups that are given this particular diet. The veterinarian wants to estimate the unknown parameter μ, the average monthly weight gain for this *hypothetical* population.

One possible *estimator* based on sample data is the sample mean, $\bar{x} = \Sigma x_i/n$. It could be used in the form of a single number or *point estimate*—for instance, 3.8 pounds—or you could use an *interval estimate* and estimate that the average weight gain will be between 2.7 and 4.9 pounds.

Both point and interval estimation procedures use information provided by the sampling distribution of the specific estimator you have chosen to use. We will begin by discussing *point estimation* and its use in estimating population means and proportions.

8.4 POINT ESTIMATION

In a practical situation, there may be several statistics that could be used as point estimators for a population parameter. To decide which of several choices is best, you need to know how the estimator behaves in repeated sampling, described by its *sampling distribution*.

By way of analogy, think of firing a revolver at a target. The parameter of interest is the bull's-eye, at which you are firing bullets. Each bullet represents a single sample estimate, fired by the revolver, which represents the estimator.

Suppose your friend fires a single shot and hits the bull's-eye. Can you conclude that he is an excellent shot? Would you stand next to the target while he fires a second shot? Probably not, because you have no measure of how well he performs in repeated trials. Does he always hit the bull's-eye, or is he consistently too high or too low? Do his shots cluster closely around the target, or do they consistently miss the target by a wide margin? Figure 8.1 shows several target configurations. Which target would you pick as belonging to the best shot?

FIGURE 8.1

Which marksman is best?

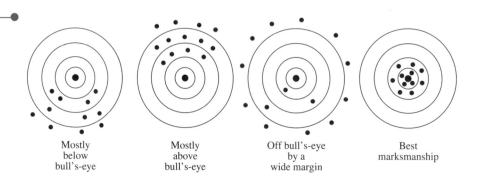

| Mostly below bull's-eye | Mostly above bull's-eye | Off bull's-eye by a wide margin | Best marksmanship |

Sampling distributions provide information that can be used to select the **best estimator.** What characteristics would be valuable? First, the **sampling distribution of the point estimator should be centered over the true value of the parameter to be estimated.** That is, the estimator should not constantly underestimate or overestimate the parameter of interest. Such an estimator is said to be **unbiased.**

Definition   An estimator of a parameter is said to be **unbiased** if the mean of its distribution is equal to the true value of the parameter. Otherwise, the estimator is said to be **biased.**

The sampling distributions for an unbiased estimator and a biased estimator are shown in Figure 8.2. The sampling distribution for the biased estimator is shifted to the right of the true value of the parameter. This biased estimator is more likely than an unbiased one to overestimate the value of the parameter.

FIGURE 8.2

Distributions for biased and unbiased estimators

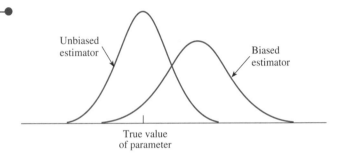

The second desirable characteristic of an estimator is that **the spread (as measured by the variance) of the sampling distribution should be as small as possible.** This ensures that, with a high probability, an individual estimate will fall close to the true value of the parameter. The sampling distributions for two unbiased estimators, one with a small variance[†] and the other with a larger variance, are shown in Figure 8.3.

FIGURE 8.3

Comparison of estimator variability

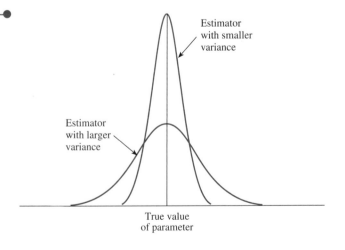

[†]Statisticians usually use the term *variance of an estimator* when in fact they mean the variance of the sampling distribution of the estimator. This contractive expression is used almost universally.

Naturally, you would prefer the estimator with the smaller variance because the estimates tend to lie closer to the true value of the parameter than in the distribution with the larger variance.

In real-life sampling situations, you may know that the sampling distribution of an estimator centers about the parameter that you are attempting to estimate, but all you have is the estimate computed from the n measurements contained in the sample. How far from the true value of the parameter will your estimate lie? How close is the marksman's bullet to the bull's-eye? The distance between the estimate and the true value of the parameter is called the **error of estimation.**

Definition The distance between an estimate and the estimated parameter is called the **error of estimation.**

In this chapter, you may assume that the sample sizes are always large and, therefore, that the *unbiased* estimators you will study have sampling distributions that can be approximated by a normal distribution (because of the Central Limit Theorem). Remember that, for any point estimator with a normal distribution, the Empirical Rule states that approximately 95% of all the point estimates will lie within two (or more exactly, 1.96) standard deviations of the mean of that distribution.

For *unbiased* estimators, this implies that the difference between the point estimator and the true value of the parameter will be less than 1.96 standard deviations or 1.96 standard errors (SE). This quantity, called the 95% **margin of error** (or simply the **"margin of error"**), provides a practical upper bound for the error of estimation (see Figure 8.4). It is possible that the error of estimation will exceed this margin of error, but that is very unlikely.

FIGURE 8.4

Sampling distribution of an unbiased estimator

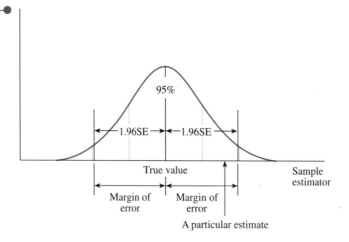

POINT ESTIMATION OF A POPULATION PARAMETER

- Point estimator: a statistic calculated using sample measurements
- 95% Margin of error: 1.96 × Standard error of the estimator

---

The sampling distributions for two unbiased point estimators were discussed in Chapter 7. It can be shown that both of these point estimators have the *minimum variability* of all unbiased estimators and are thus the *best estimators* you can find in each situation.

The variability of the estimator is measured using its standard error. However, you might have noticed that the standard error usually depends on unknown parameters such as σ or p. These parameters must be estimated using sample statistics such as s and $\hat{p}$. Although not exactly correct, experimenters generally refer to the estimated standard error as *the standard error.*

NEED TO KNOW...

How to Estimate a Population Mean or Proportion

- To estimate the population mean μ for a quantitative population, the point estimator $\bar{x}$ is *unbiased* with standard error estimated as

$$\text{SE} = \frac{s}{\sqrt{n}}^\dagger$$

The 95% margin of error when $n \geq 30$ is estimated as

$$\pm 1.96 \left(\frac{s}{\sqrt{n}} \right)$$

- To estimate the population proportion p for a binomial population, the point estimator $\hat{p} = x/n$ is *unbiased,* with standard error estimated as

$$\text{SE} = \sqrt{\frac{\hat{p}\hat{q}}{n}}$$

The 95% margin of error is estimated as

$$\pm 1.96 \sqrt{\frac{\hat{p}\hat{q}}{n}}$$

Assumptions: $n\hat{p} > 5$ and $n\hat{q} > 5$.

EXAMPLE **8.4**

An environmentalist is conducting a study of the polar bear, a species found in and around the Arctic Ocean. Their range is limited by the availability of sea ice, which they use as a platform to hunt seals, the mainstay of their diet. The destruction of its habitat on the Arctic ice, which has been attributed to global warming, threatens the bear's survival as a species; it may become extinct within the century.[1] A random sample of $n = 50$ polar bears produced an average weight of 980 pounds with a standard deviation of 105 pounds. Use this information to estimate the average weight of all Arctic polar bears.

Solution The random variable measured is weight, a quantitative random variable best described by its mean μ. The point estimate of μ, the average weight of all Arctic polar bears, is $\bar{x} = 980$ pounds. The margin of error is estimated as

$$1.96 \text{ SE} = 1.96 \left(\frac{s}{\sqrt{n}} \right) = 1.96 \left(\frac{105}{\sqrt{50}} \right) = 29.10 \approx 29 \text{ pounds}$$

---

†When you sample from a normal distribution, the statistic $(\bar{x} - \mu)/(s/\sqrt{n})$ has a t distribution, which will be discussed in Chapter 10. When the sample is *large,* this statistic is approximately normally distributed whether the sampled population is normal or nonnormal.

You can be fairly confident that the sample estimate of 980 pounds is within ±29 pounds of the population mean.

In reporting research results, investigators often attach either the sample standard deviation s (sometimes called SD) or the standard error $s/\sqrt{n}$ (usually called SE or SEM) to the estimates of population means. You should always look for an explanation somewhere in the text of the report that tells you whether the investigator is reporting $\bar{x} \pm$ SD or $\bar{x} \pm$ SE. In addition, the sample means and standard deviations or standard errors are often presented as "error bars" using the graphical format shown in Figure 8.5.

FIGURE 8.5 ●

Plot of treatment means and their standard errors

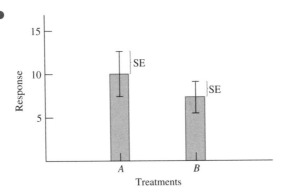

EXAMPLE 8.5

In addition to the average weight of the Arctic polar bear, the environmentalist from Example 8.4 is also interested in the opinions of adults on the subject of global warming. In particular, he wants to estimate the proportion of adults who think that global warming is a very serious problem. In a random sample of $n = 100$ adults, 73% of the sample indicated that global warming is a very serious problem. Estimate the true population proportion of adults who believe that global warming is a very serious problem, and find the margin of error for the estimate.

Solution The parameter of interest is now p, the proportion of individuals in the population who believe that global warming is a very serious problem. The best estimator of p is the sample proportion $\hat{p}$, which for this sample is $\hat{p} = .73$. In order to find the margin of error, you can approximate the value of p with its estimate $\hat{p} = .73$:

$$1.96 \; \text{SE} = 1.96 \sqrt{\frac{\hat{p}\hat{q}}{n}} = 1.96 \sqrt{\frac{.73(.27)}{100}} = .09$$

With this margin of error, you can be fairly confident that the estimate of .73 is within ±.09 of the true value of p. Hence, you can conclude that the true value of p could be as small as .64 or as large as .82. This margin of error is quite large when compared to the estimate itself and reflects the fact that large samples are required to achieve a small margin of error when estimating p.

TABLE 8.1 ● **Some Calculated Values of $\sqrt{pq}$**

p	pq	$\sqrt{pq}$	p	pq	$\sqrt{pq}$
.1	.09	.30	.6	.24	.49
.2	.16	.40	.7	.21	.46
.3	.21	.46	.8	.16	.40
.4	.24	.49	.9	.09	.30
.5	.25	.50			

Table 8.1 shows how the numerator of the standard error of $\hat{p}$ changes for various values of p. Notice that, for most values of p—especially when p is between .3 and .7—there is very little change in $\sqrt{pq}$, the numerator of SE, reaching its maximum value when $p = .5$. This means that the margin of error using the estimator $\hat{p}$ will also be a maximum when $p = .5$. Some pollsters routinely use the maximum margin of error—often called the **sampling error**—when estimating p, in which case they calculate

$$1.96 \text{ SE} = 1.96\sqrt{\frac{.5(.5)}{n}} \quad \text{or sometimes} \quad 2 \text{ SE} = 2\sqrt{\frac{.5(.5)}{n}}$$

Gallup, Harris, and Roper polls generally use sample sizes of approximately 1000, so their margin of error is

$$1.96\sqrt{\frac{.5(.5)}{1000}} = .031 \quad \text{or approximately 3\%}$$

In this case, the estimate is said to be within ± 3 percentage points of the true population proportion.

8.4 ● **EXERCISES**

BASIC TECHNIQUES

8.1 Explain what is meant by "margin of error" in point estimation.

8.2 A random sample of $n = 50$ observations from a quantitative population produced $\bar{x} = 56.4$ and $s^2 = 2.6$. Give the best point estimate for the population mean μ, and calculate the margin of error.

8.3 Calculate the margin of error in estimating a population mean μ for these values:

a. $n = 30, \sigma^2 = .2$ **b.** $n = 30, \sigma^2 = .9$
c. $n = 30, \sigma^2 = 1.5$

8.4 Suppose you are writing a questionnaire for a sample survey involving $n = 100$ individuals. The questionnaire will generate estimates for several different binomial proportions. If you want to report a single margin of error for the survey, which margin of error from Exercise 8.9 is the correct one to use?

8.5 Calculate the margin of error in estimating a population mean μ for these values:

a. $n = 50, s^2 = 4$ **b.** $n = 500, s^2 = 4$
c. $n = 5000, s^2 = 4$

8.6 Refer to Exercise 8.7. What effect does increasing the sample size have on the margin of error?

8.7 Calculate the margin of error in estimating a binomial proportion for each of the following values of n. Use $p = .5$ to calculate the standard error of the estimator.

a. $n = 30$ **b.** $n = 100$
c. $n = 400$ **d.** $n = 1000$

8.8 A random sample of $n = 75$ observations from a quantitative population produced $\bar{x} = 29.7$ and $s^2 = 10.8$. Give the best point estimate for the population mean μ and calculate the margin of error.

8.9 Calculate the margin of error in estimating a binomial proportion p using samples of size $n = 100$ and the following values for p:

a. $p = .1$ **b.** $p = .3$ **c.** $p = .5$

d. $p = .7$ **e.** $p = .9$

f. Which of the values of p produces the largest margin of error?

8.10 Refer to Exercise 8.5. What effect does an increased sample size have on the margin of error?

8.11 A random sample of $n = 900$ observations from a binomial population produced $x = 655$ successes. Estimate the binomial proportion p and calculate the margin of error.

8.12 Refer to Exercise 8.3. What effect does a larger population variance have on the margin of error?

8.13 A random sample of $n = 500$ observations from a binomial population produced $x = 450$ successes. Estimate the binomial proportion p and calculate the margin of error.

8.14 What are two characteristics of the best point estimator for a population parameter?

APPLICATIONS

8.15 The San Andreas Fault One of the most famous large fractures (cracks) in the earth's crust is the San Andreas fault in California. A geologist attempting to study the movement of the earth's crust at a particular location found many fractures in the local rock structure. In an attempt to determine the mean angle of the breaks, she sampled $n = 50$ fractures and found the sample mean and standard deviation to be 39.8° and 17.2°, respectively. Estimate the mean angular direction of the fractures and find the margin of error for your estimate.

8.16 Biomass Estimates of the earth's biomass, the total amount of vegetation held by the earth's forests, are important in determining the amount of unabsorbed carbon dioxide that is expected to remain in the earth's atmosphere.[2] Suppose a sample of 75 one-square-meter plots, randomly chosen in North America's boreal (northern) forests, produced a mean biomass of 4.2 kilograms per square meter (kg/m^2), with a standard deviation of 1.5 kg/m^2. Estimate the average biomass for the boreal forests of North America and find the margin of error for your estimate.

Source: Reprinted with permission from *Science News*, the weekly newsmagazine of *Science*, copyright 1989 by Science Services, Inc.

8.17 Consumer Confidence An increase in the rate of consumer savings is frequently tied to a lack of confidence in the economy and is said to be an indicator of a recessional tendency in the economy. A random sampling of $n = 200$ savings accounts in a local community showed a mean increase in savings account values of 7.2% over the past 12 months, with a standard deviation of 5.6%. Estimate the mean percent increase in savings account values over the past 12 months for depositors in the community. Find the margin of error for your estimate.

8.18 Multimedia Kids Do our children spend as much time enjoying the outdoors and playing with family and friends as previous generations did? Or are our children spending more and more time glued to the television, computer, and other multimedia equipment? A random sample of 250 children between the ages of 8 and 18 showed that 170 children had a TV in their bedroom and that 120 of them had a video game player in their bedroom.

a. Estimate the proportion of all 8- to 18-year-olds who have a TV in their bedroom, and calculate the margin of error for your estimate.

b. Estimate the proportion of all 8- to 18-year-olds who have a video game player in their bedroom, and calculate the margin of error for your estimate.

8.19 Illegal Immigration In a recent poll that included questions about illegal immigration into the United States, and the federal and state responses to the problem, 75% of the $n = 1004$ adults surveyed felt that the United States is not doing enough to keep illegal immigrants from coming into this country.[3]

a. What is a point estimate for the proportion of adults who feel that the United States is not doing enough to keep illegal immigrants from coming into this country? Calculate the margin of error.

b. The poll reports a margin of error of $\pm 3.5\%$. How should the reported margin of error be calculated so that it can be applied to all of the questions in the survey? Is the reported margin of error correct?

8.20 Hotel Costs Even within a particular chain of hotels, lodging during the summer months can vary substantially depending on the type of room and the amenities offered.[4] Suppose that we randomly select 50 billing statements from each of the computer databases of the Marriott, Westin, and the Doubletree hotel chains, and record the nightly room rates.

	Marriott	Westin	Doubletree
Sample Average ($)	150	165	125
Sample Standard Deviation	17.2	22.5	12.8

a. Describe the sampled population(s).

b. Find a point estimate for the average room rate for the Marriott hotel chain. Calculate the margin of error.

c. Find a point estimate for the average room rate for the Westin hotel chain. Calculate the margin of error.

d. Find a point estimate for the average room rate for the Doubletree hotel chain. Calculate the margin of error.

e. Display the results of parts b, c, and d graphically, using the form shown in Figure 8.5. Use this display to compare the average room rates for the three hotel chains.

8.21 "900" Numbers Radio and television stations often air controversial issues during broadcast time and ask viewers to indicate their agreement or disagreement with a given stand on the issue. A poll is conducted by asking those viewers who *agree* to call a certain 900 telephone number and those who *disagree* to call a second 900 telephone number. All respondents pay a fee for their calls.

a. Does this polling technique result in a random sample?

b. What can be said about the validity of the results of such a survey? Do you need to worry about a margin of error in this case?

8.22 Men On Mars? Do you think that the United States should pursue a program to send humans to Mars? An opinion poll conducted by the Associated Press indicated that 49% of the 1034 adults surveyed think that we should pursue such a program.[5]

a. Estimate the true proportion of Americans who think that the United States should pursue a program to send humans to Mars. Calculate the margin of error.

b. The question posed in part a was only one of many questions concerning our space program that were asked in the opinion poll. If the Associated Press wanted to report one sampling error that would be valid for the entire poll, what value should they report?

8.23 Hungry Rats In an experiment to assess the strength of the hunger drive in rats, 30 previously trained animals were deprived of food for 24 hours. At the end of the 24-hour period, each animal was put into a cage where food was dispensed if the animal pressed a lever. The length of time the animal continued pressing the bar (although receiving no food) was recorded for each animal. If the data yielded a sample mean of 19.3 minutes with a standard deviation of 5.2 minutes, estimate the true mean time and calculate the margin of error.

8.5

INTERVAL ESTIMATION

An *interval estimator* is a rule for calculating two numbers—say, a and b—to create an interval that you are fairly certain contains the parameter of interest. The concept of "fairly certain" means "with high probability." We measure this probability using the **confidence coefficient,** designated by $1 - \alpha$.

Definition The probability that a confidence interval will contain the estimated parameter is called the **confidence coefficient.**

NEED A TIP?

Like lariat roping:

Parameter = Fence post

Interval estimate = Lariat

For example, experimenters often construct 95% confidence intervals. This means that the confidence coefficient, or the probability that the interval will contain the estimated parameter, is .95. You can increase or decrease your amount of certainty by changing the confidence coefficient. Some values typically used by experimenters are .90, .95, .98, and .99.

Consider an analogy—this time, throwing a lariat at a fence post. The fence post represents the parameter that you wish to estimate, and the loop formed by the lariat represents the confidence interval. Each time you throw your lariat, you hope to rope the fence post; however, sometimes your lariat misses. In the same way, each time you draw a sample and construct a confidence interval for a parameter, you hope to include the parameter in your interval, but, just like the lariat, sometimes you miss. Your "success rate"—the proportion of intervals that "rope the post" in repeated sampling—is the confidence coefficient.

Constructing a Confidence Interval

When the sampling distribution of a point estimator is approximately normal, an interval estimator or **confidence interval** can be constructed using the following reasoning. For simplicity, assume that the confidence coefficient is .95 and refer to Figure 8.6.

FIGURE 8.6

Parameter ±1.96 SE

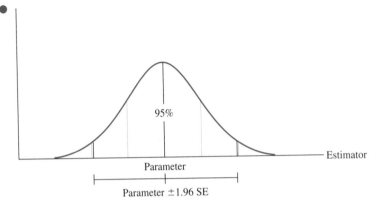

- We know that, of all possible values of the estimator that we might select, 95% of them will be in the interval

    Parameter ±1.96 SE

    shown in Figure 8.6.
- Since the value of the parameter is unknown, consider constructing the interval

    Estimator ±1.96 SE

    which has the same width as the first interval, but has a variable center.
- How often will this interval work properly and enclose the parameter of interest? Refer to Figure 8.7.

FIGURE 8.7

Some 95% confidence intervals

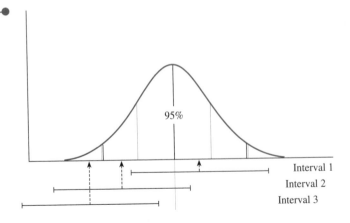

NEED A TIP?

Like a game of ring toss:
Parameter = Peg
Interval estimate = Ring

The first two intervals work properly—the parameter (marked with a light gray line) is contained within both intervals. The third interval does not work, since it fails to enclose the parameter. This happened because the value of the estimator at the center of the interval was too far away from the parameter. Fortunately, values of the estimator only fall this far away 5% of the time—our procedure will work properly 95% of the time!

You may want to change the *confidence coefficient* from $(1 - \alpha) = .95$ to another confidence level $(1 - \alpha)$. To accomplish this, you need to change the value $z = 1.96$, which locates an area .95 in the center of the standard normal curve, to a value of z that locates the area $(1 - \alpha)$ in the center of the curve, as shown in Figure 8.8.

Since the total area under the curve is 1, the remaining area in the two tails is α, and each tail contains area $\alpha/2$. The value of z that has "tail area" $\alpha/2$ to its right is called $z_{\alpha/2}$, and the area between $-z_{\alpha/2}$ and $z_{\alpha/2}$ is the confidence coefficient $(1 - \alpha)$. Values of $z_{\alpha/2}$ that are typically used by experimenters will become familiar to you as you begin to construct confidence intervals for different practical situations. Some of these values are given in Table 8.2.

FIGURE 8.8

Location of $z_{\alpha/2}$

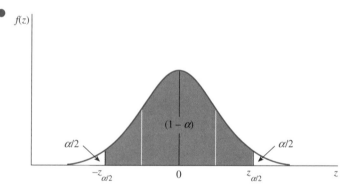

A $(1 - \alpha)$100% LARGE-SAMPLE CONFIDENCE INTERVAL

(Point estimator) $\pm z_{\alpha/2} \times$ (Standard error of the estimator)

where $z_{\alpha/2}$ is the z-value with an area $\alpha/2$ in the right tail of a standard normal distribution. This formula generates two values; the **lower confidence limit (LCL)** and the **upper confidence limit (UCL).**

TABLE 8.2 ● **Values of z Commonly Used for Confidence Intervals**

Confidence Coefficient, $(1 - \alpha)$	α	$\alpha/2$	$z_{\alpha/2}$
.90	.10	.05	1.645
.95	.05	.025	1.96
.98	.02	.01	2.33
.99	.01	.005	2.58

Large-Sample Confidence Interval for a Population Mean μ

Practical problems very often lead to the estimation of μ, the mean of a population of quantitative measurements. Here are some examples:

- The average achievement of college students at a particular university
- The average strength of a new type of steel
- The average number of deaths per age category
- The average demand for a new cosmetics product

When the sample size n is large, the sample mean $\bar{x}$ is the best point estimator for the population mean μ. Since its sampling distribution is approximately normal, it can be used to construct a confidence interval according to the general approach given earlier.

A $(1 - \alpha)100\%$ LARGE-SAMPLE CONFIDENCE INTERVAL FOR A POPULATION MEAN μ

$$\bar{x} \pm z_{\alpha/2}\frac{\sigma}{\sqrt{n}}$$

where $z_{\alpha/2}$ is the z-value corresponding to an area $\alpha/2$ in the upper tail of a standard normal z distribution, and

$n = $ Sample size
$\sigma = $ Standard deviation of the sampled population

If σ is unknown, it can be approximated by the sample standard deviation s when the sample size is large ($n \geq 30$) and the approximate confidence interval is

$$\bar{x} \pm z_{\alpha/2}\frac{s}{\sqrt{n}}$$

Another way to find the large-sample confidence interval for a population mean μ is to begin with the statistic

$$z = \frac{\bar{x} - \mu}{\sigma/\sqrt{n}}$$

which has a standard normal distribution. If you write $z_{\alpha/2}$ as the value of z with area $\alpha/2$ to its right, then you can write

$$P\left(-z_{\alpha/2} < \frac{\bar{x} - \mu}{\sigma/\sqrt{n}} < z_{\alpha/2}\right) = 1 - \alpha$$

You can rewrite this inequality as

$$-z_{\alpha/2}\frac{\sigma}{\sqrt{n}} < \bar{x} - \mu < z_{\alpha/2}\frac{\sigma}{\sqrt{n}}$$

$$-\bar{x} - z_{\alpha/2}\frac{\sigma}{\sqrt{n}} < -\mu < -\bar{x} + z_{\alpha/2}\frac{\sigma}{\sqrt{n}}$$

so that

$$P\left(\bar{x} - z_{\alpha/2}\frac{\sigma}{\sqrt{n}} < \mu < \bar{x} + z_{\alpha/2}\frac{\sigma}{\sqrt{n}}\right) = 1 - \alpha$$

Both $\bar{x} - z_{\alpha/2}(\sigma/\sqrt{n})$ and $\bar{x} + z_{\alpha/2}(\sigma/\sqrt{n})$, the lower and upper confidence limits, are actually random quantities that depend on the sample mean $\bar{x}$. Therefore, in repeated sampling, the random interval, $\bar{x} \pm z_{\alpha/2}(\sigma/\sqrt{n})$, will contain the population mean μ with probability $(1 - \alpha)$.

EXAMPLE **8.6**

A dietician selected a random sample of $n = 50$ male adults and found that their average daily intake of dairy products was $\bar{x} = 756$ grams per day with a standard deviation of $s = 35$ grams per day. Use this sample information to construct a 95% confidence interval for the mean daily intake of dairy products for men.

Solution Since the sample size of $n = 50$ is large, the distribution of the sample mean $\bar{x}$ is approximately normally distributed with mean μ and standard error estimated by $s/\sqrt{n}$. The approximate 95% confidence interval is

$$\bar{x} \pm 1.96\left(\frac{s}{\sqrt{n}}\right)$$

$$756 \pm 1.96\left(\frac{35}{\sqrt{50}}\right)$$

$$756 \pm 9.70$$

Hence, the 95% confidence interval for μ is from 746.30 to 765.70 grams per day.

 **NEED A TIP?**

A 95% confidence interval tells you that, if you were to construct many of these intervals (all of which would have slightly different endpoints), 95% of them would enclose the population mean.

Interpreting the Confidence Interval

What does it mean to say you are "95% confident" that the true value of the population mean μ is within a given interval? If you were to construct 20 such intervals, each using different sample information, your intervals might look like those shown in Figure 8.9(a). Of the 20 intervals, you might expect that 95% of them, or 19 out of 20, will perform as planned and contain μ within their upper and lower bounds. If you constructed 100 such intervals (Figure 8.9(b)), you would expect about 95 of them

FIGURE 8.9

Interpreting Confidence Intervals

MY ONLINE APPLET

Interpreting Confidence Intervals

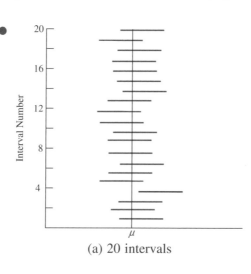

(a) 20 intervals

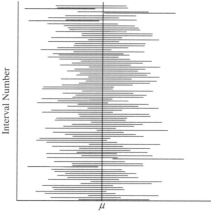

(b) 100 intervals

to perform as planned. Remember that you cannot be absolutely sure that any one particular interval contains the mean μ. You will never know whether your particular interval is one of the 19 that "worked," or whether it is the one interval that "missed." Your confidence in the estimated interval follows from the fact that when repeated intervals are calculated, 95% of these intervals will contain μ.

A good confidence interval has two desirable characteristics:

- It is as narrow as possible. The narrower the interval, the more exactly you have located the estimated parameter.
- It has a large confidence coefficient, near 1. The larger the confidence coefficient, the more likely it is that the interval will contain the estimated parameter.

EXAMPLE **8.7**

Construct a 99% confidence interval for the mean daily intake of dairy products for adult men in Example 8.6.

Solution To change the confidence level to .99, you must find the appropriate value of the standard normal z that puts area $(1 - \alpha) = .99$ in the center of the curve. This value, with tail area $\alpha/2 = .005$ to its right, is found from Table 8.2 to be $z = 2.58$ (see Figure 8.10). The 99% confidence interval is then

$$\bar{x} \pm 2.58\left(\frac{s}{\sqrt{n}}\right)$$

$$756 \pm 2.58(4.95)$$

$$756 \pm 12.77$$

or 743.23 to 768.77 grams per day. This confidence interval is *wider* than the 95% confidence interval in Example 8.6.

FIGURE 8.10

Standard normal values for a 99% confidence interval

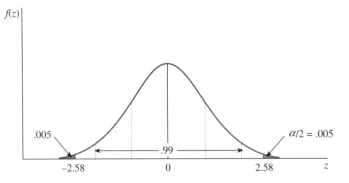

NEED A TIP?

Right Tail Area	z-Value
.05	1.645
.025	1.96
.01	2.33
.005	2.58

The increased width is necessary to increase the confidence, just as you might want a wider loop on your lariat to ensure roping the fence post! The only way to *increase the confidence* without increasing the width of the interval is to *increase the sample size, n.*

The standard error of $\bar{x}$,

$$SE = \frac{\sigma}{\sqrt{n}}$$

Exploring Confidence
Intervals

measures the variability or spread of the values of $\bar{x}$. The more variable the population data, measured by σ, the more variable will be $\bar{x}$, and the standard error will be
larger. On the other hand, if you increase the sample size n, more information is available for estimating μ. The estimates should fall closer to μ and the standard error will
be smaller.

The confidence intervals of Examples 8.6 and 8.7 are approximate because you
substituted s as an approximation for σ. That is, instead of the confidence coefficient
being .95, the value specified in the example, the true value of the coefficient may be
.92, .94, or .97. But this discrepancy is of little concern from a practical point of view;
as far as your "confidence" is concerned, there is little difference among these confidence coefficients. Most interval estimators used in statistics yield approximate confidence intervals because the assumptions upon which they are based are not satisfied
exactly. Having made this point, we will not continue to refer to confidence intervals
as "approximate." It is of little practical concern as long as the actual confidence coefficient is near the value specified.

Large-Sample Confidence Interval for a Population Proportion p

Many research experiments or sample surveys have as their objective the estimation
of the proportion of people or objects in a large group that possess a certain characteristic. Here are some examples:

- The proportion of sales that can be expected in a large number of customer
  contacts

- The proportion of seeds that germinate

- The proportion of "likely" voters who plan to vote for a particular political
  candidate

Each is a practical example of the binomial experiment, and the parameter to be estimated is the binomial proportion p.

When the sample size is large, the sample proportion,

$$\hat{p} = \frac{x}{n} = \frac{\text{Total number of successes}}{\text{Total number of trials}}$$

is the best point estimator for the population proportion p. Since its sampling distribution is approximately normal, with mean p and standard error SE $= \sqrt{pq/n}$, $\hat{p}$ can
be used to construct a confidence interval according to the general approach given in
this section.

 **NEED A TIP?**

Right Tail Area	z-Value
.05	1.645
.025	1.96
.01	2.33
.005	2.58

A $(1 - \alpha)100\%$ LARGE-SAMPLE CONFIDENCE INTERVAL FOR A POPULATION PROPORTION p

$$\hat{p} \pm z_{\alpha/2} \sqrt{\frac{pq}{n}}$$

where $z_{\alpha/2}$ is the z-value corresponding to an area $\alpha/2$ in the right tail of a standard normal z distribution. Since p and q are unknown, they are estimated using

the best point estimators: $\hat{p}$ and $\hat{q}$. The sample size is considered large when the normal approximation to the binomial distribution is adequate—namely, when $n\hat{p} > 5$ and $n\hat{q} > 5$.

EXAMPLE 8.8

A random sample of 985 "likely" voters—those who are likely to vote in the upcoming election—were polled during a phone-athon conducted by the Republican Party. Of those surveyed, 592 indicated that they intended to vote for the Republican candidate in the upcoming election. Construct a 90% confidence interval for p, the proportion of likely voters in the population who intend to vote for the Republican candidate. Based on this information, can you conclude that the candidate will win the election?

Solution   The point estimate for p is

$$\hat{p} = \frac{x}{n} = \frac{592}{985} = .601$$

and the estimated standard error is

$$\sqrt{\frac{\hat{p}\hat{q}}{n}} = \sqrt{\frac{(.601)(399)}{985}} = .016$$

The z-value for a 90% confidence interval is the value that has area $\alpha/2 = .05$ in the upper tail of the z distribution, or $z_{.05} = 1.645$ from Table 8.2. The 90% confidence interval for p is thus

$$\hat{p} \pm 1.645 \sqrt{\frac{\hat{p}\hat{q}}{n}}$$

$$.601 \pm .026$$

or $.575 < p < .627$. You estimate that the percentage of likely voters who intend to vote for the Republican candidate is between 57.5% and 62.7%. Will the candidate win the election? Assuming that she needs more than 50% of the vote to win, and since both the upper and lower confidence limits exceed this minimum value, you can say with 90% confidence that the candidate will win.

There are some problems, however, with this type of sample survey. What if the voters who consider themselves "likely to vote" do not actually go to the polls? What if a voter changes his or her mind between now and election day? What if a surveyed voter does not respond truthfully when questioned by the campaign worker? The 90% confidence interval you have constructed gives you 90% confidence only if you have selected a *random sample from the population of interest.* You can no longer be assured of "90% confidence" if your sample is biased, or if the population of voter responses changes before the day of the election!

You may have noticed that the point estimator with its 95% margin of error looks very similar to a 95% confidence interval for the same parameter. This close relationship exists for most of the parameters estimated in this book, but it is not true in general. Sometimes the best point estimator for a parameter *does not* fall in the middle of the best confidence interval; the best confidence interval may not even be a function of the best point estimator. Although this is a theoretical distinction, you should remember that there is a difference between point and interval estimation, and that the choice between the two depends on the preference of the experimenter.

8.5 EXERCISES

BASIC TECHNIQUES

8.24 Compare the confidence intervals in Exercise 8.29. What effect does each of these actions have on the width of a confidence interval?

a. Double the sample size

b. Quadruple the sample size

8.25 Find a 90% confidence interval for a population mean μ for these values:

a. $n = 125, \bar{x} = .84, s^2 = .086$

b. $n = 50, \bar{x} = 21.9, s^2 = 3.44$

c. Interpret the intervals found in parts a and b.

8.26 Suppose the number of successes observed in $n = 500$ trials of a binomial experiment is 27. Find a 95% confidence interval for p. Why is the confidence interval narrower than the confidence interval in Exercise 8.27?

8.27 A random sample of $n = 300$ observations from a binomial population produced $x = 263$ successes. Find a 90% confidence interval for p and interpret the interval.

8.28 Find a $(1 - \alpha)100\%$ confidence interval for a population mean μ for these values:

a. $\alpha = .01, n = 38, \bar{x} = 34, s^2 = 12$

b. $\alpha = .10, n = 65, \bar{x} = 1049, s^2 = 51$

c. $\alpha = .05, n = 89, \bar{x} = 66.3, s^2 = 2.48$

8.29 A random sample of n measurements is selected from a population with unknown mean μ and known standard deviation $\sigma = 10$. Calculate the width of a 95% confidence interval for μ for these values of n:

a. $n = 100$ **b.** $n = 200$ **c.** $n = 400$

8.30 Find and interpret a 95% confidence interval for a population mean μ for these values:

a. $n = 36, \bar{x} = 13.1, s^2 = 3.42$

b. $n = 64, \bar{x} = 2.73, s^2 = .1047$

8.31 Refer to Exercise 8.30.

a. Calculate the width of a 90% confidence interval for μ when $n = 100$.

b. Calculate the width of a 99% confidence interval for μ when $n = 100$.

c. Compare the widths of 90%, 95%, and 99% confidence intervals for μ. What effect does increasing the confidence coefficient have on the width of the confidence interval?

APPLICATIONS

8.32 A Chemistry Experiment In an electrolysis experiment, a class measured the amount of copper precipitated from a saturated solution of copper sulfate over a 30-minute period. The $n = 30$ students calculated a sample mean and standard deviation equal to .145 and .0051 mole, respectively. Find a 90% confidence interval for the mean amount of copper precipitated from the solution over a 30-minute period.

8.33 Acid Rain Acid rain, caused by the reaction of certain air pollutants with rainwater, is a growing problem in the United States. Pure rain falling through clean air registers a pH value of 5.7 (pH is a measure of acidity: 0 is acid; 14 is alkaline). Suppose water samples from 40 rainfalls are analyzed for pH, and $\bar{x}$ and s are equal to 3.7 and .5, respectively. Find a 99% confidence interval for the mean pH in rainfall and interpret the interval. What assumption must be made for the confidence interval to be valid?

8.34 Working Women In an *Advertising Age* white paper concerning the changing role of women as "breadwinners" in the American family, it was reported that according to their survey with JWT, working men reported doing 54 minutes of household chores a day, while working women reported tackling 72 minutes daily. But when examined more closely, Millennial men reported doing just as many household chores as the average working women, 72 minutes, compared to an average of 54 minutes among both Boomer men and Xer men.[6] The information that follows is adapted from these data and is based on random samples of 1136 men and 795 women.

	Mean	Standard Deviation	n
All Women	72	10.4	795
All Men	54	12.7	1136
Millennial	72	9.2	345
Boomers	54	13.9	475
Xers	54	10.5	316

a. Construct a 95% confidence interval for the average time all men spend doing household chores.

b. Construct a 95% confidence interval for the average time women spend doing household chores.

8.35 Hamburger Meat The meat department of a local supermarket chain packages ground beef using meat trays of two sizes: one designed to hold approximately 1 pound of meat, and one that holds approximately 3 pounds. A random sample of 35 packages in the smaller meat trays produced weight measurements with an average of 1.01 pounds and a standard deviation of .18 pound.

a. Construct a 99% confidence interval for the average weight of all packages sold in the smaller meat trays by this supermarket chain.

b. What does the phrase "99% confident" mean?

c. Suppose that the quality control department of this supermarket chain intends that the amount of ground beef in the smaller trays should be 1 pound on average. Should the confidence interval in part a concern the quality control department? Explain.

8.36 Who Killed the Electric Car? New car models with names such as the "Volt" and the "Leaf" are being hyped by automakers, as they scramble to produce electric cars that are affordable for most Americans. Still in the trial stage, BMW's *Mini E* can be leased for about $600 per month, and is claimed to be able to travel between 100 and 120 miles per battery charge.[8] Suppose that $n = 60$ field trials are conducted and that the average time between charges is 112.5 miles with a standard deviation of 4.6 miles.

a. Construct a 95% confidence interval for μ, the average time between battery charges for BMW's *Mini E*.

b. Does the confidence interval in part a confirm the claim of 100 to 120 miles per battery charge? Why or why not?

8.37 SUVs A sample survey is designed to estimate the proportion of sports utility vehicles being driven in the state of California. A random sample of 500 registrations are selected from a Department of Motor Vehicles database, and 68 are classified as sports utility vehicles.

a. Use a 95% confidence interval to estimate the proportion of sports utility vehicles in California.

b. How can you estimate the proportion of sports utility vehicles in California with a higher degree of accuracy? (HINT: There are two answers.)

8.38 Same-Sex Marriage The results of a *CBS News Poll* concerning views on same-sex marriage and gay rights given in Exercise 7.68 showed that of $n = 1082$ adults, 40% favored legal marriage, 30% favored civil unions, and 25% believed there should be no legal recognition.[7] The poll reported a margin of error of plus or minus 3%.

a. Construct a 90% confidence interval for the proportion of adults who favor the "legal marriage" position.

b. Construct a 90% confidence interval for the proportion of adults who favor the "civil unions" position.

c. How did the researchers calculate the margin of error for this survey? Confirm that their margin of error is correct.

8.39 What's Normal? What *is* normal, when it comes to people's body temperatures? A random sample of 130 human body temperatures, provided by Allen Shoemaker[9] in the *Journal of Statistical Education,* had a mean of 98.25° and a standard deviation of 0.73°.

a. Construct a 99% confidence interval for the average body temperature of healthy people.

b. Does the confidence interval constructed in part a contain the value 98.6°, the usual average temperature cited by physicians and others? If not, what conclusions can you draw?

8.40 Gonna' Vote? How likely are you to vote in the next national election? In a survey by *Pew Research,*[10] fully 77% of the registered Republican voters are *absolutely* going to vote this year while only 65% of Democrats are *absolutely* going to vote in the next election. The sample consisted of 469 registered Republicans, 490 registered Democrats, and 480 registered Independents.

a. Construct a 98% confidence interval for the proportion of registered Republicans who say they are *absolutely* going to vote in the next election. If a Republican senator predicts that at least 85% of registered Republicans will *absolutely* vote in the next election, is this figure realistic?

b. Construct a 99% confidence interval for the proportion of registered Democrats who say they are *absolutely* going to vote in the next election.

ESTIMATING THE DIFFERENCE BETWEEN TWO POPULATION MEANS

8.6

A problem equally as important as the estimation of a single population mean μ for a quantitative population is the comparison of two population means. You may want to make comparisons like these:

- The average scores on the Medical College Admission Test (MCAT) for students whose major was biochemistry and those whose major was biology
- The average yields in a chemical plant using raw materials furnished by two different suppliers
- The average stem diameters of plants grown on two different types of nutrients

For each of these examples, there are two populations: the first with mean and variance μ_1 and σ_1^2, and the second with mean and variance μ_2 and σ_2^2. A random sample of n_1 measurements is drawn from population 1, and a second random sample of size n_2 is independently drawn from population 2. Finally, the estimates of the population parameters are calculated from the sample data using the estimators $\bar{x}_1, s_1^2, \bar{x}_2$, and s_2^2 as shown in Table 8.3.

TABLE 8.3 ● **Samples From Two Quantitative Populations**

	Population 1	Population 2
Mean	μ_1	μ_2
Variance	σ_1^2	σ_2^2

	Sample 1	Sample 2
Mean	$\bar{x}_1$	$\bar{x}_2$
Variance	s_1^2	s_2^2
Sample Size	n_1	n_2

Intuitively, the difference between two sample means would provide the maximum information about the actual difference between two population means, and this is in fact the case. The best point estimator of the difference $(\mu_1 - \mu_2)$ between the population means is $(\bar{x}_1 - \bar{x}_2)$. The sampling distribution of this estimator is not difficult to derive, but we state it here without proof.

PROPERTIES OF THE SAMPLING DISTRIBUTION OF $(\bar{x}_1 - \bar{x}_2)$, THE DIFFERENCE BETWEEN TWO SAMPLE MEANS

When independent random samples of n_1 and n_2 observations have been selected from populations with means μ_1 and μ_2 and variances σ_1^2 and σ_2^2, respectively, the sampling distribution of the difference $(\bar{x}_1 - \bar{x}_2)$ has the following properties:

1. The mean of $(\bar{x}_1 - \bar{x}_2)$ is

$$\mu_1 - \mu_2$$

and the standard error is

$$SE = \sqrt{\frac{\sigma_1^2}{n_1} + \frac{\sigma_2^2}{n_2}}$$

which can be estimated as

$$SE = \sqrt{\frac{s_1^2}{n_1} + \frac{s_2^2}{n_2}} \text{ when the sample sizes are large.}$$

2. **If the sampled populations are normally distributed,** then the sampling distribution of $(\bar{x}_1 - \bar{x}_2)$ is **exactly** normally distributed, regardless of the sample size.

3. **If the sampled populations are not normally distributed,** then the sampling distribution of $(\bar{x}_1 - \bar{x}_2)$ is **approximately** normally distributed when n_1 and n_2 are both 30 or more, due to the Central Limit Theorem.

Since $(\mu_1 - \mu_2)$ is the mean of the sampling distribution, it follows that $(\bar{x}_1 - \bar{x}_2)$ is an unbiased estimator of $(\mu_1 - \mu_2)$ with an approximately normal distribution when n_1 and n_2 are large. That is, the statistic

$$z = \frac{(\bar{x}_1 - \bar{x}_2) - (\mu_1 - \mu_2)}{\sqrt{\frac{s_1^2}{n_1} + \frac{s_2^2}{n_2}}}$$

has an approximately standard normal z distribution, and the general procedures of Section 8.5 can be used to construct point and interval estimates. Although the choice between point and interval estimation depends on your personal preference, most experimenters choose to construct confidence intervals for two-sample problems. The appropriate formulas for both methods are given next.

LARGE-SAMPLE POINT ESTIMATION OF $(\mu_1 - \mu_2)$

Point estimator: $(\bar{x}_1 - \bar{x}_2)$

95% Margin of error: $\pm 1.96 \text{ SE} = \pm 1.96 \sqrt{\frac{s_1^2}{n_1} + \frac{s_2^2}{n_2}}$

A $(1 - \alpha)100\%$ LARGE-SAMPLE CONFIDENCE INTERVAL FOR $(\mu_1 - \mu_2)$

$$(\bar{x}_1 - \bar{x}_2) \pm z_{\alpha/2} \sqrt{\frac{s_1^2}{n_1} + \frac{s_2^2}{n_2}}$$

NEED A TIP?

Right Tail Area	z-Value
.05	1.645
.025	1.96
.01	2.33
.005	2.58

EXAMPLE 8.9

The wearing qualities of two types of automobile tires were compared by road-testing samples of $n_1 = n_2 = 100$ tires for each type and recording the number of miles until wearout, defined as a specific amount of tire wear. The test results are given in Table 8.4. Estimate $(\mu_1 - \mu_2)$, the difference in mean miles to wearout, using a 99% confidence interval. Is there a difference in the average wearing quality for the two types of tires?

TABLE 8.4 • **Sample Data Summary for Two Types of Tires**

Tire 1	Tire 2
$\bar{x}_1 = 26{,}400$ miles	$\bar{x}_2 = 25{,}100$ miles
$s_1^2 = 1{,}440{,}000$	$s_2^2 = 1{,}960{,}000$

Solution   The point estimate of $(\mu_1 - \mu_2)$ is

$$(\bar{x}_1 - \bar{x}_2) = 26{,}400 - 25{,}100 = 1300 \text{ miles}$$

and the standard error of $(\bar{x}_1 - \bar{x}_2)$ is estimated as

$$\text{SE} = \sqrt{\frac{s_1^2}{n_1} + \frac{s_2^2}{n_2}} = \sqrt{\frac{1{,}440{,}000}{100} + \frac{1{,}960{,}000}{100}} = 184.4 \text{ miles}$$

The 99% confidence interval is calculated as

$$(\bar{x}_1 - \bar{x}_2) \pm 2.58 \sqrt{\frac{s_1^2}{n_1} + \frac{s_2^2}{n_2}}$$

$$1300 \pm 2.58(184.4)$$

$$1300 \pm 475.8$$

or $824.2 < (\mu_1 - \mu_2) < 1775.8$. The difference in the average miles to wearout for the two types of tires is estimated to lie between LCL = 824.2 and UCL = 1775.8 miles.

Based on this confidence interval, can you conclude that there is a difference in the average miles to wearout for the two types of tires? If there were no difference in the two population means, then μ_1 and μ_2 would be equal and $(\mu_1 - \mu_2) = 0$. If you look at the confidence interval you constructed, you will see that 0 is not one of the possible values for $(\mu_1 - \mu_2)$. Therefore, it is not likely that the means are the same; you can conclude that there is a difference in the average miles to wearout for the two types of tires. The confidence interval has allowed you to *make a decision* about the equality of the two population means.

 NEED A TIP?

If 0 is not in the interval, you *can* conclude that there is a difference in the population means.

EXAMPLE **8.10**

The scientist in Example 8.6 wondered whether there was a difference in the average daily intakes of dairy products between men and women. He took a sample of $n_1 = 50$ adult men and $n_2 = 50$ adult women and recorded their daily intakes of dairy products in grams per day. A summary of his sample results is listed in Table 8.5. Construct a 95% confidence interval for the difference in the average daily intakes of dairy products for men and women. Can you conclude that there is a difference in the average daily intakes for men and women?

TABLE 8.5 • **Sample Values for Daily Intakes of Dairy Products**

	Men	Women
Sample Size	50	50
Sample Mean	756	762
Sample Standard Deviation	35	30

Solution The confidence interval is constructed using a value of z with tail area $\alpha/2 = .025$ to its right; that is, $z_{.025} = 1.96$. Using the sample standard deviations to approximate the unknown population standard deviations, the 95% confidence interval is

$$(\bar{x}_1 - \bar{x}_2) \pm 1.96\sqrt{\frac{s_1^2}{n_1} + \frac{s_2^2}{n_2}}$$

$$(756 - 762) \pm 1.96\sqrt{\frac{35^2}{50} + \frac{30^2}{50}}$$

$$-6 \pm 12.78$$

or $-18.78 < (\mu_1 - \mu_2) < 6.78$. Look at the possible values for $(\mu_1 - \mu_2)$ in the confidence interval. It is possible that the difference $(\mu_1 - \mu_2)$ could be negative (indicating that the average for women exceeds the average for men), it could be positive (indicating that men have the higher average), or it could be 0 (indicating no difference between the averages). Based on this information, you *should not be willing to conclude* that there is a difference in the average daily intakes of dairy products for men and women.

Examples 8.9 and 8.10 deserve further comment with regard to using sample estimates in place of unknown parameters. The sampling distribution of

$$\frac{(\bar{x}_1 - \bar{x}_2) - (\mu_1 - \mu_2)}{\sqrt{\frac{\sigma_1^2}{n_1} + \frac{\sigma_2^2}{n_2}}}$$

has a standard normal distribution for all sample sizes when both sampled populations are normal and an *approximate* standard normal distribution when the sampled populations are not normal but the sample sizes are large (≥ 30). When σ_1^2 and σ_2^2 are not known and are estimated by the sample estimates s_1^2 and s_2^2, the resulting statistic will still have an approximate standard normal distribution when the sample sizes are large. The behavior of this statistic when the population variances are unknown and the sample sizes are small will be discussed in Chapter 10.

8.6 EXERCISES

BASIC TECHNIQUES

8.41 Independent random samples were selected from populations 1 and 2. The sample sizes, means, and variances are as follows:

	Population	
	1	2
Sample Size	35	49
Sample Mean	12.7	7.4
Sample Variance	1.38	4.14

a. Find a 95% confidence interval for estimating the difference in the population means $(\mu_1 - \mu_2)$.

b. Based on the confidence interval in part a, can you conclude that there is a difference in the means for the two populations? Explain.

8.42 Independent random samples were selected from populations 1 and 2. The sample sizes, means, and variances are as follows:

	Population	
	1	2
Sample Size	64	64
Sample Mean	2.9	5.1
Sample Variance	0.83	1.67

a. Find a 90% confidence interval for the difference in the population means. What does the phrase "90% confident" mean?

b. Find a 99% confidence interval for the difference in the population means. Can you conclude that there is a difference in the two population means? Explain.

8.43 Independent random samples of size $n_1 = n_2 = 100$ were selected from each of two populations. The mean and standard deviations for the two samples were $\bar{x}_1 = 125.2$, $\bar{x}_2 = 123.7$, $s_1 = 5.6$, and $s_2 = 6.8$.

a. Construct a 99% confidence interval for estimating the difference in the two population means.

b. Does the confidence interval in part a provide sufficient evidence to conclude that there is a difference in the two population means? Explain.

8.44 Independent random samples of size $n_1 = n_2 = 500$ were selected from each of two populations. The mean and standard deviations for the two samples were $\bar{x}_1 = 125.2$, $\bar{x}_2 = 123.7$, $s_1 = 5.6$, and $s_2 = 6.8$.

a. Find a point estimate for the difference in the two population means. Calculate the margin of error.

b. Based on the results in part a, can you conclude that there is a difference in the two population means? Explain.

APPLICATIONS

8.45 Selenium A small amount of the trace element selenium, 50–200 micrograms (μg) per day, is considered essential to good health. Suppose that random samples of $n_1 = n_2 = 30$ adults were selected from two regions of the United States and that a day's intake of selenium was recorded for each person. The mean and standard deviation of the selenium daily intakes for the 30 adults from region 1 were $\bar{x}_1 = 167.1$ and $s_1 = 24.3$ μg, respectively. The corresponding statistics for the 30 adults from region 2 were $\bar{x}_2 = 140.9$ and $s_2 = 17.6$ μg. Find a 95% confidence interval for the difference in the mean selenium intakes for the two regions. Interpret this interval.

8.46 9-1-1 A study was conducted to compare the mean numbers of police emergency calls per 8-hour shift in two districts of a large city. Samples of 100 8-hour shifts were randomly selected from the police records for each of the two regions, and the number of emergency calls was recorded for each shift. The sample statistics are listed here:

	Region	
	1	2
Sample Size	100	100
Sample Mean	2.4	3.1
Sample Variance	1.44	2.64

Find a 90% confidence interval for the difference in the mean numbers of police emergency calls per shift between the two districts of the city. Interpret the interval.

8.47 Teaching Biology An experiment was conducted to compare a teacher-developed curriculum, "Biology: A Community Context" (BACC) that was standards-based, activity-oriented, and inquiry-centered to the traditional presentation using lecture, vocabulary, and memorized facts. The perhaps not-so-startling results when students were tested on biology concepts, published in *The American Biology Teacher*, are shown in the following table.[11]

	Mean	Sample Size	Standard Deviation
Pretest: All BACC Classes	13.38	372	5.59
Pretest: All Traditional	14.06	368	5.45
Posttest: All BACC Classes	18.5	365	8.03
Posttest: All Traditional	16.5	298	6.96

a. Find a 95% confidence interval for the mean score for the posttest for all BACC classes.

b. Find a 95% confidence interval for the mean score for the posttest for all traditional classes.

c. Find a 95% confidence interval for the difference in mean scores for the posttest BACC classes and the posttest traditional classes.

d. Does the confidence interval in c provide evidence that there is a real difference in the posttest BACC and traditional class scores? Explain.

Source: From "Performance Assessment of a Standards-Based High School Biology Curriculum," by W. Leonard, B. Speziale, and J. Pernick in The American Biology Teacher, 2001, 63(5), 310–316. Reprinted by permission of National Association of Biology Teachers.

8.48 Are You Dieting? To compare two weight reduction diets A and B, 60 dieters were randomly selected. One group of 30 dieters was placed on diet A and the other 30 on diet B, and their weight losses were recorded over a 30-day period. The means and standard deviations of the weight-loss measurements for the two groups are shown in the table. Find a 95% confidence interval for the difference in mean weight loss for the two diets. Interpret your confidence interval.

Diet A	Diet B
$\bar{x}_A = 21.3$	$\bar{x}_B = 13.4$
$s_A = 2.6$	$s_B = 1.9$

8.49 Starting Salaries As a group, students majoring in the engineering disciplines have the highest salary expectations, followed by those studying the computer science fields, according to results of NACE's *2010 Student Survey*.[12] To compare the starting salaries of college graduates majoring in engineering and computer science, random samples of 50 recent college graduates in each major were selected and the following information obtained:

Major	Mean ($)	SD
Engineering	56,202	2225
Computer science	50,657	2375

a. Find a point estimate for the difference in the average starting salaries of college students majoring in engineering and computer science. What is the margin of error for your estimate?

b. Based upon the results in part a, do you think that there is a significant difference in the average starting salaries for engineers and computer scientists? Explain.

8.50 Biology Skills Refer to Exercise 8.47. In addition to tests involving biology concepts, students were also tested on process skills. The results of pretest and posttest scores, published in *The American Biology Teacher*, are given below.[11]

	Mean	Sample Size	Standard Deviation
Pretest: All BACC Classes	10.52	395	4.79
Pretest: All Traditional	11.97	379	5.39
Posttest: All BACC Classes	14.06	376	5.65
Posttest: All Traditional	12.96	308	5.93

a. Find a 95% confidence interval for the mean score on process skills for the posttest for all BACC classes.

b. Find a 95% confidence interval for the mean score on process skills for the posttest for all traditional classes.

c. Find a 95% confidence interval for the difference in mean scores on process skills for the posttest BACC classes and the posttest traditional classes.

d. Does the confidence interval in c provide evidence that there is a real difference in the mean process skills scores between posttest BACC and traditional class scores? Explain.

Source: From "Peformance Assessment of a Standards-Based High School Biology Curriculum," by W. Leonard, B. Speziale, and J. Pernick in The American Biology Teacher, 2001, 63(5), 310–316. Reprinted by permission of National Association of Biology Teachers.

8.51 Hotel Costs Refer to Exercise 8.20. The means and standard deviations for 50 billing statements from each of the computer databases of each of the three hotel chains are given in the table:[4]

	Marriott	Westin	Doubletree
Sample Average ($)	150	165	125
Sample Standard Deviation	17.2	22.5	12.8

a. Find a 95% confidence interval for the difference in the average room rates for the Marriott and the Westin hotel chains.

b. Find a 99% confidence interval for the difference in the average room rates for the Westin and the Doubletree hotel chains.

c. Do the intervals in parts a and b contain the value $(\mu_1 - \mu_2) = 0$? Why is this of interest to the researcher?

d. Do the data indicate a difference in the average room rates between the Marriott and the Westin chains? Between the Westin and the Doubletree chains?

8.52 Noise and Stress To compare the effect of stress in the form of noise on the ability to perform a simple task, 70 subjects were divided into two groups. The first group of 30 subjects acted as a control, while the second group of 40 were the experimental group. Although each subject performed the task, the experimental group subjects had to perform the task while loud rock music was played. The time to finish the task was recorded for each subject and the following summary was obtained:

	Control	Experimental
n	30	40
$\bar{x}$	15 minutes	23 minutes
s	4 minutes	10 minutes

a. Find a 99% confidence interval for the difference in mean completion times for these two groups.

b. Based on the confidence interval in part a, is there sufficient evidence to indicate a difference in the average time to completion for the two groups? Explain.

8.53 What's Normal, continued Of the 130 people in Exercise 8.39, 65 were female and 65 were male.[9] The means and standard deviation of their temperatures are shown below.

	Men	Women
Sample Mean	98.11	98.39
Standard Deviation	0.70	0.74

Find a 95% confidence interval for the difference in the average body temperatures for males versus females. Based on this interval, can you conclude that there is a difference in the average temperatures for males versus females? Explain.

ESTIMATING THE DIFFERENCE BETWEEN TWO BINOMIAL PROPORTIONS

8.7

A simple extension of the estimation of a binomial proportion p is the estimation of the difference between two binomial proportions. You may wish to make comparisons like these:

- The proportion of defective items manufactured in two production lines
- The proportion of male and female voters who favor an equal rights amendment
- The germination rates of untreated seeds and seeds treated with a fungicide

These comparisons can be made using the difference $(p_1 - p_2)$ between two binomial proportions, p_1 and p_2. Independent random samples consisting of n_1 and n_2 trials are drawn from populations 1 and 2, respectively, and the sample estimates $\hat{p}_1$ and $\hat{p}_2$ are calculated. The unbiased estimator of the difference $(p_1 - p_2)$ is the sample difference $(\hat{p}_1 - \hat{p}_2)$.

PROPERTIES OF THE SAMPLING DISTRIBUTION OF THE DIFFERENCE $(\hat{p}_1 - \hat{p}_2)$ BETWEEN TWO SAMPLE PROPORTIONS

Assume that independent random samples of n_1 and n_2 observations have been selected from binomial populations with parameters p_1 and p_2, respectively. The sampling distribution of the difference between sample proportions

$$(\hat{p}_1 - \hat{p}_2) = \left(\frac{x_1}{n_1} - \frac{x_2}{n_2} \right)$$

has these properties:

1. The mean of $(\hat{p}_1 - \hat{p}_2)$ is

   $$p_1 - p_2$$

   and the standard error is

   $$SE = \sqrt{\frac{p_1 q_1}{n_1} + \frac{p_2 q_2}{n_2}}$$

   which is estimated as

   $$SE = \sqrt{\frac{\hat{p}_1 \hat{q}_1}{n_1} + \frac{\hat{p}_2 \hat{q}_2}{n_2}}$$

2. The sampling distribution of $(\hat{p}_1 - \hat{p}_2)$ can be approximated by a normal distribution when n_1 and n_2 are large, due to the Central Limit Theorem.

Although the range of a single proportion is from 0 to 1, the difference between two proportions ranges from -1 to 1. To use a normal distribution to approximate the distribution of $(\hat{p}_1 - \hat{p}_2)$, both $\hat{p}_1$ and $\hat{p}_2$ should be approximately normal; that is, $n_1 \hat{p}_1 > 5$, $n_1 \hat{q}_1 > 5$, and $n_2 \hat{p}_2 > 5$, $n_2 \hat{q}_2 > 5$.

The appropriate formulas for point and interval estimation are given next.

LARGE-SAMPLE POINT ESTIMATION OF $(p_1 - p_2)$

Point estimator: $(\hat{p}_1 - \hat{p}_2)$

95% Margin of error: $\pm 1.96\ \text{SE} = \pm 1.96\sqrt{\dfrac{\hat{p}_1\hat{q}_1}{n_1} + \dfrac{\hat{p}_2\hat{q}_2}{n_2}}$

A $(1 - \alpha)100\%$ LARGE-SAMPLE CONFIDENCE INTERVAL FOR $(p_1 - p_2)$

$(\hat{p}_1 - \hat{p}_2) \pm z_{\alpha/2}\sqrt{\dfrac{\hat{p}_1\hat{q}_1}{n_1} + \dfrac{\hat{p}_2\hat{q}_2}{n_2}}$

Assumption: n_1 and n_2 must be sufficiently large so that the sampling distribution of $(\hat{p}_1 - \hat{p}_2)$ can be approximated by a normal distribution—namely, if $n_1\hat{p}_1$, $n_1\hat{q}_1$, $n_2\hat{p}_2$, and $n_2\hat{q}_2$ are all greater than 5.

EXAMPLE 8.11

A bond proposal for school construction will be submitted to the voters at the next municipal election. A major portion of the money derived from this bond issue will be used to build schools in a rapidly developing section of the city, and the remainder will be used to renovate and update school buildings in the rest of the city. To assess the viability of the bond proposal, a random sample of $n_1 = 50$ residents in the developing section and $n_2 = 100$ residents from the other parts of the city were asked whether they plan to vote for the proposal. The results are tabulated in Table 8.6.

TABLE 8.6 ● **Sample Values for Opinion on Bond Proposal**

	Developing Section	Rest of the City
Sample Size	50	100
Number Favoring Proposal	38	65
Proportion Favoring Proposal	.76	.65

1. Estimate the difference in the true proportions favoring the bond proposal with a 99% confidence interval.

2. If both samples were pooled into one sample of size $n = 150$, with 103 in favor of the proposal, provide a point estimate of the proportion of city residents who will vote for the bond proposal. What is the margin of error?

Solution
1. The best point estimate of the difference $(p_1 - p_2)$ is given by

$$(\hat{p}_1 - \hat{p}_2) = .76 - .65 = .11$$

and the standard error of $(\hat{p}_1 - \hat{p}_2)$ is estimated as

$$\sqrt{\frac{\hat{p}_1\hat{q}_1}{n_1} + \frac{\hat{p}_2\hat{q}_2}{n_2}} = \sqrt{\frac{(.76)(.24)}{50} + \frac{(.65)(.35)}{100}} = .0770$$

For a 99% confidence interval, $z_{.005} = 2.58$, and the approximate 99% confidence interval is found as

$$(\hat{p}_1 - \hat{p}_2) \pm z_{.005}\sqrt{\frac{\hat{p}_1\hat{q}_1}{n_1} + \frac{\hat{p}_2\hat{q}_2}{n_2}}$$

$$.11 \pm (2.58)(.0770)$$
$$.11 \pm .199$$

or $(-.089, .309)$. Since this interval contains the value $(p_1 - p_2) = 0$, it is possible that $p_1 = p_2$, which implies that there may be no difference in the proportions favoring the bond issue in the two sections of the city.

2. If there is no difference in the two proportions, then the two samples are not really different and might well be combined to obtain an overall estimate of the proportion of the city residents who will vote for the bond issue. If both samples are pooled, then $n = 150$ and

$$\hat{p} = \frac{103}{150} = .69$$

Therefore, the point estimate of the overall value of p is .69, with a margin of error given by

$$\pm 1.96 \sqrt{\frac{(.69)(.31)}{150}} = \pm 1.96(.0378) = \pm.074$$

Notice that $.69 \pm .074$ produces the interval .62 to .76, which includes only proportions greater than .5. Therefore, if voter attitudes do not change adversely prior to the election, the bond proposal should pass by a reasonable majority.

 8.7 **EXERCISES**

BASIC TECHNIQUES

8.54 Independent random samples of $n_1 = 500$ and $n_2 = 500$ observations were selected from binomial populations 1 and 2, and $x_1 = 120$ and $x_2 = 147$ successes were observed.

a. What is the best point estimator for the difference $(p_1 - p_2)$ in the two binomial proportions?

b. Calculate the approximate standard error for the statistic used in part a.

c. What is the margin of error for this point estimate?

8.55 Independent random samples of $n_1 = 800$ and $n_2 = 640$ observations were selected from binomial populations 1 and 2, and $x_1 = 337$ and $x_2 = 374$ successes were observed.

a. Find a 90% confidence interval for the difference $(p_1 - p_2)$ in the two population proportions. Interpret the interval.

b. What assumptions must you make for the confidence interval to be valid? Are these assumptions met?

8.56 Independent random samples of $n_1 = 1265$ and $n_2 = 1688$ observations were selected from binomial populations 1 and 2, and $x_1 = 849$ and $x_2 = 910$ successes were observed.

a. Find a 99% confidence interval for the difference $(p_1 - p_2)$ in the two population proportions. What does "99% confidence" mean?

b. Based on the confidence interval in part a, can you conclude that there is a difference in the two binomial proportions? Explain.

APPLICATIONS

8.57 M&M'S Does Mars, Incorporated use the same proportion of red candies in its plain and peanut varieties? A random sample of 56 plain M&M'S contained 12 red candies, and another random sample of 32 peanut M&M'S contained 8 red candies.

a. Construct a 95% confidence interval for the difference in the proportions of red candies for the plain and peanut varieties.

b. Based on the confidence interval in part a, can you conclude that there is a difference in the proportions of red candies for the plain and peanut varieties? Explain.

8.58 Different Priorities As we approached the midterm elections, in the summer of 2010, Democrats and Republicans were split about our nation's top priorities.[13] A sample of $n = 900$ registered voters were asked the following question: "Which ONE of the following items do you think is most important for the federal government to be working on right now?" The list of options is shown in the table below. Options were rotated to reduce bias, and voters were allowed to indicate "All," "None," or "Unsure."

	All (%)	Democrats (%)	Republicans (%)	Independents (%)
Economy and Jobs	47	55	37	48
Deficit, Spending	15	8	22	16
Terrorism, Security	8	6	10	10
Iraq and Afghanistan	7	9	5	4
Immigration	5	3	7	6
All (vol.)	16	18	16	13
None/ Other (vol.)	1	-	1	2
Unsure	1	-	1	-

Suppose that there were 400 Democrats, 350 Republicans, and 150 Independents in the sample. Use a large-sample estimation procedure to compare the proportions of Republicans and Democrats who mentioned the economy and jobs as the most important item for the federal government to work on. Compare the proportions of Republicans and Independents who mentioned deficit spending as the most important item. Explain your conclusions.

8.59 Baseball Fans The first day of baseball comes in late March, ending in October with the World Series. Does fan support grow as the season goes on? Two CNN/*USA Today*/Gallup polls, one conducted in March and one in November, both involved random samples of 1001 adults aged 18 and older. In the March sample, 45% of the adults claimed to be fans of professional baseball, while 51% of the adults in the November sample claimed to be fans.[14]

a. Construct a 99% confidence interval for the difference in the proportion of adults who claim to be fans in March versus November.

b. Does the data indicate that the proportion of adults who claim to be fans increases in November, around the time of the World Series? Explain.

8.60 Auto Accidents Last year's records of auto accidents occurring on a given section of highway were classified according to whether the resulting damage was $1000 or more and to whether a physical injury resulted from the accident. The data follows:

	Under $1000	$1000 or More
Number of Accidents	32	41
Number Involving Injuries	10	23

a. Estimate the true proportion of accidents involving injuries when the damage was $1000 or more for similar sections of highway and find the margin of error.

b. Estimate the true difference in the proportion of accidents involving injuries for accidents with damage under $1000 and those with damage of $1000 or more. Use a 95% confidence interval.

8.61 Catching a Cold Do well-rounded people get fewer colds? A study on the *Chronicle of Higher Education* was conducted by scientists at Carnegie Mellon University, the University of Pittsburgh, and the University of Virginia. They found that people who have only a few social outlets get more colds than those who are involved in a variety of social activities.[16] Suppose that of the 276 healthy men and women tested, $n_1 = 96$ had only a few social outlets and $n_2 = 105$ were busy with six or more activities. When these people were exposed to a cold virus, the following results were observed:

	Few Social Outlets	Many Social Outlets
Sample Size	96	105
Percent with Colds	62%	35%

a. Construct a 99% confidence interval for the difference in the two population proportions.

b. Does there appear to be a difference in the population proportions for the two groups?

c. You might think that coming into contact with more people would lead to more colds, but the data show the opposite effect. How can you explain this unexpected finding?

8.62 Union, Yes! A sampling of political candidates—200 randomly chosen from the West and 200 from the East—was classified according to whether the candidate received backing by a national labor union and whether the candidate won. In the West, 120 winners had union backing, and in the East, 142 winners were backed by a national union. Find a 95% confidence interval for the difference between the proportions of union-backed winners in the West versus the East. Interpret this interval.

8.63 Birth Order and College Success In a study of the relationship between birth order and college success, an investigator found that 126 in a sample of 180 college graduates were firstborn or only children. In a sample of 100 nongraduates of comparable age and socioeconomic background, the number of firstborn or only children was 54. Estimate the difference between the proportions of firstborn or only children in the two populations from which these samples were drawn. Use a 90% confidence interval and interpret your results.

8.64 Generation Next Born between 1980 and 1990, Generation Next is engaged with technology, and the vast majority is dependent upon it.[17] Suppose that in a survey of 500 female and 500 male students in Generation Next, 345 of the females and 365 of the males reported that they decided to attend college in order to make more money.

a. Construct a 98% confidence interval for the difference in the proportions of female and male students who decided to attend college in order to make more money.

b. What does it mean to say that you are "98% confident"?

c. Based on the confidence interval in part a, can you conclude that there is a difference in the proportions of female and male students who decided to attend college in order to make more money?

8.65 Excedrin or Tylenol? In a study to compare the effects of two pain relievers it was found that of $n_1 = 200$ randomly selected individuals who used the first pain reliever, 93% indicated that it relieved their pain. Of $n_2 = 450$ randomly selected individuals who

used the second pain reliever, 96% indicated that it relieved their pain.

a. Find a 99% confidence interval for the difference in the proportions experiencing relief from pain for these two pain relievers.

b. Based on the confidence interval in part a, is there sufficient evidence to indicate a difference in the proportions experiencing relief for the two pain relievers? Explain.

8.66 When Bargaining Pays Off According to a national representative survey done by *Consumer Reports*, you should always try to negotiate for a better deal when shopping or paying for services.[15] Tips include researching prices at other stores and on the Internet, timing your visit late in the month when salespeople are trying to meet quotas, and talking to a manager rather than a salesperson. Suppose that random samples of 200 men and 200 women are taken, and that the men were more likely than the women to say they "always or often" bargained (30% compared with 25%).

a. Construct a 95% confidence interval for the difference in the proportion of men and women who say they "always or often" negotiate for a better deal.

b. Do the data indicate that there is a difference in the proportion of men and women who say they "always or often" negotiate for a better deal? Explain.

ONE-SIDED CONFIDENCE BOUNDS

8.8

The confidence intervals discussed in Sections 8.5 to 8.7 are sometimes called **two-sided confidence intervals** because they produce both an upper (UCL) and a lower (LCL) bound for the parameter of interest. Sometimes, however, an experimenter is interested in only one of these limits; that is, he needs only an upper bound (or possibly a lower bound) for the parameter of interest. In this case, you can construct a **one-sided confidence bound** for the parameter of interest, such as μ, p, $\mu_1 - \mu_2$, or $p_1 - p_2$.

When the sampling distribution of a point estimator is approximately normal, an argument similar to the one in Section 8.5 can be used to show that one-sided confidence bounds, constructed using the following equations *when the sample size is large,* will contain the true value of the parameter of interest $(1 - \alpha)100\%$ of the time in repeated sampling.

A $(1 - \alpha)100\%$ LOWER CONFIDENCE BOUND (LCB)

(Point estimator) $- z_\alpha \times$ (Standard error of the estimator)

A $(1 - \alpha)100\%$ UPPER CONFIDENCE BOUND (UCB)

(Point estimator) $+ z_\alpha \times$ (Standard error of the estimator)

The z-value used for a $(1 - \alpha)100\%$ one-sided confidence bound, z_α, locates an area α in a single tail of the normal distribution as shown in Figure 8.11.

FIGURE 8.11

z-value for a one-sided confidence bound

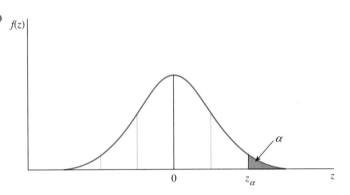

EXAMPLE 8.12 — A corporation plans to issue some short-term notes and is hoping that the interest it will have to pay will not exceed 11.5%. To obtain some information about this problem, the corporation marketed 40 notes, one through each of 40 brokerage firms. The mean and standard deviation for the 40 interest rates were 10.3% and .31%, respectively. Since the corporation is interested in only an upper limit on the interest rates, find a 95% upper confidence bound for the mean interest rate that the corporation will have to pay for the notes.

Solution Since the parameter of interest is μ, the point estimator is $\bar{x}$ with standard error SE $\approx \dfrac{s}{\sqrt{n}}$. The confidence coefficient is .95, so that $\alpha = .05$ and $z_{.05} = 1.645$. Therefore, the 95% upper confidence bound is

$$\text{UCB} = \bar{x} + 1.645\left(\frac{s}{\sqrt{n}}\right) = 10.3 + 1.645\left(\frac{.31}{\sqrt{40}}\right) = 10.3 + .0806 = 10.3806$$

Thus, you can estimate that the mean interest rate that the corporation will have to pay on its notes will be less than 10.3806%. The corporation should not be concerned about its interest rates exceeding 11.5%. How confident are you of this conclusion? Fairly confident, because intervals constructed in this manner contain μ 95% of the time.

8.9 CHOOSING THE SAMPLE SIZE

Designing an experiment is essentially a plan for buying a certain amount of information. Just as the price you pay for a video game varies depending on where and when you buy it, the price of statistical information varies depending on how and where the information is collected. As when you buy any product, you should buy as much statistical information as you can for the minimum possible cost.

The total amount of relevant information in a sample is controlled by two factors:

- The **sampling plan** or **experimental design**: the procedure for collecting the information
- The **sample size** *n*: the amount of information you collect

You can increase the amount of information you collect by *increasing* the sample size, or perhaps by *changing* the type of sampling plan or experimental design you are using. We will discuss the simplest sampling plan—random sampling from a relatively large population—and focus on ways to choose the sample size n needed to purchase a given amount of information.

A researcher makes little progress in planning an experiment before encountering the problem of sample size. **How many measurements should be included in the sample?** How much information does the researcher want to buy? In order to answer these questions, the researcher must first specify:

- The reliability he wishes to achieve, and
- The accuracy needed for his estimate

In a statistical estimation problem, the accuracy of the estimate is measured by the *margin of error* or the *width of the confidence interval,* both of which have a specified reliability. Since both of these measures are a function of the sample size, specifying the reliability and accuracy allows you to determine the necessary sample size.

For instance, suppose you want to estimate the average daily yield μ of a chemical process and you need the margin of error to be less than 4 tons. This means that, approximately 95% of the time in repeated sampling, the "reliability," the distance between the sample mean $\bar{x}$ and the population mean μ will be less than 1.96 SE. You want this quantity to be less than 4 (the "accuracy"). That is,

$$1.96 \text{ SE} < 4 \quad \text{or} \quad 1.96\left(\frac{\sigma}{\sqrt{n}}\right) < 4$$

Solving for n, you obtain

$$n > \left(\frac{1.96}{4}\right)^2 \sigma^2 \quad \text{or} \quad n > .24\sigma^2$$

If you know σ, the population standard deviation, you can substitute its value into the formula and solve for n. If σ is unknown—which is usually the case—you can use the best approximation available:

- An estimate s obtained from a previous sample
- A range estimate based on knowledge of the largest and smallest possible measurements: $\sigma \approx \text{Range}/4$

For this example, suppose that a prior study of the chemical process produced a sample standard deviation of $s = 21$ tons. Then

$$n > .24\sigma^2 = .24(21)^2 = 105.8$$

Using a sample of size $n = 106$ or larger, you could be reasonably certain (with probability approximately equal to .95) that your estimate of the average yield will be within ± 4 tons of the actual average yield.

The solution $n = 106$ is only approximate because you had to use an approximate value for σ to calculate the standard error of the mean. Although this may bother you, it is the best method available for selecting the sample size, and it is certainly better than guessing!

Sometimes researchers request a different reliability, or confidence level other than the 95% confidence specified by the margin of error. In this case, the half-width of the confidence interval provides the accuracy measure for your estimate; that is, the bound B on the error of your estimate is

$$z_{\alpha/2}\left(\frac{\sigma}{\sqrt{n}}\right) < B$$

This method for choosing the sample size can be used for all four estimation procedures presented in this chapter. The general procedure is described next.

NEED TO KNOW...

How to Choose the Sample Size

Determine the parameter to be estimated and the standard error of its point estimator. Then proceed as follows:

1. Choose B, the bound on the error of your estimate, and a confidence coefficient $(1 - \alpha)$.
2. For a one-sample problem, solve this equation for the sample size n:

   $$z_{\alpha/2} \times (\text{Standard error of the estimator}) \leq B$$

   where $z_{\alpha/2}$ is the value of z having area $\alpha/2$ to its right.
3. For a two-sample problem, set $n_1 = n_2 = n$ and solve the equation in step 2.

   [NOTE: For most estimators (all presented in this textbook), the standard error is a function of the sample size n.]

EXAMPLE 8.13

Producers of polyvinyl plastic pipe want to have a supply of pipes sufficient to meet marketing needs. They wish to survey wholesalers who buy polyvinyl pipe in order to estimate the proportion who plan to increase their purchases next year. What sample size is required if they want their estimate to be within .04 of the actual proportion with probability equal to .90?

Solution   For this particular example, the bound B on the error of the estimate is .04. Since the confidence coefficient is $(1 - \alpha) = .90$, α must equal .10 and $\alpha/2$ is .05. The z-value corresponding to an area equal to .05 in the upper tail of the z distribution is $z_{.05} = 1.645$. You then require

$$1.645 \text{ SE} = 1.645\sqrt{\frac{pq}{n}} \leq .04$$

In order to solve this equation for n, you must substitute an approximate value of p into the equation. If you want to be certain that the sample is large enough, you should use $p = .5$ (substituting $p = .5$ will yield the largest possible solution for n because the maximum value of pq occurs when $p = q = .5$). Then

$$1.645\sqrt{\frac{(.5)(.5)}{n}} \leq .04$$

or

$$\sqrt{n} \geq \frac{(1.645)(.5)}{.04} = 20.56$$

$$n \geq (20.56)^2 = 422.7$$

Therefore, the producers must survey at least 423 wholesalers if they want to estimate the proportion p correct to within .04.

EXAMPLE 8.14

A personnel director wishes to compare the effectiveness of two methods of training industrial employees to perform a certain assembly operation. Employees are to be divided into two equal groups: the first receiving training method 1 and the second training method 2. Each will perform the assembly operation, and the length of assembly time will be recorded. It is expected that the assembly times for both groups will have a range of approximately 8 minutes. For the estimate of the difference in mean times to assemble to be correct to within 1 minute with a probability equal to .95, how many workers must be included in each training group?

Solution Since you are estimating the difference between two means, the standard error of the estimate is $\sqrt{\dfrac{\sigma_1^2}{n_1} + \dfrac{\sigma_2^2}{n_2}}$. The bound is $B = 1$ minute, so that you require

$$1.96\sqrt{\frac{\sigma_1^2}{n_1} + \frac{\sigma_2^2}{n_2}} \leq 1$$

Since you want to use two equal groups, you let $n_1 = n_2 = n$ and obtain the equation

$$1.96\sqrt{\frac{\sigma_1^2}{n} + \frac{\sigma_2^2}{n}} \leq 1$$

As noted above, the variability (range) of each method of assembly is approximately the same, so that $\sigma_1^2 = \sigma_2^2 = \sigma^2$. Since the range, equal to 8 minutes, is approximately equal to 4σ, you have

$$4\sigma \approx 8 \quad \text{or} \quad \sigma \approx 2$$

Substituting this value for σ_1 and σ_2 in the earlier equation, you get

$$1.96\sqrt{\frac{(2)^2}{n} + \frac{(2)^2}{n}} \leq 1$$

$$1.96\sqrt{\frac{8}{n}} \leq 1$$

$$\sqrt{n} \geq 1.96\sqrt{8}$$

Solving, you have $n \geq 31$. Thus, each group should contain at least $n = 31$ employees.

Table 8.7 provides a summary of the formulas used to find the sample sizes required for estimation with a given bound on the error of the estimate or confidence interval width W ($W = 2B$). Notice that to estimate p, the sample size formula uses $\sigma^2 = pq$, whereas to estimate ($p_1 - p_2$), the sample size formula uses $\sigma_1^2 = p_1q_1$ and $\sigma_2^2 = p_2q_2$.

TABLE 8.7 ● **Sample Size Formulas**

Parameter	Estimator	Sample Size	Assumptions
μ	$\bar{x}$	$n \geq \dfrac{z_{\alpha/2}^2 \sigma^2}{B^2}$	
$\mu_1 - \mu_2$	$\bar{x}_1 - \bar{x}_2$	$n \geq \dfrac{z_{\alpha/2}^2(\sigma_1^2 + \sigma_2^2)}{B^2}$	$n_1 = n_2 = n$
p	$\hat{p}$	$\begin{cases} n \geq \dfrac{z_{\alpha/2}^2 pq}{B^2} \\ \text{or} \\ n \geq \dfrac{(.25)z_{\alpha/2}^2}{B^2} \end{cases}$	$p = .5$
$p_1 - p_2$	$\hat{p}_1 - \hat{p}_2$	$\begin{cases} n \geq \dfrac{z_{\alpha/2}^2(p_1 q_1 + p_2 q_2)}{B^2} \\ \text{or} \\ n \geq \dfrac{2(.25)z_{\alpha/2}^2}{B^2} \end{cases}$	$n_1 = n_2 = n$ $n_1 = n_2 = n$ and $p_1 = p_2 = .5$

8.9 EXERCISES

BASIC TECHNIQUES

8.67 Find a 90% one-sided upper confidence bound for the population mean μ for these values:

a. $n = 40$, $s^2 = 65$, $\bar{x} = 75$

b. $n = 100$, $s = 2.3$, $\bar{x} = 1.6$

8.68 Independent random samples of $n_1 = n_2 = n$ observations are to be selected from each of two populations 1 and 2. If you wish to estimate the difference between the two population means correct to within .17, with probability equal to .90, how large should n_1 and n_2 be? Assume that you know $\sigma_1^2 \approx \sigma_2^2 \approx 27.8$.

8.69 Independent random samples of size 50 are drawn from two quantitative populations, producing the sample information in the table. Find a 95% upper confidence bound for the difference in the two population means.

	Sample 1	Sample 2
Sample Size	50	50
Sample Mean	12	10
Sample Standard Deviation	5	7

8.70 Find a 99% lower confidence bound for the binomial proportion p when a random sample of $n = 400$ trials produced $x = 196$ successes.

8.71 Suppose you wish to estimate a binomial parameter p correct to within .04, with probability equal to .95. If you suspect that p is equal to some value between .1 and .3 and you want to be certain that your sample is large enough, how large should n be? (HINT: When calculating the standard error, use the value of p in the interval $.1 < p < .3$ that will give the largest sample size.)

8.72 Suppose you wish to estimate a population mean based on a random sample of n observations, and prior experience suggests that $\sigma = 12.7$. If you wish to estimate μ correct to within 1.6, with probability equal to .95, how many observations should be included in your sample?

8.73 Independent random samples of $n_1 = n_2 = n$ observations are to be selected from each of two binomial populations 1 and 2. If you wish to estimate the difference in the two population proportions correct to within .05, with probability equal to .98, how large should n be? Assume that you have no prior information on the values of p_1 and p_2, but you want to make certain that you have an adequate number of observations in the samples.

APPLICATIONS

8.74 GPAs You want to estimate the difference in grade point averages between two groups of college students accurate to within .2 grade point, with probability approximately equal to .95. If the standard deviation of the grade point measurements is approximately equal to .6, how many students must be included in each group? (Assume that the groups will be of equal size.)

8.75 Illegal Immigration Exercise 8.19 discussed a research poll conducted for *ABC News* and the *Washington Post* that included questions about illegal immigration into the United States, and the federal and state responses to the problem.[3] Suppose that you were designing a poll of this type.

a. Explain how you would select your sample. What problems might you encounter in this process?

b. If you wanted to estimate the percentage of the population who agree with a particular statement in your survey questionnaire correct to within 1% with probability .95, approximately how many people would have to be polled?

8.76 Operating Expenses A random sampling of a company's monthly operating expenses for $n = 36$ months produced a sample mean of $5474 and a standard deviation of $764. Find a 90% upper confidence bound for the company's mean monthly expenses.

8.77 Less Red Meat! As Americans become more conscious of the importance of good nutrition, some researchers believe that we may be eating less red meat. To test this theory, a researcher decides to select hospital nutritional records for subjects surveyed 10 years ago and to compare the average amount of beef consumed per year to the amounts consumed by an equal number of subjects she will interview this year. She knows that the amount of beef consumed annually by Americans ranges from 0 to approximately 104 pounds. How many subjects should the researcher select for each group if she wishes to estimate the difference in the average annual per-capita beef consumption correct to within 5 pounds with 99% confidence?

8.78 Political Corruption A questionnaire is designed to investigate attitudes about political corruption in government. The experimenter would like to survey two different groups—Republicans and Democrats—and compare the responses to various "yes/no" questions for the two groups. The experimenter requires that the sampling error for the difference in the proportion of "yes" responses for the two groups is no more than ±3 percentage points. If the two samples are the same size, how large should the samples be?

8.79 Hunting Season A wildlife service wishes to estimate the mean number of days of hunting per hunter for all hunters licensed in the state during a

given season. How many hunters must be included in the sample in order to estimate the mean with a bound on the error of estimation equal to 2 hunting days? Assume that data collected in earlier surveys have shown σ to be approximately equal to 10.

8.80 Red Meat, continued Refer to Exercise 8.77. The researcher selects two groups of 400 subjects each and collects the following sample information on the annual beef consumption now and 10 years ago:

	Ten Years Ago	This Year
Sample Mean	73	63
Sample Standard Deviation	25	28

a. The researcher would like to show that per-capita beef consumption has decreased in the last 10 years, so she needs to show that the difference in the averages is greater than 0. Find a 99% lower confidence bound for the difference in the average per-capita beef consumptions for the two groups.

b. What conclusions can the researcher draw using the confidence bound from part a?

8.81 pH in Rainfall Refer to Exercise 8.80. Suppose you wish to estimate the difference between the mean acidity for rainfalls at two different locations, one in a relatively unpolluted area and the other in an area subject to heavy air pollution. If you wish your estimate to be correct to the nearest .1 pH, with probability near .90, approximately how many rainfalls (pH values) would have to be included in each sample? (Assume that the variance of the pH measurements is approximately .25 at both locations and that the samples will be of equal size.)

8.82 Polluted Rain Suppose you wish to estimate the mean pH of rainfalls in an area that suffers heavy pollution due to the discharge of smoke from a power plant. You know that σ is approximately .5 pH, and you wish your estimate to lie within .1 of μ, with a probability near .95. Approximately how many rainfalls must be included in your sample (one pH reading per rainfall)? Would it be valid to select all of your water specimens from a single rainfall? Explain.

8.83 Selenium, again Refer to the comparison of the daily adult intake of selenium in two different regions of the United States in Exercise 8.45. Suppose you wish to estimate the difference in the mean daily intakes between the two regions correct to within 5 micrograms, with probability equal to .90. If you plan to select an equal number of adults from the two regions (i.e., $n_1 = n_2$), how large should n_1 and n_2 be?

CHAPTER REVIEW

Key Concepts and Formulas

I. Types of Estimators

1. Point estimator: a single number is calculated to estimate the population parameter.

2. Interval estimator: two numbers are calculated to form an interval that, with a certain amount of confidence, contains the parameter.

II. Properties of Good Estimators

1. Unbiased: the average value of the estimator equals the parameter to be estimated.

2. Minimum variance: of all the unbiased estimators, the best estimator has a sampling distribution with the smallest standard error.

3. The margin of error measures the maximum distance between the estimator and the true value of the parameter.

III. Large-Sample Point Estimators

To estimate one of four population parameters when the sample sizes are large, use the following point estimators with the appropriate margins of error.

Parameter	Point Estimator	95% Margin of Error
μ	$\bar{x}$	$\pm 1.96 \left(\dfrac{s}{\sqrt{n}} \right)$
p	$\hat{p} = \dfrac{x}{n}$	$\pm 1.96 \sqrt{\dfrac{\hat{p}\hat{q}}{n}}$
$\mu_1 - \mu_2$	$\bar{x}_1 - \bar{x}_2$	$\pm 1.96 \sqrt{\dfrac{s_1^2}{n_1} + \dfrac{s_2^2}{n_2}}$
$p_1 - p_2$	$(\hat{p}_1 - \hat{p}_2) = \left(\dfrac{x_1}{n_1} - \dfrac{x_2}{n_2} \right)$	$\pm 1.96 \sqrt{\dfrac{\hat{p}_1\hat{q}_1}{n_1} + \dfrac{\hat{p}_2\hat{q}_2}{n_2}}$

IV. Large-Sample Interval Estimators

To estimate one of four population parameters when the sample sizes are large, use the following interval estimators.

Parameter	$(1 - \alpha)100\%$ Confidence Interval
μ	$\bar{x} \pm z_{\alpha/2} \left(\dfrac{s}{\sqrt{n}} \right)$
p	$\hat{p} \pm z_{\alpha/2} \sqrt{\dfrac{\hat{p}\hat{q}}{n}}$
$\mu_1 - \mu_2$	$(\bar{x}_1 - \bar{x}_2) \pm z_{\alpha/2} \sqrt{\dfrac{s_1^2}{n_1} + \dfrac{s_2^2}{n_2}}$
$p_1 - p_2$	$(\hat{p}_1 - \hat{p}_2) \pm z_{\alpha/2} \sqrt{\dfrac{\hat{p}_1\hat{q}_1}{n_1} + \dfrac{\hat{p}_2\hat{q}_2}{n_2}}$

1. All values in the interval are possible values for the unknown population parameter.

2. Any values outside the interval are unlikely to be the value of the unknown parameter.

3. To compare two population means or proportions, look for the value 0 in the confidence interval. If 0 is in the interval, it is possible that the two population means or proportions are equal, and you should not declare a difference. If 0 is not in the interval, it is unlikely that the two means or proportions are equal, and you can confidently declare a difference.

V. One-Sided Confidence Bounds

Use either the upper (+) or lower (−) two-sided bound, with the critical value of z changed from $z_{\alpha/2}$ to z_α.

VI. Choosing the Sample Size

1. Determine the size of the margin of error, B, that you are willing to tolerate.

2. Choose the sample size by solving for n or $n = n_1 = n_2$ in the inequality: $z_{\alpha/2} \times \text{SE} \leq B$, where SE is a function of the sample size n.

3. For quantitative populations, estimate the population standard deviation using a previously calculated value of s or the range approximation $\sigma \approx \text{Range}/4$.

4. For binomial populations, use the conservative approach and approximate p using the value $p = .5$.

Supplementary Exercises

8.84 State the Central Limit Theorem. Of what value is the Central Limit Theorem in large-sample statistical estimation?

8.85 A random sample of $n = 64$ observations has a mean $\bar{x} = 29.1$ and a standard deviation $s = 3.9$.

a. Give the point estimate of the population mean μ and find the margin of error for your estimate.

b. Find a 90% confidence interval for μ. What does "90% confident" mean?

c. Find a 90% lower confidence bound for the population mean μ. Why is this bound different from the lower confidence limit in part b?

d. How many observations do you need to estimate μ to within .5, with probability equal to .95?

8.86 Independent random samples of $n_1 = 50$ and $n_2 = 60$ observations were selected from populations 1 and 2, respectively. The sample sizes and computed sample statistics are given in the table:

	Population	
	1	2
Sample Size	50	60
Sample Mean	100.4	96.2
Sample Standard Deviation	0.8	1.3

Find a 90% confidence interval for the difference in population means and interpret the interval.

8.87 Refer to Exercise 8.86. Suppose you wish to estimate $(\mu_1 - \mu_2)$ correct to within .2, with probability equal to .95. If you plan to use equal sample sizes, how large should n_1 and n_2 be?

8.88 A random sample of $n = 500$ observations from a binomial population produced $x = 240$ successes.

a. Find a point estimate for p, and find the margin of error for your estimator.

b. Find a 90% confidence interval for p. Interpret this interval.

8.89 Refer to Exercise 8.88. How large a sample is required if you wish to estimate p correct to within .025, with probability equal to .90?

8.90 Independent random samples of $n_1 = 40$ and $n_2 = 80$ observations were selected from binomial populations 1 and 2, respectively. The number of successes in the two samples were $x_1 = 17$ and $x_2 = 23$. Find a 99% confidence interval for the difference between the two binomial population proportions. Interpret this interval.

8.91 Refer to Exercise 8.90. Suppose you wish to estimate $(p_1 - p_2)$ correct to within .06, with probability equal to .99, and you plan to use equal sample sizes— that is, $n_1 = n_2$. How large should n_1 and n_2 be?

8.92 Ethnic Cuisine Ethnic groups in America buy differing amounts of various food products because of their ethnic cuisine. A researcher interested in market segmentation for Asian and Hispanic households would like to estimate the proportion of households that select certain brands for various products. If the researcher wishes these estimates to be within .03 with probability .95, how many households should she include in the samples? Assume that the sample sizes are equal.

8.93 Does it Pay to Haggle? In Exercise 8.60, a survey done by *Consumer Reports* indicates that you should always try to negotiate for a better deal when shopping or paying for services.[15] In fact, based on their survey, 37% of the people under age 34 were more likely to "haggle," while only 13% of those 65 and older. Suppose that this survey included 72 people under the age of 34 and 55 people who are 65 or older.

a. What are the values of $\hat{p}_1$ and $\hat{p}_2$ for the two groups in this survey?

b. Find a 95% confidence interval for the difference in the proportion of people who are more likely to "haggle" in the "under 34" versus "65 and older" age groups.

c. What conclusions can you draw regarding the groups compared in part b?

8.94 Smoking and Blood Pressure An experiment was conducted to estimate the effect of smoking on the blood pressure of a group of 35 cigarette smokers, by taking the difference in the blood pressure readings at the beginning of the experiment and again 5 years later. The sample mean increase, measured in millimeters of mercury, was $\bar{x} = 9.7$, and the sample standard deviation was $s = 5.8$. Estimate the mean increase in blood pressure that one would expect for cigarette smokers over the time span indicated by the experiment. Find the margin of error. Describe the population associated with the mean that you have estimated.

8.95 Blood Pressure, continued Using a confidence coefficient equal to .90, place a confidence interval on the mean increase in blood pressure for Exercise 8.94.

8.96 Iodine Concentration Based on repeated measurements of the iodine concentration in a solution, a chemist reports the concentration as 4.614, with an "error margin of .006."

a. How would you interpret the chemist's "error margin"?

b. If the reported concentration is based on a random sample of $n = 30$ measurements, with a sample standard deviation $s = .017$, would you agree that the chemist's "error margin" is .006?

8.97 Heights If it is assumed that the heights of men are normally distributed with a standard deviation of 2.5 inches, how large a sample should be taken to be fairly sure (probability .95) that the sample mean does not differ from the true mean (population mean) by more than .50 in absolute value?

8.98 Chicken Feed An experimenter fed different rations, A and B, to two groups of 100 chicks each. Assume that all factors other than rations are the same for both groups. Of the chicks fed ration A, 13 died, and of the chicks fed ration B, 6 died.

a. Construct a 98% confidence interval for the true difference in mortality rates for the two rations.

b. Can you conclude that there is a difference in the mortality rates for the two rations?

8.99 Antibiotics You want to estimate the mean hourly yield for a process that manufactures an antibiotic. You observe the process for 100 hourly periods chosen at random, with the results $\bar{x} = 34$ ounces per hour and $s = 3$. Estimate the mean hourly yield for the process using a 95% confidence interval.

8.100 Eating Too Much? Partly because of our addiction to fast food, the average American consumes 32.7 pounds of cheese, 14.0 pounds of ice cream, and drinks 48.8 gallons of soda each year, according to the *2010 Statistical Abstract of the United States.*[18] Suppose that we test the accuracy of these reported averages by selecting a random sample of 40 consumers, and recording the following summary statistics:

	Cheese (lbs/yr)	Ice Cream (lbs/yr)	Soda (gal/yr)
Sample Mean	33.1	11.4	49.1
Sample Standard Deviation	3.8	3.2	4.5

Use your knowledge of statistical estimation to estimate the average per-capita annual consumption for these three products. Does this sample cause you to support or to question the accuracy of the reported averages? Explain.

8.101 Fast Food! Even though we know it may not be good for us, many Americans really enjoy their fast food! A survey conducted by *Pew Research Center*[19] graphically illustrated our penchant for eating out, and in particular, eating fast food:

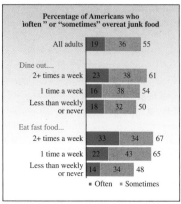

a. This survey was based on "telephone interviews conducted with a nationally representative sample of 2250 adults, ages 18 years and older, living in continental U.S. telephone households." What problems might arise with this type of sampling?

b. How accurate do you expect the percentages given in the survey to be in estimating the actual population percentages? (HINT: Find the margin of error.)

c. If you want to decrease your margin of error to be ±1%, how large a sample should you take?

8.102 Sunflowers In an article in the *Annals of Botany,* a researcher reported the basal stem diameters of two groups of dicot sunflowers: those that were left to sway freely in the wind and those that were artificially supported.[20] A similar experiment was conducted for monocot maize plants. Although the authors measured other variables in a more complicated experimental design, assume that each group consisted of 64 plants (a total of 128 sunflower and 128 maize plants). The values shown in the table are the sample means plus or minus the standard error.

	Sunflower	Maize
Free-Standing	35.3 ± .72	16.2 ± .41
Supported	32.1 ± .72	14.6 ± .40

Use your knowledge of statistical estimation to compare the free-standing and supported basal diameters for the two plants. Write a paragraph describing your conclusions, making sure to include a measure of the accuracy of your inference.

8.103 Basketball Tickets In a regular NBA season, each team plays 82 games, some at home and some on the road. Can you afford the price of a ticket? The Web site wiki.answers.com indicates that the low prices are around $10 for the high up seats while the court-side seats are around $2000 to $5000 per game and that the average price of a ticket is $75.50 a game.[21] Suppose that we test this claim by selecting a random sample of $n = 50$ ticket purchases from a computer database and find that the average ticket price is $82.50 with a standard deviation of $75.25.

a. Do you think that x, the price of an individual regular season ticket, has a mound-shaped distribution? If not, what shape would you expect?

b. If the distribution of the ticket prices is not normal, you can still use the standard normal distribution to construct a confidence interval for μ, the average price of a ticket. Why?

c. Construct a 95% confidence interval for μ, the average price of a ticket. Does this confidence interval cause you support or question the claimed average price of $75.50? Explain.

8.104 College Costs A dean of men wishes to estimate the average cost of the freshman year at a particular college correct to within $500, with a probability of .95. If a random sample of freshmen is to be selected and each asked to keep financial data, how many must be included in the sample? Assume that the dean knows only that the range of expenditures will vary from approximately $14,800 to $23,000.

8.105 Quality Control A quality-control engineer wants to estimate the fraction of defectives in a large lot of printer ink cartridges. From previous experience, he feels that the actual fraction of defectives should be somewhere around .05. How large a sample should he take if he wants to estimate the true fraction to within .01, using a 95% confidence interval?

8.106 Circuit Boards Samples of 400 printed circuit boards were selected from each of two production lines A and B. Line A produced 40 defectives, and line B produced 80 defectives. Estimate the difference in the actual fractions of defectives for the two lines with a confidence coefficient of .90.

8.107 Circuit Boards II Refer to Exercise 8.106. Suppose 10 samples of $n = 400$ printed circuit boards were tested and a confidence interval was constructed for p for each of the 10 samples. What is the probability that exactly one of the intervals will not contain the true value of p? That at least one interval will not contain the true value of p?

8.108 Ice Hockey G. Wayne Marino investigated some of the variables related to "fast starts," the acceleration and speed of a hockey player from a stopped position.[22] Sixty-nine hockey players, varsity and intramural, from the University of Illinois were included in the experiment. Each player was required to move as rapidly as possible from a stopped position to cover a distance of 6 meters. The means and standard deviations of some of the variables recorded for each of the 69 skaters are shown in the table:

	Mean	SD
Weight (kilograms)	75.270	9.470
Stride Length (meters)	1.110	.205
Stride Rate (strides/second)	3.310	.390
Average Acceleration (meters/second²)	2.962	.529
Instantaneous Velocity (meters/second)	5.753	.892
Time to Skate (seconds)	1.953	.131

a. Give the formula that you would use to construct a 95% confidence interval for one of the population means (e.g., mean time to skate the 6-meter distance).

b. Construct a 95% confidence interval for the mean time to skate. Interpret this interval.

8.109 Ice Hockey, continued Exercise 8.108 presented statistics from a study of fast starts by ice hockey skaters. The mean and standard deviation of the 69 individual average acceleration measurements over the 6-meter distance were 2.962 and .529 meters per second, respectively.

a. Find a 95% confidence interval for this population mean. Interpret the interval.

b. Suppose you were dissatisfied with the width of this confidence interval and wanted to cut the interval in half by increasing the sample size. How many skaters (total) would have to be included in the study?

8.110 Ice Hockey, continued The mean and standard deviation of the speeds of the sample of 69 skaters at the end of the 6-meter distance in Exercise 8.108 were 5.753 and .892 meters per second, respectively.

a. Find a 95% confidence interval for the mean velocity at the 6-meter mark. Interpret the interval.

b. Suppose you wanted to repeat the experiment and you wanted to estimate this mean velocity correct to within .1 second, with probability .99. How many skaters would have to be included in your sample?

8.111 School Workers In addition to teachers and administrative staff, schools also have many other employees, including bus drivers, custodians, and cafeteria workers. In Auburn, WA, the average hourly wage is $16.92 for bus drivers, $17.65 for custodians, and $12.86 for cafeteria workers.[23] Suppose that a second school district employs $n = 36$ bus drivers who earn an average of $13.45 per hour with a standard deviation of $s = $2.84. Find a 95% confidence interval for the average hourly wage of bus drivers in school districts similar to this one. Does your confidence interval enclose the Auburn, WA average of $16.92? What can you conclude about the hourly wages for bus drivers in this second school district?

8.112 Recidivism An experimental rehabilitation technique was used on released convicts. It was shown that 79 of 121 men subjected to the technique pursued useful and crime-free lives for a 3-year period following prison release. Find a 95% confidence interval for p, the probability that a convict subjected to the rehabilitation technique will follow a crime-free existence for at least 3 years after prison release.

8.113 Specific Gravity If 36 measurements of the specific gravity of aluminum had a mean of 2.705 and a standard deviation of .028, construct a 98% confidence interval for the actual specific gravity of aluminum.

8.114 Audiology Research In a study to establish the absolute threshold of hearing, 70 male college freshmen were asked to participate. Each subject was seated in a soundproof room and a 150 H tone was presented at a large number of stimulus levels in a randomized order. The subject was instructed to press a button if he detected the tone; the experimenter recorded the lowest stimulus level at which the tone was detected. The mean for the group was 21.6 db with $s = 2.1$. Estimate the mean absolute threshold for all college freshmen and calculate the margin of error.

8.115 Right- or Left-Handed A researcher classified his subjects as innately right-handed or left-handed by comparing thumbnail widths. He took a sample of 400 men and found that 80 men could be classified as left-handed according to his criterion. Estimate the proportion of all males in the population who would test to be left-handed using a 95% confidence interval.

8.116 The Citrus Red Mite An entomologist wishes to estimate the average development time of the citrus red mite correct to within .5 day. From previous experiments it is known that σ is approximately 4 days. How large a sample should the entomologist take to be 95% confident of her estimate?

8.117 The Citrus Red Mite, continued A grower believes that one in five of his citrus trees are infected with the citrus red mite mentioned in Exercise 8.116. How large a sample should be taken if the grower wishes to estimate the proportion of his trees that are infected with citrus red mite to within .08?

CASE STUDY **How Reliable Is That Poll?**
***CBS News:* How and Where America Eats**

When Americans eat out at restaurants, most choose American food; however, tastes for Mexican, Chinese, and Italian food vary from region to region of the United States. In a recent CBS telephone survey[24], it was found that 39% of families ate together 7 nights a week, slightly less than the 46% of families who reported eating together 7 nights a week in an earlier survey by CBS. Most Americans, both men and women, do some of the cooking when meals are cooked at home, as reported in the following table where we compare the number of evening meals personally cooked per week by men and women.

Number of Meals Cooked	3 or Less	4 or More
Men	76	24
Women	33	67

How often Americans eat out at restaurants is largely a function of income. "While most households earning over $50,000 got restaurant food for dinner at least once in the last week, 75% of those earning under $15,000 did not do so at all."

Income	None	1–3 Nights	4 or More Nights
All	47	49	4
Under $15,000	75	19	6
$15–$30,000	58	39	3
$30–$50,000	59	38	3
Over $50,000	31	64	5

In spite of all the negative publicity about obesity and high calories associated with burgers and fries, many Americans continue to eat fast food to save time within busy schedules.

Fast Food Nights	0	1	2–3	4+
With Kids	47	30	19	4
Without Kids	59	20	16	5

Fast Food Nights	0	1	2–3	4+
Men	46	28	20	6
Women	63	20	15	2

Fifty-three percent of families with kids ate fast food at least once last week, compared with 41% of families without kids. Furthermore, 54% of men ate fast food at least once last week, compared with only 37% of women.

The description of the survey methods that gave rise to this data was stated as follows:

> *"This poll was conducted among a nationwide random sample of 936 adults, interviewed by telephone. The error due to sampling for results based on the entire sample could be plus or minus three percentage points."*

1. Verify the margin of error of ±3 percentage points as stated for the sample of $n = 936$ adults. Suppose that the sample contained an equal number of men and women or 468 men and 468 women. What is the margin of error for men and for women?

2. Do the numbers in the tables indicate the number of people/families in the categories? If not, what do those numbers represent?

3. a. Construct a 95% confidence interval for the proportion of Americans who ate together seven nights a week.

   b. Construct a 95% confidence interval for the difference in the proportion of women and men who personally cook at least 4+ meals per week.

   c. Construct a 95% confidence interval for the proportion of Americans who eat out at restaurants at least once a week.

4. If these questions were asked today, would you expect the responses to be similar to those reported here or would you expect them to differ significantly?

Large-Sample Tests of Hypotheses

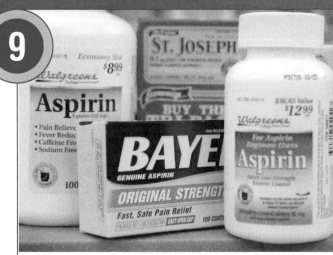

Scott Olson/Getty Images

GENERAL OBJECTIVE

In this chapter, the concept of a statistical test of hypothesis is formally introduced. The sampling distributions of statistics presented in Chapters 7 and 8 are used to construct large-sample tests concerning the values of population parameters of interest to the experimenter.

NEED TO KNOW...

Rejection Regions, *p*-Values, and Conclusions
How to Calculate β

An Aspirin a Day . . . ?

Will an aspirin a day reduce the risk of heart attack? A very large study of U.S. physicians showed that a single aspirin taken every other day reduced the risk of heart attack in men by one-half. However, 3 days later, a British study reported a completely opposite conclusion. How could this be? The case study at the end of this chapter looks at how the studies were conducted, and you will analyze the data using large-sample techniques.

TESTING HYPOTHESES ABOUT POPULATION PARAMETERS

In practical situations, statistical inference can involve either estimating a population parameter or making decisions about the value of the parameter. For example, if a pharmaceutical company is fermenting a vat of antibiotic, samples from the vat can be used to *estimate* the mean potency μ for all of the antibiotic in the vat. In contrast, suppose that the company is not concerned about the exact mean potency of the antibiotic, but is concerned only that it meet the minimum government potency standards. Then the company can use samples from the vat to decide between these two possibilities:

- The mean potency μ does not exceed the minimum allowable potency.
- The mean potency μ exceeds the minimum allowable potency.

The pharmaceutical company's problem illustrates a **statistical test of hypothesis.**

The reasoning used in a statistical test of hypothesis is similar to the process in a court trial. In trying a person for theft, the court must decide between innocence and guilt. As the trial begins, the accused person is assumed to be *innocent*. The prosecution collects and presents all available evidence in an attempt to contradict the innocent hypothesis and hence obtain a conviction.

If there is enough evidence against innocence, the court will reject the innocence hypothesis and declare the defendant *guilty*. If the prosecution does not present enough evidence to prove the defendant guilty, the court will find him *not guilty*. Notice that this does not prove that the defendant is innocent, but merely that there was not enough evidence to conclude that the defendant was guilty.

We use this same type of reasoning to explain the basic concepts of hypothesis testing. These concepts are used to test the four population parameters discussed in Chapter 8: a single population mean or proportion (μ or p) and the difference between two population means or proportions ($\mu_1 - \mu_2$ or $p_1 - p_2$). When the sample sizes are large, the point estimators for each of these four parameters have normal sampling distributions, so that all four large-sample statistical tests follow the same general pattern.

A STATISTICAL TEST OF HYPOTHESIS

A statistical test of hypothesis consists of five parts:

1. The null hypothesis, denoted by H_0
2. The alternative hypothesis, denoted by H_a
3. The test statistic and its p-value
4. The rejection region
5. The conclusion

When you specify these five elements, you define a particular test; changing one or more of the parts creates a new test. Let's look at each part of the statistical test of hypothesis in more detail.

Definition The two competing hypotheses are the **alternative hypothesis H_a,** generally the hypothesis that the researcher wishes to support, and the **null hypothesis H_0,** a contradiction of the alternative hypothesis.

As you will soon see, it is easier to show support for the alternative hypothesis by proving that the null hypothesis is false. Hence, the statistical researcher always begins by assuming that the null hypothesis H_0 is true. The researcher then uses the sample data to decide whether the evidence favors H_a rather than H_0 and draws one of these two **conclusions:**

- Reject H_0 and conclude that H_a is true.
- Accept (do not reject) H_0 as true.

EXAMPLE **9.1**

You wish to show that the average hourly wage of carpenters in the state of California is different from \$19, which is the national average. This is the alternative hypothesis, written as

2

$$H_a : \mu \neq 19$$

The null hypothesis is

1

$$H_0 : \mu = 19$$

You would like to reject the null hypothesis, thus concluding that the California mean is not equal to \$19.

EXAMPLE **9.2**

A milling process currently produces an average of 3% defectives. You are interested in showing that a simple adjustment on a machine will decrease p, the proportion of defectives produced in the milling process. Thus, the alternative hypothesis is

2

$$H_a : p < .03$$

and the null hypothesis is

1

$$H_0 : p = .03$$

If you can reject H_0, you can conclude that the adjusted process produces fewer than 3% defectives.

There is a difference in the forms of the alternative hypotheses given in Examples 9.1 and 9.2. In Example 9.1, no directional difference is suggested for the value of μ; that is, μ might be either larger or smaller than \$19 if H_a is true. This type of test is called a **two-tailed test of hypothesis.** In Example 9.2, however, you are specifically interested in detecting a directional difference in the value of p; that is, if H_a is true, the value of p is less than .03. This type of test is called a **one-tailed test of hypothesis.**

NEED A TIP?

Two-tailed ⇔ Look for a ≠ sign in H_a.

One-tailed ⇔ Look for a > or < sign in H_a.

The decision to reject or accept the null hypothesis is based on information contained in a sample drawn from the population of interest. This information takes these forms:

3

- **Test statistic:** a single number calculated from the sample data
- **p-value:** a probability calculated using the test statistic

Either or both of these measures act as decision makers for the researcher in deciding whether to reject or accept H_0.

EXAMPLE   9.3

For the test of hypothesis in Example 9.1, the average hourly wage $\bar{x}$ for a random sample of 100 California carpenters might provide a good *test statistic* for testing

$$H_0 : \mu = 19 \quad \text{versus} \quad H_a : \mu \neq 19$$

If the null hypothesis H_0 is true, then the sample mean should not be too far from the population mean $\mu = 19$. Suppose that this sample produces a sample mean $\bar{x} = 20$ with standard deviation $s = 2$. Is this sample evidence likely or unlikely to occur, if in fact H_0 is true? You can use two measures to find out. Since the sample size is large, the sampling distribution of $\bar{x}$ is approximately normal with mean $\mu = 19$ and standard error $\sigma/\sqrt{n}$, estimated as

$$SE = \frac{s}{\sqrt{n}} = \frac{2}{\sqrt{100}} = .2$$

- The **test statistic** $\bar{x} = 20$ lies

3

$$z = \frac{\bar{x} - \mu}{\sigma/\sqrt{n}} \approx \frac{20 - 19}{.2} = 5$$

standard deviations from the population mean μ.

- The **p-value** is the probability of observing a test statistic as extreme as or more extreme than the observed value, if in fact H_0 is true. For this example, we define "extreme" as far below or far above what we would have expected. That is,

$$p\text{-value} = P(z > 5) + P(z < -5) \approx 0$$

The *large value of the test statistic* and the *small p-value* mean that you have observed a very unlikely event, if indeed H_0 is true and $\mu = 19$.

4

How do you decide whether to reject or accept H_0? The entire set of values that the test statistic may assume is divided into two sets, or regions. One set, consisting of values that support the alternative hypothesis and lead to rejecting H_0, is called the **rejection region.** The other, consisting of values that support the null hypothesis, is called the **acceptance region.**

For example, in Example 9.1, you would be inclined to believe that California's average hourly wage was different from $19 if the sample mean is either much less than $19 or much greater than $19. The two-tailed rejection region consists of very small and very large values of $\bar{x}$, as shown in Figure 9.1. In Example 9.2, since you want to prove that the percentage of defectives has *decreased,* you would be inclined to reject H_0 for values of $\hat{p}$ that are much smaller than .03. Only *small* values of $\hat{p}$ belong in the left-tailed rejection region shown in Figure 9.2. When the rejection region is in the left tail of the distribution, the test is called a **left-tailed test.** A test with its rejection region in the right tail is called a **right-tailed test.**

FIGURE 9.1

Rejection and acceptance regions for Example 9.1

FIGURE 9.2 ●

Rejection and acceptance regions for Example 9.2

If the test statistic falls into the rejection region, then the null hypothesis is rejected. If the test statistic falls into the acceptance region, then either the null hypothesis is accepted or the test is judged to be inconclusive. We will clarify the different types of conclusions that are appropriate as we consider several practical examples of hypothesis tests.

Finally, how do you decide on the **critical values** that separate the acceptance and rejection regions? That is, how do you decide how much statistical evidence you need before you can reject H_0? This depends on the amount of confidence that you, the researcher, want to attach to the test conclusions and the **significance level α,** the risk you are willing to take of making an incorrect decision.

Definition   A **Type I error** for a statistical test is the error of rejecting the null hypothesis when it is true. The **level of significance (significance level)** for a statistical test of hypothesis is

$$\alpha = P(\text{Type I error}) = P(\text{falsely rejecting } H_0) = P(\text{rejecting } H_0 \text{ when it is true})$$

This value α represents the *maximum tolerable risk* of incorrectly rejecting H_0. Once this significance level is fixed, the rejection region can be set to allow the researcher to reject H_0 with a fixed degree of confidence in the decision.

In the next section, we will show you how to use a test of hypothesis to test the value of a population mean μ. As we continue, we will clarify some of the computational details and add some additional concepts to complete your understanding of hypothesis testing.

A LARGE-SAMPLE TEST ABOUT A POPULATION MEAN

Consider a random sample of n measurements drawn from a population that has mean μ and standard deviation σ. You want to test a hypothesis of the form[†]

$$H_0 : \mu = \mu_0$$

where μ_0 is some hypothesized value for μ, versus a one-tailed alternative hypothesis:

$$H_a : \mu > \mu_0$$

The subscript zero indicates the value of the parameter specified by H_0. Notice that H_0 provides an exact value for the parameter to be tested, whereas H_a gives a range of possible values for μ.

[†]Note that if the test rejects the null hypothesis $\mu = \mu_0$ in favor of the alternative hypothesis $\mu > \mu_0$, then it will certainly reject a null hypothesis that includes $\mu < \mu_0$, since this is even more contradictory to the alternative hypothesis. For this reason, in this text we state the null hypothesis for a one-tailed test as $\mu = \mu_0$ rather than $\mu \leq \mu_0$.

The Essentials of the Test

The sample mean $\bar{x}$ is the best estimate of the actual value of μ, which is presently in question. What values of $\bar{x}$ would lead you to believe that H_0 is false and μ is, in fact, greater than the hypothesized value? The values of $\bar{x}$ that are extremely *large* would imply that μ is larger than hypothesized. Hence, you should reject H_0 if $\bar{x}$ is too large.

The next problem is to define what is meant by "too large." Values of $\bar{x}$ that lie too many standard deviations to the right of the mean are not very likely to occur. Those values have very little area to their right. Hence, you can define "too large" as being too many standard deviations away from μ_0. But what is "too many"? This question can be answered using the *significance level* α, the probability of rejecting H_0 when H_0 is *true*.

Remember that the standard error of $\bar{x}$ is estimated as

$$SE = \frac{s}{\sqrt{n}}$$

Since the sampling distribution of the sample mean $\bar{x}$ is approximately normal when *n* **is large,** the number of standard deviations that $\bar{x}$ lies from μ_0 can be measured using the **standardized test statistic,**

3

$$z = \frac{\bar{x} - \mu_0}{s/\sqrt{n}}$$

which has an approximate standard normal distribution when H_0 is true and $\mu = \mu_0$. The significance level α is equal to the area under the normal curve lying above the

4 rejection region. Thus, if you want $\alpha = .01$, you will reject H_0 when $\bar{x}$ is more than 2.33 standard deviations to the right of μ_0. Equivalently, you will reject H_0 if the standardized test statistic z is greater than 2.33 (see Figure 9.3).

FIGURE 9.3

The rejection region for a right-tailed test with $\alpha = .01$

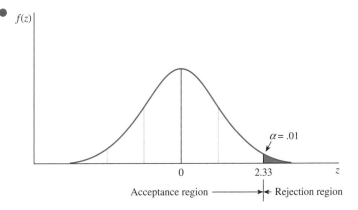

Acceptance region ⟶ ⟵ Rejection region

EXAMPLE **9.4** The average weekly earnings for female social workers is \$670. Do men in the same positions have average weekly earnings that are higher than those for women? A random sample of $n = 40$ male social workers showed $\bar{x} = \$725$ and $s = \$102$. Test the appropriate hypothesis using $\alpha = .01$.

NEED A TIP?

For one-tailed tests, look for directional words like "greater," "less than," "higher," "lower," etc.

Solution You would like to show that the average weekly earnings for men are higher than $670, the women's average. Hence, if μ is the average weekly earnings for male social workers, you can set out the formal test of hypothesis in steps:

1–2

Null and alternative hypotheses:

$$H_0: \mu = 670 \quad \text{versus} \quad H_a: \mu > 670$$

3

Test statistic: Using the sample information, with s as an estimate of the population standard deviation, calculate

$$z \approx \frac{\bar{x} - 670}{s/\sqrt{n}} = \frac{725 - 670}{102/\sqrt{40}} = 3.41$$

4

Rejection region: For this one-tailed test, values of $\bar{x}$ much larger than 670 would lead you to reject H_0; or, equivalently, values of the *standardized test statistic z* in the right tail of the standard normal distribution. To control the risk of making an incorrect decision as $\alpha = .01$, you must set the **critical value** separating the rejection and acceptance regions so that the area in the right tail is exactly $\alpha = .01$. This value is found in Table 3 of Appendix I to be $z = 2.33$, as shown in Figure 9.3. The null hypothesis will be rejected if the observed value of the test statistic, z, is greater than 2.33.

5

Conclusion: Compare the observed value of the test statistic, $z = 3.41$, with the critical value necessary for rejection, $z = 2.33$. Since the observed value of the test statistic falls in the rejection region, you can reject H_0 and conclude that the average weekly earnings for male social workers are higher than the average for female social workers. The probability that you have made an incorrect decision is $\alpha = .01$.

 NEED A TIP?

If the test is two-tailed, you will not see any directional words. The experimenter is only looking for a "difference" from the hypothesized value.

If you wish to detect departures either greater or less than μ_0, then the alternative hypothesis is *two-tailed*, written as

$$H_a : \mu \neq \mu_0$$

which implies either $\mu > \mu_0$ or $\mu < \mu_0$. Values of $\bar{x}$ that are either "too large" or "too small" in terms of their distance from μ_0 are placed in the rejection region. If you choose $\alpha = .01$, the area in the rejection region is equally divided between the two tails of the normal distribution, as shown in Figure 9.4. Using the standardized test statistic z, you can reject H_0 if $z > 2.58$ or $z < -2.58$. For different values of α, the critical values of z that separate the rejection and acceptance regions will change accordingly.

FIGURE 9.4

The rejection region for a two-tailed test with $\alpha = .01$

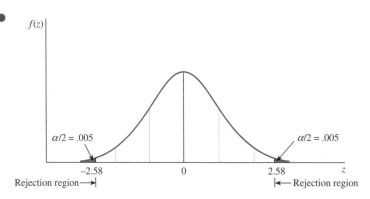

EXAMPLE 9.5

The daily yield for a local chemical plant has averaged 880 tons for the last several years. The quality control manager would like to know whether this average has changed in recent months. She randomly selects 50 days from the computer database and computes the average and standard deviation of the $n = 50$ yields as $\bar{x} = 871$ tons and $s = 21$ tons, respectively. Test the appropriate hypothesis using $\alpha = .05$.

Solution

1–2 **Null and alternative hypotheses:**

$$H_0: \mu = 880 \quad \text{versus} \quad H_a: \mu \neq 880$$

3 **Test statistic:** The point estimate for μ is $\bar{x}$. Therefore, the test statistic is

$$z \approx \frac{\bar{x} - \mu_0}{s/\sqrt{n}} = \frac{871 - 880}{21/\sqrt{50}} = -3.03$$

4 **Rejection region:** For this two-tailed test, you use values of z in both the right and left tails of the standard normal distribution. Using $\alpha = .05$, the **critical values** separating the rejection and acceptance regions cut off areas of $\alpha/2 = .025$ in the right and left tails. These values are $z = \pm 1.96$ and the null hypothesis will be rejected if $z > 1.96$ or $z < -1.96$.

5 **Conclusion:** Since $z = -3.03$, the calculated value of z, falls in the rejection region, the manager can reject the null hypothesis that $\mu = 880$ tons and conclude that it has changed. The probability of rejecting H_0 when H_0 is true is $\alpha = .05$, a fairly small probability. Hence, she is reasonably confident that the decision is correct.

LARGE-SAMPLE STATISTICAL TEST FOR μ

1. Null hypothesis: $H_0 : \mu = \mu_0$
2. Alternative hypothesis:

One-Tailed Test	**Two-Tailed Test**
$H_a : \mu > \mu_0$ (or, $H_a : \mu < \mu_0$)	$H_a : \mu \neq \mu_0$

3. Test statistic: $z = \dfrac{\bar{x} - \mu_0}{\sigma/\sqrt{n}}$ estimated as $z = \dfrac{\bar{x} - \mu_0}{s/\sqrt{n}}$

4. Rejection region: Reject H_0 when

One-Tailed Test	**Two-Tailed Test**
$z > z_\alpha$ (or $z < -z_\alpha$ when the alternative hypothesis is $H_a : \mu < \mu_0$)	$z > z_{\alpha/2}$ or $z < -z_{\alpha/2}$

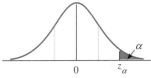
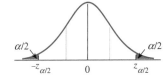

Assumptions: The n observations in the sample are randomly selected from the population and n is large—say, $n \geq 30$.

Calculating the p-Value

In the previous examples, the decision to reject or accept H_0 was made by comparing the calculated value of the test statistic with a critical value of z based on the significance level α of the test. However, different significance levels may lead to different conclusions. For example, if in a right-tailed test, the test statistic is $z = 2.03$, you can reject H_0 at the 5% level of significance because the test statistic exceeds $z = 1.645$. However, you cannot reject H_0 at the 1% level of significance, because the test statistic is less than $z = 2.33$ (see Figure 9.5). To avoid any ambiguity in their conclusions, some experimenters prefer to use a variable level of significance called the *p*-value for the test.

Definition The *p*-value or observed significance level of a statistical test is the smallest value of α for which H_0 can be rejected. It is the *actual risk* of committing a Type I error, if H_0 is rejected based on the observed value of the test statistic. The *p*-value measures the strength of the evidence against H_0.

In the right-tailed test with observed test statistic $z = 2.03$, the smallest critical value you can use and still reject H_0 is $z = 2.03$. For this critical value, the risk of an incorrect decision is

$$P(z \geq 2.03) = 1 - .9788 = .0212$$

This probability is the *p-value* for the test. *Notice that it is actually the area to the right of the calculated value of the test statistic.*

FIGURE 9.5

Variable rejection regions

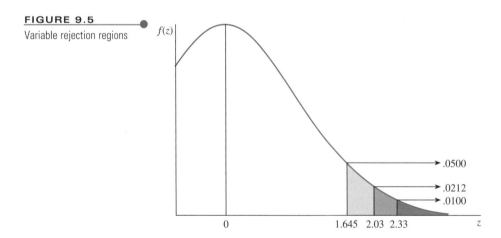

A *small p-value* indicates that the observed value of the test statistic lies far away from the hypothesized value of μ. This presents strong evidence that H_0 is false and should be rejected. *Large p-values* indicate that the observed test statistic is not far from the hypothesized mean and does not support rejection of H_0. How small does the *p*-value need to be before H_0 can be rejected?

NEED A TIP?

p-Value = Tail area (one or two tails) "beyond" the observed value of the test statistic

Definition If the p-value is less than or equal to a preassigned significance level α, then the null hypothesis can be rejected, and you can report that the results are **statistically significant** at level α.

In the previous instance, if you choose $\alpha = .05$ as your significance level, H_0 can be rejected because the p-value is less than .05. However, if you choose $\alpha = .01$ as your significance level, the p-value (.0212) is not small enough to allow rejection of H_0. The results are significant at the 5% level, but not at the 1% level. You might see these results reported in professional journals as *significant* $(p < .05)$.[†]

EXAMPLE 9.6

Refer to Example 9.5. The quality control manager wants to know whether the daily yield at a local chemical plant—which has averaged 880 tons for the last several years—has changed in recent months. A random sample of 50 days gives an average yield of 871 tons with a standard deviation of 21 tons. Calculate the p-value for this two-tailed test of hypothesis. Use the p-value to draw conclusions regarding the statistical test.

Solution The rejection region for this two-tailed test of hypothesis is found in both tails of the normal probability distribution. Since the observed value of the test statistic is $z = -3.03$, the smallest rejection region that you can use and still reject H_0 is $|z| > 3.03$. For this rejection region, the value of α is the p-value:

$$p\text{-value} = P(z > 3.03) + P(z < -3.03) = (1 - .9988) + .0012 = .0024$$

Notice that the two-tailed p-value is actually twice the tail area corresponding to the calculated value of the test statistic. If this p-value $= .0024$ is less than or equal to the preassigned level of significance α, H_0 can be rejected. For this test, you can reject H_0 at either the 1% or the 5% level of significance.

If you are reading a research report, how small should the p-value be before you decide to reject H_0? Many researchers use a "sliding scale" to classify their results.

- If the p-value is less than .01, H_0 is rejected. The results are **highly significant.**
- If the p-value is between .01 and .05, H_0 is rejected. The results are **statistically significant.**
- If the p-value is between .05 and .10, H_0 is usually not rejected. The results are only **tending toward statistical significance.**
- If the p-value is greater than .10, H_0 is not rejected. The results are **not statistically significant.**

EXAMPLE 9.7

Standards set by government agencies indicate that Americans should not exceed an average daily sodium intake of 3300 milligrams (mg). To find out whether Americans are exceeding this limit, a sample of 100 Americans is selected, and the mean and standard deviation of daily sodium intake are found to be 3400 mg and 1100 mg, respectively. Use $\alpha = .05$ to conduct a test of hypothesis.

[†]In reporting statistical significance, many researchers write $(p < .05)$ or $(P < .05)$ to mean that the p-value of the test was smaller than .05, making the results significant at the 5% level. The symbol p or P in the expression has no connection with our notation for probability or with the binomial parameter p.

Solution   The hypotheses to be tested are

$$H_0 : \mu = 3300 \quad \text{versus} \quad H_a : \mu > 3300$$

and the test statistic is

$$z \approx \frac{\bar{x} - \mu_0}{s/\sqrt{n}} = \frac{3400 - 3300}{1100/\sqrt{100}} = .91$$

The two approaches developed in this section yield the same conclusions.

NEED A TIP?

Small *p*-value ⟺ Large
z-value

Small *p*-value ⟹ Reject
H_0

How small? *p*-value ≤ α

- **The critical value approach:** Since the significance level is $\alpha = .05$ and the test is one-tailed, the rejection region is determined by a critical value with tail area equal to $\alpha = .05$; that is, H_0 can be rejected if $z > 1.645$. Since $z = .91$ is not greater than the critical value, H_0 is not rejected (see Figure 9.6).
- **The *p*-value approach:** Calculate the *p*-value, the probability that *z* is greater than or equal to $z = .91$:

$$p\text{-value} = P(z > .91) = 1 - .8186 = .1814$$

The null hypothesis can be rejected only if the *p-value is less than or equal to the specified 5% significance level.* Therefore, H_0 is not rejected and the results are *not statistically significant* (see Figure 9.6). There is not enough evidence to indicate that the average daily sodium intake exceeds 3300 mg.

FIGURE 9.6

Rejection region and
p-value for Example 9.7

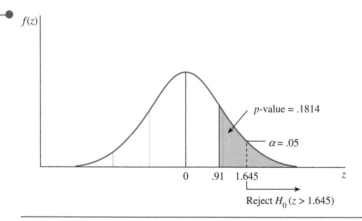

ONLINE APPLET

Large-Sample Test of a
Population Mean

Notice that these two approaches are actually the same, as shown in Figure 9.6. As soon as the calculated value of the test statistic *z* becomes *larger than* the critical value, z_α, the *p*-value becomes *smaller than* the significance level α. You can use the most convenient of the two methods; the conclusions you reach will always be the same! The *p*-value approach does have two advantages, however:

- Statistical output from computer software packages usually reports the *p*-value of the test.
- Based on the *p*-value, your test results can be evaluated using any significance level you wish to use. Many researchers report the smallest possible significance level for which their results are *statistically significant.*

Sometimes it is easy to confuse the significance level α with the *p*-value (or observed significance level). They are both probabilities calculated as areas in the tails of the sampling distribution of the test statistic. However, the significance level α is preset by the experimenter before collecting the data. The p-value is linked directly to the data and actually describes how likely or unlikely the sample results are, assuming that H_0 is true. *The smaller the p-value, the more unlikely it is that H_0 is true!*

 NEED TO KNOW...

Rejection Regions, *p*-Values, and Conclusions

The significance level, α, lets you set the risk that you are willing to take of making an incorrect decision in a test of hypothesis.

- To set a rejection region, choose a **critical value** of z so that the area in the tail(s) of the z-distribution is (are) either α for a one-tailed test or $\alpha/2$ for a two-tailed test. Use the right tail for an upper-tailed test and the left tail for a lower-tailed test. Reject H_0 when the test statistic exceeds the critical value and falls in the rejection region.

- To find a ***p*-value,** find the area in the tail "beyond" the test statistic. If the test is one-tailed, this is the p-value. If the test is two-tailed, this is only half the p-value and must be doubled. Reject H_0 when the p-value is less than α.

Two Types of Errors

You might wonder why, when H_0 was not rejected in the previous example, we did not say that H_0 was definitely true and $\mu = 3300$. This is because, if we choose to *accept H_0*, we must have a measure of the probability of error associated with this decision.

Since there are two choices in a statistical test, there are also two types of errors that can be made. In the courtroom trial, a defendant could be judged not guilty when he's really guilty, or vice versa—the same is true in a statistical test. In fact, the null hypothesis may be either true or false, regardless of the decision the experimenter makes. These two possibilities, along with the two decisions that can be made by the researcher, are shown in Table 9.1.

TABLE 9.1 ● **Decision Table**

Decision	Null Hypothesis True	False
Reject H_0	Type I error	Correct decision
Accept H_0	Correct decision	Type II error

In addition to the Type I error with probability α defined earlier in this section, it is possible to commit a second error, called a **Type II error,** which has probability β.

Definition A **Type I error** for a statistical test happens if you reject the null hypothesis when it is true. The probability of making a Type I error is denoted by the symbol α.

A **Type II error** for a statistical test happens if you accept the null hypothesis when it is false and some alternative hypothesis is true. The probability of making a Type II error is denoted by the symbol β.

Notice that the probability of a Type I error is exactly the same as the **level of significance α** and is therefore controlled by the researcher. When H_0 is rejected, you have an accurate measure of the reliability of your inference; the probability of an incorrect decision is α. However, the probability β of a Type II error is not always controlled by the experimenter. In fact, when H_0 is false and H_a is true, you may not be able to specify an exact value for μ, but only a range of values. This makes it difficult, if not impossible, to calculate β. Without a measure of reliability, it is not wise to conclude that H_0 is true. Rather than risk an incorrect decision, you should withhold judgment, concluding that you *do not have enough evidence to reject H_0.* Instead of *accepting H_0,* you should "*not reject*" or "*fail to reject*" H_0.

Keep in mind that "*accepting" a particular hypothesis means deciding in its favor.* Regardless of the outcome of a test, you are never *certain* that the hypothesis you "accept" is true. *There is always a risk of being wrong (measured by α or β).* Consequently, you never "accept" H_0 if β is unknown or its value is unacceptable to you. When this situation occurs, you should withhold judgment and collect more data.

> **NEED A TIP?**
>
> $\alpha = P(\text{reject } H_0 \text{ when } H_0 \text{ true})$
>
> $\beta = P(\text{accept } H_0 \text{ when } H_0 \text{ false})$

The Power of a Statistical Test

The goodness of a statistical test is measured by the size of the two error rates: α, the probability of rejecting H_0 when it is true; and β, the probability of accepting H_0 when H_0 is false and H_a is true. A "good" test is one for which both of these error rates are small. The experimenter begins by selecting α, the probability of a Type I error. If he or she also decides to control the value of β, the probability of accepting H_0 when H_a is true, then an appropriate sample size is chosen.

Another way of evaluating a test is to look at the complement of a Type II error—that is, rejecting H_0 when H_a is true—which has probability

$$1 - \beta = P(\text{reject } H_0 \text{ when } H_a \text{ is true})$$

The quantity $(1 - \beta)$ is called the **power** of the test because it measures the probability of taking the action that we wish to have occur—that is, rejecting the null hypothesis when it is false and H_a is true.

Definition   The **power of a statistical test,** given as

$$1 - \beta = P(\text{reject } H_0 \text{ when } H_a \text{ is true})$$

measures the ability of the test to perform as required.

A graph of $(1 - \beta)$, the probability of rejecting H_0 when in fact H_0 is false, as a function of the true value of the parameter of interest is called the **power curve** for the statistical test. Ideally, you would like α to be small and the *power* $(1 - \beta)$ to be large.

EXAMPLE **9.8**   Refer to Example 9.5. Calculate β and the power of the test $(1 - \beta)$ when μ is actually equal to 870 tons.

Solution   In Example 9.5, you assumed that H_0 was true and that $\mu = 880$. The rejection region with $\alpha = .05$ (using the right-hand curve in Figure 9.7) was set as

$$z = \frac{\bar{x} - \mu_0}{s/\sqrt{n}} > 1.96 \quad \text{or} \quad z = \frac{\bar{x} - \mu_0}{s/\sqrt{n}} < -1.96.$$

This implies that the acceptance region is

$$-1.96 < \frac{\bar{x} - 880}{21/\sqrt{50}} < 1.96 \quad \text{or} \quad 874.18 < \bar{x} < 885.82$$

shown along the horizontal axis in Figure 9.7. When H_0 is false and $\mu = 870$, the sampling distribution of $\bar{x}$ is actually represented by the left-hand curve in Figure 9.7, a normal distribution with $\mu = 870$ and SE $= 21/\sqrt{50} = 2.97$. Then β, the probability of accepting H_0 when $\mu = 870$, is the area under the left-hand normal curve located between 874.18 and 885.82 (see Figure 9.7). Calculating the z-values corresponding to 874.18 and 885.82, you get

FIGURE 9.7

Calculating β in
Example 9.8

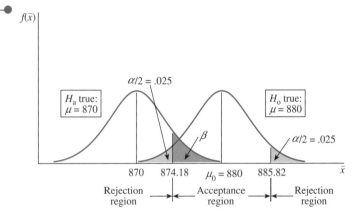

$$z_1 \approx \frac{\bar{x} - \mu}{s/\sqrt{n}} = \frac{874.18 - 870}{21/\sqrt{50}} = 1.41$$

$$z_2 \approx \frac{\bar{x} - \mu}{s/\sqrt{n}} = \frac{885.82 - 870}{21/\sqrt{50}} = 5.33$$

Then

$$\beta = P(\text{accept } H_0 \text{ when } \mu = 870) = P(874.18 < \bar{x} < 885.82 \text{ when } \mu = 870)$$
$$= P(1.41 < z < 5.33)$$

ONLINE APPLET

Power of a z-Test

You can see from Figure 9.7 that the area under the normal curve with $\mu = 870$ above $\bar{x} = 885.82$ (or $z = 5.33$) is negligible. Therefore,

$$\beta = P(z > 1.41)$$

From Table 3 in Appendix I you can find

$$\beta = 1 - .9207 = .0793$$

Hence, the power of the test is

$$1 - \beta = 1 - .0793 = .9207$$

The probability of correctly rejecting H_0, given that μ is really equal to 870, is .9207, or approximately 92 chances in 100.

Values of $(1 - \beta)$ can be calculated for various values of μ_a different from $\mu_0 = 880$ to measure the power of the test. For example, if $\mu_a = 885$,

$$\beta = P(874.18 < \bar{x} < 885.82 \text{ when } \mu = 885)$$
$$= P(-3.64 < z < .28)$$
$$= .6103 - 0 = .6103$$

and the power is $(1 - \beta) = .3897$. Table 9.2 shows the power of the test for various values of μ_a, and a power curve is graphed in Figure 9.8. Note that the power of the test increases as the distance between μ_a and μ_0 increases. The result is a U-shaped curve for this two-tailed test.

TABLE 9.2 ● **Value of $(1 - \beta)$ for Various Values of μ_a for Example 9.8**

μ_a	$(1 - \beta)$	μ_a	$(1 - \beta)$
865	.9990	883	.1726
870	.9207	885	.3897
872	.7673	888	.7673
875	.3897	890	.9207
877	.1726	895	.9990
880	.0500		

FIGURE 9.8

Power curve for Example 9.8

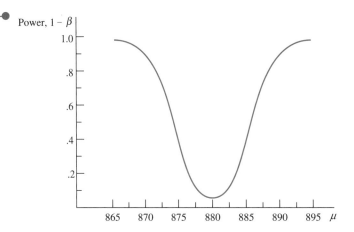

There are many important links among the two error rates, α and β, the power, $(1 - \beta)$, and the sample size, n. Look at the two curves shown in Figure 9.7.

- If α (the sum of the two tail areas in the curve on the right) is increased, the shaded area corresponding to β decreases, and vice versa.

- The only way to decrease β for a fixed α is to "buy" more information—that is, increase the sample size n.

What would happen to the area β as the curve on the left is moved closer to the curve on the right ($\mu = 880$)? With the rejection region in the right curve fixed, the value of β will *increase*. What effect does this have on the power of the test? Look at Figure 9.8.

- As the distance between the true (μ_a) and hypothesized (μ_0) values of the mean increases, the power $(1 - \beta)$ increases. The test is better at detecting *differences* when the distance is *large*.

- The closer the true value (μ_a) gets to the hypothesized value (μ_0), the less power ($1 - \beta$) the test has to detect the difference.
- The only way to increase the power ($1 - \beta$) for a fixed α is to "buy" more information—that is, increase the sample size, n.

The experimenter must decide on the values of α and β—measuring the risks of the possible errors he or she can tolerate. He or she also must decide how much power is needed to detect differences that are practically important in the experiment. Once these decisions are made, the sample size can be chosen by consulting the power curves corresponding to various sample sizes for the chosen test.

NEED TO KNOW...

How to Calculate β

1. Find the critical value or values of $\bar{x}$ used to separate the acceptance and rejection regions.
2. Using one or more values for μ consistent with the alternative hypothesis H_a, calculate the probability that the sample mean $\bar{x}$ falls in the *acceptance region*. This produces the value $\beta = P(\text{accept } H_a \text{ when } \mu = \mu_a)$.
3. Remember that the **power** of the test is ($1 - \beta$).

 EXERCISES

BASIC TECHNIQUES

9.1 Find the appropriate rejection regions for the large-sample test statistic z in these cases:

a. A right-tailed test with $\alpha = .01$

b. A two-tailed test at the 5% significance level

9.2 Find the p-value for the following large-sample z tests:

a. A right-tailed test with observed $z = 1.15$

b. A two-tailed test with observed $z = -2.78$

c. A left-tailed test with observed $z = -1.81$

9.3 Find the appropriate rejection regions for the large-sample test statistic z in these cases:

a. A left-tailed test at the 1% significance level.

b. A two-tailed test with $\alpha = .01$.

c. Suppose that the observed value of the test statistic was $z = -2.41$. For the rejection regions constructed in parts a and b, draw the appropriate conclusion for the tests. If appropriate, give a measure of the reliability of your conclusion.

9.4 Refer to Exercise 9.1. Suppose that the observed value of the test statistic was $z = 2.16$. For the rejection regions constructed in parts a and b of Exercise 9.1, draw the appropriate conclusion for the tests. If appropriate, give a measure of the reliability of your conclusion.

9.5 For the three tests given in Exercise 9.4, use the p-value to determine the significance of the results. Explain what "statistically significant" means in terms of rejecting or accepting H_0 and H_a.

9.6 A random sample of $n = 35$ observations from a quantitative population produced a mean $\bar{x} = 2.4$ and a standard deviation $s = .29$. Suppose your research objective is to show that the population mean μ exceeds 2.3.

a. Give the null and alternative hypotheses for the test.

b. Locate the rejection region for the test using a 5% significance level.

c. Find the standard error of the mean.

d. Before you conduct the test, use your intuition to decide whether the sample mean $\bar{x} = 2.4$ is likely or unlikely, assuming that $\mu = 2.3$. Now conduct the test. Do the data provide sufficient evidence to indicate that $\mu > 2.3$?

9.7 Refer to Exercise 9.6.

a. Calculate the p-value for the test statistic in part d.

b. Use the p-value to draw a conclusion at the 5% significance level.

c. Compare the conclusion in part b with the conclusion reached in part d of Exercise 9.6. Are they the same?

9.8 Refer to Exercise 9.6. You want to test $H_0 : \mu = 2.3$ against $H_a : \mu > 2.3$.

a. Find the critical value of $\bar{x}$ used for rejecting H_0.

b. Calculate $\beta = P(\text{accept } H_0 \text{ when } \mu = 2.4)$.

c. Repeat the calculation of β for $\mu = 2.3, 2.5,$ and 2.6.

d. Use the values of β from parts b and c to graph the power curve for the test.

9.9 A random sample of 100 observations from a quantitative population produced a sample mean of 26.8 and a sample standard deviation of 6.5. Use the p-value approach to determine whether the population mean is different from 28. Explain your conclusions.

APPLICATIONS

9.10 What's Normal? What *is* normal, when it comes to people's body temperatures? A random sample of 130 human body temperatures, provided by Allen Shoemaker[3] in the *Journal of Statistical Education,* had a mean of 98.25° and a standard deviation of 0.73°. Does the data indicate that the average body temperature for healthy humans is different from 98.6°, the usual average temperature cited by physicians and others? Test using both methods given in this section.

a. Use the p-value approach with $\alpha = .05$.

b. Use the critical value approach with $\alpha = .05$.

c. Compare the conclusions from parts a and b. Are they the same?

d. The 98.6 standard was derived by a German doctor in 1868, who claimed to have recorded 1 million temperatures in the course of his research.[4] What conclusions can you draw about his research in light of your conclusions in parts a and b?

9.11 Hamburger Meat Exercise 8.35 involved packages of ground beef in a small tray, intended to hold 1 pound of meat. A random sample of 35 packages in the small tray produced weight measurements with an average of 1.01 pounds and a standard deviation of .18 pound.

a. If you were the quality control manager and wanted to make sure that the average amount of ground beef was indeed 1 pound, what hypotheses would you test?

b. Find the p-value for the test and use it to perform the test in part a.

c. How would you, as the quality control manager, report the results of your study to a consumer interest group?

9.12 Airline Occupancy Rates Suppose a scheduled airline flight must average at least 60% occupancy in order to be profitable to the airline. An examination of the occupancy rate for 120 10:00 A.M. flights from Atlanta to Dallas showed a mean occupancy per flight of 58% and a standard deviation of 11%.

a. If μ is the mean occupancy per flight and if the company wishes to determine whether or not this scheduled flight is unprofitable, give the alternative and the null hypotheses for the test.

b. Does the alternative hypothesis in part a imply a one- or two-tailed test? Explain.

c. Do the occupancy data for the 120 flights suggest that this scheduled flight is unprofitable? Test using $\alpha = .05$.

9.13 Potency of an Antibiotic A drug manufacturer claimed that the mean potency of one of its antibiotics was 80%. A random sample of $n = 100$ capsules were tested and produced a sample mean of $\bar{x} = 79.7\%$ with a standard deviation of $s = .8\%$. Do the data present sufficient evidence to refute the manufacturer's claim? Let $\alpha = .05$.

a. State the null hypothesis to be tested.

b. State the alternative hypothesis.

c. Conduct a statistical test of the null hypothesis and state your conclusion.

9.14 Invasive Species In a study of the pernicious giant hogweed, Jan Pergl[1] and associates compared the density of these plants in two different sites within the Caucasus region of Russia. In its native area, the average density was found to be 5 plants/m². In an invaded area in the Czech Republic, a sample of $n = 50$ plants produced an average density of 11.17 plants/m² with a standard deviation of 3.9 plants/m².

a. Does the invaded area in the Czech Republic have an average density of giant hogweed that is different from $\mu = 5$ at the $\alpha = .05$ level of significance?

b. What is the p-value associated with the test in part a? Can you reject H_0 at the 5% level of significance using the p-value?

9.15 Does College Pay Off? An article in *Time* describing various aspects of American life indicated that higher educational achievement paid off! College grads work 7.4 hours per day, fewer than those with less than a college education.[2] Suppose that the average work day for a random sample of $n = 100$ individuals who had less than a 4-year college education was calculated to be $\bar{x} = 7.9$ hours with a standard deviation of $s = 1.9$ hours.

a. Use the p-value approach to test the hypothesis that the average number of hours worked by individuals having less than a college degree is greater than individuals having a college degree. At what level can you reject H_0?

b. If you were a college graduate, how would you state your conclusion to put yourself in the best possible light?

c. If you were not a college graduate, how might you state your conclusion?

9.16 Flextime A company was contemplating the installation of a flextime schedule in which a worker schedules his or her work hours or compresses work weeks. The company estimates that it needs a minimum mean of 7 hours per day per assembly worker in order to operate effectively. Each of a random sample of 80 of the company's assemblers was asked to submit a tentative flextime schedule. If the mean number of hours per day for Monday was 6.7 hours and the standard deviation was 2.7 hours, do the data provide sufficient evidence to indicate that the mean number of hours worked per day on Mondays, for all of the company's assemblers, will be less than 7 hours? Test using $\alpha = .05$.

9.17 Sports and Achilles Tendon Injuries Some sports that involve a significant amount of running, jumping, or hopping put participants at risk for Achilles tendinopathy (AT), an inflammation and thickening of the Achilles tendon. A study in *The American Journal of Sports Medicine* looked at the diameter (in mm) of the affected tendons for patients who participated in these types of sports activities.[5] Suppose that the Achilles tendon diameters in the general population have a mean of 5.97 millimeters (mm). When the diameters of the affected tendon were measured for a random sample of 31 patients, the average diameter was 9.80 with a standard deviation of 1.95 mm. Is there sufficient evidence to indicate that the average diameter of the tendon for patients with AT is greater than 5.97 mm? Test at the 5% level of significance.

9.4 A LARGE-SAMPLE TEST OF HYPOTHESIS FOR THE DIFFERENCE BETWEEN TWO POPULATION MEANS

In many situations, the statistical question to be answered involves a comparison of two population means. For example, the U.S. Postal Service is interested in reducing its gasoline costs by replacing gasoline-powered trucks with electric-powered trucks. To determine whether significant savings in operating costs are achieved by changing to electric-powered trucks, a pilot study should be undertaken using, say, 100 conventional gasoline-powered mail trucks and 100 electric-powered mail trucks operated under similar conditions.

The statistic that summarizes the sample information regarding the difference in population means ($\mu_1 - \mu_2$) is the difference in sample means ($\bar{x}_1 - \bar{x}_2$). Therefore, in testing whether the difference in sample means indicates that the true difference in population means differs from a specified value, ($\mu_1 - \mu_2$) = D_0, you can use the standard error of ($\bar{x}_1 - \bar{x}_2$),

$$\sqrt{\frac{\sigma_1^2}{n_1} + \frac{\sigma_2^2}{n_2}} \quad \text{estimated by} \quad SE = \sqrt{\frac{s_1^2}{n_1} + \frac{s_2^2}{n_2}}$$

in the form of a z-statistic to measure how many standard deviations the difference $(\bar{x}_1 - \bar{x}_2)$ lies from the hypothesized difference D_0. The formal testing procedure is described next.

LARGE-SAMPLE STATISTICAL TEST FOR $(\mu_1 - \mu_2)$

1. Null hypothesis: $H_0 : (\mu_1 - \mu_2) = D_0$, where D_0 is some specified difference that you wish to test. For many tests, you will hypothesize that there is no difference between μ_1 and μ_2; that is, $D_0 = 0$.

2. Alternative hypothesis:

   One-Tailed Test   **Two-Tailed Test**

   $H_a : (\mu_1 - \mu_2) > D_0$ $H_a : (\mu_1 - \mu_2) \neq D_0$
   [or $H_a : (\mu_1 - \mu_2) < D_0$]

3. Test statistic: $z \approx \dfrac{(\bar{x}_1 - \bar{x}_2) - D_0}{\text{SE}} = \dfrac{(\bar{x}_1 - \bar{x}_2) - D_0}{\sqrt{\dfrac{s_1^2}{n_1} + \dfrac{s_2^2}{n_2}}}$

4. Rejection region: Reject H_0 when

   One-Tailed Test   **Two-Tailed Test**

   $z > z_\alpha$ $z > z_{\alpha/2}$ or $z < -z_{\alpha/2}$
   [or $z < -z_\alpha$ when the
   alternative hypothesis is
   $H_a : (\mu_1 - \mu_2) < D_0$]

   or when p-value $< \alpha$

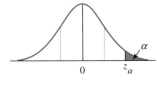   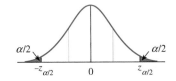

Assumptions: The samples are randomly and independently selected from the two populations and $n_1 \geq 30$ and $n_2 \geq 30$.

EXAMPLE 9.9   To determine whether car ownership affects a student's academic achievement, two random samples of 100 male students were each drawn from the student body. The grade point average for the $n_1 = 100$ nonowners of cars had an average and variance equal to $\bar{x}_1 = 2.70$ and $s_1^2 = .36$, while $\bar{x}_2 = 2.54$ and $s_2^2 = .40$ for the $n_2 = 100$ car owners. Do the data present sufficient evidence to indicate a difference in the mean achievements between car owners and nonowners of cars? Test using $\alpha = .05$.

Solution To detect a difference, if it exists, between the mean academic achievements for nonowners of cars μ_1 and car owners μ_2, you will test the null hypothesis that there is no difference between the means against the alternative hypothesis that $(\mu_1 - \mu_2) \neq 0$; that is,

$$H_0 : (\mu_1 - \mu_2) = D_0 = 0 \quad \text{versus} \quad H_a : (\mu_1 - \mu_2) \neq 0$$

Substituting into the formula for the test statistic, you get

$$z \approx \frac{(\bar{x}_1 - \bar{x}_2) - D_0}{\sqrt{\dfrac{s_1^2}{n_1} + \dfrac{s_2^2}{n_2}}} = \frac{2.70 - 2.54}{\sqrt{\dfrac{.36}{100} + \dfrac{.40}{100}}} = 1.84$$

 **NEED A TIP?**

|Test statistic| > |Critical value| ⇔ Reject H_0

- **The critical value approach:** Using a two-tailed test with significance level $\alpha = .05$, you place $\alpha/2 = .025$ in each tail of the z distribution and reject H_0 if $z > 1.96$ or $z < -1.96$. Since $z = 1.84$ does not exceed 1.96 and is not less than -1.96, H_0 cannot be rejected (see Figure 9.9). That is, there is insufficient evidence to declare a difference in the average academic achievements for the two groups. Remember that you should not be willing to *accept* H_0—declare the two means to be the same—until β is evaluated for some meaningful values of $(\mu_1 - \mu_2)$.

FIGURE 9.9

Rejection region and *p*-value for Example 9.9

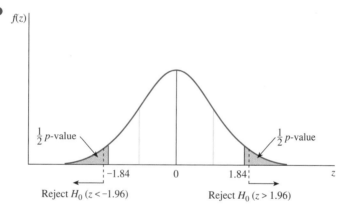

- **The *p*-value approach:** Calculate the *p*-value, the probability that z is greater than $z = 1.84$ plus the probability that z is less than $z = -1.84$, as shown in Figure 9.9:

$$p\text{-value} = P(z > 1.84) + P(z < -1.84) = (1 - .9671) + .0329 = .0658$$

The *p*-value lies between .10 and .05, so you can reject H_0 at the .10 level but not at the .05 level of significance. Since the *p*-value of .0658 exceeds the specified significance level $\alpha = .05$, H_0 cannot be rejected. Again, you should not be willing to *accept* H_0 until β is evaluated for some meaningful values of $(\mu_1 - \mu_2)$.

Hypothesis Testing and Confidence Intervals

Whether you use the critical value or the *p*-value approach for testing hypotheses about $(\mu_1 - \mu_2)$, you will always reach the same conclusion because the calculated value of the test statistic and the critical value are related *exactly* in the same way that the *p*-value and the significance level α are related. You might remember that the confidence intervals constructed in Chapter 8 could also be used to answer questions about the difference between two population means. In fact, for a two-tailed test, the $(1 - \alpha)100\%$ confidence interval for the parameter of interest can be used to test its value, just as you did informally in Chapter 8.

The value of α indicated by the confidence coefficient in the confidence interval is equivalent to the significance level α in the statistical test. For a one-tailed test, the equivalent confidence interval approach would use the one-sided confidence bounds in Section 8.8 with confidence coefficient α. In addition, by using the confidence interval approach, you gain a range of possible values for the parameter of interest, regardless of the outcome of the test of hypothesis.

- If the confidence interval you construct *contains* the value of the parameter specified by H_0, then that value is one of the likely or possible values of the parameter and H_0 should not be rejected.

- If the hypothesized value *lies outside* of the confidence limits, the null hypothesis is rejected at the α level of significance.

EXAMPLE 9.10 Construct a 95% confidence interval for the difference in average academic achievements between car owners and nonowners. Using the confidence interval, can you conclude that there is a difference in the population means for the two groups of students?

Solution Refer to Section 8.6. For the difference in two population means, the confidence interval is approximated as

$$(\bar{x}_1 - \bar{x}_2) \pm 1.96 \sqrt{\frac{s_1^2}{n_1} + \frac{s_2^2}{n_2}}$$

$$(2.70 - 2.54) \pm 1.96 \sqrt{\frac{.36}{100} + \frac{.40}{100}}$$

$$.16 \pm .17$$

or $-.01 < (\mu_1 - \mu_2) < .33$. This interval gives you a range of possible values for the difference in the population means.

Since the hypothesized difference, $(\mu_1 - \mu_2) = 0$, is contained in the confidence interval, you should not reject H_0. Look at the signs of the possible values in the confidence interval. You cannot tell from the interval whether the difference in the means is negative ($-$), positive ($+$), or zero (0)—the latter of the three would indicate that the two means are the same. Hence, you can really reach no conclusion in terms of the question posed. There is not enough evidence to indicate that there is a difference in the average achievements for car owners versus nonowners. The conclusion is the same one reached in Example 9.9.

9.4 EXERCISES

BASIC TECHNIQUES

9.18 Independent random samples of 80 measurements were drawn from two quantitative populations, 1 and 2. Here is a summary of the sample data:

	Sample 1	Sample 2
Sample Size	80	80
Sample Mean	11.6	9.7
Sample Variance	27.9	38.4

a. If your research objective is to show that μ_1 is larger than μ_2, state the alternative and the null hypotheses that you would choose for a statistical test.

b. Is the test in part a one- or two-tailed?

c. Calculate the test statistic that you would use for the test in part a. Based on your knowledge of the standard normal distribution, is this a likely or unlikely observation, assuming that H_0 is true and the two population means are the same?

d. *p-value approach:* Find the *p*-value for the test. Test for a significant difference in the population means at the 1% significance level.

e. *Critical value approach:* Find the rejection region when $\alpha = .01$. Do the data provide sufficient evidence to indicate a difference in the population means?

9.19 Independent random samples of 36 and 45 observations are drawn from two quantitative populations, 1 and 2, respectively. The sample data summary is shown here:

	Sample 1	Sample 2
Sample Size	36	45
Sample Mean	1.24	1.31
Sample Variance	.0560	.0540

Do the data present sufficient evidence to indicate that the mean for population 1 is smaller than the mean for population 2? Use one of the two methods of testing presented in this section, and explain your conclusions.

9.20 Suppose you wish to detect a difference between μ_1 and μ_2 (either $\mu_1 > \mu_2$ or $\mu_1 < \mu_2$) and, instead of running a two-tailed test using $\alpha = .05$, you use the following test procedure. You wait until you have collected the sample data and have calculated $\bar{x}_1$ and $\bar{x}_2$. If $\bar{x}_1$ is larger than $\bar{x}_2$, you choose the alternative hypothesis $H_a : \mu_1 > \mu_2$ and run a one-tailed test placing $\alpha_1 = .05$ in the upper tail of the z distribution. If, on the other hand, $\bar{x}_2$ is larger than $\bar{x}_1$, you reverse the procedure and run a one-tailed test, placing $\alpha_2 = .05$ in the lower tail of the z distribution. If you use this procedure and if μ_1 actually equals μ_2, what is the probability α that you will conclude that μ_1 is not equal to μ_2 (i.e., what is the probability α that you will incorrectly reject H_0 when H_0 is true)? This exercise demonstrates why statistical tests should be formulated *prior* to observing the data.

APPLICATIONS

9.21 Cure for the Common Cold? An experiment was planned to compare the mean time (in days) required to recover from a common cold for persons given a daily dose of 4 milligrams (mg) of vitamin C versus those who were not given a vitamin supplement. Suppose that 35 adults were randomly selected for each

treatment category and that the mean recovery times and standard deviations for the two groups were as follows:

	No Vitamin Supplement	4 mg Vitamin C
Sample Size	35	35
Sample Mean	6.9	5.8
Sample Standard Deviation	2.9	1.2

a. If your research objective is to show that the use of vitamin C reduces the mean time required to recover from a common cold and its complications, give the null and alternative hypotheses for the test. Is this a one- or a two-tailed test?

b. Conduct the statistical test of the null hypothesis in part a and state your conclusion. Test using $\alpha = .05$.

9.22 Healthy Eating As Americans become more conscious about the importance of good nutrition, researchers theorize that the consumption of red meat may have decreased over the last 10 years. A researcher selects hospital nutrition records for 400 subjects surveyed 10 years ago and compares the average amount of beef consumed per year to amounts consumed by an equal number of subjects interviewed this year. The data are given in the table.

	Ten Years Ago	This Year
Sample Mean	73	63
Sample Standard Deviation	25	28

a. Do the data present sufficient evidence to indicate that per-capita beef consumption has decreased in the last 10 years? Test at the 1% level of significance.

b. Find a 99% lower confidence bound for the difference in the average per-capita beef consumptions for the two groups. (This calculation was done as part of Exercise 8.78.) Does your confidence bound confirm your conclusions in part a? Explain. What additional information does the confidence bound give you?

9.23 Lead Levels in Drinking Water Analyses of drinking water samples for 100 homes in each of two different sections of a city gave the following means and standard deviations of lead levels (in parts per million):

	Section 1	Section 2
Sample Size	100	100
Mean	34.1	36.0
Standard Deviation	5.9	6.0

a. Calculate the test statistic and its *p*-value to test for a difference in the two population means. Use the *p*-value to evaluate the statistical significance of the results at the 5% level.

b. Use a 95% confidence interval to estimate the difference in the mean lead levels for the two sections of the city.

c. Suppose that the city environmental engineers will be concerned only if they detect a difference of more than 5 parts per million in the two sections of the city. Based on your confidence interval in part b, is the statistical significance in part a of *practical significance* to the city engineers? Explain.

9.24 Starting Salaries, again As a group, students majoring in the engineering disciplines have the highest salary expectations, followed by those studying the computer science fields, according to results of NACE's *2010 Student Survey*.[6] To compare the starting salaries of college graduates majoring in engineering and computer science (see Exercise 8.49), random samples of 50 recent college graduates in each major were selected and the following information obtained.

Major	Mean ($)	SD
Engineering	56,202	2225
Computer Science	50,657	2375

a. Do the data provide sufficient evidence to indicate a difference in average starting salaries for college graduates who majored in engineering and computer science? Test using $\alpha = .05$.

b. Compare your conclusions in part a with the results of part b in Exercise 8.49. Are they the same? Explain.

9.25 Hotel Costs In Exercise 8.20, we explored the average cost of lodging at three different hotel chains.[7] We randomly select 50 billing statements from the computer databases of the Marriott, Westin, and Doubletree hotel chains, and record the nightly room rates. A portion of the sample data is shown in the table.

	Marriott	Westin
Sample Average ($)	150	165
Sample Standard Deviation	17.2	22.5

a. Before looking at the data, would you have any preconceived idea about the direction of the difference between the average room rates for these two hotels? If not, what null and alternative hypotheses should you test?

b. Use the *critical value* approach to determine if there is a significant difference in the average room rates

for the Marriott and the Westin hotel chains. Use $\alpha = .01$.

c. Find the *p*-value for this test. Does this *p*-value confirm the results of part b?

9.26 Hotel Costs II Refer to Exercise 9.25. The table below shows the sample data collected to compare the average room rates at the Westin and Doubletree hotel chains.[7]

	Westin	Doubletree
Sample Average ($)	165	125
Sample Standard Deviation	22.5	12.8

a. Do the data provide sufficient evidence to indicate a difference in the average room rates for the Westin and the Doubletree hotel chains? Use $\alpha = .05$.

b. Construct a 95% confidence interval for the difference in the average room rates for the two chains. Does this interval confirm your conclusions in part a?

9.27 Cheaper Airfares Looking for a great airfare? *Consumer Reports*[8] has several hints about how to minimize your costs, which include being flexible about travel dates and times, checking multiple websites, and knowing when to book your flight. One suggestion involved checking fares at "secondary" airports—airports that might be slightly farther from your home, but where fares are lower. For example, the average of all domestic ticket prices at Los Angeles International Airport (LAX) was quoted as $349 compared to an average price of $287 at nearby Ontario International Airport (ONT). Suppose that these estimates were based on random samples of 1000 domestic tickets at each airport and that the standard deviation of the prices at both airports was $200.

a. Is there sufficient evidence to indicate that the mean ticket prices differ for these two airports at the $\alpha = .05$ level of significance? Use the large-sample *z*-test. What is the *p*-value of this test?

b. Construct a 95% confidence interval for $(\mu_1 - \mu_2)$. Does this interval confirm your conclusions in part a?

9.28 Noise and Stress In Exercise 8.52, you compared the effect of stress in the form of noise on the ability to perform a simple task. Seventy subjects were divided into two groups; the first group of 30 subjects acted as a control, while the second group of 40 was the experimental group. Although each subject performed the task, the experimental group subjects had to perform the task while loud rock music was played.

The time to finish the task was recorded for each subject and the following summary was obtained:

	Control	Experimental
n	30	40
$\bar{x}$	15 minutes	23 minutes
s	4 minutes	10 minutes

a. Is there sufficient evidence to indicate that the average time to complete the task was longer for the experimental "rock music" group? Test at the 1% level of significance.

b. Construct a 99% one-sided upper bound for the difference (control − experimental) in average times for the two groups. Does this interval confirm your conclusions in part a?

9.29 What's Normal II Of the 130 people in Exercise 9.16, 65 were female and 65 were male.[3] The means and standard deviations of their temperatures are shown below.

	Men	Women
Sample Mean	98.11	98.39
Standard Deviation	0.70	0.74

a. Use the p-value approach to test for a significant difference in the average temperatures for males versus females.

b. Are the results significant at the 5% level? At the 1% level?

A LARGE-SAMPLE TEST OF HYPOTHESIS FOR A BINOMIAL PROPORTION

9.5

When a random sample of n identical trials is drawn from a binomial population, the sample proportion $\hat{p}$ has an approximately normal distribution when n is large, with mean p and standard error

$$\text{SE} = \sqrt{\frac{pq}{n}}$$

When you test a hypothesis about p, the proportion in the population possessing a certain attribute, the test follows the same general form as the large-sample tests in Sections 9.3 and 9.4. To test a hypothesis of the form

$$H_0 : p = p_0$$

versus a one- or two-tailed alternative

$$H_a : p > p_0 \quad \text{or} \quad H_a : p < p_0 \quad \text{or} \quad H_a : p \neq p_0$$

the test statistic is constructed using $\hat{p}$, the best estimator of the true population proportion p. The sample proportion $\hat{p}$ is standardized, using the hypothesized mean and standard error, to form a test statistic z, which has a standard normal distribution if H_0 is true. This large-sample test is summarized next.

LARGE-SAMPLE STATISTICAL TEST FOR p

1. Null hypothesis: $H_0 : p = p_0$
2. Alternative hypothesis:

One-Tailed Test	**Two-Tailed Test**
$H_a : p > p_0$	$H_a : p \neq p_0$
(or, $H_a : p < p_0$)	

3. Test statistic: $z = \dfrac{\hat{p} - p_0}{SE} = \dfrac{\hat{p} - p_0}{\sqrt{\dfrac{p_0 q_0}{n}}}$ with $\hat{p} = \dfrac{x}{n}$

where x is the number of successes in n binomial trials.[†]

4. Rejection region: Reject H_0 when

One-Tailed Test

$z > z_\alpha$
(or $z < -z_\alpha$ when the
alternative hypothesis
is $H_a : p < p_0$)

or when p-value $< \alpha$

Two-Tailed Test

$z > z_{\alpha/2}$ or $z < -z_{\alpha/2}$

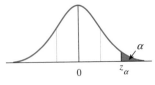

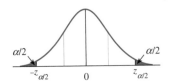

Assumption: The sampling satisfies the assumptions of a binomial experiment (see Section 5.2), and n is large enough so that the sampling distribution of $\hat{p}$ can be approximated by a normal distribution ($np_0 > 5$ and $nq_0 > 5$).

EXAMPLE 9.11

Regardless of age, about 20% of American adults participate in fitness activities at least twice a week. However, these fitness activities change as the people get older, and occasionally participants become nonparticipants as they age. In a local survey of $n = 100$ adults over 40 years old, a total of 15 people indicated that they participated in a fitness activity at least twice a week. Do these data indicate that the participation rate for adults over 40 years of age is significantly less than the 20% figure? Calculate the p-value and use it to draw the appropriate conclusions.

Solution   Assuming that the sampling procedure satisfies the requirements of a binomial experiment, you can answer the question posed using a one-tailed test of hypothesis:

$H_0 : p = .2$ versus $H_a : p < .2$

Begin by assuming that H_0 is true—that is, the true value of p is $p_0 = .2$. Then $\hat{p} = x/n$ will have an approximate normal distribution with mean p_0 and standard error $\sqrt{p_0 q_0/n}$. (NOTE: This is different from the estimation procedure in which the unknown standard error is estimated by $\sqrt{\hat{p}\hat{q}/n}$.) The observed value of $\hat{p}$ is $15/100 = .15$ and the test statistic is

NEED A TIP?

p-Value $\leq \alpha \Leftrightarrow$ Reject H_0
p-Value $> \alpha \Leftrightarrow$ Do not reject H_0

$$z = \frac{\hat{p} - p_0}{\sqrt{\dfrac{p_0 q_0}{n}}} = \frac{.15 - .20}{\sqrt{\dfrac{(.20)(.80)}{100}}} = -1.25$$

[†]An equivalent test statistic can be found by multiplying the numerator and denominator of z by n to obtain

$$z = \frac{x - np_0}{\sqrt{np_0 q_0}}$$

The *p*-value associated with this test is found as the area under the standard normal curve to the left of $z = -1.25$ as shown in Figure 9.10. Therefore,

$$p\text{-value} = P(z < -1.25) = .1056$$

FIGURE 9.10

p-value for Example 9.11

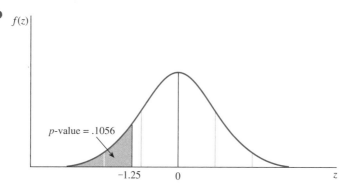

If you use the guidelines for evaluating *p*-values, then .1056 is greater than .10, and you would not reject H_0. There is insufficient evidence to conclude that the percentage of adults over age 40 who participate in fitness activities twice a week is less than 20%.

Statistical Significance and Practical Importance

It is important to understand the difference between results that are "significant" and results that are practically "important." In statistical language, the word *significant* does not necessarily mean "important," but only that the results could not have occurred by chance. For example, suppose that in Example 9.11, the researcher had used $n = 400$ adults in her experiment and had observed the same sample proportion. The test statistic is now

$$z = \frac{\hat{p} - p_0}{\sqrt{\dfrac{p_0 q_0}{n}}} = \frac{.15 - .20}{\sqrt{\dfrac{(.20)(.80)}{400}}} = -2.50$$

with

$$p\text{-value} = P(z < -2.50) = .0062$$

Now the results are *highly significant:* H_0 is rejected, and there is sufficient evidence to indicate that the percentage of adults over age 40 who participate in physical fitness activities is less than 20%. However, is this drop in activity really *important*? Suppose that physicians would be concerned only about a drop in physical activity of more than 10%. If there had been a drop of more than 10% in physical activity, this would imply that the true value of *p* was less than .10. What is the largest possible value of *p*? Using a 95% upper one-sided confidence bound, you have

$$\hat{p} + 1.645 \sqrt{\frac{\hat{p}\hat{q}}{n}}$$

$$.15 + 1.645 \sqrt{\frac{(.15)(.85)}{400}}$$

$$.15 + .029$$

or $p < .179$. The physical activity for adults aged 40 and older has dropped from 20%, but you cannot say that it has dropped below 10%. So, the results, although *statistically significant,* are not *practically important.*

In this book, you will learn how to determine whether results are statistically significant. When you use these procedures in a practical situation, however, you must also make sure the results are practically important.

9.5 EXERCISES

BASIC TECHNIQUES

9.30 A random sample of 120 observations was selected from a binomial population, and 72 successes were observed. Do the data provide sufficient evidence to indicate that p is greater than .5? Use one of the two methods of testing presented in this section, and explain your conclusions.

9.31 A random sample of $n = 1400$ observations from a binomial population produced $x = 529$.

a. If your research hypothesis is that p differs from .4, what hypotheses should you test?

b. Calculate the test statistic and its p-value. Use the p-value to evaluate the statistical significance of the results at the 1% level.

c. Do the data provide sufficient evidence to indicate that p is different from .4?

9.32 A random sample of $n = 1000$ observations from a binomial population produced $x = 279$.

a. If your research hypothesis is that p is less than .3, what should you choose for your alternative hypothesis? Your null hypothesis?

b. What is the critical value that determines the rejection region for your test with $\alpha = .05$?

c. Do the data provide sufficient evidence to indicate that p is less than .3? Use a 5% significance level.

APPLICATIONS

9.33 Childhood Obesity According to a survey in *PARADE* magazine, almost half of parents say their children's weight is fine.[9] Only 9% of parents describe their children as overweight. However, the American Obesity Association says the number of overweight children and teens is at least 15%. Suppose that you sample $n = 750$ parents and the number who describe their children as overweight is $x = 68$.

a. How would you test the hypothesis that the proportion of parents who describe their children as overweight is less than the actual proportion reported by the American Obesity Association?

b. What conclusion are you able to draw from these data at the $\alpha = .05$ level of significance?

c. What is the p-value associated with this test?

9.34 Landlines Passe? According to a new nationwide survey from the Pew Research Center's Social & Demographic Trends project, 62% of Americans consider a landline phone a necessity of life, down from 68% last year, and only 43% of Americans consider a TV a necessity.[11] Both questions were on the second of two forms used in the survey and involved $n = 1483$ individuals contacted by either a landline phone or a cell phone. Use a test of hypothesis to determine whether the $\hat{p} = .62$ figure is a significant drop from last year's value of $p = .68$ with $\alpha = .01$. What conclusions can you draw from this analysis?

9.35 Early Detection of Breast Cancer Of those women who are diagnosed to have early-stage breast cancer, one-third eventually die of the disease. Suppose a community public health department instituted a screening program to provide for the early detection of breast cancer and to increase the survival rate p of those diagnosed to have the disease. A random sample of 200 women was selected from among those who were periodically screened by the program and who were diagnosed to have the disease. Let x represent the number of those in the sample who survive the disease.

a. If you wish to determine whether the community screening program has been effective, state the alternative hypothesis that should be tested.

b. State the null hypothesis.

c. If 164 women in the sample of 200 survive the disease, can you conclude that the community screening program was effective? Test using $\alpha = .05$ and explain the practical conclusions from your test.

d. Find the p-value for the test and interpret it.

9.36 Plant Genetics A peony plant with red petals was crossed with another plant having streaky petals. A geneticist states that 75% of the offspring resulting from this cross will have red flowers. To test this claim, 100 seeds from this cross were collected and germinated, and 58 plants had red petals.

a. What hypothesis should you use to test the geneticist's claim?

b. Calculate the test statistic and its p-value. Use the p-value to evaluate the statistical significance of the results at the 1% level.

9.37 Taste Testing In a head-to-head taste test of store-brand foods versus national brands, *Consumer Reports* found that it was hard to find a taste difference in the two.[10] If the national brand is indeed better than the store brand, it should be judged as better more than 50% of the time.

a. State the null and alternative hypothesis to be tested. Is this a one- or a two-tailed test?

b. Suppose that, of the 35 food categories used for the taste test, the national brand was found to be better than the store brand in eight of the taste comparisons. Use this information to test the hypothesis in part a. Use $\alpha = .01$. What practical conclusions can you draw from the results?

9.38 Sweet Potato Whitefly Suppose that 10% of the fields in a given agricultural area are infested with the sweet potato whitefly. One hundred fields in this area are randomly selected, and 25 are found to be infested with whitefly.

a. Assuming that the experiment satisfies the conditions of the binomial experiment, do the data indicate that the proportion of infested fields is greater than expected? Use the p-value approach, and test using a 5% significance level.

b. If the proportion of infested fields is found to be significantly greater than .10, why is this of practical significance to the agronomist? What practical conclusions might she draw from the results?

9.39 A Cure for Insomnia An experimenter has prepared a drug-dose level that he claims will induce sleep for at least 80% of people suffering from insomnia. After examining the dosage we feel that his claims regarding the effectiveness of his dosage are inflated. In an attempt to disprove his claim, we administer his prescribed dosage to 50 insomniacs and observe that 37 of them have had sleep induced by the drug dose. Is there enough evidence to refute his claim at the 5% level of significance?

9.40 Taste Testing, continued In Exercise 9.37, we tried to prove that the national brand tasted better than the store brand.[10] Perhaps, however, the store brand has a better taste than the national brand! If this is true, then the store brand should be judged as better more than 50% of the time.

a. State the null and alternative hypothesis to be tested. Is this a one- or a two-tailed test?

b. Suppose that, of the 35 food categories used for the taste test, the store brand was found to be better than the national brand in six of the taste comparisons. Use this information to test the hypothesis in part a. Use $\alpha = .01$.

c. In the other 21 food comparisons in this experiment, the tasters could find no difference in taste between the store and national brands. What practical conclusions can you draw from this fact and from the two hypothesis tests in Exercises 9.37(b) and 9.38(b)?

9.41 Man's Best Friend The Humane Society reports that there are approximately 77.5 million dogs owned in the United States and that approximately 40% of all U.S. households own at least one dog.[12] In a random sample of 300 households, 114 households said that they owned at least one dog. Does this data provide sufficient evidence to indicate that the proportion of households with at least one dog is different from that reported by the Humane Society? Test using $\alpha = .05$.

A LARGE-SAMPLE TEST OF HYPOTHESIS FOR THE DIFFERENCE BETWEEN TWO BINOMIAL PROPORTIONS

9.6

When random and independent samples are selected from two *binomial* populations, the focus of the experiment may be the difference $(p_1 - p_2)$ in the proportions of individuals or items possessing a specified characteristic in the two populations. In this situation, you can use the difference in the sample proportions $(\hat{p}_1 - \hat{p}_2)$ along with its standard error,

$$SE = \sqrt{\frac{p_1 q_1}{n_1} + \frac{p_2 q_2}{n_2}}$$

in the form of a z-statistic to test for a significant difference in the two population proportions. The null hypothesis to be tested is usually of the form

$$H_0 : p_1 = p_2 \quad \text{or} \quad H_0 : (p_1 - p_2) = 0$$

versus either a one- or two-tailed alternative hypothesis. The formal test of hypothesis is summarized in the next display. In estimating the standard error for the z-statistic, you should use the fact that when H_0 is true, the two population proportions are equal to some common value—say, p. To obtain the best estimate of this common value, the sample data are "pooled" and the estimate of p is

$$\hat{p} = \frac{\text{Total number of successes}}{\text{Total number of trials}} = \frac{x_1 + x_2}{n_1 + n_2}$$

Remember that, in order for the difference in the sample proportions to have an approximately normal distribution, the sample sizes must be large and the proportions should not be too close to 0 or 1.

LARGE-SAMPLE STATISTICAL TEST FOR $(p_1 - p_2)$

1. Null hypothesis: $H_0 : (p_1 - p_2) = 0$ or equivalently $H_0 : p_1 = p_2$
2. Alternative hypothesis:

One-Tailed Test	**Two-Tailed Test**
$H_a : (p_1 - p_2) > 0$	$H_a : (p_1 - p_2) \neq 0$
[or $H_a : (p_1 - p_2) < 0$]	

3. Test statistic: $z = \dfrac{(\hat{p}_1 - \hat{p}_2) - 0}{\text{SE}} = \dfrac{\hat{p}_1 - \hat{p}_2}{\sqrt{\dfrac{p_1 q_1}{n_1} + \dfrac{p_2 q_2}{n_2}}} = \dfrac{\hat{p}_1 - \hat{p}_2}{\sqrt{\dfrac{pq}{n_1} + \dfrac{pq}{n_2}}}$

where $\hat{p}_1 = x_1/n_1$ and $\hat{p}_2 = x_2/n_2$. Since the common value of $p_1 = p_2 = p$ (used in the standard error) is unknown, it is estimated by

$$\hat{p} = \frac{x_1 + x_2}{n_1 + n_2}$$

and the test statistic is

$$z = \frac{(\hat{p}_1 - \hat{p}_2) - 0}{\sqrt{\dfrac{\hat{p}\hat{q}}{n_1} + \dfrac{\hat{p}\hat{q}}{n_2}}} \quad \text{or} \quad z = \frac{\hat{p}_1 - \hat{p}_2}{\sqrt{\hat{p}\hat{q}\left(\dfrac{1}{n_1} + \dfrac{1}{n_2}\right)}}$$

4. Rejection region: Reject H_0 when

One-Tailed Test	**Two-Tailed Test**
$z > z_\alpha$	$z > z_{\alpha/2}$ or $z < -z_{\alpha/2}$
[or $z < -z_\alpha$ when the alternative hypothesis is $H_a : (p_1 - p_2) < 0$]	

or when p-value $< \alpha$

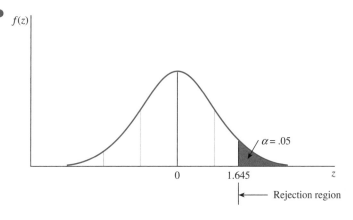

Assumptions: Samples are selected in a random and independent manner from two binomial populations, and n_1 and n_2 are large enough so that the sampling distribution of $(\hat{p}_1 - \hat{p}_2)$ can be approximated by a normal distribution. That is, $n_1\hat{p}_1$, $n_1\hat{q}_1$, $n_2\hat{p}_2$, and $n_2\hat{q}_2$ should all be greater than 5.

EXAMPLE 9.12

The records of a hospital show that 52 men in a sample of 1000 men versus 23 women in a sample of 1000 women were admitted because of heart disease. Do these data present sufficient evidence to indicate a higher rate of heart disease among men admitted to the hospital? Use $\alpha = .05$.

Solution Assume that the number of patients admitted for heart disease has an approximate binomial probability distribution for both men and women with parameters p_1 and p_2, respectively. Then, since you wish to determine whether $p_1 > p_2$, you will test the null hypothesis $p_1 = p_2$—that is, $H_0 : (p_1 - p_2) = 0$—against the alternative hypothesis $H_a : p_1 > p_2$ or, equivalently, $H_a : (p_1 - p_2) > 0$. To conduct this test, use the z-test statistic and approximate the standard error using the pooled estimate of p. Since H_a implies a one-tailed test, you can reject H_0 only for large values of z. Thus, for $\alpha = .05$, you can reject H_0 if $z > 1.645$ (see Figure 9.11).

The pooled estimate of p required for the standard error is

$$\hat{p} = \frac{x_1 + x_2}{n_1 + n_2} = \frac{52 + 23}{1000 + 1000} = .0375$$

FIGURE 9.11

Location of the rejection region in Example 9.12

and the test statistic is

$$z = \frac{\hat{p}_1 - \hat{p}_2}{\sqrt{\hat{p}\hat{q}\left(\dfrac{1}{n_1} + \dfrac{1}{n_2}\right)}} = \frac{.052 - .023}{\sqrt{(.0375)(.9625)\left(\dfrac{1}{1000} + \dfrac{1}{1000}\right)}} = 3.41$$

Since the computed value of z falls in the rejection region, you can reject the hypothesis that $p_1 = p_2$. The data present sufficient evidence to indicate that the percentage of men entering the hospital because of heart disease is higher than that of women. (NOTE: This does not imply that the *incidence* of heart disease is higher in men. Perhaps fewer women enter the hospital when afflicted with the disease!)

How *much higher* is the proportion of men than women entering the hospital with heart disease? A 95% lower one-sided confidence bound will help you find the lowest likely value for the difference.

$$(\hat{p}_1 - \hat{p}_2) - 1.645 \sqrt{\frac{\hat{p}_1\hat{q}_1}{n_1} + \frac{\hat{p}_2\hat{q}_2}{n_2}}$$

$$(.052 - .023) - 1.645 \sqrt{\frac{.052(.948)}{1000} + \frac{.023(.977)}{1000}}$$

$$.029 - .014$$

or $(p_1 - p_2) > .015$. The proportion of men is roughly 1.5% higher than women. Is this of *practical importance*? This is a question for the researcher to answer.

In some situations, you may need to test for a difference D_0 (other than 0) between two binomial proportions. If this is the case, the test statistic is modified for testing $H_0 : (p_1 - p_2) = D_0$, and a pooled estimate for a common p is no longer used in the standard error. The modified test statistic is

$$z = \frac{(\hat{p}_1 - \hat{p}_2) - D_0}{\sqrt{\frac{\hat{p}_1\hat{q}_1}{n_1} + \frac{\hat{p}_2\hat{q}_2}{n_2}}}$$

Although this test statistic is not used often, the procedure is no different from other large-sample tests you have already mastered!

9.6 EXERCISES

BASIC TECHNIQUES

9.42 Independent random samples of 280 and 350 observations were selected from binomial populations 1 and 2, respectively. Sample 1 had 132 successes, and sample 2 had 178 successes. Do the data present sufficient evidence to indicate that the proportion of successes in population 1 is smaller than the proportion in population 2? Use one of the two methods of testing presented in this section, and explain your conclusions.

9.43 Refer to Exercise 9.42. Suppose, for practical reasons, you know that p_1 cannot be larger than p_2.

a. Given this knowledge, what should you choose as the alternative hypothesis for your statistical test? The null hypothesis?

b. Does your alternative hypothesis in part a imply a one- or two-tailed test? Explain.

c. Conduct the test and state your conclusions. Test using $\alpha = .05$.

9.44 Independent random samples of $n_1 = 140$ and $n_2 = 140$ observations were randomly selected from binomial populations 1 and 2, respectively. Sample 1 had 74 successes, and sample 2 had 81 successes.

a. Suppose you have no preconceived idea as to which parameter, p_1 or p_2, is the larger, but you want only to detect a difference between the two parameters if one exists. What should you choose as the alternative hypothesis for a statistical test? The null hypothesis?

b. Calculate the standard error of the difference in the two sample proportions, $(\hat{p}_1 - \hat{p}_2)$. Make sure to use the pooled estimate for the common value of p.

c. Calculate the test statistic that you would use for the test in part a. Based on your knowledge of the standard normal distribution, is this a likely or unlikely observation, assuming that H_0 is true and the two population proportions are the same?

d. *p-value approach:* Find the *p*-value for the test. Test for a significant difference in the population proportions at the 1% significance level.

e. *Critical value approach:* Find the rejection region when $\alpha = .01$. Do the data provide sufficient evidence to indicate a difference in the population proportions?

APPLICATIONS

9.45 Treatment versus Control An experiment was conducted to test the effect of a new drug on a viral infection. After the infection was induced in 100 mice, the mice were randomly split into two groups of 50. The first group, the *control group,* received no treatment for the infection, and the second group received the drug. After a 30-day period, the proportions of survivors, $\hat{p}_1$ and $\hat{p}_2$, in the two groups were found to be .36 and .60, respectively.

a. Is there sufficient evidence to indicate that the drug is effective in treating the viral infection? Use $\alpha = .05$.

b. Use a 95% confidence interval to estimate the actual difference in the survival rates for the treated versus the control groups.

9.46 Tai Chi and Fibromyalgia A new study (Exercise 7.12) indicates that tai chi, an ancient Chinese practice of exercise and meditation, may relieve symptoms of chronic painful fibromyalgia. The study assigned 66 fibromyalgia patients to take either a 12-week tai chi class ($n_1 = 33$), or attend a wellness education class ($n_2 = 33$).[13] The results of the study are shown in the following table:

	Tai Chi	Wellness Education
Number Who Felt Better at End of Course	26	13

a. Is there a significant difference in the proportion of all fibromyalgia patients who would admit to feeling better after taking the tai chi class compared to the wellness education class? Use $\alpha = .01$.

b. Find the *p*-value for the test in part a. How would you describe the significance or nonsignificance of the test?

9.47 M&M'S In Exercise 8.57, you investigated whether Mars, Inc., uses the same proportion of red M&M'S in its plain and peanut varieties. Random samples of plain and peanut M&M'S provide the following sample data for the experiment:

	Plain	Peanut
Sample Size	56	32
Number of Red M&M'S	12	8

Use a test of hypothesis to determine whether there is a significant difference in the proportions of red candies for the two types of M&M'S. Let $\alpha = .05$ and compare your results with those of Exercise 8.57.

9.48 Hormone Therapy and Alzheimer's Disease A 4-year experiment involving 4532 women, reported in *The Press Enterprise,* was conducted at 39 medical centers to study the benefits and risks of hormone replacement therapy (HRT). Half of the women took placebos and half took Prempro, a widely prescribed type of hormone replacement therapy. There were 40 cases of dementia in the hormone group and 21 in the placebo group.[14] Is there sufficient evidence to indicate that the risk of dementia is higher for patients using Prempro? Test at the 1% level of significance.

9.49 HRT, continued Refer to Exercise 9.48. Calculate a 99% lower one-sided confidence bound for the difference in the risk of dementia for women using hormone replacement therapy versus those who do not. Would this difference be of *practical importance* to a woman considering HRT? Explain.

9.50 Clopidogrel and Aspirin A large study was conducted to test the effectiveness of clopidogrel in combination with aspirin in warding off heart attacks and strokes.[15] The trial involved more than 15,500 people 45 years of age or older from 32 countries, including the United States, who had been diagnosed with cardiovascular disease or had multiple risk factors. The subjects were randomly assigned to one of two groups. After 2 years, there was no difference in the risk of heart attack, stroke, or dying from heart disease between those who took clopidogrel and low-dose aspirin daily and those who took low-dose aspirin plus a dummy pill. The 2-drug combination actually increased the risk of dying (5.4% versus 3.8%) or dying specifically from cardiovascular disease (3.9% versus 2.2%).

a. The subjects were randomly assigned to one of the two groups. Explain how you could use the random number table to make these assignments.

b. No sample sizes were given in the article; however, let us assume that the sample sizes for each group were $n_1 = 7720$ and $n_2 = 7780$. Determine whether the risk of dying was significantly different for the two groups.

c. What do the results of the study mean in terms of *practical significance*?

9.51 Baby's Sleeping Position Does a baby's sleeping position affect the development of motor skills? In one study, published in the *Archives of Pediatric Adolescent Medicine,* 343 full-term infants were examined at their 4-month checkups for various developmental milestones, such as rolling over, grasping a rattle, reaching for an object, and so on.[16] The baby's predominant sleep position—either prone (on the stomach) or supine (on the back) or

side—was determined by a telephone interview with the parent. The sample results for 320 of the 343 infants for whom information was received are shown here:

	Prone	Supine or Side
Number of Infants	121	199
Number That Roll Over	93	119

The researcher reported that infants who slept in the side or supine position were less likely to roll over at the 4-month checkup than infants who slept primarily in the prone position ($P < .001$). Use a large-sample test of hypothesis to confirm or refute the researcher's conclusion.

9.7 SOME COMMENTS ON TESTING HYPOTHESES

A statistical test of hypothesis is a fairly clear-cut procedure that enables an experimenter to either reject or accept the null hypothesis H_0, with measured risks α and β. The experimenter can control the risk of falsely rejecting H_0 by selecting an appropriate value of α. On the other hand, the value of β depends on the sample size and the values of the parameter under test that are of practical importance to the experimenter. When this information is not available, an experimenter may decide to select an affordable sample size, in the hope that the sample will contain sufficient information to reject the null hypothesis. The chance that this decision is in error is given by α, whose value has been set in advance. If the sample does not provide sufficient evidence to reject H_0, the experimenter may wish to state the results of the test as "The data do not support the rejection of H_0" rather than accepting H_0 without knowing the chance of error β.

Some experimenters prefer to use the observed *p*-value of the test to evaluate the strength of the sample information in deciding to reject H_0. These values can usually be generated by computer and are often used in reports of statistical results:

- If the *p*-value is greater than .05, the results are reported as NS—not significant at the 5% level.

- If the *p*-value lies between .05 and .01, the results are reported as $P < .05$—significant at the 5% level.

- If the *p*-value lies between .01 and .001, the results are reported as $P < .01$—"highly significant" or significant at the 1% level.

- If the *p*-value is less than .001, the results are reported as $P < .001$—"very highly significant" or significant at the .1% level.

Still other researchers prefer to construct a confidence interval for a parameter and perform a test informally. If the value of the parameter specified by H_0 is included within the upper and lower limits of the confidence interval, then "H_0 is not rejected." If the value of the parameter specified by H_0 is not contained within the interval, then "H_0 is rejected." These results will agree with a two-tailed test; one-sided confidence bounds are used for one-tailed alternatives.

Finally, consider the choice between a one- and two-tailed test. In general, experimenters wish to know whether a treatment causes what could be a beneficial increase in a parameter or what might be a harmful decrease in a parameter. Therefore, most tests are two-tailed unless a one-tailed test is strongly dictated by practical considerations. For example, assume you will sustain a large financial loss if the mean μ is greater than μ_0 but not if it is less. You will then want to detect values larger than μ_0 with a high probability and thereby use a right-tailed test. In the same vein, if pollution levels higher than μ_0 cause critical health risks, then you will certainly wish to detect levels higher than μ_0 with a right-tailed test of hypothesis. In any case, the choice of a one- or two-tailed test should be dictated by the practical consequences that result from a decision to reject or not reject H_0 in favor of the alternative.

CHAPTER REVIEW

Key Concepts and Formulas

I. Parts of a Statistical Test

1. **Null hypothesis:** a contradiction of the alternative hypothesis

2. **Alternative hypothesis:** the hypothesis the researcher wants to support

3. **Test statistic** and its **p-value:** sample evidence calculated from the sample data

4. **Rejection region—critical values** and **significance levels:** values that lead to rejection and nonrejection of the null hypothesis

5. **Conclusion:** Reject or do not reject the null hypothesis, stating the practical significance of your conclusion

II. Errors and Statistical Significance

1. The **significance level** α is the probability of rejecting H_0 when it is in fact true.

2. The **p-value** is the probability of observing a test statistic as extreme as or more extreme than the one observed; also, the smallest value of α for which H_0 can be rejected.

3. When the **p-value** is less than the **significance level** α, the null hypothesis is rejected. This happens when the **test statistic** exceeds the **critical value.**

4. A **Type II error,** β, is the probability of accepting H_0 when it is in fact false. The **power of the test** is $(1 - \beta)$, the probability of rejecting H_0 when it is false.

III. Large-Sample Test Statistics Using the z Distribution

To test one of the four population parameters when the sample sizes are large, use the following test statistics:

Parameter	Test Statistic
μ	$z = \dfrac{\bar{x} - \mu_0}{s/\sqrt{n}}$
p	$z = \dfrac{\hat{p} - p_0}{\sqrt{\dfrac{p_0 q_0}{n}}}$
$\mu_1 - \mu_2$	$z = \dfrac{(\bar{x}_1 - \bar{x}_2) - D_0}{\sqrt{\dfrac{s_1^2}{n_1} + \dfrac{s_2^2}{n_2}}}$
$p_1 - p_2$	$z = \dfrac{\hat{p}_1 - \hat{p}_2}{\sqrt{\hat{p}\hat{q}\left(\dfrac{1}{n_1} + \dfrac{1}{n_2}\right)}}$ or $z = \dfrac{(\hat{p}_1 - \hat{p}_2) - D_0}{\sqrt{\dfrac{\hat{p}_1 \hat{q}_1}{n_1} + \dfrac{\hat{p}_2 \hat{q}_2}{n_2}}}$

Supplementary Exercises

Starred (*) exercises are optional.

9.52 Generation Next Born between 1980 and 1990, Generation Next was the topic of Exercise 8.64.[17] In a survey of 500 female and 500 male students in Generation Next, 345 of the females and 365 of the males reported that they decided to attend college in order to make more money.

a. Is there a significant difference in the population proportions of female and male students who decided to attend college in order to make more money? Use $\alpha = .01$.

b. Can you think of any reason why a statistically significant difference in these population proportions might be of *practical importance*? To whom might this difference be important?

9.53 What is the *p*-value for a test of hypothesis? How is it calculated for a large-sample test?

9.54 What conditions must be met so that the *z* test can be used to test a hypothesis concerning a population mean μ?

9.55 Define the power of a statistical test. As the alternative value of μ gets farther from μ_0, how is the power affected?

9.56 Acidity in Rainfall Refer to Exercise 8.33 and the collection of water samples to estimate the mean acidity (in pH) of rainfalls. As noted, the pH for pure rain falling through clean air is approximately 5.7. The sample of $n = 40$ rainfalls produced pH readings with $\bar{x} = 3.7$ and $s = .5$. Do the data provide sufficient evidence to indicate that the mean pH for rainfalls is more acidic ($H_a : \mu < 5.7$ pH) than pure rainwater? Test using $\alpha = .05$. Note that this inference is appropriate only for the area in which the rainwater specimens were collected.

9.57 Washing Machine Colors A manufacturer of automatic washers provides a particular model in one of three colors. Of the first 1000 washers sold, it is noted that 400 were of the first color. Can you conclude that more than one-third of all customers have a preference for the first color?

a. Find the *p*-value for the test.

b. If you plan to conduct your test using $\alpha = .05$, what will be your test conclusions?

9.58 a. Define α and β for a statistical test of hypothesis.

b. For a fixed sample size *n,* if the value of α is decreased, what is the effect on β?

c. In order to decrease both α and β for a particular alternative value of μ, how must the sample size change?

9.59 Bass Fishing The pH factor is a measure of the acidity or alkalinity of water. A reading of 7.0 is neutral; values in excess of 7.0 indicate alkalinity; those below 7.0 imply acidity. Loren Hill states that the best chance of catching bass occurs when the pH of the water is in the range 7.5 to 7.9.[18] Suppose you suspect that acid rain is lowering the pH of your favorite fishing spot and you wish to determine whether the pH is less than 7.5.

a. State the alternative and null hypotheses that you would choose for a statistical test.

b. Does the alternative hypothesis in part a imply a one- or a two-tailed test? Explain.

c. Suppose that a random sample of 30 water specimens gave pH readings with $\bar{x} = 7.3$ and $s = .2$. Just glancing at the data, do you think that the difference $\bar{x} - 7.5 = -.2$ is large enough to indicate that the mean pH of the water samples is less than 7.5? (Do *not* conduct the test.)

d. Now conduct a statistical test of the hypotheses in part a and state your conclusions. Test using $\alpha = .05$. Compare your statistically based decision with your intuitive decision in part c.

9.60 Boomers, Xers, and Millennial Men In Exercise 8.34, an *Advertising Age* white paper reported the number of minutes spent doing household chores for women when compared to men. However, there may be a difference among men, depending on the generation to which they belong.[19] The information that follows is adapted from these data and is based on random samples of 1136 men and 795 women.

	Mean	Standard Deviation	n
All Women	72	10.4	795
All Men	54	12.7	1136
Millennial	72	9.2	345
Boomers	54	13.9	475
Xers	54	10.5	316

a. Is there a significant difference in the average number of minutes spent doing household chores for men and women? Use $\alpha = .01$.

b. Is there a significant difference in the average number of minutes spent doing household chores for Millennial men and men classified as Boomers? Use $\alpha = .01$.

c. Is there a significant difference in the average number of minutes spent doing household chores for men classified as Xers and men classified as Boomers? Use $\alpha = .01$.

d. Write a conclusion which explains the practical conclusions which can be drawn from parts a, b, and c.

9.61 White-Tailed Deer In an article entitled "A Strategy for Big Bucks," Charles Dickey discusses studies of the habits of white-tailed deer that indicate that they live and feed within very limited ranges—approximately 150 to 205 acres.[20] To determine whether there was a difference between the ranges of deer located in two different geographic areas, 40 deer were caught, tagged, and fitted with small radio transmitters. Several months later, the deer were tracked and identified, and the distance x from the release point was recorded. The mean and standard deviation of the distances from the release point were as follows:

	Location 1	Location 2
Sample Size	40	40
Sample Mean (ft)	2980	3205
Sample Standard Deviation (ft)	1140	963

a. If you have no preconceived reason for believing one population mean is larger than another, what would you choose for your alternative hypothesis? Your null hypothesis?

b. Does your alternative hypothesis in part a imply a one- or a two-tailed test? Explain.

c. Do the data provide sufficient evidence to indicate that the mean distances differ for the two geographic locations? Test using $\alpha = .05$.

9.62 Biomass Exercise 7.65 reported that the biomass for tropical woodlands, thought to be about 35 kilograms per square meter (kg/m^2), may in fact be too high and that tropical biomass values vary regionally—from about 5 to 55 kg/m^2.[21] Suppose you measure the tropical biomass in 400 randomly selected square-meter plots and obtain $\bar{x} = 31.75$ and $s = 10.5$. Do the data present sufficient evidence to indicate that scientists are overestimating the mean biomass for tropical woodlands and that the mean is in fact lower than estimated?

a. State the null and alternative hypotheses to be tested.

b. Locate the rejection region for the test with $\alpha = .01$.

c. Conduct the test and state your conclusions.

9.63 Bolts Random samples of 200 bolts manufactured by a type A machine and 200 bolts manufactured by a type B machine showed 16 and 8 defective bolts,

respectively. Do these data present sufficient evidence to suggest a difference in the performance of the machine types? Use $\alpha = .05$.

9.64 Female Models In a study to assess various effects of using a female model in automobile advertising, 100 men were shown photographs of two automobiles matched for price, color, and size, but of different makes. One of the automobiles was shown with a female model to 50 of the men (group A), and both automobiles were shown without the model to the other 50 men (group B). In group A, the automobile shown with the model was judged as more expensive by 37 men; in group B, the same automobile was judged as the more expensive by 23 men. Do these results indicate that using a female model influences the perceived cost of an automobile? Use a one-tailed test with $\alpha = .05$.

9.65 Adolescents and Social Stress In a study to compare ethnic differences in adolescents' social stress, researchers recruited subjects from three middle schools in Houston, Texas.[22] A tabulation of student responses to a question regarding their socioeconomic status (SES) compared with other families in which the students chose one of five responses (much worse off, somewhat worse off, about the same, better off, or much better off) resulted in the tabulation that follows.

	European American	African American	Hispanic American	Asian American
Sample Size	144	66	77	19
About the Same	68	42	48	8

a. Do these data support the hypothesis that the proportion of adolescent African Americans who state that their SES is "about the same" exceeds that for adolescent Hispanic Americans?

b. Find the p-value for the test.

c. If you plan to test using $\alpha = .05$, what is your conclusion?

9.66* Adolescents and Social Stress, continued Refer to Exercise 9.65. Some thought should have been given to designing a test for which β is tolerably low when p_1 exceeds p_2 by an important amount. For example, find a common sample size n for a test with $\alpha = .05$ and $\beta \leq .20$ when in fact p_1 exceeds p_2 by 0.1. (HINT: The maximum value of $p(1 - p) = .25$.)

9.67 Losing Weight In a comparison of the mean 1-month weight losses for women aged 20–30 years, these sample data were obtained for each of two diets:

	Diet I	Diet II
Sample Size n	40	40
Sample Mean $\bar{x}$(lb)	10	8
Sample Variance s^2	4.3	5.7

Do the data provide sufficient evidence to indicate that diet I produces a greater mean weight loss than diet II? Use $\alpha = .05$.

9.68 Increased Yield An agronomist has shown experimentally that a new irrigation/fertilization regimen produces an increase of 2 bushels per quadrat (significant at the 1% level) when compared with the regimen currently in use. The cost of implementing and using the new regimen will not be a factor if the increase in yield exceeds 3 bushels per quadrat. Is statistical significance the same as practical importance in this situation? Explain.

9.69 Breaking Strengths of Cables A test of the breaking strengths of two different types of cables was conducted using samples of $n_1 = n_2 = 100$ pieces of each type of cable.

Cable I	Cable II
$\bar{x}_1 = 1925$	$\bar{x}_2 = 1905$
$s_1 = 40$	$s_2 = 30$

Do the data provide sufficient evidence to indicate a difference between the mean breaking strengths of the two cables? Use $\alpha = .05$.

9.70 Breaststroke Swimmers How much training time does it take to become a world-class breaststroke swimmer? A survey published in *The American Journal of Sports Medicine* reported the number of meters per week swum by two groups of swimmers—those who competed only in breaststroke and those who competed in the individual medley (which includes breaststroke). The number of meters per week practicing the breaststroke swim was recorded and the summary statistics are shown below.[26]

	Breaststroke	Individual Medley
Sample Size	130	80
Sample Mean	9017	5853
Sample Standard Deviation	7162	1961

Is there sufficient evidence to indicate a difference in the average number of meters swum by these two groups of swimmers? Test using $\alpha = .01$.

9.71 Spraying Fruit Trees A fruit grower wants to test a new spray that a manufacturer claims will *reduce* the loss due to insect damage. To test the claim, the grower sprays 200 trees with the new spray and 200 other trees with the standard spray. The following data were recorded:

	New Spray	Standard Spray
Mean Yield per Tree $\bar{x}$ (lb)	240	227
Variance s^2	980	820

a. Do the data provide sufficient evidence to conclude that the mean yield per tree treated with the new spray exceeds that for trees treated with the standard spray? Use $\alpha = .05$.

b. Construct a 95% confidence interval for the difference between the mean yields for the two sprays.

9.72 Put on the Brakes The braking ability was compared for two 2012 automobile models. Random samples of 64 automobiles were tested for each type. The recorded measurement was the distance (in feet) required to stop when the brakes were applied at 50 miles per hour. These are the computed sample means and variances:

Model I	Model II
$\bar{x}_1 = 118$	$\bar{x}_2 = 109$
$s_1^2 = 102$	$s_2^2 = 87$

Do the data provide sufficient evidence to indicate a difference between the mean stopping distances for the two models?

9.73 SAT Scores How do California high school students compare to students nationwide in their college readiness, as measured by their SAT scores? The national average scores for the class of 2010 were 501 on the critical reading portion, 516 on the math portion, and 492 on the writing portion.[23] Suppose that 100 California students from the class of 2010 were randomly selected and their SAT scores recorded in the following table:

	Critical Reading	Math	Writing
Sample Average	499	514	490
Sample Standard Deviation	98	96	92

a. Do the data provide sufficient evidence to indicate that the average critical reading score for all California students in the class of 2010 is different from the national average? Test using $\alpha = .05$.

b. Do the data provide sufficient evidence to indicate that the average math score for all California students in the class of 2010 is different from the national average? Test using $\alpha = .05$.

c. Could you use this data to determine if there is a difference between the average math and critical reading scores for all California students in the class of 2010? Explain your answer.

9.74 Actinomycin D A biologist hypothesizes that high concentrations of actinomycin D inhibit RNA synthesis in cells and hence the production of proteins as well. An experiment conducted to test this theory compared the RNA synthesis in cells treated with two concentrations of actinomycin D: .6 and .7 microgram per milliliter. Cells treated with the lower concentration (.6) of actinomycin D showed that 55 out of 70 developed normally, whereas only 23 out of 70 appeared to develop normally for the higher concentration (.7). Do these data provide sufficient evidence to indicate a difference between the rates of normal RNA synthesis for cells exposed to the two different concentrations of actinomycin D?

a. Find the *p*-value for the test.

b. If you plan to conduct your test using $\alpha = .05$, what will be your test conclusions?

9.75 PCBs Polychlorinated biphenyls (PCBs) have been found to be dangerously high in some game birds found along the marshlands of the southeastern coast of the United States. The Federal Drug Administration (FDA) considers a concentration of PCBs higher than 5 parts per million (ppm) in these game birds to be dangerous for human consumption. A sample of 38 game birds produced an average of 7.2 ppm with a standard deviation of 6.2 ppm. Is there sufficient evidence to indicate that the mean ppm of PCBs in the population of game birds exceeds the FDA's recommended limit of 5 ppm? Use $\alpha = .01$.

9.76* PCBs, continued Refer to Exercise 9.75.

a. Calculate β and $1 - \beta$ if the true mean ppm of PCBs is 6 ppm.

b. Calculate β and $1 - \beta$ if the true mean ppm of PCBs is 7 ppm.

c. Find the power, $1 - \beta$, when $\mu = 8, 9, 10$, and 12. Use these values to construct a power curve for the test in Exercise 9.75.

d. For what values of μ does this test have power greater than or equal to .90?

9.77 Goin' to the Chapel? If you choose to marry, what type of wedding site will you pick? A *USA Today* snapshot claims that 43% of all brides-to-be choose a house of worship for their wedding site.[24]

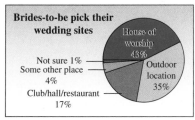

Brides-to-be pick their wedding sites
House of worship 43%
Outdoor location 35%
Club/hall/restaurant 17%
Some other place 4%
Not sure 1%

By Michelle Healey and Veronica Salazar, *USA TODAY*
Source: 'BRIDES' Magazine

In a follow-up study, a random sample of 100 brides-to-be found that 46 of those sampled had chosen or would choose a house of worship for their wedding site. Does this sample contradict the reported 43% figure? Test at the $\alpha = .05$ level of significance.

9.78 A Maze Experiment In a maze running study, a rat is run in a T maze and the result of each run recorded. A reward in the form of food is always placed at the right exit. If learning is taking place, the rat will choose the right exit more often than the left. If no learning is taking place, the rat should randomly choose either exit. Suppose that the rat is given $n = 100$ runs in the maze and that he chooses the right exit $x = 64$ times. Would you conclude that learning is taking place? Use the *p*-value approach, and make a decision based on this *p*-value.

9.79 English as a Second Language The state of California monitors the progress of elementary-aged students whose native language is not English using the California English Language Development Test.[25] The test results for two school districts in Riverside County for the 2009–2010 school year follow. The proportion given is the proportion of 6th grade students judged as Advanced or Early-Advanced in English proficiency.

District	Riverside Unified	Palm Springs
Number Tested	602	459
Proportion	.54	.31

Does the data provide sufficient evidence to indicate that there is a difference in the proportion of students who are Advanced or Early-Advanced in English proficiency for these two school districts? Test using $\alpha = .01$.

9.80 Heights and Gender It is a well-accepted fact that males are taller on the average than females. But how much taller? The genders of 105 biomedical students (Exercise 1.54) were also recorded and the data are summarized below:

	Males	Females
Sample Size	48	77
Sample Mean	69.58	64.43
Sample Standard Deviation	2.62	2.58

a. Perform a test of hypothesis to either confirm or refute our initial claim that males are taller on the average than females? Use $\alpha = .01$.

b. If the results of part a show that our claim was correct, construct a 99% confidence one-sided lower confidence bound for the average difference in heights between male and female college students. How much taller are males than females?

9.81 Breaststroke, continued Refer to Exercise 9.80.

a. Construct a 99% confidence interval for the difference in the average number of meters swum by breaststroke versus individual medley swimmers.

b. How much longer do pure breaststroke swimmers practice that stroke than individual medley swimmers? What is the practical reason for this difference?

CASE STUDY

An Aspirin a Day . . . ?

On Wednesday, January 27, 1988, the front page of the *New York Times* read, "Heart attack risk found to be cut by taking aspirin: Lifesaving effects seen." A very large study of U.S. physicians showed that a single aspirin tablet taken every other day reduced by one-half the risk of heart attack in men.[27] Three days later, a headline in the *Times* read, "Value of daily aspirin disputed in British study of heart attacks." How could two seemingly similar studies, both involving doctors as participants, reach such opposite conclusions?

The U.S. physicians' health study consisted of two randomized clinical trials in one. The first tested the hypothesis that 325 milligrams (mg) of aspirin taken every other day reduces mortality from cardiovascular disease. The second tested whether 50 mg of β-carotene taken on alternate days decreases the incidence of cancer. From names on an American Medical Association computer tape, 261,248 male physicians between the ages of 40 and 84 were invited to participate in the trial. Of those who responded, 59,285 were willing to participate. After the exclusion of those physicians who had a history of medical disorders, or who were currently taking aspirin or had negative reactions to aspirin, 22,071 physicians were randomized into one of four treatment groups: (1) buffered aspirin and β-carotene, (2) buffered aspirin and a β-carotene placebo, (3) aspirin placebo and β-carotene, and (4) aspirin placebo and β-carotene placebo. Thus, half were assigned to receive aspirin and half to receive β-carotene.

The study was conducted as a double-blind study, in which neither the participants nor the investigators responsible for following the participants knew to which group a participant belonged. The results of the American study concerning myocardial infarctions (the technical name for heart attacks) are given in the following table:

	American Study	
	Aspirin ($n = 11{,}037$)	Placebo ($n = 11{,}034$)
Myocardial Infarction		
Fatal	5	18
Nonfatal	99	171
Total	104	189

The objective of the British study was to determine whether 500 mg of aspirin taken daily would reduce the incidence of and mortality from cardiovascular disease. In 1978 all male physicians in the United Kingdom were invited to participate. After the usual exclusions, 5139 doctors were randomly allocated to take aspirin, unless some problem developed, and one-third were randomly allocated to *avoid* aspirin. Placebo tablets were not used, so the study was not blind! The results of the British study are given here:

	British Study	
	Aspirin ($n = 3429$)	Control ($n = 1710$)
Myocardial Infarction		
Fatal	89 (47.3)	47 (49.6)
Nonfatal	80 (42.5)	41 (43.3)
Total	169 (89.8)	88 (92.9)

To account for unequal sample sizes, the British study reported rates per 10,000 subject-years alive (given in parentheses).

1. Test whether the American study does in fact indicate that the rate of heart attacks for physicians taking 325 mg of aspirin every other day is significantly different from the rate for those on the placebo. Is the American claim justified?

2. Repeat the analysis using the data from the British study in which one group took 500 mg of aspirin every day and the control group took none. Based on their data, is the British claim justified?

3. Can you think of some possible reasons why the results of these two studies, which were alike in some respects, produced such different conclusions?

10

Inference from Small Samples

ZouZou/Shutterstock.com

GENERAL OBJECTIVE

The basic concepts of large-sample statistical estimation and hypothesis testing for practical situations involving population means and proportions were introduced in Chapters 8 and 9. Because all of these techniques rely on the Central Limit Theorem to justify the normality of the estimators and test statistics, they apply only when the samples are large. This chapter supplements the large-sample techniques by presenting small-sample tests and confidence intervals for population means and variances. Unlike their large-sample counterparts, these small-sample techniques require the sampled populations to be normal, or approximately so.

CHAPTER INDEX

NEED TO KNOW...
How to Decide Which Test to Use

School Accountability Study—How is Your School Doing?

Schools are being held responsible for guidelines that have been set forth by states and the federal government in an attempt to quantify students' progress. It has been said that these mandates simply detract from the time that teachers actually teach. In California, one such accountability report is based upon the API, the California state Academic Performance Index. See how the inference methods for small samples can be used to examine this index.

INTRODUCTION

Suppose you need to run an experiment to estimate a population mean or the difference between two means. The process of collecting the data may be very expensive or very time-consuming. If you cannot collect a *large sample,* the estimation and test procedures of Chapters 8 and 9 are of no use to you.

This chapter introduces some equivalent statistical procedures that can be used when the *sample size is small.* The estimation and testing procedures involve these familiar parameters:

- A single population mean, μ
- The difference between two population means, $(\mu_1 - \mu_2)$
- A single population variance, σ^2
- The comparison of two population variances, σ_1^2 and σ_2^2

Small-sample tests and confidence intervals for binomial proportions will be omitted from our discussion.[†]

STUDENT'S *t* DISTRIBUTION

In conducting an experiment to evaluate a new but very costly process for producing synthetic diamonds, you are able to study only six diamonds generated by the process. How can you use these six measurements to make inferences about the average weight μ of diamonds from this process?

In discussing the sampling distribution of $\bar{x}$ in Chapter 7, we made these points:

- When the original sampled population is normal, $\bar{x}$ and $z = (\bar{x} - \mu)/(\sigma/\sqrt{n})$ both have normal distributions, *for any sample size.*
- When the original sampled population is *not* normal, $\bar{x}$, $z = (\bar{x} - \mu)/(\sigma/\sqrt{n})$, and $z \approx (\bar{x} - \mu)/(s/\sqrt{n})$ all have approximately normal distributions, if the sample size is *large.*

When $n < 30$, the Central Limit Theorem will not guarantee that

$$\frac{\bar{x} - \mu}{s/\sqrt{n}}$$

is approximately normal.

Unfortunately, when the sample size n is small, the statistic $(\bar{x} - \mu)/(s/\sqrt{n})$ *does not* have a normal distribution. Therefore, all the critical values of z that you used in Chapters 8 and 9 are no longer correct. For example, you *cannot say* that $\bar{x}$ will lie within 1.96 standard errors of μ 95% of the time.

This problem is not new; it was studied by statisticians and experimenters in the early 1900s. To find the sampling distribution of this statistic, there are two ways to proceed:

- Use an empirical approach. Draw repeated samples and compute $(\bar{x} - \mu)/(s/\sqrt{n})$ for each sample. The relative frequency distribution that you construct using these values will approximate the shape and location of the sampling distribution.
- Use a mathematical approach to derive the actual density function or curve that describes the sampling distribution.

[†]A small-sample test for the binomial parameter *p* will be presented in Chapter 15.

This second approach was used by an Englishman named W.S. Gosset in 1908. He derived a complicated formula for the density function of

$$t = \frac{\bar{x} - \mu}{s/\sqrt{n}}$$

for random samples of size n from a normal population, and he published his results under the pen name "Student." Ever since, the statistic has been known as **Student's t.** It has the following characteristics:

- It is mound-shaped and symmetric about $t = 0$, just like z.

- It is more variable than z, with "heavier tails"; that is, the t curve does not approach the horizontal axis as quickly as z does. This is because the t statistic involves two random quantities, $\bar{x}$ and s, whereas the z statistic involves only the sample mean, $\bar{x}$. You can see this phenomenon in Figure 10.1.

- The shape of the t distribution depends on the sample size n. As n increases, the variability of t decreases because the estimate s of σ is based on more and more information. Eventually, when n is infinitely large, the t and z distributions are identical!

FIGURE 10.1 ●

Standard normal z and the t distribution with 5 degrees of freedom

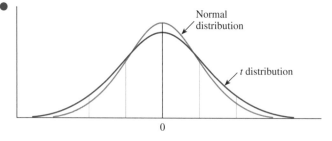

The divisor $(n - 1)$ in the formula for the sample variance s^2 is called the **number of degrees of freedom (df) associated with s^2.** It determines the *shape* of the t distribution. The origin of the term *degrees of freedom* is theoretical and refers to the number of independent squared deviations in s^2 that are available for estimating σ^2. These degrees of freedom may change for different applications and, since they specify the correct t distribution to use, you need to remember to calculate the correct degrees of freedom for each application.

The table of probabilities for the standard normal z distribution is no longer useful in calculating critical values or p-values for the t statistic. Instead, you will use Table 4 in Appendix I, which is partially reproduced in Table 10.1. When you index a particular number of degrees of freedom, the table records t_α, a value of t that has tail area a to its right, as shown in Figure 10.2.

NEED A TIP?

For a one-sample t, $df = n - 1$.

FIGURE 10.2 ●

Tabulated values of Student's t

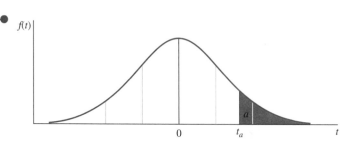

TABLE 10.1 ● **Format of the Student's *t* Table from Table 4 in Appendix I**

ONLINE APPLET

Student's *t* Probabilities
Comparing *t* and *z*

df	$t_{.100}$	$t_{.050}$	$t_{.025}$	$t_{.010}$	$t_{.005}$	df
1	3.078	6.314	12.706	31.821	63.657	1
2	1.886	2.920	4.303	6.965	9.925	2
3	1.638	2.353	3.182	4.541	5.841	3
4	1.533	2.132	2.776	3.747	4.604	4
5	1.476	2.015	2.571	3.365	4.032	5
6	1.440	1.943	2.447	3.143	3.707	6
7	1.415	1.895	2.365	2.998	3.499	7
8	1.397	1.860	2.306	2.896	3.355	8
9	1.383	1.833	2.262	2.821	3.250	9
.	.	.	.	.	.	.
.	.	.	.	.	.	.
.	.	.	.	.	.	.
26	1.315	1.706	2.056	2.479	2.779	26
27	1.314	1.703	2.052	2.473	2.771	27
28	1.313	1.701	2.048	2.467	2.763	28
29	1.311	1.699	2.045	2.462	2.756	29
inf.	1.282	1.645	1.960	2.326	2.576	inf.

EXAMPLE 10.1

For a *t* distribution with 5 degrees of freedom, the value of *t* that has area .05 to its right is found in row 5 in the column marked $t_{.050}$. For this particular *t* distribution, the area to the right of $t = 2.015$ is .05; only 5% of all values of the *t* statistic will exceed this value.

EXAMPLE 10.2

Suppose you have a sample of size $n = 10$ from a normal distribution. Find a value of *t* such that only 1% of all values of *t* will be smaller.

Solution The degrees of freedom that specify the correct *t* distribution are $df = n - 1 = 9$, and the necessary *t*-value must be in the lower portion of the distribution, with area .01 to its left, as shown in Figure 10.3. Since the *t* distribution is symmetric about 0, this value is simply the negative of the value on the right-hand side with area .01 to its right, or $-t_{.01} = -2.821$.

FIGURE 10.3 ●

t Distribution for
Example 10.2

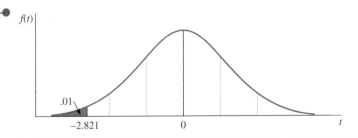

You might wonder why the degrees of freedom (*df*) in Table 10.1 jump from $df = 29$ to $df = inf.$ (infinity). The critical values of *t* for various degrees of freedom between 29 and 300 are given in Figure 10.4. You will notice that the value of *t* for the same right-tail area decreases as the degrees of freedom increase. When the degrees of freedom become infinitely large (*inf.*) the value *t* equals the value of *z* which is given in the last row of Figure 10.4.

FIGURE 10.4

Critical Values of Student's *t* for Degrees of Freedom between *df* = 29 and *df* = *infinity*

	Right-Tail Area		
df	0.05	0.025	0.01
29	1.699	2.045	2.462
49	1.677	2.010	2.405
69	1.667	1.995	2.382
100	1.660	1.984	2.364
200	1.653	1.972	2.345
300	1.650	1.968	2.339
inf.	1.645	1.96	2.326

At the same time, as the degrees of freedom increase, the shape of the *t* distribution becomes less variable until it ultimately looks like (and is) the standard normal distribution. Notice that when the degrees of freedom with *t* are *df* = 300, there is almost no difference. When *df* = 29 and *n* = 30 the critical values of *t* are quite close to their normal counterparts; this may explain why we use the arbitrary dividing line between large and small sample as *n* = 30. Rather than produce a *t*-table with many more critical values, the critical values of *z* are sufficient when *n* reaches 30.

Assumptions behind Student's *t* Distribution

NEED A TIP?

Assumptions for one-sample *t*:
- Random sample
- Normal distribution

The critical values of *t* allow you to make reliable inferences *only if* you follow all the rules; that is, your sample must meet these requirements specified by the *t* distribution:

- The sample must be randomly selected.
- The population from which you are sampling must be normally distributed.

These requirements may seem quite restrictive. How can you possibly know the shape of the probability distribution for the entire population if you have only a sample? If this were a serious problem, however, the *t* statistic could be used in only very limited situations. Fortunately, the shape of the *t* distribution is not affected very much as long as the sampled population has an *approximately mound-shaped* distribution. Statisticians say that the *t* statistic is **robust,** meaning that the distribution of the statistic does not change significantly when the normality assumption is violated.

How can you tell whether your sample is from a normal population? Although there are statistical procedures designed for this purpose, the easiest and quickest way to check for normality is to use the graphical techniques of Chapter 2: Draw a dotplot or construct a stem and leaf plot. As long as your plot tends to "mound up" in the center, you can be fairly safe in using the *t* statistic for making inferences.

The random sampling requirement, on the other hand, is quite critical if you want to produce reliable inferences. If the sample is not random, or if it does not at *least behave as* a random sample, then your sample results may be affected by some unknown factor and your conclusions may be incorrect. When you design an experiment or read about experiments conducted by others, look critically at the way the data have been collected!

SMALL-SAMPLE INFERENCES CONCERNING A POPULATION MEAN

10.3

As with large-sample inference, small-sample inference can involve either **estimation** or **hypothesis testing,** depending on the preference of the experimenter. We explained the basics of these two types of inference in the earlier chapters, and we use them again now, with a different sample statistic, $t = (\bar{x} - \mu)/(s/\sqrt{n})$, and a different sampling distribution, the Student's t, with $(n - 1)$ degrees of freedom.

SMALL-SAMPLE HYPOTHESIS TEST FOR μ

1. Null hypothesis: $H_0 : \mu = \mu_0$
2. Alternative hypothesis:

One-Tailed Test	**Two-Tailed Test**
$H_a : \mu > \mu_0$ (or, $H_a : \mu < \mu_0$)	$H_a : \mu \neq \mu_0$

3. Test statistic: $t = \dfrac{\bar{x} - \mu_0}{s/\sqrt{n}}$

4. Rejection region: Reject H_0 when

One-Tailed Test	**Two-Tailed Test**
$t > t_\alpha$ (or $t < -t_\alpha$ when the alternative hypothesis is $H_a : \mu < \mu_0$)	$t > t_{\alpha/2}$ or $t < -t_{\alpha/2}$

or when p-value $< \alpha$

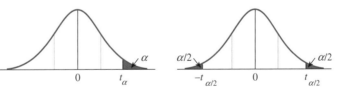

The critical values of t, t_α, and $t_{\alpha/2}$ are based on $(n - 1)$ degrees of freedom. These tabulated values can be found using Table 4 of Appendix I.

Assumption: The sample is randomly selected from a normally distributed population.

SMALL-SAMPLE $(1 - \alpha)100\%$ CONFIDENCE INTERVAL FOR μ

$$\bar{x} \pm t_{\alpha/2}\frac{s}{\sqrt{n}}$$

where $s/\sqrt{n}$ is the estimated standard error of $\bar{x}$, often referred to as the **standard error of the mean.**

EXAMPLE 10.3

A new process for producing synthetic diamonds can be operated at a profitable level only if the average weight of the diamonds is greater than .5 karat. To evaluate the profitability of the process, six diamonds are generated, with recorded weights .46, .61, .52, .48, .57, and .54 karat. Do the six measurements present sufficient evidence to indicate that the average weight of the diamonds produced by the process is in excess of .5 karat?

Solution The population of diamond weights produced by this new process has mean μ, and you can set out the formal test of hypothesis in steps, as you did in Chapter 9:

1–2 **Null and alternative hypotheses:**

$$H_0: \mu = .5 \quad \text{versus} \quad H_a: \mu > .5$$

3 **Test statistic:** You can use your calculator to verify that the mean and standard deviation for the six diamond weights are .53 and .0559, respectively. The test statistic is a t statistic, calculated as

$$t = \frac{\bar{x} - \mu_0}{s/\sqrt{n}} = \frac{.53 - .5}{.0559/\sqrt{6}} = 1.32$$

As with the large-sample tests, the test statistic provides evidence for either rejecting or accepting H_0 depending on how far from the center of the t distribution it lies.

4 **Rejection region:** If you choose a 5% level of significance ($\alpha = .05$), the right-tailed rejection region is found using the critical values of t from Table 4 of Appendix I. With $df = n - 1 = 5$, you can reject H_0 if $t > t_{.05} = 2.015$, as shown in Figure 10.5.

5 **Conclusion:** Since the calculated value of the test statistic, 1.32, does not fall in the rejection region, you cannot reject H_0. The data do not present sufficient evidence to indicate that the mean diamond weight exceeds .5 karat.

FIGURE 10.5

Rejection region for Example 10.3

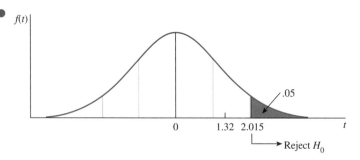

As in Chapter 9, the conclusion to *accept* H_0 would require the difficult calculation of β, the probability of a Type II error. To avoid this problem, we choose to *not reject* H_0. We can then calculate the lower bound for μ using a small-sample lower one-sided confidence bound. This bound is similar to the large-sample one-sided confidence bound, except that the critical z_α is replaced by a critical t_α from Table 4. For this example, a 95% lower one-sided confidence bound for μ is:

$$\bar{x} - t_\alpha \frac{s}{\sqrt{n}}$$

$$.53 - 2.015\frac{.0559}{\sqrt{6}}$$

$$.53 - .046$$

The 95% lower bound for μ is $\mu > .484$. The range of possible values includes mean diamond weights both smaller and greater than .5; this confirms the failure of our test to show that μ exceeds .5.

Remember from Chapter 9 that there are two ways to conduct a test of hypothesis:

- **The critical value approach:** Set up a rejection region based on the critical values of the statistic's sampling distribution. If the test statistic falls in the rejection region, you can reject H_0.

- **The p-value approach:** Calculate the p-value based on the observed value of the test statistic. If the p-value is smaller than the significance level, α, you can reject H_0. If there is no *preset* significance level, use the guidelines in Section 9.3 to judge the statistical significance of your sample results.

We used the first approach in the solution to Example 10.3. We use the second approach to solve Example 10.4.

EXAMPLE **10.4**

Labels on 1-gallon cans of paint usually indicate the drying time and the area that can be covered in one coat. Most brands of paint indicate that, in one coat, a gallon will cover between 250 and 500 square feet, depending on the texture of the surface to be painted. One manufacturer, however, claims that a gallon of its paint will cover 400 square feet of surface area. To test this claim, a random sample of ten 1-gallon cans of white paint were used to paint 10 identical areas using the same kind of equipment. The actual areas (in square feet) covered by these 10 gallons of paint are given here:

310	311	412	368	447
376	303	410	365	350

Do the data present sufficient evidence to indicate that the average coverage differs from 400 square feet? Find the p-value for the test, and use it to evaluate the statistical significance of the results.

NEED A TIP?

Remember from Chapter 2 how to calculate $\bar{x}$ and s using the data entry method on your calculator.

Solution   To test the claim, the hypotheses to be tested are

$$H_0 : \mu = 400 \quad \text{versus} \quad H_a : \mu \neq 400$$

The sample mean and standard deviation for the recorded data are

$$\bar{x} = 365.2 \qquad s = 48.417$$

and the test statistic is

$$t = \frac{\bar{x} - \mu_0}{s/\sqrt{n}} = \frac{365.2 - 400}{48.417/\sqrt{10}} = -2.27$$

The p-value for this test is the probability of observing a value of the t statistic as contradictory to the null hypothesis as the one observed for this set of data—namely, $t = -2.27$. Since this is a two-tailed test, the p-value is the probability that either $t \leq -2.27$ or $t \geq 2.27$.

Unlike the z-table, the table for t gives the values of t corresponding to upper-tail areas equal to .100, .050, .025, .010, and .005. Consequently, you can only approximate the upper-tail area that corresponds to the probability that $t > 2.27$. Since the t statistic for this test is based on 9 df, we refer to the row corresponding to $df = 9$ in Table 4. The five critical values for various tail areas are shown in Figure 10.6, an enlargement of the tail of the t distribution with 9 degrees of freedom. The value $t = 2.27$ falls between $t_{.025} = 2.262$ and $t_{.010} = 2.821$. Therefore, the right-tail area corresponding to the probability that $t > 2.27$ lies between .01 and .025. Since this area represents only half of the p-value, you can write

$$.01 < \frac{1}{2}(p\text{-value}) < .025 \quad \text{or} \quad .02 < p\text{-value} < .05$$

FIGURE 10.6

Calculating the p-value for Example 10.4 (shaded area $= \frac{1}{2}\,p$-value)

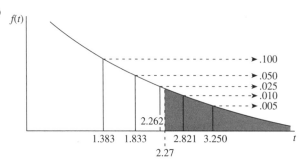

What does this tell you about the significance of the statistical results? For you to reject H_0, the p-value must be less than the specified significance level, α. Hence, you could reject H_0 at the 5% level, but not at the 2% or 1% level. Therefore, the p-value for this test would typically be reported by the experimenter as

p-value $< .05$ (or sometimes $P < .05$)

Small Sample Test of a Population Mean

For this test of hypothesis, H_0 is rejected at the 5% significance level. There is sufficient evidence to indicate that the average coverage differs from 400 square feet.

Within what limits does this average coverage *really* fall? A 95% confidence interval gives the upper and lower limits for μ as

$$\bar{x} \pm t_{\alpha/2}\left(\frac{s}{\sqrt{n}}\right)$$

$$365.2 \pm 2.262\left(\frac{48.417}{\sqrt{10}}\right)$$

$$365.2 \pm 34.63$$

Thus, you can estimate that the average area covered by 1 gallon of this brand of paint lies in the interval 330.6 to 399.8. A more precise interval estimate (a shorter interval) can generally be obtained by increasing the sample size. Notice that the upper limit of this interval is very close to the value of 400 square feet, the coverage claimed on the label. This coincides with the fact that the observed value of $t = -2.27$ is just slightly less than the left-tail critical value of $t_{.025} = -2.262$, making the p-value just slightly less than .05.

Many statistical computing packages contain programs that will implement the Student's *t*-test or construct a confidence interval for μ. Although *MS Excel* does not have a single command to implement these procedures, you can use the function tool in *Excel* to find the test statistic, the *p*-value, and/or the upper and lower confidence limits yourself. *MINITAB*, however, calculates and reports all of these values with one set of commands, allowing you to quickly and accurately draw conclusions about the statistical significance of the results. The results of the *MINITAB* one-sample *t*-test and confidence interval procedures are given in Figure 10.7. Besides the observed value of $t = -2.27$ and the confidence interval (330.6, 399.8), the output gives the sample mean, the sample standard deviation, the standard error of the mean (SE Mean $= s/\sqrt{n}$), and the exact *p*-value of the test (P = .049). This is consistent with the range for the *p*-value that we found using Table 4 in Appendix I:

$$.02 < p\text{-value} < .05$$

You will find instructions for generating this *MINITAB* output in the "Technology Today" section at the end of this chapter.

FIGURE 10.7
MINITAB output for
Example 10.4

```
One-Sample T: Area
Test of mu = 400  vs  not = 400

Variable      N      Mean        StDev        SE Mean
Area         10     365.2         48.4           15.3

Variable     95% CI                     T          P
Area        (330.6,  399.8)          -2.27      0.049
```

You can see the value of using the computer output to evaluate statistical results:

- The exact *p*-value eliminates the need for tables and critical values.
- All of the numerical calculations are done for you.

The most important job—which is left for the experimenter—is to *interpret* the results in terms of their practical significance!

10.3 EXERCISES

BASIC TECHNIQUES

10.1 Find the following *t*-values in Table 4 of Appendix I:

a. $t_{.05}$ for 5 *df*

b. $t_{.025}$ for 8 *df*

c. $t_{.10}$ for 18 *df*

d. $t_{.025}$ for 30 *df*

10.2 Find the critical value(s) of *t* that specify the rejection region in these situations:

a. A two-tailed test with $\alpha = .01$ and 12 *df*

b. A right-tailed test with $\alpha = .05$ and 16 *df*

c. A two-tailed test with $\alpha = .05$ and 25 *df*

d. A left-tailed test with $\alpha = .01$ and 7 *df*

10.3 Use Table 4 in Appendix I to approximate the *p*-value for the *t* statistic in each situation:

a. A two-tailed test with $t = 2.43$ and 12 *df*

b. A right-tailed test with $t = 3.21$ and 16 *df*

c. A two-tailed test with $t = -1.19$ and 25 *df*

d. A left-tailed test with $t = -8.77$ and 7 *df*

 Data set

EX1004

10.4 Test Scores The test scores on a 100-point test were recorded for 20 students:

71	93	91	86	75
73	86	82	76	57
84	89	67	62	72
77	68	65	75	84

a. Can you reasonably assume that these test scores have been selected from a normal population? Use a stem and leaf plot to justify your answer.

b. Calculate the mean and standard deviation of the scores.

c. If these students can be considered a random sample from the population of all students, find a 95% confidence interval for the average test score in the population.

10.5 The following $n = 10$ observations are a sample from a normal population:

7.4   7.1   6.5   7.5   7.6   6.3   6.9   7.7   6.5   7.0

a. Find the mean and standard deviation of these data.

b. Find a 99% upper one-sided confidence bound for the population mean μ.

c. Test $H_0 : \mu = 7.5$ versus $H_a : \mu < 7.5$. Use $\alpha = .01$.

d. Do the results of part b support your conclusion in part c?

APPLICATIONS

Data set **10.6 Tuna Fish** Is there a difference in the
EX1006   prices of tuna, depending on the method of packaging? *Consumer Reports* gives the estimated average price for a 6-ounce can or a 7.06-ounce pouch of tuna, based on prices paid nationally in supermarkets.[1] These prices are recorded for a variety of different brands of tuna.

Light Tuna in Water		White Tuna in Oil	White Tuna in Water	Light Tuna in Oil	
.99	.53	1.27	1.49	2.56	.62
1.92	1.41	1.22	1.29	1.92	.66
1.23	1.12	1.19	1.27	1.30	62
.85	.63	1.22	1.35	1.79	.65
.65	.67		1.29	1.23	.60
.69	.60		1.00		.67
.60	.66		1.27		
			1.28		

Source: Case Study "Pricing of Tuna" Copyright 2001 by Consumers Union of U.S., Inc., Yonkers, NY 10703-1057, a nonprofit organization. Reprinted with permission from the June 2001 issue of *Consumer Reports*® for educational purposes only. No commercial use or reproduction permitted. www.ConsumerReports.org®.

Assume that the tuna brands included in this survey represent a random sample of all tuna brands available in the United States.

a. Find a 95% confidence interval for the average price for light tuna in water. Interpret this interval. That is, what does the "95%" refer to?

b. Find a 95% confidence interval for the average price for white tuna in oil. How does the width of this interval compare to the width of the interval in part a? Can you explain why?

c. Find 95% confidence intervals for the other two samples (white tuna in water and light tuna in oil). Plot the four treatment means and their standard errors in a two-dimensional plot similar to Figure 8.5. What kind of broad comparisons can you make about the four treatments? (We will discuss the procedure for comparing more than two population means in Chapter 11.)

10.7 Dissolved O$_2$ Content Industrial wastes and sewage dumped into our rivers and streams absorb oxygen and thereby reduce the amount of dissolved oxygen available for fish and other forms of aquatic life. One state agency requires a minimum of 5 parts per million (ppm) of dissolved oxygen in order for the oxygen content to be sufficient to support aquatic life. Six water specimens taken from a river at a specific location during the low-water season (July) gave readings of 4.9, 5.1, 4.9, 5.0, 5.0, and 4.7 ppm of dissolved oxygen. Do the data provide sufficient evidence to indicate that the dissolved oxygen content is less than 5 ppm? Test using $\alpha = .05$.

10.8 Lobsters In a study of the infestation of the *Thenus orientalis* lobster by two types of barnacles, *Octolasmis tridens* and *O. lowei*, the carapace lengths (in millimeters) of 10 randomly selected lobsters caught in the seas near Singapore are measured:[2]

78   66   65   63   60   60   58   56   52   50

Find a 95% confidence interval for the mean carapace length of the *T. orientalis* lobsters.

Data set **10.9 Smoking and Lung Capacity** In a
EX1009   study of the effect of cigarette smoking on the carbon monoxide diffusing capacity (DL) of the lung, researchers found that current smokers had DL readings significantly lower than those of either exsmokers or nonsmokers. The carbon monoxide diffusing capacities for a random sample of $n = 20$ current smokers are listed here:

103.768	88.602	73.003	123.086	91.052
92.295	61.675	90.677	84.023	76.014
100.615	88.017	71.210	82.115	89.222
102.754	108.579	73.154	106.755	90.479

a. Do these data indicate that the mean DL reading for current smokers is significantly lower than 100 DL, the average for nonsmokers? Use $\alpha = .01$.

b. Find a 99% upper one-sided confidence bound for the mean DL reading for current smokers. Does this bound confirm your conclusions in part a?

Data set **10.10 Ben Roethlisberger #7** The number
EX1010 of passes completed by Ben Roethlisberger, quarterback for the Pittsburgh Steelers, was recorded for each of the 12 regular season games in which he played during the fall of 2010 (www.ESPN.com):[3]

16	19	17	17	30	18
20	22	21	23	22	15

a. A stem and leaf plot of the $n = 12$ observations is shown below:

Stem-and-Leaf Display: Roethlisberger
```
Stem-and-leaf of Roethlisberger   N = 12
Leaf Unit = 1.0
  1   1   5
  4   1   677
  6   1   89
  6   2   01
  4   2   223
  1   2
  1   2
  1   2
  1   3   0
```

Based on this plot, is it reasonable to assume that the underlying population is approximately normal, as required for the one-sample *t*-test? Explain.

b. Calculate the mean and standard deviation for Ben Roethlisberger's per game pass completions.

c. Construct a 95% confidence interval to estimate the average pass completions per game for Ben Roethlisberger.

10.11 Purifying Organic Compound Organic chemists often purify organic compounds by a method known as fractional crystallization. One chemist prepared ten 4.85-g quantities of aniline and purified it to acetanilide. The following dry yields were recorded:

3.85	3.80	3.88	3.85	3.90
3.36	3.62	4.01	3.72	3.83

Estimate the mean grams of acetanilide that can be recovered from an initial amount of 4.85 g of aniline. Use a 95% confidence interval.

10.12 Organic Compounds, continued Refer to Exercise 10.11. Approximately how many 4.85-g specimens of aniline are required if you wish to estimate the mean number of grams of acetanilide correct to within .06 g with probability equal to .95?

10.13 Bulimia In a study to determine which factors predict who will benefit from treatment for bulimia nervosa, an article in the *British Journal of Clinical Psychology* indicates that self-esteem was one of these important predictors.[4] The table gives the mean and standard deviation of self-esteem scores prior to treatment, at posttreatment, and during a follow-up:

	Pretreatment	Posttreatment	Follow-up
Sample Mean $\bar{x}$	20.3	26.6	27.7
Standard Deviation s	5.0	7.4	8.2
Sample Size n	21	21	20

a. Use a test of hypothesis to determine whether there is sufficient evidence to conclude that the true pretreatment mean is less than 25.

b. Construct a 95% confidence interval for the true posttreatment mean.

c. In Section 10.4, we will introduce small-sample techniques for making inferences about the difference between two population means. Without the formality of a statistical test, what are you willing to conclude about the differences among the three sampled population means represented by the results in the table?

Data set **10.14 RBC Counts** Here are the red blood
EX1014 cell counts (in 10^6 cells per microliter) of a healthy person measured on each of 15 days:

5.4	5.2	5.0	5.2	5.5
5.3	5.4	5.2	5.1	5.3
5.3	4.9	5.4	5.2	5.2

Find a 95% confidence interval estimate of μ, the true mean red blood cell count for this person during the period of testing.

Data set **10.15 Hamburger Meat** These data are the
EX1015 weights (in pounds) of 27 packages of ground beef in a supermarket meat display:

1.08	.99	.97	1.18	1.41	1.28	.83
1.06	1.14	1.38	.75	.96	1.08	.87
.89	.89	.96	1.12	1.12	.93	1.24
.89	.98	1.14	.92	1.18	1.17	

a. Interpret the accompanying *MINITAB* printouts for the one-sample test and estimation procedures.

MINITAB output for Exercise 10.15

One-Sample T: Weight
```
Test of mu = 1 vs not = 1

Variable        N        Mean     StDev   SE Mean
Weight         27      1.0522    0.1657    0.0319

Variable            95% CI               T       P
Weight      (0.9867, 1.1178)          1.64   0.113
```

b. Verify the calculated values of *t* and the upper and lower confidence limits.

10.16 Cholesterol The serum cholesterol levels of 50 subjects randomly selected from the L.A. Heart Data, data from an epidemiological heart disease study on Los Angeles County employees,[5] follow.

148	304	300	240	368	139	203	249	265	229
303	315	174	209	253	169	170	254	212	255
262	284	275	229	261	239	254	222	273	299
278	227	220	260	221	247	178	204	250	256
305	225	306	184	242	282	311	271	276	248

a. Construct a histogram for the data. Are the data approximately mound-shaped?

b. Use a *t*-distribution to construct a 95% confidence interval for the average serum cholesterol levels for L.A. County employees.

10.17 Cholesterol, continued Refer to Exercise 10.16. Since $n > 30$, use the methods of Chapter 8 to create a large-sample 95% confidence interval for the average serum cholesterol level for L.A. County employees. Compare the two intervals. (HINT: The two intervals should be quite similar. This is the reason we choose to approximate the sample distribution of $\dfrac{\bar{x} - \mu}{s/\sqrt{n}}$ with a *z*-distribution when $n > 30$.)

SMALL-SAMPLE INFERENCES FOR THE DIFFERENCE BETWEEN TWO POPULATION MEANS: INDEPENDENT RANDOM SAMPLES

10.4

The physical setting for the problem considered in this section is the same as the one in Section 8.6, except that the sample sizes are no longer large. Independent random samples of n_1 and n_2 measurements are drawn from two populations, with means and variances μ_1, σ_1^2, μ_2, and σ_2^2, and your objective is to make inferences about $(\mu_1 - \mu_2)$, the difference between the two population means.

When the sample sizes are small, you can no longer rely on the Central Limit Theorem to ensure that the sample means will be normal. If the original populations *are normal*, however, then the sampling distribution of the difference in the sample means, $(\bar{x}_1 - \bar{x}_2)$, will be normal (even for small samples) with mean $(\mu_1 - \mu_2)$ and standard error

$$\sqrt{\frac{\sigma_1^2}{n_1} + \frac{\sigma_2^2}{n_2}}$$

NEED A TIP?

Assumptions for the two-sample (independent) *t*-test:

• Random independent samples

• Normal distributions

• $\sigma_1 = \sigma_2$

In Chapters 7 and 8, you used the sample variances, s_1^2 and s_2^2, to calculate an *estimate* of the standard error, which was then used to form a large-sample confidence interval or a test of hypothesis based on the large-sample *z* statistic:

$$z \approx \frac{(\bar{x}_1 - \bar{x}_2) - (\mu_1 - \mu_2)}{\sqrt{\dfrac{s_1^2}{n_1} + \dfrac{s_2^2}{n_2}}}$$

Unfortunately, when the sample sizes are small, this statistic does not have an approximately normal distribution—nor does it have a Student's *t* distribution. In order to form a statistic with a sampling distribution that can be derived theoretically, you must make one more assumption.

Suppose that the variability of the measurements in the two normal populations is the same and can be measured by a common variance σ^2. That is, *both populations have exactly the same shape,* and $\sigma_1^2 = \sigma_2^2 = \sigma^2$. Then the standard error of the difference in the two sample means is

$$\sqrt{\frac{\sigma_1^2}{n_1} + \frac{\sigma_2^2}{n_2}} = \sqrt{\sigma^2\left(\frac{1}{n_1} + \frac{1}{n_2}\right)}$$

It can be proven mathematically that, if you use the appropriate sample estimate s^2 for the population variance σ^2, then the resulting test statistic,

$$t = \frac{(\bar{x}_1 - \bar{x}_2) - (\mu_1 - \mu_2)}{\sqrt{s^2\left(\frac{1}{n_1} + \frac{1}{n_2}\right)}}$$

has a *Student's t distribution.* The only remaining problem is to find the sample estimate s^2 and the appropriate number of *degrees of freedom* for the t statistic.

Remember that the population variance σ^2 describes the shape of the normal distributions from which your samples come, so that either s_1^2 or s_2^2 would give you an estimate of σ^2. But why use just one when information is provided by both? A better procedure is to combine the information in both sample variances using a *weighted average,* in which the weights are determined by the relative amount of information (the number of measurements) in each sample. For example, if the first sample contained twice as many measurements as the second, you might consider giving the first sample variance twice as much weight. To achieve this result, use this formula:

$$s^2 = \frac{(n_1 - 1)s_1^2 + (n_2 - 1)s_2^2}{n_1 + n_2 - 2}$$

Remember from Section 10.3 that the degrees of freedom for the one-sample t statistic are $(n - 1)$, the denominator of the sample estimate s^2. Since s_1^2 has $(n_1 - 1)$ df and s_2^2 has $(n_2 - 1)$ df, the total number of degrees of freedom is the sum

$$(n_1 - 1) + (n_2 - 1) = n_1 + n_2 - 2$$

shown in the denominator of the formula for s^2.

CALCULATION OF s^2

- If you have a scientific calculator, calculate each of the two sample standard deviations s_1 and s_2 separately, using the data entry procedure for your particular calculator. These values are squared and used in this formula:

$$s^2 = \frac{(n_1 - 1)s_1^2 + (n_2 - 1)s_2^2}{n_1 + n_2 - 2}$$

 NEED A TIP?

For the two-sample (independent) t-test, $df = n_1 + n_2 - 2$

It can be shown that s^2 is an unbiased estimator of the common population variance σ^2. If s^2 is used to estimate σ^2 and if the samples have been randomly and independently drawn from normal populations with a common variance, then the statistic

$$t = \frac{(\bar{x}_1 - \bar{x}_2) - (\mu_1 - \mu_2)}{\sqrt{s^2\left(\frac{1}{n_1} + \frac{1}{n_2}\right)}}$$

has a Student's t distribution with $(n_1 + n_2 - 2)$ degrees of freedom. The small-sample estimation and test procedures for the difference between two means are given next.

TEST OF HYPOTHESIS CONCERNING THE DIFFERENCE BETWEEN TWO MEANS: INDEPENDENT RANDOM SAMPLES

1. Null hypothesis: $H_0 : (\mu_1 - \mu_2) = D_0$, where D_0 is some specified difference that you wish to test. For many tests, you will hypothesize that there is no difference between μ_1 and μ_2; that is, $D_0 = 0$.

2. Alternative hypothesis:

One-Tailed Test	**Two-Tailed Test**
$H_a : (\mu_1 - \mu_2) > D_0$	$H_a : (\mu_1 - \mu_2) \neq D_0$
[or $H_a : (\mu_1 - \mu_2) < D_0$]	

3. Test statistic: $t = \dfrac{(\bar{x}_1 - \bar{x}_2) - D_0}{\sqrt{s^2\left(\dfrac{1}{n_1} + \dfrac{1}{n_2}\right)}}$ where

$$s^2 = \frac{(n_1 - 1)s_1^2 + (n_2 - 1)s_2^2}{n_1 + n_2 - 2}$$

4. Rejection region: Reject H_0 when

One-Tailed Test	**Two-Tailed Test**
$t > t_\alpha$	$t > t_{\alpha/2}$ or $t < -t_{\alpha/2}$
[or $t < -t_\alpha$ when the alternative hypothesis is $H_a : (\mu_1 - \mu_2) < D_0$]	

or when p-value $< \alpha$

The critical values of t, t_α, and $t_{\alpha/2}$ are based on $(n_1 + n_2 - 2)$ df. The tabulated values can be found using Table 4 of Appendix I.

Assumptions: The samples are randomly and independently selected from normally distributed populations. The variances of the populations σ_1^2 and σ_2^2 are equal.

SMALL-SAMPLE $(1 - \alpha)100\%$ CONFIDENCE INTERVAL FOR $(\mu_1 - \mu_2)$ BASED ON INDEPENDENT RANDOM SAMPLES

$$(\bar{x}_1 - \bar{x}_2) \pm t_{\alpha/2}\sqrt{s^2\left(\frac{1}{n_1} + \frac{1}{n_2}\right)}$$

where s^2 is the pooled estimate of σ^2.

EXAMPLE 10.5

A course can be taken for credit either by attending lecture sessions at fixed times and days, or by doing online sessions that can be done at the student's own pace and at those times the student chooses. The course coordinator wants to determine if these two ways of taking the course resulted in a significant difference in achievement as measured by the final exam for the course. Table 10.2 gives the scores on an examination with 45 possible points for one group of $n_1 = 9$ students who took the course online, and a second group of $n_2 = 9$ students who took the course with conventional lectures. Do these data present sufficient evidence to indicate that the average grade for students who take the course online is significantly higher than for those who attend a conventional class?

TABLE 10.2

Test Scores for Online and Classroom Presentations

Online	Classroom
32	35
37	31
35	29
28	25
41	34
44	40
35	27
31	32
34	31

Solution Let μ_1 and μ_2 be the mean scores for the online group and the classroom group, respectively. Then, since you seek evidence to support the theory that $\mu_1 > \mu_2$, you can test the null hypothesis

$$H_0 : \mu_1 = \mu_2 \qquad [\text{or } H_0 : (\mu_1 - \mu_2) = 0]$$

versus the alternative hypothesis

$$H_a : \mu_1 > \mu_2 \qquad [\text{or } H_a : (\mu_1 - \mu_2) > 0]$$

To conduct the t-test for these two independent samples, you must assume that the sampled populations are both normal and have the same variance σ^2. Is this reasonable? Stem and leaf plots of the data in Figure 10.8 show at least a "mounding" pattern, so that the assumption of normality is not unreasonable.

FIGURE 10.8

Stem and leaf plots for Example 10.5

Online		Classroom	
2	8	2	579
3	124	3	1124
3	557	3	5
4	14	4	0

Furthermore, the standard deviations of the two samples, calculated as

 NEED A TIP?

Stem and leaf plots can help you decide if the normality assumption is reasonable.

$$s_1 = 4.9441 \quad \text{and} \quad s_2 = 4.4752$$

are not different enough for us to doubt that the two distributions may have the same shape. If you make these two assumptions and calculate (using full accuracy) the pooled estimate of the common variance as

$$s^2 = \frac{(n_1 - 1)s_1^2 + (n_2 - 1)s_2^2}{n_1 + n_2 - 2} = \frac{8(4.9441)^2 + 8(4.4752)^2}{9 + 9 - 2} = 22.2361$$

you can then calculate the test statistic,

$$t = \frac{\bar{x}_1 - \bar{x}_2}{\sqrt{s^2\left(\dfrac{1}{n_1} + \dfrac{1}{n_2}\right)}} = \frac{35.22 - 31.56}{\sqrt{22.2361\left(\dfrac{1}{9} + \dfrac{1}{9}\right)}} = 1.65$$

NEED a tip? NEED A TIP?

If you are using a calculator, don't round off until the final step!

The alternative hypothesis $H_a : \mu_1 > \mu_2$ or, equivalently, $H_a : (\mu_1 - \mu_2) > 0$ implies that you should use a one-tailed test in the upper tail of the t distribution with $(n_1 + n_2 - 2) = 16$ degrees of freedom. You can find the appropriate critical value for a rejection region with $\alpha = .05$ in Table 4 of Appendix I, and H_0 will be rejected if $t > 1.746$. Comparing the observed value of the test statistic $t = 1.65$ with the critical value $t_{.05} = 1.746$, you cannot reject the null hypothesis (see Figure 10.9). There is insufficient evidence to indicate that the average online course grade is higher than the average conventional course grade at the 5% level of significance.

FIGURE 10.9

Rejection region for Example 10.5

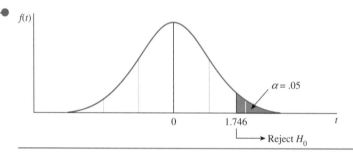

EXAMPLE 10.6

Find the p-value that would be reported for the statistical test in Example 10.5.

Solution The observed value of t for this one-tailed test is $t = 1.65$. Therefore,

$$p\text{-value} = P(t > 1.65)$$

MY ONLINE APPLET

Two-Sample t-Test: Independent Samples

for a t statistic with 16 degrees of freedom. Remember that you cannot obtain this probability directly from Table 4 in Appendix I; you can only *bound* the p-value using the critical values in the table. Since the observed value, $t = 1.65$, lies between $t_{.100} = 1.337$ and $t_{.050} = 1.746$, the tail area to the right of 1.65 is between .05 and .10. The p-value for this test would be reported as

$$.05 < p\text{-value} < .10$$

Because the p-value is greater than .05, most researchers would report the results as *not significant*.

EXAMPLE 10.7

Use a lower 95% confidence bound to estimate the difference $(\mu_1 - \mu_2)$ in Example 10.5. Does the lower confidence bound indicate that the online average is significantly higher than the classroom average?

Solution The lower confidence bound takes a familiar form—the point estimator $(\bar{x}_1 - \bar{x}_2)$ minus an amount equal to t_α times the standard error of the estimator. Substituting into the formula, you can calculate the 95% lower confidence bound:

$$(\bar{x}_1 - \bar{x}_2) - t_\alpha \sqrt{s^2\left(\frac{1}{n_1} + \frac{1}{n_2}\right)}$$

$$(35.22 - 31.56) - 1.746 \sqrt{22.2361\left(\frac{1}{9} + \frac{1}{9}\right)}$$

$$3.66 - 3.88$$

or $(\mu_1 - \mu_2) > -.22$. Since the value $(\mu_1 - \mu_2) = 0$ is included in the confidence interval, it is possible that the two means are equal. There is insufficient evidence to indicate that the online average is higher than the classroom average.

NEED A TIP?

Larger s^2/Smaller $s^2 < 3$

$\Leftrightarrow$

Variance assumption is reasonable

The two-sample procedure that uses a pooled estimate of the common variance σ^2 relies on four important assumptions:

- The samples must be *randomly selected*. Samples not randomly selected may introduce bias into the experiment and thus alter the significance levels you are reporting.
- The samples must be *independent*. If not, this is not the appropriate statistical procedure. We discuss another procedure for dependent samples in Section 10.5.
- The populations from which you sample must be *normal*. However, moderate departures from normality do not seriously affect the distribution of the test statistic, especially if the sample sizes are nearly the same.
- The population *variances should be equal* or nearly equal to ensure that the procedures are valid.

If the population variances are far from equal, there is an alternative procedure for estimation and testing that has an *approximate t* distribution in repeated sampling. As a rule of thumb, you should use this procedure if the ratio of the two sample variances,

$$\frac{\text{Larger } s^2}{\text{Smaller } s^2} > 3$$

Since the population variances are not equal, the pooled estimator s^2 is no longer appropriate, and each population variance must be estimated by its corresponding sample variance. The resulting test statistic is

$$\frac{(\bar{x}_1 - \bar{x}_2) - D_0}{\sqrt{\dfrac{s_1^2}{n_1} + \dfrac{s_2^2}{n_2}}}$$

When the sample sizes are *small*, critical values for this statistic are found using degrees of freedom approximated by the formula

$$df \approx \frac{\left(\dfrac{s_1^2}{n_1} + \dfrac{s_2^2}{n_2}\right)^2}{\dfrac{(s_1^2/n_1)^2}{(n_1 - 1)} + \dfrac{(s_2^2/n_2)^2}{(n_2 - 1)}}$$

The degrees of freedom are taken to be the integer part of this result.

Computer packages such as *MINITAB* and *MS Excel* can be used to implement this procedure, sometimes called *Satterthwaite's approximation,* as well as the *pooled method* described earlier. In fact, some experimenters choose to analyze their data using *both* methods. As long as both analyses lead to the same conclusions, you need not concern yourself with the equality or inequality of variances.

The *MINITAB* and *Excel* outputs resulting from the pooled method of analysis for the data of Example 10.5 are shown in Figure 10.10(a) and (b). Notice that the ratio of the two sample variances, $(4.94/4.48)^2 = 1.22$, is less than 3, which makes the pooled method appropriate. The calculated value of $t = 1.65$ and the exact *p*-value $= .059$ with 16 degrees of freedom are shown in both of the outputs. The exact *p*-value makes it quite easy for you to determine the significance or nonsignificance of the sample results. You will find instructions for generating this output in the section "Technology Today" at the end of this chapter.

FIGURE 10.10(a)

MINITAB output for Example 10.5

Two-Sample T-Test and CI: Online, Classroom

```
Two-sample T for Online vs Classroom
                N      Mean     StDev    SE Mean
Online          9     35.22      4.94        1.6
Classroom       9     31.56      4.48        1.5

Difference = mu (Online) - mu (Classroom)
Estimate for difference: 3.67
95% lower bound for difference: -0.21
T-Test of difference = 0 (vs >): T-Value = 1.65  P-Value = 0.059  DF = 16
Both use Pooled StDev = 4.7155
```

FIGURE 10.10(b)

Excel output for Example 10.5

D	E	F
t-Test: Two-Sample Assuming Equal Variances		
	Online	Classroom
Mean	35.222	31.556
Variance	24.444	20.028
Observations	9	9
Pooled Variance	22.236	
Hypothesized Mean Difference	0	
df	16	
t Stat	1.649	
P(T<=t) one-tail	0.059	
t Critical one-tail	1.746	
P(T<=t) two-tail	0.119	
t Critical two-tail	2.120	

If there is reason to believe that the normality assumptions have been violated, you can test for a shift in location of two population distributions using the nonparametric Wilcoxon rank sum test of Chapter 15. This test procedure, which requires fewer assumptions concerning the nature of the population probability distributions, is almost as sensitive in detecting a difference in population means when the conditions necessary for the *t*-test are satisfied. It may be more sensitive when the normality assumption is not satisfied.

10.4 EXERCISES

BASIC TECHNIQUES

10.18 Two independent random samples of sizes $n_1 = 4$ and $n_2 = 5$ are selected from each of two normal populations:

Population 1	12	3	8	5	
Population 2	14	7	7	9	6

a. Calculate s^2, the pooled estimator of σ^2.

b. Find a 90% confidence interval for $(\mu_1 - \mu_2)$, the difference between the two population means.

c. Test $H_0 : (\mu_1 - \mu_2) = 0$ against $H_a : (\mu_1 - \mu_2) < 0$ for $\alpha = .05$. State your conclusions.

10.19 Calculate s^2, the pooled estimator for σ^2, in these cases:

a. $n_1 = 10$, $n_2 = 4$, $s_1^2 = 3.4$, $s_2^2 = 4.9$

b. $n_1 = 12$, $n_2 = 21$, $s_1^2 = 18$, $s_2^2 = 23$

10.20 Give the number of degrees of freedom for s^2, the pooled estimator of σ^2, in these cases:

a. $n_1 = 16$, $n_2 = 8$

b. $n_1 = 10$, $n_2 = 12$

c. $n_1 = 15$, $n_2 = 3$

10.21 Independent random samples of $n_1 = 16$ and $n_2 = 13$ observations were selected from two normal populations with equal variances:

	Population	
	1	2
Sample Size	16	13
Sample Mean	34.6	32.2
Sample Variance	4.8	5.9

a. Suppose you wish to detect a difference between the population means. State the null and alternative hypotheses for the test.

b. Find the rejection region for the test in part a for $\alpha = .01$.

c. Find the value of the test statistic.

d. Find the approximate p-value for the test.

e. Conduct the test and state your conclusions.

10.22 Refer to Exercise 10.21. Find a 99% confidence interval for $(\mu_1 - \mu_2)$.

10.23 The *MINITAB* printout shows a test for the difference in two population means.

MINITAB output for Exercise 10.23

Two-Sample T-Test and CI: Sample 1, Sample 2
```
Two-sample T for Sample 1 vs Sample 2
          N     Mean    StDev   SE Mean
Sample 1  6    29.00     4.00       1.6
Sample 2  7    28.86     4.67       1.8

Difference = mu (Sample 1) - mu (Sample 2)
Estimate for difference: 0.14
95% CI for difference: (-5.2, 5.5)
T-Test of difference = 0 (vs not =):
T-Value = 0.06  P-Value = 0.95  DF = 11
Both use Pooled StDev = 4.38
```

a. Do the two sample standard deviations indicate that the assumption of a common population variance is reasonable?

b. What is the observed value of the test statistic? What is the p-value associated with this test?

c. What is the pooled estimate s^2 of the population variance?

d. Use the answers to part b to draw conclusions about the difference in the two population means.

e. Find the 95% confidence interval for the difference in the population means. Does this interval confirm your conclusions in part d?

10.24 The *MS Excel* printout shows a test for the difference in two population means.

MS Excel output for Exercise 10.24

D	E	F
t-Test: Two-Sample Assuming Equal Variances		
	Sample 1	Sample 2
Mean	28.667	28.286
Variance	5.067	2.238
Observations	6	7
Pooled Variance	3.524	
Hypothesized Mean Difference	0	
df	11	
t Stat	0.365	
P(T<=t) one-tail	0.361	
t Critical one-tail	1.796	
P(T<=t) two-tail	0.722	
t Critical two-tail	2.201	

a. Do the two sample variances indicate that the assumption of a common population variance is reasonable?

b. What is the observed value of the test statistic? If this is a two-tailed test, what is the p-value associated with the test?

c. What is the pooled estimate s^2 of the population variance?

d. Use the answers to part b to draw conclusions about the difference in the two population means.

e. Use the information in the printout to construct a 95% confidence interval for the difference in the population means. Does this interval confirm your conclusions in part d?

APPLICATIONS

10.25 Healthy Teeth Jan Lindhe conducted a study on the effect of an oral antiplaque rinse on plaque buildup on teeth.[6] Fourteen people whose teeth were thoroughly cleaned and polished were randomly assigned to two groups of seven subjects each. Both groups were assigned to use oral rinses (no brushing) for a 2-week period. Group 1 used a rinse that contained an antiplaque agent. Group 2, the control group, received a similar rinse except that, unknown to the subjects, the rinse contained no antiplaque agent. A plaque index x, a measure of plaque buildup, was recorded at 14 days with means and standard deviations for the two groups shown in the table.

	Control Group	Antiplaque Group
Sample Size	7	7
Mean	1.26	.78
Standard Deviation	.32	.32

a. State the null and alternative hypotheses that should be used to test the effectiveness of the antiplaque oral rinse.

b. Do the data provide sufficient evidence to indicate that the oral antiplaque rinse is effective? Test using $\alpha = .05$.

c. Find the approximate p-value for the test.

Data set **10.26 Tuna, again** In Exercise 10.6 we
EX1026 presented data on the estimated average price for a 6-ounce can or a 7.06-ounce pouch of tuna, based on prices paid nationally in supermarkets. A portion of the data is reproduced in the table below. Use the *MINITAB* printout to answer the questions.

Light Tuna in Water		Light Tuna in Oil	
.99	.53	2.56	.62
1.92	1.41	1.92	.66
1.23	1.12	1.30	.62
.85	.63	1.79	.65
.65	.67	1.23	.60
.69	.60		.67
.60	.66		

MINITAB output for Exercise 10.26

Two-Sample T-Test and CI: Water, Oil

```
Two-sample T for Water vs Oil
            N      Mean    StDev   SE Mean
Water      14     0.896    0.400     0.11
Oil        11     1.147    0.679     0.20

Difference = mu (Water) - mu (Oil)
Estimate for difference: -0.251
95% CI for difference: (-0.700, 0.198)
T-Test of difference = 0 (vs not =):
T-Value = -1.16  P-Value = 0.260  DF = 23
Both use Pooled StDev = 0.5389
```

a. Do the data in the table present sufficient evidence to indicate a difference in the average prices of light tuna in water versus oil? Test using $\alpha = .05$.

b. What is the p-value for the test?

c. The *MINITAB* analysis uses the pooled estimate of σ^2. Is the assumption of equal variances reasonable? Why or why not?

10.27 Runners and Cyclists Chronic anterior compartment syndrome is a condition characterized by exercise-induced pain in the lower leg. Swelling and impaired nerve and muscle function also accompany this pain, which is relieved by rest. Susan Beckham and colleagues conducted an experiment involving 10 healthy runners and 10 healthy cyclists to determine whether there are significant differences in pressure measurements within the anterior muscle compartment for runners and cyclists.[7] The data summary—compartment pressure in millimeters of mercury (Hg)—is as follows:

	Runners		Cyclists	
Condition	Mean	Standard Deviation	Mean	Standard Deviation
Rest	14.5	3.92	11.1	3.98
80% maximal O_2 consumption	12.2	3.49	11.5	4.95
Maximal O_2 consumption	19.1	16.9	12.2	4.47

a. Test for a significant difference in the average compartment pressure between runners and cyclists under the resting condition. Use $\alpha = .05$.

b. Construct a 95% confidence interval estimate of the difference in means for runners and cyclists under the condition of exercising at 80% of maximal oxygen consumption.

c. To test for a significant difference in the average compartment pressures at maximal oxygen consumption, should you use the pooled or unpooled t-test? Explain.

10.28 Disinfectants An experiment published in *The American Biology Teacher* studied the efficacy of using 95% ethanol or 20% bleach as a disinfectant in removing bacterial and fungal contamination when culturing plant tissues. The experiment was repeated 15 times with each disinfectant, using eggplant as the plant tissue being cultured.[8] Five cuttings per plant were placed on a petri dish for each disinfectant and stored at 25°C for 4 weeks. The observation reported was the number of uncontaminated eggplant cuttings after the 4-week storage.

Disinfectant	95% Ethanol	20% Bleach
Mean	3.73	4.80
Variance	2.78095	.17143
n	15	15
Pooled variance 1.47619		

a. Are you willing to assume that the underlying variances are equal?

b. Using the information from part a, are you willing to conclude that there is a significant difference in the mean numbers of uncontaminated eggplants for the two disinfectants tested?

Data set **10.29 Titanium** A geologist collected 20 different ore samples, all of the same weight, and randomly divided them into two groups. The titanium contents of the samples, found using two different methods, are listed in the table:

Method 1					Method 2				
.011	.013	.013	.015	.014	.011	.016	.013	.012	.015
.013	.010	.013	.011	.012	.012	.017	.013	.014	.015

a. Use an appropriate method to test for a significant difference in the average titanium contents using the two different methods.

b. Determine a 95% confidence interval estimate for $(\mu_1 - \mu_2)$. Does your interval estimate substantiate your conclusion in part a? Explain.

Data set **10.30 Raisins** The numbers of raisins in each of 14 miniboxes (1/2-ounce size) were counted for a generic brand and for Sunmaid® brand raisins:

Generic Brand				Sunmaid			
25	26	25	28	25	29	24	24
26	28	28	27	28	24	28	22
26	27	24	25	25	28	30	27
26	26			28	24		

a. Although counts cannot have a normal distribution, do these data have approximately normal distributions? (HINT: Use a histogram or stem and leaf plot.)

b. Are you willing to assume that the underlying population variances are equal? Why?

c. Use the p-value approach to determine whether there is a significant difference in the mean numbers of raisins per minibox. What are the implications of your conclusion?

10.31 Dissolved O_2 Content, continued Refer to Exercise 10.7, in which we measured the dissolved oxygen content in river water to determine whether a stream had sufficient oxygen to support aquatic life. A pollution control inspector suspected that a river community was releasing amounts of semitreated sewage into a river. To check his theory, he drew five randomly selected specimens of river water at a location above the town, and another five below. The dissolved oxygen readings (in parts per million) are as follows:

Above Town	4.8	5.2	5.0	4.9	5.1
Below Town	5.0	4.7	4.9	4.8	4.9

a. Do the data provide sufficient evidence to indicate that the mean oxygen content below the town is less than the mean oxygen content above? Test using $\alpha = .05$.

b. Suppose you prefer estimation as a method of inference. Estimate the difference in the mean dissolved oxygen contents for locations above and below the town. Use a 95% confidence interval.

Data set **10.32 Freestyle Swimmers** In an effort to compare the average swimming times for two swimmers, each swimmer was asked to swim freestyle for a distance of 100 yards at randomly selected times. The swimmers were thoroughly rested between laps and did not race against each other, so that each sample of times was an independent random sample. The times for each of 10 trials are shown for the two swimmers.

Swimmer 1		Swimmer 2	
59.62	59.74	59.81	59.41
59.48	59.43	59.32	59.63
59.65	59.72	59.76	59.50
59.50	59.63	59.64	59.83
60.01	59.68	59.86	59.51

Suppose that swimmer 2 was last year's winner when the two swimmers raced. Does it appear that the average time for swimmer 2 is still faster than the average time for swimmer 1 in the 100-yard freestyle? Find the approximate p-value for the test and interpret the results.

10.33 Freestyle Swimmers, continued Refer to Exercise 10.32. Construct a lower 95% one-sided confidence bound for the difference in the average times for the two swimmers. Does this interval confirm your conclusions in Exercise 10.32?

Data set **10.34 Comparing NFL Quarterbacks** How does Aaron Rodgers, quarterback for the 2011 Super Bowl Champion Green Bay Packers, compare to Drew Brees, quarterback for the 2010 Super Bowl winners, the New Orleans Saints? The table below shows the number of completed passes for each athlete during the 2010 NFL football season.[9] Use the *Excel* printout to answer the questions that follow.

Aaron Rodgers			Drew Brees		
19	21	7	27	37	25
19	15	25	28	34	29
34	27	19	30	27	35
12	22		33	29	22
27	26		24	23	
18	21		21	24	

E	F	G
t-Test: Two-Sample Assuming Equal Variances		
	Rodgers	Brees
Mean	20.800	28.000
Variance	44.029	23.333
Observations	15.000	16.000
Pooled Variance	33.324	
Hypothesized Mean Difference	0.000	
df	29.000	
t Stat	-3.470	
P(T<=t) one-tail	0.001	
t Critical one-tail	1.699	
P(T<=t) two-tail	0.002	
t Critical two-tail	2.045	

a. The *Excel* analysis uses the pooled estimate of σ^2. Is the assumption of equal variances reasonable? Why or why not?

b. Do the data indicate that there is a difference in the average number of completed passes for the two quarterbacks? Test using $\alpha = .05$.

c. What is the p-value for the test?

d. Use the information given in the printout to construct a 95% confidence interval for the difference in the average number of completed passes for the two quarterbacks. Does the confidence interval confirm your conclusion in part b? Explain.

Data set **10.35 An Archeological Find** An article in
EX1035 *Archaeometry* involved an analysis of 26 samples of Romano-British pottery, found at four different kiln sites in the United Kingdom.[10] The samples were analyzed to determine their chemical composition and the percentage of aluminum oxide in each of 10 samples at two sites is shown below.

Island Thorns	Ashley Rails
18.3	17.7
15.8	18.3
18.0	16.7
18.0	14.8
20.8	19.1

Does the data provide sufficient information to indicate that there is a difference in the average percentage of aluminum oxide at the two sites? Test at the 5% level of significance.

SMALL-SAMPLE INFERENCES FOR THE DIFFERENCE BETWEEN TWO MEANS: A PAIRED-DIFFERENCE TEST

10.5

To compare the wearing qualities of two types of automobile tires, A and B, a tire of type A and one of type B are randomly assigned and mounted on the rear wheels of each of five automobiles. The automobiles are then operated for a specified number of miles, and the amount of wear is recorded for each tire. These measurements appear in Table 10.3. Do the data present sufficient evidence to indicate a difference in the average wear for the two tire types?

TABLE 10.3 ● **Average Wear for Two Types of Tires**

Automobile	Tire A	Tire B
1	10.6	10.2
2	9.8	9.4
3	12.3	11.8
4	9.7	9.1
5	8.8	8.3
	$\bar{x}_1 = 10.24$	$\bar{x}_2 = 9.76$
	$s_1 = 1.316$	$s_2 = 1.328$

Table 10.3 shows a difference of $(\bar{x}_1 - \bar{x}_2) = (10.24 - 9.76) = .48$ between the two sample means, while the standard deviations of both samples are approximately 1.3. Given the variability of the data and the small number of measurements, this is a rather

small difference, and you would probably not suspect a difference in the average wear for the two types of tires. Let's check your suspicions using the methods of Section 10.4.

Look at the *MINITAB* analysis in Figure 10.11. The two-sample *pooled t*-test is used for testing the difference in the means based on two independent random samples. The calculated value of t used to test the null hypothesis $H_0 : \mu_1 = \mu_2$ is $t = .57$ with *p*-value $= .582$, a value that is not nearly small enough to indicate a significant difference in the two population means. The corresponding 95% confidence interval, given as

$$-1.448 < (\mu_1 - \mu_2) < 2.408$$

is quite wide and also does not indicate a significant difference in the population means.

FIGURE 10.11

MINITAB output using
t-test for independent
samples for the tire data

Two-Sample T-Test and CI: Tire A, Tire B

```
Two-sample T for Tire A vs Tire B
          N      Mean     StDev    SE Mean
Tire A    5     10.24      1.32       0.59
Tire B    5      9.76      1.33       0.59

Difference = mu (Tire A) - mu (Tire B)
Estimate for difference: 0.480
95% CI for difference: (-1.448, 2.408)
T-Test of difference = 0 (vs not =): T-Value = 0.57  P-Value = 0.582  DF = 8
Both use Pooled StDev = 1.3221
```

Take a second look at the data and you will notice that the wear measurement for type A is greater than the corresponding value for type B for *each* of the five automobiles. Wouldn't this be unlikely, if there's really no difference between the two tire types?

Consider a simple intuitive test, based on the binomial distribution of Chapter 5. If there is no difference in the mean tire wear for the two types of tires, then it is just as likely as not that tire A shows more wear than tire B. The five automobiles then correspond to five binomial trials with $p = P$(tire A shows more wear than tire B) $= .5$.

Is the observed value of $x = 5$ positive differences shown in Table 10.4 unusual? The probability of observing $x = 5$ or the equally unlikely value $x = 0$ can be found in Table 1 in Appendix I to be $2(.031) = .062$, which is quite small compared to the likelihood of the more powerful *t*-test, which had a *p*-value of .58. Isn't it peculiar that the *t*-test, which uses more information (the actual sample measurements) than the binomial test, fails to supply sufficient information for rejecting the null hypothesis?

There is an explanation for this inconsistency. The *t*-test described in Section 10.4 is *not* the proper statistical test to be used for our example. The statistical test procedure of Section 10.4 requires that the two samples be *independent and random*. Certainly, the independence requirement is violated by the manner in which the experiment was conducted.

The (pair of) measurements, an A and a B tire, for a particular automobile are definitely related. A glance at the data shows that the readings have approximately the same magnitude for a particular automobile but vary markedly from one automobile to another. This, of course, is exactly what you might expect. Tire wear is largely determined by driver habits, the balance of the wheels, and the road surface. Since each automobile has a different driver, you would expect a large amount of variability in the data from one automobile to another.

In designing the tire wear experiment, the experimenter realized that the measurements would vary greatly from automobile to automobile. If the tires (five of type A and five of type B) were randomly assigned to the 10 wheels, resulting in *independent* random samples, this variability would result in a large standard error and make it difficult to detect a difference in the means. Instead, he chose to "pair" the measurements, comparing the wear for type A and type B tires on each of the five automobiles. This experimental design, sometimes called a **paired-difference** or **matched pairs** design, allows us to eliminate the car-to-car variability by looking at only the five difference measurements shown in Table 10.4. These five differences form a single random sample of size $n = 5$.

TABLE 10.4 ● **Differences in Tire Wear, Using the Data of Table 10.3**

Automobile	A	B	$d = A - B$
1	10.6	10.2	.4
2	9.8	9.4	.4
3	12.3	11.8	.5
4	9.7	9.1	.6
5	8.8	8.3	.5
			$\bar{d} = .48$

Notice that in Table 10.4 the sample mean of the differences, $d = A - B$, is calculated as

$$\bar{d} = \frac{\Sigma d_i}{n} = .48$$

and is exactly the same as the difference of the sample means: $(\bar{x}_1 - \bar{x}_2) = (10.24 - 9.76) = .48$. It should not surprise you that this can be proven to be true in general, and also that the same relationship holds for the population means. That is, the average of the population differences is

$$\mu_d = (\mu_1 - \mu_2)$$

Because of this fact, you can use the sample differences to test for a significant difference in the two population means, $(\mu_1 - \mu_2) = \mu_d$. The test is a single-sample *t*-test of the difference measurements to test the null hypothesis

$$H_0 : \mu_d = 0 \qquad [\text{or } H_0 : (\mu_1 - \mu_2) = 0]$$

versus the alternative hypothesis

$$H_a : \mu_d \neq 0 \qquad [\text{or } H_a : (\mu_1 - \mu_2) \neq 0]$$

The test procedures take the same form as the procedures used in Section 10.3 and are described next.

PAIRED-DIFFERENCE TEST OF HYPOTHESIS FOR $(\mu_1 - \mu_2) = \mu_d$: DEPENDENT SAMPLES

1. Null hypothesis: $H_0 : \mu_d = 0$
2. Alternative hypothesis:

One-Tailed Test	Two-Tailed Test
$H_a : \mu_d > 0$ (or $H_a : \mu_d < 0$)	$H_a : \mu_d \neq 0$

3. Test statistic: $t = \dfrac{\bar{d} - 0}{s_d / \sqrt{n}} = \dfrac{\bar{d}}{s_d / \sqrt{n}}$

where n = Number of paired differences
$\bar{d}$ = Mean of the sample differences
s_d = Standard deviation of the sample differences

$$= \sqrt{\frac{\Sigma(d_i - \bar{d})^2}{n - 1}} = \sqrt{\frac{\Sigma d_i^2 - \dfrac{(\Sigma d_i)^2}{n}}{n - 1}}$$

4. Rejection region: Reject H_0 when

One-Tailed Test	Two-Tailed Test
$t > t_\alpha$ (or $t < -t_\alpha$ when the alternative hypothesis is $H_a : \mu_d < 0$)	$t > t_{\alpha/2}$ or $t < -t_{\alpha/2}$

or when p-value $< \alpha$

The critical values of t, t_α, and $t_{\alpha/2}$ are based on $(n - 1)$ df. These tabulated values can be found using Table 4 in Appendix I.

$(1 - \alpha)100\%$ SMALL-SAMPLE CONFIDENCE INTERVAL FOR $(\mu_1 - \mu_2) = \mu_d$, BASED ON A PAIRED-DIFFERENCE EXPERIMENT

$$\bar{d} \pm t_{\alpha/2} \left(\frac{s_d}{\sqrt{n}} \right)$$

Assumptions: The experiment is designed as a paired-difference test so that the n differences represent a random sample from a normal population.

EXAMPLE   10.8   Do the data in Table 10.3 provide sufficient evidence to indicate a difference in the mean wear for tire types A and B? Test using $\alpha = .05$.

Solution   You can verify using your calculator that the average and standard deviation of the five difference measurements are

$$\bar{d} = .48 \quad \text{and} \quad s_d = \sqrt{\dfrac{1.18 - \dfrac{(2.4)^2}{5}}{4}} = .0837$$

Then

$$H_0 : \mu_d = 0 \quad \text{and} \quad H_a : \mu_d \neq 0$$

and

$$t = \frac{\bar{d} - 0}{s_d/\sqrt{n}} = \frac{.48}{.0837/\sqrt{5}} = 12.8$$

The critical value of t for a two-tailed statistical test, $\alpha = .05$ and 4 df, is 2.776. Certainly, the observed value of $t = 12.8$ is extremely large and highly significant. Hence, you can conclude that there is a difference in the mean wear for tire types A and B.

EXAMPLE 10.9

Find a 95% confidence interval for $(\mu_1 - \mu_2) = \mu_d$ using the data in Table 10.3.

Solution A 95% confidence interval for the difference between the mean levels of wear is

$$\bar{d} \pm t_{\alpha/2}\left(\frac{s_d}{\sqrt{n}}\right)$$

$$.48 \pm 2.776\left(\frac{.0837}{\sqrt{5}}\right)$$

$$.48 \pm .10$$

 NEED A TIP?

Confidence intervals are always interpreted in the same way! In repeated sampling, intervals constructed in this way enclose the true value of the parameter $100(1 - \alpha)$% of the time.

or $.38 < (\mu_1 - \mu_2) < .58$. How does the width of this interval compare with the width of an interval you might have constructed *if* you had designed the experiment in an unpaired manner? It probably would have been of the same magnitude as the interval calculated in Figure 10.11, where the observed data were *incorrectly* analyzed using the unpaired analysis. This interval, $-1.448 < (\mu_1 - \mu_2) < 2.408$, is much wider than the paired interval, which indicates that the paired difference design increased the accuracy of our estimate, and we have gained valuable information by using this design.

The *paired-difference test* or *matched pairs design* used in the tire wear experiment is a simple example of an experimental design called a **randomized block design.** When there is a great deal of variability among the experimental units, even before any experimental procedures are implemented, the effect of this variability can be minimized by **blocking**—that is, comparing the different procedures within groups of relatively similar experimental units called **blocks.** In this way, the "noise" caused by the large variability does not mask the true differences between the procedures. We will discuss randomized block designs in more detail in Chapter 11.

It is important for you to remember that the *pairing* or *blocking* occurs when the experiment is planned, and not after the data are collected. An experimenter may choose to use pairs of identical twins to compare two learning methods. A physician may record a patient's blood pressure before and after a particular medication is given. Once you have used a paired design for an experiment, you no longer have the option of using the unpaired analysis of Section 10.4. The independence assumption has been purposely violated, and your only choice is to use the paired analysis described here!

NEED A TIP?

Paired difference test:

$df = n - 1$

Although pairing was very beneficial in the tire wear experiment, this may not always be the case. In the paired analysis, the degrees of freedom for the t-test are cut in half—from $(n + n - 2) = 2(n - 1)$ to $(n - 1)$. This reduction *increases* the critical value of t for rejecting H_0 and also increases the width of the confidence interval for the difference in the two means. If pairing is not effective, this increase is not offset by a *decrease* in the variability, and you may in fact lose rather than gain information by pairing. This, of course, did not happen in the tire experiment—the large reduction in the standard error more than compensated for the loss in degrees of freedom.

Except for notation, the paired-difference analysis is the same as the single-sample analysis presented in Section 10.3. However, both *MINITAB* and *MS Excel* provide a single procedure (**Paired *t*** in *MINITAB* and ***t*-Test: Paired Two Sample for Means** in *MS Excel*) to analyze the differences. The *MINITAB* output, shown in Figure 10.12, shows the *p*-value for the paired analysis, .000, indicating a *highly significant* difference in the means. You will find instructions for generating both the *MINITAB* and *MS Excel* outputs in the "Technology Today" section at the end of this chapter.

FIGURE 10.12 ●

MINITAB output for paired-difference analysis of tire wear data

Paired T-Test and CI: Tire A, Tire B

Paired T for Tire A - Tire B

	N	Mean	StDev	SE Mean
Tire A	5	10.240	1.316	0.589
Tire B	5	9.760	1.328	0.594
Difference	5	0.4800	0.0837	0.0374

95% CI for mean difference: (0.3761, 0.5839)
T-Test of mean difference = 0 (vs not = 0): T-Value = 12.83 P-Value = 0.000

10.5 EXERCISES

BASIC TECHNIQUES

10.36 A paired-difference experiment was conducted to compare the means of two populations:

Population	Pairs				
	1	2	3	4	5
1	1.3	1.6	1.1	1.4	1.7
2	1.2	1.5	1.1	1.2	1.8

a. Do the data provide sufficient evidence to indicate that μ_1 differs from μ_2? Test using $\alpha = .05$.

b. Find the approximate *p*-value for the test and interpret its value.

c. Find a 95% confidence interval for $(\mu_1 - \mu_2)$. Compare your interpretation of the confidence interval with your test results in part a.

d. What assumptions must you make for your inferences to be valid?

10.37 A paired-difference experiment consists of $n = 18$ pairs, $\bar{d} = 5.7$, and $s_d^2 = 256$. Suppose you wish to detect $\mu_d > 0$.

a. Give the null and alternative hypotheses for the test.

b. Conduct the test and state your conclusions.

10.38 A paired-difference experiment was conducted using $n = 10$ pairs of observations.

a. Test the null hypothesis $H_0 : (\mu_1 - \mu_2) = 0$ against $H_a : (\mu_1 - \mu_2) \neq 0$ for $\alpha = .05$, $\bar{d} = .3$, and $s_d^2 = .16$. Give the approximate *p*-value for the test.

b. Find a 95% confidence interval for $(\mu_1 - \mu_2)$.

c. How many pairs of observations do you need if you want to estimate $(\mu_1 - \mu_2)$ correct to within .1 with probability equal to .95?

APPLICATIONS

 10.39 Auto Insurance In Exercise 2.4, we EX1039 presented the annual 2010 premium for a male, licensed for 6–8 years, who drives a Honda Accord 12,600 to 15,000 miles per year and has no violations or accidents.[11]

City	GEICO ($)	21st Century ($)
Long Beach	2780	2352
Pomona	2411	2462
San Bernardino	2261	2284
Moreno Valley	2263	2520

Source: www.insurance.ca.gov

a. Why would you expect these pairs of observations to be dependent?

b. Do the data provide sufficient evidence to indicate that there is a difference in the average annual premiums between GEICO and 21st Century insurance? Test using $\alpha = .01$.

c. Find the approximate *p*-value for the test and interpret its value.

d. Find a 99% confidence interval for the difference in the average annual premiums for GEICO and 21st Century insurance.

e. Can we use the information in the table to make valid comparisons between GEICO and 21st

Century insurance throughout the United States? Why or why not?

10.40 Runners and Cyclists II Refer to Exercise 10.27. In addition to the compartment pressures, the level of creatine phosphokinase (CPK) in blood samples, a measure of muscle damage, was determined for each of 10 runners and 10 cyclists before and after exercise.[7] The data summary—CPK values in units/liter—is as follows:

Condition	Runners Mean	Runners Standard Deviation	Cyclists Mean	Cyclists Standard Deviation
Before Exercise	255.63	115.48	173.8	60.69
After Exercise	284.75	132.64	177.1	64.53
Difference	29.13	21.01	3.3	6.85

a. Test for a significant difference in mean CPK values for runners and cyclists before exercise under the assumption that $\sigma_1^2 \neq \sigma_2^2$; use $\alpha = .05$. Find a 95% confidence interval estimate for the corresponding difference in means.

b. Test for a significant difference in mean CPK values for runners and cyclists after exercise under the assumption that $\sigma_1^2 \neq \sigma_2^2$; use $\alpha = .05$. Find a 95% confidence interval estimate for the corresponding difference in means.

c. Test for a significant difference in mean CPK values for runners before and after exercise.

d. Find a 95% confidence interval estimate for the difference in mean CPK values for cyclists before and after exercise. Does your estimate indicate that there is no significant difference in mean CPK levels for cyclists before and after exercise?

Data set
EX1041
10.41 America's Market Basket An advertisement for a popular supermarket chain claims that it has had consistently lower prices than one of its competitors. As part of a survey conducted by an independent price-checking company, the average weekly total, based on the prices of approximately 95 items, is given for this chain and for its competitor recorded during four consecutive weeks in a particular month.

Week	Advertiser ($)	Competitor ($)
1	254.26	256.03
2	240.62	255.65
3	231.90	255.12
4	234.13	261.18

a. Is there a significant difference in the average prices for these two different supermarket chains?

b. What is the approximate p-value for the test conducted in part a?

c. Construct a 99% confidence interval for the difference in the average prices for the two supermarket chains. Interpret this interval.

Data set
EX1042
10.42 No Left Turn An experiment was conducted to compare the mean reaction times to two types of traffic signs: prohibitive (No Left Turn) and permissive (Left Turn Only). Ten drivers were included in the experiment. Each driver was presented with 40 traffic signs, 20 prohibitive and 20 permissive, in random order. The mean time to reaction (in milliseconds) was recorded for each driver and is shown here.

Driver	Prohibitive	Permissive
1	824	702
2	866	725
3	841	744
4	770	663
5	829	792
6	764	708
7	857	747
8	831	685
9	846	742
10	759	610

MS Excel printout for Exercise 10.42

D	E	F
t-Test: Paired Two Sample for Means		
	Prohibitive	Permissive
Mean	818.7	711.8
Variance	1573.344444	2596.4
Observations	10	10
Pearson Correlation	0.693852753	
Hypothesized Mean Difference	0	
df	9	
t Stat	9.14983257	
P(T<=t) one-tail	3.72891E-06	
t Critical one-tail	1.833112933	
P(T<=t) two-tail	7.45782E-06	
t Critical two-tail	2.262157163	

a. Explain why this is a paired-difference experiment and give reasons why the pairing should be useful in increasing information on the difference between the mean reaction times to prohibitive and permissive traffic signs.

b. Use the *Excel* printout to determine whether there is a significant difference in mean reaction times to prohibitive and permissive traffic signs. Use the p-value approach.

10.43 Healthy Teeth II Exercise 10.25 describes a dental experiment conducted to investigate the effectiveness of an oral rinse used to inhibit the growth of plaque on teeth. Subjects were divided into two groups: One group used a rinse with an antiplaque ingredient, and the control group used a rinse containing inactive ingredients. Suppose that the plaque growth on each person's teeth was measured after using the rinse after 4 hours and then again after 8 hours. If you wish to estimate the difference in plaque growth from 4 to 8 hours, should you use a confidence interval based on a paired or an unpaired analysis? Explain.

10.44 Ground or Air? The earth's temperature can be measured using either ground-based sensors or infrared-sensing devices mounted in aircraft or space satellites. Ground-based sensing is very accurate but tedious, while infrared-sensing appears to introduce a bias into the temperature readings—that is, the average temperature reading may not be equal to the average obtained by ground-based sensing. To determine the bias, readings were obtained at five different locations using both ground- and air-based temperature sensors. The readings (in degrees Celsius) are listed here:

Location	Ground	Air
1	46.9	47.3
2	45.4	48.1
3	36.3	37.9
4	31.0	32.7
5	24.7	26.2

a. Do the data present sufficient evidence to indicate a bias in the air-based temperature readings? Explain.

b. Estimate the difference in mean temperatures between ground- and air-based sensors using a 95% confidence interval.

c. How many paired observations are required to estimate the difference between mean temperatures for ground- versus air-based sensors correct to within .2°C, with probability approximately equal to .95?

10.45 Red Dye To test the comparative EX1045 brightness of two red dyes, nine samples of cloth were taken from a production line and each sample was divided into two pieces. One of the two pieces in each sample was randomly chosen and red dye 1 applied; red dye 2 was applied to the remaining piece. The following data represent a "brightness score" for each piece. Is there sufficient evidence to indicate a difference in mean brightness scores for the two dyes? Use $\alpha = .05$.

Sample	1	2	3	4	5	6	7	8	9
Dye 1	10	12	9	8	15	12	9	10	15
Dye 2	8	11	10	6	12	13	9	8	13

10.46 Tax Assessors In response to a complaint that a particular tax assessor (A) was EX1046 biased, an experiment was conducted to compare the assessor named in the complaint with another tax assessor (B) from the same office. Eight properties were selected, and each was assessed by both assessors. The assessments (in thousands of dollars) are shown in the table.

Property	Assessor 1	Assessor 2
1	276.3	275.1
2	288.4	286.8
3	280.2	277.3
4	294.7	290.6
5	268.7	269.1
6	282.8	281.0
7	276.1	275.3
8	279.0	279.1

Use the *MINITAB* printout to answer the questions that follow.

MINITAB output for Exercise 10.46

Paired T-Test and CI: Assessor A, Assessor B
```
Paired T for Assessor A - Assessor B
                 N       Mean      StDev    SE Mean
Assessor A   8     280.78       7.99       2.83
Assessor B   8     279.29       6.85       2.42
Difference   8      1.487       1.491      0.527

95% lower bound for mean difference: 0.489
T-Test of mean difference = 0 (vs > 0):
T-Value = 2.82  P-value = 0.013
```

a. Do the data provide sufficient evidence to indicate that assessor A tends to give higher assessments than assessor B?

b. Estimate the difference in mean assessments for the two assessors.

c. What assumptions must you make in order for the inferences in parts a and b to be valid?

d. Suppose that assessor A had been compared with a more stable standard—say, the average $\bar{x}$ of the assessments given by four assessors selected from the tax office. Thus, each property would be assessed by A and also by each of the four other assessors and $(x_A - \bar{x})$ would be calculated. If the test in part a is valid, can you use the paired-difference t-test to test the hypothesis that the bias, the mean difference between A's assessments and the mean of the assessments of the four assessors, is equal to 0? Explain.

10.47 Memory Experiments A psychology EX1047 class performed an experiment to compare whether a recall score in which instructions to form images of 25 words were given is better than an initial recall score for which no imagery instructions were given. Twenty students participated in the experiment with the following results:

Student	With Imagery	Without Imagery	Student	With Imagery	Without Imagery
1	20	5	11	17	8
2	24	9	12	20	16
3	20	5	13	20	10
4	18	9	14	16	12
5	22	6	15	24	7
6	19	11	16	22	9
7	20	8	17	25	21
8	19	11	18	21	14
9	17	7	19	19	12
10	21	9	20	23	13

Does it appear that the average recall score is higher when imagery is used?

10.48 Music in the Workplace Before contracting to have stereo music piped into each of his suites of offices, an executive had his office manager randomly select seven offices in which to have the system installed. The average time (in minutes) spent outside these offices per excursion among the employees involved was recorded before and after the music system was installed with the following results.

Office Number	1	2	3	4	5	6	7
No Music	8	9	5	6	5	10	7
Music	5	6	7	5	6	7	8

Would you suggest that the executive proceed with the installation? Conduct an appropriate test of hypothesis. Find the approximate *p*-value and interpret your results.

10.6 INFERENCES CONCERNING A POPULATION VARIANCE

You have seen in the preceding sections that an estimate of the population variance σ^2 is usually needed before you can make inferences about population means. Sometimes, however, the population variance σ^2 is the primary objective in an experimental investigation. It may be *more* important to the experimenter than the population mean! Consider these examples:

- Scientific measuring instruments must provide unbiased readings with a very small error of measurement. An aircraft altimeter that measures the correct altitude on the *average* is fairly useless if the measurements are in error by as much as 1000 feet above or below the correct altitude.

- Machined parts in a manufacturing process must be produced with minimum variability in order to reduce out-of-size and hence defective parts.

- Aptitude tests must be designed so that scores *will* exhibit a reasonable amount of variability. For example, an 800-point test is not very discriminatory if all students score between 601 and 605.

In previous chapters, you have used

$$s^2 = \frac{\Sigma(x_i - \bar{x})^2}{n - 1}$$

as an unbiased estimator of the population variance σ^2. This means that, in repeated sampling, the average of all your sample estimates will equal the target parameter, σ^2. But how close or far from the target is your estimator s^2 likely to be? To answer this question, we use the sampling distribution of s^2, which describes its behavior in repeated sampling.

Consider the distribution of s^2 based on repeated *random* sampling from a *normal* distribution with a specified mean and variance. We can show theoretically that the distribution begins at $s^2 = 0$ (since the variance cannot be negative) with a mean equal to σ^2. Its shape is *nonsymmetric* and changes with each different sample size and each different value of σ^2. Finding critical values for the sampling distribution of s^2 would be quite difficult and would require separate tables for each population variance. Fortunately, we can simplify the problem by *standardizing*, as we did with the z distribution.

Definition   The standardized statistic

$$\chi^2 = \frac{(n-1)s^2}{\sigma^2}$$

is called a **chi-square variable** and has a sampling distribution called the **chi-square probability distribution,** with $n - 1$ degrees of freedom.

The equation of the density function for this statistic is quite complicated to look at, but it traces the curve shown in Figure 10.13.

FIGURE 10.13 ●

A chi-square distribution

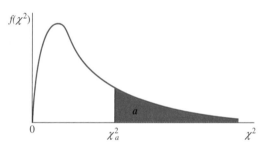

Certain critical values of the chi-square statistic, which are used for making inferences about the population variance, have been tabulated by statisticians and appear in Table 5 of Appendix I. Since the shape of the distribution varies with the sample size n or, more precisely, the degrees of freedom, $n - 1$, associated with s^2, Table 5, partially reproduced in Table 10.5, is constructed in exactly the same way as the t table, with the degrees of freedom in the first and last columns. The symbol χ_a^2 indicates that the tabulated χ^2-value has an area a to its right (see Figure 10.13).

TABLE 10.5 ●   **Format of the Chi-Square Table from Table 5 in Appendix I**

df	$\chi^2_{.995}$	$\cdots$	$\chi^2_{.950}$	$\chi^2_{.900}$	$\chi^2_{.100}$	$\chi^2_{.050}$	$\cdots$	$\chi^2_{.005}$	df
1	.0000393		.0039321	.0157908	2.70554	3.84146		7.87944	1
2	.0100251		.102587	.210720	4.60517	5.99147		10.5966	2
3	.0717212		.351846	.584375	6.25139	7.81473		12.8381	3
4	.206990		.710721	1.063623	7.77944	9.48773		14.8602	4
5	.411740		1.145476	1.610310	9.23635	11.0705		16.7496	5
6	.0675727		1.63539	2.204130	10.6446	12.5916		18.5476	6
.	.		.	.	.	.		.	.
.	.		.	.	.	.		.	.
.	.		.	.	.	.		.	.
15	4.60094		7.26094	8.54675	22.3072	24.9958		32.8013	15
16	5.14224		7.96164	9.31223	23.5418	26.2962		34.2672	16
17	5.69724		8.67176	10.0852	24.7690	27.5871		35.7185	17
18	6.26481		9.39046	10.8649	25.9894	28.8693		37.1564	18
19	6.84398		10.1170	11.6509	27.2036	30.1435		38.5822	19
.	.		.	.	.	.		.	.
.	.		.	.	.	.		.	.
.	.		.	.	.	.		.	.

NEED A TIP?

Testing one variance:
$df = n - 1$

ONLINE APPLET

Chi-Square Probabilities

You can see in Table 10.5 that, because the distribution is nonsymmetric and starts at 0, both upper and lower tail areas must be tabulated for the chi-square statistic. For example, the value $\chi^2_{.95}$ is the value that has 95% of the area under the curve to its right and 5% of the area to its left. This value cuts off an area equal to .05 in the lower tail of the chi-square distribution.

EXAMPLE 10.10

Check your ability to use Table 5 in Appendix I by verifying the following statements:

1. The probability that χ^2, based on $n = 16$ measurements ($df = 15$), exceeds 24.9958 is .05.

2. For a sample of $n = 6$ measurements, 95% of the area under the χ^2 distribution lies to the right of 1.145476.

These values are shaded in Table 10.5.

The statistical test of a null hypothesis concerning a population variance

$$H_0 : \sigma^2 = \sigma_0^2$$

uses the test statistic

$$\chi^2 = \frac{(n-1)s^2}{\sigma_0^2}$$

Notice that when H_0 is true, s^2/σ_0^2 should be near 1, so χ^2 should be close to $(n - 1)$, the degrees of freedom. If σ^2 is really greater than the hypothesized value σ_0^2, the test statistic will tend to be larger than $(n - 1)$ and will probably fall toward the upper tail of the distribution. If $\sigma^2 < \sigma_0^2$, the test statistic will tend to be smaller than $(n - 1)$ and will probably fall toward the lower tail of the chi-square distribution. As in other testing situations, you may use either a one- or a two-tailed statistical test, depending on the alternative hypothesis. This test of hypothesis and the $(1 - \alpha)100\%$ confidence interval for σ^2 are both based on the chi-square distribution and are described next.

TEST OF HYPOTHESIS CONCERNING A POPULATION VARIANCE

1. Null hypothesis: $H_0 : \sigma^2 = \sigma_0^2$
2. Alternative hypothesis:

One-Tailed Test	**Two-Tailed Test**
$H_a : \sigma^2 > \sigma_0^2$ (or $H_a : \sigma^2 < \sigma_0^2$)	$H_a : \sigma^2 \neq \sigma_0^2$

3. Test statistic: $\chi^2 = \dfrac{(n-1)s^2}{\sigma_0^2}$

4. Rejection region: Reject H_0 when

One-Tailed Test	**Two-Tailed Test**
$\chi^2 > \chi_\alpha^2$ (or $\chi^2 < \chi_{(1-\alpha)}^2$ when the alternative hypothesis is $H_a : \sigma^2 < \sigma_0^2$), where χ_α^2 and $\chi_{(1-\alpha)}^2$ are, respectively, the upper- and lower-tail values of χ^2 that place α in the tail areas	$\chi^2 > \chi_{\alpha/2}^2$ or $\chi^2 < \chi_{(1-\alpha/2)}^2$, where $\chi_{\alpha/2}^2$ and $\chi_{(1-\alpha/2)}^2$ are, respectively, the upper- and lower-tail values of χ^2 that place $\alpha/2$ in the tail areas

or when p-value $< \alpha$

The critical values of χ^2 are based on $(n - 1)$ *df*. These tabulated values can be found using Table 5 of Appendix I.

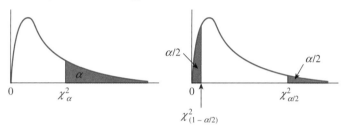

$(1 - \alpha)100\%$ CONFIDENCE INTERVAL FOR σ^2

$$\frac{(n - 1)s^2}{\chi_{\alpha/2}^2} < \sigma^2 < \frac{(n - 1)s^2}{\chi_{(1-\alpha/2)}^2}$$

where $\chi_{\alpha/2}^2$ and $\chi_{(1-\alpha/2)}^2$ are the upper and lower χ^2-values, which locate one-half of α in each tail of the chi-square distribution.
Assumption: The sample is randomly selected from a normal population.

EXAMPLE **10.11**

A cement manufacturer claims that concrete prepared from his product has a relatively stable compressive strength and that the strength measured in kilograms per square centimeter (kg/cm^2) lies within a range of 40 kg/cm^2. A sample of $n = 10$ measurements produced a mean and variance equal to, respectively,

$$\bar{x} = 312 \quad \text{and} \quad s^2 = 195$$

Do these data present sufficient evidence to reject the manufacturer's claim?

Solution In Section 2.5, you learned that the range of a set of measurements should be approximately four standard deviations. The manufacturer's claim that the range of the strength measurements is within 40 kg/cm^2 must mean that the standard deviation of the measurements is roughly 10 kg/cm^2 or less. To test his claim, the appropriate hypotheses are

$$H_0 : \sigma^2 = 10^2 = 100 \quad \text{versus} \quad H_a : \sigma^2 > 100$$

If the sample variance is much larger than the hypothesized value of 100, then the test statistic

$$\chi^2 = \frac{(n - 1)s^2}{\sigma_0^2} = \frac{1755}{100} = 17.55$$

will be unusually large, favoring rejection of H_0 and acceptance of H_a. There are two ways to use the test statistic to make a decision for this test.

- **The critical value approach:** The appropriate test requires a one-tailed rejection region in the right tail of the χ^2 distribution. The critical value for $\alpha = .05$ and $(n - 1) = 9$ df is $\chi^2_{.05} = 16.9190$ from Table 5 in Appendix I. Figure 10.14 shows the rejection region; you can reject H_0 if the test statistic exceeds 16.9190. Since the observed value of the test statistic is $\chi^2 = 17.55$, you can conclude that the null hypothesis is false and that the range of concrete strength measurements exceeds the manufacturer's claim.

FIGURE 10.14

Rejection region and p-value (shaded) for Example 10.11

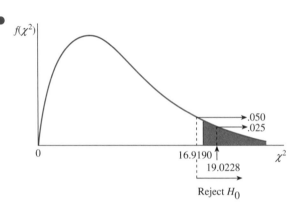

- **The p-value approach:** The p-value for a statistical test is the smallest value of α for which H_0 can be rejected. It is calculated, as in other one-tailed tests, as the area in the tail of the χ^2 distribution to the right of the observed value, $\chi^2 = 17.55$. Although computer packages allow you to calculate this area exactly, Table 5 in Appendix I allows you only to bound the p-value. Since the value 17.55 lies between $\chi^2_{.050} = 16.9190$ and $\chi^2_{.025} = 19.0228$, the p-value lies between .025 and .05. Most researchers would reject H_0 and report these results as significant at the 5% level, or $P < .05$. Again, you can reject H_0 and conclude that the range of measurements exceeds the manufacturer's claim.

EXAMPLE 10.12

An experimenter is convinced that her measuring instrument had a variability measured by standard deviation $\sigma = 2$. During an experiment, she recorded the measurements 4.1, 5.2, and 10.2. Do these data confirm or disprove her assertion? Test the appropriate hypothesis, and construct a 90% confidence interval to estimate the true value of the population variance.

Solution Since there is no preset level of significance, you should choose to use the p-value approach in testing these hypotheses:

$$H_0 : \sigma^2 = 4 \quad \text{versus} \quad H_a : \sigma^2 \neq 4$$

Use your scientific calculator to verify that the sample variance is $s^2 = 10.57$ and the test statistic is

$$\chi^2 = \frac{(n - 1)s^2}{\sigma_0^2} = \frac{2(10.57)}{4} = 5.285$$

Since this is a two-tailed test, the rejection region is divided into two parts, half in each tail of the χ^2 distribution. If you approximate the area to the right of the observed test statistic, $\chi^2 = 5.285$, you will have only *half* of the p-value for the test. Since an equally unlikely value of χ^2 might occur in the lower tail of the distribution, with equal probability, you must *double* the upper area to obtain the p-value. With 2 *df*, the observed value, 5.29, falls between $\chi^2_{.10}$ and $\chi^2_{.05}$ so that

$$.05 < \frac{1}{2}(p\text{-value}) < .10 \quad \text{or} \quad .10 < p\text{-value} < .20$$

Since the p-value is greater than .10, the results are not statistically significant. There is insufficient evidence to reject the null hypothesis $H_0 : \sigma^2 = 4$.

The corresponding 90% confidence interval is

$$\frac{(n-1)s^2}{\chi^2_{\alpha/2}} < \sigma^2 < \frac{(n-1)s^2}{\chi^2_{(1-\alpha/2)}}$$

The values of $\chi^2_{(1-\alpha/2)}$ and $\chi^2_{\alpha/2}$ are

$$\chi^2_{(1-\alpha/2)} = \chi^2_{.95} = .102587$$

$$\chi^2_{\alpha/2} = \chi^2_{.05} = 5.99147$$

Substituting these values into the formula for the interval estimate, you get

$$\frac{2(10.57)}{5.99147} < \sigma^2 < \frac{2(10.57)}{.102587} \quad \text{or} \quad 3.53 < \sigma^2 < 206.07$$

Thus, you can estimate the population variance to fall into the interval 3.53 to 206.07. This very wide confidence interval indicates how little information on the population variance is obtained from a sample of only three measurements. Consequently, it is not surprising that there is insufficient evidence to reject the null hypothesis $\sigma^2 = 4$. To obtain more information on σ^2, the experimenter needs to increase the sample size.

Although *MS Excel* does not have a single command to implement these procedures, you can use the function tool in *Excel* to find the test statistic, the p-value and/or the upper and lower confidence limits yourself. If you use *MINITAB*, the command **Stat ▶ Basic Statistics ▶ 1 Variance** allows you to enter either raw data or summary statistics to perform the chi-square test for a single variance, and calculate a confidence interval. The pertinent part of the *MINITAB 16* printout for Example 10.12 is shown in Figure 10.15.

FIGURE 10.15 ●

MINITAB output for Example 10.12

Test and CI for One Variance: Measurements

```
Null hypothesis          Sigma-squared = 4
Alternative hypothesis   Sigma-squared not = 4

Statistics
Variable       N   StDev   Variance
Measurements   3   3.25      10.6

90% Confidence Intervals
CI for         CI for
Variable       Method        StDev           Variance
Measurements   Chi-Square    (1.88, 14.36)   (3.5, 206.1)

Test
Variable       Method        Statistic   DF   P-Value
Measurements   Chi-Square         5.28    2    0.142
```

(10.6) **EXERCISES**

BASIC TECHNIQUES

10.49 A random sample of $n = 25$ observations from a normal population produced a sample variance equal to 21.4. Do these data provide sufficient evidence to indicate that $\sigma^2 > 15$? Test using $\alpha = .05$.

10.50 A random sample of $n = 15$ observations was selected from a normal population. The sample mean and variance were $\bar{x} = 3.91$ and $s^2 = .3214$. Find a 90% confidence interval for the population variance σ^2.

10.51 A random sample of size $n = 7$ from a normal population produced these measurements: 1.4, 3.6, 1.7, 2.0, 3.3, 2.8, 2.9.

a. Calculate the sample variance, s^2.

b. Construct a 95% confidence interval for the population variance, σ^2.

c. Test $H_0 : \sigma^2 = .8$ versus $H_a : \sigma^2 \neq .8$ using $\alpha = .05$. State your conclusions.

d. What is the approximate p-value for the test in part c?

APPLICATIONS

10.52 Instrument Precision A precision instrument is guaranteed to read accurately to within 2 units. A sample of four instrument readings on the same object yielded the measurements 353, 351, 351, and 355.

a. Test the null hypothesis that $\sigma = .7$ against the alternative $\sigma > .7$. Use $\alpha = .05$.

b. Find a 90% confidence interval for the population variance.

10.53 Drug Potency To properly treat patients, drugs prescribed by physicians must not only have a mean potency value as specified on the drug's container, but also the variation in potency values must be small. Otherwise, pharmacists would be distributing drug prescriptions that could be harmfully potent or have a low potency and be ineffective. A drug manufacturer claims that his drug has a potency of $5 \pm .1$ milligram per cubic centimeter (mg/cc). A random sample of four containers gave potency readings equal to 4.94, 5.09, 5.03, and 4.90 mg/cc.

a. Do the data present sufficient evidence to indicate that the mean potency differs from 5 mg/cc?

b. Do the data present sufficient evidence to indicate that the variation in potency differs from the error limits specified by the manufacturer? (HINT: It is sometimes difficult to determine exactly what is

meant by limits on potency as specified by a manufacturer. Since he implies that the potency values will fall into the interval $5 \pm .1$ mg/cc with very high probability—the implication is almost *always*—let us assume that the range .2; or 4.9 to 5.1, represents 6σ, as suggested by the Empirical Rule).

10.54 Drug Potency, continued Refer to Exercise 10.53. Testing of 60 additional randomly selected containers of the drug gave a sample mean and variance equal to 5.04 and .0063 (for the total of $n = 64$ containers). Using a 95% confidence interval, estimate the variance of the manufacturer's potency measurements.

10.55 Hard Hats A manufacturer of hard safety hats for construction workers is concerned about the mean and the variation of the forces helmets transmit to wearers when subjected to a standard external force. The manufacturer desires the mean force transmitted by helmets to be 800 pounds (or less), well under the legal 1000-pound limit, and σ to be less than 40. A random sample of $n = 40$ helmets was tested, and the sample mean and variance were found to be equal to 825 pounds and 2350 pounds^2, respectively.

a. If $\mu = 800$ and $\sigma = 40$, is it likely that any helmet, subjected to the standard external force, will transmit a force to a wearer in excess of 1000 pounds? Explain.

b. Do the data provide sufficient evidence to indicate that when the helmets are subjected to the standard external force, the mean force transmitted by the helmets exceeds 800 pounds?

10.56 Hard Hats, continued Refer to Exercise 10.55. Do the data provide sufficient evidence to indicate that σ exceeds 40?

 10.57 Light Bulbs A manufacturer of industrial light bulbs likes its bulbs to have a mean length of life that is acceptable to its customers and a variation in length of life that is relatively small. A sample of 20 bulbs tested produced the following lengths of life (in hours):

| 2100 | 2302 | 1951 | 2067 | 2415 | 1883 | 2101 | 2146 | 2278 | 2019 |
| 1924 | 2183 | 2077 | 2392 | 2286 | 2501 | 1946 | 2161 | 2253 | 1827 |

The manufacturer wishes to control the variability in length of life so that σ is less than 150 hours. Do the data provide sufficient evidence to indicate that the manufacturer is achieving this goal? Test using $\alpha = .01$.

COMPARING TWO POPULATION VARIANCES

Just as a single population variance is sometimes important to an experimenter, you might also need to compare two population variances. You might need to compare the precision of one measuring device with that of another, the stability of one manufacturing process with that of another, or even the variability in the grading procedure of one college professor with that of another.

One way to compare two population variances, σ_1^2 and σ_2^2, is to use the ratio of the sample variances, s_1^2/s_2^2. If s_1^2/s_2^2 is nearly equal to 1, you will find little evidence to indicate that σ_1^2 and σ_2^2 are unequal. On the other hand, a very large or very small value for s_1^2/s_2^2 provides evidence of a difference in the population variances.

How large or small must s_1^2/s_2^2 be for sufficient evidence to exist to reject the following null hypothesis?

$$H_0 : \sigma_1^2 = \sigma_2^2$$

The answer to this question may be found by studying the distribution of s_1^2/s_2^2 in repeated sampling.

When independent random samples are drawn from two *normal* populations with *equal variances*—that is, $\sigma_1^2 = \sigma_2^2$—then s_1^2/s_2^2 has a probability distribution in repeated sampling that is known to statisticians as an **F distribution.** The equation of the density function for this statistic is quite complicated to look at, but it traces the curve shown in Figure 10.16.

FIGURE 10.16 ●

An *F* distribution with $df_1 = 10$ and $df_2 = 10$

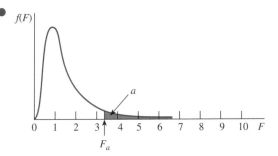

ASSUMPTIONS FOR s_1^2/s_2^2 TO HAVE AN *F* DISTRIBUTION

- Random and independent samples are drawn from each of two normal populations.
- The variability of the measurements in the two populations is the same and can be measured by a common variance, σ^2; that is, $\sigma_1^2 = \sigma_2^2 = \sigma^2$.

ONLINE APPLET

F Probabilities

It is not important for you to know the complex equation of the density function for *F*. For your purposes, you need only to use the well-tabulated critical values of *F* given in Table 6 in Appendix I.

Like the χ^2 distribution, the shape of the *F* distribution is nonsymmetric and depends on the number of degrees of freedom associated with s_1^2 and s_2^2, represented as

NEED A TIP?

Testing two variances:
$df_1 = n_1 - 1$ and
$df_2 = n_2 - 1$

$df_1 = (n_1 - 1)$ and $df_2 = (n_2 - 1)$, respectively. This complicates the tabulation of critical values of the F distribution because a table is needed for each different combination of df_1, df_2, and a.

In Table 6 in Appendix I, critical values of F for right-tailed areas corresponding to $a = .100, .050, .025, .010$, and $.005$ are tabulated for various combinations of df_1 numerator degrees of freedom and df_2 denominator degrees of freedom. A portion of Table 6 is reproduced in Table 10.6. The numerator degrees of freedom df_1 are listed across the top margin, and the denominator degrees of freedom df_2 are listed along the side margin. The values of a are listed in the second column. For a fixed combination of df_1 and df_2, the appropriate critical values of F are found in the line indexed by the value of a required.

EXAMPLE 10.13

Check your ability to use Table 6 in Appendix I by verifying the following statements:

1. The value of F with area .05 to its right for $df_1 = 6$ and $df_2 = 9$ is 3.37.
2. The value of F with area .05 to its right for $df_1 = 5$ and $df_2 = 10$ is 3.33.
3. The value of F with area .01 to its right for $df_1 = 6$ and $df_2 = 9$ is 5.80.

These values are shaded in Table 10.6.

TABLE 10.6 ● **Format of the F Table from Table 6 in Appendix I**

df_2	a	1	2	3	4	5	6
				df_1			
1	.100	39.86	49.50	53.59	55.83	57.24	58.20
	.050	161.4	199.5	215.7	224.6	230.2	234.0
	.025	647.8	799.5	864.2	899.6	921.8	937.1
	.010	4052	4999.5	5403	5625	5764	5859
	.005	16211	20000	21615	22500	23056	23437
2	.100	8.53	9.00	9.16	9.24	9.29	9.33
	.050	18.51	19.00	19.16	19.25	19.30	19.33
	.025	38.51	39.00	39.17	39.25	39.30	39.33
	.010	98.50	99.00	99.17	99.25	99.30	99.33
	.005	198.5	199.0	199.2	199.2	199.3	199.3
3	.100	5.54	5.46	5.39	5.34	5.31	5.28
	.050	10.13	9.55	9.28	9.12	9.01	8.94
	.025	17.44	16.04	15.44	15.10	14.88	14.73
	.010	34.12	30.82	29.46	28.71	28.24	27.91
	.005	55.55	49.80	47.47	46.19	45.39	44.84
⋮	⋮			⋮			⋮
9	.100	3.36	3.01	2.81	2.69	2.61	2.55
	.050	5.12	4.26	3.86	3.63	3.48	3.37
	.025	7.21	5.71	5.08	4.72	4.48	4.32
	.010	10.56	8.02	6.99	6.42	6.06	5.80
	.005	13.61	10.11	8.72	7.96	7.47	7.13
10	.100	3.29	2.92	2.73	2.61	2.52	2.46
	.050	4.96	4.10	3.71	3.48	3.33	3.22
	.025	6.94	5.46	4.83	4.47	4.24	4.07
	.010	10.04	7.56	6.55	5.99	5.64	5.39
	.005	12.83	9.43	8.08	7.34	6.87	6.54

The statistical test of the null hypothesis

$$H_0 : \sigma_1^2 = \sigma_2^2$$

uses the test statistic

$$F = \frac{s_1^2}{s_2^2}$$

When the alternative hypothesis implies a one-tailed test—that is,

$$H_a : \sigma_1^2 > \sigma_2^2$$

you can find the right-tailed critical value for rejecting H_0 directly from Table 6 in Appendix I. However, when the alternative hypothesis requires a two-tailed test—that is,

$$H_0 : \sigma_1^2 \neq \sigma_2^2$$

the rejection region is divided between the upper and lower tails of the F distribution. These left-tailed critical values are *not given* in Table 6 for the following reason: You are free to decide which of the two populations you want to call "Population 1." If you always choose to call the population with the *larger* sample variance "Population 1," then the observed value of your test statistic will always be in the right tail of the F distribution. Even though half of the rejection region, the area $\alpha/2$ to its left, will be in the lower tail of the distribution, you will never need to use it! Remember these points, though, for a two-tailed test:

- The area in the right tail of the rejection region is only $\alpha/2$.

- The area to the right of the observed test statistic is only (p-value)/2.

The formal procedures for a test of hypothesis and a $(1 - \alpha)100\%$ confidence interval for two population variances are shown next.

TEST OF HYPOTHESIS CONCERNING THE EQUALITY OF TWO POPULATION VARIANCES

1. Null hypothesis: $H_0 : \sigma_1^2 = \sigma_2^2$
2. Alternative hypothesis:

One-Tailed Test	**Two-Tailed Test**
$H_a : \sigma_1^2 > \sigma_2^2$ (or $H_a : \sigma_1^2 < \sigma_2^2$)	$H_a : \sigma_1^2 \neq \sigma_2^2$

3. Test statistic:

One-Tailed Test	**Two-Tailed Test**
$F = \dfrac{s_1^2}{s_2^2}$	$F = \dfrac{s_1^2}{s_2^2}$

   where s_1^2 is the larger sample variance.
4. Rejection region: Reject H_0 when

One-Tailed Test	**Two-Tailed Test**
$F > F_\alpha$	$F > F_{\alpha/2}$

   or when p-value $< \alpha$

(continued)

TEST OF HYPOTHESIS CONCERNING
THE EQUALITY OF TWO POPULATION VARIANCES
(continued)

The critical values of F_α and $F_{\alpha/2}$ are based on $df_1 = (n_1 - 1)$ and $df_2 = (n_2 - 1)$. These tabulated values, for $a = .100, .050, .025, .010,$ and $.005$, can be found using Table 6 in Appendix I.

Assumptions: The samples are randomly and independently selected from normally distributed populations.

CONFIDENCE INTERVAL FOR σ_1^2/σ_2^2

$$\left(\frac{s_1^2}{s_2^2}\right)\frac{1}{F_{df_1,df_2}} < \frac{\sigma_1^2}{\sigma_2^2} < \left(\frac{s_1^2}{s_2^2}\right)F_{df_2,df_1}$$

where $df_1 = (n_1 - 1)$ and $df_2 = (n_2 - 1)$. F_{df_1,df_2} is the tabulated critical value of F corresponding to df_1 and df_2 degrees of freedom in the numerator and denominator of F, respectively, with area $\alpha/2$ to its right.
Assumptions: The samples are randomly and independently selected from normally distributed populations.

EXAMPLE 10.14

An experimenter is concerned that the variability of responses using two different experimental procedures may not be the same. Before conducting his research, he conducts a prestudy with random samples of 10 and 8 responses and gets $s_1^2 = 7.14$ and $s_2^2 = 3.21$, respectively. Do the sample variances present sufficient evidence to indicate that the population variances are unequal?

Solution Assume that the populations have probability distributions that are reasonably mound-shaped and hence satisfy, for all practical purposes, the assumption that the populations are normal. You wish to test these hypotheses:

$$H_0 : \sigma_1^2 = \sigma_2^2 \quad \text{versus} \quad H_a : \sigma_1^2 \neq \sigma_2^2$$

Using Table 6 in Appendix I for $\alpha/2 = .025$, you can reject H_0 when $F > 4.82$ with $\alpha = .05$. The calculated value of the test statistic is

$$F = \frac{s_1^2}{s_2^2} = \frac{7.14}{3.21} = 2.22$$

Because the test statistic does not fall into the rejection region, you cannot reject $H_0 : \sigma_1^2 = \sigma_2^2$. Thus, there is insufficient evidence to indicate a difference in the population variances.

EXAMPLE ❨10.15❩ Refer to Example 10.14 and find a 90% confidence interval for σ_1^2/σ_2^2.

Solution The 90% confidence interval for σ_1^2/σ_2^2 is

$$\left(\frac{s_1^2}{s_2^2}\right)\frac{1}{F_{df_1,df_2}} < \frac{\sigma_1^2}{\sigma_2^2} < \left(\frac{s_1^2}{s_2^2}\right)F_{df_2,df_1}$$

where

$$s_1^2 = 7.14 \qquad\qquad s_2^2 = 3.21$$
$$df_1 = (n_1 - 1) = 9 \qquad df_2 = (n_2 - 1) = 7$$
$$F_{9,7} = 3.68 \qquad\qquad F_{7,9} = 3.29$$

Substituting these values into the formula for the confidence interval, you get

$$\left(\frac{7.14}{3.21}\right)\frac{1}{3.68} < \frac{\sigma_1^2}{\sigma_2^2} < \left(\frac{7.14}{3.21}\right)3.29 \quad \text{or} \quad .60 < \frac{\sigma_1^2}{\sigma_2^2} < 7.32$$

The calculated interval estimate .60 to 7.32 includes 1.0, the value hypothesized in H_0. This indicates that it is quite possible that $\sigma_1^2 = \sigma_2^2$ and therefore agrees with the test conclusions. Do not reject $H_0 : \sigma_1^2 = \sigma_2^2$.

The *Excel* function called **F.TEST (FTEST** in *Excel 2007* and earlier versions) can be used to perform the *F*-test for the equality of variances when you have entered the raw data into the spreadsheet. The *MINITAB* command **Stat ▶ Basic Statistics ▶ 2 Variances** is a little more flexible, since it allows you to enter either raw data or summary statistics to perform the *F*-test. In addition, *MINITAB 16* calculates confidence intervals for the ratio of two variances or two standard deviations (which we have not discussed). The *MINITAB 16* printout for Example 10.14, containing the *F* statistic and its *p*-value, is shown in Figure 10.17.

FIGURE 10.17

MINITAB output for Example 10.14

Test and CI for Two Variances

```
Null hypothesis          Sigma(1) / Sigma(2) = 1
Alternative hypothesis   Sigma(1) / Sigma(2) not = 1
Significance level       Alpha = 0.05

Statistics
Sample    N   StDev    Variance
1         10  2.672     7.140
2          8  1.792     3.210
Ratio of standard deviations = 1.491
Ratio of variances = 2.224


95% Confidence Intervals
CI for
Distribution    CI for StDev       Variance
of Data             Ratio             Ratio
Normal          (0.679, 3.055)   (0.461, 9.335)
Test
Method          DF1  DF2   Statistic   P-Value
F Test (normal)   9    7       2.22      0.304
```

FIGURE 10.18 ●

Distributions of impurity measurements for two production lines

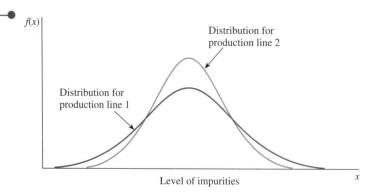

EXAMPLE 10.16

The variability in the amount of impurities present in a batch of chemical used for a particular process depends on the length of time the process is in operation. A manufacturer using two production lines, 1 and 2, has made a slight adjustment to line 2, hoping to reduce the variability as well as the average amount of impurities in the chemical. Samples of $n_1 = 25$ and $n_2 = 25$ measurements from the two batches yield these means and variances:

$$\bar{x}_1 = 3.2 \qquad s_1^2 = 1.04$$
$$\bar{x}_2 = 3.0 \qquad s_2^2 = .51$$

Do the data present sufficient evidence to indicate that the process variability is less for line 2?

Solution The experimenter believes that the average levels of impurities are the same for the two production lines but that her adjustment may have decreased the variability of the levels for line 2, as illustrated in Figure 10.18. This adjustment would be good for the company because it would decrease the probability of producing shipments of the chemical with unacceptably high levels of impurities.

To test for a decrease in variability, the test of hypothesis is

$$H_0 : \sigma_1^2 = \sigma_2^2 \quad \text{versus} \quad H_a : \sigma_1^2 > \sigma_2^2$$

and the observed value of the test statistic is

$$F = \frac{s_1^2}{s_2^2} = \frac{1.04}{.51} = 2.04$$

Using the p-value approach, you can bound the one-tailed p-value using Table 6 in Appendix I with $df_1 = df_2 = (25 - 1) = 24$. The observed value of F falls between $F_{.050} = 1.98$ and $F_{.025} = 2.27$, so that $.025 < p\text{-value} < .05$. The results are judged significant at the 5% level, and H_0 is rejected. You can conclude that the variability of line 2 is less than that of line 1.

The F-test for the difference in two population variances completes the battery of tests you have learned in this chapter for making inferences about population parameters under these conditions:

- The sample sizes are small.
- The sample or samples are drawn from normal populations.

You will find that the F and χ^2 distributions, as well as the Student's t distribution, are very important in other applications in the chapters that follow. They will be used for different estimators designed to answer different types of inferential questions, but the basic techniques for making inferences remain the same.

In the next section, we review the assumptions required for all of these inference tools, and discuss options that are available when the assumptions do not seem to be reasonably correct.

10.7 EXERCISES

BASIC TECHNIQUES

10.58 Independent random samples from two normal populations produced the variances listed here:

Sample Size	Sample Variance
16	55.7
20	31.4

a. Do the data provide sufficient evidence to indicate that σ_1^2 differs from σ_2^2? Test using $\alpha = .05$.

b. Find the approximate p-value for the test and interpret its value.

10.59 Refer to Exercise 10.58 and find a 95% confidence interval for σ_1^2/σ_2^2.

10.60 Independent random samples from two normal populations produced the given variances:

Sample Size	Sample Variance
13	18.3
13	7.9

a. Do the data provide sufficient evidence to indicate that $\sigma_1^2 > \sigma_2^2$? Test using $\alpha = .05$.

b. Find the approximate p-value for the test and interpret its value.

APPLICATIONS

10.61 SAT Scores The SAT subject tests in chemistry and physics[12] for two groups of 15 students each electing to take these tests are given below.

Chemistry	Physics
$\bar{x} = 644$	$\bar{x} = 658$
$s = 114$	$s = 103$
$n = 15$	$n = 15$

To use the two-sample t-test with a pooled estimate of σ^2, you must assume that the two population variances are equal. Test this assumption using the F-test for equality of variances. What is the approximate p-value for the test?

10.62 Tuna III In Exercise 10.26 and dataset EX1026, you conducted a test to detect a difference in the average prices of light tuna in water versus light tuna in oil.

a. What assumption had to be made concerning the population variances so that the test would be valid?

b. Do the data present sufficient evidence to indicate that the variances violate the assumption in part a? Test using $\alpha = .05$.

 10.63 Roethlisberger and Rodgers Quarterbacks not only need to have a good passing EX1063 percentage, but they need to be consistent. That is, the variability in the number of passes completed per game should be small. The table below gives the number of passes completed for Ben Roethlisberger and Aaron Rodgers, quarterbacks for the Pittsburgh Steelers and Green Bay Packers, respectively, during the 2010 NFL season.[13]

Aaron Rodgers			Ben Roethlisberger	
19	21	7	16	20
19	15	25	19	22
34	27	19	17	21
12	22		17	23
27	26		30	22
18	21		18	15

a. Does the data indicate that there is a difference in the variability in the number of passes completed for the two quarterbacks? Use $\alpha = .01$.

b. If you were going to test for a difference in the two population means, would it be appropriate to use the two-sample t-test that assumes equal variances? Explain.

10.64 Impurities A pharmaceutical manufacturer purchases a particular material from two different suppliers. The mean level of impurities in the raw material is approximately the same for both suppliers, but the manufacturer is concerned about the variability of the impurities from shipment to shipment. To compare the variation in percentage impurities for the two suppliers, the manufacturer selects 10 shipments from each of the two suppliers and measures the percentage of impurities in the raw material for each shipment. The sample means and variances are shown in the table.

Supplier A	Supplier B
$\bar{x}_1 = 1.89$	$\bar{x}_2 = 1.85$
$s_1^2 = .273$	$s_2^2 = .094$
$n_1 = 10$	$n_2 = 10$

a. Do the data provide sufficient evidence to indicate a difference in the variability of the shipment impurity levels for the two suppliers? Test using $\alpha = .01$. Based on the results of your test, what recommendation would you make to the pharmaceutical manufacturer?

b. Find a 99% confidence interval for σ_2^2 and interpret your results.

10.65 Runners and Cyclists III Refer to Exercise 10.27. Susan Beckham and colleagues conducted an experiment involving 10 healthy runners and 10 healthy cyclists to determine if there are significant differences in pressure measurements within the anterior muscle compartment for runners and cyclists.[7] The data—compartment pressure, in millimeters of mercury (Hg)—are reproduced here:

	Runners		Cyclists	
Condition	Mean	Standard Deviation	Mean	Standard Deviation
Rest	14.5	3.92	11.1	3.98
80% maximal O₂ consumption	12.2	3.49	11.5	4.95
Maximal O₂ consumption	19.1	16.9	12.2	4.47

For each of the three variables measured in this experiment, test to see whether there is a significant difference in the variances for runners versus cyclists. Find the approximate p-values for each of these tests. Will a two-sample t-test with a pooled estimate of σ^2 be appropriate for all three of these variables? Explain.

10.66 Lithium Batteries The stability of measurements on a manufactured product is important in maintaining product quality. A manufacturer of lithium batteries, such as the ones used for digital cameras, suspected that one of the production lines was producing batteries with a wide variation in length of life. To test this theory, he randomly selected $n = 50$ batteries from the suspect line and $n = 50$ from a line that was judged to be "in control." He then measured the length of time (in hours) until depletion to 0.85V with a 5-Ohm load for both samples. The sample means and variances for the two samples were as follows:

Suspect Line	Line "in Control"
$\bar{x}_1 = 9.40$	$\bar{x}_2 = 9.25$
$s_1 = .25$	$s_2 = .12$

a. Do the data provide sufficient evidence to indicate that batteries produced by the "suspect line" have a larger variance in length of life than those produced by the line that is assumed to be in control? Test using $\alpha = .05$.

b. Find the approximate p-value for the test and interpret its value.

c. Construct a 90% confidence interval for the variance ratio.

NEED TO KNOW...

How to Decide Which Test to Use

Are you interested in testing means? If the design involves:

a. One random sample, use the one-sample t statistic.
b. Two independent random samples, are the population variances equal?
    i. If equal, use the two-sample t statistic with pooled s^2.
    ii. If unequal, use the unpooled t with estimated df.
c. Two paired samples with random pairs, use a one-sample t for analyzing differences.

Are you interested in testing variances? If the design involves:

a. One random sample, use the χ^2 test for a single variance.
b. Two independent random samples, use the F-test to compare two variances.

REVISITING THE SMALL-SAMPLE ASSUMPTIONS

All of the tests and estimation procedures discussed in this chapter require that the data satisfy certain conditions in order that the error probabilities (for the tests) and the confidence coefficients (for the confidence intervals) be equal to the values you have specified. For example, if you construct what you believe to be a 95% confidence interval, you want to be certain that, in repeated sampling, 95% (and not 85% or 75% or less) of all such intervals will contain the parameter of interest. These conditions are summarized in these assumptions:

ASSUMPTIONS

1. For all tests and confidence intervals described in this chapter, it is assumed that **samples are randomly selected from normally distributed populations.**

2. When two samples are selected, it is assumed that they are **selected in an independent manner** except in the case of the paired-difference experiment.

3. For tests or confidence intervals concerning the difference between two population means μ_1 and μ_2 based on independent random samples, it is assumed that $\sigma_1^2 = \sigma_2^2$.

In reality, you will never know everything about the sampled population. If you did, there would be no need for sampling or statistics. It is also highly unlikely that a population will *exactly* satisfy the assumptions given in the box. Fortunately, the procedures presented in this chapter give good inferences even when the data exhibit moderate departures from the necessary conditions.

A statistical procedure that is not sensitive to departures from the conditions on which it is based is said to be **robust.** The Student's t-tests are quite robust for moderate departures from normality. Also, as long as the sample sizes are nearly equal, there is not much difference between the pooled and unpooled t statistics for the difference in two population means. However, if the sample sizes are not clearly equal, and if the population variances are unequal, the pooled t statistic provides inaccurate conclusions.

If you are concerned that your data do not satisfy the assumptions, other options are available:

• If you can select relatively large samples, you can use one of the large-sample procedures of Chapters 8 and 9, which do not rely on the normality or equal variance assumptions.

• You may be able to use a *nonparametric test* to answer your inferential questions. These tests have been developed specifically so that few or no distributional assumptions are required for their use. Tests that can be used to compare the locations or variability of two populations are presented in Chapter 15.

CHAPTER REVIEW

Key Concepts and Formulas

I. Experimental Designs for Small Samples

1. **Single random sample:** The sampled population must be normal.

2. **Two independent random samples:** Both sampled populations must be normal.

   a. Populations have a common variance σ^2.

   b. Populations have different variances: σ_1^2 and σ_2^2.

3. **Paired-difference** or **matched pairs design:** The samples are not independent.

II. Statistical Tests of Significance

1. Based on the t, F, and χ^2 distributions

2. Use the same procedure as in Chapter 9

3. **Rejection region—critical values** and **significance levels:** based on the t, F, or χ^2 distributions with the appropriate degrees of freedom

4. **Tests of population parameters:** a single mean, the difference between two means, a single variance, and the ratio of two variances

III. Small-Sample Test Statistics

To test one of the population parameters when the sample sizes are small, use the following test statistics:

Parameter	Test Statistic	Degrees of Freedom
μ	$t = \dfrac{\bar{x} - \mu_0}{s/\sqrt{n}}$	$n - 1$
$\mu_1 - \mu_2$ (equal variances)	$t = \dfrac{(\bar{x}_1 - \bar{x}_2) - (\mu_1 - \mu_2)}{\sqrt{s^2\left(\dfrac{1}{n_1} + \dfrac{1}{n_2}\right)}}$	$n_1 + n_2 - 2$
$\mu_1 - \mu_2$ (unequal variances)	$t \approx \dfrac{(\bar{x}_1 - \bar{x}_2) - (\mu_1 - \mu_2)}{\sqrt{\dfrac{s_1^2}{n_1} + \dfrac{s_2^2}{n_2}}}$	Satterthwaite's approximation
$\mu_1 - \mu_2$ (paired samples)	$t = \dfrac{\bar{d} - \mu_d}{s_d/\sqrt{n}}$	$n - 1$
σ^2	$\chi^2 = \dfrac{(n - 1)s^2}{\sigma_0^2}$	$n - 1$
σ_1^2/σ_2^2	$F = s_1^2/s_2^2$	$n_1 - 1$ and $n_2 - 1$

 TECHNOLOGY TODAY

Small-Sample Testing—*Microsoft Excel*

The tests of hypotheses for two population means based on the Student's t distribution and the F-test for the ratio of two variances can be found using the *Microsoft Excel* command **Data ▶ Data Analysis.** Remember that you need to have loaded the *Excel* add-ins called **Analysis ToolPak** (see the instructions in the "Technology Today." section of Chapter 1). You will find three choices for the two-sample t-tests and one F-test in the list of "Analysis Tools." To choose the proper t-tests, you must first decide whether the samples are independent or paired; for the independent samples test, you must decide whether or not the population variances can be assumed equal.

 **(Two-Sample t-Test Assuming Equal Variances)** The test scores on the same algebra test were recorded for nine students randomly selected from a classroom taught by teacher A and eight students randomly selected from a classroom taught by Teacher B. Is there a difference in the average scores for students taught by these two teachers?

Teacher A	65	88	93	95	80	76	79	77	
Teacher B	91	85	70	82	92	68	86	87	75

Enter the data into columns A and B of an *Excel* spreadsheet.

1. Use **Data ▶ Data ▶ Analysis ▶ Descriptive Statistics** or 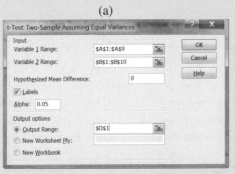 **▶ Statistical ▶ STDEV.S** (**STDEV** in *Excel 2007* and earlier versions) to find the standard deviations for the two samples, $s_1 = 9.913$ and $s_2 = 8.800$. Since the ratio of the two variances is $s_1^2/s_2^2 = 1.27$ (less than 3), you are safe in assuming that the population variances are the same.

2. Select **Data ▶ Data Analysis ▶ t-Test: Two-Sample Assuming Equal Variances** to generate the Dialog box in Figure 10.19(a). Highlight or type the **Variable 1 Range** and **Variable 2 Range** (the data in the first and second columns) into the first two boxes. In the box marked "Hypothesized Mean Difference" type **0** (since we are testing $H_0: \mu_1 - \mu_2 = 0$) and check "Labels" if necessary.

3. The default significance level is $\alpha = .05$ in *Excel.* Change this significance level if necessary. Enter a cell location for the **Output Range** and click **OK.** The output will appear in the selected cell location, and should be adjusted using **Format ▶ AutoFit Column Width** on the **Home** tab in the **Cells** group while it is still highlighted. You can decrease the decimal accuracy if you like, using ⊕ on the **Home** tab in the **Number** group.

4. The observed value of the test statistic $t = -.0337$ is found in Figure 10.19(b) in the row labeled "*t Stat*" followed by the one-tailed *p*-value "*P(T<=t) one-tail*" and the critical value marking the rejection region for a one-tailed test with $\alpha = .05$. The last two rows of output give the *p*-value and critical *t*-value for a two-tailed test.

5. For this example, the *p*-value $= .9736$ indicates that there is no significant difference in the average scores for students taught by the two teachers.

FIGURE 10.19

(a)

t-Test: Two-Sample Assuming Equal Variances

Input	
Variable 1 Range:	$A$1:$A$9
Variable 2 Range:	$B$1:$B$10
Hypothesized Mean Difference:	0
✓ Labels	
Alpha:	0.05
Output options	
⦿ Output Range:	$D$1
○ New Worksheet Ply:	
○ New Workbook	

OK Cancel Help

(b)

D	E	F
t-Test: Two-Sample Assuming Equal Variances		
	A	B
Mean	81.625	81.778
Variance	98.268	77.444
Observations	8	9
Pooled Variance	87.162	
Hypothesized Mean Difference	0	
df	15	
t Stat	-0.0337	
P(T<=t) one-tail	0.4868	
t Critical one-tail	1.7531	
P(T<=t) two-tail	0.9736	
t Critical two-tail	2.1314	

6. In Section 10.7, we presented a formal test of hypothesis for the equality of two variances using the *F*-test. To implement this test using *Excel,* select **Data ▶ Data Analysis ▶ F-Test: Two-Sample for Variances.** Follow the directions for the Equal Variances *t*-test, but replace the "Alpha" value with 0.025, and you will generate the output in Figure 10.20.

FIGURE 10.20

F-Test Two-Sample for Variances		
	A	B
Mean	81.625	81.78
Variance	98.268	77.44
Observations	8	9
df	7	8
F	1.2689	
P(F<=f) one-tail	0.3701	
F Critical one-tail	4.5286	

Notice that only the one-tailed p-value and critical value are given in the output, which is why we specified the single tail to be 0.025. Hence, for our two-tailed test, $\alpha = 0.05$ and p-value $= .7402$. There is no significant difference in the two variances.

(**Two-Sample t-Test Assuming Unequal Variances**)

1. Refer to Example 10.17. If the ratio of the two sample variances had been so large that you could not assume equal variances (we use "greater than 3" as a rule of thumb), you should select **Data ▶ Data Analysis ▶ t-Test: Two-Sample Assuming Unequal Variances.**

2. Follow the directions for the Equal Variances t-test, and you will generate similar output. If we use this test on the data from Example 10.17, the following output results (Figure 10.21).

FIGURE 10.21 ●

H	I	J
t-Test: Two-Sample Assuming Unequal Variances		
	A	B
Mean	81.625	81.778
Variance	98.268	77.444
Observations	8	9
Hypothesized Mean Difference	0	
df	14	
t Stat	-0.0334	
P(T<=t) one-tail	0.4869	
t Critical one-tail	1.7613	
P(T<=t) two-tail	0.9738	
t Critical two-tail	2.1448	

3. You will see slight differences in the observed value of the test statistic, the degrees of freedom and the p-values for the test, but the conclusions did not change.

4. NOTE: When calculating the degrees of freedom for *Satterthwaite's Approximation,* the **Data Analysis Tool** in *Excel* rounds to the nearest integer. An alternative *Excel* function for calculating the p-value for this test (**f_x ▶ Statistical ▶ T.TEST**) uses the exact value of *df* given by Satterthwaite's formula. Because of these different approaches to determining the degrees of freedom, the results of **T.TEST** and the t-Test tool will differ slightly in the Unequal Variances case, and will *also* differ slightly from the *MINITAB* output.

(**Paired t-Test**) Refer to the tire wear data from Table 10.3.

1. To perform a paired-difference test for these dependent samples, enter the data into the first two columns of an *Excel* spreadsheet and select **Data ▶ Data Analysis ▶ t-Test: Paired Two Sample for Means.**

2. Follow the directions for the Equal Variances t-test, and you will generate similar output. For the data in Table 10.3, you obtain the output in Figure 10.22. Again, you can decrease the decimal accuracy if you like, using ⇥.₀.₀₀ on the **Home** tab in the **Number** group.

3. Using the observed value of the test statistic ($t = 12.83$) with two-tailed p-value $= .0002$, there is strong evidence to indicate a difference in the two population means.

FIGURE 10.22

	D	E	F
t-Test: Paired Two Sample for Means			
		Tire A	Tire B
Mean		10.24	9.76
Variance		1.733	1.763
Observations		5	5
Pearson Correlation		0.998	
Hypothesized Mean Difference		0	
df		4	
t Stat		12.8285	
P(T<=t) one-tail		0.0001	
t Critical one-tail		2.1318	
P(T<=t) two-tail		0.0002	
t Critical two-tail		2.7764	

Small-Sample Testing and Estimation—*MINITAB*

The tests of hypotheses for two population means based on the Student's t distribution and the F-test for the ratio of two variances can be found using the *MINITAB* command **Stat ▶ Basic Statistics.** You will find choices for **1-Sample t, 2-Sample t, Paired t,** and **2 Variances,** which will perform the tests and estimation procedures of Sections 10.3, 10.4, 10.5, and 10.7. To choose the proper two sample t-tests, you must first decide whether the samples are independent or paired; for the independent samples test, you must decide whether or not the population variances can be assumed equal.

EXAMPLE  **(One Sample *t*-Test)** Refer to Example 10.3, in which the average weight of diamonds using a new process was compared to an average weight of .5 karat.

1. Enter the six recorded weights—.46, .61, .52, .48, .57, .54—in column C1 and name them "Weights." Use **Stat ▶ Basic Statistics ▶ 1-Sample t** to generate the Dialog boxes in Figure 10.23.

FIGURE 10.23

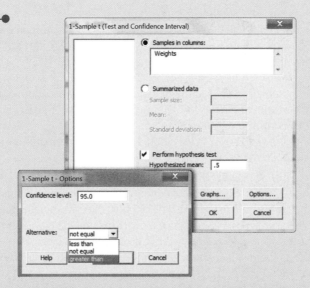

2. To test $H_0: \mu = .5$ versus $H_a: \mu > .5$, use the list on the left to select "Weights" for the box marked "Samples in Columns." Check the box marked "Perform hypothesis test." Then, place your cursor in the box marked "Hypothesized mean:" and enter **.5** as the test value. Finally, use **Options** and the drop-down menu marked "Alternative" to select "greater than." You can change the default confidence coefficient of **.95** if you wish. Click **OK** twice to obtain the output in Figure 10.24.

FIGURE 10.24

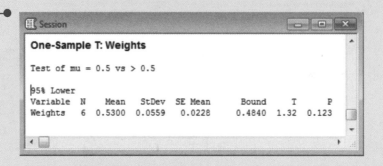

```
One-Sample T: Weights

Test of mu = 0.5 vs > 0.5

                                    95% Lower
Variable   N    Mean    StDev   SE Mean    Bound     T      P
Weights    6   0.5300  0.0559   0.0228    0.4840   1.32  0.123
```

3. Notice that *MINITAB* produces a one- or a two-sided confidence interval for the single population mean, consistent with the alternative hypothesis you have chosen.

EXAMPLE  **10.21**

(Two Sample *t*-Test) The test scores on the same algebra test were recorded for nine students randomly selected from a classroom taught by teacher A and eight students randomly selected from a classroom taught by Teacher B. Is there a difference in the average scores for students taught by these two teachers?

Teacher A	65	88	93	95	80	76	79	77	
Teacher B	91	85	70	82	92	68	86	87	75

1. The data can be entered into the worksheet in one of three ways:
   - Enter measurements from both samples into a single column and enter letters (A or B) in a second column to identify the sample from which the measurement comes.
   - Enter the samples in two separate columns.
   - If you do not have the raw data, but rather have summary statistics, *MINITAB 16* will allow you to use these values by selecting "Summarized data" and entering the appropriate values in the boxes.

2. Use the second method and enter the data into two columns of the worksheet. Use **Stat ▶ Basic Statistics ▶ Display Descriptive Statistics** to find the standard deviations for the two samples, $s_1 = 9.91$ and $s_2 = 8.80$. Since the ratio of the two variances is $s_1^2/s_2^2 = 1.27$ (less than 3), you are safe in assuming that the population variances are the same.

3. Select **Stat ▶ Basic Statistics ▶ 2-Sample t** to generate the Dialog box in Figure 10.25(a). Check "samples in different columns," selecting the appropriate columns from the list at the left. Check the "Assume equal variances" box and select the proper alternative in the **Options** box. The two-sample output when you click **OK** twice automatically contains a 95% one- or two-sided confidence interval as well as the test statistic and *p*-value (you can change the confidence coefficient if you wish). The output is shown in Figure 10.25(b).

FIGURE 10.25

(a)

(b)

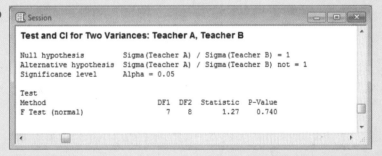

4. The observed value of the test statistic $t = -.03$ is labeled "*T-Value*" followed by the two-tailed "*P-Value*." For this example, the *p*-value $= .974$ indicates that there is no significant difference in the average scores for students taught by the two teachers.

5. In Section 10.7, we presented a formal test of hypothesis for the equality of two variances using the *F*-test. To implement this test using *MINITAB*, select **Stat ▶ Basic Statistics ▶ 2 Variances.** In the drop-down list, select "Samples in two columns" and enter the appropriate columns from the list on the left. The pertinent portion of the output is shown in Figure 10.26. For our two-tailed test with $\alpha = 0.05$, the test statistic is $F = 1.27$ and the *p*-value $= .740$. There is no significant difference in the two variances.

FIGURE 10.26

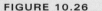

```
Session

Test and CI for Two Variances: Teacher A, Teacher B

Null hypothesis          Sigma(Teacher A) / Sigma(Teacher B) = 1
Alternative hypothesis   Sigma(Teacher A) / Sigma(Teacher B) not = 1
Significance level       Alpha = 0.05

Test
Method                      DF1  DF2  Statistic  P-Value
F Test (normal)               7    8       1.27    0.740
```

EXAMPLE 10.22 — (**Two-Sample *t*-Test Assuming Unequal Variances**)

1. Refer to Example 10.21. If the ratio of the two sample variances had been so large that you could not assume equal variances (we use "greater than 3" as a rule of thumb), you should select **Stat ▶ Basic Statistics ▶ 2-Sample t,** but DO NOT check the box marked "Assume Equal Variances." If we use this test on the data from Example 10.17, the following output results (Figure 10.27).

FIGURE 10.27

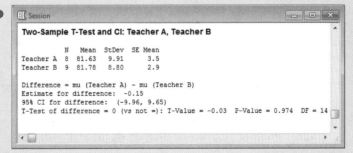

```
Session

Two-Sample T-Test and CI: Teacher A, Teacher B

           N   Mean  StDev  SE Mean
Teacher A  8  81.63   9.91      3.5
Teacher B  9  81.78   8.80      2.9

Difference = mu (Teacher A) - mu (Teacher B)
Estimate for difference:  -0.15
95% CI for difference:  (-9.96, 9.65)
T-Test of difference = 0 (vs not =): T-Value = -0.03  P-Value = 0.974  DF = 14
```

2. You will see slight differences in the degrees of freedom, there is no "Pooled StDev" listed, but the conclusions did not change.

3. NOTE: When calculating the degrees of freedom for *Satterthwaite's Approximation*, *MINITAB* uses the integer part of the calculated value, which is different from the procedures used in *MS Excel*. Because of these different approaches to determining the degrees of freedom, the results of the outputs from *MS Excel* and *MINITAB* will differ slightly.

EXAMPLE 10.23 — **(Paired-Difference Test)** Refer to the tire wear data from Table 10.3.

1. To perform a paired-difference test for these dependent samples, enter the data into the first two columns of a *MINITAB* worksheet and select **Stat ▶ Basic Statistics ▶ Paired t.**

2. Follow the directions for the independent samples *t*-test, and you will generate similar output. For the data in Table 10.3, you obtain the output in Figure 10.28.

3. Using the observed value of the test statistic ($t = 12.83$) with two-tailed *p*-value = .000, there is strong evidence to indicate a difference in the two population means.

FIGURE 10.28 ●

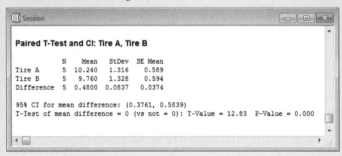

```
Session

Paired T-Test and CI: Tire A, Tire B

              N    Mean   StDev   SE Mean
Tire A        5   10.240  1.316   0.589
Tire B        5    9.760  1.328   0.594
Difference    5   0.4800  0.0837  0.0374

95% CI for mean difference: (0.3761, 0.5839)
T-Test of mean difference = 0 (vs not = 0): T-Value = 12.83  P-Value = 0.000
```

Supplementary Exercises

10.67 What assumptions are made when Student's *t*-test is used to test a hypothesis concerning a population mean?

10.68 What assumptions are made about the populations from which random samples are obtained when the *t* distribution is used in making small-sample inferences concerning the difference in population means?

10.69 Why use paired observations to estimate the difference between two population means rather than estimation based on independent random samples selected from the two populations? Is a paired experiment always preferable? Explain.

10.70 Use Table 4 in Appendix I to find the following critical values:
a. An upper one-tailed rejection region with $\alpha = .05$ and 11 *df*.

b. A two-tailed rejection region with $\alpha = .05$ and 7 *df*.

c. A lower one-tailed rejection region with $\alpha = .01$ and 15 *df*.

10.71 Use Table 4 in Appendix I to bound the following *p*-values:
a. $P(t > 1.2)$ with 5 *df*
b. $P(t > 2) + P(t < -2)$ with 10 *df*
c. $P(t < -3.3)$ with 8 *df*
d. $P(t > 0.6)$ with 12 *df*

10.72 A random sample of $n = 12$ observations from a normal population produced $\bar{x} = 47.1$ and $s^2 = 4.7$. Test the hypothesis $H_0: \mu = 48$ against $H_0: \mu \neq 48$ at the 5% level of significance.

10.73 Impurities II A manufacturer can tolerate a small amount (.05 milligrams per liter (mg/l)) of impurities in a raw material needed for manufacturing its product. Because the laboratory test for the impurities is subject to experimental error, the manufacturer tests each batch 10 times. Assume that the mean value of the experimental error is 0 and hence that the mean value of the 10 test readings is an unbiased estimate of the true amount of the impurities in the batch. For a particular batch of the raw material, the mean of the 10 test readings is .058 mg/l, with a standard deviation of .012 mg/l. Do the data provide sufficient evidence to indicate that the amount of impurities in the batch exceeds .05 mg/l? Find the p-value for the test and interpret its value.

10.74 Red Pine The main stem growth measured for a sample of seventeen 4-year-old red pine trees produced a mean and standard deviation equal to 11.3 and 3.4 inches, respectively. Find a 90% confidence interval for the mean growth of a population of 4-year-old red pine trees subjected to similar environmental conditions.

10.75 Sodium Hydroxide The object of a general chemistry experiment is to determine the amount (in milliliters) of sodium hydroxide (NaOH) solution needed to neutralize 1 gram of a specified acid. This will be an exact amount, but when the experiment is run in the laboratory, variation will occur as the result of experimental error. Three titrations are made using phenolphthalein as an indicator of the neutrality of the solution (pH equals 7 for a neutral solution). The three volumes of NaOH required to attain a pH of 7 in each of the three titrations are as follows: 82.10, 75.75, and 75.44 milliliters. Use a 99% confidence interval to estimate the mean number of milliliters required to neutralize 1 gram of the acid.

10.76 Sodium Chloride Measurements of water intake, obtained from a sample of 17 rats that had been injected with a sodium chloride solution, produced a mean and standard deviation of 31.0 and 6.2 cubic centimeters (cm^3), respectively. Given that the average water intake for noninjected rats observed over a comparable period of time is 22.0 cm^3, do the data indicate that injected rats drink more water than noninjected rats? Test at the 5% level of significance. Find a 90% confidence interval for the mean water intake for injected rats.

10.77 Sea Urchins An experimenter was interested in determining the mean thickness of the cortex of the sea urchin egg. The thickness was measured for

$n = 10$ sea urchin eggs. These measurements were obtained:

4.5	6.1	3.2	3.9	4.7
5.2	2.6	3.7	4.6	4.1

Estimate the mean thickness of the cortex using a 95% confidence interval.

10.78 Fabricating Systems A production plant has two complex fabricating systems, both of which are maintained at 2-week intervals. However, one system is twice as old as the other. The number of finished products fabricated daily by each of the systems is recorded for 30 working days, with the results given in the table. Do these data present sufficient evidence to conclude that the variability in daily production warrants increased maintenance of the older fabricating system? Use the p-value approach.

New System	Old System
$\bar{x}_1 = 246$	$\bar{x}_2 = 240$
$s_1 = 15.6$	$s_2 = 28.2$

Data set **EX1079** **10.79 Fossils** The data in the table are the diameters and heights of 10 fossil specimens of a species of small shellfish, *Rotularia (Annelida) fallax,* that were unearthed in a mapping expedition near the Antarctic Peninsula.[14] The table gives an identification symbol for the fossil specimen, the fossil's diameter and height in millimeters, and the ratio of diameter to height.

Specimen	Diameter	Height	D/H
OSU 36651	185	78	2.37
OSU 36652	194	65	2.98
OSU 36653	173	77	2.25
OSU 36654	200	76	2.63
OSU 36655	179	72	2.49
OSU 36656	213	76	2.80
OSU 36657	134	75	1.79
OSU 36658	191	77	2.48
OSU 36659	177	69	2.57
OSU 36660	199	65	3.06
$\bar{x}$:	184.5	73	2.54
s:	21.5	5	.37

a. Find a 95% confidence interval for the mean diameter of the species.

b. Find a 95% confidence interval for the mean height of the species.

c. Find a 95% confidence interval for the mean ratio of diameter to height.

d. Compare the three intervals constructed in parts a, b, and c. Is the average of the ratios the same as the ratio of the average diameter to average height?

10.80 Fossils, continued Refer to Exercise 10.79 and data set EX1079. Suppose you want to estimate the mean diameter of the fossil specimens correct to within 5 millimeters with probability equal to .95. How many fossils do you have to include in your sample?

Data set **10.81 Alcohol and Reaction Times** To test EX1081 the effect of alcohol in increasing the reaction time to respond to a given stimulus, the reaction times of seven people were measured. After consuming 3 ounces of 40% alcohol, the reaction time for each of the seven people was measured again. Do the following data indicate that the mean reaction time after consuming alcohol was greater than the mean reaction time before consuming alcohol? Use $\alpha = .05$.

Person	1	2	3	4	5	6	7
Before	4	5	5	4	3	6	2
After	7	8	3	5	4	5	5

Data set **10.82 Cheese, Please** Here are the prices per EX1082 ounce of $n = 13$ different brands of individually wrapped cheese slices:

29.0	24.1	23.7	19.6	27.5
28.7	28.0	23.8	18.9	23.9
21.6	25.9	27.4		

Construct a 95% confidence interval estimate of the underlying average price per ounce of individually wrapped cheese slices.

10.83 Drug Absorption An experiment was conducted to compare the mean lengths of time required for the bodily absorption of two drugs A and B. Ten people were randomly selected and assigned to receive one of the drugs. The length of time (in minutes) for the drug to reach a specified level in the blood was recorded, and the data summary is given in the table:

Drug A	Drug B
$\bar{x}_1 = 27.2$	$\bar{x}_2 = 33.5$
$s_1^2 = 16.36$	$s_2^2 = 18.92$

a. Do the data provide sufficient evidence to indicate a difference in mean times to absorption for the two drugs? Test using $\alpha = .05$.

b. Find the approximate p-value for the test. Does this value confirm your conclusions?

c. Find a 95% confidence interval for the difference in mean times to absorption. Does the interval confirm your conclusions in part a?

10.84 Drug Absorption, continued Refer to Exercise 10.83. Suppose you wish to estimate the difference in mean times to absorption correct to within 1 minute with probability approximately equal to .95.

a. Approximately how large a sample is required for each drug (assume that the sample sizes are equal)?

b. If conducting the experiment using the sample sizes of part a will require a large amount of time and money, can anything be done to reduce the sample sizes and still achieve the 1-minute margin of error for estimation?

Data set **10.85 Ring-Necked Pheasants** The weights EX1085 in grams of 10 males and 10 female juvenile ring-necked pheasants are given below.

Males		Females	
1384	1672	1073	1058
1286	1370	1053	1123
1503	1659	1038	1089
1627	1725	1018	1034
1450	1394	1146	1281

a. Use a statistical test to determine if the population variance of the weights of the male birds differs from that of the females.

b. Test whether the average weight of juvenile male ring-necked pheasants exceeds that of the females by more than 300 grams. (HINT: The procedure that you use should take into account the results of the analysis in part a.)

Data set **10.86 Bees** Insects hovering in flight expend EX1086 enormous amounts of energy for their size and weight. The data shown here were taken from a much larger body of data collected by T.M. Casey and colleagues.[15] They show the wing stroke frequencies (in hertz) for two different species of bees, $n_1 = 4$ *Euglossa mandibularis* Friese and $n_2 = 6$ *Euglossa imperialis* Cockerell.

E. mandibularis Friese	E. imperialis Cockerell
235	180
225	169
190	180
188	185
	178
	182

a. Based on the observed ranges, do you think that a difference exists between the two population variances?

b. Use an appropriate test to determine whether a difference exists.

c. Explain why a Student's t-test with a pooled estimator s^2 is unsuitable for comparing the mean wing stroke frequencies for the two species of bees.

Data set **10.87 Calcium** The calcium (Ca) content of
EX1087 a powdered mineral substance was analyzed 10 times with the following percent compositions recorded:

.0271	.0282	.0279	.0281	.0268
.0271	.0281	.0269	.0275	.0276

a. Find a 99% confidence interval for the true calcium content of this substance.

b. What does the phrase "99% confident" mean?

c. What assumptions must you make about the sampling procedure so that this confidence interval will be valid? What does this mean to the chemist who is performing the analysis?

10.88 Sun or Shade? Karl Niklas and T.G. Owens examined the differences in a particular plant, *Plantago Major L.,* when grown in full sunlight versus shade conditions.[16] In this study, shaded plants received direct sunlight for less than 2 hours each day, whereas full-sun plants were never shaded. A partial summary of the data based on $n_1 = 16$ full-sun plants and $n_2 = 15$ shade plants is shown here:

	Full Sun		Shade	
	$\bar{x}$	s	$\bar{x}$	s
Leaf Area (cm²)	128.00	43.00	78.70	41.70
Overlap Area (cm²)	46.80	2.21	8.10	1.26
Leaf Number	9.75	2.27	6.93	1.49
Thickness (mm)	.90	.03	.50	.02
Length (cm)	8.70	1.64	8.91	1.23
Width (cm)	5.24	.98	3.41	.61

a. What assumptions are required in order to use the small-sample procedures given in this chapter to compare full-sun versus shade plants? From the summary presented, do you think that any of these assumptions have been violated?

b. Do the data present sufficient evidence to indicate a difference in mean leaf area for full-sun versus shade plants?

c. Do the data present sufficient evidence to indicate a difference in mean overlap area for full-sun versus shade plants?

10.89 Orange Juice A comparison of the precisions of two machines developed for extracting juice from oranges is to be made using the following data:

Machine A	Machine B
$s^2 = 3.1$ ounces²	$s^2 = 1.4$ ounces²
$n = 25$	$n = 25$

a. Is there sufficient evidence to indicate that there is a difference in the precision of the two machines at the 5% level of significance?

b. Find a 95% confidence interval for the ratio of the two population variances. Does this interval confirm your conclusion from part a? Explain.

Data set **10.90 At Home or at Preschool?** Four sets
EX1090 of identical twins (pairs A, B, C, and D) were selected at random from a computer database of identical twins. One child was selected at random from each pair to form an "experimental group." These four children were sent to preschool. The other four children were kept at home as a control group. At the end of the year, the following IQ scores were obtained:

Pair	Experimental Group	Control Group
A	110	111
B	125	120
C	139	128
D	142	135

Does this evidence justify the conclusion that lack of preschool experience has a depressing effect on IQ scores? Use the p-value approach.

Data set **10.91 Dieting** Eight obese persons were
EX1091 placed on a diet for 1 month, and their weights, at the beginning and at the end of the month, were recorded:

	Weights	
Subjects	Initial	Final
1	310	263
2	295	251
3	287	249
4	305	259
5	270	233
6	323	267
7	277	242
8	299	265

Estimate the mean weight loss for obese persons when placed on the diet for a 1-month period. Use a 95% confidence interval and interpret your results. What assumptions must you make so that your inference is valid?

10.92 Price Wars Many seniors are ordering their drugs online to take advantage of lower costs for these pharmacies. A random sample of nine online pharmacies was selected and the cost of a 10-mg Buspar (Buspirone) tablet recorded, as given in the following table.[17]

Pharmacy	Brand ($)	Generic ($)
CanadaDrugStop.com	1.33	.79
CanadaDrugCenter	1.33	.79
Big Mountain Drugs	1.16	.74
Blue Sky Drugs	1.17	.75
CanadaDrugsPharmacy	1.33	.79
Canada Drugs Online	1.11	.75
PharmStore.com	1.13	.59
Buy Low Drugs	1.45	.45
Planetdrugsdirect.com	1.14	.59

a. Test the hypothesis of no difference in costs between brand and generic Buspar 10-mg tablets at the $\alpha = .05$ level of significance. (HINT: the observations are paired by the online pharmacies.)

b. Find the estimated savings per tablet by purchasing the generic as opposed to the brand-name tablets with a 95% confidence interval.

10.93 Breathing Patterns Research psychologists measured the baseline breathing patterns—the total ventilation (in liters of air per minute) adjusted for body size—for each of $n = 30$ patients, so that they could estimate the average total ventilation for patients before any experimentation was done. The data, along with some *MINITAB* output, are presented here:

5.23	5.72	5.77	4.99	5.12	4.82
5.54	4.79	5.16	5.84	4.51	5.14
5.92	6.04	5.83	5.32	6.19	5.70
4.72	5.38	5.48	5.37	4.96	5.58
4.67	5.17	6.34	6.58	4.35	5.63

MINITAB output for Exercise 10.93

Stem-and-Leaf Display: Ltrs/min
```
Stem-and-leaf of Ltrs/min  N = 30
Leaf Unit = 0.10

   1    4  3
   2    4  5
   5    4  677
   8    4  899
  12    5  1111
  (4)   5  2333
  14    5  455
  11    5  6777
   7    5  889
   4    6  01
   2    6  3
   1    6  5
```

Descriptive Statistics: Ltrs/min
```
Variable    N   N*    Mean   SE Mean    StDev
Ltrs/min    30   0  5.3953   0.0997   0.5462

Minimum      Q1   Median      Q3   Maximum
 4.3500  4.9825   5.3750  5.7850   6.5800
```

a. What information does the stem and leaf plot give you about the data? Why is this important?

b. Use the output to construct a 99% confidence interval for the average total ventilation for patients.

10.94 Reaction Times A comparison of reaction times (in seconds) for two different stimuli in a psychological word-association experiment produced the following results when applied to a random sample of 16 people:

Stimulus 1	1	3	2	1	2	1	3	2
Stimulus 2	4	2	3	3	1	2	3	3

Do the data present sufficient evidence to indicate a difference in mean reaction times for the two stimuli? Test using $\alpha = .05$.

10.95 Reaction Times II Refer to Exercise 10.94. Suppose that the word-association experiment is conducted using eight people as blocks and making a comparison of reaction times within each person; that is, each person is subjected to both stimuli in a random order. The reaction times (in seconds) for the experiment are as follows:

Person	Stimulus 1	Stimulus 2
1	3	4
2	1	2
3	1	3
4	2	1
5	1	2
6	2	3
7	3	3
8	2	3

Do the data present sufficient evidence to indicate a difference in mean reaction times for the two stimuli? Test using $\alpha = .05$.

10.96 Refer to Exercises 10.94 and 10.95. Calculate a 95% confidence interval for the difference in the two population means for each of these experimental designs. Does it appear that blocking increased the amount of information available in the experiment?

Data set **10.97 Impact Strength** The following data
EX1097 are readings (in foot-pounds) of the impact
strengths of two kinds of packaging material:

A	B
1.25	.89
1.16	1.01
1.33	.97
1.15	.95
1.23	.94
1.20	1.02
1.32	.98
1.28	1.06
1.21	.98

MS Excel output for Exercise 10.97

D	E	F
t-Test: Two-Sample Assuming Equal Variances		
	A	B
Mean	1.2367	0.9778
Variance	0.0042	0.0024
Observations	9	9
Pooled Variance	0.0033	
Hypothesized Mean Difference	0	
df	16	
t Stat	9.5641	
P(T<=t) one-tail	0.0000	
t Critical one-tail	1.7459	
P(T<=t) two-tail	0.0000	
t Critical two-tail	2.1199	

a. Use the *MS Excel* printout to determine whether
there is evidence of a difference in the mean
strengths for the two kinds of material.

b. Are there practical implications to your results?

Data set **10.98 Cake Mixes** An experiment was con-
EX1098 ducted to compare the densities (in ounces per
cubic inch) of cakes prepared from two different cake
mixes. Six cake pans were filled with batter A, and six
were filled with batter B. Expecting a variation in oven
temperature, the experimenter placed a pan filled with
batter A and another with batter B *side by side* at six
different locations in the oven. The six paired observations of densities are as follows:

Location	1	2	3	4	5	6
Batter A	.135	.102	.098	:141	.131	.144
Batter B	.129	.120	.112	.152	.135	.163

a. Do the data present sufficient evidence to indicate
a difference between the average densities of cakes
prepared using the two types of batter?

b. Construct a 95% confidence interval for the difference between the average densities for the two
mixes.

10.99 Under what assumptions can the F distribution
be used in making inferences about the ratio of population variances?

10.100 Got Milk? A dairy is in the market for a
new container-filling machine and is considering two
models, manufactured by company A and company B.
Ruggedness, cost, and convenience are comparable in
the two models, so the deciding factor is the variability
of fills. The model that produces fills with the smaller
variance is preferred. If you obtain samples of fills for
each of the two models, an F-test can be used to test
for the equality of population variances. Which type of
rejection region would be most favored by each of
these individuals?

a. The manager of the dairy—Why?

b. A sales representative for company A—Why?

c. A sales representative for company B—Why?

10.101 Got Milk II Refer to Exercise 10.100.
Wishing to demonstrate that the variability of fills is
less for her model than for her competitor's, a sales
representative for company A acquired a sample of
30 fills from her company's model and a sample of
10 fills from her competitor's model. The sample
variances were $s_A^2 = .027$ and $s_B^2 = .065$, respectively. Does this result provide statistical support at
the .05 level of significance for the sales representative's claim?

10.102 Chemical Purity A chemical manufacturer
claims that the purity of his product never varies
by more than 2%. Five batches were tested and
given purity readings of 98.2, 97.1, 98.9, 97.7, and
97.9%.

a. Do the data provide sufficient evidence to contradict the manufacturer's claim? (HINT: To be generous, let a range of 2% equal 4σ.)

b. Find a 90% confidence interval for σ^2.

10.103 16-Ounce Cans? A cannery prints "weight
16 ounces" on its label. The quality control supervisor
selects nine cans at random and weighs them. She
finds $\bar{x} = 15.7$ and $s = .5$. Do the data present sufficient evidence to indicate that the mean weight is less
than that claimed on the label?

10.104 Reaction Time III A psychologist wishes to
verify that a certain drug increases the reaction time to
a given stimulus. The following reaction times (in

tenths of a second) were recorded before and after injection of the drug for each of four subjects:

	Reaction Time	
Subject	Before	After
1	7	13
2	2	3
3	12	18
4	12	13

Test at the 5% level of significance to determine whether the drug significantly increases reaction time.

10.105 Food Production At a time when energy conservation is so important, some scientists think closer scrutiny should be given to the cost (in energy) of producing various forms of food. Suppose you wish to compare the mean amount of oil required to produce 1 acre of corn versus 1 acre of cauliflower. The readings (in barrels of oil per acre), based on 20-acre plots, seven for each crop, are shown in the table.

Corn	Cauliflower
5.6	15.9
7.1	13.4
4.5	17.6
6.0	16.8
7.9	15.8
4.8	16.3
5.7	17.1

a. Use these data to find a 90% confidence interval for the difference between the mean amounts of oil required to produce these two crops.

b. Based on the interval in part a, is there evidence of a difference in the average amount of oil required to produce these two crops? Explain.

10.106 Alcohol and Altitude The effect of alcohol consumption on the body appears to be much greater at high altitudes than at sea level. To test this theory, a scientist randomly selects 12 subjects and randomly divides them into two groups of six each. One group is put into a chamber that simulates conditions at an altitude of 12,000 feet, and each subject ingests a drink containing 100 cubic centimeters (cc) of alcohol. The second group receives the same drink in a chamber that simulates conditions at sea level. After 2 hours, the amount of alcohol in the blood (grams per 100 cc) for each subject is measured. The data are shown in the table. Do the data provide sufficient evidence to support the theory that average amount of alcohol in the blood after 2 hours is greater at high altitudes?

Sea Level	12,000 Feet
.07	.13
.10	.17
.09	.15
.12	.14
.09	.10
.13	.14

10.107 Stock Risks The closing prices of two common stocks were recorded for a period of 15 days. The means and variances are

$$\bar{x}_1 = 40.33 \quad \bar{x}_2 = 42.54$$
$$s_1^2 = 1.54 \quad s_2^2 = 2.96$$

a. Do these data present sufficient evidence to indicate a difference between the variabilities of the closing prices of the two stocks for the populations associated with the two samples? Give the p-value for the test and interpret its value.

b. Construct a 99% confidence interval for the ratio of the two population variances.

10.108 Auto Design An experiment is conducted to compare two new automobile designs. Twenty people are randomly selected, and each person is asked to rate each design on a scale of 1 (poor) to 10 (excellent). The resulting ratings will be used to test the null hypothesis that the mean level of approval is the same for both designs against the alternative hypothesis that one of the automobile designs is preferred. Do these data satisfy the assumptions required for the Student's t-test of Section 10.4? Explain.

10.109 Safety Programs The data shown here were collected on lost-time accidents (the figures given are mean work-hours lost per month over a period of 1 year) before and after an industrial safety program was put into effect. Data were recorded for six industrial plants. Do the data provide sufficient evidence to indicate whether the safety program was effective in reducing lost-time accidents? Test using $\alpha = .01$.

	Plant Number					
	1	2	3	4	5	6
Before Program	38	64	42	70	58	30
After Program	31	58	43	65	52	29

10.110 Two Different Entrees To compare the demand for two different entrees, the manager of a cafeteria recorded the number of purchases of each entree on seven consecutive days. The data are shown in the table. Do the data provide sufficient evidence to indicate a greater mean demand for one of the entrees? Use the *Excel* printout.

Day	A	B
Monday	420	391
Tuesday	374	343
Wednesday	434	469
Thursday	395	412
Friday	637	538
Saturday	594	521
Sunday	679	625

MS Excel output for Exercise 10.110

E	F	G
t-Test: Paired Two Sample for Means		
	A	*B*
Mean	504.714	471.286
Variance	16191.238	9495.571
Observations	7	7
Pearson Correlation	0.945	
Hypothesized Mean Difference	0	
df	6	
t Stat	1.862	
P(T<=t) one-tail	0.056	
t Critical one-tail	1.943	
P(T<=t) two-tail	0.112	
t Critical two-tail	2.447	

10.111 Pollution Control The EPA limit on the allowable discharge of suspended solids into rivers and streams is 60 milligrams per liter (mg/l) per day. A study of water samples selected from the discharge at a phosphate mine shows that over a long period, the mean daily discharge of suspended solids is 48 mg/l, but day-to-day discharge readings are variable. State inspectors measured the discharge rates of suspended solids for $n = 20$ days and found $s^2 = 39$ (mg/l)^2. Find a 90% confidence interval for σ^2. Interpret your results.

Data set **10.112 Enzymes** Two methods were used to
EX10112 measure the specific activity (in units of enzyme activity per milligram of protein) of an enzyme. One unit of enzyme activity is the amount that catalyzes the formation of 1 micromole of product per minute under specified conditions. Use an appropriate test or estimation procedure to compare the two methods of measurement. Comment on the validity of any assumptions you need to make.

Method 1	125	137	130	151	142
Method 2	137	143	151	156	149

10.113 Connector Rods A producer of machine parts claimed that the diameters of the connector rods produced by his plant had a variance of at most .03 inch^2. A random sample of 15 connector rods from his plant produced a sample mean and variance of .55 inch and .053 inch^2, respectively.

a. Is there sufficient evidence to reject his claim at the $\alpha = .05$ level of significance?

b. Find a 95% confidence interval for the variance of the rod diameters.

10.114 Sleep and the College Student How much sleep do you get on a typical school night? A group of 10 college students were asked to report the number of hours that they slept on the previous night with the following results:

7, 6, 7.25, 7, 8.5, 5, 8, 7, 6.75, 6

a. Find a 99% confidence interval for the average number of hours that college students sleep.

b. What assumptions are required in order for this confidence interval to be valid?

Data set **10.115 Arranging Objects** The following
EX10115 data are the response times in seconds for $n = 25$ first graders to arrange three objects by size.

5.2	3.8	5.7	3.9	3.7
4.2	4.1	4.3	4.7	4.3
3.1	2.5	3.0	4.4	4.8
3.6	3.9	4.8	5.3	4.2
4.7	3.3	4.2	3.8	5.4

Find a 95% confidence interval for the average response time for first graders to arrange three objects by size. Interpret this interval.

Data set **10.116 The NBA Finals** Want to attend a
EX10116 pro-basketball finals game? The average prices for the NBA rematch of the Boston Celtics and the LA Lakers in 2010 compared to the average ticket prices in 2008 are given in the table that follows.[18]

Game	2008 ($)	2010 ($)
1	593	532
2	684	855
3	727	541
4	907	458
5	769	621
6	753	681
7	533	890

a. If we were to assume that the prices given in the table have been randomly selected, test for a significant difference between the 2008 and 2010 prices. Use $\alpha = .01$.

b. Find a 98% confidence interval for the mean difference, $\mu_d = \mu_{08} - \mu_{10}$. Does this estimate confirm the results of part a?

10.117 Mall Rats An article in *American Demographics* investigated consumer habits at the mall. We tend to spend the most money shopping on the weekends, and, in particular, on Sundays from 4 to 6 P.M. Wednesday morning shoppers spend the least![19] Suppose that a random sample of 20 weekend shoppers and a random sample of 20 weekday shoppers were selected, and the amount spent per trip to the mall was recorded.

	Weekends	Weekdays
Sample Size	20	20
Sample Mean ($)	78	67
Sample Standard Deviation ($)	22	20

a. Is it reasonable to assume that the two population variances are equal? Use the F-test to test this hypothesis with $\alpha = .05$.

b. Based on the results of part a, use the appropriate test to determine whether there is a difference in the average amount spent per trip on weekends versus weekdays. Use $\alpha = .05$.

10.118 Books or iPads? As part of a larger pilot study, students at a Riverside, California middle school, will compare the learning of algebra by students using iPads versus students using the traditional algebra textbook with the same author and publisher.[20] To remove teacher-to-teacher variation, the same teacher will teach both classes, and the iPad and textbook material are both provided by the same author and publisher. Suppose that after 1 month, 10 students were selected from each class and their scores on an algebra advancement test recorded. The summarized data follow.

	iPad	Textbook
Mean	86.4	79.7
Standard Deviation	8.95	10.7
Sample Size	10	10

a. Use the summary data to test for a significant difference in advancement scores for the two groups using $\alpha = .05$.

b. Find a 95% confidence interval for the difference in mean scores for the two groups.

c. In light of parts a and b, what can we say about the efficacy of using an iPad versus a traditional textbook in learning algebra at the middle school level?

 API
Riverside
School

School Accountability Study—How Is Your School Doing?

If you are a California resident, the API (Academic Performance Index) may be published by your local paper. The 2009–2010 accountability progress report for Riverside County, California, was published in the *Press-Enterprise* listing the API and Growth (difference in 2010 and 2009 API) for all school districts in the county, together with whether each school has met the federal Adequate Yearly Progress guidelines.[21] The first two of these measures are given for a sample of elementary schools from the two school districts in Riverside, California.

Alvord Schools	API	Growth	Riverside Unified Schools	API	Growth
Stokoe	786	−5	Lake Matthews	891	18
Orrenmaa	760	−2	Beatty	778	8
Collett	768	8	Alcott	827	10
La Granada	745	40	Emerson	789	23
Terrace	731	−6	Harrison	805	28
McAuliffe	815	0	Highgrove	762	6
Valley View	690	−16	Hyatt	798	35
			Jefferson	796	34
			Longfellow	731	−4
			Magnolia	818	28
			Mountain View	769	3
			Victoria	815	7
			Woodcrest	841	8
			Franklin	876	3

1. To study these data statistically, begin by finding the means and standard deviations for the API and the Growth scores for each of the two districts.

2. Test for a significant growth in API scores for both the Alvord and Riverside Unified school districts. Use $\alpha = .05$.

3. Test for a significant difference in API for Alvord and Riverside Unified school districts using $\alpha = .05$.

4. Summarize your results on the progress of these two school districts in the form of a report.

chirapbogdan/Shutterstock.com

11

The Analysis of Variance

GENERAL OBJECTIVE

The quantity of information contained in a sample is affected by various factors that the experimenter may or may not be able to control. This chapter introduces three different *experimental designs*, two of which are direct extensions of the unpaired and paired designs of Chapter 10. A new technique called the *analysis of variance* is used to determine how the different experimental factors affect the average response.

CHAPTER INDEX

 NEED TO KNOW...

How to Determine Whether My Calculations Are Accurate

How to Save Money on Groceries!

Canning or freezing produce that you buy in bulk will almost always save you money compared to buying in supermarkets. You can save more than 75% by canning—and more than 80% by freezing—produce purchased in bulk. The case study at the end of this chapter investigates the costs of purchasing in bulk, canning, and freezing using the analysis of variance procedures presented in this chapter.

THE DESIGN OF AN EXPERIMENT

The way that a sample is selected is called the *sampling plan* or *experimental design* and determines the amount of information in the sample. Some research involves an **observational study,** in which the researcher does not actually produce the data but only *observes* the characteristics of data that already exist. Most sample surveys, in which information is gathered with a questionnaire, fall into this category. The researcher forms a plan for collecting the data—called the *sampling plan*—and then uses the appropriate statistical procedures to draw conclusions about the population or populations from which the sample comes.

Other research involves **experimentation.** The researcher may deliberately impose one or more experimental conditions on the experimental units in order to determine their effect on the response. Here are some new terms we will use to discuss the design of a statistical experiment.

Definition   An **experimental unit** is the object on which a measurement (or measurements) is taken.

A **factor** is an independent variable whose values are controlled and varied by the experimenter.

A **level** is the intensity setting of a factor.

A **treatment** is a specific combination of factor levels.

The **response** is the variable being measured by the experimenter.

EXAMPLE   11.1

A group of people is randomly divided into an experimental and a control group. The control group is given an aptitude test after having eaten a full breakfast. The experimental group is given the same test without having eaten any breakfast. What are the factors, levels, and treatments in this experiment?

Solution   The *experimental units* are the people on which the *response* (test score) is measured. The *factor* of interest could be described as "meal" and has two *levels:* "breakfast" and "no breakfast." Since this is the only factor controlled by the experimenter, the two levels—"breakfast" and "no breakfast"—also represent the *treatments* of interest in the experiment.

EXAMPLE   11.2

Suppose that the experimenter in Example 11.1 began by randomly selecting 20 men and 20 women for the experiment. These two groups were then randomly divided into 10 each for the experimental and control groups. What are the factors, levels, and treatments in this experiment?

Solution   Now there are two *factors* of interest to the experimenter, and each factor has two *levels:*

- "Gender" at two levels: men and women
- "Meal" at two levels: breakfast and no breakfast

In this more complex experiment, there are four *treatments,* one for each specific combination of factor levels: men without breakfast, men with breakfast, women without breakfast, and women with breakfast.

In this chapter, we will concentrate on experiments that have been designed in three different ways, and we will use a technique called the *analysis of variance* to judge the effects of various factors on the experimental response. Two of these *experimental designs* are extensions of the unpaired and paired designs from Chapter 10.

11.2 WHAT IS AN ANALYSIS OF VARIANCE?

The responses that are generated in an experimental situation always exhibit a certain amount of *variability*. In an **analysis of variance,** you divide the total variation in the response measurements into portions that may be attributed to various *factors* of interest to the experimenter. If the experiment has been properly designed, these portions can then be used to answer questions about the effects of the various factors on the response of interest.

You can better understand the logic underlying an analysis of variance by looking at a simple experiment. Consider two sets of samples randomly selected from populations 1 (◆) and 2 (○), each with identical pairs of means, $\bar{x}_1$ and $\bar{x}_2$. The two sets are shown in Figure 11.1. Is it easier to detect the difference in the two means when you look at set A or set B? You will probably agree that set A shows the difference much more clearly. In set A, the variability of the measurements *within* the groups (◆s and ○s) is much smaller than the variability *between* the two groups. In set B, there is more variability *within* the groups (◆s and ○s), causing the two groups to "mix" together and making it more difficult to see the *identical* difference in the means.

FIGURE 11.1

Two sets of samples with the same means

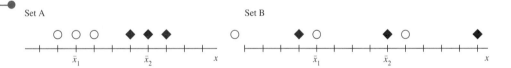

The comparison you have just done intuitively is formalized by the analysis of variance. Moreover, the analysis of variance can be used not only to compare two means but also to make comparisons of *more than two* population means and to determine the effects of various factors in more complex experimental designs. The analysis of variance relies on statistics with sampling distributions that are modeled by the *F* distribution of Section 10.7.

11.3 THE ASSUMPTIONS FOR AN ANALYSIS OF VARIANCE

The assumptions required for an analysis of variance are similar to those required for the Student's *t* and *F* statistics of Chapter 10. Regardless of the experimental design used to generate the data, you must assume that the observations within each treatment group are **normally distributed** with a **common variance** σ^2. As in Chapter 10, the analysis of variance procedures are fairly **robust** when the sample sizes are equal and when the data are fairly mound-shaped. Violating the assumption of a common variance is more serious, especially when the sample sizes are not nearly equal.

ASSUMPTIONS FOR ANALYSIS OF VARIANCE TEST AND ESTIMATION PROCEDURES

- The observations within each population are normally distributed with a common variance σ^2.
- Assumptions regarding the sampling procedure are specified for each design in the sections that follow.

This chapter describes the analysis of variance for three different experimental designs. The first design is based on independent random sampling from several populations and is an extension of the *unpaired t-test* of Chapter 10. The second is an extension of the *paired-difference* or *matched pairs* design and involves a random assignment of treatments within matched sets of observations. The third is a design that allows you to judge the effect of two experimental factors on the response. The sampling procedures necessary for each design will be restated in their respective sections.

THE COMPLETELY RANDOMIZED DESIGN: A ONE-WAY CLASSIFICATION

11.4

One of the simplest experimental designs is the **completely randomized design,** in which random samples are selected independently from each of k populations. This design involves only one *factor,* the population from which the measurement comes—hence the designation as a **one-way classification.** There are k different *levels* corresponding to the k populations, which are also the *treatments* for this one-way classification. Are the k population means all the same, or is at least one mean different from the others?

Why do you need a new procedure, the *analysis of variance,* to compare the population means when you already have the Student's *t*-test available? In comparing $k = 3$ means, you could test each of three pairs of hypotheses:

$$H_0 : \mu_1 = \mu_2 \qquad H_0 : \mu_1 = \mu_3 \qquad H_0 : \mu_2 = \mu_3$$

to find out where the differences lie. However, you must remember that each test you perform is subject to the possibility of error. To compare $k = 4$ means, you would need six tests, and you would need 10 tests to compare $k = 5$ means. The more tests you perform on a set of measurements, the more likely it is that at least one of your conclusions will be incorrect. The analysis of variance procedure provides one overall test to judge the equality of the k population means. Once you have determined whether there is *actually* a difference in the means, you can use another procedure to find out where the differences lie.

How can you select these k random samples? Sometimes the populations actually exist in fact, and you can use a computerized random number generator or a random number table to randomly select the samples. For example, in a study to compare the average sizes of health insurance claims in four different states, you could use a computer database provided by the health insurance companies to select random samples from the four states. In other situations, the populations may be *hypothetical,* and responses can be generated only after the experimental treatments have been applied.

EXAMPLE **11.3**

A researcher is interested in the effects of five types of insecticides for use in controlling the boll weevil in cotton fields. Explain how to implement a completely randomized design to investigate the effects of the five insecticides on crop yield.

Solution The only way to generate the equivalent of five random samples from the hypothetical populations corresponding to the five insecticides is to use a method called a **randomized assignment.** A fixed number of cotton plants are chosen for treatment, and each is assigned a random number. Suppose that each sample is to have an equal number of measurements. Using a randomization device, you can assign the first n_1 plants chosen to receive insecticide 1, the second n_2 plants to receive insecticide 2, and so on, until all five treatments have been assigned.

Whether by *random selection* or *random assignment,* both of these examples result in a completely randomized design, or one-way classification, for which the analysis of variance is used.

THE ANALYSIS OF VARIANCE FOR A COMPLETELY RANDOMIZED DESIGN

11.5

Suppose you want to compare k population means, $\mu_1, \mu_2, \ldots, \mu_k$, based on independent random samples of size $n_1, n_2, \ldots, n_k$ from normal populations with a common variance σ^2. That is, each of the normal populations has the same shape, but their locations might be different, as shown in Figure 11.2.

FIGURE 11.2 ●

Normal populations with a common variance but different means

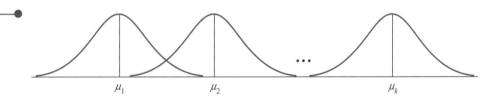

Partitioning the Total Variation in an Experiment

Let x_{ij} be the jth measurement ($j = 1, 2, \ldots, n_i$) in the ith sample. The analysis of variance procedure begins by considering the total variation in the experiment, which is measured by a quantity called the **total sum of squares (TSS):**

$$\text{Total SS} = \Sigma(x_{ij} - \bar{x})^2 = \Sigma x_{ij}^2 - \frac{(\Sigma x_{ij})^2}{n}$$

This is the familiar numerator in the formula for the sample variance for the entire set of $n = n_1 + n_2 + \cdots + n_k$ measurements. The second part of the calculational formula is sometimes called the **correction for the mean (CM).** If we let G represent the *grand total* of all n observations, then

$$\text{CM} = \frac{(\Sigma x_{ij})^2}{n} = \frac{G^2}{n}$$

This Total SS is partitioned into two components. The first component, called the **sum of squares for treatments (SST),** measures the variation among the k sample means:

$$\text{SST} = \Sigma n_i(\bar{x}_i - \bar{x})^2 = \Sigma \frac{T_i^2}{n_i} - \text{CM}$$

where T_i is the total of the observations for treatment i. The second component, called the **sum of squares for error (SSE)**, is used to measure the pooled variation within the k samples:

$$\text{SSE} = (n_1 - 1)s_1^2 + (n_2 - 1)s_2^2 + \cdots + (n_k - 1)s_k^2$$

This formula is a direct extension of the numerator in the formula for the pooled estimate of σ^2 from Chapter 10. We can show algebraically that, in the analysis of variance,

Total SS = SST + SSE

Therefore, you need to calculate only two of the three sums of squares—Total SS, SST, and SSE—and the third can be found by subtraction.

Each of the sources of variation, when divided by its appropriate **degrees of freedom**, provides an estimate of the variation in the experiment. Since Total SS involves n squared observations, its degrees of freedom are $df = (n - 1)$. Similarly, the sum of squares for treatments involves k squared observations, and its degrees of freedom are $df = (k - 1)$. Finally, the sum of squares for error, a direct extension of the pooled estimate in Chapter 10, has

$$df = (n_1 - 1) + (n_2 - 1) + \cdots + (n_k - 1) = n - k$$

Notice that the degrees of freedom for treatments and error are additive—that is,

$$df(\text{total}) = df(\text{treatments}) + df(\text{error})$$

These two sources of variation and their respective degrees of freedom are combined to form the **mean squares** as MS = SS/df. The total variation in the experiment is then displayed in an **analysis of variance** (or **ANOVA**) **table**.

ANOVA TABLE FOR k INDEPENDENT RANDOM SAMPLES: COMPLETELY RANDOMIZED DESIGN

Source	df	SS	MS	F
Treatments	$k - 1$	SST	MST = SST/$(k - 1)$	MST/MSE
Error	$n - k$	SSE	MSE = SSE/$(n - k)$	
Total	$n - 1$	Total SS		

where

$$\text{Total SS} = \Sigma x_{ij}^2 - \text{CM}$$
$$= (\text{Sum of squares of all } x\text{-values}) - \text{CM}$$

with

$$\text{CM} = \frac{(\Sigma x_{ij})^2}{n} = \frac{G^2}{n}$$

$$\text{SST} = \Sigma \frac{T_i^2}{n_i} - \text{CM} \qquad \text{MST} = \frac{\text{SST}}{k - 1}$$

$$\text{SSE} = \text{Total SS} - \text{SST} \qquad \text{MSE} = \frac{\text{SSE}}{n - k}$$

and

G = Grand total of all n observations
T_i = Total of all observations in sample i
n_i = Number of observations in sample i
$n = n_1 + n_2 + \cdots + n_k$

EXAMPLE 11.4 In an experiment to determine the effect of nutrition on the attention spans of elementary school students, a group of 15 students were randomly assigned to each of three meal plans: no breakfast, light breakfast, and full breakfast. Their attention spans (in minutes) were recorded during a morning reading period and are shown in Table 11.1. Construct the analysis of variance table for this experiment.

TABLE 11.1 ● **Attention Spans of Students After Three Meal Plans**

No Breakfast	Light Breakfast	Full Breakfast
8	14	10
7	16	12
9	12	16
13	17	15
10	11	12
$T_1 = 47$	$T_2 = 70$	$T_3 = 65$

Solution To use the calculational formulas, you need the $k = 3$ treatment totals together with $n_1 = n_2 = n_3 = 5$, $n = 15$, and $\Sigma x_{ij} = 182$. Then

$$CM = \frac{(182)^2}{15} = 2208.2667$$

Total SS $= (8^2 + 7^2 + \cdots + 12^2) - CM = 2338 - 2208.2667 = 129.7333$
with $(n - 1) = (15 - 1) = 14$ degrees of freedom,

$$SST = \frac{47^2 + 70^2 + 65^2}{5} - CM = 2266.8 - 2208.2667 = 58.5333$$

with $(k - 1) = (3 - 1) = 2$ degrees of freedom, and by subtraction,

SSE $=$ Total SS $-$ SST $= 129.7333 - 58.5333 = 71.2$

with $(n - k) = (15 - 3) = 12$ degrees of freedom. These three sources of variation, their degrees of freedom, sums of squares, and mean squares are shown in the shaded area of the ANOVA tables generated by both *MS Excel* and *MINITAB* and given in Figure 11.3. You will find instructions for generating this output in the "Technology Today" section at the end of this chapter.

FIGURE 11.3(a)
MINITAB output for
Example 11.4

One-way ANOVA: Span versus Meal

```
Source   DF      SS     MS      F      P
Meal      2   58.53  29.27   4.93  0.027
Error    12   71.20   5.93
Total    14  129.73

S = 2.436   R-Sq = 45.12%   R-Sq(adj) = 35.97%
                            Individual 95% CIs For Mean
                            Based on Pooled StDev
Level   N    Mean   StDev   --+---------+---------+---------+-------
1       5   9.400   2.302   (---------*--------)
2       5  14.000   2.550                       (--------*--------)
3       5  13.000   2.449                 (--------*--------)
                            --+---------+---------+---------+-------
                            7.5      10.0      12.5      15.0

Pooled StDev = 2.436
```

FIGURE 11.3(b)

MS Excel output for Example 11.4

Anova: Single Factor

SUMMARY

Groups	Count	Sum	Average	Variance
None	5	47	9.4	5.3
Light	5	70	14	6.5
Full	5	65	13	6

ANOVA

Source of Variation	SS	df	MS	F	P-value	F crit
Between Groups	58.533	2	29.267	4.933	0.027	3.885
Within Groups	71.2	12	5.933			
Total	129.733	14				

The computer outputs give some additional information about the variation in the experiment. The lower section in *MINITAB* and the upper section in *MS Excel* show the means and standard deviations (or variances) for the three meal plans. More important, you can see in the upper section in *MINITAB* and the lower section in *MS Excel* two columns marked "F" and "P" ("*F-*" and "*P*-value" in *Excel*). We can use these values to test a hypothesis concerning the equality of the three treatment means.

Testing the Equality of the Treatment Means

The *mean squares* in the analysis of variance table can be used to test the null hypothesis

NEED a tip? **NEED A TIP?**

MS = SS/*df*

$$H_0 : \mu_1 = \mu_2 = \cdots = \mu_k$$

versus the alternative hypothesis

H_a : At least one of the means is different from the others

using the following theoretical argument:

- Remember that σ^2 is the common variance for all k populations. The quantity

$$\text{MSE} = \frac{\text{SSE}}{n - k}$$

is a pooled estimate of σ^2, a weighted average of all k sample variances, whether or not H_0 is true.

- If H_0 is true, then the variation in the sample means, measured by MST = [SST/($k - 1$)], also provides an unbiased estimate of σ^2. However, if H_0 is false and the population means are different, then MST—which measures the variation in the sample means—will be unusually *large*, as shown in Figure 11.4.

FIGURE 11.4

Sample means drawn from identical versus different populations

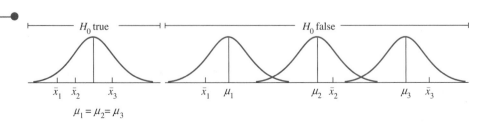

- The test statistic

$$F = \frac{\text{MST}}{\text{MSE}}$$

tends to be larger than usual if H_0 is false. Hence, you can reject H_0 for large values of F, using a *right-tailed* statistical test. When H_0 is true, this test statistic has an F distribution with $df_1 = (k - 1)$ and $df_2 = (n - k)$ degrees of freedom, and *right-tailed* critical values of the F distribution (from Table 6 in Appendix I) or computer-generated p-values can be used to draw statistical conclusions about the equality of the population means.

F-TEST FOR COMPARING k POPULATION MEANS

1. Null hypothesis: $H_0 : \mu_1 = \mu_2 = \cdots = \mu_k$
2. Alternative hypothesis: H_a : One or more pairs of population means differ
3. Test statistic: $F = \text{MST}/\text{MSE}$, where F is based on $df_1 = (k - 1)$ and $df_2 = (n - k)$
4. Rejection region: Reject H_0 if $F > F_\alpha$, where F_α lies in the upper tail of the F distribution (with $df_1 = k - 1$ and $df_2 = n - k$) or if the p-value $< \alpha$.

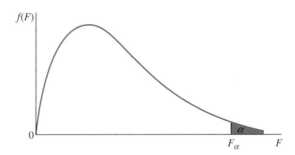

Assumptions:

- The samples are randomly and independently selected from their respective populations.
- The populations are normally distributed with means $\mu_1, \mu_2, \ldots, \mu_k$ and equal variances, $\sigma_1^2 = \sigma_2^2 = \cdots = \sigma_k^2 = \sigma^2$.

EXAMPLE 11.5

Do the data in Example 11.4 provide sufficient evidence to indicate a difference in the average attention spans depending on the type of breakfast eaten by the student?

Solution To test $H_0 : \mu_1 = \mu_2 = \mu_3$ versus the alternative hypothesis that the average attention span is different for at least one of the three treatments, you use the analysis of variance F statistic, calculated as

$$F = \frac{\text{MST}}{\text{MSE}} = \frac{29.2667}{5.9333} = 4.93$$

and shown in the columns marked "F" in Figure 11.3(a) and "*F*" in Figure 11.3(b). It will not surprise you to know that the value in the column marked "P" in Figure 11.3(a) and "*P*-value" in Figure 11.3(b) is the exact *p*-value for this statistical test.

The test statistic MST/MSE calculated above has an *F* distribution with $df_1 = 2$ and $df_2 = 12$ degrees of freedom. Using the critical value approach with $\alpha = .05$, you can reject H_0 if $F > F_{.05} = 3.89$ from Table 6 in Appendix I (see Figure 11.5). Since the observed value, $F = 4.93$, exceeds the critical value, you reject H_0. There is sufficient evidence to indicate that at least one of the three average attention spans is different from at least one of the others.

FIGURE 11.5

Rejection region for
Example 11.5

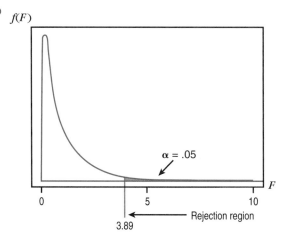

NEED A TIP?

Computer printouts give
the exact *p*-value—use
the *p*-value to make your
decision.

You could have reached this same conclusion using the exact *p*-value, .027, given in Figure 11.3. Since the *p*-value is less than $\alpha = .05$, the results are statistically significant at the 5% level. You still conclude that at least one of the three average attention spans is different from at least one of the others.

Estimating Differences in the Treatment Means

The next obvious question you might ask involves the nature of the differences in the population means. Which means are different from the others? How can you estimate the difference, or possibly the individual means for each of the three treatments? In Section 11.6, we will present a procedure that you can use to compare all possible pairs of treatment means simultaneously. However, if you have a special interest in a particular mean or pair of means, you can construct confidence intervals using the small-sample procedures of Chapter 10, based on the Student's *t* distribution. For a single population mean, μ_i, the confidence interval is

$$\bar{x}_i \pm t_{\alpha/2}\left(\frac{s}{\sqrt{n_i}}\right)$$

where $\bar{x}_i$ is the sample mean for the *i*th treatment. Similarly, for a comparison of two population means—say, μ_i and μ_j—the confidence interval is

$$(\bar{x}_i - \bar{x}_j) \pm t_{\alpha/2}\sqrt{s^2\left(\frac{1}{n_i} + \frac{1}{n_j}\right)}$$

Before you can use these confidence intervals, however, two questions remain:

- How do you calculate s or s^2, the best estimate of the common variance σ^2?
- How many degrees of freedom are used for the critical value of t?

To answer these questions, remember that in an analysis of variance, the mean square for error, MSE, always provides an unbiased estimator of σ^2 and uses information from the entire set of measurements. Hence, it is the best available estimator of σ^2, regardless of what test or estimation procedure you are using. You should *always* use

$$s^2 = \text{MSE} \qquad \text{with } df = (n - k)$$

to estimate σ^2! You can find the positive square root of this estimator, $s = \sqrt{\text{MSE}}$, on the last line of Figure 11.3(a) labeled "Pooled StDev."

NEED A TIP?

Degrees of freedom for confidence intervals are the *df* for **error**.

COMPLETELY RANDOMIZED DESIGN: $(1 - \alpha)100\%$ CONFIDENCE INTERVALS FOR A SINGLE TREATMENT MEAN AND THE DIFFERENCE BETWEEN TWO TREATMENT MEANS

Single treatment mean:

$$\bar{x}_i \pm t_{\alpha/2}\left(\frac{s}{\sqrt{n_i}}\right)$$

Difference between two treatment means:

$$(\bar{x}_i - \bar{x}_j) \pm t_{\alpha/2}\sqrt{s^2\left(\frac{1}{n_i} + \frac{1}{n_j}\right)}$$

with

$$s = \sqrt{s^2} = \sqrt{\text{MSE}} = \sqrt{\frac{\text{SSE}}{n - k}}$$

where $n = n_1 + n_2 + \cdots + n_k$ and $t_{\alpha/2}$ is based on $(n - k)$ *df*.

EXAMPLE 11.6

The researcher in Example 11.4 believes that students who have no breakfast will have significantly shorter attention spans but that there may be no difference between those who eat a light or a full breakfast. Find a 95% confidence interval for the average attention span for students who eat no breakfast, as well as a 95% confidence interval for the difference in the average attention spans for light versus full breakfast eaters.

Solution For $s^2 = \text{MSE} = 5.9333$ so that $s = \sqrt{5.9333} = 2.436$ with $df = (n - k) = 12$, you can calculate the two confidence intervals:

- For no breakfast:

$$\bar{x}_1 \pm t_{\alpha/2}\left(\frac{s}{\sqrt{n_1}}\right)$$

$$9.4 \pm 2.179\left(\frac{2.436}{\sqrt{5}}\right)$$

$$9.4 \pm 2.37$$

or between 7.03 and 11.77 minutes.

• For light versus full breakfast:

$$(\bar{x}_2 - \bar{x}_3) \pm t_{\alpha/2}\sqrt{s^2\left(\frac{1}{n_2} + \frac{1}{n_3}\right)}$$

$$(14 - 13) \pm 2.179\sqrt{5.9333\left(\frac{1}{5} + \frac{1}{5}\right)}$$

$$1 \pm 3.36$$

a difference of between -2.36 and 4.36 minutes.

You can see that the second confidence interval does not indicate a difference in the average attention spans for students who ate light versus full breakfasts, as the researcher suspected. If the researcher, because of prior beliefs, wishes to test the other two possible pairs of means—none versus light breakfast, and none versus full breakfast—the methods given in Section 11.6 should be used for testing all three pairs.

Some computer programs have graphics options that provide a powerful visual description of data and the k treatment means. One such option in the *MINITAB* program is shown in Figure 11.6. The treatment means are indicated by the symbol ⊕ and are connected with straight lines. Notice that the "no breakfast" mean appears to be somewhat different from the other two means, as the researcher suspected, although there is a bit of overlap in the box plots. In the next section, we present a formal procedure for testing the significance of the differences between all pairs of treatment means.

FIGURE 11.6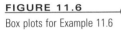

Box plots for Example 11.6

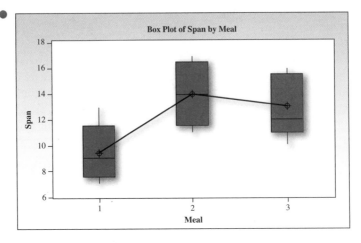

NEED TO KNOW...

How to Determine Whether My Calculations Are Accurate

The following suggestions apply to all the analyses of variance in this chapter:

1. When calculating sums of squares, be certain to carry at least six significant figures before performing subtractions.

2. Remember, sums of squares can never be negative. If you obtain a negative sum of squares, you have made a mistake in arithmetic.

3. Always check your analysis of variance table to make certain that the degrees of freedom sum to the total degrees of freedom $(n - 1)$ and that the sums of squares sum to Total SS.

11.5 EXERCISES

BASIC TECHNIQUES

11.1 Suppose you wish to compare the means of six populations based on independent random samples, each of which contains 10 observations. Insert, in an ANOVA table, the sources of variation and their respective degrees of freedom.

11.2 The values of Total SS and SSE for the experiment in Exercise 11.1 are Total SS = 21.4 and SSE = 16.2.

a. Complete the ANOVA table for Exercise 11.1.

b. How many degrees of freedom are associated with the F statistic for testing $H_0 : \mu_1 = \mu_2 = \cdots = \mu_6$?

c. Give the rejection region for the test in part b for $\alpha = .05$.

d. Do the data provide sufficient evidence to indicate differences among the population means?

e. Estimate the p-value for the test. Does this value confirm your conclusions in part d?

11.3 The sample means corresponding to populations 1 and 2 in Exercise 11.1 are $\bar{x}_1 = 3.07$ and $\bar{x}_2 = 2.52$.

a. Find a 95% confidence interval for μ_1.

b. Find a 95% confidence interval for the difference $(\mu_1 - \mu_2)$.

11.4 Suppose you wish to compare the means of four populations based on independent random samples, each of which contains six observations. Insert, in an ANOVA table, the sources of variation and their respective degrees of freedom.

11.5 The values of Total SS and SST for the experiment in Exercise 11.4 are Total SS = 473.2 and SST = 339.8.

a. Complete the ANOVA table for Exercise 11.4.

b. How many degrees of freedom are associated with the F statistic for testing $H_0 : \mu_1 = \mu_2 = \mu_3 = \mu_4$?

c. Give the rejection region for the test in part b for $\alpha = .05$.

d. Do the data provide sufficient evidence to indicate differences among the population means?

e. Approximate the p-value for the test. Does this confirm your conclusions in part d?

11.6 The sample means corresponding to populations 1 and 2 in Exercise 11.4 are $\bar{x}_1 = 88.0$ and $\bar{x}_2 = 83.9$.

a. Find a 90% confidence interval for μ_1.

b. Find a 90% confidence interval for the difference $(\mu_1 - \mu_2)$.

 11.7 These data are observations collected EX1107 using a completely randomized design:

Sample 1	Sample 2	Sample 3
3	4	2
2	3	0
4	5	2
3	2	1
2	5	

a. Calculate CM and Total SS.

b. Calculate SST and MST.

c. Calculate SSE and MSE.

d. Construct an ANOVA table for the data.

e. State the null and alternative hypotheses for an analysis of variance F-test.

f. Use the p-value approach to determine whether there is a difference in the three population means.

11.8 Refer to Exercise 11.7 and data set EX1107. Do the data provide sufficient evidence to indicate a difference between μ_2 and μ_3? Test using the t-test of Section 10.4 with $\alpha = .05$.

11.9 Refer to Exercise 11.7 and data set EX1107.

a. Find a 90% confidence interval for μ_1.

b. Find a 90% confidence interval for the difference $(\mu_1 - \mu_3)$.

APPLICATIONS

11.10 Reducing Hostility A clinical psychologist wished to compare three methods for reducing hostility levels in university students using a certain psychological test (HLT). High scores on this test were taken to indicate great hostility, and 11 students who got high and nearly equal scores were used in the experiment. Five were selected at random from among the 11 students and treated by method A, three were taken at random from the remaining six students and treated by method B, and the other three students were treated by method C. All treatments continued throughout a semester, when the HLT test was given again. The results are shown in the table.

Method	Scores on the HLT Test				
A	73	83	76	68	80
B	54	74	71		
C	79	95	87		

a. Perform an analysis of variance for this experiment.

b. Do the data provide sufficient evidence to indicate a difference in mean student scores after treatment for the three methods?

11.11 Hostility, continued Refer to Exercise 11.10. Let μ_A and μ_B, respectively, denote the mean scores at the end of the semester for the populations of extremely hostile students who were treated throughout that semester by method A and method B.

a. Find a 95% confidence interval for μ_A.

b. Find a 95% confidence interval for μ_B.

c. Find a 95% confidence interval for $(\mu_A - \mu_B)$.

d. Is it correct to claim that the confidence intervals found in parts a, b, and c are jointly valid?

11.12 Assembling Electronic Equipment An experiment was conducted to compare the effectiveness of three training programs, A, B, and C, in training assemblers of a piece of electronic equipment. Fifteen employees were randomly assigned, five each, to the three programs. After completion of the program, each person was required to assemble four pieces of the equipment, and the average léngth of time required to complete the assembly was recorded. Several of the employees resigned during the course of the program; the remainder were evaluated, producing the data shown in the accompanying table. Use the *Excel* printout to answer the questions.

Training Program	Average Assembly Time (min)				
A	59	64	57	62	
B	52	58	54		
C	58	65	71	63	64

a. Do the data provide sufficient evidence to indicate a difference in mean assembly times for people trained by the three programs? Give the p-value for the test and interpret its value.

b. Find a 99% confidence interval for the difference in mean assembly times between persons trained by programs A and B.

c. Find a 99% confidence interval for the mean assembly times for persons trained by program A.

d. Do you think the data will satisfy (approximately) the assumption that they have been selected from normal populations? Why?

MS Excel output for Exercise 11.12

Anova: Single Factor

SUMMARY

Groups	Count	Sum	Average	Variance		
A	4	242	60.5	9.667		
B	3	164	54.667	9.333		
C	5	321	64.2	21.7		

ANOVA

Source of Variation	SS	df	MS	F	P-value	F crit
Between Groups	170.45	2	85.225	5.704	0.0251	4.256
Within Groups	134.467	9	14.941			
Total	304.917	11				

11.13 Swampy Sites An ecological study was conducted to compare the rates of growth of vegetation at four swampy undeveloped sites and to determine the cause of any differences that might be observed. Part of the study involved measuring the leaf lengths of a particular plant species on a preselected date in May. Six plants were randomly selected at each of the four sites to be used in the comparison. The data in the table are the mean leaf length per plant (in centimeters) for a random sample of 10 leaves per plant.

The *MINITAB* analysis of variance computer printout for these data is also provided.

Location	Mean Leaf Length (cm)					
1	5.7	6.3	6.1	6.0	5.8	6.2
2	6.2	5.3	5.7	6.0	5.2	5.5
3	5.4	5.0	6.0	5.6	4.9	5.2
4	3.7	3.2	3.9	4.0	3.5	3.6

MINITAB output for Exercise 11.13

```
One-way ANOVA: Length versus Location

Source      DF       SS        MS       F        P
Location     3    19.740    6.580    57.38    0.000
Error       20     2.293    0.115
Total       23    22.033
S = 0.3386    R-Sq = 89.59%    R-Sq(adj) = 88.03%
                                    Individual 95% CIs For Mean
                                    Based on Pooled StDev
Level   N     Mean     StDev    --------+---------+---------+---------+-
1       6   6.0167    0.2317                             (--*---)
2       6   5.6500    0.3937                     (---*--)
3       6   5.3500    0.4087                  (---*--)
4       6   3.6500    0.2881    (---*--)
                               --------+---------+---------+---------+-
Pooled StDev =   0.3386            4.00     4.80     5.60     6.40
```

a. You will recall that the test and estimation procedures for an analysis of variance require that the observations be selected from normally distributed (at least, roughly so) populations. Why might you feel reasonably confident that your data satisfy this assumption?

b. Do the data provide sufficient evidence to indicate a difference in mean leaf length among the four locations? What is the p-value for the test?

c. Suppose, prior to seeing the data, you decided to compare the mean leaf lengths of locations 1 and 4. Test the null hypothesis $\mu_1 = \mu_4$ against the alternative $\mu_1 \neq \mu_4$.

d. Refer to part c. Construct a 99% confidence interval for $(\mu_1 - \mu_4)$.

e. Rather than use an analysis of variance F-test, it would seem simpler to examine one's data, select the two locations that have the smallest and largest sample mean lengths, and then compare these two means using a Student's t-test. If there is evidence to indicate a difference in these means, there is clearly evidence of a difference among the four. (If you were to use this logic, there would be no need for the analysis of variance F-test.) Explain why this procedure is invalid.

11.14 Dissolved O_2 Content Water samples EX1114 were taken at four different locations in a river to determine whether the quantity of dissolved oxygen, a measure of water pollution, varied from one location to another. Locations 1 and 2 were selected above an industrial plant, one near the shore and the other in

midstream; location 3 was adjacent to the industrial water discharge for the plant; and location 4 was slightly downriver in midstream. Five water specimens were randomly selected at each location, but one specimen, corresponding to location 4, was lost in the laboratory. The data and an *MS Excel* analysis of variance computer printout are provided here (the greater the pollution, the lower the dissolved oxygen readings).

Location	Mean Dissolved Oxygen Content				
1	5.9	6.1	6.3	6.1	6.0
2	6.3	6.6	6.4	6.4	6.5
3	4.8	4.3	5.0	4.7	5.1
4	6.0	6.2	6.1	5.8	

MS Excel output for Exercise 11.14

Anova: Single Factor

SUMMARY

Groups	Count	Sum	Average	Variance		
1	5	30.4	6.08	0.022		
2	5	32.2	6.44	0.013		
3	5	23.9	4.78	0.097		
4	4	24.1	6.025	0.0292		

ANOVA

Source of Variation	SS	df	MS	F	P-value	F crit
Between Groups	7.8361	3	2.6120	63.656	9E-09	3.287
Within Groups	0.6155	15	0.0410			
Total	8.4516	18				

a. Do the data provide sufficient evidence to indicate a difference in the mean dissolved oxygen contents for the four locations?

b. Compare the mean dissolved oxygen content in midstream above the plant with the mean content adjacent to the plant (location 2 versus location 3). Use a 95% confidence interval.

11.15 Calcium The calcium content of a EX1115 powdered mineral substance was analyzed five times by each of three methods, with similar standard deviations:

Method	Percent Calcium				
1	.0279	.0276	.0270	.0275	.0281
2	.0268	.0274	.0267	.0263	.0267
3	.0280	.0279	.0282	.0278	.0283

Use an appropriate test to compare the three methods of measurement. Comment on the validity of any assumptions you need to make.

11.16 Tuna Fish In Exercise 10.6, we EX1116 reported the estimated average prices for a 6-ounce can or a 7.06-ounce pouch of tuna fish, based on prices paid nationally for a variety of different brands of tuna.[1]

Light Tuna in Water		White Tuna in Oil	White Tuna in Water	Light Tuna in Oil	
.99	.53	1.27	1.49	2.56	.62
1.92	1.41	1.22	1.29	1.92	.66
1.23	1.12	1.19	1.27	1.30	.62
.85	.63	1.22	1.35	1.79	.65
.65	.67		1.29	1.23	.60
.69	.60		1.00		.67
.60	.66		1.27		
			1.28		

Source: From "Pricing of Tuna" Copyright 2001 by Consumers Union of U.S., Inc., Yonkers, NY 10703-1057, a nonprofit organization. Reprinted with permission from the June 2001 issue of *Consumer Reports*® for educational purposes only. No commercial use or reproduction permitted. www.ConsumerReports.org®.

a. Use an analysis of variance for a completely randomized design to determine if there are significant differences in the prices of tuna packaged in these four different ways. Can you reject the hypothesis of no difference in average price for these packages at the $\alpha = .05$ level of significance? At the $\alpha = .01$ level of significance?

b. Find a 95% confidence interval estimate of the difference in price between light tuna in water and light tuna in oil. Does there appear to be a significant difference in the price of these two kinds of packaged tuna?

c. Find a 95% confidence interval estimate of the difference in price between white tuna in water and white tuna in oil. Does there appear to be a significant difference in the price of these two kinds of packaged tuna?

d. What other confidence intervals might be of interest to the researcher who conducted this experiment?

Data set **11.17 The Cost of Lumber** A national home EX1117 builder wants to compare the prices per 1000 board feet of standard or better grade green Douglas fir framing lumber. He randomly selects five suppliers in each of the four states where the builder is planning to begin construction. The prices are given in the table.

State			
1 ($)	2 ($)	3 ($)	4 ($)
261	236	250	265
255	240	245	270
258	225	255	258
267	233	248	275
270	240	260	275

a. What type of experimental design has been used?

b. Construct the analysis of variance table for this data.

c. Do the data provide sufficient evidence to indicate that the average price per 1000 board feet of Douglas fir differs among the four states? Test using $\alpha = .05$.

Data set **11.18 Good at Math?** Twenty third graders EX1118 were randomly separated into four equal groups, and each group was taught a mathematical concept using a different teaching method. At the end of the teaching period, progress was measured by a unit test. The scores are shown below (one child in group 3 was absent on the day that the test was administered).

Group			
1	2	3	4
112	111	140	101
92	129	121	116
124	102	130	105
89	136	106	126
97	99		119

a. What type of design has been used in this experiment?

b. Construct an ANOVA table for the experiment.

c. Do the data present sufficient evidence to indicate a difference in the average scores for the four teaching methods? Test using $\alpha = .05$.

RANKING POPULATION MEANS

11.6

Many experiments are exploratory in nature. You have no preconceived notions about the results and have not decided (before conducting the experiment) to make specific treatment comparisons. Rather, you want to rank the treatment means, determine which means differ, and identify sets of means for which no evidence of difference exists.

One option might be to order the sample means from the smallest to the largest and then to conduct *t*-tests for adjacent means in the ordering. If two means differ by more than

$$t_{\alpha/2}\sqrt{s^2\left(\frac{1}{n_1}+\frac{1}{n_2}\right)}$$

you conclude that the pair of population means differ. The problem with this procedure is that the probability of making a Type I error—that is, concluding that two means differ when, in fact, they are equal—is α for each test. If you compare a large number of pairs of means, the probability of detecting at least one difference in means, when in fact none exists, is quite large.

A simple way to avoid the high risk of declaring differences when they do not exist is to use the **studentized range,** the difference between the smallest and the largest in a set of k sample means, as the yardstick for determining whether there is a difference in a pair of population means. This method, often called **Tukey's method for paired comparisons,** makes the probability of declaring that a difference exists between at least one pair in a set of k treatment means, when no difference exists, equal to α.

Tukey's method for making paired comparisons is based on the usual analysis of variance assumptions. **In addition, it assumes that the sample means are independent and based on samples of equal size.** The yardstick that determines whether a difference exists between a pair of treatment means is the quantity ω (Greek lowercase omega), which is presented next.

YARDSTICK FOR MAKING PAIRED COMPARISONS

$$\omega = q_\alpha(k, df)\left(\frac{s}{\sqrt{n_t}}\right)$$

where

k = Number of treatments
s^2 = MSE = Estimator of the common variance σ^2 and $s = \sqrt{s^2}$
df = Number of degrees of freedom for s^2
n_t = Common sample size—that is, the number of observations in each of the k treatment means
$q_\alpha(k, df)$ = Tabulated value from Tables 11(a) and 11(b) in Appendix I, for $\alpha = .05$ and .01, respectively, and for various combinations of k and df

Rule: Two population means are judged to differ if the corresponding sample means differ by ω or more.

Tables 11(a) and 11(b) in Appendix I list the values of $q_\alpha(k, df)$ for $\alpha = .05$ and .01, respectively. To illustrate the use of the tables, refer to the portion of Table 11.2 reproduced in Table 11.2. Suppose you want to make pairwise comparisons of $k = 5$ means with $\alpha = .05$ for an analysis of variance, where s^2 possesses 9 df. The tabulated value for $k = 5$, $df = 9$, and $\alpha = .05$, shaded in Table 11.2, is $q_{.05}(5, 9) = 4.76$.

TABLE 11.2 ●
A Partial Reproduction of Table 11(a) in Appendix I;
Upper 5% Points

df	2	3	4	5	6	7	8	9	10	11	12
1	17.97	26.98	32.82	37.08	40.41	43.12	45.40	47.36	49.07	50.59	51.96
2	6.08	8.33	9.80	10.88	11.74	12.44	13.03	13.54	13.99	14.39	14.75
3	4.50	5.91	6.82	7.50	8.04	8.48	8.85	9.18	9.46	9.72	9.95
4	3.93	5.04	5.76	6.29	6.71	7.05	7.35	7.60	7.83	8.03	8.21
5	3.64	4.60	5.22	5.67	6.03	6.33	6.58	6.80	6.99	7.17	7.32
6	3.46	4.34	4.90	5.30	5.63	5.90	6.12	6.32	6.49	6.65	6.79
7	3.34	4.16	4.68	5.06	5.36	5.61	5.82	6.00	6.16	6.30	6.43
8	3.26	4.04	4.53	4.89	5.17	5.40	5.60	5.77	5.92	6.05	6.18
9	3.20	3.95	4.41	4.76	5.02	5.24	5.43	5.59	5.74	5.87	5.98
10	3.15	3.88	4.33	4.65	4.91	5.12	5.30	5.46	5.60	5.72	5.83
11	3.11	3.82	4.26	4.57	4.82	5.03	5.20	5.35	5.49	5.61	5.71
12	3.08	3.77	4.20	4.51	4.75	4.95	5.12	5.27	5.39	5.51	5.61

EXAMPLE ● 11.7

Refer to Example 11.4, in which you compared the average attention spans for students given three different "meal" treatments in the morning: no breakfast, a light breakfast, or a full breakfast. The ANOVA F-test in Example 11.5 indicated a significant difference in the population means. Use Tukey's method for paired comparisons to determine which of the three population means differ from the others.

Solution   For this example, there are $k = 3$ treatment means, with $s = \sqrt{\text{MSE}} = 2.436$. Tukey's method can be used, with each of the three samples containing $n_t = 5$ measurements and $(n - k) = 12$ degrees of freedom. Consult Table 11 in Appendix I to find $q_{.05}(k, df) = q_{.05}(3, 12) = 3.77$ and calculate the "yardstick" as

$$\omega = q_{.05}(3, 12)\left(\frac{s}{\sqrt{n_t}}\right) = 3.77\left(\frac{2.436}{\sqrt{5}}\right) = 4.11$$

The three treatment means are arranged in order from the smallest, 9.4, to the largest, 14.0, in Figure 11.7. The next step is to check the difference between every pair of means. The only difference that exceeds $\omega = 4.11$ is the difference between no breakfast and a light breakfast. These two treatments are thus declared significantly different. You cannot declare a difference between the other two pairs of treatments. To indicate this fact visually, Figure 11.7 shows a line under those pairs of means that are not significantly different.

FIGURE 11.7 ●

Ranked means for
Example 11.7

None	Full	Light
9.4	13.0	14.0

The results here may seem confusing. However, it usually helps to think of ranking the means and interpreting nonsignificant differences as our inability to distinctly rank those means underlined by the same line. For this example, the light breakfast definitely ranked higher than no breakfast, but the full breakfast could not be ranked higher than no breakfast, or lower than the light breakfast. The probability that we make at least one error among the three comparisons is at most $\alpha = .05$.

NEED A TIP?

If zero is not in the interval, there is evidence of a difference between the two methods.

Many computer programs provide an option to perform **paired comparisons,** including Tukey's method, although *MS Excel* does not. The *MINITAB* output in Figure 11.8 shows its form of Tukey's test, which differs slightly from the method we have presented. The three intervals that you see in the printout marked "Lower" and "Upper" represent the difference in the two sample means plus or minus the yardstick ω. If the interval contains the value 0, the two means are judged to be not significantly different. You can see that only means 1 and 2 (none versus light) show a significant difference.

FIGURE 11.8

MINITAB output for Example 11.7

```
Tukey's 95% Simultaneous Confidence Intervals
All Pairwise Comparisons among Levels of Meal
Individual confidence level = 97.94%

Meal = 1 subtracted from:

Meal  Lower   Center  Upper   -----+---------+---------+---------+----
2     0.493   4.600   8.707                      (-----------*-----------)
3    -0.507   3.600   7.707                   (----------*-----------)
                              -----+---------+---------+---------+----
                               -3.5      0.0       3.5       7.0
Meal = 2 subtracted from:

Meal  Lower   Center  Upper   -----+---------+---------+---------+----
3    -5.107  -1.000   3.107   (-----------*-----------)
                              -----+---------+---------+---------+----
                               -3.5      0.0       3.5       7.0
```

As you study two more experimental designs in the next sections of this chapter, remember that, once you have found a factor to be significant, you should use Tukey's method or another method of paired comparisons to find out exactly where the differences lie!

11.6 EXERCISES

BASIC TECHNIQUES

11.19 Suppose you wish to use Tukey's method of paired comparisons to rank a set of population means. In addition to the analysis of variance assumptions, what other property must the treatment means satisfy?

11.20 Consult Tables 11(a) and 11(b) in Appendix I and find the values of $q_\alpha(k, df)$ for these cases:

a. $\alpha = .05$, $k = 5$, $df = 7$

b. $\alpha = .05$, $k = 3$, $df = 10$

c. $\alpha = .01$, $k = 4$, $df = 8$

d. $\alpha = .01$, $k = 7$, $df = 5$

11.21 If the sample size for each treatment is n_t and if s^2 is based on 12 *df*, find ω in these cases:

a. $\alpha = .05$, $k = 4$, $n_t = 5$

b. $\alpha = .01$, $k = 6$, $n_t = 8$

11.22 An independent random sampling design was used to compare the means of six treatments based on samples of four observations per treatment. The pooled estimator of σ^2 is 9.12, and the sample means follow:

$\bar{x}_1 = 101.6$ $\bar{x}_2 = 98.4$ $\bar{x}_3 = 112.3$
$\bar{x}_4 = 92.9$ $\bar{x}_5 = 104.2$ $\bar{x}_6 = 113.8$

a. Give the value of ω that you would use to make pairwise comparisons of the treatment means for $\alpha = .05$.

b. Rank the treatment means using pairwise comparisons.

APPLICATIONS

11.23 Swamp Sites, again Refer to Exercise 11.13 and data set EX1113. Rank the mean leaf growth for the four locations. Use $\alpha = .01$.

11.24 Calcium Refer to Exercise 11.15 and data set EX1115. The paired comparisons option in *MINITAB* generated the output provided here. What do these results tell you about the differences in the population means? Does this confirm your conclusions in Exercise 11.15?

MINITAB output for Exercise 11.24

```
Tukey's 95% Simultaneous Confidence Intervals
All Pairwise Comparisons among Levels of Method
Individual confidence level = 97.94%

Method = 1 subtracted from:

Method       Lower       Center       Upper
2        -0.0014377  -0.0008400  -0.0002423
3        -0.0001777   0.0004200   0.0010177
Method   --------+---------+---------+---------+
2            (-----*-----)
3                       (-----*-----)
         --------+---------+---------+---------+
             -0.0010    0.0000    0.0010    0.0020

Method = 2 subtracted from:

Method       Lower       Center       Upper
3         0.0006623   0.0012600   0.0018577
Method   --------+---------+---------+---------+
3                           (-----*-----)
         --------+---------+---------+---------+
             -0.0010    0.0000    0.0010    0.0020
```

11.25 Glucose Tolerance Physicians depend on laboratory test results when managing medical problems such as diabetes or epilepsy. In a uniformity test for glucose tolerance, three different laboratories were each sent $n_t = 5$ identical blood samples from a person who had drunk 50 milligrams (mg) of glucose dissolved in water. The laboratory results (in mg/dl) are listed here:

Lab 1	Lab 2	Lab 3
120.1	98.3	103.0
110.7	112.1	108.5
108.9	107.7	101.1
104.2	107.9	110.0
100.4	99.2	105.4

a. Do the data indicate a difference in the average readings for the three laboratories?

b. Use Tukey's method for paired comparisons to rank the three treatment means. Use $\alpha = .05$.

11.26 The Cost of Lumber, continued The analysis of variance *F*-test in Exercise 11.17 (and data set EX1117) determined that there was indeed a difference in the average cost of lumber for the four states. The following information from Exercise 11.17 is given in the table:

Sample Means	$\bar{x}_1 = 262.2$	MSE	41.25
	$\bar{x}_2 = 234.8$	Error *df*:	16
	$\bar{x}_3 = 251.6$	n_i:	5
	$\bar{x}_4 = 268.6$	*k*:	4

Use Tukey's method for paired comparisons to determine which means differ significantly from the others at the $\alpha = .01$ level.

11.27 GRE Scores The quantitative reasoning scores on the Graduate Record Examination (GRE)[2] were recorded for students admitted to three different graduate programs at a local university.

Graduate Program					
Life Sciences		Physical Sciences		Social Sciences	
630	660	660	760	440	540
640	660	640	670	330	450
470	480	720	700	670	570
600	650	690	710	570	530
580	710	530	450	590	630

a. Do these data provide sufficient evidence to indicate a difference in the mean GRE scores for applicants admitted to the three programs?

b. Find a 95% confidence interval for the difference in mean GRE scores for Life Sciences and Physical Sciences.

c. If you find a significant difference in the average GRE scores for the three programs, use Tukey's method for paired comparisons to determine which means differ significantly from the others. Use $\alpha = .05$.

11.7 THE RANDOMIZED BLOCK DESIGN: A TWO-WAY CLASSIFICATION

The *completely randomized design* introduced in Section 11.4 is a generalization of the *two independent samples* design presented in Section 10.4. It is meant to be used when the experimental units are quite similar or *homogeneous* in their makeup and when there is only one factor—the *treatment*—that might influence the response.

Any other variation in the response is due to random variation or *experimental error.* Sometimes it is clear to the researcher that the experimental units are *not homogeneous.* Experimental subjects or animals, agricultural fields, days of the week, and other experimental units often add their own variability to the response. Although the researcher is not really interested in this source of variation, but rather in some *treatment* he chooses to apply, he may be able to increase the information by isolating this source of variation using the **randomized block design**—a direct extension of the *matched pairs* or *paired-difference design* in Section 10.5.

In a randomized block design, the experimenter is interested in comparing k treatment means. The design uses *blocks* of k experimental units that are relatively similar, or *homogeneous,* with one unit within each block *randomly* assigned to each treatment. If the randomized block design involves k treatments within each of b blocks, then the total number of observations in the experiment is $n = bk$.

A production supervisor wants to compare the mean times for assembly-line operators to assemble an item using one of three methods: A, B, or C. Expecting variation in assembly times from operator to operator, the supervisor uses a randomized block design to compare the three methods. Five assembly-line operators are selected to serve as blocks, and each is assigned to assemble the item three times, once for each of the three methods. Since the sequence in which the operator uses the three methods may be important (fatigue or increasing dexterity may be factors affecting the response), each operator should be assigned a random sequencing of the three methods. For example, operator 1 might be assigned to perform method C first, followed by A and B. Operator 2 might perform method A first, then C and B, and so on.

To compare four different teaching methods, a group of students might be divided into blocks of size 4, so that the groups are most nearly *matched* according to academic achievement. To compare the average costs for three different cellular phone companies, costs might be compared at each of three usage levels: low, medium, and high. To compare the average yields for three species of fruit trees when a variation in yield is expected because of the field in which the trees are planted, a researcher uses five fields. She divides each field into three *plots* on which the three species of fruit trees are planted.

Matching or *blocking* can take place in many different ways. Comparisons of treatments are often made within blocks of time, within blocks of people, or within similar external environments. The purpose of blocking is to remove or isolate the *block-to-block* variability that might otherwise hide the effect of the treatments. You will find more examples of the use of the randomized block design in the exercises at the end of the next section.

NEED A TIP?

b = blocks
k = treatments
$n = bk$

THE ANALYSIS OF VARIANCE FOR A RANDOMIZED BLOCK DESIGN

11.8

The randomized block design identifies two factors: **treatments** and **blocks**—both of which affect the response.

Partitioning the Total Variation in the Experiment

Let x_{ij} be the response when the ith treatment ($i = 1, 2, \ldots, k$) is applied in the jth block ($j = 1, 2, \ldots, b$). The total variation in the $n = bk$ observations is

$$\text{Total SS} = \Sigma(x_{ij} - \overline{x})^2 = \Sigma x_{ij}^2 - \frac{(\Sigma x_{ij})^2}{n}$$

This is partitioned into *three* (rather than two) parts in such a way that

Total SS = SSB + SST + SSE

where

- SSB (sum of squares for blocks) measures the variation among the block means.
- SST (sum of squares for treatments) measures the variation among the treatment means.
- SSE (sum of squares for error) measures the variation of the differences among the treatment observations *within* blocks, which measures the experimental error.

The calculational formulas for the four sums of squares are similar in form to those you used for the completely randomized design in Section 11.5. Although you can simplify your work by using a computer program to calculate these sums of squares, the formulas are given next.

CALCULATING THE SUMS OF SQUARES FOR A RANDOMIZED BLOCK DESIGN, *k* TREATMENTS IN *b* BLOCKS

$$CM = \frac{G^2}{n}$$

where

$$G = \Sigma x_{ij} = \text{Total of all } n = bk \text{ observations}$$

$$\text{Total SS} = \Sigma x_{ij}^2 - CM$$

$$= (\text{Sum of squares of all } x\text{-values}) - CM$$

$$SST = \Sigma \frac{T_i^2}{b} - CM$$

$$SSB = \Sigma \frac{B_j^2}{k} - CM$$

$$SSE = \text{Total SS} - SST - SSB$$

with

$$T_i = \text{Total of all observations receiving treatment } i, \ i = 1, 2, \ldots, k$$
$$B_j = \text{Total of all observations in block } j, \ j = 1, 2, \ldots, b$$

NEED A TIP?

Total SS = SST + SSB + SSE

Each of the three **sources of variation,** when divided by the appropriate **degrees of freedom,** provides an estimate of the variation in the experiment. Since Total SS involves $n = bk$ squared observations, its degrees of freedom are $df = (n - 1)$. Similarly, SST involves k squared totals, and its degrees of freedom are $df = (k - 1)$, while SSB involves b squared totals and has $(b - 1)$ degrees of freedom. Finally, since the degrees of freedom are additive, the remaining degrees of freedom associated with SSE can be shown algebraically to be $df = (b - 1)(k - 1)$.

These three sources of variation and their respective degrees of freedom are combined to form the **mean squares** as MS = SS/*df*, and the total variation in the experiment is then displayed in an **analysis of variance** (or **ANOVA**) **table** as shown here:

ANOVA TABLE FOR A RANDOMIZED BLOCK DESIGN, k TREATMENTS AND b BLOCKS

Source	df	SS	MS	F
Treatments	$k - 1$	SST	MST = SST/$(k - 1)$	MST/MSE
Blocks	$b - 1$	SSB	MSB = SSB/$(b - 1)$	MSB/MSE
Error	$(b - 1)(k - 1)$	SSE	MSE = SSE/$(b - 1)(k - 1)$	
Total	$n - 1 = bk - 1$			

EXAMPLE **11.8**

The cell phone industry is involved in a fierce battle for customers, with each company devising its own complex pricing plan to lure customers. Since the cost of a cell phone minute varies drastically depending on the number of minutes per month used by the customer, a consumer watchdog group decided to compare the average costs for four cellular phone companies using three different usage levels as blocks. The monthly costs (in dollars) computed by the cell phone companies for peak-time callers at low (20 minutes per month), middle (150 minutes per month), and high (1000 minutes per month) usage levels are given in Table 11.3. Construct the analysis of variance table for this experiment.

TABLE 11.3 ● **Monthly Phone Costs of Four Companies at Three Usage Levels**

	Company				
Usage Level	A	B	C	D	Totals
Low	27	24	31	23	$B_1 = 105$
Middle	68	76	65	67	$B_2 = 276$
High	308	326	312	300	$B_3 = 1246$
Totals	$T_1 = 403$	$T_2 = 426$	$T_3 = 408$	$T_4 = 390$	$G = 1627$

Solution The experiment is designed as a *randomized block design* with $b = 3$ usage levels (blocks) and $k = 4$ companies (treatments), so there are $n = bk = 12$ observations and $G = 1627$. Then

$$CM = \frac{G^2}{n} = \frac{1627^2}{12} = 220{,}594.0833$$

$$\text{Total SS} = (27^2 + 24^2 + \cdots + 300^2) - CM = 189{,}798.9167$$

$$SST = \frac{403^2 + \cdots + 390^2}{3} - CM = 222.25$$

$$SSB = \frac{105^2 + 276^2 + 1246^2}{4} - CM = 189{,}335.1667$$

and by subtraction,

$$SSE = \text{Total SS} - SST - SSB = 241.5$$

These four sources of variation, their degrees of freedom, sums of squares, and mean squares are shown in the shaded areas of the analysis of variance tables, generated by *MINITAB* and *MS Excel* and given in Figures 11.9(a) and 11.9(b). You will find instructions for generating this output in the section "Technology Today" at the end of this chapter.

Two-way ANOVA: Dollars versus Usage, Company

```
Source    DF      SS        MS         F        P
Usage      2   189335   94667.6    2351.99   0.000
Company    3      222      74.1       1.84   0.240
Error      6      242      40.3
Total     11   189799
```

S = 6.344   R-Sq = 99.87%   R-Sq(adj) = 99.77%

Anova: Two-Factor Without Replication

SUMMARY	Count	Sum	Average	Variance
Low	4	105	26.25	12.917
Middle	4	276	69	23.333
High	4	1246	311.5	118.333
A	3	403	134.333	23040.333
B	3	426	142	26068
C	3	408	136	23521
D	3	390	130	22159

ANOVA

Source of Variation	SS	df	MS	F	P-value	F crit
Usage	189335.167	2	94667.583	2351.990	0.000	5.143
Company	222.25	3	74.083	1.841	0.240	4.757
Error	241.5	6	40.25			
Total	189798.917	11				

Notice that both the ANOVA tables show two different F statistics and p-values. It will not surprise you to know that these statistics are used to test hypotheses concerning the equality of both the *treatment* and *block* means.

Testing the Equality of the Treatment and Block Means

The *mean squares* in the analysis of variance table can be used to test the null hypotheses

H_0 : No difference among the k treatment means

or

H_0 : No difference among the b block means

versus the alternative hypothesis

H_a : At least one of the means is different from at least one other

using a theoretical argument similar to the one we used for the completely randomized design.

- Remember that σ^2 is the common variance for the observations in all bk block-treatment combinations. The quantity

$$\text{MSE} = \frac{\text{SSE}}{(b-1)(k-1)}$$

is an unbiased estimate of σ^2, whether or not H_0 is true.

- The two mean squares, MST and MSB, estimate σ^2 only if H_0 is true and tend to be unusually *large* if H_0 is false and either the treatment or block means are different.
- The test statistics

$$F = \frac{MST}{MSE} \quad \text{and} \quad F = \frac{MSB}{MSE}$$

are used to test the equality of treatment and block means, respectively. Both statistics tend to be larger than usual if H_0 is false. Hence, you can reject H_0 for large values of F, using *right-tailed* critical values of the F distribution with the appropriate degrees of freedom (see Table 6 in Appendix I) or computer-generated p-values to draw statistical conclusions about the equality of the population means.

TESTS FOR A RANDOMIZED BLOCK DESIGN

For comparing treatment means:

1. Null hypothesis: H_0 : The treatment means are equal
2. Alternative hypothesis: H_a : At least two of the treatment means differ
3. Test statistic: $F = MST/MSE$, where F is based on $df_1 = (k - 1)$ and $df_2 = (b - 1)(k - 1)$
4. Rejection region: Reject if $F > F_\alpha$, where F_α lies in the upper tail of the F distribution (see the figure), or when the p-value $< \alpha$

For comparing block means:

1. Null hypothesis: H_0 : The block means are equal
2. Alternative hypothesis: H_a : At least two of the block means differ
3. Test statistic: $F = MSB/MSE$, where F is based on $df_1 = (b - 1)$ and $df_2 = (b - 1)(k - 1)$
4. Rejection region: Reject if $F > F_\alpha$, where F_α lies in the upper tail of the F distribution (see the figure), or when the p-value $< \alpha$

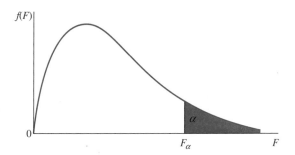

EXAMPLE 11.9

Do the data in Example 11.8 provide sufficient evidence to indicate a difference in the average monthly cell phone cost depending on the company the customer uses?

Solution The cell phone companies represent the *treatments* in this randomized block design, and the differences in their average monthly costs are of primary interest

to the researcher. To test

H_0 : No difference in the average cost among companies

versus the alternative that the average cost is different for at least one of the four companies, you use the analysis of variance F statistic, calculated as

$$F = \frac{\text{MST}}{\text{MSE}} = \frac{74.1}{40.3} = 1.84$$

and shown in the column marked "F" and the row marked "Company" in Figures 11.9(a) and 11.9(b). The exact p-value for this statistical test is also given in Figures 11.9(a) and 11.9(b) as .240, which is too large to allow rejection of H_0. The results do not show a significant difference in the treatment means. That is, there is insufficient evidence to indicate a difference in the average monthly costs for the four companies.

The researcher in Example 11.9 was fairly certain in using a *randomized block design* that there would be a significant difference in the block means—that is, a significant difference in the average monthly costs depending on the usage level. This suspicion is justified by looking at the test of equality of block means. Notice that the observed test statistic is $F = 2351.99$ with p-value $= .000$, showing a highly significant difference, as expected, in the block means.

Identifying Differences in the Treatment and Block Means

Once the overall F-test for equality of the treatment or block means has been performed, what more can you do to identify the nature of any differences you have found? As in Section 11.5, you can use Tukey's method of paired comparisons to determine which pairs of treatment or block means are significantly different from one another. However, if the F-test does not indicate a significant difference in the means, there is no reason to use Tukey's procedure. If you have a special interest in a particular *pair* of treatment or block means, you can estimate the difference using a $(1 - \alpha)100\%$ confidence interval.[†] The formulas for these procedures, shown next, follow a pattern similar to the formulas for the completely randomized design. Remember that MSE always provides an unbiased estimator of σ^2 and uses information from the entire set of measurements. Hence, it is the best available estimator of σ^2, regardless of what test or estimation procedure you are using. You will again use

 NEED A TIP?

Degrees of freedom for Tukey's test and for confidence intervals are **error *df*.**

$$s^2 = \text{MSE} \qquad \text{with } df = (b - 1)(k - 1)$$

to estimate σ^2 in comparing the treatment and block means.

COMPARING TREATMENT AND BLOCK MEANS

Tukey's yardstick for comparing block means:

$$\omega = q_\alpha(b, df)\left(\frac{s}{\sqrt{k}}\right)$$

Tukey's yardstick for comparing treatment means:

$$\omega = q_\alpha(k, df)\left(\frac{s}{\sqrt{b}}\right)$$

[†]You cannot construct a confidence interval for a single mean unless the blocks have been randomly selected from among the population of all blocks. The procedure for constructing intervals for single means is beyond the scope of this book.

$(1 - \alpha)100\%$ confidence interval for the difference in two block means:

$$(\bar{B}_i - \bar{B}_j) \pm t_{\alpha/2} \sqrt{s^2 \left(\frac{1}{k} + \frac{1}{k} \right)}$$

where $\bar{B}_i$ is the average of all observations in block i

$(1 - \alpha)100\%$ confidence interval for the difference in two treatment means:

$$(\bar{T}_i - \bar{T}_j) \pm t_{\alpha/2} \sqrt{s^2 \left(\frac{1}{b} + \frac{1}{b} \right)}$$

where $\bar{T}_i$ is the average of all observations in treatment i.

Note: The values $q_\alpha(*, df)$ from Table 11 in Appendix I, $t_{\alpha/2}$ from Table 4 in Appendix I, and $s^2 = $ MSE all depend on $df = (b - 1)(k - 1)$ degrees of freedom.

EXAMPLE 11.10

Identify the nature of any differences you found in the average monthly cell phone costs from Example 11.8.

Solution   Since the F-test did not show any significant differences in the average costs for the four companies, there is no reason to use Tukey's method of paired comparisons. Suppose, however, that you are an executive for company B and your major competitor is company C. Can you claim a significant difference in the two average costs? Using a 95% confidence interval, you can calculate

$$(\bar{T}_2 - \bar{T}_3) \pm t_{.025} \sqrt{\text{MSE} \left(\frac{2}{b} \right)}$$

$$\left(\frac{426}{3} - \frac{408}{3} \right) \pm 2.447 \sqrt{40.3 \left(\frac{2}{3} \right)}$$

$$6 \pm 12.68$$

so the difference between the two average costs is estimated as between $-\$6.68$ and $\$18.68$. Since 0 is contained in the interval, you do not have evidence to indicate a significant difference in your average costs. Sorry!

NEED a tip? NEED A TIP?

You **cannot** form a confidence interval or test an hypothesis about a single treatment mean in a randomized block design!

Some Cautionary Comments on Blocking

Here are some important points to remember:

- A randomized block design should not be used when treatments and blocks both correspond to **experimental** factors of interest to the researcher. In designating one factor as a *block,* you may assume that the effect of the treatment will be the same, regardless of which block you are using. If this is *not* the case, the two factors—blocks and treatments—are said to **interact,** and your analysis could lead to incorrect conclusions regarding the relationship between the treatments and the response. When an *interaction* is suspected between two factors, you should analyze the data as a **factorial experiment,** which is introduced in the next section.

- Remember that blocking may not always be beneficial. When SSB is removed from SSE, the number of degrees of freedom associated with SSE gets smaller. For blocking to be beneficial, the information gained by isolating the block variation must outweigh the loss of degrees of freedom for error. Usually, though, if you suspect that the experimental units are not homogeneous and you can group the units into blocks, it pays to use the *randomized block design!*

* Finally, remember that you cannot construct confidence intervals for individual treatment means unless it is reasonable to assume that the b blocks have been randomly selected from a population of blocks. If you construct such an interval, the sample treatment mean will be biased by the positive and negative effects that the blocks have on the response.

11.8 EXERCISES

BASIC TECHNIQUES

11.28 A randomized block design was used to compare the means of three treatments within six blocks. Construct an ANOVA table showing the sources of variation and their respective degrees of freedom.

11.29 Suppose that the analysis of variance calculations for Exercise 11.28 are SST = 11.4, SSB = 17.1, and Total SS = 42.7. Complete the ANOVA table, showing all sums of squares, mean squares, and pertinent F-values.

11.30 Do the data of Exercise 11.28 provide sufficient evidence to indicate that blocking increased the amount of information in the experiment about the treatment means? Justify your answer.

11.31 Refer to Exercise 11.28. Find a 95% confidence interval for the difference between a pair of treatment means A and B if $\bar{x}_A = 21.9$ and $\bar{x}_B = 24.2$.

11.32 Do the data of Exercise 11.28 provide sufficient evidence to indicate differences among the treatment means? Test using $\alpha = .05$.

11.33 The data that follow are observations collected from an experiment that compared four treatments, A, B, C, and D, within each of three blocks, using a randomized block design.

Block	A	B	C	D	Total
			Treatment		
1	6	10	8	9	33
2	4	9	5	7	25
3	12	15	14	14	55
Total	22	34	27	30	113

a. Do the data present sufficient evidence to indicate differences among the treatment means? Test using $\alpha = .05$.

b. Do the data present sufficient evidence to indicate differences among the block means? Test using $\alpha = .05$.

c. Rank the four treatment means using Tukey's method of paired comparisons with $\alpha = .01$.

d. Find a 95% confidence interval for the difference in means for treatments A and B.

e. Does it appear that the use of a randomized block design for this experiment was justified? Explain.

11.34 The data shown here are observations collected from an experiment that compared three treatments, A, B, and C, within each of five blocks, using a randomized block design:

Treatment	1	2	3	4	5	Total
			Block			
A	2.1	2.6	1.9	3.2	2.7	12.5
B	3.4	3.8	3.6	4.1	3.9	18.8
C	3.0	3.6	3.2	3.9	3.9	17.6
Total	8.5	10.0	8.7	11.2	10.5	48.9

MS Excel output for Exercise 11.34

Anova: Two-Factor Without Replication

SUMMARY	Count	Sum	Average	Variance
A	5	12.5	2.5	0.265
B	5	18.8	3.76	0.073
C	5	17.6	3.52	0.167
1	3	8.5	2.833	0.443
2	3	10	3.333	0.413
3	3	8.7	2.9	0.79
4	3	11.2	3.733	0.223
5	3	10.5	3.5	0.48

ANOVA

Source of Variation	SS	df	MS	F	P-value	F crit
Rows	4.476	2	2.238	79.929	0.000	4.459
Columns	1.796	4	0.449	16.036	0.001	3.838
Error	0.224	8	0.028			
Total	6.496	14				

Use the *Excel* ouput to analyze the experiment. Investigate possible differences in the block and/or treatment means and, if any differences exist, use an appropriate method to specifically identify where the differences lie. Has blocking been effective in this experiment? Present your results in the form of a report.

11.35 The partially completed ANOVA table for a randomized block design is presented here:

Source	df	SS	MS	F
Treatments	4	14.2		
Blocks		18.9		
Error	24			
Total	34	41.9		

a. How many blocks are involved in the design?
b. How many observations are in each treatment total?
c. How many observations are in each block total?
d. Fill in the blanks in the ANOVA table.
e. Do the data present sufficient evidence to indicate differences among the treatment means? Test using $\alpha = .05$.
f. Do the data present sufficient evidence to indicate differences among the block means? Test using $\alpha = .05$.

APPLICATIONS

11.36 Gas Mileage A study was conducted EX1136 to compare automobile gasoline mileage for three formulations of gasoline. Four automobiles, all of the same make and model, were used in the experiment, and each formulation was tested in each automobile. Using each formulation in the same automobile has the effect of eliminating (blocking out) automobile-to-automobile variability. The data (in miles per gallon) follow.

	Automobile			
Formulation	1	2	3	4
A	25.7	27.0	27.3	26.1
B	27.2	28.1	27.9	27.7
C	26.1	27.5	26.8	27.8

a. Do the data provide sufficient evidence to indicate a difference in mean mileage per gallon for the three gasoline formulations?
b. Is there evidence of a difference in mean mileage for the four automobiles?

c. Suppose that *prior to looking at the data,* you had decided to compare the mean mileage per gallon for formulations A and B. Find a 90% confidence interval for this difference.
d. Use an appropriate method to identify the pairwise differences, if any, in the average mileages for the three formulations.

11.37 Water Resistance in Textiles An EX1137 experiment was conducted to compare the effects of four different chemicals, A, B, C, and D, in producing water resistance in textiles. A strip of material, randomly selected from a bolt, was cut into four pieces, and the four pieces were randomly assigned to receive one of the four chemicals, A, B, C, or D. This process was replicated three times, thus producing a randomized block design. The design, with moisture-resistance measurements, is as shown in the figure (low readings indicate low moisture penetration). Analyze the experiment using a method appropriate for this randomized block design. Identify the blocks and treatments, and investigate any possible differences in treatment means. If any differences exist, use an appropriate method to specifically identify where the differences lie. What are the practical implications for the chemical producers? Has blocking been effective in this experiment? Present your results in the form of a report.

Illustration for Exercise 11.37

Blocks (bolt samples)

1	2	3
C 9.9	D 13.4	B 12.7
A 10.1	B 12.9	D 12.9
B 11.4	A 12.2	C 11.4
D 12.1	C 12.3	A 11.9

11.38 Glare in Rearview Mirrors An experiment was conducted to compare the glare characteristics of four types of automobile rearview mirrors. Forty drivers were randomly selected to participate in the experiment. Each driver was exposed to the glare produced by a headlight located 30 feet behind the rear window of the experimental automobile. The driver then rated the glare produced by the rearview mirror on a scale of 1 (low) to 10 (high). Each of the four mirrors was tested by each driver; the mirrors were assigned to a driver in random order. An analysis of variance of the data produced this ANOVA table:

Source	df	SS	MS	F
Mirrors		46.98		
Drivers			8.42	
Error				
Total		638.61		

a. Fill in the blanks in the ANOVA table.

b. Do the data present sufficient evidence to indicate differences in the mean glare ratings of the four rearview mirrors? Calculate the approximate p-value and use it to make your decision.

c. Do the data present sufficient evidence to indicate that the level of glare perceived by the drivers varied from driver to driver? Use the p-value approach.

d. Based on the results of part b, what are the practical implications of this experiment for the manufacturers of the rearview mirrors?

Data set **11.39 Slash Pine Seedings** An experiment EX1139 was conducted to determine the effects of three methods of soil preparation on the first-year growth of slash pine seedlings. Four locations (state forest lands) were selected, and each location was divided into three plots. Since it was felt that soil fertility within a location was more homogeneous than between locations, a randomized block design was employed using locations as blocks. The methods of soil preparation were A (no preparation), B (light fertilization), and C (burning). Each soil preparation was randomly applied to a plot within each location. On each plot, the same number of seedlings were planted and the average first-year growth of the seedlings was recorded on each plot. Use the *MINITAB* printout to answer the questions.

Soil Preparation	Location 1	2	3	4
A	11	13	16	10
B	15	17	20	12
C	10	15	13	10

a. Conduct an analysis of variance. Do the data provide evidence to indicate a difference in the mean growths for the three soil preparations?

b. Is there evidence to indicate a difference in mean rates of growth for the four locations?

c. Use Tukey's method of paired comparisons to rank the mean growths for the three soil preparations. Use $\alpha = .05$.

d. Use a 95% confidence interval to estimate the difference in mean growths for methods A and B.

MINITAB output for Exercise 11.39

Two-way ANOVA: Growth versus Soil Prep, Location

```
Source      DF        SS        MS       F       P
Soil Prep    2    38.000   19.0000   10.06   0.012
Location     3    61.667   20.5556   10.88   0.008
Error        6    11.333    1.8889
Total       11   111.000

S = 1.374   R-Sq = 89.79%   R-Sq(adj) = 81.28%

                          Individual 95% CIs For Mean Based on
                          Pooled StDev
Soil Prep   Mean     ---------+---------+---------+---------+--
1           12.5        (-------*-------)
2           16.0                            (-------*-------)
3           12.0     (-------*-------)
                     ---------+---------+---------+---------+-
                        12.0      14.0      16.0      18.0

                          Individual 95% CIs For Mean Based on
                          Pooled StDev
Location    Mean     ------+---------+---------+---------+-----
1           12.0000     (-------*-------)
2           15.0000                 (-------*-------)
3           16.3333                       (------*-------)
4           10.6667  (-------*------)
                     ------+---------+---------+---------+-----
                        10.0      12.5      15.0      17.5
```

Data set **11.40 Digitalis and Calcium Uptake** A study EX1140 was conducted to compare the effects of three levels of digitalis on the levels of calcium in the heart muscles of dogs. Because general level of calcium uptake varies from one animal to another, the tissue for a heart muscle was regarded as a block, and comparisons of the three digitalis levels (treatments) were made within a given animal. The calcium uptakes for the three levels of digitalis, A, B, and C, were compared based on the heart muscles of four dogs and the results are given in the table. Use the *Excel* printout to answer the questions.

	Dogs			
	1	2	3	4
	A	C	B	A
	1342	1698	1296	1150
	B	B	A	C
	1608	1387	1029	1579
	C	A	C	B
	1881	1140	1549	1319

a. How many degrees of freedom are associated with SSE?

b. Do the data present sufficient evidence to indicate a difference in the mean uptakes of calcium for the three levels of digitalis?

c. Use Tukey's method of paired comparisons with $\alpha = .01$ to rank the mean calcium uptakes for the three levels of digitalis.

d. Do the data indicate a difference in the mean uptakes of calcium for the four heart muscles?

e. Use Tukey's method of paired comparisons with $\alpha = .01$ to rank the mean calcium uptakes for the heart muscles of the four dogs used in the experiment. Are these results of any practical value to the researcher?

f. Give the standard error of the difference between the mean calcium uptakes for two levels of digitalis.

g. Find a 95% confidence interval for the difference in mean responses between treatments A and B.

MS Excel output for Exercise 11.40

Anova: Two-Factor Without Replication

SUMMARY	Count	Sum	Average	Variance
A	4	4661	1165.25	16891.583
B	4	5610	1402.5	20261.667
C	4	6707	1676.75	72681.583
1	3	4831	1610.333	72634.333
2	3	4225	1408.333	78182.333
3	3	3874	1291.333	67616.333
4	3	4048	1349.333	46700.333

ANOVA						
Source of Variation	SS	df	MS	F	P-value	F crit
Digitalis	524177.167	2	262088.583	258.237	0.000	5.143
Dogs	173415	3	57805	56.955	0.000	4.757
Error	6089.5	6	1014.917			
Total	703681.667	11				

11.41 Bidding on Construction Jobs A building contractor employs three construction engineers, A, B, and C, to estimate and bid on jobs. To determine whether one tends to be a more conservative (or liberal) estimator than the others, the contractor selects four projected construction jobs and has each estimator independently estimate the cost (in dollars per square foot) of each job. The data are shown in the table:

Estimator	Construction Job				Total
	1	2	3	4	
A	35.10	34.50	29.25	31.60	130.45
B	37.45	34.60	33.10	34.40	139.55
C	36.30	35.10	32.45	32.90	136.75
Total	108.85	104.20	94.80	98.90	406.75

Analyze the experiment using the appropriate methods. Identify the blocks and treatments, and investigate any possible differences in treatment means. If any differences exist, use an appropriate method to specifically identify where the differences lie. Has blocking been effective in this experiment? What are the practical implications of the experiment? Present your results in the form of a report.

11.42 Premium Equity? The cost of auto insurance varies by coverage, location, and the driving record of the driver. The following are estimates of the annual cost for standard coverage as of January 1, 2011 for a male driver with 6–8 years of experience, driving a Honda Accord with no accidents or violations.[3] (These are quotes and not premiums.)

Location	21st Century	Geico	AAA	Fireman's Fund	State Farm
West Hollywood	3922	4073	3663	4075	3876
Laguna Beach	2378	2512	2478	3056	2508
Redlands	2560	2476	2549	2756	2614
Riverside	2584	2759	2494	2940	2714

Source: www.insurance.ca.gov

a. What type of design was used in collecting these data?

b. Is there sufficient evidence to indicate that insurance premiums for the same type of coverage differs from company to company?

c. Is there sufficient evidence to indicate that insurance premiums vary from location to location?

d. Use Tukey's procedure to determine which insurance companies listed here differ from others in the premiums they charge for this typical client. Use $\alpha = .05$.

e. Summarize your findings.

11.43 Where to Shop? Do you shop at the grocery store closest to home or do you look for the store that has the best prices? We compared the regular prices at four different grocery stores for eight items purchased on the same day.

	Stores			
Items	Vons	Ralphs	Stater Bros	WinCo
Salad mix, 12 oz. bag	3.99	2.79	1.99	1.78
Hillshire Farm® Beef Smoked Sausage, 14 oz.	4.29	4.29	3.99	2.50
Kellogg's Raisin Bran®, 25.5 oz.	4.49	5.49	4.49	3.15
Kraft® Philadelphia® Cream Cheese, 8 oz.	2.99	3.19	2.79	1.48
Kraft® Ranch Dressing, 16 oz.	3.19	3.49	3.49	1.48
Treetop® Apple Juice, 64 oz.	2.99	3.49	3.49	1.58
Dial® Bar Soap, Gold, 8–4 oz.	5.99	6.49	5.79	5.14
Jif® Peanut Butter, Creamy, 28 oz.	5.15	5.49	4.79	4.34

a. What are the blocks and treatments in this experiment?

b. Do the data provide evidence to indicate that there are significant differences in prices from store to store? Support your answer statistically using the ANOVA printout that follows.

c. Are there significant differences from block to block? Was blocking effective?

Two-way ANOVA: Price versus Item, Store

```
Source   DF        SS        MS        F       P
Item      7   40.2184   5.74548   29.99   0.000
Store     3   14.6695   4.88982   25.53   0.000
Error    21    4.0230   0.19157
Total    31   58.9108

S = 0.4377  R-Sq = 93.17%  R-Sq(adj) = 89.92%
```

11.44 Where to Shop?, continued Refer to
Exercise 11.43. The printout that follows provides the
average costs of the selected items for the $k = 4$ stores.

Store	Mean
Vons	4.1350
Ralphs	4.3400
Stater	3.8525
WinCo	2.5975

a. What is the appropriate value of $q_{.05}(k, df)$ for testing for differences among stores?

b. What is the value of $\omega = q_{.05}(k, df)\sqrt{\dfrac{MSE}{b}}$?

c. Use Tukey's pairwise comparison test among stores used to determine which stores differ significantly in average prices of the selected items.

THE $a \times b$ FACTORIAL EXPERIMENT: A TWO-WAY CLASSIFICATION

11.9

Suppose the manager of a manufacturing plant suspects that the output (in number of units produced per shift) of a production line depends on two factors:

- Which of two supervisors is in charge of the line
- Which of three shifts—day, swing, or night—is being measured

That is, the manager is interested in two *factors:* "supervisor" at two levels and "shift" at three levels. Can you use a randomized block design, designating one of the two factors as a block factor? In order to do this, you would need to assume that the effect of the two supervisors is the same, regardless of which shift you are considering. This may not be the case; maybe the first supervisor is most effective in the morning, and the second is more effective at night. You cannot generalize and say that one supervisor is better than the other or that the output of one particular shift is best. You need to investigate not only the average output for the two supervisors and the average output for the three shifts, but also the **interaction** or relationship between the two factors. Consider two different examples that show the effect of *interaction* on the responses in this situation.

EXAMPLE 11.11

Suppose that the two supervisors are each observed on three randomly selected days for each of the three different shifts. The average outputs for the three shifts are shown in Table 11.4 for each of the supervisors. Look at the relationship between the two factors in the line chart for these means, shown in Figure 11.10. Notice that supervisor 2 always produces a higher output, regardless of the shift. The two factors behave *independently;* that is, the output is always about 100 units higher for supervisor 2, no matter which shift you look at.

Now consider another set of data for the same situation, shown in Table 11.5. There is a definite difference in the results, depending on which shift you look at, and the *interaction* can be seen in the crossed lines of the chart in Figure 11.11.

TABLE 11.4 • **Average Outputs for Two Supervisors on Three Shifts**

Supervisor	Day	Shift Swing	Night
1	487	498	550
2	602	602	637

FIGURE 11.10

Interaction plot for means in Table 11.4

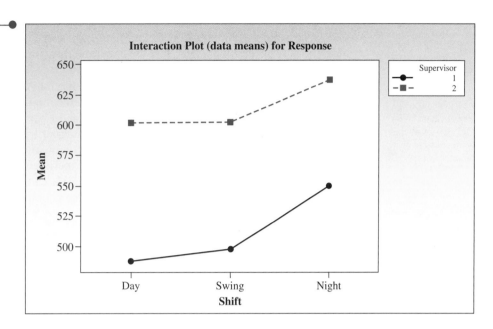

TABLE 11.5 ● **Average Outputs for Two Supervisors on Three Shifts**

Supervisor	Shift		
	Day	Swing	Night
1	602	498	450
2	487	602	657

FIGURE 11.11

Interaction plot for means in Table 11.5

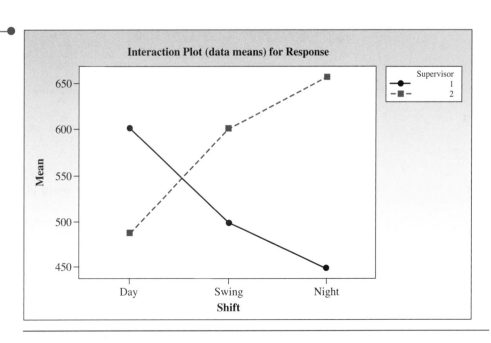

This situation is an example of a **factorial experiment** in which there are a total of 2×3 possible combinations of the levels for the two factors. These $2 \times 3 = 6$ combinations form the *treatments,* and the experiment is called a **2 × 3 factorial experiment.** This type of experiment can actually be used to investigate the effects of three or more factors on a response and to explore the interactions between the factors. However, we confine our discussion to two factors and their interaction.

When you compare treatment means for a factorial experiment (or for any other experiment), you will need more than one observation per treatment. For example, if you obtain two observations for each of the factor combinations of a complete factorial experiment, you have two **replications** of the experiment. In the next section on the analysis of variance for a factorial experiment, you can assume that each treatment or combination of factor levels is replicated the same number of times r.

THE ANALYSIS OF VARIANCE FOR AN *a* × *b* FACTORIAL EXPERIMENT

An analysis of variance for a two-factor factorial experiment replicated r times follows the same pattern as the previous designs. If the letters A and B are used to identify the two factors, the total variation in the experiment

$$\text{Total SS} = \Sigma(x - \bar{x})^2 = \Sigma x^2 - \text{CM}$$

is partitioned into *four* parts in such a way that

$$\text{Total SS} = \text{SSA} + \text{SSB} + \text{SS(AB)} + \text{SSE}$$

where

- SSA (sum of squares for factor A) measures the variation among the factor A means.
- SSB (sum of squares for factor B) measures the variation among the factor B means.
- SS(AB) (sum of squares for interaction) measures the variation *among* the different combinations of factor levels.
- SSE (sum of squares for error) measures the variation of the differences among the observations *within* each combination of factor levels—the experimental error.

Sums of squares SSA and SSB are often called the **main effect** sums of squares, to distinguish them from the **interaction** sum of squares. Although you can simplify your work by using a computer program to calculate these sums of squares, the calculational formulas are given next. You can assume that there are:

- a levels of factor A
- b levels of factor B
- r replications of each of the ab factor combinations
- A total of $n = abr$ observations

CALCULATING THE SUMS OF SQUARES FOR A TWO-FACTOR FACTORIAL EXPERIMENT

$$CM = \frac{G^2}{n} \qquad \text{Total SS} = \Sigma x^2 - CM$$

$$SSA = \Sigma \frac{A_i^2}{br} - CM \qquad SSB = \Sigma \frac{B_j^2}{ar} - CM$$

$$SS(AB) = \Sigma \frac{(AB)_{ij}^2}{r} - CM - SSA - SSB$$

where

G = Sum of all $n = abr$ observations
A_i = Total of all observations at the ith level of factor A,
  $i = 1, 2, \ldots, a$
B_j = Total of all observations at the jth level of factor B,
  $j = 1, 2, \ldots, b$
$(AB)_{ij}$ = Total of the r observations at the ith level of factor A and the jth level of factor B

Each of the four **sources of variation,** when divided by the appropriate **degrees of freedom,** provides an estimate of the variation in the experiment. These estimates are called **mean squares**—MS = SS/df—and are displayed along with their respective sums of squares and df in the **analysis of variance** (or **ANOVA**) **table.**

ANOVA TABLE FOR r REPLICATIONS OF A TWO-FACTOR FACTORIAL EXPERIMENT: FACTOR A AT a LEVELS AND FACTOR B AT b LEVELS

Source	df	SS	MS	F
A	$a - 1$	SSA	$MSA = \dfrac{SSA}{a-1}$	$\dfrac{MSA}{MSE}$
B	$b - 1$	SSB	$MSB = \dfrac{SSB}{b-1}$	$\dfrac{MSB}{MSE}$
AB	$(a-1)(b-1)$	SS(AB)	$MS(AB) = \dfrac{SS(AB)}{(a-1)(b-1)}$	$\dfrac{MS(AB)}{MSE}$
Error	$ab(r-1)$	SSE	$MSE = \dfrac{SSE}{ab(r-1)}$	
Total	$abr - 1$	Total SS		

Finally, the equality of means for various levels of the factor combinations (the interaction effect) and for the levels of both main effects, A and B, can be tested using the ANOVA F-tests, as shown next.

TESTS FOR A FACTORIAL EXPERIMENT

- **For interaction:**
  1. Null hypothesis: H_0 : Factors A and B do not interact
  2. Alternative hypothesis: H_a : Factors A and B interact
  3. Test statistic: $F = $ MS(AB)/MSE, where F is based on $df_1 = (a - 1)(b - 1)$ and $df_2 = ab(r - 1)$
  4. Rejection region: Reject H_0 when $F > F_\alpha$, where F_α lies in the upper tail of the F distribution (see the figure), or when the p-value $< \alpha$

- **For main effects, factor A:**
  1. Null hypothesis: H_0 : There are no differences among the factor A means
  2. Alternative hypothesis: H_a : At least two of the factor A means differ
  3. Test statistic: $F = $ MSA/MSE, where F is based on $df_1 = (a - 1)$ and $df_2 = ab(r - 1)$
  4. Rejection region: Reject H_0 when $F > F_\alpha$ (see the figure) or when the p-value $< \alpha$

- **For main effects, factor B:**
  1. Null hypothesis: H_0 : There are no differences among the factor B means
  2. Alternative hypothesis: H_a : At least two of the factor B means differ
  3. Test statistic: $F = $ MSB/MSE, where F is based on $df_1 = (b - 1)$ and $df_2 = ab(r - 1)$
  4. Rejection region: Reject H_0 when $F > F_\alpha$ (see the figure) or when the p-value $< \alpha$

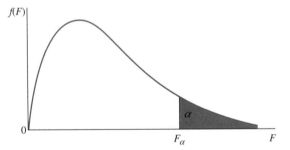

EXAMPLE 11.12

Table 11.6 shows the original data used to generate Table 11.5 in Example 11.11. That is, the two supervisors were each observed on three randomly selected days for each of the three different shifts, and the production outputs were recorded. Analyze these data using the appropriate analysis of variance procedure.

TABLE 11.6 ● **Outputs for Two Supervisors on Three Shifts**

Supervisor	Day	Swing	Night
		Shift	
1	571	480	470
	610	474	430
	625	540	450
2	480	625	630
	516	600	680
	465	581	661

Solution The computer output in Figure 11.12(a) was generated using the two-way analysis of variance procedure in the *MINITAB* software package. A similar print-out generated by *Excel* is shown in Figure 11.12(b). You can verify the quantities in the ANOVA table using the calculational formulas presented earlier, or you may choose just to use the results and interpret their meaning.

FIGURE 11.12(a) ●

MINITAB output for Example 11.12

Two-way ANOVA: Output versus Supervisor, Shift

```
Source        DF        SS        MS        F        P
Supervisor     1     19208   19208.0    26.68    0.000
Shift          2       247     123.5     0.17    0.844
Interaction    2     81127   40563.5    56.34    0.000
Error         12      8640     720.0
Total         17    109222

S = 26.83    R-Sq = 92.09%    R-Sq(adj) = 88.79%

                           Individual 95% CIs For Mean Based on
                           Pooled StDev
Supervisor       Mean    ----+---------+---------+---------+-----
1             516.667    (-------*------)
2             582.000                         (-------*-------)
                         ----+---------+---------+---------+-----
                          510        540       570       600

                           Individual 95% CIs For Mean Based on
                           Pooled StDev
Shift            Mean    ---+---------+---------+---------+--------
Day            544.5     (---------------*---------------)
Swing          550.0       (---------------*--------------)
Night          553.5          (---------------*--------------)
                         ---+---------+---------+---------+--------
                          525        540       555       570
```

FIGURE 11.12(b) ●

MS Excel output for Example 11.12

ANOVA: Two-Factor With Replication

SUMMARY	Day	Swing	Night	Total
1				
Count	3	3	3	9
Sum	1806	1494	1350	4650
Average	602	498	450	516.667
Variance	777	1332	400	5155.25
2				
Count	3	3	3	9
Sum	1461	1806	1971	5238
Average	487	602	657	582
Variance	687	487	637	6096.5
Total				
Count	6	6	6	
Sum	3267	3300	3321	
Average	544.5	550	553.5	
Variance	4553.1	3972.4	13269.5	

ANOVA

Source of Variation	SS	df	MS	F	P-value	F crit
Supervisor	19208	1	19208	26.678	0.000	4.747
Shift	247	2	123.5	0.172	0.844	3.885
Interaction	81127	2	40563.5	56.338	0.000	3.885
Within	8640	12	720			
Total	109222	17				

NEED A TIP?

If the interaction is **not significant,** test each of the factors individually.

At this point, you have undoubtedly discovered the familiar pattern in testing the significance of the various experimental factors with the F statistic and its p-value. The small p-value (.000) in the row marked "Supervisor" means that there is sufficient evidence to declare a difference in the mean levels for factor A—that is, a difference in mean outputs per supervisor. This fact is visually apparent in the nonoverlapping confidence intervals for the supervisor means shown in the printout. But this is overshadowed by the fact that there is strong evidence ($P = .000$) of an *interaction* between factors A and B. This means that the average output for a given shift depends on the supervisor on duty. You saw this effect clearly in Figure 11.10. The three largest mean outputs occur when supervisor 1 is on the day shift and when supervisor 2 is on either the swing or night shift. As a practical result, the manager should schedule supervisor 1 for the day shift and supervisor 2 for the night shift.

If the interaction effect *is* significant, the differences in the treatment means can be further studied, *not* by comparing the means for factor A or B individually but rather by looking at comparisons for the 2×3 (AB) factor level combinations. If the interaction effect *is not significant,* then the significance of the main effect means should be investigated, first with the overall F-test and next with Tukey's method for paired comparisons and/or specific confidence intervals. Remember that these analysis of variance procedures always use $s^2 = \text{MSE}$ as the best estimator of σ^2 with degrees of freedom equal to $df = ab(r - 1)$.

For example, using Tukey's yardstick to compare the average outputs for the two supervisors on each of the three shifts, you could calculate

$$\omega = q_{.05}(6, 12)\left(\frac{s}{\sqrt{r}}\right) = 4.75\left(\frac{\sqrt{720}}{\sqrt{3}}\right) = 73.59$$

Since all three pairs of means—602 and 487 on the day shift, 498 and 602 on the swing shift, and 450 and 657 on the night shift—differ by more than ω, our practical conclusions have been confirmed statistically.

11.10 EXERCISES

BASIC TECHNIQUES

11.45 Suppose you were to conduct a two-factor factorial experiment, factor A at four levels and factor B at five levels, with three replications per treatment.

a. How many treatments are involved in the experiment?

b. How many observations are involved?

c. List the sources of variation and their respective degrees of freedom.

11.46 The analysis of variance table for a 3×4 factorial experiment, with factor A at three levels and fac-

tor B at four levels, and with two observations per treatment, is shown here:

Source	df	SS	MS	F
	2	5.3		
	3	9.1		
	6			
	12	24.5		
Total	23	43.7		

a. Fill in the missing items in the table.

b. Do the data provide sufficient evidence to indicate that factors A and B interact? Test using $\alpha = .05$. What are the practical implications of your answer?

c. Do the data provide sufficient evidence to indicate that factors A and B affect the response variable x? Explain.

11.47 Refer to Exercise 11.46. The means of two of the factor level combinations—say, A_1B_1 and A_2B_1—are $\bar{x}_1 = 8.3$ and $\bar{x}_2 = 6.3$, respectively. Find a 95% confidence interval for the difference between the two corresponding population means.

Data set EX1148 **11.48** The table gives data for a 3×3 factorial experiment, with two replications per treatment:

Levels of Factor B	Levels of Factor A		
	1	2	3
1	5, 7	9, 7	4, 6
2	8, 7	12, 13	7, 10
3	14, 11	8, 9	12, 15

a. Perform an analysis of variance for the data, and present the results in an analysis of variance table.

b. What do we mean when we say that factors A and B interact?

c. Do the data provide sufficient evidence to indicate interaction between factors A and B? Test using $\alpha = .05$.

d. Find the approximate p-value for the test in part c.

e. What are the practical implications of your results in part c? Explain your results using a line graph similar to the one in Figure 11.11.

Data set EX1149 **11.49 2 × 2 Factorial** The table gives data for a 2×2 factorial experiment, with four replications per treatment.

Levels of Factor B	Levels of Factor A	
	1	2
1	2.1, 2.7, 2.4, 2.5	3.7, 3.2, 3.0, 3.5
2	3.1, 3.6, 3.4, 3.9	2.9, 2.7, 2.2, 2.5

a. The accompanying graph was generated by *MINITAB*. Verify that the four points that connect the two lines are the means of the four observations within each factor level combination. What does the graph tell you about the interaction between factors A and B?

MINITAB interaction plot for Exercise 11.49

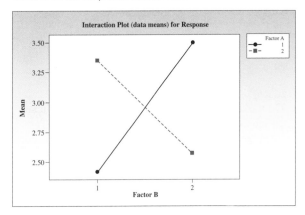

b. Use the *MINITAB* output to test for a significant interaction between A and B. Does this confirm your conclusions in part a?

MINITAB output for Exercise 11.49

Two-way ANOVA: Response versus Factor A, Factor B

Source	DF	SS	MS	F	P
Factor A	1	0.0000	0.00000	0.00	1.000
Factor B	1	0.0900	0.09000	1.00	0.338
Interaction	1	3.4225	3.42250	37.85	0.000
Error	12	1.0850	0.09042		
Total	15	4.5975			

S = 0.3007   R-Sq = 76.40%   R-Sq(adj) = 70.50%

c. Considering your results in part b, how can you explain the fact that neither of the main effects is significant?

d. If a significant interaction is found, is it necessary to test for significant main effect differences? Explain.

e. Write a short paragraph summarizing the results of this experiment.

APPLICATIONS

Data set EX1150 **11.50 Demand for Diamonds** A chain of jewelry stores conducted an experiment to investigate the effect of price markup and location on the demand for its diamonds. Six small-town stores were selected for the study, as well as six stores located in large suburban malls. Two stores in each of these locations were assigned to each of three item percentage markups. The percentage gain (or loss) in sales for each store was recorded at the end of 1 month. The data are shown in the accompanying table.

	Markup		
Location	1	2	3
Small towns	10	−3	−10
	4	7	−24
Suburban malls	14	8	−4
	18	3	3

a. Do the data provide sufficient evidence to indicate an interaction between markup and location? Test using $\alpha = .05$.

b. What are the practical implications of your test in part a?

c. Draw a line graph similar to Figure 11.11 to help visualize the results of this experiment. Summarize the results.

d. Find a 95% confidence interval for the difference in mean change in sales for stores in small towns versus those in suburban malls if the stores are using price markup 3.

11.51 Terrain Visualization
A study was conducted to determine the effect of two factors on terrain visualization training for soldiers.[4] During the training programs, participants viewed contour maps of various terrains and then were permitted to view a computer reconstruction of the terrain as it would appear from a specified angle. The two factors investigated in the experiment were the participants' spatial abilities (abilities to visualize in three dimensions) and the viewing procedures (active or passive). Active participation permitted participants to view the computer-generated reconstructions of the terrain from any and all angles. Passive participation gave the participants a set of preselected reconstructions of the terrain. Participants were tested according to spatial ability, and from the test scores 20 were categorized as possessing high spatial ability, 20 medium, and 20 low. Then 10 participants within each of these groups were assigned to each of the two training modes, active or passive. The accompanying tables are the ANOVA table computed by the researchers and the table of the treatment means.

Source	df	MS	Error df	F	p
Main effects:					
Training condition	1	103.7009	54	3.66	.0610
Ability	2	760.5889	54	26.87	.0005
Interaction:					
Training condition × Ability	2	124.9905	54	4.42	.0167
Within cells	54	28.3015			

	Training Condition	
Spatial Ability	Active	Passive
High	17.895	9.508
Medium	5.031	5.648
Low	1.728	1.610

Note: Maximum score = 36.

a. Explain how the authors arrived at the degrees of freedom shown in the ANOVA table.

b. Are the F-values correct?

c. Interpret the test results. What are their practical implications?

d. Use Table 6 in Appendix I to approximate the p-values for the F statistics shown in the ANOVA table.

Source: H.F. Barsam and Z.M. Simutis, "Computer-Based Graphics for Terrain Visualization Training," Human Factors, no. 26, 1984. Copyright 1984 by the Human Factors Society, Inc. Reproduced by permission.

Data set **11.52 Animation Helps?** To explore ways
EX1152 to increase the educational experience using animation versus static images in a learning environment, Cyril Rebetez and colleagues[5] ran a factorial experiment that measured retention of information under four factorial conditions: with animation or without animation; and reinforcement through snapshots or without snapshots of the major frames in the animation. It was expected that the animation would lead to better retention of information and that having the snapshots available would also help with retention of information. The following data are based on the results of their experiment:

	Learning Setting			
	Static		Animated	
Snapshots	Without	With	Without	With
	58.9	42.0	57.7	64.3
	48.9	53.9	55.9	66.4
	51.8	54.4	57.2	63.1
	53.0	47.6	65.1	55.8
	51.3	50.5	59.3	57.9
	49.8	50.2	65.7	61.5
	61.5	47.0	60.8	61.2
	47.8	52.4	59.3	61.9

Use the *MINITAB* output that follows to analyze the experiment with the appropriate method. Identify the two factors, and investigate any possible effects due to their interaction or the main effects. What are the practical implications of these results? Why do the interaction plots seem counterintuitive to the analysis? If the interaction effect is real, what might you do as an experimenter to show that the interaction is, in fact, significant? Explain your conclusions in the form of a report.

MINITAB output for Exercise 11.52

ANOVA: Retention versus Setting, Snapshots
```
Factor        Type    Levels   Values
Setting       fixed      2     Animated, Static
Snapshots     fixed      2     With, Without
Analysis of Variance for Retention

Source        DF      SS      MS
Setting        1   722.95  722.95      F       P
Snapshots      1     6.04    6.04    44.56   0.000
Setting*                              0.37    0.547
  Snapshots    1    40.73   40.73
Error         28   454.28   16.22     2.51   0.124
Total         31  1223.99
S = 4.02793    R-Sq = 62.89%    R-Sq(adj) = 58.91%
```

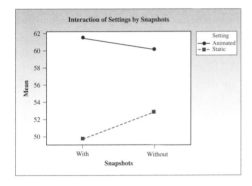

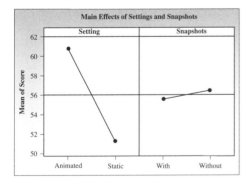

11.53 Fourth-Grade Test Scores A local school board was interested in comparing test scores on a standardized reading test for fourth-grade students in their district. They selected a random sample of five male and five female fourth grade students at each of four different elementary schools in the

district and recorded the test scores. The results are shown in the table below.

Gender	School 1	School 2	School 3	School 4
Male	631	642	651	350
	566	710	611	565
	620	649	755	543
	542	596	693	509
	560	660	620	494
Female	669	722	709	505
	644	769	545	498
	600	723	657	474
	610	649	722	470
	559	766	711	463

a. What type of experimental design is this? What are the experimental units? What are the factors and levels of interest to the school board?

b. Perform the appropriate analysis of variance for this experiment.

c. Do the data indicate that effect of gender on the average test score is different depending on the student's school? Test the appropriate hypothesis using $\alpha = .05$.

d. Plot the average scores using an interaction plot. How would you describe the effect of gender and school on the average test scores?

e. Do the data indicate that either of the main effects is significant? If the main effect is significant, use Tukey's method of paired comparisons to examine the differences in detail. Use $\alpha = .01$.

11.54 Management Training An experiment was conducted to investigate the effect of management training on the decision-making abilities of supervisors in a large corporation. Sixteen supervisors were selected, and eight were randomly chosen to receive managerial training. Four trained and four untrained supervisors were then randomly selected to function in a situation in which a standard problem arose. The other eight supervisors were presented with an emergency situation in which standard procedures could not be used. The response was a management behavior rating for each supervisor as assessed by a rating scheme devised by the experimenter.

a. What are the experimental units in this experiment?

b. What are the two factors considered in the experiment?

c. What are the levels of each factor?

d. How many treatments are there in the experiment?

e. What type of experimental design has been used?

11.55 Management Training, continued
EX1155 Refer to Exercise 11.54. The data for this experiment are shown in the table.

| | Training (A) | | |
Situation (B)	Trained	Not Trained	Totals
Standard	85	53	519
	91	49	
	80	38	
	78	45	
Emergency	76	40	473
	67	52	
	82	46	
	71	39	
Totals	630	362	992

a. Construct the ANOVA table for this experiment.

b. Is there a significant interaction between the presence or absence of training and the type of decision-making situation? Test at the 5% level of significance.

c. Do the data indicate a significant difference in behavior ratings for the two types of situations at the 5% level of significance?

d. Do behavior ratings differ significantly for the two types of training categories at the 5% level of significance.

e. Plot the average scores using an interaction plot. How would you describe the effect of training and emergency situation on the decision-making abilities of the supervisors?

REVISITING THE ANALYSIS OF VARIANCE ASSUMPTIONS

11.11

In Section 11.3, you learned that the assumptions and test procedures for the analysis of variance are similar to those required for the t and F-tests in Chapter 10—namely, that observations within a treatment group must be normally distributed with common variance σ^2. You also learned that the analysis of variance procedures are fairly robust when the sample sizes are equal and the data are fairly mound-shaped. If this is the case, one way to protect yourself from inaccurate conclusions is to try when possible to select samples of equal sizes!

There are some quick and simple ways to check the data for violation of assumptions. Look first at the type of response variable you are measuring. You might immediately see a problem with either the normality or common variance assumption. It may be that the data you have collected cannot be measured *quantitatively*. For example, many responses, such as product preferences, can be ranked only as "A is better than B" or "C is the least preferable." Data that are *qualitative* cannot have a normal distribution. If the response variable is *discrete* and can assume only three values—say, 0, 1, or 2—then it is again unreasonable to assume that the response variable is normally distributed.

Suppose that the response variable is binomial—say, the proportion p of people who favor a particular type of investment (see Section 7.6). Although binomial data can be approximately mound-shaped under certain conditions, they violate the equal variance assumption. The variance of a sample proportion is

$$\sigma^2 = \frac{pq}{n} = \frac{p(1-p)}{n}$$

so that the variance changes depending on the value of p. As the treatment means change, the value of p changes and so does the variance σ^2. A similar situation occurs when the response variable is a Poisson random variable—say, the number of industrial accidents per month in a manufacturing plant (see Section 5.3). Since the variance of a Poisson random variable is $\sigma^2 = \mu$, the variance changes exactly as the treatment mean changes.

If you cannot see any flagrant violations in the type of data being measured, look at the range of the data within each treatment group. If these ranges are nearly the same, then the common variance assumption is probably reasonable. To check for normality, you might make a quick dotplot or stem and leaf plot for a particular treatment group. However, quite often you do not have enough measurements to obtain a reasonable plot.

If you are using a computer program to analyze your experiment, there are some valuable **diagnostic tools** you can use. These procedures are too complicated to be performed using hand calculations, but they are easy to use when the computer does all the work!

Residual Plots

In the analysis of variance, the total variation in the data is partitioned into several parts, depending on the factors identified as important to the researcher. Once the effects of these sources of variation have been removed, the "leftover" variability in each observation is called the **residual** for that data point. These residuals represent **experimental error,** the basic variability in the experiment, and should have an approximately *normal distribution* with a mean of 0 and the *same variation* for each treatment group. Many computer packages will provide options for plotting these residuals:

- The **normal probability plot of residuals** is a graph that plots the residuals for each observation against the expected value of that residual *had it come from a normal distribution.* If the residuals are approximately normal, the plot will closely resemble a *straight line,* sloping upward to the right.

- The **plot of residuals versus fit** or **residuals versus variables** is a graph that plots the residuals against the expected value of that observation *using the experimental design we have used.* If no assumptions have been violated and there are no "leftover" sources of variation other than experimental error, this plot should show a *random scatter* of points around the horizontal "zero error line" for each treatment group, with approximately the same vertical spread.

EXAMPLE 11.13

The data from Example 11.4 involving the attention spans of three groups of elementary students were analyzed using *MINITAB*. The graphs in Figure 11.13, generated by *MINITAB*, are the normal probability plot and the residuals versus fit plot for this experiment. Look at the straight-line pattern in the normal probability plot, which indicates a normal distribution in the residuals. In the other plot, the residuals are plotted against the estimated expected values, which are the sample averages for each of the three treatments in the completely randomized design. The random scatter around the horizontal "zero error line" and the constant spread indicate *no violations* in the constant variance assumption.

FIGURE 11.13

MINITAB diagnostic plots for Example 11.13

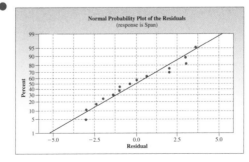
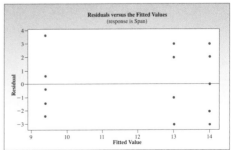

EXAMPLE **11.14**

A company plans to promote a new product by using one of three advertising campaigns. To investigate the extent of product recognition from these three campaigns, 15 market areas were selected and five were randomly assigned to each advertising plan. At the end of the ad campaigns, random samples of 400 adults were selected in each area and the proportions who were familiar with the new product were recorded, as in Table 11.7. Have any of the analysis of variance assumptions been violated in this experiment?

TABLE 11.7 ● **Proportions of Product Recognition for Three Advertising Campaigns**

Campaign 1	Campaign 2	Campaign 3
.33	.28	.21
.29	.41	.30
.21	.34	.26
.32	.39	.33
.25	.27	.31

Solution   The experiment is designed as a *completely randomized design,* but the response variable is a binomial sample proportion. This indicates that both the normality and the common variance assumptions might be invalid. Look at the normal probability plot of the residuals and the plot of residuals versus fit shown in Figure 11.14. The curved pattern in the normal probability plot indicates that the residuals *do not have a normal distribution.* In the residual versus fit plot, you can see three vertical lines of residuals, one for each of the three ad campaigns. Notice that two of the lines (campaigns 1 and 3) are close together and have similar spread. However, the third line (campaign 2) is farther to the right, which indicates a larger sample proportion and consequently a *larger variance* in this group. Both analysis of variance assumptions are suspect in this experiment.

FIGURE 11.14 ●

MINITAB diagnostic plots for Example 11.14

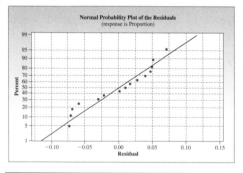
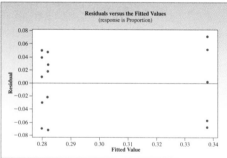

What can you do when the ANOVA assumptions are not satisfied? The *constant variance* assumption can often be remedied by **transforming** the response measurements. That is, instead of using the original measurements, you might use their square roots, logarithms, or some other function of the response. Transformations that tend to stabilize the variance of the response also tend to make their distributions more nearly normal.

When nothing can be done to *even approximately* satisfy the ANOVA assumptions or if the data are rankings, you should use **nonparametric** testing and estimation procedures, presented in Chapter 15. We have mentioned these procedures before; they are almost as powerful in detecting treatment differences as the tests presented in this chapter when the data are normally distributed. When the parametric ANOVA assumptions are violated, the nonparametric tests are generally more powerful.

11.12 A BRIEF SUMMARY

We presented three different experimental designs in this chapter, each of which can be analyzed using the analysis of variance procedure. The objective of the analysis of variance is to detect differences in the mean responses for experimental units that have received different treatments—that is, different combinations of the experimental factor levels. Once an overall test of the differences is performed, the nature of these differences (if any exist) can be explored using methods of paired comparisons and/or interval estimation procedures.

The three designs presented in this chapter represent only a brief introduction to the subject of analyzing designed experiments. Designs are available for experiments that involve several design variables, as well as more than two treatment factors and other more complex designs. Remember that **design variables** are factors whose effect you want to control and hence remove from experimental error, whereas **treatment variables** are factors whose effect you want to investigate. If your experiment is properly designed, you will be able to analyze it using the analysis of variance. Experiments in which the levels of a variable are *measured experimentally* rather than *controlled* or *preselected* ahead of time may be analyzed using **linear** or **multiple regression analysis**—the subject of Chapters 12 and 13.

CHAPTER REVIEW

Key Concepts and Formulas

I. Experimental Designs

1. Experimental units, factors, levels, treatments, response variables.

2. Assumptions: Observations within each treatment group must be normally distributed with a common variance σ^2.

3. One-way classification—completely randomized design: Independent random samples are selected from each of k populations.

4. Two-way classification—randomized block design: k treatments are compared within b relatively homogeneous groups of experimental units called *blocks*.

5. Two-way classification—$a \times b$ factorial experiment: Two factors, A and B, are compared at several levels. Each factor level combination is replicated r times to allow for the investigation of an interaction between the two factors.

II. Analysis of Variance

1. The total variation in the experiment is divided into variation (sums of squares) explained by the various experimental factors and variation due to experimental error (unexplained).

2. If there is an effect due to a particular factor, its mean square (MS = SS/df) is usually large and F = MS(factor)/MSE is large.

3. Test statistics for the various experimental factors are based on F statistics, with appropriate degrees of freedom (df_2 = Error degrees of freedom).

III. Interpreting an Analysis of Variance

1. For the completely randomized and randomized block design, each factor is tested for significance.

2. For the factorial experiment, first test for a significant interaction. If the interaction is significant,

main effects need not be tested. The nature of the differences in the factor level combinations should be further examined.

3. If a significant difference in the population means is found, Tukey's method of pairwise comparisons or a similar method can be used to further identify the nature of the differences.

4. If you have a special interest in one population mean or the difference between two population means, you can use a confidence interval estimate. (For a randomized block design, confidence intervals do not provide unbiased estimates for single population means.)

IV. Checking the Analysis of Variance Assumptions

1. To check for normality, use the normal probability plot for the residuals. The residuals should exhibit a straight-line pattern, increasing upwards toward the right.

2. To check for equality of variance, use the residuals versus fit plot. The plot should exhibit a random scatter, with the same vertical spread around the horizontal "zero error line."

 TECHNOLOGY TODAY

Analysis of Variance Procedures—*Microsoft Excel*

The statistical procedures to perform the analysis of variance for the three experimental designs in this chapter can be found using the *Microsoft Excel* command **Data ▶ Data Analysis.** You will see choices for **Single Factor, Two-Factor Without Replication,** and **Two-Factor With Replication** that will generate Dialog boxes used for the completely randomized, randomized block, and factorial designs, respectively.

 **(Completely Randomized Design)** Refer to the breakfast study in Example 11.4, in which the effect of nutrition on attention span was studied.

No Breakfast	Light Breakfast	Full Breakfast
8	14	10
7	16	12
9	12	16
13	17	15
10	11	12

Enter the data into columns A, B, and C of an *Excel* spreadsheet with one sample per column.

1. Use **Data ▶ Data Analysis ▶ Anova: Single Factor** to generate the Dialog box in Figure 11.15(a). Highlight or type the **Input Range** (the data in the first three columns) into the first box. In the section marked "Grouped by" choose the radio button for **Columns** and check "Labels" if necessary.

2. The default significance level is $\alpha = .05$ in *Excel.* Change this significance level if necessary. Enter a cell location for the **Output Range** and click **OK.** The output will appear in the selected cell location, and should be adjusted using **Format ▶ AutoFit Column Width** on the **Home** tab in the **Cells** group while it is still highlighted. You can decrease the decimal accuracy if you like, using $\begin{smallmatrix}*.0\\.00\end{smallmatrix}$ on the **Home** tab in the **Number** group (see Figure 11.15(b)).

3. The observed value of the test statistic $F = 4.93$ is found in the row labeled "*Between Groups*" followed by the "*P-value*" and the critical value marking the rejection region for a one-tailed test with $\alpha = .05$. For this example, the p-value $= .0273$ indicates

that there is a significant difference in the average attention spans depending on the type of breakfast.

FIGURE 11.15

(a)

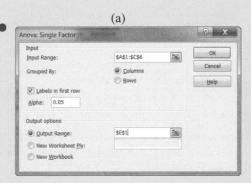

(b)

	E	F	G	H	I	J	K
Anova: Single Factor							
SUMMARY							
Groups		*Count*	*Sum*	*Average*	*Variance*		
None		5	47	9.4	5.3		
Light		5	70	14	6.5		
Full		5	65	13	6		
ANOVA							
Source of Variation		*SS*	*df*	*MS*	*F*	*P-value*	*F crit*
Between Groups		58.533	2	29.2667	4.93258	0.0273	3.89
Within Groups		71.2	12	5.93333			
Total		129.73	14				

EXAMPLE 11.16 — **(Randomized Block Design)** Refer to the cell phone study in Example 11.8, in which the effect of usage level on cost was studied for four different companies.

		Company		
Usage Level	A	B	C	D
Low	27	24	31	23
Middle	68	76	65	67
High	308	326	312	300

Enter the data into columns A–E of an *Excel* spreadsheet, using column A for usage labels and row 1 for company labels, just as shown in the table above.

1. Use **Data ▶ Data Analysis ▶ Anova: Two-Factor Without Replication** to generate the Dialog box in Figure 11.16(a). Highlight or type the **Input Range** (the data in the first five columns) into the first box and check "Labels" if necessary. Change the significance level if needed, and click **OK.** You can adjust the output, possibly changing the labels "Rows" and "Columns" to "Usage" and "Company," as shown in Figure 11.16(b).

2. The observed value of the test statistic for treatments (companies) is $F = 1.84$ with p-value $= .240$ indicating that there is no significant difference among the four companies. The test for blocks (usage) is highly significant, with p-value $= .000$.

FIGURE 11.16

(a)

(b)

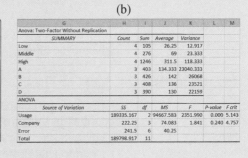

	G	H	I	J	K	L	M
Anova: Two-Factor Without Replication							
	SUMMARY	*Count*	*Sum*	*Average*	*Variance*		
Low		4	105	26.25	12.917		
Middle		4	276	69	23.333		
High		4	1246	311.5	118.333		
A		3	403	134.333	23040.333		
B		3	426	142	26068		
C		3	408	136	23521		
D		3	390	130	22159		
ANOVA							
	Source of Variation	*SS*	*df*	*MS*	*F*	*P-value*	*F crit*
Usage		189335.167	2	94667.583	2351.990	0.000	5.143
Company		222.25	3	74.083	1.841	0.240	4.757
Error		241.5	6	40.25			
Total		189798.917	11				

EXAMPLE 11.17 — **(Factorial Experiment)** Refer to the production output study in Example 11.12, in which the effect of supervisor and shift on production output was studied.

Supervisor	Shift		
	Day	Swing	Night
1	571	480	470
	610	474	430
	625	540	450
2	480	625	630
	516	600	680
	465	581	661

Enter the data into columns A–D of an *Excel* spreadsheet, using column A for supervisor labels and row 1 for shift labels, just as shown in the table above.

1. Use **Data ▶ Data Analysis ▶ Anova: Two-Factor With Replication** to generate the Dialog box in Figure 11.17(a). Highlight or type the **Input Range** (the data in the first four columns) into the first box. Enter the number of replications (3) into "Rows per Sample," change the significance level if needed, and click **OK.** You can adjust the output, possibly changing the labels "Sample" and "Columns" to "Supervisor" and "Shift," as shown in Figure 11.17(b).

2. Refer to the ANOVA table at the bottom of the printout. There is a significant interaction between shift and supervisor (*p*-value = .000). The differences in the treatment means can now be studied by looking at comparisons for the 3 × 2 = 6 factor level combinations.

FIGURE 11.17 ●

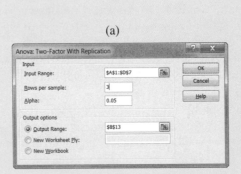

(a)

(b)

G	H	I	J	K	L	M
Anova: Two-Factor With Replication						
SUMMARY		Day	Swing	Night	Total	
1						
Count		3	3	3	9	
Sum		1806	1494	1350	4650	
Average		602	498	450	516.667	
Variance		777	1332	400	5155.25	
2						
Count		3	3	3	9	
Sum		1461	1806	1971	5238	
Average		487	602	657	582	
Variance		687	487	637	6096.5	
Total						
Count		6	6	6		
Sum		3267	3300	3321		
Average		544.5	550	553.5		
Variance		4553.1	3972.4	13269.5		
ANOVA						
Source of Variation	SS	df	MS	F	P-value	F crit
Supervisor	19208	1	19208	26.678	0.000	4.747
Shift	247	2	123.5	0.172	0.844	3.885
Interaction	81127	2	40563.5	56.338	0.000	3.885
Within	8640	12	720			
Total	109222	17				

(NOTE: *MS Excel* does not provide options for performing Tukey's test or for generating diagnostic plots.)

Analysis of Variance Procedures—*MINITAB*

The statistical procedures to perform the analysis of variance for the three experimental designs in this chapter can be found using the *MINITAB* command **Stat ▶ ANOVA.** You will see choices for **One-Way, One-Way (Unstacked),** and **Two-Way** that will generate Dialog boxes used for the completely randomized, randomized block, and factorial designs, respectively.

(**Completely Randomized Design**) Refer to the breakfast study in Example 11.4, in which the effect of nutrition on attention span was studied.

No Breakfast	Light Breakfast	Full Breakfast
8	14	10
7	16	12
9	12	16
13	17	15
10	11	12

1. Enter the 15 recorded attention spans in column C1 of a *MINITAB* worksheet and name them "Span." Next, enter the integers 1, 2, and 3 into a second column C2 to identify the meal assignment (*treatment*) for each observation. You can let *MINITAB* set this pattern for you using **Calc ▶ Make Patterned Data ▶ Simple Set of Numbers** and entering the appropriate numbers, as shown in Figure 11.18.

FIGURE 11.18 ●

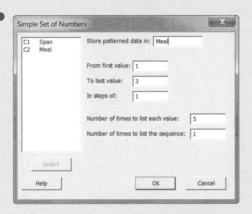

1. Use **Stat ▶ ANOVA ▶ One-Way** to generate the Dialog box in Figure 11.19(a).[†] Select the column of observations for the "Response" box and the column of treatment indicators for the "Factor" box.

2. Now you have several options. Under **Comparisons,** you can select "Tukey's family error rate" (which has a default level of 5%) to obtain paired comparisons output. Under **Graphs,** you can select individual value plots and/or box plots to compare the three meal assignments, and you can generate residual plots (use "Normal plot of residuals" and/or "Residuals versus fits") to verify the validity of the ANOVA assumptions. Click **OK** from the main Dialog box to obtain the output in Figure 11.19(b).

FIGURE 11.19 ●

(a)

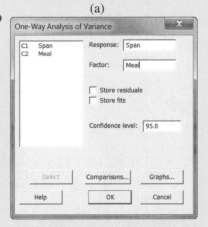

(b)

[†]If you had entered each of the three samples into separate columns, the proper command would have been Stat ▶ ANOVA ▶ One-Way (Unstacked).

3. The observed value of the test statistic $F = 4.93$ is found in the row labeled "*Meal*" followed by the *p*-value $= .0273$. With $\alpha = .05$, there is a significant difference in the average attention spans depending on the type of breakfast.

The **Stat ▶ ANOVA ▶ Two-Way** command can be used for both the randomized block and the factorial designs. You must first enter all of the observations into a single column and then integers or descriptive names to indicate either of these cases:

- The *block* and *treatment* for each of the measurements in a randomized block design.
- The levels of *factors A and B* for the factorial experiment.

MINITAB will recognize a number of replications within each factor level combination in the factorial experiment and will break out the sum of squares for interaction as long as you do not check the box "Fit additive model." Since these two designs involve the same sequence of commands, we will use the data from Example 11.12 to generate the ANOVA.

EXAMPLE 11.19

(Two-Way Classification) Refer to the production output study in Example 11.12, in which the effect of supervisor and shift on production output was studied.

Supervisor	Shift		
	Day	Swing	Night
1	571	480	470
	610	474	430
	625	540	450
2	480	625	630
	516	600	680
	465	581	661

1. Enter the data into the worksheet as shown in Figure 11.20(a). See if you can use the **Calc ▶ Make Patterned Data ▶ Simple Set of Numbers** to enter the data in columns C2–C3.

2. Use **Stat ▶ ANOVA ▶ Two-Way** to generate the Dialog box in Figure 11.20(b). Choose "Output" for the "Response" box, and "Supervisor" and "Shift" for the "Row Factor" and "Column Factor," respectively. You may choose to display the main effect means along with 95% confidence intervals by checking "Display means," and you may select residual plots if you wish. Click OK to obtain the ANOVA printout in Figure 11.12(a).

FIGURE 11.20

(a)

Data Display

Row	Output	Supervisor	Shift
1	571	1	1
2	610	1	1
3	625	1	1
4	480	2	1
5	516	2	1
6	465	2	1
7	480	1	2
8	474	1	2
9	540	1	2
10	625	2	2
11	600	2	2
12	581	2	2
13	470	1	3
14	430	1	3
15	450	1	3
16	630	2	3
17	680	2	3
18	661	2	3

(b)

Two-Way Analysis of Variance

C1 Output
C2 Supervisor
C3 Shift

Response: Output

Row factor: Supervisor ☑ Display means

Column factor: Shift ☑ Display means

☐ Store residuals
☐ Store fits

Confidence level: 95.0

☐ Fit additive model

Select   Help   OK   Cancel   Graphs...

3. Since the interaction between supervisors and shifts is highly significant, you may want to explore the nature of this interaction by plotting the average output for each supervisor at each of the three shifts. Use **Stat ▶ ANOVA ▶ Interaction Plot** and choose the appropriate response and factor variables. The plot is shown in Figure 11.21. You can see the strong difference in the behaviors of the mean outputs for the two supervisors, indicating a strong interaction between the two factors.

FIGURE 11.21

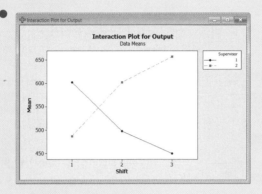

Supplementary Exercises

11.56 Reaction Times versus Stimuli

EX1156 Twenty-seven people participated in an experiment to compare the effects of five different stimuli on reaction time. The experiment was run using a completely randomized design, and, regardless of the results of the analysis of variance, the experimenters wanted to compare stimuli A and D. The results of the experiment are given here. Use the *MINITAB* printout to complete the exercise.

Stimulus	Reaction Time (sec)							Total	Mean
A	.8	.6	.6	.5				2.5	.625
B	.7	.8	.5	.5	.6	.9	.7	4.7	.671
C	1.2	1.0	.9	1.2	1.3	.8		6.4	1.067
D	1.0	.9	.9	1.1	.7			4.6	.920
E	.6	.4	.4	.7	.3			2.4	.480

MINITAB output for Exercise 11.56

One-way ANOVA: Time versus Stimulus

```
Source     DF      SS       MS      F       P
Stimulus    4   1.2118   0.3030  11.67   0.000
Error      22   0.5711   0.0260
Total      26   1.7830

S = 0.1611   R-Sq = 67.97%   R-Sq(adj) = 62.14%

                         Individual 95% CIs For Mean Based on
                         Pooled StDev
Level   N    Mean    StDev  -------+---------+---------+---------+--
A       4   0.6250  0.1258  (-----*------)
B       7   0.6714  0.1496     (----*----)
C       6   1.0667  0.1966                    (-----*----)
D       5   0.9200  0.1483                (-----*-----)
E       5   0.4800  0.1643  (-----*----)
                            -------+---------+---------+---------+--
Pooled StDev =  0.1611            0.50      0.75      1.00      1.25
```

a. Conduct an analysis of variance and test for a difference in the mean reaction times due to the five stimuli.

b. Compare stimuli A and D to see if there is a difference in mean reaction times.

11.57 Refer to Exercise 11.56. Use this *MINITAB* output to identify the differences in the treatment means.

MINITAB output for Exercise 11.57

```
Tukey 95% Simultaneous Confidence Intervals
All Pairwise Comparisons among Levels of Stimulus

Individual confidence level = 99.29%

Stimulus = A subtracted from:

Stimulus    Lower    Center   Upper  --------+---------+---------+---------+-
B         -0.2535   0.0464   0.3463         (-----*-----)
C          0.1328   0.4417   0.7505              (-----*-----)
D         -0.0260   0.2950   0.6160            (-----*-----)
E         -0.4660  -0.1450   0.1760     (-----*-----)
                                       --------+---------+---------+---------+-
                                          -0.50     0.00      0.50      1.00

Stimulus = B subtracted from:

Stimulus    Lower    Center   Upper  --------+---------+---------+---------+-
C          0.1290   0.3952   0.6615              (-----*-----)
D         -0.0316   0.2486   0.5288            (-----*-----)
E         -0.4716  -0.1914   0.0888     (-----*-----)
                                       --------+---------+---------+---------+-
                                          -0.50     0.00      0.50      1.00

Stimulus = C subtracted from:

Stimulus    Lower    Center   Upper  --------+---------+---------+---------+-
D         -0.4364  -0.1467   0.1431        (-----*-----)
E         -0.8764  -0.5867  -0.2969  (-----*-----)
                                     --------+---------+---------+---------+-
                                        -0.50     0.00      0.50      1.00

Stimulus = D subtracted from:

Stimulus    Lower    Center   Upper  --------+---------+---------+---------+-
E         -0.7426  -0.4400  -0.1374  (-----*-----)
                                     --------+---------+---------+---------+-
                                        -0.50     0.00      0.50      1.00
```

11.58 Refer to Exercise 11.56. What do the normal probability plot and the residuals versus fit plot tell you about the validity of your analysis of variance results?

MINITAB diagnostic plots for Exercise 11.58

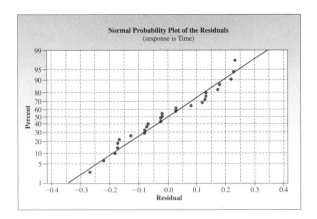

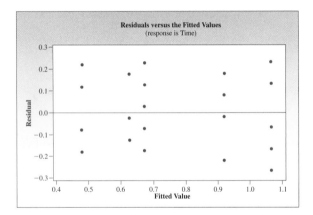

Excel output for Exercise 11.59

ANOVA: Two-Factor Without Replication

SUMMARY	Count	Sum	Average	Variance
A	4	2.8	0.7	0.02
B	4	3.2	0.8	0.0267
C	4	4.2	1.05	0.0167
D	4	4.1	1.025	0.0025
E	4	2.1	0.525	0.009

ANOVA

Source of Variation	SS	df	MS	F	P-value	F crit
Rows	0.14	3	0.046667	6.588	0.007	3.490
Columns	0.787	4	0.196750	27.776	0.000	3.259
Error	0.085	12	0.007083			
Total	1.012	19				

a. Use the *Excel* printout to analyze the data and test for differences in treatment means.

b. Use Tukey's method of paired comparisons to identify the significant pairwise differences in the stimuli.

c. Does it appear that blocking was effective in this experiment?

11.60 Heart Rate and Exercise An experiment was conducted to examine the effect of age on heart rate when a person is subjected to a specific amount of exercise. Ten male subjects were randomly selected from four age groups: 10–19, 20–39, 40–59, and 60–69. Each subject walked on a treadmill at a fixed grade for a period of 12 minutes, and the increase in heart rate, the difference before and after exercise, was recorded (in beats per minute):

	10–19	20–39	40–59	60–69
	29	24	37	28
	33	27	25	29
	26	33	22	34
	27	31	33	36
	39	21	28	21
	35	28	26	20
	33	24	30	25
	29	34	34	24
	36	21	27	33
	22	32	33	32
Total	309	275	295	282

Use an appropriate computer program to answer these questions:

a. Do the data provide sufficient evidence to indicate a difference in mean increase in heart rate among the four age groups? Test by using $\alpha = .05$.

b. Find a 90% confidence interval for the difference in mean increase in heart rate between age groups 10–19 and 60–69.

11.59 Reaction Times II The experiment in Exercise 11.56 might have been conducted more effectively using a randomized block design with people as blocks, since you would expect mean reaction time to vary from one person to another. Hence, four people were used in a new experiment, and each person was subjected to each of the five stimuli in a random order. The reaction times (in seconds) are listed here:

Subject	Stimulus				
	A	B	C	D	E
1	.7	.8	1.0	1.0	.5
2	.6	.6	1.1	1.0	.6
3	.9	1.0	1.2	1.1	.6
4	.6	.8	.9	1.0	.4

c. Find a 90% confidence interval for the mean increase in heart rate for the age group 20–39.

d. Approximately how many people would you need in each group if you wanted to be able to estimate a group mean correct to within two beats per minute with probability equal to .95?

Data set **EX1161**

11.61 Learning to Sell A company wished to study the effects of four training programs on the sales abilities of their sales personnel. Thirty-two people were randomly divided into four groups of equal size, and each group was then subjected to one of the different sales training programs. Because there were some dropouts during the training programs due to illness, vacations, and so on, the number of trainees completing the programs varied from group to group. At the end of the training programs, each salesperson was randomly assigned a sales area from a group of sales areas that were judged to have equivalent sales potentials. The sales made by each of the four groups of salespeople during the first week after completing the training program are listed in the table:

	Training Program			
	1	2	3	4
	78	99	74	81
	84	86	87	63
	86	90	80	71
	92	93	83	65
	69	94	78	86
	73	85		79
		97		73
		91		70
Total	482	735	402	588

Analyze the experiment using the appropriate method. Identify the treatments or factors of interest to the researcher and investigate any significant effects. What are the practical implications of this experiment? Write a paragraph explaining the results of your analysis.

11.62 Refer to Exercise 11.63. The means of all observations, at the factor A levels A_1 and A_2 are $\bar{x}_1 = 3.7$ and $\bar{x}_2 = 1.4$, respectively. Find a 95% confidence interval for the difference in mean response for factor levels A_1 and A_2.

11.63 2 × 3 Factorial The analysis of variance for a 2×3 factorial experiment, factor A at two levels and factor B at three levels, with five observations per treatment, is shown in the table.

Source	df	SS	MS	F
A		1.14		
B		2.58		
AB		.49		
Error				
Total		8.41		

a. Do the data provide sufficient evidence to indicate an interaction between factors A and B? Test using $\alpha = .05$. What are the practical implications of your answer?

b. Give the approximate p-value for the test in part a.

c. Do the data provide sufficient evidence to indicate that factor A affects the response? Test using $\alpha = .05$.

d. Do the data provide sufficient evidence to indicate that factor B affects the response? Test using $\alpha = .05$.

11.64 4 × 2 Factorial Suppose you were to conduct a two-factor factorial experiment, factor A at four levels and factor B at two levels, with r replications per treatment.

a. How many treatments are involved in the experiment?

b. How many observations are involved?

c. List the sources of variation and their respective degrees of freedom.

11.65 The Whitefly in California The whitefly, which causes defoliation of shrubs and trees and a reduction in salable crop yields, has emerged as a pest in Southern California. In a study to determine factors that affect the life cycle of the whitefly, an experiment was conducted in which whiteflies were placed on two different types of plants at three different temperatures. The observation of interest was the total number of eggs laid by caged females under one of the six possible treatment combinations. Each treatment combination was run using five cages.

Plant	Temperature		
	70°F	77°F	82°F
Cotton	37	34	46
	21	54	32
	36	40	41
	43	42	36
	31	16	38
Cucumber	50	59	43
	53	53	62
	25	31	71
	37	69	49
	48	51	59

Excel output for Exercise 11.65

ANOVA

Source of Variation	SS	df	MS	F	P-value	F crit
Plant	1512.3	1	1512.3	12.293	0.002	4.260
Temperature	487.4667	2	243.733	1.981	0.160	3.403
Interaction	111.2	2	55.6	0.452	0.642	3.403
Within	2952.4	24	123.017			
Total	5063.367	29				

a. What type of experimental design has been used?

b. Do the data provide sufficient evidence to indicate that the effect of temperature on the number of eggs laid is different depending on the type of plant? Use the printout to test the appropriate hypothesis.

c. Plot the treatment means for cotton as a function of temperature. Plot the treatment means for cucumber as a function of temperature. Comment on the similarity or difference in these two plots.

d. Find the mean number of eggs laid on cotton and cucumber based on 15 observations each. Calculate a 95% confidence interval for the difference in the underlying population means.

11.66 Pollution from Chemical Plants EX1166 Four chemical plants, producing the same product and owned by the same company, discharge effluents into streams in the vicinity of their locations. To check on the extent of the pollution created by the effluents and to determine whether this varies from plant to plant, the company collected random samples of liquid waste, five specimens for each of the four plants. The data are shown in the table:

Plant	Polluting Effluents (lb/gal of waste)				
A	1.65	1.72	1.50	1.37	1.60
B	1.70	1.85	1.46	2.05	1.80
C	1.40	1.75	1.38	1.65	1.55
D	2.10	1.95	1.65	1.88	2.00

a. Do the data provide sufficient evidence to indicate a difference in the mean amounts of effluents discharged by the four plants?

b. If the maximum mean discharge of effluents is 1.5 lb/gal, do the data provide sufficient evidence to indicate that the limit is exceeded at plant A?

c. Estimate the difference in the mean discharge of effluents between plants A and D, using a 95% confidence interval.

11.67 America's Market Basket Exercise EX1167 10.41 examined an advertisement for a popular supermarket chain that claimed it has had consistently lower prices than one of its competitors. As part of a survey conducted by an independent price-checking company, the average weekly total, based on the prices of approximately 95 items, is given for this chain and for its competitor recorded during four consecutive weeks in a particular month.[6]

Week	Advertiser ($)	Competitor ($)
1	254.26	256.03
2	240.62	255.65
3	231.90	255.12
4	234.13	261.18

a. What type of design has been used in this experiment?

b. Conduct an analysis of variance for the data.

c. Is there sufficient evidence to indicate that there is a difference in the average weekly totals for the two supermarkets? Use $\alpha = .05$.

11.68 Yield of Wheat The yields of wheat EX1168 (in bushels per acre) were compared for five different varieties, A, B, C, D, and E, at six different locations. Each variety was randomly assigned to a plot at each location. The results of the experiment are shown in the accompanying table, along with a *MINITAB* printout of the analysis of variance. Analyze the experiment using the appropriate method. Identify the treatments or factors of interest to the researcher and investigate any effects that exist. Use the diagnostic plots to comment on the validity of the analysis of variance assumptions. What are the practical implications of this experiment? Write a paragraph explaining the results of your analysis.

Variety	Location					
	1	2	3	4	5	6
A	35.3	31.0	32.7	36.8	37.2	33.1
B	30.7	32.2	31.4	31.7	35.0	32.7
C	38.2	33.4	33.6	37.1	37.3	38.2
D	34.9	36.1	35.2	38.3	40.2	36.0
E	32.4	28.9	29.2	30.7	33.9	32.1

MINITAB output for Exercise 11.68

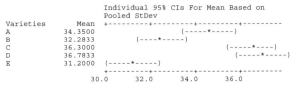

```
Two-way ANOVA: Yield versus Varieties, Location

Source      DF        SS         MS        F        P
Varieties    4    142.670    35.6675    18.61    0.000
Locations    5     68.142    13.6283     7.11    0.001
Error       20     38.303     1.9165
Total       29    249.142

S = 1.384   R-Sq = 84.62%   R-Sq(adj) = 77.69%

                        Individual 95% CIs For Mean Based on
                        Pooled StDev
Varieties     Mean    +---------+---------+---------+---------
A          34.3500                        (-----*-----)
B          32.2833              (----*-----)
C          36.3000                              (-----*----)
D          36.7833                               (-----*-----)
E          31.2000    (-----*-----)
                      +---------+---------+---------+---------
                     30.0      32.0      34.0      36.0
```

MINITAB diagnostic plots for Exercise 11.68

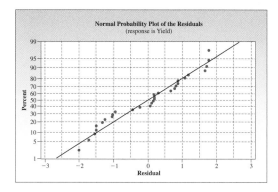

Normal Probability Plot of the Residuals (response is Yield)

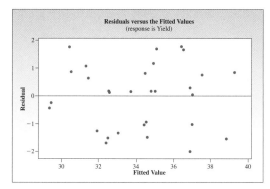

Residuals versus the Fitted Values (response is Yield)

Data set
EX1169

11.69 Physical Fitness Researchers Russell R. Pate and colleagues analyzed the results of the National Health and Nutrition Examination Survey to assess cardiorespiratory fitness levels in youth aged 12 to 19 years.[6] Estimated maximum oxygen uptake (VO_{2max}) was used to measure a person's cardiorespiratory level. The focus of our study investigates the relationship between levels of physical activity (more than others, same as others, or less than others) and gender on VO_{2max}. The data that follows are based on this study.

	Physical Activity		
	More	Same	Less
Males	50.1	45.7	40.9
	47.2	44.2	41.3
	49.7	46.8	39.2
	50.4	44.9	40.9
Females	41.2	37.2	36.5
	39.8	39.4	35.0
	41.5	38.6	37.2
	38.2	37.8	35.4

a. Is this a factorial experiment or a randomized block design? Explain.

b. Is there a significant interaction between levels of physical activity and gender? Are there significant differences between males and females? Levels of physical activity?

c. If the interaction is significant, use Tukey's pairwise procedure to investigate differences among the six cell means. Comment on the results found using this procedure. Use $\alpha = .05$.

Data set
EX1170

11.70 Professor's Salaries In a study of starting salaries of assistant professors,[7] five male and five female beginning assistant professors at each of two types of institutions granting doctoral degrees were polled and their initial starting salaries were recorded. The results of the survey in thousands of dollars are given in the following table.

	Type	
Gender	Public ($)	Not-For-Profit ($)
Male	64.6	75.2
	63.8	75.7
	63.3	75.0
	64.1	75.1
	64.9	75.8
Female	59.7	71.6
	56.9	71.6
	58.4	71.2
	60.1	73.0
	59.8	69.5

a. What type of design was used in collecting these data?

b. Use an analysis of variance to test if there are significant differences in gender, in type of institution, and to test for a significant interaction of gender $\times$ type of institution.

c. Find a 95% confidence interval estimate for the difference in starting salaries for male assistant professors and female assistant professors. Interpret this interval in terms of a gender difference in starting salaries.

d. Summarize the results of your analysis.

Data set
EX1171

11.71 Pottery in the United Kingdom An article in *Archaeometry* involved an analysis of 26 samples of Romano-British pottery, found at four different kiln sites in the United Kingdom.[8] Since one site only yielded two samples, consider the samples found at the other three sites. The samples were analyzed to determine their chemical composition and the percentage of iron oxide is shown next.

Llanederyn		Island Thorns	Ashley Rails
7.00	5.78	1.28	1.12
7.08	5.49	2.39	1.14
7.09	6.92	1.50	.92
6.37	6.13	1.88	2.74
7.06	6.64	1.51	1.64
6.26	6.69		
4.26	6.44		

a. What type of experimental design is this?

b. Use an analysis of variance to determine if there is a difference in the average percentage of iron oxide at the three sites. Use $\alpha = .01$.

c. If you have access to a computer program, generate the diagnostic plots for this experiment. Does it appear that any of the analysis of variance assumptions have been violated? Explain.

Data set **11.72 Smart Phones** A **smart phone** is a
EX1172 mobile phone that offers more advanced computing ability and connectivity than a contemporary basic "feature phone." The data that follow are the ratings for six smart phones from each of the four suppliers, three of which cost $150 or more and three of which cost less than $150.[9] The ratings have a maximum value of 100 and a minimum of 0.

	Supplier			
	AT&T	Sprint	T-Mobile	Verizon
Cost ≥ $150	76	74	72	75
	74	69	71	73
	69	68	71	73
Cost < $150	69	69	71	72
	67	64	71	71
	64	60	70	70

a. What type of experiment was used to evaluate these smart phones? What are the factors? How many levels of each factor are used in the experiment?

b. Produce an analysis of variance table appropriate for this design, specifying the sources of variation, degrees of freedom, sums of squares, mean squares, and the appropriate values of F used in testing.

c. Is there significant interaction between the two factors?

d. Is there a main effect due to suppliers? What is its p-value?

e. Is there a main effect due to cost? What is its p-value?

f. Summarize the results of parts c–e.

11.73 Smart Phones, continued Refer to Exercise 11.72. The diagnostic plots for this experiment are shown below. Does it appear that any of the analysis of variance assumptions have been violated? Explain.

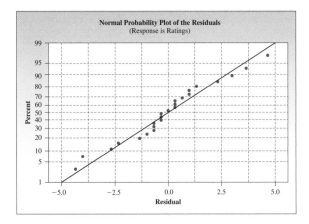

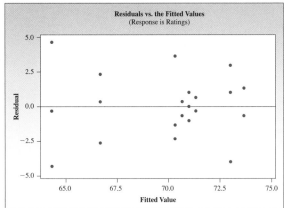

Data set **11.74 Professors' Salaries II** The U.S.
EX1174 Department of Education[7] reports the salaries of professors at universities and colleges in the United States. The following data (in thousands of dollars) is adapted from the report for full-time faculty on 9-month contracts at not-for-profit institutions offering doctoral programs. Ten samples were taken from each of the three professorial levels for both male and females.

	Rank					
Gender	Assistant Professor		Associate Professor		Full Professor	
Male	75.4	76.5	94.8	92.6	133.1	132.0
	75.6	75.1	96.4	95.0	147.7	122.1
	76.6	74.8	89.7	72.6	143.3	132.7
	76.2	74.9	103.1	100.2	158.1	142.0
	76.1	76.3	91.5	88.3	126.3	127.6
Female	74.8	71.1	86.3	74.7	161.1	122.9
	70.6	70.3	91.0	77.6	141.4	130.8
	72.4	71.1	88.3	71.8	130.8	145.4
	70.5	71.3	83.0	80.3	145.0	127.9
	71.1	69.3	72.5	80.9	127.5	122.9

a. Identify the design used in this survey.

b. Use the appropriate analysis of variance for these data.

c. Do the data indicate that the salary at the different ranks vary by gender?

d. If there is no interaction, determine whether there are differences in salaries by rank, and whether there are differences by gender. Discuss your results.

e. Plot the average salaries using an interaction plot. If the main effect of ranks is significant, use Tukey's method of pairwise comparisons to determine if there are significant differences among the ranks. Use $\alpha = .01$.

CASE STUDY

 Groceries

How to Save Money on Groceries!

Canning or freezing produce that you buy in bulk will almost always save you money compared to buying in supermarkets. You can save more than 75% by canning—and more than 80% by freezing—produce purchased in bulk. The following prices are found in "Save Money on Groceries," an article by Roberta R. Bailey and Craig Idlebrook on the website www.MotherEarthNews.com.[10]

Produce	Bulk Cost/lb ($)	Canned		Frozen	
		Home 16 oz ($)	Store 16 oz ($)	Home 16 oz ($)	Store 16 oz ($)
Green beans	1.00	1.00	1.31	1.00	1.99*
Sweet corn	2.00 (doz)	0.83	1.31	0.83	1.99*
Shell peas	2.00	6.00	1.31	6.00	1.99*
Whole tomatoes	1.00	1.50	1.31*	1.50	N/A
Beets	1.00	1.00	1.31	1.00	N/A
Broccoli	1.50	N/A	N/A	1.50	2.29
Spinach	4.00	N/A	N/A	4.00	4.62*
Pears	1.00	1.00	2.59	1.00	3.59
Blueberries	.50	.50	2.19	.50	3.59*
Peaches	1.00	1.00	1.99*	1.00	3.59

*The lowest price from a range is reported here.

There are some produce that are not canned and others that are not frozen. You may eliminate those entries for the analysis that follows.

1. Does this layout correspond to any of the designs studied in this chapter? If so, identify the rows and/or columns as they relate to that design.

2. Since the table is incomplete, consider deleting the rows corresponding to whole tomatoes, beets, broccoli, and spinach prior to analysis. Is the design given in part 1 still valid?

3. Use an appropriate analysis of variance procedure to analyze this data. If you find significance, use Tukey's procedure to identify real differences in prices.

4. Summarize your results in the form of a report. Can you really save money by buying produce in bulk? Explain.

AP Photo/David Zalubowski

Linear Regression and Correlation

GENERAL OBJECTIVES

In this chapter, we consider the situation in which the mean value of a random variable y is related to another variable x. By measuring both y and x for each experimental unit, thereby generating bivariate data, you can use the information provided by x to estimate the average value of y and to predict values of y for preassigned values of x.

CHAPTER INDEX

 NEED TO KNOW...

How to Make Sure That My Calculations Are Correct

Is Your Car "Made in the U.S.A."?

The phrase "made in the U.S.A." has become a battle cry in the past few years as American workers try to protect their jobs from overseas competition. In the case study at the end of this chapter, we explore the changing attitudes of American consumers toward automobiles made outside the United States, using a simple linear regression analysis.

INTRODUCTION

High school seniors, freshmen entering college, their parents, and a university administration are concerned about the academic achievement of a student after he or she has enrolled in a university. Can you estimate or predict a student's grade point average (GPA) at the end of the freshman year before the student enrolls in the university? At first glance this might seem like a difficult problem. However, you would expect highly motivated students who have graduated with a high class rank from a high school with superior academic standards to achieve a high GPA at the end of the college freshman year. On the other hand, students who lack motivation or who have achieved only moderate success in high school are not expected to do so well. You would expect the college achievement of a student to be a function of several variables:

- Rank in high school class
- High school's overall rating
- High school GPA
- SAT scores

This problem is of a fairly general nature. You are interested in a random variable y (college GPA) that is related to a number of independent variables. The objective is to create a *prediction equation* that expresses y as a function of these independent variables. Then, if you can measure the independent variables, you can substitute these values into the prediction equation and obtain the prediction for y—the student's college GPA in our example. But which variables should you use as predictors? How strong is their relationship to y? How do you construct a good prediction equation for y as a function of the selected predictor variables? We will answer these questions in the next two chapters.

In this chapter, we restrict our attention to the simple problem of predicting y as a linear function of a single predictor variable x. This problem was originally addressed in Chapter 3 in the discussion of *bivariate data.* Remember that we used the equation of a straight line to describe the relationship between x and y and we described the strength of the relationship using the correlation coefficient r. We rely on some of these results as we revisit the subject of linear regression and correlation.

A SIMPLE LINEAR PROBABILISTIC MODEL

Consider the problem of trying to predict the value of a response y based on the value of an independent variable x. The best-fitting line of Chapter 3,

$$y = a + bx$$

was based on a *sample* of n bivariate observations drawn from a larger *population* of measurements. The line that describes the relationship between y and x in the *population* is similar to, but not the same as, the best-fitting line from the *sample*. How can you construct a **population model** to describe the relationship between a random variable y and a related independent variable x?

You begin by assuming that the variable of interest, y, is *linearly* related to an independent variable x. To describe the linear relationship, you can use the **deterministic model**

$$y = \alpha + \beta x$$

where α is the y-intercept—the value of y when $x = 0$—and β is the slope of the line, defined as the change in y for a one-unit change in x, as shown in Figure 12.1. This model describes a deterministic relationship between the variable of interest y, sometimes called the **response variable,** and the independent variable x, often called the **predictor variable.** That is, the linear equation determines an exact value of y when the value of x is given. Is this a realistic model for an experimental situation? Consider the following example.

FIGURE 12.1

The y-intercept and slope for a line

NEED A TIP?

Slope = Change in y for a 1-unit change in x

y-Intercept = Value of y when $x = 0$

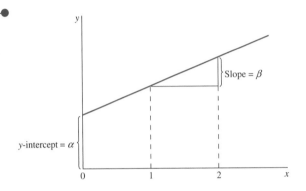

ONLINE APPLET

Building a Scatterplot

Table 12.1 displays the mathematics achievement test scores for a random sample of $n = 10$ college freshmen, along with their final calculus grades. A bivariate plot of these scores and grades is given in Figure 12.2. Notice that the points *do not lie exactly on a line* but rather seem to be deviations about an underlying line. A simple way to modify the deterministic model is to add a **random error component** to explain the deviations of the points about the line. A particular response y is described using the **probabilistic model**

$$y = \alpha + \beta x + \epsilon$$

TABLE 12.1

Mathematics Achievement Test Scores and Final Calculus Grades for College Freshmen

Student	Mathematics Achievement Test Score	Final Calculus Grade
1	39	65
2	43	78
3	21	52
4	64	82
5	57	92
6	47	89
7	28	73
8	75	98
9	34	56
10	52	75

FIGURE 12.2

Scatterplot of the data in
Table 12.1

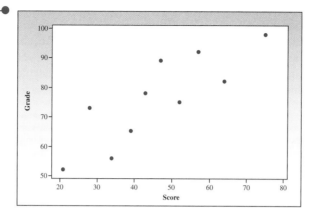

The first part of the equation, $\alpha + \beta x$—called the **line of means**—describes the average value of y for a given value of x. The error component ϵ allows each individual response y to deviate from the line of means by a small amount.

In order to use this *probabilistic model* for making inferences, you need to be more specific about this "small amount," ϵ.

ASSUMPTIONS ABOUT THE RANDOM ERROR ϵ

Assume that the values of ϵ satisfy these conditions:

- Are independent in the probabilistic sense
- Have a mean of 0 and a common variance equal to σ^2
- Have a normal probability distribution

These assumptions about the random error ϵ are shown in Figure 12.3 for three fixed values of x—say, x_1, x_2, and x_3. Notice the similarity between these assumptions and the assumptions necessary for the tests in Chapters 10 and 11. We will revisit these assumptions later in this chapter and provide some diagnostic tools for you to use in checking their validity.

FIGURE 12.3

Linear probabilistic model

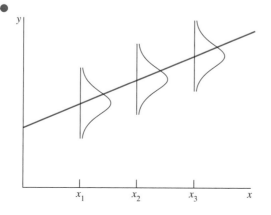

Remember that this model is created for a population of measurements that is generally unknown to you. However, you can use sample information to estimate the values of α and β, which are the coefficients of the line of means, $E(y) = \alpha + \beta x$. These estimates are used to form the best-fitting line for a given set of data, called the **least squares line** or **regression line.** We review how to calculate the intercept and the slope of this line in the next section.

THE METHOD OF LEAST SQUARES

12.3

The statistical procedure for finding the best-fitting line for a set of bivariate data does mathematically what you do visually when you move a ruler until you think you have minimized the vertical distances, or deviations, from the ruler to a set of points. The formula for the best-fitting line is

NEED A TIP?

Slope = Coefficient of x

y-Intercept = Constant term

$$\hat{y} = a + bx$$

where a and b are the estimates of the intercept and slope parameters α and β, respectively. The fitted line for the data in Table 12.1 is shown in Figure 12.4. The vertical lines drawn from the prediction line $(x_i, \hat{y}_i)$ to each point (x_i, y_i) represent the deviations of the points from the line—that is, $(y_i, - \hat{y}_i)$.

FIGURE 12.4

Method of Least Squares

ONLINE APPLET

Method of Least Squares

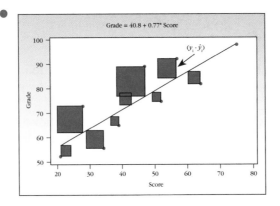

Notice that some points are below the prediction line, and hence $(y_i - \hat{y}_i)$ will be negative. To avoid the positive and negative distances from "cancelling each other out," we choose to minimize the distances from the points to the fitted line, using the **principle of least squares.**

PRINCIPLE OF LEAST SQUARES

The line that minimizes the sum of squares of the deviations of the observed values of y from those predicted is the **best-fitting line.** The sum of squared deviations is commonly called the **sum of squares for error** (SSE) and defined as

$$\text{SSE} = \Sigma(y_i - \hat{y}_i)^2 = \Sigma(y_i - a - bx_i)^2$$

Look at the regression line and the data points in Figure 12.4. SSE is the sum of the squared distances represented by the area of the blue squares.

Finding the values of a and b, the estimates of α and β, uses differential calculus, which is beyond the scope of this text. Rather than derive their values, we will simply present formulas for calculating the values of a and b—called the **least-squares estimators** of α and β. We will use notation that is based on the **sums of squares** for the variables in the regression problem, which are similar in form to the sums of squares used in Chapter 11. These formulas look different from the formulas presented in Chapter 3, but they are in fact algebraically identical!

You should use the data entry method for your scientific calculator to enter the sample data.

- If your calculator has only a one-variable statistics function, you can still save some time in finding the necessary sums and sums of squares.

- If your calculator has a two-variable statistics function, or if you have a graphing calculator, the calculator will automatically store all of the sums and sums of squares as well as the values of a, b, and the correlation coefficient r.

- Make sure you consult your calculator manual to find the easiest way to obtain the least-squares estimators.

LEAST-SQUARES ESTIMATORS OF α AND β

$$b = \frac{S_{xy}}{S_{xx}} \quad \text{and} \quad a = \bar{y} - b\bar{x}$$

where the quantities S_{xy} and S_{xx} are defined as

$$S_{xy} = \Sigma(x_i - \bar{x})(y_i - \bar{y}) = \Sigma x_i y_i - \frac{(\Sigma x_i)(\Sigma y_i)}{n}$$

and

$$S_{xx} = \Sigma(x_i - \bar{x})^2 = \Sigma x_i^2 - \frac{(\Sigma x_i)^2}{n}$$

Notice that the sum of squares of the x-values is found using the computing formula given in Section 2.3 and the sum of the cross-products is the numerator of the *covariance* defined in Section 3.4.

EXAMPLE 12.1 Find the least-squares prediction line for the calculus grade data in Table 12.1.

Solution   Use the data in Table 12.2 and the data entry method in your scientific calculator to find the following sums of squares:

$$S_{xx} = \Sigma x_i^2 - \frac{(\Sigma x_i)^2}{n} = 23{,}634 - \frac{(460)^2}{10} = 2474$$

$$S_{xy} = \Sigma x_i y_i - \frac{(\Sigma x_i)(\Sigma y_i)}{n} = 36{,}854 - \frac{(460)(760)}{10} = 1894$$

$$\bar{y} = \frac{\Sigma y_i}{n} = \frac{760}{10} = 76 \qquad \bar{x} = \frac{\Sigma x_i}{n} = \frac{460}{10} = 46$$

TABLE 12.2 **Calculations for the Data in Table 12.1**

y_i	x_i	x_i^2	$x_i y_i$	y_i^2
65	39	1521	2535	4225
78	43	1849	3354	6084
52	21	441	1092	2704
82	64	4096	5248	6724
92	57	3249	5244	8464
89	47	2209	4183	7921
73	28	784	2044	5329
98	75	5625	7350	9604
56	34	1156	1904	3136
75	52	2704	3900	5625
Sum 760	460	23,634	36,854	59,816

Then

$$b = \frac{S_{xy}}{S_{xx}} = \frac{1894}{2474} = .76556 \quad \text{and} \quad a = \bar{y} - b\bar{x} = 76 - (.76556)(46) = 40.78424$$

The least-squares regression line is then

$$\hat{y} = a + bx = 40.78424 + .76556x$$

The graph of this line is shown in Figure 12.4. It can now be used to predict y for a given value of x—either by referring to Figure 12.4 or by substituting the proper value of x into the equation. For example, if a freshman scored $x = 50$ on the achievement test, the student's predicted calculus grade is (using full decimal accuracy)

$$\hat{y} = a + b(50) = 40.78424 + (.76556)(50) = 79.06$$

> **NEED A TIP?**
>
> You can predict y for a given value of x by substituting x into the equation to find $\hat{y}$.

 NEED TO KNOW...

How to Make Sure That My Calculations Are Correct

- Be careful of rounding errors. Carry at least six significant figures, and round off only in reporting the end result.
- Use a scientific or graphing calculator to do all the work for you. Most of these calculators will calculate the values for a and b if you enter the data properly.
- Use a computer software program if you have access to one.
- Always plot the data and graph the line. If the line does not fit through the points, you have probably made a mistake!

AN ANALYSIS OF VARIANCE FOR LINEAR REGRESSION

12.4

In Chapter 11, you used the analysis of variance procedures to divide the total variation in the experiment into portions attributed to various factors of interest to the experimenter. In a regression analysis, the response y is related to the independent variable x. Hence, the total variation in the response variable y, given by

$$\text{Total SS} = S_{yy} = \Sigma(y_i - \bar{y})^2 = \Sigma y_i^2 - \frac{(\Sigma y_i)^2}{n}$$

is divided into two portions:

- SSR (sum of squares for regression) measures the amount of variation explained by using the regression line with one independent variable x

- SSE (sum of squares for error) measures the "residual" variation in the data that is not explained by the independent variable x

so that

$$\text{Total SS} = \text{SSR} + \text{SSE}$$

For a particular value of the response y_i, you can visualize this breakdown in the variation using the vertical distances illustrated in Figure 12.5. You can see that SSR is the sum of the squared deviations of the differences between the estimated response without using x ($\bar{y}$) and the estimated response using x (the regression line, $\hat{y}$); SSE is the sum of the squared differences between the regression line ($\hat{y}$) and the point y.

FIGURE 12.5

Deviations from the fitted line

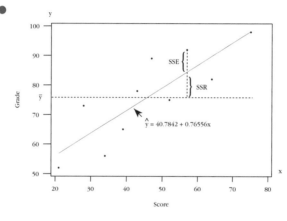

It is not too hard to show algebraically that

$$\text{SSR} = \Sigma(\hat{y}_i - \bar{y})^2 = \Sigma(a + bx_i - \bar{y})^2 = \Sigma(\bar{y} - b\bar{x} + bx_i - \bar{y})^2 = b^2\Sigma(x_i - \bar{x})^2$$

$$= \left(\frac{S_{xy}}{S_{xx}}\right)^2 S_{xx} = \frac{(S_{xy})^2}{S_{xx}}$$

Since Total SS = SSR + SSE, you can complete the partition by calculating

$$\text{SSE} = \text{Total SS} - \text{SSR} = S_{yy} - \frac{(S_{xy})^2}{S_{xx}}$$

Remember from Chapter 11 that each of the various sources of variation, when divided by the appropriate **degrees of freedom,** provides an estimate of the variation in the experiment. These estimates are called **mean squares**—MS = SS/df—and are displayed in an ANOVA table.

In examining the degrees of freedom associated with each of these sums of squares, notice that the total degrees of freedom for n measurements is $(n - 1)$. Since estimating the regression line, $\hat{y} = a + bx_i = \bar{y} - b\bar{x} + bx_i$, involves estimating *one additional* parameter β, there is *one* degree of freedom associated with SSR, leaving $(n - 2)$ degrees of freedom with SSE.

As with all ANOVA tables we have discussed, the mean square for error,

$$\text{MSE} = s^2 = \frac{\text{SSE}}{n-2}$$

is an unbiased estimator of the underlying variance σ^2. The analysis of variance table is shown in Table 12.3.

TABLE 12.3 ● **Analysis of Variance for Linear Regression**

Source	df	SS	MS
Regression	1	$\dfrac{(S_{xy})^2}{S_{xx}}$	MSR
Error	$n-2$	$S_{yy} - \dfrac{(S_{xy})^2}{S_{xx}}$	MSE
Total	$n-1$	S_{yy}	

For the data in Table 12.1, you can calculate

$$\text{Total SS} = S_{yy} = \Sigma y_i^2 - \frac{(\Sigma y_i)^2}{n} = 59{,}816 - \frac{(760)^2}{10} = 2056$$

$$\text{SSR} = \frac{(S_{xy})^2}{S_{xx}} = \frac{(1894)^2}{2474} = 1449.9741$$

so that

$$\text{SSE} = \text{Total SS} - \text{SSR} = 2056 - 1449.9741 = 606.0259$$

and

$$\text{MSE} = \frac{\text{SSE}}{n-2} = \frac{606.0259}{8} = 75.7532$$

The analysis of variance table, part of the *linear regression output* generated by *MINITAB*, is the lower shaded section in the printout in Figure 12.6(a). The first two lines give the equation of the least-squares line, $\hat{y} = 40.8 + .766x$. The least-squares estimates a and b are given with greater accuracy in the column labeled "Coef." The

FIGURE 12.6(a) ●
MINITAB output for the data of Table 12.1

Regression Analysis: y versus x

The regression equation is
y = 40.8 + 0.766 x

Predictor	Coef	SE Coef	T	P
Constant	40.784	8.507	4.79	0.001
x	0.7656	0.1750	4.38	0.002

S = 8.70363   R-Sq = 70.5%   R-Sq(adj) = 66.8%

Analysis of Variance

Source	DF	SS	MS	F	P
Regression	1	1450.0	1450.0	19.14	0.002
Residual Error	8	606.0	75.8		
Total	9	2056.0			

FIGURE 12.6(b)

MS Excel output for the data of Table 12.1

SUMMARY OUTPUT							
Regression Statistics							
Multiple R	0.8398						
R Square	0.7052						
Adjusted R Square	0.6684						
Standard Error	8.7036						
Observations	10						

ANOVA							
	df	*SS*	*MS*	*F*	*Significance F*		
Regression	1	1449.974	1449.974	19.141	0.002		
Residual	8	606.026	75.753				
Total	9	2056					

	Coefficients	*Standard Error*	*t Stat*	*P-value*	*Lower 95%*	*Upper 95%*	*Lower 95.0%*
Intercept	40.784	8.507	4.794	0.001	21.167	60.401	21.167
Score	0.766	0.175	4.375	0.002	0.362	1.169	0.362

MS Excel output for the same data is shown in Figure 12.6(b). The ANOVA table is in the middle of the shaded output; the least squares estimates are found at the bottom of the shaded output in the column labeled "Coefficients." You can find instructions for generating this output in the section "Technology Today" at the end of this chapter.

The computer outputs also give some information about the variation in the experiment. Each of the least-squares estimates, *a* and *b,* has an associated standard error, labeled "SE Coef" in Figure 12.6(a) and "Standard Error" in Figure 12.6(b). In the middle of the *MINITAB* output, you will find the best unbiased estimate of σ—$s = \sqrt{\text{MSE}} = \sqrt{75.7532} = 8.70363$—which measures the **residual error,** the unexplained or "leftover" variation in the experiment. This same measure is found in the top portion of the *MS Excel* output labeled "Standard Error." It will not surprise you to know that the *t* and *F* statistics and their *p*-values found in the printout are used to test statistical hypotheses. We explain these entries in the next section.

 NEED A TIP?

Look for *a* and *b* in the column called "Coef" or "Coefficients."

12.4 EXERCISES

BASIC TECHNIQUES

12.1 Graph the line corresponding to the equation $y = 2x + 1$ by graphing the points corresponding to $x = 0, 1,$ and 2. Give the *y*-intercept and slope for the line.

12.2 You are given five points with these coordinates:

x	−2	−1	0	1	2
y	1	1	3	5	5

a. Use the data entry method on your scientific or graphing calculator to enter the $n = 5$ observations. Find the sums of squares and cross-products, $S_{xx}, S_{xy},$ and $S_{yy}.$

b. Find the least-squares line for the data.

c. Plot the five points and graph the line in part b. Does the line appear to provide a good fit to the data points?

d. Construct the ANOVA table for the linear regression.

12.3 Give the equation and graph for a line with *y*-intercept equal to 3 and slope equal to −1.

12.4 Graph the line corresponding to the equation $y = -2x + 1$ by graphing the points corresponding to $x = 0, 1,$ and 2. Give the *y*-intercept and slope for the line. How is this line related to the line $y = 2x + 1$ of Exercise 12.1?

12.5 What is the difference between deterministic and probabilistic mathematical models?

12.6 Give the equation and graph for a line with *y*-intercept equal to −3 and slope equal to 1.

12.7 Six points have these coordinates:

x	1	2	3	4	5	6
y	5.6	4.6	4.5	3.7	3.2	2.7

a. Find the least-squares line for the data.

b. Plot the six points and graph the line. Does the line appear to provide a good fit to the data points?

c. Use the least-squares line to predict the value of *y* when $x = 3.5.$

d. Fill in the missing entries in the *MINITAB* analysis of variance table.

MINITAB ANOVA table for Exercise 12.7

```
Analysis of Variance

Source          DF        SS        MS
Regression      *        ***       5.4321
Residual Error  *      0.1429       ***
Total           *      5.5750
```

12.8 Six points have these coordinates:

x	1	2	3	4	5	6
y	9.7	6.5	6.4	4.1	2.1	1.0

a. Find the least-squares line for the data.

b. Plot the six points and graph the line. Does the line appear to provide a good fit to the data points?

c. Use the least-squares line to predict the value of y when $x = 3.5$.

d. Fill in the missing entries in the *MS Excel* analysis of variance table.

ANOVA

	df	SS	MS
Regression	*	***	49.7286
Residual	*	1.7848	***
Total	*	51.5133	

APPLICATIONS

12.9 Professor Asimov Professor Isaac Asimov was one of the most prolific writers of all time. Prior to his death, he wrote nearly 500 books during a 40-year career. In fact, as his career progressed, he became even more productive in terms of the number of books written within a given period of time.[1] The data give the time in months required to write his books in increments of 100:

Number of Books, x	100	200	300	400	490
Time in Months, y	237	350	419	465	507

a. Assume that the number of books x and the time in months y are linearly related. Find the least-squares line relating y to x.

b. Plot the time as a function of the number of books written using a scatterplot, and graph the least-squares line on the same paper. Does it seem to provide a good fit to the data points?

c. Construct the ANOVA table for the linear regression.

12.10 A Chemical Experiment Using a chemical procedure called *differential pulse polarography,* a chemist measured the peak current generated (in microamperes) when a solution containing a given amount of nickel (in parts per billion) is added to a buffer:[2]

x = Ni (ppb)	y = Peak Current (mA)
19.1	.095
38.2	.174
57.3	.256
76.2	.348
95	.429
114	.500
131	.580
150	.651
170	.722

a. Use the data entry method for your calculator to calculate the preliminary sums of squares and cross-products, S_{xx}, S_{yy}, and S_{xy}.

b. Calculate the least-squares regression line.

c. Plot the points and the fitted line. Does the assumption of a linear relationship appear to be reasonable?

d. Use the regression line to predict the peak current generated when a solution containing 100 ppb of nickel is added to the buffer.

e. Construct the ANOVA table for the linear regression.

12.11 Sleep Deprivation A study was conducted to determine the effects of sleep deprivation on people's ability to solve problems without sleep. A total of 10 subjects participated in the study, two at each of five sleep deprivation levels—8, 12, 16, 20, and 24 hours. After his or her specified sleep deprivation period, each subject was administered a set of simple addition problems, and the number of errors was recorded. These results were obtained:

Number of Errors, y	8, 6	6, 10	8, 14
Number of Hours without Sleep, x	8	12	16
Number of Errors, y	14, 12	16, 12	
Number of Hours without Sleep, x	20	24	

a. How many pairs of observations are in the experiment?

b. What are the total number of degrees of freedom?

c. Complete the *MINITAB* printout.

MINITAB output for Exercise 12.11

Regression Analysis: y versus x

```
The regression equation is
y = 3.00 + 0.475 x

Predictor       Coef     SE Coef         T       P
Constant       3.000       2.127      1.41   0.196
x                ***      0.1253      3.79   0.005

S = 2.24165    R-Sq = 64.2%    R-Sq(adj) = 59.8%

Analysis of Variance
Source          DF        SS        MS       F       P
Regression      **    72.200    72.200   14.37   0.005
Residual Error  **       ***     5.025
Total           **       ***
```

d. What is the least-squares prediction equation?

e. Use the prediction equation to predict the number of errors for a person who has not slept for 10 hours.

12.12 Sleep Deprivation II Refer to the data given in the sleep deprivation experiment in Exercise 12.11. Answer the questions posed in parts a, b, d, and e of that exercise by completing the following *MS Excel* printout:

ANOVA

	df	SS	MS	F	Significance F
Regression	**	72.2	72.2	14.36816	0.005308
Residual	**	***	5.025		
Total	**	***			

	Coefficients	Standard Error	t Stat	P-value
Intercept	3	2.126617	1.410691	0.196016
x	0.475	0.125312	3.790535	0.005308

 12.13 Achievement Tests The Academic EX1213 Performance Index (API) is a measure of school achievement based on the results of the Stanford 9 Achievement test. Scores range from 200 to 1000, with 800 considered a long-range goal for schools. The following table shows the API for eight elementary schools in Riverside County, California, along with the percent of students at that school who are considered "English Learners" (EL).[3]

School	1	2	3	4	5	6	7	8
API	745	808	798	791	854	688	801	751
EL	71	18	24	50	17	71	11	57

a. Which of the two variables is the independent variable and which is the dependent variable? Explain your choice.

b. Use a scatterplot to plot the data. Is the assumption of a linear relationship between x and y reasonable?

c. Assuming that x and y are linearly related, calculate the least-squares regression line.

d. Plot the line on the scatterplot in part b. Does the line fit through the data points?

12.14 How Long Is It? How good are you at EX1214 estimating? To test a subject's ability to estimate sizes, he was shown 10 different objects and asked to estimate their length or diameter. The object was then measured, and the results were recorded in the table below.

Object	Estimated (inches)	Actual (inches)
Pencil	7.00	6.00
Dinner plate	9.50	10.25
Book 1	7.50	6.75
Cell phone	4.00	4.25
Photograph	14.50	15.75
Toy	3.75	5.00
Belt	42.00	41.50
Clothespin	2.75	3.75
Book 2	10.00	9.25
Calculator	3.50	4.75

a. Find the least-squares regression line for predicting the actual measurement as a function of the estimated measurement.

b. Plot the points and the fitted line. Does the assumption of a linear relationship appear to be reasonable?

12.15 Test Interviews Of two personnel EX1215 evaluation techniques available, the first requires a two-hour test interview while the second can be completed in less than an hour. The scores for each of the 15 individuals who took both tests are given in the next table.

Applicant	Test 1 (x)	Test 2 (y)
1	75	38
2	89	56
3	60	35
4	71	45
5	92	59
6	105	70
7	55	31
8	87	52
9	73	48
10	77	41
11	84	51
12	91	58
13	75	45
14	82	49
15	76	47

a. Construct a scatterplot for the data. Does the assumption of linearity appear to be reasonable?

b. Find the least-squares line for the data.

c. Use the regression line to predict the score on the second test for an applicant who scored 85 on Test 1.

12.16 Test Interviews, continued Refer to Exercise 12.15. Construct the ANOVA table for the linear regression relating y, the score on Test 2, to x, the score on Test 1.

Data set **12.17 Armspan and Height** Leonardo EX1217 da Vinci (1452–1519) drew a sketch of a man, indicating that a person's armspan (measuring across the back with your arms outstretched to make a "T") is roughly equal to the person's height. To test this claim, we measured eight people with the following results:

Person	1	2	3	4
Armspan (inches)	68	62.25	65	69.5
Height (inches)	69	62	65	70

Person	5	6	7	8
Armspan (inches)	68	69	62	60.25
Height (inches)	67	67	63	62

a. Draw a scatterplot for armspan and height. Use the same scale on both the horizontal and vertical axes. Describe the relationship between the two variables.

b. If da Vinci is correct, and a person's armspan is roughly the same as the person's height, what should the slope of the regression line be?

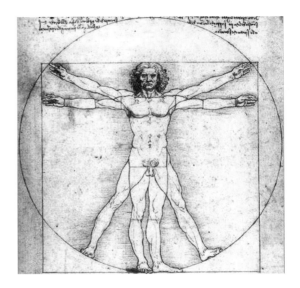

c. Calculate the regression line for predicting height based on a person's armspan. Does the value of the slope b confirm your conclusions in part b?

d. If a person has an armspan of 62 inches, what would you predict the person's height to be?

Data set **12.18 Strawberries** The following data were EX1218 obtained in an experiment relating the dependent variable, y (texture of strawberries), with x (coded storage temperature).

x	−2	−2	0	2	2
y	4.0	3.5	2.0	0.5	0.0

a. Find the least-squares line for the data.

b. Plot the data points and graph the least-squares line as a check on your calculations.

c. Construct the ANOVA table.

TESTING THE USEFULNESS OF THE LINEAR REGRESSION MODEL

In considering linear regression, you may ask two questions:

- Is the independent variable x useful in predicting the response variable y?
- If so, how well does it work?

This section examines several statistical tests and measures that will help you reach some answers. Once you have determined that the model is working, you can then use the model for predicting the response y for a given value of x.

Inferences Concerning β, the Slope of the Line of Means

Is the least-squares regression line useful? That is, is the regression equation that uses information provided by x substantially better than the simple predictor $\bar{y}$ that does not rely on x? If the independent variable x is *not useful* in the population model $y = \alpha + \beta x + \epsilon$, then the value of y does not change for different values of x. The only way that this happens for all values of x is when the slope β of the line of means equals 0. This would indicate that the relationship between y and x is not linear, so that the initial question about the usefulness of the independent variable x can be restated as: Is there a linear relationship between x and y?

You can answer this question by using either a test of hypothesis or a confidence interval for β. Both of these procedures are based on the sampling distribution of b, the sample estimator of the slope β. It can be shown that, if the assumptions about the random error ϵ are valid, then the estimator b has a normal distribution in repeated sampling with mean

$$E(b) = \beta$$

and standard error given by

$$\text{SE} = \sqrt{\frac{\sigma^2}{S_{xx}}}$$

where σ^2 is the variance of the random error ϵ. Since the value of σ^2 is estimated with $s^2 = \text{MSE}$, you can base inferences on the statistic given by

$$t = \frac{b - \beta}{\sqrt{\text{MSE}/S_{xx}}}$$

which has a t distribution with $df = (n - 2)$, the degrees of freedom associated with MSE.

TEST OF HYPOTHESIS CONCERNING THE SLOPE OF A LINE

1. Null hypothesis: $H_0 : \beta = \beta_0$
2. Alternative hypothesis:

One-Tailed Test	Two-Tailed Test
$H_a : \beta > \beta_0$ (or $\beta < \beta_0$)	$H_a : \beta \neq \beta_0$

3. Test statistic: $t = \dfrac{b - \beta_0}{\sqrt{\text{MSE}/S_{xx}}}$

When the assumptions given in Section 12.2 are satisfied, the test statistic will have a Student's t distribution with $(n - 2)$ degrees of freedom.

TEST OF HYPOTHESIS CONCERNING THE SLOPE OF A LINE (*Continued*)

4. Rejection region: Reject H_0 when

One-Tailed Test

$t > t_\alpha$
(or $t < -t_\alpha$ when the alternative hypothesis is $H_a : \beta < \beta_0$)

Two-Tailed Test

$t > t_{\alpha/2}$ or $t < -t_{\alpha/2}$

or when *p*-value $< \alpha$

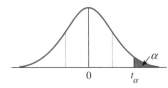 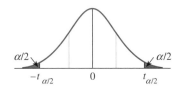

The values of t_α and $t_{\alpha/2}$ corresponding to $(n - 2)$ degrees of freedom can be found using Table 4 in Appendix I.

EXAMPLE 12.2

Determine whether there is a significant linear relationship between the calculus grades and test scores listed in Table 12.1. Test at the 5% level of significance.

Solution The hypotheses to be tested are

$$H_0 : \beta = 0 \quad \text{versus} \quad H_a : \beta \neq 0$$

MY ONLINE APPLET

t-Test for the Slope & Student's Probabilities

and the observed value of the test statistic is calculated as

$$t = \frac{b - 0}{\sqrt{\text{MSE}/S_{xx}}} = \frac{.7656 - 0}{\sqrt{75.7532/2474}} = 4.38$$

with $(n - 2) = 8$ degrees of freedom. With $\alpha = .05$, you can reject H_0 when $t > 2.306$ or $t < -2.306$. Since the observed value of the test statistic falls into the rejection region, H_0 is rejected and you can conclude that there is a significant linear relationship between the calculus grades and the test scores for the population of college freshmen.

Another way to make inferences about the value of β is to construct a confidence interval for β and examine the range of possible values for β.

A $(1 - \alpha)100\%$ CONFIDENCE INTERVAL FOR β

$$b \pm t_{\alpha/2}(\text{SE})$$

where $t_{\alpha/2}$ is based on $(n - 2)$ degrees of freedom, $s^2 = \text{MSE}$, and

$$\text{SE} = \sqrt{\frac{s^2}{S_{xx}}} = \sqrt{\frac{\text{MSE}}{S_{xx}}}$$

EXAMPLE 12.3

Find a 95% confidence interval estimate of the slope β for the calculus grade data in Table 12.1.

Solution Substituting previously calculated values into

$$b \pm t_{.025}\sqrt{\frac{\text{MSE}}{S_{xx}}}$$

you have

$$.766 \pm 2.306 \sqrt{\frac{75.7532}{2474}}$$

$$.766 \pm .404$$

The resulting 95% confidence interval is .362 to 1.170. Since the interval does not contain 0, you can conclude that the true value of β is not 0, and you can reject the null hypothesis $H_0 : \beta = 0$ in favor of $H_a : \beta \neq 0$, a conclusion that agrees with the findings in Example 12.2. Furthermore, the confidence interval estimate indicates that there is an increase from as little as .4 to as much as 1.2 points in a calculus test score for each 1-point increase in the achievement test score.

If you are using computer software to perform the regression analysis, you will find the t statistic and its p-value on the printout. In the second section of the *MINITAB* output in Figure 12.7(a), you will find the least-squares estimate b of the slope in the line marked "x," along with its standard error "SE Coef," the calculated value of the test statistic "T" used for testing the hypothesis $H_0 : \beta = 0$ and its p-value "P." You will find the same information in the last line of the *MS Excel* output in Figure 12.7(b), along with the upper and lower confidence limits of a 95% confidence interval for β. The t-test for significant regression, $H_0 : \beta = 0$, has a p-value of $P = .002$, and the null hypothesis is rejected, as in Example 12.2. There is a significant linear relationship between x and y.

FIGURE 12.7(a)

MINITAB output for the calculus grade data

NEED A TIP?

Look for the standard error of *b* in the column marked "SE Coef" on the *MINITAB* output and "Standard Error" on the *MS Excel* output.

Regression Analysis: y versus x

```
The regression equation is
y = 40.8 + 0.766 x

Predictor      Coef      SE Coef        T         P
Constant     40.784        8.507      4.79     0.001
x             0.7656       0.1750     4.38     0.002

S = 8.70363    R-Sq = 70.5%    R-Sq(adj) = 66.8%

Analysis of Variance
Source            DF        SS        MS        F         P
Regression         1     1450.0    1450.0    19.14     0.002
Residual Error     8      606.0      75.8
Total              9     2056.0
```

FIGURE 12.7(b)

MS Excel output for the calculus grade data

SUMMARY OUTPUT

Regression Statistics	
Multiple R	0.8398
R Square	0.7052
Adjusted R Square	0.6684
Standard Error	8.7036
Observations	10

ANOVA

	df	SS	MS	F	Significance F
Regression	1	1449.974	1449.974	19.141	0.002
Residual	8	606.026	75.753		
Total	9	2056			

	Coefficients	Standard Error	t Stat	P-value	Lower 95%	Upper 95%
Intercept	40.784	8.507	4.794	0.001	21.167	60.401
x	0.766	0.175	4.375	0.002	0.362	1.169

The Analysis of Variance F-Test

ANOVA *F*-tests are always one-tailed (upper-tail).

The analysis of variance portion of the printout in Figures 12.7(a) and 12.7(b) shows an F statistic given by

$$F = \frac{\text{MSR}}{\text{MSE}} = 19.14$$

with 1 numerator degree of freedom and $(n - 2) = 8$ denominator degrees of freedom. This is an *equivalent test statistic* that can also be used for testing the hypothesis $H_0 : \beta = 0$. Notice that, within rounding error, the value of F is equal to t^2 with the identical p-value. In this case, if you use five-decimal-place accuracy prior to rounding, you find that $t^2 = (.76556/.17498)^2 = (4.37513)^2 = 19.14175 \approx 19.14 = F$ as given in the printout. This is no accident and results from the fact that the square of a t statistic with df degrees of freedom has the same distribution as an F-statistic with 1 numerator and df denominator degrees of freedom. The F-test is a more general test of the usefulness of the model and can be used when the model has more than one independent variable.

Measuring the Strength of the Relationship: The Coefficient of Determination

How well does the regression model fit? To answer this question, you can use a measure related to the *correlation coefficient r,* introduced in Chapter 3. Remember that

$$r = \frac{s_{xy}}{s_x s_y} = \frac{S_{xy}}{\sqrt{S_{xx} S_{yy}}} \qquad \text{for } -1 \le r \le 1$$

where s_{xy}, s_x, and s_y were defined in Chapter 3 and the various sums of squares were defined in Section 12.4.

The sum of squares for regression, SSR, in the analysis of variance measures the portion of the total variation, Total SS $= S_{yy}$, that can be explained by the regression of y on x. The remaining portion, SSE, is the "unexplained" variation attributed to random error. One way to measure the strength of the relationship between the response variable y and the predictor variable x is to calculate the **coefficient of determination**—the proportion of the total variation that is explained by the linear regression of y on x. For the calculus grade data, this proportion is equal to

On computer printouts, r^2 is often given as a **percentage** rather than a proportion.

$$\frac{\text{SSR}}{\text{Total SS}} = \frac{1450}{2056} = .705 \quad \text{or} \quad 70.5\%$$

Since Total SS $= S_{yy}$ and SSR $= \dfrac{(S_{xy})^2}{S_{xx}}$, you can write

$$\frac{\text{SSR}}{\text{Total SS}} = \frac{(S_{xy})^2}{S_{xx} S_{yy}} = \left(\frac{S_{xy}}{\sqrt{S_{xx} S_{yy}}}\right)^2 = r^2$$

Therefore, the coefficient of determination, which was calculated as SSR/Total SS, is simply the square of the correlation coefficient r. It is the entry labeled "R-Sq" in Figure 12.7(a) and "R Square" in Figure 12.7(b).

Remember that the analysis of variance table isolates the variation due to regression (SSR) from the total variation in the experiment. Doing so reduces the amount of *random variation* in the experiment, now measured by SSE rather than Total SS. In this context, the **coefficient of determination, r^2,** can be defined as follows:

NEED A TIP?

r^2 is called "R-Sq" on the *MINITAB* printout and "R Square" on the *Excel* printout.

Definition The **coefficient of determination** r^2 can be interpreted as the percent reduction in the total variation in the experiment obtained by using the regression line $\hat{y} = a + bx$, instead of ignoring x and using the sample mean $\bar{y}$ to predict the response variable y.

For the calculus grade data, a reduction of $r^2 = .705$ or 70.5% is substantial. The regression model is working very well!

Interpreting the Results of a Significant Regression

Once you have performed the t-test or F-test to determine the significance of the linear regression, you must interpret your results carefully. The slope β of the line of means is estimated based on data from only a particular region of observation. Even if you do not reject the null hypothesis that the slope of the line equals 0, it does not necessarily mean that y and x are unrelated. It may be that you have committed a Type II error—falsely declaring that the slope is 0 and that x and y are unrelated.

Fitting the Wrong Model

It may happen that y and x are perfectly related in a nonlinear way, as shown in Figure 12.8. Here are three possibilities:

FIGURE 12.8

Curvilinear relationship

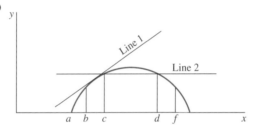

- If observations were taken only within the interval $b < x < c$, the relationship would appear to be linear with a positive slope.

- If observations were taken only within the interval $d < x < f$, the relationship would appear to be linear with a negative slope.

- If the observations were taken over the interval $c < x < d$, the line would be fitted with a slope close to 0, indicating no linear relationship between y and x.

For the example shown in Figure 12.8, no straight line accurately describes the true relationship between x and y, which is really a *curvilinear relationship*. In this case, we have chosen the *wrong model* to describe the relationship. Sometimes this type of mistake can be detected using residual plots, the subject of Section 12.7.

 NEED A TIP?

It is dangerous to try to predict values of y outside of the range of the fitted data.

Extrapolation

One serious problem is to apply the results of a linear regression analysis to values of x that are *not included* within the range of the fitted data. This is called **extrapolation** and can lead to serious errors in prediction, as shown for line 1 in Figure 12.8. Prediction results would be good over the interval $b < x < c$ but would seriously overestimate the values of y for $x > c$.

Causality

When there is a significant regression of y and x, it is tempting to conclude that x *causes* y. However, it is possible that one or more unknown variables that you have not even measured and that are not included in the analysis may be causing the observed relationship. In general, the statistician reports the results of an analysis but leaves conclusions concerning causality to scientists and investigators who are experts in these areas. These experts are better prepared to make such decisions!

12.5 EXERCISES

BASIC TECHNIQUES

12.19 Refer to Exercise 12.6. The data are reproduced below.

x	−2	−1	0	1	2
y	1	1	3	5	5

a. Do the data present sufficient evidence to indicate that y and x are linearly related? Test the hypothesis that $\beta = 0$ at the 5% level of significance.

b. Use the ANOVA table from Exercise 12.6 to calculate $F = MSR/MSE$. Verify that the square of the t statistic used in part a is equal to F.

c. Compare the two-tailed critical value for the t-test in part a with the critical value for F with $\alpha = .05$. What is the relationship between the critical values?

12.20 Refer to Exercise 12.19. Find a 95% confidence interval for the slope of the line. What does the phrase "95% confident" mean?

12.21 Refer to Exercise 12.7. The data, along with the *MINITAB* analysis of variance table are reproduced below.

x	1	2	3	4	5	6
y	5.6	4.6	4.5	3.7	3.2	2.7

MINITAB ANOVA table for Exercise 12.21

Regression Analysis: y versus x

```
Analysis of Variance
Source          DF        SS        MS        F        P
Regression       1    5.4321    5.4321   152.10    0.000
Residual Error   4    0.1429    0.0357
Total            5    5.5750
```

a. Do the data provide sufficient evidence to indicate that y and x are linearly related? Use the information in the *MINITAB* printout to answer this question at the 1% level of significance.

b. Calculate the coefficient of determination r^2. What information does this value give about the usefulness of the linear model?

12.22 Refer to Exercise 12.8. The data, along with the *MS Excel* analysis of variance table are reproduced below:

x	1	2	3	4	5	6
y	9.7	6.5	6.4	4.1	2.1	1.0

MS Excel ANOVA table for Exercise 12.22

ANOVA

	df	SS	MS	F	Significance F
Regression	1	49.72857	49.72857	111.45	0.000
Residual	4	1.78476	0.4462		
Total	5	51.51333			

a. Do the data provide sufficient evidence to indicate that y and x are linearly related? Use the information in the printout to answer this question at the 5% level of significance.

b. Calculate the coefficient of determination r^2. What information does this value give about the usefulness of the linear model?

APPLICATIONS

12.23 Chirping Crickets In Exercise 3.18, EX1223 we found that male crickets chirp by rubbing their front wings together, and their chirping is temperature dependent. The table below shows the number of chirps per second for a cricket, recorded at 10 different temperatures:

Chirps per Second	20	16	19	18	18	16	14	17	15	16
Temperature	88	73	91	85	82	75	69	82	69	83

a. Use the formulas given in this chapter to find the least-squares regression line relating the number of chirps to temperature. Compare to the results obtained in Exercise 3.18.

b. Do the data provide sufficient evidence to indicate that there is a linear relationship between number of chirps and temperature?

c. Calculate r^2. What does this value tell you about the effectiveness of the linear regression analysis?

Data set **12.24 Gestation Times and Longevity** The
EX1224 table below, a subset of the data given in Exercise
3.33, shows the gestation time in days and the average
longevity in years for a variety of mammals in captivity.[4]

Animal	Gestation (days)	Avg Longevity (yrs)
Baboon	187	20
Bear (black)	219	18
Bison	285	15
Cat (domestic)	63	12
Elk	250	15
Fox (red)	52	7
Goat (domestic)	151	8
Gorilla	258	20
Horse	330	20
Monkey (rhesus)	166	15
Mouse (meadow)	21	3
Pig (domestic)	112	10
Puma	90	12
Sheep (domestic)	154	12
Wolf (maned)	63	5

a. If you want to estimate the average longevity of an
animal based on its gestation time, which variable
is the response variable and which is the independent predictor variable?

b. Assume that there is a linear relationship between
gestation time and longevity. Calculate the least-
squares regression line describing longevity as a
linear function of gestation time.

c. Plot the data points and the regression line. Does it
appear that the line fits the data?

d. Use the appropriate statistical tests and measures to
explain the usefulness of the regression model for
predicting longevity.

12.25 Professor Asimov, continued Refer to the
data in Exercise 12.9, relating x, the number of books
written by Professor Isaac Asimov, to y, the number of
months he took to write his books (in increments of
100). The data are reproduced below.

Number of Books, x	100	200	300	400	490
Time in Months, y	237	350	419	465	507

a. Do the data support the hypothesis that $\beta = 0$? Use
the p-value approach, bounding the p-value using
Table 4 of Appendix I. Explain your conclusions in
practical terms.

b. Use the ANOVA table in Exercise 12.9, part c,
to calculate the coefficient of determination r^2.
What percentage reduction in the total variation

is achieved by using the linear regression
model?

c. Plot the data or refer to the plot in Exercise 12.9,
part b. Do the results of parts a and b indicate that
the model provides a good fit for the data? Are
there any assumptions that may have been violated
in fitting the linear model?

12.26 Refer to the sleep deprivation experiment
described in Exercises 12.11 and 12.12 and data set
EX1211. The data and the *MINITAB* and *MS Excel*
printout are reproduced here.

Number of Errors, y	8, 6	6, 10	8, 14
Number of Hours without Sleep, x	8	12	16

Number of Errors, y	14, 12	16, 12
Number of Hours without Sleep, x	20	24

MINITAB output for Exercise 12.26

Regression Analysis: y versus x

```
The regression equation is
y = 3.00 + 0.475 x

Predictor     Coef    SE Coef       T       P
Constant     3.000     2.127     1.41   0.196
x            0.4750    0.1253     3.79   0.005

S = 2.24165   R-Sq = 64.2%   R-Sq(adj) = 59.8%

Analysis of Variance

Source          DF       SS       MS       F      P
Regression       1   72.200   72.200   14.37  0.005
Residual Error   8   40.200    5.025
Total            9  112.400
```

MS Excel output for Exercise 12.26

ANOVA

	df	SS	MS	F	Significance F
Regression	1	72.2	72.2	14.368	0.005
Residual	8	40.2	5.025		
Total	9	112.4			

	Coefficients	Standard Error	t Stat	P-value	Lower 95%	Upper 95%
Intercept	3	2.1266	1.4107	0.1960	-1.9040	7.9040
x	0.475	0.1253	3.7905	0.0053	0.1860	0.7640

a. Do the data present sufficient evidence to indicate
that the number of errors is linearly related to the
number of hours without sleep? Identify the two
test statistics in the printout that can be used to
answer this question.

b. Would you expect the relationship between y and x
to be linear if x varied over a wider range (say,
$x = 4$ to $x = 48$)?

c. How do you describe the strength of the relationship between y and x?

d. What is the best estimate of the common population variance σ^2?

e. Find a 95% confidence interval for the slope of the line.

12.27 Strawberries II The following data (Exercise 12.18 and data set EX1218) were obtained in an experiment relating the dependent variable, y (texture of strawberries), with x (coded storage temperature). Use the information from Exercise 12.18 to answer the following questions:

x	−2	−2	0	2	2
y	4.0	3.5	2.0	0.5	0.0

a. What is the best estimate of σ^2, the variance of the random error ε?

b. Do the data indicate that texture and storage temperature are linearly related? Use $\alpha = .05$.

c. Calculate the coefficient of determination, r^2.

d. Of what value is the *linear* model in increasing the accuracy of prediction as compared to the predictor, $\bar{y}$?

Data set **12.28 Laptops and Learning** In Exercise
EX1228 1.61 we described an informal experiment conducted at McNair Academic High School in Jersey City, New Jersey. Two freshman algebra classes were studied, one of which used laptop computers at school and at home, while the other class did not. In each class, students were given a survey at the beginning and end of the semester, measuring his or her technological level. The scores were recorded for the end of semester survey (x) and the final examination (y) for the laptop group.[5] The data and the *MINITAB* printout are shown here.

Student	Posttest	Final Exam	Student	Posttest	Final Exam
1	100	98	11	88	84
2	96	97	12	92	93
3	88	88	13	68	57
4	100	100	14	84	84
5	100	100	15	84	81
6	96	78	16	88	83
7	80	68	17	72	84
8	68	47	18	88	93
9	92	90	19	72	57
10	96	94	20	88	83

MINITAB output for Exercise 12.28

Regression Analysis: y versus x

```
The regression equation is
y = -26.8 + 1.26 x

Predictor      Coef    SE Coef       T      P
Constant     -26.82      14.76   -1.82  0.086
x            1.2617      0.1685    7.49  0.000

S = 7.61912   R-Sq = 75.7%   R-Sq(adj) = 74.3%

Analysis of Variance
Source           DF       SS      MS      F      P
Regression        1   3254.0  3254.0  56.05  0.000
Residual Error   18   1044.9    58.1
Total            19   4299.0
```

a. Construct a scatterplot for the data. Does the assumption of linearity appear to be reasonable?

b. What is the equation of the regression line used for predicting final exam score as a function of the posttest score?

c. Do the data present sufficient evidence to indicate that final exam score is linearly related to the posttest score? Use $\alpha = .01$.

d. Find a 99% confidence interval for the slope of the regression line.

12.29 Laptops and Learning, continued Refer to Exercise 12.28.

a. Use the *MINITAB* printout to find the value of the coefficient of determination, r^2. Show that $r^2 = $ SSR/Total SS.

b. What percentage reduction in the total variation is achieved by using the linear regression model?

12.30 Armspan and Height II In Exercise 12.17 (data set EX1217), we measured the armspan and height of eight people with the following results:

Person	1	2	3	4
Armspan (inches)	68	62.25	65	69.5
Height (inches)	69	62	65	70

Person	5	6	7	8
Armspan (inches)	68	69	62	60.25
Height (inches)	67	67	63	62

a. Does the data provide sufficient evidence to indicate that there is a linear relationship between armspan and height? Test at the 5% level of significance.

b. Construct a 95% confidence interval for the slope of the line of means, β.

c. If Leonardo da Vinci is correct, and a person's armspan is roughly the same as the person's height, the slope of the regression line is approximately equal to 1. Is this supposition confirmed by the confidence interval constructed in part b? Explain.

DIAGNOSTIC TOOLS FOR CHECKING THE REGRESSION ASSUMPTIONS

Even though you have determined—using the *t*-test for the slope (or the ANOVA *F*-test) and the value of r^2—that *x* is useful in predicting the value of *y*, the results of a regression analysis are valid only when the data satisfy the necessary regression assumptions.

REGRESSION ASSUMPTIONS

- The relationship between *y* and *x* must be linear, given by the model

$$y = \alpha + \beta x + \epsilon$$

- The values of the random error term ϵ (1) are independent, (2) have a mean of 0 and a common variance σ^2, independent of *x*, and (3) are normally distributed.

Since these assumptions are quite similar to those presented in Chapter 11 for an analysis of variance, it should not surprise you to find that the **diagnostic tools** for checking these assumptions are the same as those we used in that chapter. These tools involve the analysis of the **residual error,** the unexplained variation in each observation once the variation explained by the regression model has been removed.

Dependent Error Terms

The error terms are often dependent when the observations are collected at regular time intervals. When this is the case, the observations make up a **time series** whose error terms are correlated. This in turn causes bias in the estimates of model parameters. Time series data should be analyzed using time series methods.

Residual Plots

The other regression assumptions can be checked using **residual plots,** which are fairly complicated to construct by hand but easy to use once a computer has graphed them for you!

In simple linear regression, you can use the **plot of residuals versus fit** to check for a constant variance as well as to make sure that the linear model is in fact adequate. This plot should be free of any patterns. It should appear as a random scatter of points about 0 on the vertical axis with approximately the same vertical spread for all values of $\hat{y}$. One property of the residuals is that they sum to 0 and therefore have a sample mean of 0. The plot of the residuals versus fit for the calculus grade example is shown in Figure 12.9. There are no apparent patterns in this residual plot, which indicates that the model assumptions appear to be satisfied for these data.

FIGURE 12.9

Plot of the residuals versus $\hat{y}$ for Example 12.1

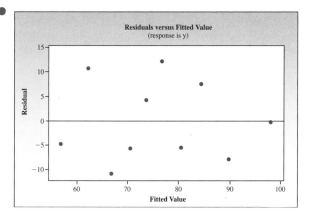

Recall from Chapter 11 that the **normal probability plot** is a graph that plots the residuals against the expected value of that residual if it had come from a normal distribution. When the residuals are normally distributed or approximately so, the plot should appear as a straight line, sloping upward. The normal probability plot for the residuals in Example 12.1 is given in Figure 12.10. With the exception of the fourth and fifth plotted points, the remaining points appear to lie approximately on a straight line. This plot is not unusual and does not indicate underlying nonnormality. The most serious violations of the normality assumption usually appear in the tails of the distribution because this is where the normal distribution differs most from other types of distributions with a similar mean and measure of spread. Hence, curvature in either or both of the two ends of the normal probability plot is indicative of nonnormality.

NEED A TIP?

Residuals versus fits ⇔ Random scatter

Normal plot ⇔ Straight line, sloping up

FIGURE 12.10

Normal probability plot of residuals for Example 12.1

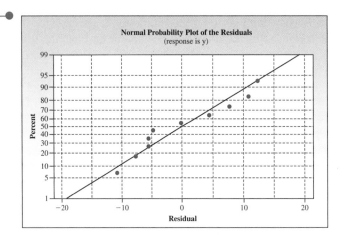

12.6 **EXERCISES**

BASIC TECHNIQUES

12.31 What diagnostic plot can you use to determine whether the data satisfy the normality assumption? What should the plot look like for normal residuals?

12.32 What diagnostic plot can you use to determine whether the incorrect model has been used? What should the plot look like if the correct model has been used?

12.33 What diagnostic plot can you use to determine whether the assumption of equal variance has been violated? What should the plot look like when the variances are equal for all values of x?

12.34 Refer to the data in Exercise 12.7. The normal probability plot and the residuals versus fitted values plots generated by *MINITAB* are shown here. Does it appear that any regression assumptions have been violated? Explain.

Diagnostic plots for Exercise 12.34

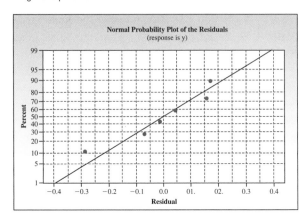

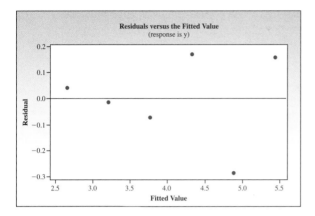

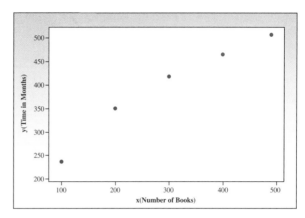

APPLICATIONS

12.35 Chirping Crickets Refer to Exercise 12.23, in which the number of chirps per second for a cricket was recorded at 10 different temperatures. Use the *MINITAB* diagnostic plots to comment on the validity of the regression assumptions.

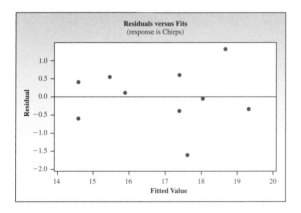

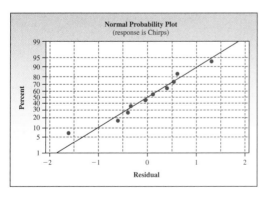

12.36 Professor Asimov, again Refer to Exercise 12.9, in which the number of books x written by Isaac Asimov are related to the number of months y he took to write them. A plot of the data is shown.

a. Can you see any pattern other than a linear relationship in the original plot?

b. The value of r^2 for these data is .959. What does this tell you about the fit of the regression line?

c. Look at the accompanying diagnostic plots for these data. Do you see any pattern in the residuals? Does this suggest that the relationship between number of months and number of books written is something other than linear?

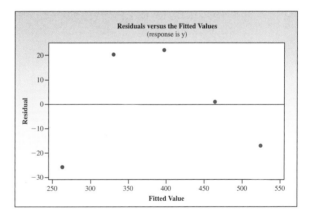

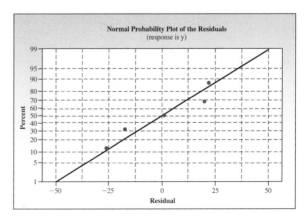

12.37 Laptops and Learning, again Refer to the data given in Exercise 12.28. The *MINITAB* printout is reproduced here.

Regression Analysis: y versus x

```
The regression equation is
y = -26.8 + 1.26 x

Predictor      Coef    SE Coef       T       P
Constant     -26.82      14.76   -1.82   0.086
x            1.2617      0.1685    7.49   0.000

S = 7.61912    R-Sq = 75.7%    R-Sq(adj) = 74.3%

Analysis of Variance
Source          DF       SS      MS       F       P
Regression       1   3254.0  3254.0   56.05   0.000
Residual Error  18   1044.9    58.1
Total           19   4299.0
```

a. What assumptions must be made about the distribution of the random error, ε?

b. What is the best estimate of σ^2, the variance of the random error, ε?

c. Use the diagnostic plots for these data to comment on the validity of the regression assumptions.

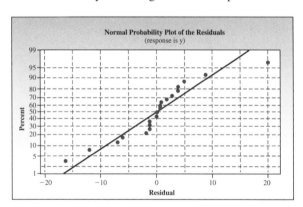

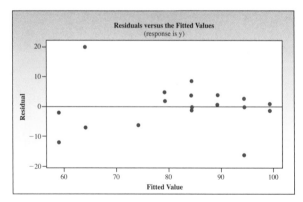

Data set **12.38 How to Choose a TV** In Exercise
EX1238 3.19, *Consumer Reports*[6] gave the prices and screen sizes for the top 10 LCD TVs in the 46-inch

and higher categories. Does the price of an LCD TV depend on the size of the screen?

Brand	Price ($)	Size
Sony Bravia KDL-52NX800	2340	52
Samsung LN55C650	1600	55
Vizio VF550M	1330	55
Sony Bravia KDL-60EX700	2700	60
Sharp Aquos LED LC-52LE700UN	1620	52
Sony Bravia KDL-46XBR10	2500	46
Samsung UN46C8000	2200	46
Vizio SV472XVT	1400	47
Samsung UN46C7000	2100	46
LG 47LD450	900	47

a. Suppose that we assume that the relationship between size and price is linear, and perform a linear regression, resulting in a value of $r^2 = .027$. What does the value of r^2 tell you about the strength of the relationship between price and screen size?

b. The diagnostic plots for this data are shown below. Does it appear that either the normality or equal variance assumptions have been violated?

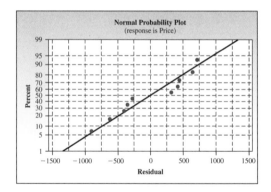

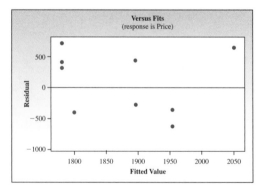

c. Use a scatterplot to plot price versus screen size for the 10 LCD TVs. Based on the information in part a, which assumption for the linear regression model has been violated?

ESTIMATION AND PREDICTION USING THE FITTED LINE

12.7

Now that you have

- tested the fitted regression line, $\hat{y} = a + bx$, to make sure that it is useful for prediction and
- used the diagnostic tools to make sure that none of the regression assumptions have been violated

you are ready to use the line for one of its two purposes:

- Estimating the average value of y for a given value of x
- Predicting a particular value of y for a given value of x

The sample of n pairs of observations have been chosen from a population in which the *average* value of y is related to the value of the predictor variable x by the **line of means,**

$$E(y) = \alpha + \beta x$$

an unknown line, shown as a broken line in Figure 12.11. Remember that for a fixed value of x—say, x_0—the *particular* values of y deviate from the line of means. These values of y are assumed to have a normal distribution with mean equal to $\alpha + \beta x_0$ and variance σ^2, as shown in Figure 12.11.

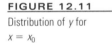

FIGURE 12.11

Distribution of y for $x = x_0$

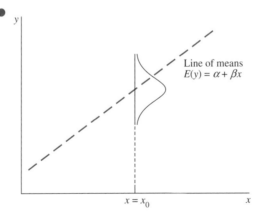

Since the computed values of a and b vary from sample to sample, each new sample produces a different regression line $\hat{y} = a + bx$, which can be used either to estimate the line of means or to predict a particular value of y. Figure 12.12 shows one of the possible configurations of the fitted line (blue), the unknown line of means (gray), and a particular value of y (the blue dot).

FIGURE 12.12

Error in estimating $E(y)$ and in predicting y

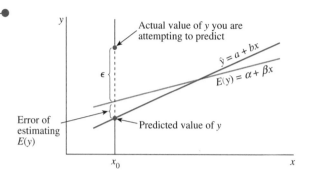

How far will our estimator $\hat{y} = a + bx_0$ be from the quantity to be estimated or predicted? This depends, as always, on the variability in our estimator, measured by its **standard error.** It can be shown that

$$\hat{y} = a + bx_0$$

the estimated value of y when $x = x_0$, is an unbiased estimator of the line of means, $\alpha + \beta x_0$, and that $\hat{y}$ is normally distributed with the standard error of $\hat{y}$ estimated by

$$SE(\hat{y}) = \sqrt{MSE\left(\frac{1}{n} + \frac{(x_0 - \bar{x})^2}{S_{xx}}\right)}$$

Estimation and testing are based on the statistic

$$t = \frac{\hat{y} - E(y)}{SE(\hat{y})}$$

which has a t distribution with $(n - 2)$ degrees of freedom.

To form a $(1 - \alpha)100\%$ confidence interval for the average value of y when $x = x_0$, measured by the line of means, $\alpha + \beta x_0$, you can use the usual form for a confidence interval based on the t distribution:

$$\hat{y} \pm t_{\alpha/2}SE(\hat{y})$$

NEED A TIP?

For a given value of x, the prediction interval is always wider than the confidence interval.

If you choose to predict a *particular* value of y when $x = x_0$, however, there is some additional error in the prediction because of the deviation of y from the line of means. If you examine Figure 12.12, you can see that the error in prediction has two components:

- The error in using the fitted line to estimate the line of means
- The error caused by the deviation of y from the line of means, measured by σ^2

The variance of the difference between y and $\hat{y}$ is the sum of these two variances and forms the basis for the standard error of $(y - \hat{y})$ used for prediction:

$$SE(y - \hat{y}) = \sqrt{MSE\left[1 + \frac{1}{n} + \frac{(x_0 - \bar{x})^2}{S_{xx}}\right]}$$

and the $(1 - \alpha)100\%$ prediction interval is formed as

$$\hat{y} \pm t_{\alpha/2}SE(y - \hat{y})$$

$(1 - \alpha)100\%$ CONFIDENCE AND PREDICTION INTERVALS

- For estimating the average value of y when $x = x_0$:

$$\hat{y} \pm t_{\alpha/2}\sqrt{\text{MSE}\left[\frac{1}{n} + \frac{(x_0 - \bar{x})^2}{S_{xx}}\right]}$$

- For predicting a particular value of y when $x = x_0$:

$$\hat{y} \pm t_{\alpha/2}\sqrt{\text{MSE}\left[1 + \frac{1}{n} + \frac{(x_0 - \bar{x})^2}{S_{xx}}\right]}$$

where $t_{\alpha/2}$ is the value of t with $(n - 2)$ degrees of freedom and area $\alpha/2$ to its right.

EXAMPLE 12.4

Use the information in Example 12.1 to estimate the average calculus grade for students whose achievement score is 50, with a 95% confidence interval.

Solution The point estimate of $E(y|x_0 = 50)$, the average calculus grade for students whose achievement score is 50, is

$$\hat{y} = 40.78424 + .76556(50) = 79.06$$

The standard error of $\hat{y}$ is

$$\sqrt{\text{MSE}\left[\frac{1}{n} + \frac{(x_0 - \bar{x})^2}{S_{xx}}\right]} = \sqrt{75.7532\left[\frac{1}{10} + \frac{(50 - 46)^2}{2474}\right]} = 2.840$$

and the 95% confidence interval is

$$79.06 \pm 2.306(2.840)$$

$$79.06 \pm 6.55$$

Our results indicate that the average calculus grade for students who score 50 on the achievement test will lie between 72.51 and 85.61.

EXAMPLE 12.5

A student took the achievement test and scored 50 but has not yet taken the calculus test. Using the information in Example 12.1, predict the calculus grade for this student with a 95% prediction interval.

Solution The predicted value of y is $\hat{y} = 79.06$, as in Example 12.4. However, the error in prediction is measured by $\text{SE}(y - \hat{y})$, and the 95% prediction interval is

$$79.06 \pm 2.306\sqrt{75.7532\left[1 + \frac{1}{10} + \frac{(50 - 46)^2}{2474}\right]}$$

$$79.06 \pm 2.306(9.155)$$

$$79.06 \pm 21.11$$

or from 57.95 to 100.17. The prediction interval is *wider* than the confidence interval in Example 12.4 because of the extra variability in predicting the actual value of the response y.

One particular point on the line of means is often of interest to experimenters, the **y-intercept α**—the average value of y when $x_0 = 0$.

EXAMPLE 12.6

Prior to fitting a line to the calculus grade-achievement score data, you may have thought that a score of 0 on the achievement test would predict a grade of 0 on the calculus test. This implies that we should fit a model with α equal to 0. Do the data support the hypothesis of a 0 intercept?

Solution You can answer this question by constructing a 95% confidence interval for the y-intercept α, which is the average value of y when $x = 0$. The estimate of α is

$$\hat{y} = 40.784 + .76556(0) = 40.784 = a$$

and the 95% confidence interval is

$$\hat{y} \pm t_{\alpha/2} \sqrt{\text{MSE}\left[\frac{1}{n} + \frac{(x_0 - \bar{x})^2}{S_{xx}}\right]}$$

$$40.784 \pm 2.306 \sqrt{75.7532\left[\frac{1}{10} + \frac{(0 - 46)^2}{2474}\right]}$$

$$40.784 \pm 19.617$$

or from 21.167 to 60.401, an interval that does not contain the value $\alpha = 0$. Hence, it is unlikely that the y-intercept is 0. You should include a nonzero intercept in the model $y = \alpha + \beta x + \epsilon$.

For this special situation in which you are interested in testing or estimating the y-intercept α for the line of means, the inferences involve the sample estimate a. The test for a 0 intercept is given in Figure 12.13 in the shaded line labeled "Constant." The coefficient given as 40.784 is a, with standard error given in the column labeled "SE Coef" as 8.507, which agrees with the value calculated in Example 12.6. The value of $t = 4.79$ is found by dividing a by its standard error with p-value $= .001$.

FIGURE 12.13

Portion of the *MINITAB* output for Example 12.6

Predictor	Coef	SE Coef	T	P
Constant	40.784	8.507	4.79	0.001
x	0.7656	0.1750	4.38	0.002

You can see that it is quite time-consuming to calculate these estimation and prediction intervals by hand. Moreover, it is difficult to maintain accuracy in your calculations. Fortunately, computer programs can perform these calculations for you. The *MINITAB* regression command provides an option for either estimation or prediction when you specify the necessary value(s) of x. The printout in Figure 12.14 gives the values of $\hat{y} = 79.06$ labeled "Fit," the standard error of $\hat{y}$, SE($\hat{y}$), labeled "SE Fit," the *confidence interval* for the average value of y when $x = 50$, labeled "95.0% CI," and the prediction interval for y when $x = 50$, labeled "95.0% PI."

FIGURE 12.14

MINITAB option for estimation and prediction

```
Predicted Values for New Observations
New Obs    Fit    SE Fit       95% CI            95% PI
1         79.06    2.84   (72.51,  85.61)   (57.95,  100.17)

Values of Predictors for New Observations
New Obs        x
1           50.0
```

The confidence bands and prediction bands generated by *MINITAB* for the calculus grades data are shown in Figure 12.15. Notice that in general the confidence bands are narrower than the prediction bands for every value of the achievement test score x. Certainly you would expect predictions for an individual value to be much more variable than estimates of the average value. Also notice that the bands seem to get wider as the value of x_0 gets farther from the mean $\bar{x}$. This is because the standard errors used in the confidence and prediction intervals contain the term $(x_0 - \bar{x})^2$, which gets larger as the two values diverge. In practice, this means that estimation and prediction are more accurate when x_0 is near the center of the range of the x-values. You can locate the calculated confidence and prediction intervals when $x = 50$ in Figure 12.15.

FIGURE 12.15 ●

Confidence and prediction intervals for the data in Table 12.1

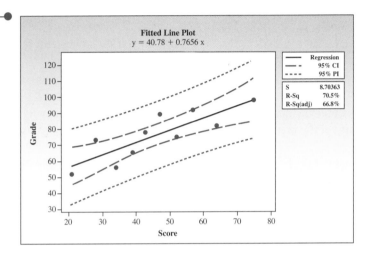

(12.7) **EXERCISES**

BASIC TECHNIQUES

12.39 Refer to Exercise 12.6.

a. Estimate the average value of y when $x = 1$, using a 90% confidence interval.

b. Find a 90% prediction interval for some value of y to be observed in the future when $x = 1$.

12.40 Refer to Exercise 12.7. Portions of the *MINITAB* printout are shown here.

MINITAB output for Exercise 12.40

```
Regression Analysis: y versus x

The regression equation is
y = 6.00 - 0.557 x

Predictor      Coef    SE Coef        T       P
Constant     6.0000     0.1759    34.10   0.000
x           -0.55714    0.04518   -12.33   0.000

Predicted Values for New Observations
New Obs    Fit    SE Fit       95.0% CI            95.0% PI
1       4.8857   0.1027   (4.6006,  5.1708)   (4.2886,  5.4829)
2       1.5429   0.2174   (0.9392,  2.1466)   (0.7430,  2.3427) X
X  denotes a point that is an outlier in the predictors.

Values of Predictors for New Observations
New Obs      x
1         2.00
2         8.00
```

a. Find a 95% confidence interval for the average value of y when $x = 2$.

b. Find a 95% prediction interval for some value of y to be observed in the future when $x = 2$.

c. The last line in the third section of the printout indicates a problem with one of the fitted values. What value of x corresponds to the fitted value $\hat{y} = 1.5429$? What problem has the *MINITAB* program detected?

APPLICATIONS

Data set

EX1241

12.41 What to Buy? A marketing research experiment was conducted to study the relationship between the length of time necessary for a buyer to reach a decision and the number of alternative package designs of a product presented. Brand names were eliminated from the packages and the buyers made their selections using the manufacturer's product descriptions on the packages as the only buying guide. The length of time necessary to reach a decision was recorded for 15 participants in the marketing research study.

Length of Decision Time, y (sec)	5, 8, 8, 7, 9	7, 9, 8, 9, 10	10, 11, 10, 12, 9
Number of Alternatives, x	2	3	4

a. Find the least-squares line appropriate for these data.

b. Plot the points and graph the line as a check on your calculations.

c. Calculate s^2.

d. Do the data present sufficient evidence to indicate that the length of decision time is linearly related to the number of alternative package designs? (Test at the $\alpha = .05$ level of significance.)

e. Find the approximate p-value for the test and interpret its value.

f. If they are available, examine the diagnostic plots to check the validity of the regression assumptions.

g. Estimate the average length of time necessary to reach a decision when three alternatives are presented, using a 95% confidence interval.

12.42 Housing Prices The data in the table give the square footages and sales prices of $n = 12$ houses randomly selected from those sold in a small city. Use the *MINITAB* printout to answer the questions.

Square Feet, x	Price, y	Square Feet, x	Price, y
1460	$288,700	1977	$305,400
2108	309,300	1610	297,000
1743	301,400	1530	292,400
1499	291,100	1759	298,200
1864	302,400	1821	304,300
2391	314,900	2216	311,700

Plot of data for Exercise 12.42

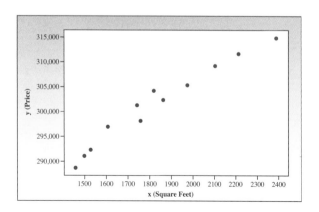

MINITAB output for Exercise 12.42

Regression Analysis: y versus x

```
The regression equation is
y = 251206 + 27.4 x

Predictor        Coef     SE Coef         T        P
Constant       251206        3389     74.13    0.000
x              27.406       1.828     14.99    0.000

S = 1792.72     R-Sq = 95.7%     R-Sq(adj) = 95.3%

Predicted Values for New Observations
New Obs     Fit   SE Fit        95.0% CI            95.0% PI
1        299989      526   (298817,  301161)   (295826,  304151)
2        306018      602   (304676,  307360)   (301804,  310232)

Values of Predictors for New Observations
New Obs       x
1          1780
2          2000
```

a. Can you see any pattern other than a linear relationship in the original plot?

b. The value of r^2 for these data is .957. What does this tell you about the fit of the regression line?

c. Look at the accompanying diagnostic plots for these data. Do you see any pattern in the residuals? Does this suggest that the relationship between price and square feet is something other than linear?

Diagnostic plots for Exercise 12.42

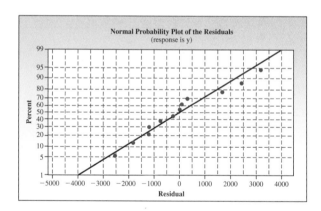

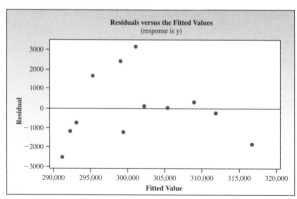

12.43 Housing Prices II Refer to Exercise 12.42 and data set EX1242.

a. Estimate the average increase in the price for an increase of 1 square foot for houses sold in the city. Use a 99% confidence interval. Interpret your estimate.

b. A real estate salesperson needs to estimate the average sales price of houses with a total of 2000 square feet of heated space. Use a 95% confidence interval and interpret your estimate.

c. Calculate the price per square foot for each house and then calculate the sample mean. Why is this estimate of the average cost per square foot not equal to the answer in part a? Should it be? Explain.

d. Suppose that a house with 1780 square feet of heated floor space is offered for sale. Construct a 95% prediction interval for the price at which the house will sell.

12.44 Strawberries III The following data (Exercises 12.18 and 12.27) were obtained in an experiment relating the dependent variable, y (texture of strawberries), with x (coded storage temperature).

x	−2	−2	0	2	2
y	4.0	3.5	2.0	0.5	0.0

a. Estimate the expected strawberry texture for a coded storage temperature of $x = -1$. Use a 99% confidence interval.

b. Predict the particular value of y when $x = 1$ with a 99% prediction interval.

c. At what value of x will the width of the prediction interval for a particular value of y be a minimum, assuming n remains fixed?

12.45 Drew Brees The number of passes
EX1245 completed and the total number of passing yards for Drew Brees, quarterback for the New Orleans Saints, were recorded for the 16 regular games in the 2010 football season.[7] Week 10 was a bye and no data was reported.

Week	Completions	Total Yards
1	27	237
2	28	254
3	30	365
4	33	275
5	24	279
6	21	263
7	37	356
8	34	305
9	27	253
10	–	–
11	29	382
12	23	352
13	24	313
14	25	221
15	29	267
16	25	302
17	22	196

a. What is the least-squares line relating the total passing yards to the number of pass completions for Drew Brees?

b. What proportion of the total variation is explained by the regression of total passing yards (y) on the number of pass completions (x)?

c. If they are available, examine the diagnostic plots to check the validity of the regression assumptions.

12.46 Drew Brees, continued Refer to Exercise 12.45.

a. Estimate the average number of passing yards for games in which Brees throws 20 completed passes using a 95% confidence interval.

b. Predict the actual number of passing yards for games in which Brees throws 20 completed passes using a 95% confidence interval.

c. Would it be advisable to use the least-squares line from Exercise 12.45 to predict Brees' total number of passing yards for a game in which he threw only five completed passes? Explain.

12.8 CORRELATION ANALYSIS

In Chapter 3, we introduced the *correlation coefficient* as a measure of the strength of the linear relationship between two variables. The correlation coefficient, r—formally called the **Pearson product moment sample coefficient of correlation**—is defined next.

PEARSON PRODUCT MOMENT COEFFICIENT OF CORRELATION

$$r = \frac{S_{xy}}{s_x s_y} = \frac{S_{xy}}{\sqrt{S_{xx} S_{yy}}} \qquad \text{for } -1 \le r \le 1$$

NEED A TIP?

r is always between
-1 and $+1$.

The variances and covariance can be found by direct calculation, by using a calculator with a two-variable statistics capacity, or by using a statistical package such as *MINITAB* or *MS Excel.* The variances and covariance are calculated as

$$s_{xy} = \frac{S_{xy}}{n-1} \qquad s_x^2 = \frac{S_{xx}}{n-1} \qquad s_y^2 = \frac{S_{yy}}{n-1}$$

and use S_{xy}, S_{xx}, and S_{yy}, the same quantities used in regression analysis earlier in this chapter. In general, when a sample of n individuals or experimental units is selected and two variables are measured on each individual or unit so that *both variables are random,* the correlation coefficient r is the appropriate measure of linearity for use in this situation.

EXAMPLE 12.7

The heights and weights of $n = 10$ offensive backfield football players are randomly selected from a county's football all-stars. Calculate the correlation coefficient for the heights (in inches) and weights (in pounds) given in Table 12.4.

TABLE 12.4 • **Heights and Weights of n = 10 Backfield All-Stars**

Player	Height, x	Weight, y
1	73	185
2	71	175
3	75	200
4	72	210
5	72	190
6	75	195
7	67	150
8	69	170
9	71	180
10	69	175

Solution You should use the appropriate data entry method of your scientific calculator to verify the calculations for the sums of squares and cross-products,

$$S_{xy} = 328 \qquad S_{xx} = 60.4 \qquad S_{yy} = 2610$$

using the calculational formulas given earlier in this chapter. Then

$$r = \frac{328}{\sqrt{(60.4)(2610)}} = .8261$$

or $r = .83$. This value of r is fairly close to 1, the largest possible value of r, which indicates a fairly strong positive linear relationship between height and weight.

There is a direct relationship between the calculational formulas for the correlation coefficient r and the slope of the regression line b. Since the numerator of both quantities is S_{xy}, both r and b have the same sign. Therefore, the correlation coefficient has these general properties:

NEED A TIP?

The sign of r is always the **same** as the sign of the slope b.

- When $r = 0$, the slope is $b = 0$, and there is no linear relationship between x and y.

- When r is positive, so is b, and there is a positive linear relationship between x and y.

- When r is negative, so is b, and there is a negative linear relationship between x and y.

In Section 12.5, we showed that

$$r^2 = \frac{\text{SSR}}{\text{Total SS}} = \frac{\text{Total SS} - \text{SSE}}{\text{Total SS}}$$

Exploring Correlation

In this form, you can see that r^2 can never be greater than 1, so that $-1 \le r \le 1$. Moreover, you can see the relationship between the random variation (measured by SSE) and r^2.

- If there is no random variation and all the points fall on the regression line, then SSE $= 0$ and $r^2 = 1$.
- If the points are randomly scattered and there is no variation explained by regression, then SSR $= 0$ and $r^2 = 0$.

Figure 12.16 shows four typical scatterplots and their associated correlation coefficients. Notice that in scatterplot (d) there appears to be a curvilinear relationship between x and y, but r is approximately 0, which reinforces the fact that r is a measure of a *linear* (not *curvilinear*) relationship between two variables.

FIGURE 12.16

Some typical scatterplots with approximate values of r

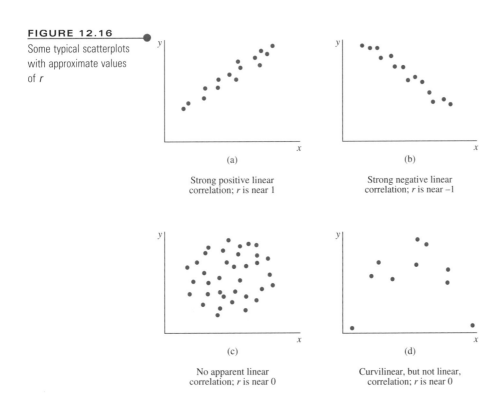

Consider a population generated by measuring two random variables on each experimental unit. In this *bivariate* population, the **population correlation coefficient** ρ (Greek lowercase rho) is calculated and interpreted as it is in the sample. In this situation, the experimenter can test the hypothesis that there is no correlation between the variables x and y using a test statistic that is *exactly equivalent* to the test of the slope β in Section 12.5. The test procedure is shown next.

TEST OF HYPOTHESIS CONCERNING THE CORRELATION COEFFICIENT ρ

1. Null hypothesis: $H_0 : \rho = 0$
2. Alternative hypothesis:

One-Tailed Test	Two-Tailed Test
$H_a : \rho > 0$	$H_a : \rho \neq 0$
(or $\rho < 0$)	

3. Test statistic: $t = r\sqrt{\dfrac{n-2}{1-r^2}}$

   When the assumptions given in Section 12.2 are satisfied, the test statistic will have a Student's t distribution with $(n - 2)$ degrees of freedom.

4. Rejection region: Reject H_0 when

One-Tailed Test	Two-Tailed Test
$t > t_\alpha$	$t > t_{\alpha/2}$ or $t < -t_{\alpha/2}$
(or $t < -t_\alpha$ when the alternative hypothesis is $H_a : \rho < 0$)	

   or when p-value $< \alpha$

   The values of t_α and $t_{\alpha/2}$ corresponding to $(n - 2)$ degrees of freedom can be found using Table 4 in Appendix I.

NEED A TIP?

You can prove that

$t = r\sqrt{\dfrac{n-2}{1-r^2}}$

$= \dfrac{b - 0}{\sqrt{MSE/S_{xx}}}$.

EXAMPLE 12.8

Refer to the height and weight data in Example 12.7. The correlation of height and weight was calculated to be $r = .8261$. Is this correlation significantly different from 0?

Solution To test the hypotheses

$$H_0 : \rho = 0 \quad \text{versus} \quad H_a : \rho \neq 0$$

the value of the test statistic is

NEED A TIP?

The t-value and p-value for testing $H_0 : \rho = 0$ will be identical to the t- and p-value for testing $H_0 : \beta = 0$.

$$t = r\sqrt{\frac{n-2}{1-r^2}} = .8261\sqrt{\frac{10-2}{1-(.8261)^2}} = 4.15$$

which for $n = 10$ has a t distribution with 8 degrees of freedom. Since this value is greater than $t_{.005} = 3.355$, the two-tailed p-value is less than $2(.005) = .01$, and the correlation is declared significant at the 1% level ($P < .01$). The value $r^2 = .8261^2 = .6824$ means that about 68% of the variation in one of the variables is explained by the other. The *MINITAB* printout in Figure 12.17 displays the correlation r and the exact p-value for testing its significance.

FIGURE 12.17

MINITAB output for Example 12.8

Correlations: x, y

```
Pearson correlation of x and y = 0.826
P-Value = 0.003
```

If the linear coefficients of correlation between y and each of two variables x_1 and x_2 are calculated to be .4 and .5, respectively, it does not follow that a predictor using both variables will account for $[(.4)^2 + (.5)^2] = .41$, or a 41% reduction in the sum of squares of deviations. Actually, x_1 and x_2 might be highly correlated and therefore contribute virtually the same information for the prediction of y.

Finally, remember that r is a measure of **linear correlation** and that x and y could be perfectly related by some **nonlinear** function when the observed value of r is equal to 0. The problem of estimating or predicting y using information given by several independent variables, $x_1, x_2, \ldots, x_k$, is the subject of Chapter 13.

12.8 EXERCISES

BASIC TECHNIQUES

12.47 How does the coefficient of correlation measure the strength of the linear relationship between two variables y and x?

12.48 Describe the significance of the algebraic sign and the magnitude of r.

12.49 What value does r assume if all the data points fall on the same straight line in these cases?

a. The line has positive slope.

b. The line has negative slope.

12.50 You are given these data:

x	−2	−1	0	1	2
y	2	2	3	4	4

a. Plot the data points. Based on your graph, what will be the sign of the sample correlation coefficient?

b. Calculate r and r^2 and interpret their values.

12.51 You are given these data:

x	1	2	3	4	5	6
y	7	5	5	3	2	0

a. Plot the six points on graph paper.

b. Calculate the sample coefficient of correlation r and interpret.

c. By what percentage was the sum of squares of deviations reduced by using the least-squares predictor $\hat{y} = a + bx$ rather than $\bar{y}$ as a predictor of y?

12.52 Reverse the slope of the line in Exercise 12.51 by reordering the y observations, as follows:

x	1	2	3	4	5	6
y	0	2	3	5	5	7

Repeat the steps of Exercise 12.51. Notice the change in the sign of r and the relationship between the values of r^2 of Exercise 12.51 and this exercise.

APPLICATIONS

12.53 Lobster The table gives the numbers of *Octolasmis tridens* and *O. lowei* barnacles on each of 10 lobsters.[8] Does it appear that the barnacles compete for space on the surface of a lobster?

Lobster Field Number	O. tridens	O. lowei
A061	645	6
A062	320	23
A066	401	40
A070	364	9
A067	327	24
A069	73	5
A064	20	86
A068	221	0
A065	3	109
A063	5	350

a. If they do compete, do you expect the number x of *O. tridens* and the number y of *O. lowei* barnacles to be positively or negatively correlated? Explain.

b. If you want to test the theory that the two types of barnacles compete for space by conducting a test of the null hypothesis "the population correlation coefficient ρ equals 0," what is your alternative hypothesis?

c. Conduct the test in part b and state your conclusions.

12.54 Social Skills Training A social skills training program was implemented with seven mildly challenged students in a study to determine whether the program caused improvement in pre/post measures and behavior ratings. For one such test, the pre- and posttest scores for the seven students are given in the table.[9]

Subject	Pretest	Posttest
Earl	101	113
Ned	89	89
Jasper	112	121
Charlie	105	99
Tom	90	104
Susie	91	94
Lori	89	99

a. What type of correlation, if any, do you expect to see between the pre- and posttest scores? Plot the data. Does the correlation appear to be positive or negative?

b. Calculate the correlation coefficient, r. Is there a significant positive correlation?

12.55 Hockey G. W. Marino investigated the variables related to a hockey player's ability to make a fast start from a stopped position.[10] In the experiment, each skater started from a stopped position and attempted to move as rapidly as possible over a 6-meter distance. The correlation coefficient r between a skater's stride rate (number of strides per second) and the length of time to cover the 6-meter distance for the sample of 69 skaters was $-.37$.

a. Do the data provide sufficient evidence to indicate a correlation between stride rate and time to cover the distance? Test using $\alpha = .05$.

b. Find the approximate p-value for the test.

c. What are the practical implications of the test in part a?

12.56 Hockey II Refer to Exercise 12.55. Marino calculated the sample correlation coefficient r for the stride rate and the average acceleration rate for the 69 skaters to be .36. Do the data provide sufficient evidence to indicate a correlation between stride rate and average acceleration for the skaters? Use the p-value approach.

Data set **12.57 Geothermal Power** Geothermal EX1257 power is an important source of energy. Since the amount of energy contained in 1 pound of water is a function of its temperature, you might wonder whether water obtained from deeper wells contains

more energy per pound. The data in the table are reproduced from an article on geothermal systems by A.J. Ellis.[11]

Location of Well	Average (max.) Drill Hole Depth (m)	Average (max.) Temperature (°C)
El Tateo, Chile	650	230
Ahuachapan, El Salvador	1000	230
Namafjall, Iceland	1000	250
Larderello (region), Italy	600	200
Matsukawa, Japan	1000	220
Cerro Prieto, Mexico	800	300
Wairakei, New Zealand	800	230
Kizildere, Turkey	700	190
The Geysers, United States	1500	250

Is there a significant positive correlation between average maximum drill hole depth and average maximum temperature?

Data set **12.58 Ice Cream, Anyone?** As much as EX1258 Americans try to avoid high fat, high calorie foods, the demand for a cold, creamy ice cream cone on a hot day is hard to resist. The popular ice cream franchise *Coldstone Creamery* posted the nutritional information for its ice cream offerings in three serving sizes—"Like it", "Love it", and "Gotta Have it"—on their website.[12] A portion of that information for the "Like it" serving size is shown in the table.

Flavor	Calories	Total Fat (grams)
Cake Batter	340	19
Cinnamon Bun	370	21
French Toast	330	19
Mocha	320	20
OREO® Crème	440	31
Peanut Butter	370	24
Strawberry Cheesecake	320	21

a. Should you use the methods of linear regression analysis or correlation analysis to analyze the data? Explain.

b. Analyze the data to determine the nature of the relationship between total fat and calories in *Coldstone Creamery* ice cream.

Data set **12.59 Body Temperature and Heart Rate** EX1259 Is there any relationship between these two variables? To find out, we randomly selected 12 people from a data set constructed by Allen Shoemaker (*Journal of Statistics Education*) and recorded their body temperature and heart rate.[13]

Person	1	2	3	4	5	6
Temperature (degrees)	96.3	97.4	98.9	99.0	99.0	96.8
Heart Rate (beats per minute)	70	68	80	75	79	75

Person	7	8	9	10	11	12
Temperature (degrees)	98.4	98.4	98.8	98.8	99.2	99.3
Heart Rate (beats per minute)	74	84	73	84	66	68

a. Find the correlation coefficient r, relating body temperature to heart rate.

b. Is there sufficient evidence to indicate that there is a correlation between these two variables? Test at the 5% level of significance.

Data set **12.60 Baseball Stats** Does a team's batting
EX1260 average depend in any way on the number of home runs hit by the team? The data in the table show the number of team home runs and the overall team batting average for eight selected major league teams for the 2010 season.[14]

Team	Total Home Runs	Team Batting Average
Atlanta Braves	139	.258
Baltimore Orioles	133	.259
Boston Red Sox	211	.268
Chicago White Sox	177	.268
Houston Astros	108	.247
LA Dodgers	120	.252
Philadelphia Phillies	166	.260
Seattle Mariners	101	.236

Source: ESPN.com

a. Plot the points using a scatterplot. Does it appear that there is any relationship between total home runs and team batting average?

b. Is there a significant positive correlation between total home runs and team batting average? Test at the 5% level of significance.

c. Do you think that the relationship between these two variables would be different if we had looked at the entire set of major league franchises?

CHAPTER REVIEW

Key Concepts and Formulas

I. A Linear Probabilistic Model

1. When the data exhibit a linear relationship, the appropriate model is $y = \alpha + \beta x + \epsilon$.

2. The random error ϵ has a normal distribution with mean 0 and variance σ^2.

II. Method of Least Squares

1. Estimates a and b, for α and β, are chosen to minimize SSE, the sum of squared deviations about the regression line, $\hat{y} = a + bx$.

2. The least-squares estimates are $b = S_{xy}/S_{xx}$ and $a = \bar{y} - b\bar{x}$.

III. Analysis of Variance

1. Total SS = SSR + SSE, where Total SS = S_{yy} and SSR = $(S_{xy})^2/S_{xx}$.

2. The best estimate of σ^2 is MSE = SSE/$(n - 2)$.

IV. Testing, Estimation, and Prediction

1. A test for the significance of the linear regression—$H_0 : \beta = 0$—can be implemented using one of two test statistics:

$$t = \frac{b}{\sqrt{\text{MSE}/S_{xx}}} \quad \text{or} \quad F = \frac{\text{MSR}}{\text{MSE}}$$

2. The strength of the relationship between x and y can be measured using

$$r^2 = \frac{\text{SSR}}{\text{Total SS}}$$

which gets closer to 1 as the relationship gets stronger.

3. Use residual plots to check for nonnormality, inequality of variances, or an incorrectly fitted model.

4. Confidence intervals can be constructed to estimate the intercept α and slope β of the regression line and to estimate the average value of y, $E(y)$, for a given value of x.

5. Prediction intervals can be constructed to predict a particular observation, y, for a given value of x. For a given x, prediction intervals are always wider than confidence intervals.

V. Correlation Analysis

1. Use the correlation coefficient to measure the relationship between x and y when both variables are random:

$$r = \frac{S_{xy}}{\sqrt{S_{xx}S_{yy}}}$$

2. The sign of r indicates the direction of the relationship; r near 0 indicates no linear relationship, and r near 1 or -1 indicates a strong linear relationship.

3. A test of the significance of the correlation coefficient uses the statistic

$$t = r\sqrt{\frac{n-2}{1-r^2}}$$

and is identical to the test of the slope β.

 TECHNOLOGY TODAY

Linear Regression Procedures—*Microsoft Excel*

In Chapter 3, we used some of the linear regression procedures available in *Microsoft Excel* to obtain a scatterplot of the data and the least-squares regression line and to calculate the correlation coefficient r for a bivariate data set. Now that you have studied the testing and estimation techniques for a simple linear regression analysis, more *MS Excel* options are available to you.

 EXAMPLE 12.9

Refer to Table 12.1, in which the relationship between x = mathematics achievement test score and y = final calculus grade was studied.

Student	Mathematics Achievement Test Score (x)	Final Calculus Grade (y)
1	39	65
2	43	78
3	21	52
4	64	82
5	57	92
6	47	89
7	28	73
8	75	98
9	34	56
10	52	75

Enter the values for x and y into columns A and B of an *Excel* spreadsheet.

1. Use **Data ▶ Data Analysis ▶ Regression** to generate the Dialog box in Figure 12.18(a). Highlight or type in the cell ranges for the x and y values and check "Labels" if necessary.

2. If you click "Confidence Level," *Excel* will calculate confidence intervals for the regression estimates, a and b. Enter a cell location for the **Output Range** and click **OK**.

3. The output will appear in the selected cell location, and should be adjusted using **Format ▶ AutoFit Column Width** on the **Home** tab in the **Cells** group while it is still highlighted. You can decrease the decimal accuracy if you like, using ⌄⌄⌄ on the **Home** tab in the **Number** group (see Figure 12.18(b)).

4. The output in Figure 12.18(b) can also be found in Figures 12.6(b) and 12.7(b), with its interpretation found in Sections 12.4 and 12.5 of the text.

FIGURE 12.18

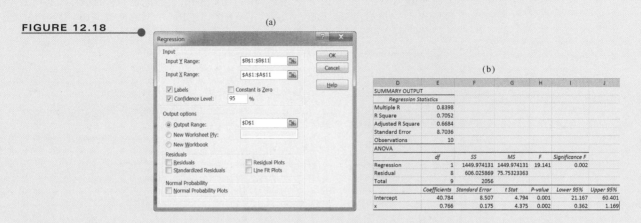

(a)

(b)

NOTE: *MS Excel* does not provide options for estimation and prediction or for the test of significant correlation in Section 12.8. The diagnostic plots which can be generated in *Excel* are not the same plots as we have discussed in Section 12.6 and will not be discussed in this section.

Linear Regression Procedures—*MINITAB*

In Chapter 3, we used some of the linear regression procedures available in *MINITAB* to obtain a graph of the best-fitting least-squares regression line and to calculate the correlation coefficient *r* for a bivariate data set. Now that you have studied the testing and estimation techniques for a simple linear regression analysis, more *MINITAB* options are available to you.

EXAMPLE 12.10

Refer to Table 12.1, in which the relationship between x = mathematics achievement test score and y = final calculus grade was studied.

Student	Mathematics Achievement Test Score (x)	Final Calculus Grade (y)
1	39	65
2	43	78
3	21	52
4	64	82
5	57	92
6	47	89
7	28	73
8	75	98
9	34	56
10	52	75

Enter the values for x and y into the first two columns of a *MINITAB* worksheet.

1. The main tools for linear regression analysis are generated using **Stat ▶ Regression ▶ Regression.** (You will use this same sequence of commands in Chapter 13 when you study *multiple regression analysis*.) The Dialog box for the Regression command is shown in Figure 12.19(a).

2. Select **y** for the "Response" variable and **x** for the "Predictor" variable. You can now generate some residual plots to check the validity of your regression assumptions before you use the model for estimation or prediction. Choose **Graphs** to display the Dialog box in Figure 12.19(b). We have used **Regular** residual plots, checking the boxes for "Normal plot of residuals" and "Residuals versus fits." Click **OK** to return to the main Dialog box.

FIGURE 12.19

(a)

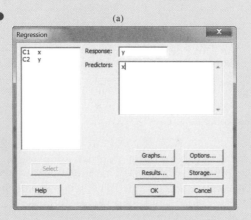

(b)

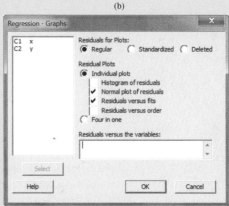

3. If you now choose **Options,** you can obtain confidence and prediction intervals for either of these cases:

- A single value of x (typed in the box marked "Prediction intervals for new observations").
- Several values of x stored in a column (say, C3) of the worksheet.

4. Enter the value **50** in Figure 12.20(a) to match the output given in Figure 12.14. When you click **OK** twice, the regression output is generated as shown in Figure 12.20(b). The two diagnostic plots will appear in separate graphics windows.

FIGURE 12.20

(b)

(a)

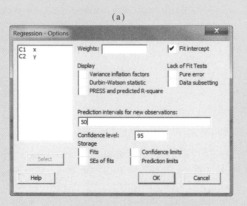

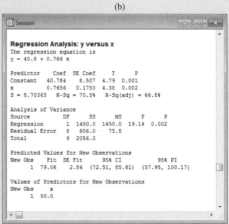

5. If you wish, you can now plot the data points, the regression line, and the upper and lower confidence and prediction limits (see Figure 12.15) using **Stat ▶ Regression ▶ Fitted Line Plot.** Select y and x for the response and predictor variables and click "Display confidence interal" and "Display prediction interval" in the **Options** Dialog box. Make sure that **Linear** is selected as the "Type of Regression Model," so that you will obtain a linear fit to the data.

6. Recall that in Chapter 3, we used the command **Stat ▶ Basic Statistics ▶ Correlation** to obtain the value of the correlation coefficient r. Make sure that the box marked "Display p-values" is checked. The output for this command (using the test/grade data) is shown in Figure 12.21. Notice that the p-value for the test of $H_0: \rho = 0$ is identical to the p-value for the test of $H_0: \beta = 0$ because the tests are exactly equivalent!

FIGURE 12.21

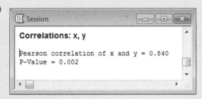

```
Correlations: x, y

Pearson correlation of x and y = 0.840
P-Value = 0.002
```

Supplementary Exercises

12.61 Potency of an Antibiotic An experiment was conducted to observe the effect of an increase in temperature on the potency of an antibiotic. Three 1-ounce portions of the antibiotic were stored for equal lengths of time at each of these temperatures: 30°, 50°, 70°, and 90°. The potency readings observed at each temperature of the experimental period are listed here:

Potency Readings, y	38, 43, 29	32, 26, 33	19, 27, 23	14, 19, 21
Temperature, x	30°	50°	70°	90°

Use an appropriate computer program to answer these questions:

a. Find the least-squares line appropriate for these data.

b. Plot the points and graph the line as a check on your calculations.

c. Construct the ANOVA table for linear regression.

d. If they are available, examine the diagnostic plots to check the validity of the regression assumptions.

e. Estimate the change in potency for a 1-unit change in temperature. Use a 95% confidence interval.

f. Estimate the average potency corresponding to a temperature of 50°. Use a 95% confidence interval.

g. Suppose that a batch of the antibiotic was stored at 50° for the same length of time as the experimental period. Predict the potency of the batch at the end of the storage period. Use a 95% prediction interval.

12.62 Plant Science An experiment was conducted to determine the effect of various levels of phosphorus on the inorganic phosphorus levels in a particular plant. The data in the table represent the levels of inorganic phosphorus in micromoles (μmol) per gram dry weight of Sudan grass roots grown in the greenhouse for 28 days, in the absence of zinc. Use the *MINITAB* output to answer the questions.

Phosphorus Applied, x	Phosphorus in Plant, y
.50 μmol	204
	195
	247
	245
.25 μmol	159
	127
	95
	144
.10 μmol	128
	192
	84
	71

a. Plot the data. Do the data appear to exhibit a linear relationship?

b. Find the least-squares line relating the plant phosphorus levels y to the amount of phosphorus applied to the soil x. Graph the least-squares line as a check on your answer.

c. Do the data provide sufficient evidence to indicate that the amount of phosphorus present in the plant is linearly related to the amount of phosphorus applied to the soil?

d. Estimate the mean amount of phosphorus in the plant if .20 μmol of phosphorus is applied to the soil, in the absence of zinc. Use a 90% confidence interval.

MINITAB output for Exercise 12.62

Regression Analysis: y versus x

```
The regression equation is
y = 80.9 + 271 x

Predictor       Coef     SE Coef        T        P
Constant       80.85       22.40     3.61    0.005
x             270.82       68.31     3.96    0.003

S = 39.0419    R-Sq = 61.1%     R-Sq(adj) = 57.2%

Predicted Values for New Observations
New Obs     Fit    SE Fit       90.0% CI           90.0% PI
1         135.0      12.6   (112.1,   157.9)   (60.6,   209.4)

Values of Predictors for New Observations
New Obs        x
1          0.200
```

Data set **12.63 Track Stats!** An experiment was conducted to investigate the effect of a training program on the time to complete the 100-yard dash. Nine students were placed in the program. The reduction y in time to complete the race was measured for three students at the end of 2 weeks, for three at the end of 4 weeks, and for three at the end of 6 weeks of training. The data are given in the table.

EX1263

Reduction in Time, y (sec)	1.6, .8, 1.0	2.1, 1.6, 2.5	3.8, 2.7, 3.1
Length of Training, x (wk)	2	4	6

Use an appropriate computer software package to analyze these data. State any conclusions you can draw.

Data set **12.64 Nematodes** Some varieties of nematodes, roundworms that live in the soil and feed on the roots of lawn grasses and other plants, can be treated by the application of nematicides. Data collected on the percent kill of nematodes for various rates of application (dosages given in pounds per acre of active ingredient) are as follows:

EX1264

Rate of Application, x	2	3	4	5
Percent Kill, y	50, 56, 48	63, 69, 71	86, 82, 76	94, 99, 97

MINITAB diagnostic plots for Exercise 12.64

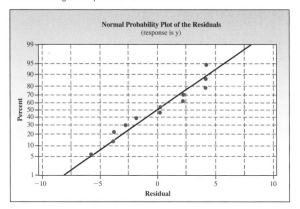

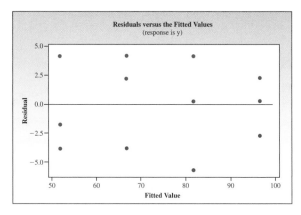

Use an appropriate computer printout to answer these questions:

a. Calculate the coefficient of correlation r between rates of application x and percent kill y.

b. Calculate the coefficient of determination r^2 and interpret.

c. Fit a least-squares line to the data.

d. Suppose you wish to estimate the mean percent kill for an application of 4 pounds of the nematicide per acre. What do the diagnostic plots tell you about the validity of the regression assumptions? Which assumptions may have been violated? Can you explain why?

12.65 Knee Injuries Athletes and others suffering the same type of knee injury often require anterior and posterior ligament reconstruction. In order to determine the proper length of bone grafts, experiments were done using three imaging techniques, and these results were compared to the actual length required. A summary of the results of a simple linear regression analysis for each of these three methods is given in the following table.[15]

Imaging Technique	Coefficient of Determination, r^2	Intercept	Slope	p-value
Radiographs	0.80	−3.75	1.031	<0.0001
Standard MRI	0.43	20.29	0.497	0.011
3-Dimensional MRI	0.65	1.80	0.977	<0.0001

a. What can you say about the significance of each of the three regression analyses?

b. How would you rank the effectiveness of the three regression analyses? What is the basis of your decision?

c. How do the values of r^2 and the p-values compare in determining the best predictor of actual graft lengths of ligament required?

Data set **12.66 Achievement Tests II** Refer to Exercise 12.13 and data set EX1213 regarding the EX1266 relationship between the Academic Performance Index (API), a measure of school achievement based on the results of the Stanford 9 Achievement test, and the percentage of students who are considered English Learners (EL). The following table shows the API for eight elementary schools in Riverside County, California, along with the percentage of students at that school who are considered "English Learners" (EL).[3]

School	1	2	3	4	5	6	7	8
API	745	808	798	791	854	688	801	751
EL	71	18	24	50	17	71	11	57

a. Use an appropriate program to analyze the relationship between API and EL.

b. Explain all pertinent details of your analysis.

12.67 How Long Is It? Refer to Exercise 12.14 and data set EX1214 regarding a subject's ability to estimate

sizes. The table that follows gives the actual and estimated lengths of the specified objects.

Object	Estimated (inches)	Actual (inches)
Pencil	7.00	6.00
Dinner plate	9.50	10.25
Book 1	7.50	6.75
Cell phone	4.00	4.25
Photograph	14.50	15.75
Toy	3.75	5.00
Belt	42.00	41.50
Clothespin	2.75	3.75
Book 2	10.00	9.25
Calculator	3.50	4.75

a. Use an appropriate program to analyze the relationship between the actual and estimated lengths of the listed objects.

b. Explain all pertinent details of your analysis.

Data set **12.68 Tennis, Anyone?** If you play tennis, EX1268 you know that tennis racquets vary in their physical characteristics. The data in the accompanying table give measures of bending stiffness and twisting stiffness as measured by engineering tests for 12 tennis racquets:

Racquet	Bending Stiffness, x	Twisting Stiffness, y
1	419	227
2	407	231
3	363	200
4	360	211
5	257	182
6	622	304
7	424	384
8	359	194
9	346	158
10	556	225
11	474	305
12	441	235

a. If a racquet has bending stiffness, is it also likely to have twisting stiffness? Do the data provide evidence that x and y are positively correlated?

b. Calculate the coefficient of determination r^2 and interpret its value.

Data set **12.69 Avocado Research** Movement of avocados into the United States from certain areas is EX1269 prohibited because of the possibility of bringing fruit flies into the country with the avocado shipments. However, certain avocado varieties supposedly are resistant to fruit fly infestation before they soften as a result of ripening. The data in the table resulted from an experiment in which avocados ranging from 1 to 9 days after harvest were exposed to Mediterranean fruit flies. Penetrability of the avocados was measured on the day of exposure, and the percentage of the avocado fruit infested was assessed.

Days after Harvest	Penetrability	Percentage Infected
1	.91	30
2	.81	40
4	.95	45
5	1.04	57
6	1.22	60
7	1.38	75
9	1.77	100

Use the *MINITAB* printout of the regression of percentage infected (y) on days after harvest (x) to analyze the relationship between these two variables. Explain all pertinent parts of the printout and interpret the results of any tests.

MINITAB output for Exercise 12.69

Regression Analysis: Percent versus x

```
The regression equation is
Percent = 18.4 + 8.18 x

Predictor      Coef     SE Coef        T       P
Constant     18.427       5.110     3.61   0.015
x           8.1768       0.9285     8.81   0.000

S = 6.35552     R-Sq = 93.9%     R-Sq(adj) = 92.7%

Analysis of Variance
Source        DF        SS        MS      F       P
Regression     1    3132.9    3132.9  77.56   0.000
Residual Error 5     202.0      40.4
Total          6    3334.9
```

12.70 Avocados II Refer to Exercise 12.69. Suppose the experimenter wants to examine the relationship between the penetrability and the number of days after harvest. Does the method of linear regression discussed in this chapter provide an appropriate method of analysis? If not, what assumptions have been violated? Use the diagnostic plots provided.

Diagnostic plots for Exercise 12.70

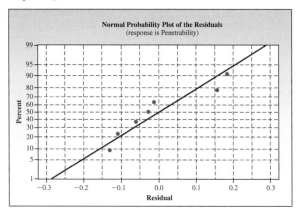

Normal Probability Plot of the Residuals
(response is Penetrability)

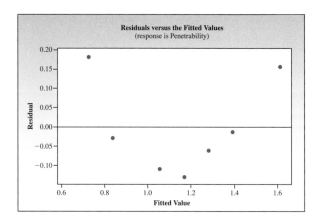

Residuals versus the Fitted Values
(response is Penetrability)

Data set
EX1271 **12.71 Metabolism and Weight Gain** Why is it that one person may tend to gain weight, even if he eats no more and exercises no less than a slim friend? Recent studies suggest that the factors that control metabolism may depend on your genetic makeup. One study involved 11 pairs of identical twins fed about 1000 calories per day more than needed to maintain initial weight. Activities were kept constant, and exercise was minimal. At the end of 100 days, the changes in body weight (in kilograms) were recorded for the 22 twins.[16] Is there a significant positive correlation between the changes in body weight for the twins? Can you conclude that this similarity is caused by genetic similarities? Explain.

Pair	Twin A	Twin B
1	4.2	7.3
2	5.5	6.5
3	7.1	5.7
4	7.0	7.2
5	7.8	7.9
6	8.2	6.4
7	8.2	6.5
8	9.1	8.2
9	11.5	6.0
10	11.2	13.7
11	13.0	11.0

Data set
EX1272 **12.72 Movie Reviews** How many weeks can a movie run and still make a reasonable profit? The data that follow show the number of weeks in release (x) and the gross to date (y) for the top 12 movies during a recent week.[17]

Movie	Gross to Date ($ millions)	Weeks in Release
1. *The Town*	33.1	1
2. *Easy A*	22.1	1
3. *Resident Evil: Afterlife 3D*	47.1	2
4. *Devil*	15.3	1
5. *Alpha and Omega 3D*	10.4	1
6. *The American*	33.7	3
7. *Takers*	53.3	4
8. *Eat Pray Love*	78.3	6
9. *Inception*	285.8	10
10. *Machete*	25.1	3
11. *The Other Guys*	116.0	7
12. *Going the Distance*	17.3	3

a. Plot the points in a scatterplot. Does it appear that the relationship between x and y is linear? How would you describe the direction and strength of the relationship?

b. Calculate the value of r^2. What percentage of the overall variation is explained by using the linear model rather than $\bar{y}$ to predict the response variable y?

c. What is the regression equation? Do the data provide evidence to indicate that x and y are linearly related? Test using a 5% significance level.

d. Given the results of parts b and c, is it appropriate to use the regression line for estimation and prediction? Explain your answer.

12.73 In addition to increasingly large bounds on error, why should an experimenter refrain from predicting y for values of x outside the experimental region?

12.74 If the experimenter stays within the experimental region, when will the error in predicting a particular value of y be maximum?

12.75 Oatmeal, Anyone? An agricultural EX1275 experimenter, investigating the effect of the amount of nitrogen x applied in 100 pounds per acre on the yield of oats y measured in bushels per acre, collected the following data:

x	1	2	3	4
y	22	38	57	68
	19	41	54	65

a. Find the least-squares line for the data.

b. Construct the ANOVA table.

c. Is there sufficient evidence to indicate that the yield of oats is linearly related to the amount of nitrogen applied? Use $\alpha = .05$.

d. Predict the expected yield of oats with 95% confidence if 250 pounds of nitrogen per acre are applied.

e. Estimate the average increase in yield for an increase of 100 pounds of nitrogen per acre with 99% confidence.

f. Calculate r^2 and explain its significance in terms of predicting y, the yield of oats.

12.76 Fresh Roses A horticulturalist EX1276 devised a scale to measure the freshness of roses that were packaged and stored for varying periods of time before transplanting. The freshness measurement y and the length of time in days that the rose is packaged and stored before transplanting x are given below.

x	5	10	15	20	25
y	15.3	13.6	9.8	5.5	1.8
	16.8	13.8	8.7	4.7	1.0

a. Fit a least-squares line to the data.

b. Construct the ANOVA table.

c. Is there sufficient evidence to indicate that freshness is linearly related to storage time? Use $\alpha = .05$.

d. Estimate the mean rate of change in freshness for a 1-day increase in storage time using a 98% confidence interval.

e. Estimate the expected freshness measurement for a storage time of 14 days with a 95% confidence interval.

f. Of what value is the linear model when compared to $\bar{y}$ in predicting freshness?

12.77 Lexus, Inc. The Lexus GX is a mid- EX1277 size sport utility vehicle (SUV) sold in North American and Eurasian markets by *Lexus*. The GX 470 was introduced in 2002 (as a 2003 model) and was later upgraded with a new off-road suspension system. The sales of the Lexus GX 470 from its inception until 2009 are given in the table:[18]

Generation	Model(s)	Calendar Year	Total Sales (U.S.)
		2002	2,190
		2003	31,376
		2004	35,420
UZJ120	GX 470	2005	34,339
		2006	25,454
		2007	23,035
		2008	15,759
		2009	6,235

a. Plot the data using a scatterplot. How would you describe the relationship between year and sales of the Lexus GX 470?

b. Even though the scatterplot in part a might indicate differently, assume that the relationship between year and sales is linear. Find the least-squares regression line relating the sales of the Lexus GX 470 to the year being measured.

c. Is there sufficient evidence to indicate that sales are linearly related to year? Use $\alpha = .05$.

d. Examine the diagnostic plots shown below. What can you conclude about the validity of the regression assumptions?

Diagnostic plots for Exercise 12.77

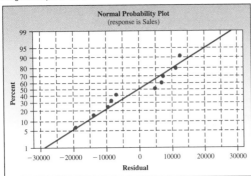

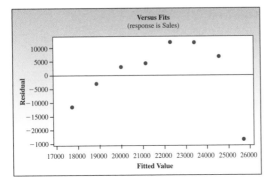

e. Based on your conclusions in part d, is it advisable to predict the 2010 sales using the regression line from part b? Explain.

12.78 Starbucks Here is some nutritional data for a sampling of Starbucks 16-ounce Espresso beverages, made with 2% milk. The nutritional information for all of Starbucks products can be found on the company website, www.starbucks.com.[19]

Product	Calories	Fat (g)	Carb. (g)	Fiber (g)	Protein (g)
Caffe Latte	190	7	18	0	12
Caffe Mocha	260	8	41	2	13
Cappuccino	120	4	12	0	8
Caramel Macchiato	240	7	34	0	10
Cinnamon Dolce Latte	260	6	40	0	11
Flavored Latte	250	6	36	0	12
Iced Caffe Latte	130	4.5	13	0	8
Iced Caffe Mocha	200	6	35	2	9
Iced Caramel Macchiato	230	6	33	0	10
Iced Cinnamon Dolce Latte	200	4	34	0	7
Iced Flavored Latte	250	6	36	0	12
Iced Peppermint Mocha	260	6	52	2	8
Iced Peppermint White Chocolate Mocha	400	9	72	0	10
Iced Pumpkin Spice Latte	250	4	44	0	10
Iced Skinny Flavored Latte	110	4	12	0	7
Iced Toffee Mocha	280	3.5	51	2	12
Iced White Chocolate Mocha	340	9	55	0	10
Peppermint Mocha	330	8	57	2	12
Peppermint White Chocolate Mocha	470	12	78	0	14
Pumpkin Spice Latte	310	6	49	0	14
Skinny Cinnamon Dolce Latte	180	6	18	0	12
Skinny Flavored Latte	180	6	18	0	12
Toffee Mocha	350	7	58	2	17
White Chocolate Mocha	400	11	61	0	15

Use the appropriate statistical methods to analyze the relationships between some of the nutritional variables given in the table. Write a summary report explaining any conclusions that you can draw from your analysis.

CASE STUDY

Is Your Car "Made in the U.S.A."?

Foreign Cars

The phrase "made in the U.S.A." has become a familiar battle cry as U.S. workers try to protect their jobs from overseas competition. For the past few decades, a major trade imbalance in the United States has been caused by a flood of imported goods that enter the country and are sold at lower cost than comparable American-made goods. One prime concern is the automotive industry, in which the number of imported cars steadily increased during the 1970s and 1980s. The U.S. automobile industry has been besieged with complaints about product quality, worker layoffs, and high prices, and has spent billions in advertising and research to produce an American-made car that will satisfy consumer demands. Have they been successful in stopping the flood of imported cars purchased by American consumers? The data in the table represent the numbers of imported cars y sold in the United States (in millions) for the years 1969–2009.[20] To simplify the analysis, we have coded the year using the coded variable $x = $ Year $- 1969$.

Year	(Year − 1969), x	Number of Imported Cars, y	Year	(Year − 1969), x	Number of Imported Cars, y
1969	0	1.1	1989	20	2.7
1970	1	1.3	1990	21	2.4
1971	2	1.6	1991	22	2.0
1972	3	1.6	1992	23	1.9
1973	4	1.8	1993	24	1.8
1974	5	1.4	1994	25	1.7
1975	6	1.6	1995	26	1.5
1976	7	1.5	1996	27	1.3
1977	8	2.1	1997	28	1.4
1978	9	2.0	1998	29	1.4
1979	10	2.3	1999	30	1.7
1980	11	2.4	2000	31	2.0
1981	12	2.3	2001	32	2.1
1982	13	2.2	2002	33	2.2
1983	14	2.4	2003	34	2.1
1984	15	2.4	2004	35	2.1
1985	16	2.8	2005	36	2.2
1986	17	3.2	2006	37	2.3
1987	18	3.1	2007	38	2.4
1988	19	3.0	2008	39	2.3
			2009	40	1.8

1. Using a scatterplot, plot the data for the years 1969–1988. Does there appear to be a linear relationship between the number of imported cars and the year?

2. Use a computer software package to find the least-squares line for predicting the number of imported cars as a function of year for the years 1969–1988.

3. Is there a significant linear relationship between the number of imported cars and the year?

4. Use the computer program to predict the number of cars that will be imported using 95% prediction intervals for each of the years 2007, 2008, and 2009.

5. Now look at the actual data points for the years 2007–2009. Do the predictions obtained in step 4 provide accurate estimates of the *actual* values observed in these years? Explain.

6. Add the data for 1989–2009 to your database, and recalculate the regression line. What effect have the new data points had on the slope? What is the effect on SSE?

7. Given the form of the scatterplot for the years 1969–2009, does it appear that a straight line provides an accurate model for the data? What other type of model might be more appropriate? (Use residual plots to help answer this question.)

Multiple Regression Analysis

© Will & Deni McIntyre/CORBIS

GENERAL OBJECTIVES

In this chapter, we extend the concepts of linear regression and correlation to a situation where the average value of a random variable y is related to several independent variables—x_1, x_2, . . . , x_k—in models that are more flexible than the straight-line model of Chapter 12. With *multiple regression analysis*, we can use the information provided by the independent variables to fit various types of models to the sample data, to evaluate the usefulness of these models, and finally to estimate the average value of y or predict the actual value of y for given values of x_1, x_2, . . . , x_k.

CHAPTER INDEX

"Made in the U.S.A."— Another Look

In Chapter 12, we used simple linear regression analysis to try to predict the number of cars imported into the United States over a period of years. Unfortunately, the number of imported cars does not really follow a linear trend pattern, and our predictions were far from accurate. We reexamine the same data at the end of this chapter, using the methods of multiple regression analysis.

INTRODUCTION

Multiple linear regression is an extension of simple linear regression to allow for more than one independent variable. That is, instead of using only a single independent variable x to explain the variation in y, you can simultaneously use several independent (or predictor) variables. By using more than one independent variable, you should do a better job of explaining the variation in y and hence be able to make more accurate predictions.

For example, a company's regional sales y of a product might be related to three factors:

- x_1—the amount spent on television advertising
- x_2—the amount spent on newspaper advertising
- x_3—the number of sales representatives assigned to the region

A researcher would collect data measuring the variables y, x_1, x_2, and x_3, and then use these sample data to construct a prediction equation relating y to the three predictor variables. Of course, several questions arise, just as they did with simple linear regression:

- How well does the model fit?
- How strong is the relationship between y and the predictor variables?
- Have any important assumptions been violated?
- How good are estimates and predictions?

The methods of **multiple regression analysis**—which are almost always done with a computer software program—can be used to answer these questions. This chapter provides a brief introduction to multiple regression analysis and the difficult task of model building—that is, choosing the correct model for a practical application.

THE MULTIPLE REGRESSION MODEL

The **general linear model** for a multiple regression analysis describes a particular response y using the model given next.

GENERAL LINEAR MODEL AND ASSUMPTIONS

$$y = \beta_0 + \beta_1 x_1 + \beta_2 x_2 + \cdots + \beta_k x_k + \epsilon$$

where

- y is the **response variable** that you want to predict.
- $\beta_0, \beta_1, \beta_2, \ldots, \beta_k$ are unknown constants.
- $x_1, x_2, \ldots, x_k$ are independent **predictor variables** that are measured without error.
- ϵ is the random error, which allows each response to deviate from the average value of y by the amount ϵ. You must assume that the values of ϵ (1) are independent; (2) have a mean of 0 and a common variance σ^2 for any set $x_1, x_2, \ldots, x_k$; and (3) are normally distributed.

When these assumptions about ϵ are met, the *average* value of y for a given set of values $x_1, x_2, \ldots, x_k$ is equal to the *deterministic* part of the model:

$$E(y) = \beta_0 + \beta_1 x_1 + \beta_2 x_2 + \cdots + \beta_k x_k$$

You will notice that the multiple regression model and assumptions are *very similar* to the model and assumptions used for linear regression. It will probably not surprise you that the testing and estimation procedures are also extensions of those used in Chapter 12.

Multiple regression models are very flexible and can take many forms, depending on the way in which the independent variables x_1, x_2, . . . , x_k are entered into the model. We begin with a simple multiple regression model, explaining the basic concepts and procedures with an example. As you become more familiar with the multiple regression procedures, we increase the complexity of the examples, and you will see that the same procedures can be used for models of different forms, depending on the particular application.

EXAMPLE ● 13.1

Suppose you want to relate a random variable y to two independent variables x_1 and x_2. The multiple regression model is

$$y = \beta_0 + \beta_1 x_1 + \beta_2 x_2 + \epsilon$$

with the mean value of y given as

$$E(y) = \beta_0 + \beta_1 x_1 + \beta_2 x_2$$

This equation is a three-dimensional extension of the **line of means** from Chapter 12 and traces a **plane** in three-dimensional space (see Figure 13.1). The constant β_0 is called the **intercept**—the average value of y when x_1 and x_2 are both 0. The coefficients β_1 and β_2 are called the **partial slopes** or **partial regression coefficients.** The partial slope β_i (for $i = 1$ or 2) measures the change in y for a one-unit change in x_i when *all other independent variables are held constant.* The value of the partial regression coefficient—say, β_1—with x_1 and x_2 in the model is generally *not* the same as the slope when you fit a line with x_1 alone. These coefficients are the unknown constants, which must be estimated using sample data to obtain the prediction equation.

 NEED A TIP?

Instead of *x* and *y* plotted in two-dimensional space, *y* and x_1, x_2, . . . , x_k have to be plotted in $(k + 1)$ dimensions.

FIGURE 13.1

Plane of means for Example 13.1

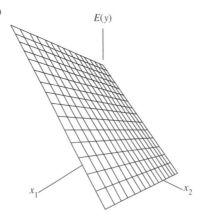

A MULTIPLE REGRESSION ANALYSIS

13.3

A multiple regression analysis involves estimation, testing, and diagnostic procedures designed to fit the multiple regression model

$$E(y) = \beta_0 + \beta_1 x_1 + \beta_2 x_2 + \cdots + \beta_k x_k$$

to a set of data. Because of the complexity of the calculations involved, these procedures are almost always implemented with a regression program from one of several computer software packages. All give similar output in slightly different forms. We follow the basic patterns set in simple linear regression, beginning with an outline of the general procedures and illustrated with an example.

The Method of Least Squares

The prediction equation

$$\hat{y} = b_0 + b_1 x_1 + b_2 x_2 + \cdots + b_k x_k$$

is the line that minimizes SSE, the sum of squares of the deviations of the observed values y from the predicted values $\hat{y}$. These values are calculated using a regression program.

EXAMPLE 13.2 How do real estate agents decide on the asking price for a newly listed condominium? A computer database in a small community contains the listed selling price y (in thousands of dollars), the amount of living area x_1 (in hundreds of square feet), and the numbers of floors x_2, bedrooms x_3, and bathrooms x_4, for $n = 15$ randomly selected condos currently on the market. The data are shown in Table 13.1.

TABLE 13.1 ● **Data on 15 Condominiums**

Observation	List Price, y	Living Area, x_1	Floors, x_2	Bedrooms, x_3	Baths, x_4
1	169.0	6	1	2	1
2	218.5	10	1	2	2
3	216.5	10	1	3	2
4	225.0	11	1	3	2
5	229.9	13	1	3	1.7
6	235.0	13	2	3	2.5
7	239.9	13	1	3	2
8	247.9	17	2	3	2.5
9	260.0	19	2	3	2
10	269.9	18	1	3	2
11	234.9	13	1	4	2
12	255.0	18	1	4	2
13	269.9	17	2	4	3
14	294.5	20	2	4	3
15	309.9	21	2	4	3

The multiple regression model is

$$E(y) = \beta_0 + \beta_1 x_1 + \beta_2 x_2 + \beta_3 x_3 + \beta_4 x_4$$

which can be fit using either the *MINITAB* or *Microsoft Excel* software packages. You can find instructions for generating this output in the section "Technology Today" at the end of this chapter. The first portion of the *MINITAB* regression output is shown in Figure 13.2(a). You will find the fitted regression equation in the first two lines of the printout:

$$\hat{y} = 119 + 6.27 x_1 - 16.2 x_2 - 2.67 x_3 + 30.3 x_4$$

The partial regression coefficients are shown with slightly more accuracy in the second section of the *MINITAB* printout; a similar output generated by *MS Excel* is shown in Figure 13.2(b). The columns list the name given to each independent predictor variable, its estimated regression coefficient, its standard error, and the t- and p-values that are used to test its significance *in the presence of all the other predictor variables*. We explain these tests in more detail in a later section.

FIGURE 13.2(a)

A portion of the *MINITAB* printout for Example 13.2

Regression Analysis: List Price versus Square Feet, Number of Floors, Bedrooms, Baths
```
The regression equation is
List Price = 119 + 6.27 Square Feet - 16.2 Number of Floors
                 - 2.67 Bedrooms + 30.3 Baths

Predictor              Coef      SE Coef       T        P
Constant            118.763        9.207    12.90    0.000
Square Feet          6.2698       0.7252     8.65    0.000
Number of Floors    -16.203        6.212    -2.61    0.026
Bedrooms             -2.673        4.494    -0.59    0.565
Baths                30.271        6.849     4.42    0.001
```

FIGURE 13.2(b)

A portion of the *MS Excel* printout for Example 13.2

	Coefficients	Standard Error	t Stat	P-value	Lower 95%	Upper 95%
Intercept	118.763	9.207	12.899	0.000	98.248	139.279
Square Feet	6.270	0.725	8.645	0.000	4.654	7.886
Number of						
Floors	-16.203	6.212	-2.608	0.026	-30.045	-2.362
Bedrooms	-2.673	4.494	-0.595	0.565	-12.686	7.340
Baths	30.271	6.849	4.420	0.001	15.011	45.530

The Analysis of Variance for Multiple Regression

The analysis of variance divides the total variation in the response variable y,

$$\text{Total SS} = \Sigma y_i^2 - \frac{(\Sigma y_i)^2}{n}$$

into two portions:

- SSR (sum of squares for regression) measures the amount of variation explained by using the regression equation.
- SSE (sum of squares for error) measures the residual variation in the data that is not explained by the independent variables.

so that

$$\text{Total SS} = \text{SSR} + \text{SSE}$$

The **degrees of freedom** for these sums of squares are found using the following argument. There are $(n - 1)$ total degrees of freedom. Estimating the regression line requires estimating k unknown coefficients—$\beta_1, \beta_2, \ldots, \beta_k$; the constant b_0 (which estimates β_0) is a function of $\bar{y}$ and the other estimates. Hence, there are k regression degrees of freedom, leaving $(n - 1) - k$ degrees of freedom for error. As in previous chapters, the mean squares are calculated as $\text{MS} = \text{SS}/df$.

The ANOVA table for the real estate data in Table 13.1 is shown in the second portion of the *MINITAB* printout and the lower section of the *Excel* printout in Figure 13.3. There are $n = 15$ observations and $k = 4$ independent predictor variables. You can verify that the total degrees of freedom, $(n - 1) = 14$, is divided into $k = 4$ for regression and $(n - k - 1) = 10$ for error.

FIGURE 13.3(a)

A portion of the *MINITAB* printout for Example 13.2

```
S = 6.84930        R-Sq = 97.1%     R-Sq(adj) = 96.0%

Analysis of Variance
Source            DF          SS          MS        F        P
Regression         4      15913.0      3978.3    84.80    0.000
Residual Error    10        469.1        46.9
Total             14      16382.2

Source            DF      Seq SS
Square Feet        1     14829.3
Number of Floors   1         0.9
Bedrooms           1       166.4
Baths              1       916.5
```

FIGURE 13.3(b)

A portion of the *MS Excel* printout for Example 13.2

SUMMARY OUTPUT

Regression Statistics	
Multiple R	0.986
R Square	0.971
Adjusted R Square	0.960
Standard Error	6.849
Observations	15

ANOVA

	df	SS	MS	F	Significance F
Regression	4	15913.048	3978.262	84.801	0.000
Residual	10	469.129	46.913		
Total	14	16382.177			

The best estimate of the random variation σ^2 in the experiment—the variation that is unexplained by the predictor variables—is as usual given by

$$s^2 = \text{MSE} = \frac{\text{SSE}}{n - k - 1} = 46.913$$

from the ANOVA table. The first line of Figure 13.3(a) and the fourth line in Figure 13.3(b) also shows $s = \sqrt{s^2} = 6.849$. The computer uses these values internally to produce test statistics, confidence intervals, and prediction intervals, which we discuss in subsequent sections.

The last section of Figure 13.3(a) shows a decomposition of SSR = 15,913.0 in which the conditional contribution of each predictor variable *given the variables already entered into the model* is shown for the order of entry that you specify in your regression program. For the real estate example, the *MINITAB* program entered the variables in this order: square feet, then numbers of floors, bedrooms, and baths. These conditional or **sequential sums of squares** each account for one of the $k = 4$ regression degrees of freedom. It is interesting to notice that the predictor variable x_1 alone accounts for 14,829.3/15,913.0 = .932 or 93.2% of the total variation explained by the regression model. However, if you change the order of entry, another variable may account for the major part of the regression sum of squares!

Testing the Usefulness of the Regression Model

Recall in Chapter 12 that you tested to see whether y and x were linearly related by testing $H_0 : \beta = 0$ with either a t-test or an equivalent F-test. In multiple regression, there is more than one *partial slope*—the *partial regression coefficients*. The t- and F-tests are no longer equivalent.

The Analysis of Variance F-Test

Is the regression equation that uses information provided by the predictor variables $x_1, x_2, \ldots, x_k$ substantially better than the simple predictor $\bar{y}$ that does not rely on any of the x-values? This question is answered using an overall F-test with the hypotheses:

NEED A TIP?

The overall F-test (for the significance of the model) in multiple regression is one-tailed.

$$H_0 : \beta_1 = \beta_2 = \cdots = \beta_k = 0$$

versus

$$H_a : \text{At least one of } \beta_1, \beta_2, \ldots, \beta_k \text{ is not } 0$$

The test statistic is found in the ANOVA table (Figure 13.3) as

$$F = \frac{\text{MSR}}{\text{MSE}} = \frac{3978.3}{46.9} = 84.80$$

which has an F distribution with $df_1 = k = 4$ and $df_2 = (n - k - 1) = 10$. Since the exact p-value, $P = .000$, is given in the printout, you can declare the regression to be highly significant. That is, at least one of the predictor variables is contributing significant information for the prediction of the response variable y.

The Coefficient of Determination, R^2

NEED A TIP?

MINITAB printouts report R^2 as a percentage rather than a proportion.

How well does the regression model fit? The regression printout provides a statistical measure of the strength of the model in the **coefficient of determination, R^2**—the proportion of the total variation that is explained by the regression of y on $x_1, x_2, \ldots, x_k$—defined as

$$R^2 = \frac{\text{SSR}}{\text{Total SS}} = \frac{15{,}913.0}{16{,}382.2} = .971 \quad \text{or } 97.1\%$$

NEED A TIP?

R^2 is the multivariate equivalent of r^2, used in linear regression.

The coefficient of determination is sometimes called **multiple R^2** and is found in the first line of Figure 13.3(a), labeled "R-Sq." and in the second line of Figure 13.3(b), labeled "R Square." Hence, for the real estate example, 97.1% of the total variation has been explained by the regression model. The model fits very well!

It may be helpful to know that the value of the F statistic is related to R^2 by the formula

$$F = \frac{R^2/k}{(1 - R^2)/(n - k - 1)}$$

so that when R^2 is large, F is large, and vice versa.

Interpreting the Results of a Significant Regression

NEED A TIP?

You can show that
$$F = \frac{\text{MSR}}{\text{MSE}} = \frac{R^2/k}{(1 - R^2)/(n - k - 1)}$$

Testing the Significance of the Partial Regression Coefficients

Once you have determined that the model is useful for predicting y, you should explore the nature of the "usefulness" in more detail. Do all of the predictor variables add important information for prediction *in the presence of other predictors already in the model?* The individual t-tests in the first section of the regression printout are designed to test the hypotheses

$$H_0 : \beta_i = 0 \quad \text{versus} \quad H_a : \beta_i \neq 0$$

for each of the partial regression coefficients, *given that the other predictor variables are already in the model.* These tests are based on the Student's t statistic given by

$$t = \frac{b_i - \beta_i}{SE(b_i)}$$

which has $df = (n - k - 1)$ degrees of freedom. The procedure is identical to the one used to test a hypothesis about the slope β in the simple linear regression model.[†]

NEED A TIP?

Test for the significance of the individual coefficient β_i, using t-tests.

Figure 13.4 shows the t-test and p-values from the upper portion of the *MINITAB* printout and the lower section of the *MS Excel* printout. By examining the p-values in the last column, you can see that all the variables *except* x_3, the number of bedrooms, add very significant information for predicting y, **even with all the other independent variables already in the model.** Could the model be any better? It may be that x_3 is an unnecessary predictor variable. One option is to remove this variable and re-fit the model with a new set of data!

FIGURE 13.4(a)

A portion of the *MINITAB* printout for Example 13.2

Predictor	Coef	SE Coef	T	P
Constant	118.763	9.207	12.90	0.000
Square Feet	6.2698	0.7252	8.65	0.000
Number of Floors	-16.203	6.212	-2.61	0.026
Bedrooms	-2.673	4.494	-0.59	0.565
Baths	30.271	6.849	4.42	0.001

FIGURE 13.4(b)

A portion of the *MS Excel* printout for Example 13.2

	Coefficients	Standard Error	t Stat	P-value
Intercept	118.763	9.207	12.899	0.000
Square Feet	6.270	0.725	8.645	0.000
Number of Floors	-16.203	6.212	-2.608	0.026
Bedrooms	-2.673	4.494	-0.595	0.565
Baths	30.271	6.849	4.420	0.001

The Adjusted Value of R^2

Notice from the definition of $R^2 = \text{SSR/Total SS}$ that its value can never decrease with the addition of more variables into the regression model. Hence, R^2 can be artificially inflated by the inclusion of more and more predictor variables.

An alternative measure of the strength of the regression model is adjusted for degrees of freedom by using mean squares rather than sums of squares:

$$R^2(\text{adj}) = \left(1 - \frac{\text{MSE}}{\text{Total SS}/(n - 1)} \right) 100\%$$

NEED A TIP?

Use $R^2(\text{adj})$ for comparing one or more possible models.

For the real estate data in Figure 13.3,

$$R^2(\text{adj}) = \left(1 - \frac{46.9}{16{,}382.2/14} \right) 100\% = 96.0\%$$

[†]Some packages use the t statistic just described, whereas others use the equivalent F statistic ($F = t^2$), since the square of a t statistic with v degrees of freedom is equal to an F statistic with 1 df in the numerator and v degrees of freedom in the denominator.

is found in the first line of the *MINITAB* printout and the third line of the *Excel* printout. The value "R-Sq(adj) = 96.0%" or "Adjusted R Square = 0.960" represents the percentage of variation in the response *y* explained by the independent variables, corrected for degrees of freedom. The adjusted value of R^2 is mainly used to compare two or more regression models that use different numbers of independent predictor variables.

Checking the Regression Assumptions

Before using the regression model for its main purpose—estimation and prediction of *y*—you should look at computer-generated **residual plots** to make sure that all the regression assumptions are valid. The *normal probability plot* and the *plot of residuals versus fit* are shown in Figure 13.5 for the real estate data. There appear to be three observations that do not fit the general pattern. You can see them as outliers in both graphs. These three observations should probably be investigated; however, they do not provide strong evidence that the assumptions are violated.

FIGURE 13.5

Diagnostic plots

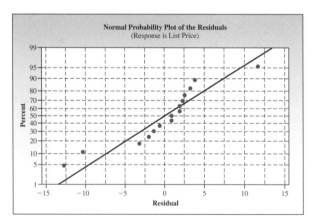

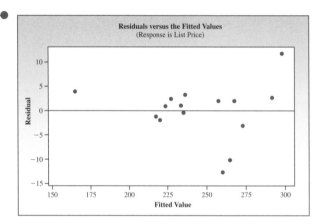

Using the Regression Model for Estimation and Prediction

Finally, once you have determined that the model is effective in describing the relationship between *y* and the predictor variables $x_1, x_2, \ldots, x_k$, the model can be used for these purposes:

NEED A TIP?

For given values of x_1, $x_2, \ldots, x_k$, the prediction interval will **always** be wider than the confidence interval.

- Estimating the average value of *y*—$E(y)$—for given values of $x_1, x_2, \ldots, x_k$
- Predicting a particular value of *y* for given values of $x_1, x_2, \ldots, x_k$

The values of $x_1, x_2, \ldots, x_k$ are entered into the computer, and the computer generates the fitted value $\hat{y}$ together with its estimated standard error and the confidence and prediction intervals. Remember that the prediction interval is *always wider* than the confidence interval.

Let's see how well our prediction works for the real estate data, using another house from the computer database—a house with 1000 square feet of living area, one floor, three bedrooms, and two baths, which was listed at $221,500. The printout in Figure 13.6 shows the confidence and prediction intervals for these values. The actual value falls within both intervals, which indicates that the model is working very well!

FIGURE 13.6 ●

Confidence and prediction intervals for Example 13.2

```
Predicted Values for New Observations
New Obs      Fit     SE Fit       95% CI              95% PI
1          217.78      3.11   (210.86,  224.70)   (201.02,  234.54)

Values of Predictors for New Observations
New      Square    Number of
Obs       Feet      Floors      Bedrooms      Baths
1         10.0        1.00          3.00       2.00
```

A POLYNOMIAL REGRESSION MODEL

In Section 13.3, we explained in detail the various portions of the multiple regression printout. When you perform a multiple regression analysis, you should use a step-by-step approach:

1. Obtain the fitted prediction model.
2. Use the analysis of variance F-test and R^2 to determine how well the model fits the data.
3. Check the t-tests for the partial regression coefficients to see which ones are contributing significant information in the presence of the others.
4. If you choose to compare several different models, use $R^2(\text{adj})$ to compare their effectiveness.
5. Use computer-generated residual plots to check for violation of the regression assumptions.

Once all of these steps have been taken, you are ready to use your model for estimation and prediction.

The predictor variables $x_1, x_2, \ldots, x_k$ used in the general linear model do not have to represent *different* predictor variables. For example, if you suspect that one independent variable x affects the response y, but that the relationship is *curvilinear* rather than *linear,* then you might choose to fit a **quadratic model:**

$$y = \beta_0 + \beta_1 x + \beta_2 x^2 + \epsilon$$

NEED A TIP?

A quadratic equation is $y = a + bx + cx^2$. The graph forms a **parabola**.

The quadratic model is an example of a **second-order model** because it involves a term whose exponents sum to 2 (in this case, x^2).[†] It is also an example of a **polynomial model**—a model that takes the form

$$y = a + bx + cx^2 + dx^3 + \cdots$$

To fit this type of model using the multiple regression program, observed values of y, x, and x^2 are entered into the computer, and the printout can be generated as in Section 13.3.

[†]The *order* of a term is determined by the sum of the exponents of variables making up that term. Terms involving x_1 or x_2 are first-order. Terms involving x_1^2, x_2^2, or $x_1 x_2$ are second-order.

EXAMPLE 13.3

In a study of variables that affect productivity in the retail grocery trade, W.S. Good uses value added per work-hour to measure the productivity of retail grocery outlets.[1] He defines "value added" as "the surplus [money generated by the business] available to pay for labor, furniture and fixtures, and equipment." Data consistent with the relationship between value added per work-hour y and the size x of a grocery outlet described in Good's article are shown in Table 13.2 for 10 fictitious grocery outlets. Choose a model to relate y to x.

Data on Store Size and Value Added

Store	Value Added per Work-Hour, y ($)	Size of Store (thousand square feet), x
1	4.08	21.0
2	3.40	12.0
3	3.51	25.2
4	3.09	10.4
5	2.92	30.9
6	1.94	6.8
7	4.11	19.6
8	3.16	14.5
9	3.75	25.0
10	3.60	19.1

Solution You can investigate the relationship between y and x by looking at the plot of the data points in Figure 13.7. The graph suggests that productivity, y, increases as the size of the grocery outlet, x, increases until an optimal size is reached. Above that size, productivity tends to decrease. The relationship appears to be *curvilinear*, and a quadratic model,

$$E(y) = \beta_0 + \beta_1 x + \beta_2 x^2$$

FIGURE 13.7

Plot of store size x and value added y for Example 13.3

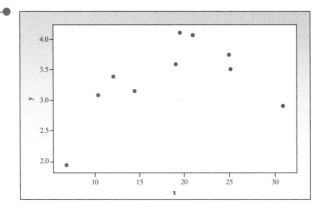

may be appropriate. Remember that, in choosing to use this model, we are not saying that the true relationship is quadratic, but only that it may provide more accurate estimations and predictions than, say, a linear model.

EXAMPLE 13.4 Refer to the data on grocery retail outlet productivity and outlet size in Example 13.3. *MINITAB* was used to fit a quadratic model to the data and to graph the quadratic prediction curve, along with the plotted data points. Discuss the adequacy of the fitted model.

Solution From the printout in Figure 13.8, you can see that the regression equation is

$$\hat{y} = -.159 + .392x - .00949x^2$$

The graph of this quadratic equation together with the data points is shown in Figure 13.9.

FIGURE 13.8

MINITAB printout for Example 13.4

NEED A TIP?

Look at the computer printout and find the labels for "Predictor." This will tell you what variables have been used in the model.

Regression Analysis: y versus x, x-sq

```
The regression equation is
y = - 0.159 + 0.392 x - 0.00949 x-sq

Predictor        Coef      St Coef          T         P
Constant      -0.1594       0.5006      -0.32     0.760
x             0.39193       0.05801       6.76     0.000
x-sq         -0.009495      0.001535     -6.19     0.000

S = 0.250298    R-Sq = 87.9%    R-Sq(adj) = 84.5%

Analysis of Variance
Source          DF        SS        MS        F        P
Regression       2    3.1989    1.5994    25.53    0.001
Residual Error   7    0.4385    0.0626
Total            9    3.6374

Source          DF     Seq SS
x                1     0.8003
x-sq             1     2.3986
```

FIGURE 13.9

Fitted quadratic regression line for Example 13.4

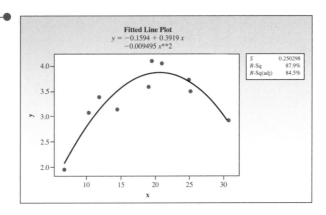

To assess the adequacy of the quadratic model, the test of

$$H_0 : \beta_1 = \beta_2 = 0$$

versus

$$H_a : \text{Either } \beta_1 \text{ or } \beta_2 \text{ is not } 0$$

is given in the printout as

$$F = \frac{\text{MSR}}{\text{MSE}} = 25.53$$

with p-value $= .001$. Hence, the overall fit of the model is highly significant. Quadratic regression accounts for $R^2 = 87.9\%$ of the variation in y [R^2(adj) $= 84.5\%$].

From the t-tests for the individual variables in the model, you can see that both b_1 and b_2 are highly significant, with p-values equal to .000. Notice from the sequential sum of squares section that the sum of squares for linear regression is .8003, with an additional sum of squares of 2.3986 when the quadratic term is added. It is apparent that the simple linear regression model is inadequate in describing the data.

One last look at the residual plots in Figure 13.10 ensures that the regression assumptions are valid. Notice the relatively linear appearance of the normal plot and the relative scatter of the residuals versus fits. The quadratic model provides accurate predictions for values of x that lie *within the range of the sampled values of x*.

FIGURE 13.10

Diagnostic plots for
Example 13.4

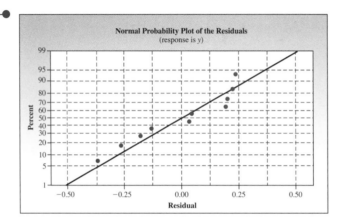

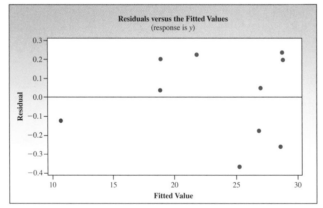

(13.4) **EXERCISES**

BASIC TECHNIQUES

13.1 Suppose that $E(y)$ is related to two predictor variables, x_1 and x_2, by the equation

$$E(y) = 3 + x_1 - 2x_2$$

a. Graph the relationship between $E(y)$ and x_1 when $x_2 = 2$. Repeat for $x_2 = 1$ and for $x_2 = 0$.

b. What relationship do the lines in part a have to one another?

13.2 Refer to Exercise 13.1.

a. Graph the relationship between $E(y)$ and x_2 when $x_1 = 0$. Repeat for $x_1 = 1$ and for $x_1 = 2$.

b. What relationship do the lines in part a have to one another?

c. Suppose, in a practical situation, you want to model the relationship between $E(y)$ and two predictor variables x_1 and x_2. What is the implication of using the first-order model $E(y) = \beta_0 + \beta_1 x_1 + \beta_2 x_2$?

13.3 Suppose that you fit the model

$$E(y) = \beta_0 + \beta_1 x_1 + \beta_2 x_2 + \beta_3 x_3$$

to 15 data points and found F equal to 57.44.

a. Do the data provide sufficient evidence to indicate that the model contributes information for the prediction of y? Test using a 5% level of significance.

b. Use the value of F to calculate R^2. Interpret its value.

13.4 The computer output for the multiple regression analysis for Exercise 13.3 provides this information:

$$b_0 = 1.04 \qquad\qquad b_1 = 1.29$$
$$\qquad\qquad\qquad SE(b_1) = .42$$
$$b_2 = 2.72 \qquad\qquad b_3 = .41$$
$$SE(b_2) = .65 \qquad\qquad SE(b_3) = .17$$

a. Which, if any, of the independent variables x_1, x_2, and x_3 contribute information for the prediction of y?

b. Give the least-squares prediction equation.

c. On the same sheet of graph paper, graph y versus x_1 when $x_2 = 1$ and $x_3 = 0$ and when $x_2 = 1$ and $x_3 = .5$. What relationship do the two lines have to each other?

d. What is the practical interpretation of the parameter β_1?

13.5 Suppose that you fit the model

$$E(y) = \beta_0 + \beta_1 x + \beta_2 x^2$$

to 20 data points and obtained the accompanying *MINITAB* printout.

MINITAB output for Exercise 13.5

Regression Analysis: y versus x, x-sq

```
The regression equation is
y = 10.6 + 4.44 x - 0.648 x-sq

Predictor      Coef    SE Coef       T      P
Constant    10.5638     0.6951   15.20  0.000
x            4.4366     0.5150    8.61  0.000
x-sq        -0.64754    0.07988  -8.11  0.000

S = 1.191    R-Sq = 81.5%    R-Sq(adj) = 79.3%

Analysis of Variance
Source           DF       SS       MS       F      P
Regression        2  106.072   53.036   37.37  0.000
Residual Error   17   24.128    1.419
Total            19  130.200
```

a. What type of model have you chosen to fit the data?

b. How well does the model fit the data? Explain.

c. Do the data provide sufficient evidence to indicate that the model contributes information for the prediction of y? Use the p-value approach.

13.6 Refer to Exercise 13.5.

a. What is the prediction equation?

b. Graph the prediction equation over the interval $0 \leq x \leq 6$.

13.7 Refer to Exercise 13.5.

a. What is your estimate of the average value of y when $x = 0$?

b. Do the data provide sufficient evidence to indicate that the average value of y differs from 0 when $x = 0$?

13.8 Refer to Exercise 13.5.

a. Suppose that the relationship between $E(y)$ and x is a straight line. What would you know about the value of β_2?

b. Do the data provide sufficient evidence to indicate curvature in the relationship between y and x?

13.9 Refer to Exercise 13.5. Suppose that y is the profit for some business and x is the amount of capital invested, and you know that the rate of increase in profit for a unit increase in capital invested can only decrease as x increases. You want to know whether the data provide sufficient evidence to indicate a decreasing rate of increase in profit as the amount of capital invested increases.

a. The circumstances described imply a one-tailed statistical test. Why?

b. Conduct the test at the 1% level of significance. State your conclusions.

APPLICATIONS

Data set **13.10 College Textbooks** A publisher of EX1310 college textbooks conducted a study to relate profit per text y to cost of sales x over a 6-year period when its sales force (and sales costs) were growing rapidly. These inflation-adjusted data (in thousands of dollars) were collected:

Profit per Text, y	16.5	22.4	24.9	28.8	31.5	35.8
Sales Cost per Text, x	5.0	5.6	6.1	6.8	7.4	8.6

Expecting profit per book to rise and then plateau, the publisher fitted the model $E(y) = \beta_0 + \beta_1 x + \beta_2 x^2$ to the data.

Excel output for Exercise 13.10

```
SUMMARY OUTPUT
    Regression Statistics
Multiple R          0.9978
R Square            0.9955
Adjusted R Square   0.9925
Standard Error      0.5944
Observations             6
```

ANOVA

	df	SS	MS	F	Significance F
Regression	2	234.995	117.478	332.528	0.000
Residual	3	1.060	0.353		
Total	5	236.015			

	Coefficients	Standard Error	t Stat	P-value
Intercept	-44.192	8.287	-5.333	0.013
x	16.334	2.490	6.560	0.007
x-sq	-0.820	0.182	-4.494	0.021

a. Plot the data points. Does it look as though the quadratic model is necessary?

b. Find *s* on the printout. Confirm that

$$s = \sqrt{\frac{\text{SSE}}{n - k - 1}}$$

c. Do the data provide sufficient evidence to indicate that the model contributes information for the prediction of *y*? What is the *p*-value for this test, and what does it mean?

d. What sign would you expect the actual value of β_2 to have? Find the value of β_2 in the printout. Does this value confirm your expectation?

e. Do the data indicate a significant curvature in the relationship between *y* and *x*? Test at the 5% level of significance.

f. What conclusions can you draw from the accompanying residual plots?

Diagnostic plots for Exercise 13.10

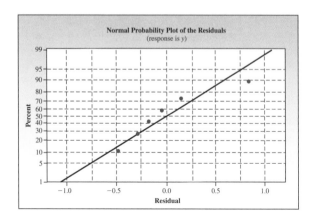

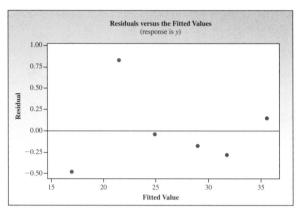

13.11 College Textbooks II Refer to Exercise 13.10.

a. Use the values of SSR and Total SS to calculate R^2. Compare this value with the value given in the printout.

b. Calculate R^2(adj). When would it be appropriate to use this value rather than R^2 to assess the fit of the model?

c. The value of R^2(adj) was 95.66% when a simple linear model was fit to the data. Does the linear or the quadratic model fit better?

Data set **13.12 Choosing a Good Camera** Cameras
EX1312 come with many options, and it appears that the more that you want, the higher the cost of the camera. *Consumer Reports*[2] has rated $n = 20$ cameras on qualities that we consumers are looking for. Variables that may relate to the cost of a camera are given in the following table where y = overall customer score, x_1 = megapixels, x_2 = weight (oz), x_3 = optical zoom, x_4 = widest angle, and x_5 = battery life (shots).

Camera	Cost	y	x_1	x_2	x_3	x_4	x_5
Nikon Coolpix S8000	300	65	14	6	10.0	30	210
Canon PowerShot SD1400 IS Elph	250	61	14	5	4.0	28	230
Sony Cyber-shot DSC-HX5V	350	60	10	7	10.0	25	310
Sony Cyber-shot DSC-TX7	400	59	10	5	4.0	25	230
Panasonic Lumix DMC-FPI	150	59	12	6	4.0	35	300
Canon PowerShot A3100 IS	180	58	12	6	4.0	35	240
Sony Cyber-shot DSC-W380	140	57	14	4	5.0	24	220
Nikon Coolpix S70	300	57	12	6	5.0	28	200
Pentax Optio I-10	300	56	12	6	5.0	28	250
Olympus Stylus 5010	200	55	14	5	5.0	26	120
Olympus Stylus 7040	250	55	14	5	7.0	28	120
Sony Cyber-shot DSC-W350	200	55	14	4	4.0	26	220
Canon PowerShot A490	110	55	10	7	3.0	37	150
Canon PowerShot A945	130	54	10	7	3.0	37	150
Casio Exilim EX-G1	250	54	12	5	3.0	38	300
Samsung TL210	230	54	12	5	5.0	27	200
Olympus Stylus Tough-3000	230	48	12	6	3.6	28	160
Sony Cyber-shot DSC-W310	150	47	12	5	4.0	28	220
Olympus FE-47	120	47	14	7	5.0	36	120
Sony Cyber-shot DSC-S2100	120	46	12	7	3.0	35	170

The *MINITAB* printout below shows the regression of y on the predictor variables x_1 through x_5.

MINITAB output for Exercise 13.12

Regression Analysis: y versus x1, x2, x3, x4, x5

```
The regression equation is
y = 45.6 - 0.0583 x1 - 0.83 x2 + 1.18 x3 + 1.129 x4 + 0.0277 x5

Predictor       Coef    SE Coef       T       P
Constant      45.584      8.664    5.26   0.000
x1          -0.05832    0.04527   -1.29   0.219
x2            -0.835      1.393   -0.60   0.559
x3            1.1773     0.5994    1.96   0.070
x4            0.1286     0.3040    0.42   0.679
x5           0.02773    0.01664    1.67   0.118

S = 4.09203    R-Sq = 50.3%    R-Sq(adj) = 32.6%

Analysis of Variance
Source          DF      SS      MS      F      P
Regression       5  237.37   47.47   2.84  0.057
Residual Error  14  234.43   16.74
Total           19  471.80

Source   DF    Seq SS
x1        1     86.68
x2        1      1.69
x3        1     95.43
x4        1      7.05
x5        1     46.51
```

a. Write a multiple regression model using each of the x-variables as independent variables and y as the response variable.

b. Comment on the fit of the model using the statistical test for the overall fit and the coefficient of determination, R^2.

c. If you were to refit the model, eliminating one of the predictor variables, which one would you choose? Why?

13.13 Choosing a Good Camera II Refer to Exercise 13.12. A command in the *MINITAB* regression menu provides output in which R^2 and R^2(adj) are calculated for all possible subsets of the five independent variables. The printout is provided here.

MINITAB output for Exercise 13.13

Best Subsets Regression: y versus x1, x2, x3, x4, x5

```
Response is y
                R-Sq   Mallows              x x x x x
Vars  R-Sq     (adj)    C-p       S         1 2 3 4 5
  1   26.4     22.3     4.7    4.3928       X
  1   20.7     16.3     6.3    4.5577               X
  2   41.0     34.1     2.6    4.0449           X   X
  2   37.4     30.1     3.6    4.1674       X   X
  3   49.0     39.5     2.4    3.8766       X   X   X
  3   43.5     32.9     3.9    4.0807         X X   X
  4   49.7     36.3     4.2    3.9784       X X X   X
  4   49.0     35.4     4.4    4.0037       X   X X X
```

a. If you had to compare these models and choose the best one, which model would you choose? Explain.

b. Comment on the usefulness of the model you chose in part a. Is your model valuable in predicting the overall score based on the chosen predictor variables?

13.14 Lexus, Inc. In Exercise 12.77 we presented sales data for the Lexus GX, a mid-size sport utility vehicle (SUV) sold in North American and Eurasian markets by *Lexus*. The sales of the Lexus GX 470 from its inception until 2009 are given in the table:[3]

Generation	Model(s)	Calendar Year	Total Sales (United States)
		2002	2,190
		2003	31,376
		2004	35,420
UZJ120	GX 470	2005	34,339
		2006	25,454
		2007	23,035
		2008	15,759
		2009	6,235

a. Plot the data. What model would you expect to provide the best fit to the data? Write the equation of that model.

b. Use a computer software package to fit the model from part a.

c. Find the least-squares prediction equation relating the sales of the Lexus GX 470 to the year of production.

d. Does the model contribute significant information for the prediction of sales based on the year of production? Use the appropriate p-value to make your decision.

e. Find R^2 on the printout. What does this value tell you about the effectiveness of the multiple regression analysis?

13.15 Corporate Profits In order to study the relationship of advertising and capital investment with corporate profits, the following data, recorded in units of $100,000, were collected for 10 medium-sized firms in the same year. The variable y represents profit for the year, x_1 represents capital investment, and x_2 represents advertising expenditures.

y	x_1	x_2	y	x_1	x_2
15	25	4	1	20	0
16	1	5	16	12	4
2	6	3	18	15	5
3	30	1	13	6	4
12	29	2	2	16	2

a. Using the model

$$E(y) = \beta_0 + \beta_1 x_1 + \beta_2 x_2$$

and an appropriate computer software package, find the least-squares prediction equation for these data.

b. Use the overall F-test to determine whether the model contributes significant information for the prediction of y. Use $\alpha = .01$.

c. Does advertising expenditure x_2 contribute significant information for the prediction of y, given that x_1 is already in the model? Use $\alpha = .01$.

d. Calculate the coefficient of determination, R^2. What percentage of the overall variation is explained by the model?

Data set **13.16 The New Route 66?** One of the most
EX1316 famous national highways from its beginnings in the 1920s until its demise around 1970 was *Route 66*. On its way from Chicago to Los Angeles, one of the last stops was San Bernardino, California. Now, a new transportation hub—the San Bernardino International Airport—is in the planning stages. Although no agreements with any air carriers have been signed, officials forecast 2 million passengers by 2030, as shown in the following table.[4]

Year	2010	2015	2020	2025	2030
Coded Year, x	1	2	3	4	5
Number of Travelers, y	500,000	750,000	1,000,000	1,500,000	2,000,000

Linear and quadratic fitted plots for these data follow.

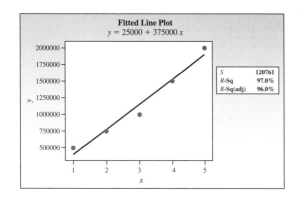

Fitted Line Plot
$y = 25000 + 375000\,x$

S	120761
R-Sq	97.0%
R-Sq(adj)	96.0%

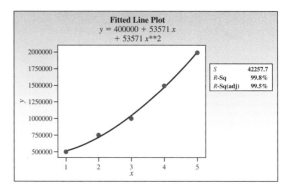

Fitted Line Plot
$y = 400000 + 53571\,x + 53571\,x^{**}2$

S	42257.7
R-Sq	99.8%
R-Sq(adj)	99.5%

a. Based on the summary statistics in the line plots, which of the two models better fits the data?

b. Write the theoretical equation for the quadratic model.

c. Use the following printout to determine if the quadratic model contributes significant information to the prediction of y.

d. Use the following printout to determine if the quadratic term contributes significant information to the prediction of y, in the presence of the linear term.

Regression Analysis: y versus x, x-sq
```
The regression equation is
y = 400000 + 53571 x + 53571 x-sq

Predictor     Coef    SE Coef      T      P
Constant    400000      90633   4.41  0.048
x            53571      69068   0.78  0.519
x-sq         53571      11294   4.74  0.042

S = 42257.7   R-Sq = 99.8%    R-Sq(adj) = 99.5%

Analysis of Variance
Source        DF          SS           MS       F      P
Regression     2  1.44643E+12  7.23214E+11  405.00  0.002
Residual Error 2   3571428571   1785714286
Total          4  1.45000E+12
```

USING QUANTITATIVE AND QUALITATIVE PREDICTOR VARIABLES IN A REGRESSION MODEL

13.5

One reason multiple regression models are very flexible is that they allow for the use of both *qualitative* and *quantitative* predictor variables. For the multiple regression methods used in this chapter, the response variable *y must be quantitative,* measuring a numerical random variable that has a normal distribution (according to the assumptions of Section 13.2). However, each independent predictor variable can be either a quantitative variable *or* a qualitative variable, whose levels represent qualities or characteristics and can only be categorized.

Quantitative and qualitative variables enter the regression model in different ways. To make things more complicated, we can allow a combination of different types of variables in the model, *and* we can allow the variables to *interact,* a concept that may be familiar to you from the *factorial experiment* of Chapter 11. We consider these options one at a time.

A **quantitative variable** x can be entered as a linear term, x, or to some higher power such as x^2 or x^3, as in the quadratic model in Example 13.3. When more than one quantitative variable is necessary, the interpretation of the possible models becomes more complicated. For example, with two quantitative variables x_1 and x_2, you could use a **first-order model** such as

$$E(y) = \beta_0 + \beta_1 x_1 + \beta_2 x_2$$

which traces a plane in three-dimensional space (see Figure 13.1). However, it may be that one of the variables—say, x_2—is not related to y in the same way when $x_1 = 1$ as it is when $x_1 = 2$. To allow x_2 to behave differently depending on the value of x_1, we add an **interaction term,** $x_1 x_2$, and allow the two-dimensional plane to *twist.* The model is now a **second-order model:**

$$E(y) = \beta_0 + \beta_1 x_1 + \beta_2 x_2 + \beta_3 x_1 x_2$$

The models become complicated quickly when you allow curvilinear relationships *and* interaction for the two variables. One way to decide on the type of model you need is to plot some of the data—perhaps y versus x_1, y versus x_2, and y versus x_2 for various values of x_1.

In contrast to quantitative predictor variables, **qualitative predictor variables** are entered into a regression model through **dummy** or **indicator variables.** For example, in a model that relates the mean salary of a group of employees to a number of predictor variables, you may want to include the employee's ethnic background. If each employee included in your study belongs to one of three ethnic groups—say, A, B, or C—you can enter the qualitative variable "ethnicity" into your model using two *dummy variables:*

$$x_1 = \begin{cases} 1 & \text{if group B} \\ 0 & \text{if not} \end{cases} \qquad x_2 = \begin{cases} 1 & \text{if group C} \\ 0 & \text{if not} \end{cases}$$

Look at the effect these two variables have on the model $E(y) = \beta_0 + \beta_1 x_1 + \beta_2 x_2$: For employees in group A,

$$E(y) = \beta_0 + \beta_1(0) + \beta_2(0) = \beta_0$$

for employees in group B,

$$E(y) = \beta_0 + \beta_1(1) + \beta_2(0) = \beta_0 + \beta_1$$

and for those in group C,

$$E(y) = \beta_0 + \beta_1(0) + \beta_2(1) = \beta_0 + \beta_2$$

The model allows a different average response for each group. β_1 measures the difference in the average responses between groups B and A, while β_2 measures the difference between groups C and A.

When a qualitative variable involves k categories or levels, $(k - 1)$ dummy variables should be added to the regression model. This model may contain other predictor variables—quantitative or qualitative—as well as cross products (**interactions**) of the dummy variables with other variables that appear in the model. As you can see, the process of model building—deciding on the appropriate terms to enter into the regression model—can be quite complicated. However, you can become more proficient

NEED a tip? **NEED A TIP?**

Enter **quantitative** variables as

• a single x

• a higher power, x^2 or x^3

• an interaction with another variable

NEED a tip? **NEED A TIP?**

Qualitative variables are entered as dummy variables—one fewer than the number of categories or levels.

at model building, gaining experience with the chapter exercises. The next example involves one quantitative and one qualitative variable that interact.

EXAMPLE **13.5**

A study was conducted to examine the relationship between university salary y, the number of years of experience of the faculty member, and the gender of the faculty member. If you expect a straight-line relationship between mean salary and years of experience for both men and women, write the model that relates mean salary to the two predictor variables: years of experience (quantitative) and gender of the professor (qualitative).

Solution Since you may suspect the mean salary lines for women and men to be different, your model for mean salary $E(y)$ may appear as shown in Figure 13.11. A straight-line relationship between $E(y)$ and years of experience x_1 implies the model

$$E(y) = \beta_0 + \beta_1 x_1 \qquad \text{(graphs as a straight line)}$$

FIGURE 13.11

Hypothetical relationship for mean salary $E(y)$, years of experience (x_1), and gender (x_2) for Example 13.5

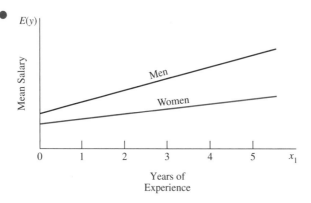

The qualitative variable "gender" involves $k = 2$ categories, men and women. Therefore, you need $(k - 1) = 1$ dummy variable, x_2, defined as

$$x_2 = \begin{cases} 1 & \text{if a man} \\ 0 & \text{if a woman} \end{cases}$$

and the model is expanded to become

$$E(y) = \beta_0 + \beta_1 x_1 + \beta_2 x_2 \qquad \text{(graphs as two parallel lines)}$$

The fact that the slopes of the two lines may differ means that the two predictor variables **interact;** that is, the change in $E(y)$ corresponding to a change in x_1 depends on whether the professor is a man or a woman. To allow for this interaction (difference in slopes), the interaction term $x_1 x_2$ is introduced into the model. The complete model that characterizes the graph in Figure 13.11 is

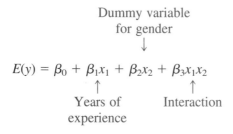

Dummy variable
for gender
↓

$$E(y) = \beta_0 + \beta_1 x_1 + \beta_2 x_2 + \beta_3 x_1 x_2$$

↑   ↑

Years of   Interaction
experience

where

$$x_1 = \text{Years of experience}$$
$$x_2 = \begin{cases} 1 & \text{if a man} \\ 0 & \text{if a woman} \end{cases}$$

You can see how the model works by assigning values to the dummy variable x_2. When the faculty member is a woman, the model is

$$E(y) = \beta_0 + \beta_1 x_1 + \beta_2(0) + \beta_3 x_1(0) = \beta_0 + \beta_1 x_1$$

which is a straight line with slope β_1 and intercept β_0. When the faculty member is a man, the model is

$$E(y) = \beta_0 + \beta_1 x_1 + \beta_2(1) + \beta_3 x_1(1) = (\beta_0 + \beta_2) + (\beta_1 + \beta_3) x_1$$

which is a straight line with slope $(\beta_1 + \beta_3)$ and intercept $(\beta_0 + \beta_2)$. The two lines have *different slopes and different intercepts,* which allows the relationship between salary y and years of experience x_1 to behave differently for men and women.

EXAMPLE 13.6

Random samples of six female and six male assistant professors were selected from among the assistant professors in a college of arts and sciences. The data on salary and years of experience are shown in Table 13.3. Note that each of the two samples (male and female) contained two professors with 3 years of experience, but no male professor had 2 years of experience. Interpret the output of the *MS Excel* regression printout and graph the predicted salary lines.

TABLE 13.3

Salary versus Gender and Years of Experience

Years of Experience, x_1	Salary for Men, y ($)	Salary for Women, y ($)
1	60,710	59,510
2	—	60,440
3	63,160	61,340
3	63,210	61,760
4	64,140	62,750
5	65,760	63,200
5	65,590	—

Solution The *Excel* regression printout for the data in Table 13.3 is shown in Figure 13.12. You can use a step-by-step approach to interpret this regression analysis, beginning with the fitted prediction equation, $\hat{y} = 58{,}593 + 969x_1 + 866.71x_2 + 260.13x_1x_2$. By substituting $x_2 = 0$ or 1 into this equation, you get two straight lines— one for women and one for men—to predict the value of y for a given x_1. These lines are

Women: $\hat{y} = 58{,}593 + 969x_1$

Men: $\hat{y} = 59{,}459.71 + 1229.13x_1$

and are graphed in Figure 13.13.

FIGURE 13.12

MS Excel output for Example 13.6

SUMMARY OUTPUT

Regression Statistics	
Multiple R	0.9962
R Square	0.9924
Adjusted R Square	0.9895
Standard Error	201.3438
Observations	12

ANOVA

	df	SS	MS	F	Significance F
Regression	3	42108777.03	14036259.01	346.238	0.000
Residual	8	324314.64	40539.330		
Total	11	42433091.67			

	Coefficients	Standard Error	t Stat	P-value
Intercept	58593	207.9470	281.7689	0.000
x1	969	63.6705	15.2190	0.000
x2	866.710	305.2568	2.8393	0.022
x1x2	260.130	87.0580	2.9880	0.017

Next, consider the overall fit of the model using the analysis of variance F-test. Since the observed test statistic in the ANOVA portion of the printout is $F = 346.238$ with p-value ("Significance F") equal to .000, you can conclude that at least one of the predictor variables is contributing information for the prediction of y. The strength of this model is further measured by the coefficient of determination, $R^2 = 99.24\%$. You can see that the model appears to fit very well.

FIGURE 13.13

A graph of the faculty salary prediction lines for Example 13.6

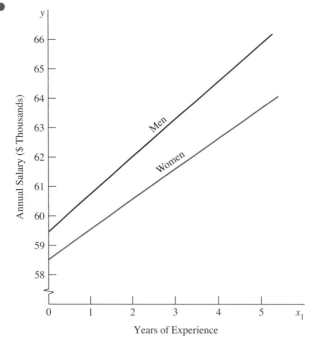

To explore the effect of the predictor variables in more detail, look at the individual t-tests for the three predictor variables. The p-values for these tests—.000, .022, and .017, respectively—are all significant, which means that all of the predictor variables add significant information to the prediction *with the other two variables already in*

the model. Finally, check the diagnostic plots to make sure that there are no strong violations of the regression assumptions. These plots, which behave as expected for a properly fit model, are shown in Figure 13.14.

FIGURE 13.14 ●

Diagnostic plots for Example 13.6

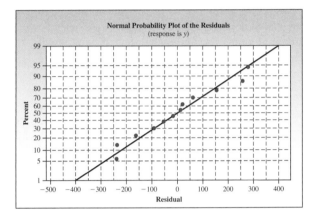

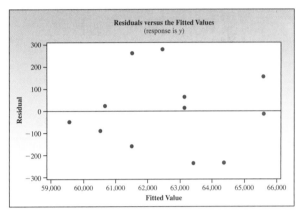

EXAMPLE (13.7)

Refer to Example 13.6. Do the data provide sufficient evidence to indicate that the annual rate of increase in male junior faculty salaries exceeds the annual rate of increase in female junior faculty salaries? That is, do the data provide sufficient evidence to indicate that the slope of the men's faculty salary line is greater than the slope of the women's faculty salary line?

Solution   Since β_3 measures the difference in slopes, the slopes of the two lines will be identical if $\beta_3 = 0$. Therefore, you want to test the null hypothesis

$$H_0 : \beta_3 = 0$$

—that is, the slopes of the two lines are identical—versus the alternative hypothesis

$$H_a : \beta_3 > 0$$

—that is, the slope of the men's faculty salary line is greater than the slope of the women's faculty salary line.

The calculated value of t corresponding to β_3, shown in the row labeled "x1x2" in Figure 13.12, is 2.988. Since the *Excel* regression output provides p-values for two-tailed significance tests, the p-value in the printout, .017, is *twice* what it would be for a one-tailed test. For this one-tailed test, the p-value is .017/2 = .0085, and the null hypothesis is rejected. There is sufficient evidence to indicate that the annual rate of increase in men's faculty salaries exceeds the rate for women.[†]

[†]If you want to determine whether the data provide sufficient evidence to indicate that male faculty members start at higher salaries, you would test $H_0 : \beta_2 = 0$ versus the alternative hypothesis $H_a : \beta_2 > 0$.

13.5 **EXERCISES**

BASIC TECHNIQUES

13.17 Production Yield Suppose you wish to predict production yield y as a function of several independent predictor variables. Indicate whether each of the following independent variables is qualitative or quantitative. If qualitative, define the appropriate dummy variable(s).

a. The prevailing interest rate in the area

b. The price per pound of one item used in the production process

c. The plant (A, B, or C) at which the production yield is measured

d. The length of time that the production machine has been in operation

e. The shift (night or day) in which the yield is measured

13.18 Suppose $E(y)$ is related to two predictor variables x_1 and x_2 by the equation

$$E(y) = 3 + x_1 - 2x_2 + x_1x_2$$

a. Graph the relationship between $E(y)$ and x_1 when $x_2 = 0$. Repeat for $x_2 = 2$ and for $x_2 = -2$.

b. Repeat the instructions of part a for the model

$$E(y) = 3 + x_1 - 2x_2$$

c. Note that the equation for part a is exactly the same as the equation in part b except that we have added the term x_1x_2. How does the addition of the x_1x_2 term affect the graphs of the three lines?

d. What flexibility is added to the first-order model $E(y) = \beta_0 + \beta_1x_1 + \beta_2x_2$ by the addition of the term $\beta_3x_1x_2$, using the model $E(y) = \beta_0 + \beta_1x_1 + \beta_2x_2 + \beta_3x_1x_2$?

13.19 A multiple linear regression model involving one qualitative and one quantitative independent variable produced this prediction equation:

$$\hat{y} = 12.6 + .54x_1 - 1.2x_1x_2 + 3.9x_2^2$$

a. Which of the two variables is the quantitative variable? Explain.

b. If x_1 can take only the values 0 or 1, find the two possible prediction equations for this experiment.

c. Graph the two equations found in part b. Compare the shapes of the two curves.

APPLICATIONS

Data set **13.20 Less Red Meat!** One desirable dietary
EX1320 change if you want to "eat right," is to reduce the intake of red meat and to substitute poultry or fish. Researchers tracked beef and chicken consumption, y (in annual pounds per person), and found the consumption of beef declining and the consumption of chicken increasing over a period of 7 years. A summary of their data is shown in the table.

Year	Beef	Chicken
1	85	37
2	89	36
3	76	47
4	76	47
5	68	62
6	67	74
7	60	79

Consider fitting the following model, which allows for simultaneously fitting two simple linear regression lines:

$$E(y) = \beta_0 + \beta_1x_1 + \beta_2x_2 + \beta_3x_1x_2$$

where y is the annual meat (either beef or chicken) consumption per person per year,

$$x_1 = \begin{cases} 1 & \text{if beef} \\ 0 & \text{if chicken} \end{cases} \quad \text{and} \quad x_2 = \text{Year}$$

MINITAB output for Exercise 13.20

Regression Analysis: y versus x1, x2, x1x2

```
The regression equation is
y = 23.6 + 69.0 x1 + 7.75 x2 - 12.3 x1x2

Predictor      Coef     SE Coef       T        P
Constant     23.571       3.522     6.69    0.000
x1           69.000       4.981    13.85    0.000
x2            7.7500      0.7875     9.84    0.000
x1x2        -12.286       1.114   -11.03    0.000

S = 4.16705    R-Sq = 95.4%     R-Sq(adj) = 94.1%

Analysis of Variance
Source          DF       SS       MS       F        P
Regression       3   3637.9   1212.6   69.83    0.000
Residual Error  10    173.6     17.4
Total           13   3811.5

Source    DF    Seq SS
x1         1    1380.1
x2         1     144.6
x1x2       1    2113.1

Predicted Values for New Observations
New
Obs    Fit   SE Fit       95% CI              95% PI
1    56.29     3.52   (48.44,  64.13)   (44.13,  68.44)

Values of Predictors for New Observations
New Obs       x1        x2      x1x2
1           1.00      8.00      8.00
```

Diagnostic plots for Exercise 13.20

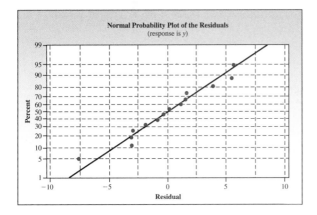

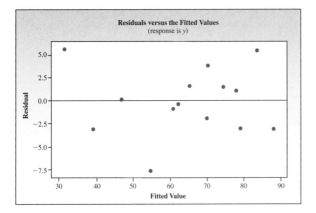

a. How well does the model fit? Use any relevant statistics and diagnostic tools from the printout to answer this question.

b. Write the equations of the two straight lines that describe the trend in consumption over the period of 7 years for beef and for chicken.

c. Use the prediction equation to find a point estimate of the average per-person beef consumption in year 8. Compare this value with the value labeled "Fit" in the printout.

d. Use the printout to find a 95% confidence interval for the average per-person beef consumption in year 8. What is the 95% prediction interval for the per-person beef consumption in year 8? Is there any problem with the validity of the 95% confidence level for these intervals?

Data set **13.21 Cotton versus Cucumber** In Exer-
EX1321 cise 11.65, you used the analysis of variance procedure to analyze a 2 × 3 factorial experiment in

which each factor–level combination was replicated five times. The experiment involved the number of eggs laid by caged female whiteflies on two different plants at three different temperature levels. Suppose that several of the whiteflies died before the experiment was completed, so that the number of replications was no longer the same for each treatment. The analysis of variance formulas of Chapter 11 can no longer be used, but the experiment *can* be analyzed using a multiple regression analysis. The results of this **2 × 3 factorial experiment with unequal replications** are shown in the table.

Cotton			Cucumber		
70°	77°	82°	70°	77°	82°
37	34	46	50	59	43
21	54	32	53	53	62
36	40	41	25	31	71
43	42		37	69	49
31			48	51	

a. Write a model to analyze this experiment. Make sure to include a term for the interaction between plant and temperature.

b. Use a computer software package to perform the multiple regression analysis.

c. Do the data provide sufficient evidence to indicate that the effect of temperature on the number of eggs laid is *different* depending on the type of plant?

d. Based on the results of part c, do you suggest refitting a different model? If so, rerun the regression analysis using the new model and analyze the printout.

e. Write a paragraph summarizing the results of your analyses.

Data set **13.22 Achievement Scores III** The Aca-
EX1322 demic Performance Index (API), described in Exercise 12.13, is a measure of school achievement based on the results of the Stanford 9 Achievement Test. The API scores for 12 elementary schools in Riverside County, California, are shown below, along with several other independent variables.[5]

School	API Score, y	EL(%), x_1	Free/ Reduced Lunch(%), x_2	Avg Parent Education Level, x_3	Gifted and Talented(%), x_4	Previous Year's API, x_5
1	745	71	89	1.70	4	705
2	808	18	51	2.91	16	809
3	798	24	79	2.21	10	763
4	791	50	76	2.19	5	786
5	854	17	56	2.84	7	839
6	688	71	27	1.70	6	673

School	API Score, y	EL(%), x_1	Free/ Reduced Lunch(%), x_2	Avg Parent Education Level, x_3	Gifted and Talented(%), x_4	Previous Year's API, x_5
7	801	11	39	2.79	7	804
8	751	57	87	1.72	1	750
9	778	34	81	2.14	6	770
10	846	9	31	3.22	22	841
11	690	53	78	2.14	3	706
12	685	77	28	1.46	8	665

The variables are defined as

y = API score in 2010

x_1 = % of students who are "English learners"

x_2 = % of students who receive a free or reduced cost lunch

x_3 = Average parent education level (with 1 = Not a high school graduate, 2 = High school graduate, 3 = Some college, 4 = College graduate, 5 = Graduate school)

x_4 = % of students in Gifted and Talented Education Program

x_5 = API score in 2009

The *MINITAB* printout for a first-order regression model is given below.

Regression Analysis: y versus x1, x2, x3, x4, x5
```
The regression equation is
y = 15 - 0.306 x1 + 0.076 x2 - 48.1 x3
    + 1.93 x4 + 1.13 x5

Predictor      Coef     SE Coef       T       P
Constant       14.9       179.7    0.08   0.936
x1           -0.3060     0.6483   -0.47   0.654
x2            0.0763     0.2885    0.26   0.800
x3           -48.06       34.33   -1.40   0.211
x4            1.927        1.450    1.33   0.232
x5            1.1269      0.2503    4.50   0.004

S = 17.0180    R-Sq = 95.4%      R-Sq(adj) = 91.6%

Analysis of Variance
Source          DF        SS       MS       F       P
Regression       5    36121.2   7224.2   24.94   0.001
Residual Error   6     1737.7    289.6
Total           11    37858.9

Source    DF    Seq SS
x1         1   28748.4
x2         1     701.0
x3         1     591.6
x4         1     212.3
x5         1    5868.0
```

a. What is the model that has been fit to this data? What is the least-squares prediction equation?

b. How well does the model fit? Use any relevant statistics from the printout to answer this question.

c. Which, if any, of the independent variables are useful in predicting the API, given the other independent variables already in the model? Explain.

d. Use the values of R^2 and R^2(adj) in the following printout to choose the best model for prediction.

Would you be confident in using the chosen model for predicting the API score for next year based on a model containing similar variables? Explain.

Best Subsets Regression: y versus x1, x2, x3, x4, x5
```
Response is y
                             R-Sq   Mallows           x x x x x
Vars   R-Sq   (adj)          Cp         S      1 2 3 4 5
   1   93.2   92.5          0.9    16.095              X
   1   75.9   73.5         23.5    30.184      X
   2   93.8   92.5          2.0    16.090          X     X
   2   93.3   91.8          2.8    16.789            X X
   3   95.2   93.3          2.3    15.132          X X X
   3   94.0   91.7          3.9    16.858      X   X   X
   4   95.4   92.7          4.1    15.847      X   X X X
   4   95.2   92.5          4.2    16.045        X X X X
   5   95.4   91.6          6.0    17.018      X X X X X
```

13.23 Particle Board A quality control engineer is interested in predicting the strength of particle board y as a function of the size of the particles x_1 and two types of bonding compounds. If the basic response is expected to be a quadratic function of particle size, write a linear model that incorporates the qualitative variable "bonding compound" into the predictor equation.

13.24 Construction Projects In a study to
EX1324 examine the relationship between the time required to complete a construction project and several pertinent independent variables, an analyst compiled a list of four variables that might be useful in predicting the time to completion. These four variables were size of the contract, x_1 (in $1000 unit), number of work-days adversely affected by the weather x_2, number of subcontractors involved in the project x_4, and a variable x_3 that measured the presence ($x_3 = 1$) or absence ($x_3 = 0$) of a workers' strike during the construction. Fifteen construction projects were randomly chosen, and each of the four variables as well as the time to completion were measured.

y	x_1	x_2	x_3	x_4
29	60	7	0	7
15	80	10	0	8
60	100	8	1	10
10	50	14	0	5
70	200	12	1	11
15	50	4	0	3
75	500	15	1	12
30	75	5	0	6
45	750	10	0	10
90	1200	20	1	12
7	70	5	0	3
21	80	3	0	6
28	300	8	0	8
50	2600	14	1	13
30	110	7	0	4

An analysis of these data using a first-order model in x_1, x_2, x_3, and x_4 produced the following printout. Give a complete analysis of the printout and interpret your results. What can you say about the apparent contribution of x_1 and x_2 in predicting y?

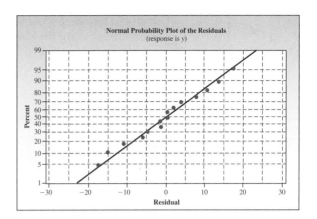

SUMMARY OUTPUT

Regression Statistics	
Multiple R	0.9204
R Square	0.8471
Adjusted R Square	0.7859
Standard Error	11.8450
Observations	15

ANOVA

	df	SS	MS	F	Significance F
Regression	4	7770.297	1942.574	13.846	0.000
Residual	10	1403.036	140.304		
Total	14	9173.333			

	Coefficients	Standard Error	t Stat	P-value
Intercept	-1.589	11.656	-0.136	0.894
x1	-0.008	0.006	-1.259	0.237
x2	0.675	1.000	0.675	0.515
x3	28.013	11.371	2.463	0.033
x4	3.489	1.935	1.803	0.102

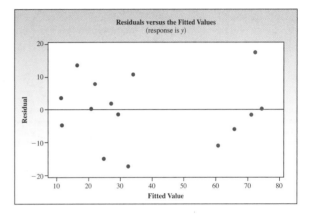

TESTING SETS OF REGRESSION COEFFICIENTS

13.6

In the preceding sections, you have tested the complete set of partial regression coefficients using the F-test for the overall fit of the model, and you have tested the partial regression coefficients individually using the Student's t-test. Besides these two important tests, you might want to test hypotheses about some subsets of these regression coefficients.

For example, suppose a company suspects that the demand y for some product could be related to as many as five independent variables, x_1, x_2, x_3, x_4, and x_5. The cost of obtaining measurements on the variables x_3, x_4, and x_5 is very high. If, in a small pilot study, the company could show that these three variables contribute little or no information for the prediction of y, they can be eliminated from the study at great savings to the company.

If all five variables, x_1, x_2, x_3, x_4, and x_5, are used to predict y, the regression model would be written as

$$y = \beta_0 + \beta_1 x_1 + \beta_2 x_2 + \beta_3 x_3 + \beta_4 x_4 + \beta_5 x_5 + \epsilon$$

However, if x_3, x_4, and x_5 contribute no information for the prediction of y, then they would not appear in the model—that is, $\beta_3 = \beta_4 = \beta_5 = 0$—and the reduced model would be

$$y = \beta_0 + \beta_1 x_1 + \beta_2 x_2 + \epsilon$$

Hence, you want to test the null hypothesis

$$H_0 : \beta_3 = \beta_4 = \beta_5 = 0$$

—that is, the independent variables x_3, x_4, and x_5 contribute no information for the prediction of y—versus the alternative hypothesis

$$H_a : \text{At least one of the parameters } \beta_3, \beta_4, \text{ or } \beta_5 \text{ differs from } 0$$

—that is, at least one of the variables x_3, x_4, or x_5 contributes information for the prediction of y. Thus, in deciding whether the complete model is preferable to the reduced model in predicting demand, you are led to a test of hypothesis about a set of three parameters, β_3, β_4, and β_5.

A test of hypothesis concerning a set of model parameters involves two models:

Model 1 (reduced model)

$$E(y) = \beta_0 + \beta_1 x_1 + \beta_2 x_2 + \cdots + \beta_r x_r$$

Model 2 (complete model)

$$E(y) = \underbrace{\beta_0 + \beta_1 x_1 + \beta_2 x_2 + \cdots + \beta_r x_r}_{\text{terms in model 1}} + \underbrace{\beta_{r+1} x_{r+1} + \beta_{r+2} x_{r+2} + \cdots \beta_k x_k}_{\text{additional terms in model 2}}$$

Suppose you fit both models to the data set and calculated the sum of squares for error for both regression analyses. If model 2 contributes more information for the prediction of y than model 1, then the errors of prediction for model 2 should be smaller than the corresponding errors for model 1, and SSE_2 should be smaller than SSE_1. In fact, the greater the difference between SSE_1 and SSE_2, the greater is the evidence to indicate that model 2 contributes more information for the prediction of y than model 1.

The test of the null hypothesis

$$H_0 : \beta_{r+1} = \beta_{r+2} = \cdots = \beta_k = 0$$

versus the alternative hypothesis

$$H_a : \text{At least one of the parameters } \beta_{r+1}, \beta_{r+2}, \ldots, \beta_k \text{ differs from } 0$$

uses the test statistic

$$F = \frac{(SSE_1 - SSE_2)/(k - r)}{MSE_2}$$

where F is based on $df_1 = (k - r)$ and $df_2 = n - (k + 1)$. Note that the $(k - r)$ parameters involved in H_0 are those added to model 1 to obtain model 2. The numerator degrees of freedom df_1 always equals $(k - r)$, the number of parameters involved in H_0. The denominator degrees of freedom df_2 is the number of degrees of freedom associated with the sum of squares for error, SSE_2, for the complete model.

The rejection region for the test is identical to the rejection region for all of the analysis of variance F-tests—namely,

$$F > F_\alpha$$

EXAMPLE 13.8 — Refer to the real estate data of Example 13.2 that relate the listed selling price y to the square feet of living area x_1, the number of floors x_2, the number of bedrooms x_3, and the number of bathrooms, x_4. The realtor suspects that the square footage of living area is the most important predictor variable and that the other variables might be eliminated from the model without loss of much prediction information. Test this claim with $\alpha = .05$.

Solution The hypothesis to be tested is

$$H_0 : \beta_2 = \beta_3 = \beta_4 = 0$$

versus the alternative hypothesis that at least one of β_2, β_3, or β_4 is different from 0. The **complete model 2,** given as

$$y = \beta_0 + \beta_1 x_1 + \beta_2 x_2 + \beta_3 x_3 + \beta_4 x_4 + \epsilon$$

was fitted in Example 13.2. A portion of the *MINITAB* printout from Figure 13.3 is reproduced in Figure 13.15 along with a portion of the *MINITAB* printout for the simple linear regression analysis of the **reduced model 1,** given as

$$y = \beta_0 + \beta_1 x_1 + \epsilon$$

FIGURE 13.15 ●

Portions of the *MINITAB* regression printouts for (a) complete and (b) reduced models for Example 13.8

Regression Analysis: (a) List Price versus Square Feet, Number of Floors, Bedrooms and Baths

```
S = 6.84930     R-Sq = 97.1%     R-Sq(adj) = 96.0%

Analysis of Variance
Source          DF        SS        MS        F        P
Regression       4    15913.0    3978.3    84.80    0.000
Residual Error  10      469.1      46.9
Total           14    16382.2
```

Regression Analysis: (b) List Price versus Square Feet

```
S = 10.9294     R-Sq = 90.5%     R-Sq(adj) = 89.8%

Analysis of Variance
Source          DF        SS        MS        F        P
Regression       1     14829     14829    124.14    0.000
Residual Error  13      1553       119
Total           14     16382
```

Then $SSE_1 = 1553$ from Figure 13.15(b) and $SSE_2 = 469.1$ and $MSE_2 = 46.9$ from Figure 13.15(a). The test statistic is

$$F = \frac{(SSE_1 - SSE_2)/(k - r)}{MSE_2}$$

$$= \frac{(1553 - 469.1)/(4 - 1)}{46.9} = 7.70$$

The critical value of F with $\alpha = .05$, $df_1 = 3$, and $df_2 = n - (k + 1) = 15 - (4 + 1) = 10$ is $F_{.05} = 3.71$. Hence, H_0 is rejected. There is evidence to indicate that at least one of the three variables—number of floors, bedrooms, or bathrooms—is contributing significant information for predicting the listed selling price.

INTERPRETING RESIDUAL PLOTS

13.7

Once again, you can use residual plots to discover possible violations in the assumptions required for a regression analysis. There are several common patterns you should recognize because they occur frequently in practical applications.

The variance of some types of data changes as the mean changes:

- Poisson data exhibit variation that *increases* with the mean.
- Binomial data exhibit variation that *increases* for values of p from .0 to .5, and then *decreases* for values of p from .5 to 1.0.

Residual plots for these types of data have a pattern similar to that shown in Figure 13.16.

FIGURE 13.16

Plots of residuals against $\hat{y}$

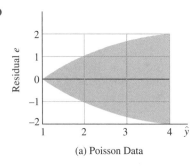 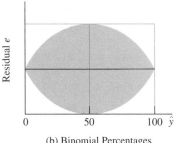

(a) Poisson Data   (b) Binomial Percentages

If the range of the residuals increases as $\hat{y}$ increases and you know that the data are measurements on Poisson variables, you can stabilize the variance of the response by running the regression analysis on $y^* = \sqrt{y}$. Or if the percentages are calculated from binomial data, you can use the arcsin transformation, $y^* = \sin^{-1}\sqrt{y}$.[†]

Even if you are not sure why the range of the residuals increases as $\hat{y}$ increases, you can still use a transformation of y that affects larger values of y more than smaller values—say, $y^* = \sqrt{y}$ or $y^* = \ln y$. These transformations have a tendency both to stabilize the variance of y^* and to make the distribution of y^* more nearly normal when the distribution of y is highly skewed.

Plots of the residuals versus the fits $\hat{y}$ or versus the individual predictor variables often show a pattern that indicates you have chosen an incorrect model. For example, if $E(y)$ and a single independent variable x are linearly related—that is,

$$E(y) = \beta_0 + \beta_1 x$$

and you fit a straight line to the data, then the observed y-values should vary in a random manner about $\hat{y}$, and a plot of the residuals against x will appear as shown in Figure 13.17.

FIGURE 13.17

Residual plot when the model provides a good approximation to reality

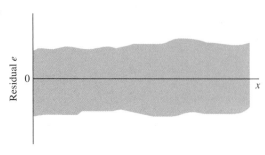

[†]In Chapter 11 and earlier chapters, we represented the response variable by the symbol x. In the chapters on regression analysis, Chapters 12 and 13, the response variable is represented by the symbol y.

In Example 13.3, you fit a quadratic model relating productivity y to store size x. If you had incorrectly used a linear model to fit these data, the residual plot in Figure 13.18 would show that the unexplained variation exhibits a curved pattern, which suggests that there is a quadratic effect that has not been included in the model.

FIGURE 13.18

Residual plot for linear fit of store size and productivity data in Example 13.3

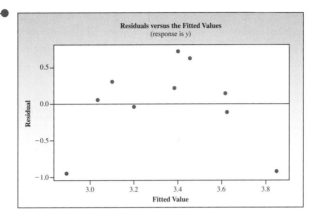

For the data in Example 13.6, the residuals of a linear regression of salary with years of experience x_1 without including gender, x_2, would show one distinct set of positive residuals corresponding to the men and a set of negative residuals corresponding to the women (see Figure 13.19). This pattern signals that the "gender" variable was not included in the model.

FIGURE 13.19

Residual plot for linear fit of salary data in Example 13.6

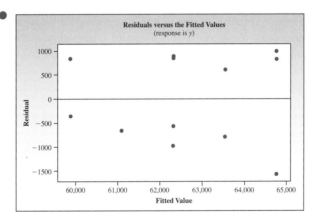

Unfortunately, not all residual plots give such a clear indication of the problem. You should examine the residual plots carefully, looking for nonrandomness in the pattern of residuals. If you can find an explanation for the behavior of the residuals, you may be able to modify your model to eliminate the problem.

13.8 STEPWISE REGRESSION ANALYSIS

Sometimes there are a large number of independent predictor variables that *might* have an effect on the response variable y. For example, try to list all the variables that might affect a college freshman's GPA:

- Grades in high school courses, high school GPA, SAT score, ACT score
- Major, number of units carried, number of courses taken
- Work schedule, marital status, commute or live on campus

Which of this large number of independent variables should be included in the model? Since the number of terms could quickly get unmanageable, you might choose to use a procedure called a **stepwise regression analysis,** which is implemented by computer and is available in most statistical packages.

A stepwise regression analysis fits a variety of models to the data, adding and deleting variables as their significance in the presence of the other variables is either *significant* or *nonsignificant,* respectively. Once the program has performed a sufficient number of iterations and no more variables are significant when added to the model, and none of the variables in the model are nonsignificant when removed, the procedure stops.

A stepwise regression analysis is an easy way to locate some variables that contribute information for predicting *y,* but it is not foolproof. Since these programs always fit first-order models of the form

$$E(y) = \beta_0 + \beta_1 x_1 + \beta_2 x_2 + \cdots + \beta_k x_k$$

they are not helpful in detecting *curvature* or *interaction* in the data. The stepwise regression analysis is best used as a preliminary tool for identifying which of a large number of variables should be considered in your model. You must then decide how to enter these variables into the actual model you will use for prediction.

MISINTERPRETING A REGRESSION ANALYSIS

13.9

Several misinterpretations of the output of a regression analysis are common. We have already mentioned the importance of model selection. If a model does not fit a set of data, it does not mean that the variables included in the model contribute little or no information for the prediction of *y.* The variables may be very important contributors of information, but you may have entered the variables into the model in the wrong way. For example, a second-order model in the variables might provide a very good fit to the data when a first-order model appears to be completely useless in describing the response variable *y.*

Causality

You must be careful not to conclude that changes in *x cause* changes in *y.* This type of **causal relationship** can be detected only with a *carefully designed experiment.* For example, if you randomly assign experimental units to each of two levels of a variable *x*—say, $x = 5$ and $x = 10$—and the data show that the mean value of *y* is larger when $x = 10$, then you can say that the change in the level of *x* caused a change in the mean value of *y.* But in most regression analyses, in which the experiments are not designed, there is no guarantee that an important predictor variable—say, x_1—caused *y* to change. It is quite possible that some variable that is not even in the model causes *both* y and x_1 to change.

Multicollinearity

Neither the size of a regression coefficient nor its *t*-value indicates the importance of the variable as a contributor of information. For example, suppose you intend to predict *y,* a college student's calculus grade, based on x_1 = high school mathematics average and x_2 = score on mathematics aptitude test. Since these two variables contain much of the same or **shared information,** it will not surprise you to learn that, once one of the variables is entered into the model, the other contributes very little additional information. The individual t-value is small. If the variables were entered in the reverse order, however, you would see the size of the t-values reversed.

The situation described above is called **multicollinearity,** and it occurs when two or more of the predictor variables are highly correlated with one another. When multicollinearity is present in a regression problem, it can have these effects on the analysis:

- The estimated regression coefficients will have large standard errors, causing imprecision in confidence and prediction intervals.

- Adding or deleting a predictor variable may cause significant changes in the values of the other regression coefficients.

How can you tell whether a regression analysis exhibits multicollinearity? Look for these clues:

- The value of R^2 is large, indicating a good fit, but the individual t-tests are nonsignificant.

- The signs of the regression coefficients are contrary to what you would intuitively expect the contributions of those variables to be.

- A matrix of correlations, generated by computer, shows you which predictor variables are highly correlated with each other and with the response y.

Figure 13.20 displays the matrix of correlations generated for the real estate data from Example 13.2. The first column of the matrix shows the correlations of each predictor variable with the response variable y. They are all significantly nonzero, but the first variable, $x_1 = $ living area, is the most highly correlated. The last three columns of the matrix show significant correlations between all but one pair of predictor variables. This is a strong indication of multicollinearity. If you try to eliminate one of the variables in the model, it may drastically change the effects of the other three! Another clue can be found by examining the coefficients of the prediction line,

```
ListPrice = 119 + 6.27  Square Feet - 16.2  Number of Floors
          - 2.67 Bedrooms + 30.3 Baths
```

Correlations: List Price, Square Feet, Number of Floors, Bedrooms, Baths

```
                 ListPrice   SqFeet   Numflrs    Bdrms
Square Feet         0.951
                    0.000

Number of Fl        0.605    0.630
                    0.017    0.012

Bedrooms            0.746    0.711    0.375
                    0.001    0.003    0.168

Baths               0.834    0.720    0.760    0.675
                    0.000    0.002    0.001    0.006

Cell Contents: Pearson Correlation
               P-Value
```

You would expect more floors and bedrooms to increase the list price, but their coefficients are negative.

Since multicollinearity exists to some extent in all regression problems, you should think of the individual terms as *information contributors,* rather than try to measure the practical importance of each term. The primary decision to be made is whether a term contributes sufficient information to justify its inclusion in the model.

STEPS TO FOLLOW WHEN BUILDING A MULTIPLE REGRESSION MODEL

13.10

The ultimate objective of a multiple regression analysis is to develop a model that will accurately predict y as a function of a set of predictor variables $x_1, x_2, \ldots, x_k$. The step-by-step procedure for developing this model was presented in Section 13.4 and is restated next with some additional detail. If you use this approach, what may appear to be a complicated problem can be made simpler. As with any statistical procedure, your confidence will grow as you gain experience with multiple regression analysis in a variety of practical situations.

1. Select the predictor variables to be included in the model. Since some of these variables may contain shared information, you can reduce the list by running a stepwise regression analysis (see Section 13.8). Keep the number of predictors small enough to be effective yet manageable. Be aware that the number of observations in your data set must exceed the number of terms in your model; the greater the excess, the better!

2. Write a model using the selected predictor variables. If the variables are qualitative, it is best to begin by including interaction terms. If the variables are quantitative, it is best to start with a second-order model. Unnecessary terms can be deleted later. Obtain the fitted prediction model.

3. Use the analysis of variance F-test and R^2 to determine how well the model fits the data.

4. Check the t-tests for the partial regression coefficients to see which ones are contributing significant information in the presence of the others. If some terms appear to be nonsignificant, consider deleting them. If you choose to compare several different models, use $R^2(\text{adj})$ to compare their effectiveness.

5. Use computer-generated residual plots to check for violation of the regression assumptions.

CHAPTER REVIEW

Key Concepts and Formulas

I. The General Linear Model

1. $y = \beta_0 + \beta_1 x_1 + \beta_2 x_2 + \cdots + \beta_k x_k + \epsilon$

2. The random error ϵ has a normal distribution with mean 0 and variance σ^2.

II. Method of Least Squares

1. Estimates $b_0, b_1, \ldots, b_k$, for $\beta_0, \beta_1, \ldots, \beta_k$, are chosen to minimize SSE, the sum of squared deviations about the regression line, $\hat{y} = b_0 + b_1 x_1 + b_2 x_2 + \cdots + b_k x_k$.

2. Least-squares estimates are produced by computer.

III. Analysis of Variance

1. Total SS = SSR + SSE, where Total SS = S_{yy}. The ANOVA table is produced by computer.

2. Best estimate of σ^2 is

$$\text{MSE} = \frac{\text{SSE}}{n - k - 1}$$

IV. Testing, Estimation, and Prediction

1. A test for the significance of the regression, $H_0 : \beta_1 = \beta_2 = \cdots = \beta_k = 0$, can be implemented using the analysis of variance F-test:

$$F = \frac{\text{MSR}}{\text{MSE}}$$

2. The strength of the relationship between x and y can be measured using

$$R^2 = \frac{\text{SSR}}{\text{Total SS}}$$

which gets closer to 1 as the relationship gets stronger.

3. Use residual plots to check for nonnormality, inequality of variances, and an incorrectly fit model.

4. Significance tests for the partial regression coefficients can be performed using the Student's t-test:

$$t = \frac{b_i - \beta_i}{\text{SE}(b_i)} \qquad \text{with error } df = n - k - 1$$

5. Confidence intervals can be generated by computer to estimate the average value of y, $E(y)$, for given values of $x_1, x_2, \ldots, x_k$. Computer-generated prediction intervals can be used to predict a particular observation y for given values of $x_1, x_2, \ldots, x_k$. For given $x_1, x_2, \ldots, x_k$, prediction intervals are always wider than confidence intervals.

V. Model Building

1. The number of terms in a regression model cannot exceed the number of observations in the data set and should be considerably less!

2. To account for a curvilinear effect in a *quantitative* variable, use a second-order polynomial model. For a cubic effect, use a third-order polynomial model.

3. To add a *qualitative* variable with k categories, use $(k - 1)$ dummy or indicator variables.

4. There may be interactions between two quantitative variables or between a quantitative and qualitative variable. Interaction terms are entered as $\beta x_i x_j$.

5. Compare models using $R^2(\text{adj})$.

 TECHNOLOGY TODAY

Multiple Regression Procedures—*Microsoft Excel*

The procedure for performing a multiple regression analysis in *MS Excel* is identical to the linear regression procedure described in the "Technology Today" section in Chapter 12, except that the range of the x-variables will cover more than one column. You might want to review this section before continuing.

EXAMPLE 13.9

Suppose that a response variable y is related to four predictor variables, x_1, x_2, x_3, and x_4, so that $k = 4$.

1. Enter the observed values of y and each of the $k = 4$ predictor variables into the first $(k + 1)$ columns of an *Excel* spreadsheet. (NOTE: In order for the multiple regression analysis to work properly, there must be a column of values for each independent predictor variable x_i in your model, and the x columns **must be adjacent to each other.**)

2. Use **Data ▶ Data Analysis ▶ Regression** to generate the Dialog box, highlighting or typing in the cell ranges for the x_i and y values and check "Labels" if necessary.

3. If you click "Confidence Level," *Excel* will calculate confidence intervals for the regression estimates, $b_0, b_1, b_2, b_3,$ and b_4. Enter a cell location for the **Output Range** and click **OK** to generate the regression output.

NOTE: *MS Excel* does not provide options for estimation and prediction. Also, the diagnostic plots which can be generated in *Excel* are not the same plots as we have discussed in Section 13.3 and will not be discussed in this section.

Multiple Regression Procedures—*MINITAB*

The procedure for performing a multiple regression analysis in *MINITAB* is similar to the linear regression procedure described in the "Technology Today" section in Chapter 12, except that the range of the *x*-variables will cover more than one column. You might want to review this section before continuing.

EXAMPLE  13.10

Suppose that a response variable y is related to four predictor variables, x_1, x_2, x_3, and x_4, so that $k = 4$.

1. Enter the observed values of y and each of the $k = 4$ predictor variables into the first $(k + 1)$ columns of a *MINITAB* worksheet. Once this is done, the main inferential tools for multiple regression analysis are generated using **Stat ▶ Regression ▶ Regression.** Th e Dialog box for the **Regression** command is shown in Figure 13.21(a).

2. Select **y** for the Response variable and $\mathbf{x_1}$, $\mathbf{x_2}$, ..., $\mathbf{x_k}$ for the Predictor variables. You may now choose to generate some residual plots to check the validity of your regression assumptions before you use the model for estimation or prediction. Choose **Graphs** to display the Dialog box for residual plots, and choose the appropriate diagnostic plot.

FIGURE 13.21 ●

(a)

(b)

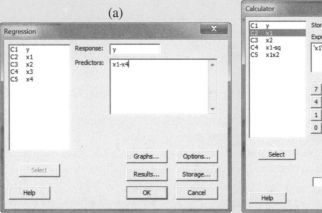

3. Once you have verified the appropriateness of your multiple regression model, you can choose **Options** and obtain confidence and prediction intervals for either of these cases:

   • A single set of values x_1, x_2, ..., x_k (typed in the box marked "Prediction intervals for new observations")

   • Several sets of values x_1, x_2, ..., x_k stored in k columns of the worksheet

   When you click **OK** twice, the regression output is generated.

4. The only difficulty in performing the multiple regression analysis using *MINITAB* might be properly entering the data for your particular model. If the model involves polynomial terms or interaction terms, the **Calc ▶ Calculator** command will help you. For example, suppose you want to fit the model

$$E(y) = \beta_0 + \beta_1 x_1 + \beta_2 x_2 + \beta_3 x_1^2 + \beta_4 x_1 x_2$$

You will need to enter the observed values of y, x_1, and x_2 into the first three columns of the *MINITAB* worksheet. Name column C4 "x1-sq" and name C5 "x1x2." You can now use the calculator Dialog box shown in Figure 13.21(b) to generate these two columns. In the **Expression** box, select **x1 * x1** or **x1 ** 2** and store the results in **C4** (x1-sq). Click **OK.** Similarly, to obtain the data for C5, select **x1 * x2** and store the results in **C5** (x1x2). Click **OK.** You are now ready to perform the multiple regression analysis.

5. If you are fitting either a quadratic or a cubic model in one variable x, you can now plot the data points, the polynomial regression curve, and the upper and lower confidence and prediction limits using **Stat ▶ Regression ▶ Fitted line Plot.** Select y and x for the Response and Predictor variables, and click "Display confidence interval" and "Display prediction interval" in the **Options** Dialog box. Make sure that **Quadratic** or **Cubic** is selected as the "Type of Regression Model," so that you will get the proper fit to the data.

6. Recall that in Chapter 12, you used **Stat ▶ Basic Statistics ▶ Correlation** to obtain the value of the correlation coefficient r. In multiple regression analysis, the same command will generate a matrix of correlations, one for each pair of variables in the set y, x_1, x_2, . . . , x_k. Make sure that the box marked "Display p-values" is checked. The p-values will provide information on the significant correlation between a particular pair, in the presence of all the other variables in the model, and they are identical to the p-values for the individual t-tests of the regression coefficients.

Supplementary Exercises

Data set
EX1325
13.25 Biotin Intake in Chicks Groups of 10-day-old chicks were randomly assigned to seven treatment groups in which a basal diet was supplemented with 0, 50, 100, 150, 200, 250, or 300 micrograms/kilogram ($\mu g/kg$) of biotin. The table gives the average biotin intake (x) in micrograms per day and the average weight gain (y) in grams per day.[6]

Added Biotin	Biotin Intake, x	Weight Gain, y
0	.14	8.0
50	2.01	17.1
100	6.06	22.3
150	6.34	24.4
200	7.15	26.5
250	9.65	23.4
300	12.50	23.3

In the *MINITAB* printout, the second-order polynomial model

$$E(y) = \beta_0 + \beta_1 x + \beta_2 x^2$$

is fitted to the data. Use the printout to answer the questions.

a. What is the fitted least-squares line?

b. Find R^2 and interpret its value.

c. Do the data provide sufficient evidence to conclude that the model contributes significant information for predicting y?

d. Find the results of the test of $H_0 : \beta_2 = 0$. Is there sufficient evidence to indicate that the quadratic model provides a better fit to the data than a simple linear model does?

e. Do the residual plots indicate that any of the regression assumptions have been violated? Explain.

MINITAB output for Exercise 13.25

Regression Analysis: y versus x, x-sq

```
The regression equation is
y = 8.59 + 3.82 x - 0.217 x-sq

Predictor        Coef      SE Coef         T         P
Constant        8.585        1.641      5.23     0.006
x              3.8208       0.5683      6.72     0.003
x-sq         -0.21663      0.04390     -4.93     0.008

S = 1.83318     R-Sq = 94.4%     R-Sq(adj) = 91.5%

Analysis of Variance
Source          DF          SS        MS       F        P
Regression       2      224.75    112.37   33.44    0.003
Residual Error   4       13.44      3.36
Total            6      238.19

Source          DF      Seq SS
x                1      142.92
x-sq             1       81.83
```

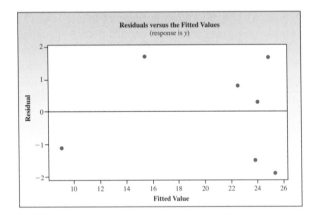

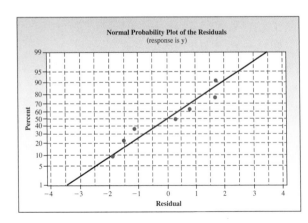

Data set **13.26 Advertising and Sales** A department
EX1326 store conducted an experiment to investigate the effects of advertising expenditures on the weekly sales for its men's wear, children's wear, and women's wear

departments. Five weeks for observation were randomly selected from each department, and an advertising budget x_1 (in hundreds of dollars) was assigned for each. The weekly sales (in thousands of dollars) are shown in the accompanying table for each of the 15 one-week sales periods. If we expect weekly sales $E(y)$ to be linearly related to advertising expenditure x_1, and if we expect the slopes of the lines corresponding to the three departments to differ, then an appropriate model for $E(y)$ is

$$E(y) = \beta_0 + \underbrace{\beta_1 x_1}_{\substack{\text{quantitative} \\ \text{variable} \\ \text{"advertising} \\ \text{expenditure"}}} + \underbrace{\beta_2 x_2 + \beta_3 x_3}_{\substack{\text{dummy variables} \\ \text{used to introduce} \\ \text{the qualitative} \\ \text{variable "department"} \\ \text{into the model}}} + \underbrace{\beta_4 x_1 x_2 + \beta_5 x_1 x_3}_{\substack{\text{interaction terms that} \\ \text{introduce differences} \\ \text{in slopes}}}$$

where

$x_1 =$ Advertising expenditure

$$x_2 = \begin{cases} 1 & \text{if children's wear department B} \\ 0 & \text{if not} \end{cases}$$

$$x_3 = \begin{cases} 1 & \text{if women's wear department C} \\ 0 & \text{if not} \end{cases}$$

	Advertising Expenditure (hundreds of dollars)				
Department	1	2	3	4	5
Men's wear A	5.2	5.9	7.7	7.9	9.4
Children's wear B	8.2	9.0	9.1	10.5	10.5
Women's wear C	10.0	10.3	12.1	12.7	13.6

a. Find the equation of the line relating $E(y)$ to advertising expenditure x_1 for the men's wear department A. [HINT: According to the coding used for the dummy variables, the model represents mean sales $E(y)$ for the men's wear department A when $x_2 = x_3 = 0$. Substitute $x_2 = x_3 = 0$ into the equation for $E(y)$ to find the equation of this line.]

b. Find the equation of the line relating $E(y)$ to x_1 for the children's wear department B. [HINT: According to the coding, the model represents $E(y)$ for the children's wear department when $x_2 = 1$ and $x_3 = 0$.]

c. Find the equation of the line relating $E(y)$ to x_1 for the women's wear department C.

d. Find the difference between the intercepts of the $E(y)$ lines corresponding to the children's wear B and men's wear A departments.

e. Find the difference in slopes between $E(y)$ lines corresponding to the women's wear C and men's wear A departments.

f. Refer to part e. Suppose you want to test the null hypothesis that the slopes of the lines corresponding to the three departments are equal. Express this as a test of hypothesis about one or more of the model parameters.

13.27 Advertising and Sales, continued Refer to Exercise 13.26. Use a computer software package to perform the multiple regression analysis and obtain diagnostic plots if possible.

a. Comment on the fit of the model, using the analysis of variance F-test, R^2, and the diagnostic plots to check the regression assumptions.

b. Find the prediction equation, and graph the three department sales lines.

c. Examine the graphs in part b. Do the slopes of the lines corresponding to the children's wear B and men's wear A departments appear to differ? Test the null hypothesis that the slopes do not differ $(H_0 : \beta_4 = 0)$ versus the alternative hypothesis that the slopes are different.

d. Are the interaction terms in the model significant? Use the methods described in Section 13.5 to test $H_0 : \beta_4 = \beta_5 = 0$. Do the results of this test suggest that the fitted model should be modified?

e. Write a short explanation of the practical implications of this regression analysis.

13.28 Demand for Utilities A short-term
EX1328 study was conducted to investigate the effect of mean monthly daily temperature x_1 and cost per kilowatt-hour x_2 on the mean daily consumption of electricity (in kilowatt-hours, kWh) per household. The investigators expected the demand for electricity to rise in cold weather (due to heating), fall when the weather was moderate, and rise again when the temperature rose and there was need for air-conditioning. They expected demand to decrease as the cost per kilowatt-hour increased, reflecting greater attention to conservation. Data were available for 2 years, a period in which the cost per kilowatt-hour x_2

increased because of the increasing cost of fuel. The company fitted the model

$$E(y) = \beta_0 + \beta_1 x_1 + \beta_2 x_1^2 + \beta_3 x_2 + \beta_4 x_1 x_2 + \beta_5 x_1^2 x_2$$

to the data shown in the table. The *Excel* printout for this multiple regression problem is also provided.

Price per kWh, x_2	Daily Temperature and Consumption	Mean Daily Consumption (kWh) per Household					
8¢	Mean daily temperature (°F), x_1	31 62	34 66	39 68	42 71	47 75	56 78
	Mean daily consumption, y	55 41	49 46	46 44	47 51	40 62	43 73
10¢	Mean daily temperature, x_1	32 62	36 66	39 68	42 72	48 75	56 79
	Mean daily consumption, y	50 39	44 44	42 40	42 44	38 50	40 55

Excel output for Exercise 13.28

SUMMARY OUTPUT

Regression Statistics	
Multiple R	0.948
R Square	0.898
Adjusted R Square	0.870
Standard Error	2.908
Observations	24

ANOVA

	df	SS	MS	F	Significance F
Regression	5	1346.448	269.290	31.852	0.000
Residual	18	152.177	8.454		
Total	23	1498.625			

	Coefficients	Standard Error	t Stat	P-value
Intercept	325.606	83.064	3.920	0.001
x1	-11.383	3.239	-3.515	0.002
x1-sq	0.113	0.029	3.854	0.001
x2	-21.699	9.224	-2.352	0.030
x1x2	0.873	0.359	2.433	0.026
x1sqx2	-0.009	0.003	-2.723	0.014

a. Do the data provide sufficient evidence to indicate that the model contributes information for the prediction of mean daily kilowatt-hour consumption per household? Test at the 5% level of significance.

b. Graph the curve depicting $\hat{y}$ as a function of temperature x_1 when the cost per kilowatt-hour is $x_2 = 8$¢. Construct a similar graph for the case when $x_2 = 10$¢ per kilowatt-hour. Are the consumption curves different?

c. If cost per kilowatt-hour is unimportant in predicting use, then you do not need the terms involving x_2 in the model. Therefore, the null hypothesis

$H_0 : x_2$ does not contribute information for the prediction of y

is equivalent to the null hypothesis $H_0 : \beta_3 = \beta_4 = \beta_5 = 0$ (if $\beta_3 = \beta_4 = \beta_5 = 0$, the terms involving x_2 disappear from the model). The *MINITAB* printout, obtained by fitting the reduced model

$$E(y) = \beta_0 + \beta_1 x_1 + \beta_2 x_1^2$$

to the data, is shown here. Use the methods of Section 13.5 to determine whether price per kilowatt-hour x_2 contributes significant information for the prediction of y.

Excel output for Exercise 13.28

SUMMARY OUTPUT

Regression Statistics	
Multiple R	0.8304
R Square	0.6896
Adjusted R Square	0.6601
Standard Error	4.7063
Observations	24

ANOVA

	df	SS	MS	F	Significance F
Regression	2	1033.490	516.745	23.330	0.000
Residual	21	465.135	22.149		
Total	23	1498.625			

	Coefficients	Standard Error	t Stat	P-value
Intercept	130.009	14.876	8.740	0.000
x1	-3.502	0.579	-6.049	0.000
x1-sq	0.033	0.005	6.349	0.000

d. Compare the values of $R^2(\text{adj})$ for the two models fit in this exercise. Which of the two models would you recommend?

Data set **13.29 Mercury Concentration in Dolphins**
EX1329 Because dolphins (and other large marine mammals) are considered to be the top predators in the marine food chain, the heavy metal concentrations in striped dolphins were measured as part of a marine pollution study. The concentration of mercury, the heavy metal reported in this study, is expected to differ in males and females because the mercury in a female is apparently transferred to her offspring during gestation and nursing. This study involved 28 males between the ages of .21 and 39.5 years, and 17 females between the ages of .80 and 34.5 years. For the data in the table,

$x_1 = $ Age of the dolphin (in years)

$$x_2 = \begin{cases} 0 & \text{if female} \\ 1 & \text{if male} \end{cases}$$

$y = $ Mercury concentration (in micrograms/gram) in the liver

y	x_1	x_2	y	x_1	x_2
1.70	.21	1	481.00	22.50	1
1.72	.33	1	485.00	24.50	1
8.80	2.00	1	221.00	24.50	1
5.90	2.20	1	406.00	25.50	1
101.00	8.50	1	252.00	26.50	1
85.40	11.50	1	329.00	26.50	1
118.00	11.50	1	316.00	26.50	1
183.00	13.50	1	445.00	26.50	1
168.00	16.50	1	278.00	27.50	1
218.00	16.50	1	286.00	28.50	1
180.00	17.50	1	315.00	29.50	1
264.00	20.50	1			

y	x_1	x_2	y	x_1	x_2
241.00	31.50	1	142.00	17.50	0
397.00	31.50	1	180.00	17.50	0
209.00	36.50	1	174.00	18.50	0
314.00	37.50	1	247.00	19.50	0
318.00	39.50	1	223.00	21.50	0
2.50	.80	0	167.00	21.50	0
9.35	1.58	0	157.00	25.50	0
4.01	1.75	0	177.00	25.50	0
29.80	5.50	0	475.00	32.50	0
45.30	7.50	0	342.00	34.50	0
101.00	8.05	0			
135.00	11.50	0			

a. Write a second-order model relating y to x_1 and x_2. Allow for curvature in the relationship between age and mercury concentration, and allow for an interaction between gender and age.

Use a computer software package to perform the multiple regression analysis. Refer to the printout to answer these questions.

b. Comment on the fit of the model, using relevant statistics from the printout.

c. What is the prediction equation for predicting the mercury concentration in a female dolphin as a function of her age?

d. What is the prediction equation for predicting the mercury concentration in a male dolphin as a function of his age?

e. Does the quadratic term in the prediction equation for females contribute significantly to the prediction of the mercury concentration in a female dolphin?

f. Are there any other important conclusions that you feel were not considered regarding the fitted prediction equation?

Data set **13.30 GRE Scores** The quantitative reasoning scores on the Graduate Record Examination (GRE)[7] were recorded for students admitted to three different graduate programs at a local university. This data was analyzed in Exercise 11.27 using the analysis of variance for a completely randomized design.
EX1330

Graduate Program					
Life Sciences		Physical Sciences		Social Sciences	
630	660	660	760	440	540
640	660	640	670	330	450
470	480	720	700	670	570
600	650	690	710	570	530
580	710	530	450	590	630

a. Write the theoretical model relating the GRE score to the qualitative variable "graduate program" using two dummy (indicator) variables to represent the three graduate programs.

b. Use a computer package to analyze the data with a multiple regression analysis. Is there sufficient evidence to indicate a difference in the average scores between the students who have been admitted to the three graduate programs? Use $\alpha = .05$.

c. Comment on the fit of the model and any regression assumptions that may have been violated. Summarize your results in a report, including printouts and graphs if possible.

Data set **13.31 On the Road Again** Until recently, performance tires were fitted mostly on sporty or luxury vehicles. Now they come standard on many everyday sedans. Increased levels of handling and grip have come at the expense of tread wear. The data that follows is abstracted from a report on all-season tires by *Consumer Reports*[8] in which several aspects of performance were evaluated for $n = 26$ different tires where
EX1331

y = overall score	x_1 = dry braking
x_2 = wet braking	x_3 = handling
x_4 = roll resistance	x_5 = tread life

Brand and Model	Price ($)	y	x_1	x_2	x_3	x_4	x_5
Michelin HydroEdge	116	84	4	5	4	3	5
Continental ProContact ECOPLUS	90	82	4	4	3	5	3
Michelin Energy Saver A/S	120	82	4	4	3	5	3

Brand and Model	Price ($)	y	x_1	x_2	x_3	x_4	x_5
Hankook Optimo H727	96	82	4	4	3	3	3
Pirelli P4 Four Seasons	98	80	4	5	3	2	2
Goodyear Assurance TripleTred	121	80	4	4	3	3	3
Pirelli Cinturato P5	98	78	3	5	3	2	3
Kumho Solus KR21	77	78	4	5	3	3	4
Maxxis Escapade MA-T1	77	76	3	4	4	3	4
Toyo Extensa A/S	99	76	3	4	3	3	3
Cooper GFE	83	76	3	4	3	4	2
Toyo Versado LX	91	76	4	4	3	3	5
BFGoodrich Traction T/A T	90	74	4	4	4	2	3
General Altimax RT	77	74	3	5	3	3	3
Yokohama Avid TRZ	81	74	4	5	3	2	3
Dayton Quadra LE	74	74	3	4	3	3	3
Cooper CS4 Touring	86	72	3	5	3	2	3
Uniroyal Tiger Paw Tour SR	77	72	4	4	2	5	2
Yokohama Avid Touring-S	70	70	4	4	3	4	4
Cooper Lifeliner GLS	78	68	3	4	3	4	2
Yokohama Avid T4	85	66	3	4	3	3	2
Bridgestone Turanze EL400	97	66	3	4	2	3	2
Falken Sincera SN828	82	64	3	5	3	3	1
Dunlop SP 60	79	64	3	3	2	2	4
Sumitomo HTR T4	67	64	3	4	3	4	2
Firestone FR710	80	60	3	5	3	3	4

The variables x_1 through x_5 are coded using the scale 5 = excellent, 4 = very good, 3 = good, 2 = fair, and 1 = poor.

a. Use a program of your choice to find the correlation matrix for the variables under study including price. Is price significantly correlated with any of the study variables? Which variables appear to be highly correlated with y, the overall score?

b. Write a model to describe y, overall score, as a function of the variables x_1 = dry braking, x_2 = wet braking, x_3 = handling, x_4 = roll resistance, and x_5 = tread life.

c. Use a regression program of your choice to fit the full model using all of the predictors. What proportion of the variation in y is explained by regression? Does this convey the impression that the model adequately explains the inherent variability in y?

d. Which variable or variables appear to be good predictors of y? How might you refine the model in light of these results? Use these variables in refitting the model. What proportion of the variation is explained by this refitted model? Comment on the adequacy of this reduced model in comparison to the full model.

Data set **13.32 Tuna Fish** The tuna fish data from Exercise 11.16 were analyzed as a completely randomized design with four treatments. However, we could also view the experimental design as a 2×2
EX1332

factorial experiment with unequal replications. The data are shown below.[9]

	Oil		Water	
Light Tuna	2.56	.62	.99	1.12
	1.92	.66	1.92	.63
	1.30	.62	1.23	.67
	1.79	.65	.85	.69
	1.23	.60	.65	.60
		.67	.53	.60
			1.41	.66
White Tuna	1.27		1.49	1.29
	1.22		1.29	1.00
	1.19		1.27	1.27
	1.22		1.35	1.28

Source: Case Study "Tuna Goes Upscale" Copyright 2001 by Consumers Union of U.S., Inc., Yonkers, NY 10703-1057, a nonprofit organization. Reprinted with permission from the June 2001 issue of *Consumer Reports®* for educational purposes only. No commercial use or reproduction permitted. www.ConsumerReports.org.

The data can be analyzed using the model

$$y = \beta_0 + \beta_1 x_1 + \beta_2 x_2 + \beta_3 x_1 x_2 + \epsilon$$

where

$x_1 = 0$ if oil, 1 if water

$x_2 = 0$ if light tuna, 1 if white tuna

a. Show how you would enter the data into a computer spreadsheet, entering the data into columns for y, x_1, x_2, and $x_1 x_2$.

b. The printout generated by *MINITAB* is shown below. What is the least-squares prediction equation?

MINITAB output for Exercise 13.32
Regression Analysis: y versus x1, x2, x1x2
The regression equation is
y = 1.15 - 0.251 x1 + 0.078 x2 + 0.306 x1x2

Predictor	Coef	SE Coef	T	P
Constant	1.1473	0.1370	8.38	0.000
x1	-0.2508	0.1830	-1.37	0.180
x2	0.0777	0.2652	0.29	0.771
x1x2	0.3058	0.3330	0.92	0.365

S = 0.454287   R-Sq = 11.9%   R-Sq(adj) = 3.9%

Analysis of Variance

Source	DF	SS	MS	F	P
Regression	3	0.9223	0.3074	1.49	0.235
Residual Error	33	6.8104	0.2064		
Total	36	7.7328			

c. Is there an interaction between type of tuna and type of packing liquid?

d. Which, if any, of the main effects (type of tuna and type of packing liquid) contribute significant information for the prediction of y?

e. How well does the model fit the data? Explain.

13.33 Tuna, continued Refer to Exercise 13.32. The hypothesis tested in Chapter 11—that the average prices for the four types of tuna are the same—is equivalent to saying that $E(y)$ will not change as x_1 and x_2 change. This can only happen when $\beta_1 = \beta_2 = \beta_3 = 0$. Use the *MINITAB* printout for the one-way ANOVA shown below to perform the test for equality of treatment means. Verify that this test is identical to the test for significant regression in Exercise 13.32.

MINITAB output for Exercise 13.33
One-Way ANOVA: Light Water, White Oil, White Water, Light Oil

Source	DF	SS	MS	F	P
Factor	3	0.922	0.307	1.49	0.235
Error	33	6.810	0.206		
Total	36	7.733			

S = 0.4543   R-Sq = 11.93%   R-Sq(adj) = 3.92%

13.34 Quality Control A manufacturer EX1334 recorded the number of defective items (y) produced on a given day by each of 10 machine operators and also recorded the average output per hour (x_1) for each operator and the time in weeks from the last machine service (x_2).

y	x_1	x_2
13	20	3.0
1	15	2.0
11	23	1.5
2	10	4.0
20	30	1.0
15	21	3.5
27	38	0
5	18	2.0
26	24	5.0
1	16	1.5

The printout that follows resulted when these data were analyzed using the *MINITAB* package using the model:

$$E(y) = \beta_0 + \beta_1 x_1 + \beta_2 x_2$$

MINITAB output for Exercise 13.34
Regression Analysis: y versus x1, x2
The regression equation is
y = -28.4 + 1.46 x1 + 3.84 x2

Predictor	Coef	SE Coef	T	P
Constant	-28.3906	0.8273	-34.32	0.000
x1	1.46306	0.02699	-54.20	0.000
x2	3.8446	0.1426	26.97	0.000

S = 0.548433   R-Sq = 99.8%   R-Sq(adj) = 99.7%

Analysis of Variance

Source	DF	SS	MS	F	P
Regression	2	884.79	442.40	1470.84	0.000
Residual Error	7	2.11	0.30		
Total	9	886.90			

Source	DF	Seq SS
x1	1	666.04
x2	1	218.76

a. Interpret R^2 and comment on the fit of the model.

b. Is there evidence to indicate that the model contributes significantly to the prediction of y at the $\alpha = .01$ level of significance?

c. What is the prediction equation relating $\hat{y}$ and x_1 when $x_2 = 4$?

d. Use the fitted prediction equation to predict the number of defective items produced for an operator whose average output per hour is 25 and whose machine was serviced 3 weeks ago.

e. What do the residual plots tell you about the validity of the regression assumptions?

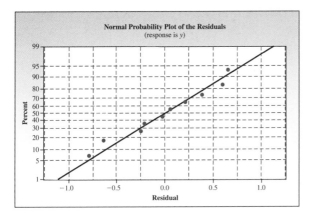

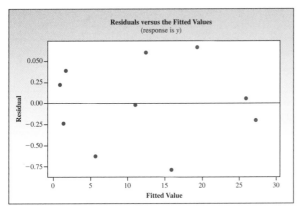

Data set **13.35 Metal Corrosion and Soil Acids** In
EX1335   an investigation to determine the relationship
between the degree of metal corrosion and the length
of time the metal is exposed to the action of soil acids,
the percentage of corrosion and exposure time were
measured weekly.

y	0.1	0.3	0.5	0.8	1.2	1.8	2.5	3.4
x	1	2	3	4	5	6	7	8

The data were fitted using the quadratic model, $E(y) = \beta_0 + \beta_1 x + \beta_2 x^2$, with the following results.

Excel output for Exercise 13.35

SUMMARY OUTPUT

Regression Statistics	
Multiple R	0.9993
R Square	0.9985
Adjusted R Square	0.9979
Standard Error	0.0530
Observations	8

ANOVA

	df	SS	MS	F	Significance F
Regression	2	9.421	4.710	1676.610	0.000
Residual	5	0.014	0.003		
Total	7	9.435			

	Coefficients	Standard Error	t Stat	P-value
Intercept	0.196	0.074	2.656	0.045
x	-0.100	0.038	-2.652	0.045
x-sq	0.062	0.004	15.138	0.000

a. What percentage of the total variation is explained by the quadratic regression of y on x?

b. Is the regression on x and x^2 significant at the $\alpha = .05$ level of significance?

c. Is the linear regression coefficient significant when x^2 is in the model?

d. Is the quadratic regression coefficient significant when x is in the model?

e. The data were fitted to a linear model without the quadratic term with the results that follow. What can you say about the contribution of the quadratic term when it is included i   n the model?

Excel output for Exercise 13.35
SUMMARY OUTPUT

Regression Statistics	
Multiple R	0.9645
R Square	0.9303
Adjusted R Square	0.9187
Standard Error	0.3311
Observations	8

ANOVA

	df	SS	MS	F	Significance F
Regression	1	8.777	8.777	80.052	0.000
Residual	6	0.658	0.110		
Total	7	9.435			

	Coefficients	Standard Error	t Stat	P-value
Intercept	-0.732	0.258	-2.838	0.030
x	0.457	0.051	8.947	0.000

f. The plot of the residuals from the linear regression model in part e shows a specific pattern. What is the term in the model that seems to be missing?

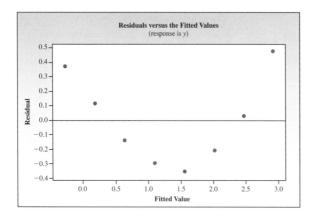

Residuals versus the Fitted Values
(response is y)

The following computer printout resulted when the data were analyzed.

Regression Analysis: y versus x1, x2, x3

```
The regression equation is
y = -3.11 + 0.503 x1 - 1.61 x2 - 1.15 x3

Predictor      Coef    SE Coef       T       P
Constant     -3.112      3.600   -0.86   0.421
x1          0.50314    0.07670    6.56   0.001
x2          -1.6126     0.6579   -2.45   0.050
x3           -1.155      1.791   -0.64   0.543

S = 1.89646    R-Sq = 92.2%    R-Sq(adj) = 88.4%

Analysis of Variance
Source           DF       SS      MS      F      P
Regression        3  256.621  85.540  23.78  0.001
Residual Error    6   21.579   3.597
Total             9  278.200

Source   DF    Seq SS
x1        1   229.113
x2        1    26.012
x3        1     1.496
```

13.36 Managing your Money A particular
EX1336 savings and loan corporation is interested in determining how well the amount of money in family savings accounts can be predicted using the three independent variables—annual income, number in the family unit, and area in which the family lives. Suppose that there are two specific areas of interest to the corporation. The following data were collected, where

y = Amount in all savings accounts

x_1 = Annual income

x_2 = Number in family unit

x_3 = 0 if in Area 1; 1 if not

Both y and x_1 were recorded in units of $1000.

y	x_1	x_2	x_3
0.5	19.2	3	0
0.3	23.8	6	0
1.3	28.6	5	0
0.2	15.4	4	0
5.4	30.5	3	1
1.3	20.3	2	1
12.8	34.7	2	1
1.5	25.2	4	1
0.5	18.6	3	1
15.2	45.8	2	1

a. Interpret R^2 and comment on the fit of the model.

b. Test for a significant regression of y on x_1, x_2, and x_3 at the 5% level of significance.

c. Test the hypothesis $H_0 : \beta_3 = 0$ against $H_a : \beta_3 \neq 0$ using $\alpha = .05$. Comment on the results of your test.

d. What can be said about the utility of x_3 as a predictor variable in this problem?

CASE STUDY **"Made in the U.S.A."—Another Look**

Foreign Cars

The case study in Chapter 12 examined the effect of foreign competition in the automotive industry as the number of imported cars steadily increased during the 1970s and 1980s.[10] The U.S. automobile industry has been besieged with complaints about product quality, worker layoffs, and high prices and has spent billions in advertising and research to produce an American-made car that will satisfy consumer demands. Have they been successful in stopping the flood of imported cars purchased by American consumers? The data shown in the table give the number of imported cars (y) sold in the United States (in millions) for the years 1969–2009. To simplify the analysis, we have coded the year using the coded variable x = Year − 1969.

Year	(Year–1969), x	Number of Imported Cars, y	Year	(Year–1969), x	Number of Imported Cars, y
1969	0	1.1	1989	20	2.7
1970	1	1.3	1990	21	2.4
1971	2	1.6	1991	22	2.0
1972	3	1.6	1992	23	1.9
1973	4	1.8	1993	24	1.8
1974	5	1.4	1994	25	1.7
1975	6	1.6	1995	26	1.5
1976	7	1.5	1996	27	1.3
1977	8	2.1	1997	28	1.4
1978	9	2.0	1998	29	1.4
1979	10	2.3	1999	30	1.7
1980	11	2.4	2000	31	2.0
1981	12	2.3	2001	32	2.1
1982	13	2.2	2002	33	2.2
1983	14	2.4	2003	34	2.1
1984	15	2.4	2004	35	2.1
1985	16	2.8	2005	36	2.2
1986	17	3.2	2006	37	2.3
1987	18	3.1	2007	38	2.4
1988	19	3.0	2008	39	2.3
			2009	40	1.8

By examining a scatterplot of these data, you will find that the number of imported cars does not appear to follow a linear relationship over time, but rather exhibits a curvilinear response. The question, then, is to decide whether a second-, third-, or higher-order model adequately describes the data.

1. Plot the data and sketch what you consider to be the best-fitting linear, quadratic, and cubic models.

2. Find the residuals using the fitted linear regression model. Does there appear to be any pattern in the residuals when plotted against x? What model do the residuals indicate would produce a better fit?

3. What is the increase in R^2 when you fit a quadratic rather than a linear model? Is the coefficient of the quadratic term significant? Is the fitted quadratic model significantly better than the fitted linear model? Plot the residuals from the fitted quadratic model. Does there seem to be any apparent pattern in the residuals when plotted against x?

4. What is the increase in R^2 when you compare the fitted cubic with the fitted quadratic model? Is the fitted cubic model significantly better than the fitted quadratic? Are there any patterns in a plot of the residuals versus x? What proportion of the variation in the response y is not accounted for by fitting a cubic model? Should any higher-order polynomial model be considered? Why or why not?

Analysis of Categorical Data

© rubberball/Jupiter Images

GENERAL OBJECTIVES

Many types of surveys and experiments result in qualitative rather than quantitative response variables, so that the responses can be classified but not quantified. Data from these experiments consist of the count or number of observations that fall into each of the response categories included in the experiment. In this chapter, we are concerned with methods for analyzing categorical data.

CHAPTER INDEX

NEED TO KNOW...
How to Determine the Appropriate Number of Degrees of Freedom

Who is the Primary Breadwinner in Your Family?

How have the roles of the working women of America changed? How many of the 130.2 million jobs in America are held by women? How has advertising refocused their ads to influence the 31% of women who are the primary breadwinners in their family? The case study at the end of this chapter examines some of these issues using the statistical techniques presented in this chapter.

A DESCRIPTION OF THE EXPERIMENT

14.1

Many experiments result in measurements that are *qualitative* or *categorical* rather than *quantitative*; that is, a *quality* or *characteristic* (rather than a numerical value) is measured for each experimental unit. You can summarize this type of data by creating a list of the categories or characteristics and reporting a **count** of the number of measurements that fall into each category. Here are a few examples:

- People can be classified into five income brackets.
- A mouse can respond in one of three ways to a stimulus.
- An M&M'S candy can have one of six colors.
- An industrial process manufactures items that can be classified as "acceptable," "second quality," or "defective."

These are some of the many situations in which the data set has characteristics appropriate for the **multinomial experiment.**

THE MULTINOMIAL EXPERIMENT

- The experiment consists of n identical trials.
- The outcome of each trial falls into one of k categories.
- The probability that the outcome of a single trial falls into a particular category—say, category i—is p_i and remains constant from trial to trial. This probability must be between 0 and 1, for each of the k categories, and the sum of all k probabilities is $\Sigma p_i = 1$.
- The trials are independent.
- The experimenter counts the *observed* number of outcomes in each category, written as $O_1, O_2, \ldots, O_k$, with $O_1 + O_2 + \cdots + O_k = n$.

You can visualize the multinomial experiment by thinking of k boxes or **cells** into which n balls are tossed. The n tosses are independent, and on each toss the chance of hitting the ith box is the same. However, this chance can vary from box to box; it might be easier to hit box 1 than box 3 on each toss. Once all n balls have been tossed, the number in each box or **cell**—$O_1, O_2, \ldots, O_k$—is counted.

You have probably noticed the similarity between the *multinomial experiment* and the *binomial experiment* introduced in Chapter 5. In fact, when there are $k = 2$ categories, the two experiments are identical, except for notation. Instead of p and q, we write p_1 and p_2 to represent the probabilities for the two categories, "success" and "failure." Instead of x and $(n - x)$, we write O_1 and O_2 to represent the observed number of "successes" and "failures."

When we presented the binomial random variable, we made inferences about the binomial parameter p (and by default, $q = 1 - p$) using large-sample methods based on the z statistic. In this chapter, we extend this idea to make inferences about the *multinomial parameters,* $p_1, p_2, \ldots, p_k$, using a different type of statistic. This statistic, whose approximate sampling distribution was derived by a British statistician named Karl Pearson in 1900, is called the **chi-square** (or sometimes **Pearson's chi-square**) **statistic.**

 NEED A TIP?

The multinomial experiment is an extension of the *binomial experiment.* For a binomial experiment, $k = 2$.

PEARSON'S CHI-SQUARE STATISTIC

Suppose that $n = 100$ balls are tossed at the cells (boxes) and you know that the probability of a ball falling into the first box is $p_1 = .1$. How many balls would you *expect* to fall into the first box? Intuitively, you would expect to see $100(.1) = 10$ balls in the first box. This should remind you of the average or expected number of successes, $\mu = np$, in the binomial experiment. In general, the expected number of balls that fall into cell i—written as E_i—can be calculated using the formula

$$E_i = np_i$$

for any of the cells $i = 1, 2, \ldots, k$.

Now suppose that you *hypothesize* values for each of the probabilities $p_1, p_2, \ldots, p_k$ and calculate the expected number for each category or cell. If your hypothesis is correct, the actual *observed cell counts, O_i*, should not be too different from the *expected cell counts, $E_i = np_i$*. The larger the differences, the more likely it is that the hypothesis is incorrect. The *Pearson chi-square statistic* uses the differences $(O_i - E_i)$ by first squaring these differences to eliminate negative contributions, and then forming a *weighted* average of the squared differences.

PEARSON'S CHI-SQUARE TEST STATISTIC

$$X^2 = \Sigma \frac{(O_i - E_i)^2}{E_i}$$

summed over all k cells, with $E_i = np_i$.

 NEED A TIP?

The Pearson's chi-square tests are always upper-tailed tests.

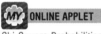 **ONLINE APPLET**

Chi-Square Probabilities

Although the mathematical proof is beyond the scope of this book, it can be shown that when n is large, X^2 has an approximate **chi-square probability distribution** in repeated sampling. If the hypothesized expected cell counts are correct, the differences $(O_i - E_i)$ are small and X^2 is close to 0. But, if the hypothesized probabilities are incorrect, large differences $(O_i - E_i)$ result in a *large* value of X^2. You should use a **right-tailed statistical test** and look for an unusually large value of the test statistic.

The chi-square distribution was used in Chapter 10 to make inferences about a single population variance σ^2. Like the F distribution, its shape is not symmetric and depends on a specific number of **degrees of freedom.** Once these degrees of freedom are specified, you can use Table 5 in Appendix I to find critical values or to bound the p-value for a particular chi-square statistic.

The appropriate degrees of freedom for the chi-square statistic vary depending on the particular application you are using. Although we will specify the appropriate degrees of freedom for the applications presented in this chapter, you should use the general rule given next for determining degrees of freedom for the chi-square statistic.

 NEED TO KNOW...

How to Determine the Appropriate Number of Degrees of Freedom

1. Start with the number of *categories* or cells in the experiment.
2. Subtract one degree of freedom for each linear restriction on the cell probabilities. You will always lose one *df* because $p_1 + p_2 + \cdots + p_k = 1$.

3. Sometimes the expected cell counts cannot be calculated directly but must be estimated using the sample data. Subtract one degree of freedom for every independent population parameter that must be estimated to obtain the estimated values of E_i.

We begin with the simplest applications of the chi-square test statistic—the **goodness-of-fit** test.

TESTING SPECIFIED CELL PROBABILITIES: THE GOODNESS-OF-FIT TEST

14.3

The simplest hypothesis concerning the cell probabilities specifies a numerical value for each cell. The expected cell counts are easily calculated using the hypothesized probabilities, $E_i = np_i$, and are used to calculate the observed value of the X^2 test statistic. For a multinomial experiment consisting of k categories or cells, the test statistic has an approximate χ^2 distribution with $df = (k - 1)$.

EXAMPLE **14.1**

A researcher designs an experiment in which a rat is attracted to the end of a ramp that divides, leading to doors of three different colors. The researcher sends the rat down the ramp $n = 90$ times and observes the choices listed in Table 14.1. Does the rat have (or acquire) a preference for one of the three doors?

TABLE 14.1 ● **Rat's Door Choices**

	Door		
	Green	Red	Blue
Observed Count (O_i)	20	39	31

Solution If the rat has no preference in the choice of a door, you would expect in the long run that the rat would choose each door an equal number of times. That is, the null hypothesis is

$$H_0 : p_1 = p_2 = p_3 = \frac{1}{3}$$

versus the alternative hypothesis

$$H_a : \text{At least one } p_i \text{ is different from } \frac{1}{3}$$

where p_i is the probability that the rat chooses door i, for $i = 1, 2,$ and 3. The expected cell counts are the same for each of the three categories—namely, $np_i = 90(1/3) = 30$. The chi-square test statistic can now be calculated as

NEED NEED A TIP?
a tip?

The rejection region and
p-value are in the upper
tail of the chi-square
distribution.

$$X^2 = \Sigma\frac{(O_i - E_i)^2}{E_i}$$

$$= \frac{(20 - 30)^2}{30} + \frac{(39 - 30)^2}{30} + \frac{(31 - 30)^2}{30} = 6.067$$

For this example, the test statistic has $(k - 1) = 2$ degrees of freedom because the only linear restriction on the cell probabilities is that they must sum to 1. Hence, you can use Table 5 in Appendix I to find bounds for the right-tailed p-value. Since the observed value, $X^2 = 6.067$, lies between $\chi^2_{.050} = 5.99$ and $\chi^2_{.025} = 7.38$, the p-value is between .025 and .050. The researcher would report the results as significant at the 5% level ($P < .05$), meaning that the null hypothesis of no preference is rejected. There is sufficient evidence to indicate that the rat has a preference for one of the three doors.

What more can you say about the experiment once you have determined statistically that the rat has a preference? Look at the data to see where the differences lie.

The blue door was chosen only a little more than one-third of the time:

$$\frac{31}{90} = .344$$

However, the sample proportions for the other two doors are quite different from one-third. The rat chooses the green door least often—only 22% of the time:

$$\frac{20}{90} = .222$$

The rat chooses the red door most often—43% of the time:

$$\frac{39}{90} = .433$$

You would summarize the results of the experiment by saying that the rat has a preference for the red door. Can you conclude that the preference is *caused* by the door color? The answer is no—the cause could be some other physiological or psychological factor that you have not yet explored. Avoid declaring a *causal* relationship between color and preference!

EXAMPLE 14.2

The proportions of blood phenotypes A, B, AB, and O in the population of all Caucasians in the United States are .41, .10, .04, and .45, respectively. To determine whether or not the actual population proportions fit this set of reported probabilities, a random sample of 200 Americans were selected and their blood phenotypes were recorded. The observed and expected cell counts are shown in Table 14.2. The expected cell counts are calculated as $E_i = 200p_i$. Test the goodness-of-fit of these blood phenotype proportions.

TABLE 14.2 ● **Counts of Blood Phenotypes**

	A	B	AB	O
Observed (O_i)	89	18	12	81
Expected (E_i)	82	20	8	90

Solution The hypothesis to be tested is determined by the model probabilities:

$$H_0 : p_1 = .41; p_2 = .10; p_3 = .04; p_4 = .45$$

versus

$$H_a : \text{At least one of the four probabilities is different from the specified value}$$

NEED A TIP?

Degrees of freedom for a simple goodness-of-fit test: $df = k - 1$

Then

$$X^2 = \Sigma \frac{(O_i - E_i)^2}{E_i}$$

$$= \frac{(89 - 82)^2}{82} + \cdots + \frac{(81 - 90)^2}{90} = 3.70$$

From Table 5 in Appendix I, indexing $df = (k - 1) = 3$, you can find that the observed value of the test statistic is less than $\chi^2_{.100} = 6.25$, so that the p-value is greater than .10. You do not have sufficient evidence to reject H_0; that is, you cannot declare that the blood phenotypes for American Caucasians are *different* from those reported earlier. The results are nonsignificant (NS).

You can find instructions in the "Technology Today" section at the end of this chapter that allow you to use *MINITAB* (versions 15 or 16) to perform the chi-square goodness-of-fit test and generate the results.

Notice the difference in the goodness-of-fit hypothesis compared to other hypotheses that you have tested. In the goodness-of-fit test, the researcher uses the null hypothesis to specify the model he believes to be *true*, rather than a model he hopes to prove false! When you could not reject H_0 in the blood type example, the results were as expected. Be careful, however, when you report your results for goodness-of-fit tests. You cannot declare with confidence that the model is absolutely correct without reporting the value of β for some practical alternatives.

14.3 EXERCISES

BASIC TECHNIQUES

14.1 List the characteristics of a multinomial experiment.

14.2 Use Table 5 in Appendix I to find the value of χ^2 with the following area α to its right:

a. $\alpha = .05, df = 3$ **b.** $\alpha = .01, df = 8$

14.3 Give the rejection region for a chi-square test of specified probabilities if the experiment involves k categories in these cases:

a. $k = 7, \alpha = .05$ **b.** $k = 10, \alpha = .01$

14.4 Use Table 5 in Appendix I to bound the p-value for a chi-square test:

a. $X^2 = 4.29, df = 5$ **b.** $X^2 = 20.62, df = 6$

14.5 Suppose that a response can fall into one of $k = 5$ categories with probabilities $p_1, p_2, \ldots, p_5$ and that $n = 300$ responses produced these category counts:

Category	1	2	3	4	5
Observed Count	47	63	74	51	65

a. Are the five categories equally likely to occur? How would you test this hypothesis?

b. If you were to test this hypothesis using the chi-square statistic, how many degrees of freedom would the test have?

c. Find the critical value of χ^2 that defines the rejection region with $\alpha = .05$.

d. Calculate the observed value of the test statistic.

e. Conduct the test and state your conclusions.

14.6 Suppose that a response can fall into one of $k = 3$ categories with probabilities $p_1 = .4, p_2 = .3$, and $p_3 = .3$, and $n = 300$ responses produce these category counts:

Category	1	2	3
Observed Count	130	98	72

Do the data provide sufficient evidence to indicate that the cell probabilities are different from those specified for the three categories? Find the approximate p-value and use it to make your decision.

APPLICATIONS

14.7 Your Favorite Lane A freeway with four lanes in each direction was studied to see whether drivers prefer to drive on the inside lanes. A total of 1000 automobiles were observed during heavy early-morning traffic, and the number of cars in each lane was recorded:

Lane	1	2	3	4
Observed Count	294	276	238	192

Do the data present sufficient evidence to indicate that some lanes are preferred over others? Test using $\alpha = .05$. If there are any differences, discuss the nature of the differences.

14.8 Peonies A peony plant with red petals was crossed with another plant having streaky petals. A geneticist states that 75% of the offspring from this cross will have red flowers. To test this claim, 100 seeds from this cross were collected and germinated, and 58 plants had red petals. Use the chi-square goodness-of-fit test to determine whether the sample data confirm the geneticist's prediction.

14.9 Heart Attacks on Mondays Researchers from Germany have concluded that the risk of a heart attack for a working person may be as much as 50% greater on Monday than on any other day.[1] The researchers kept track of heart attacks and coronary arrests over a period of 5 years among 330,000 people who lived near Augsburg, Germany. In an attempt to verify their claim, you survey 200 working people who had recently had heart attacks and recorded the day on which their heart attacks occurred:

Day	Observed Count
Sunday	24
Monday	36
Tuesday	27
Wednesday	26
Thursday	32
Friday	26
Saturday	29

Do the data present sufficient evidence to indicate that there is a difference in the incidence of heart attacks depending on the day of the week? Test using $\alpha = .05$.

14.10 Mortality Statistics Medical statistics show that deaths due to four major diseases—call them A, B, C, and D—account for 15%, 21%, 18%, and 14%, respectively, of all nonaccidental deaths. A study of the causes of 308 nonaccidental deaths at a hospital gave the following counts:

Disease	A	B	C	D	Other
Deaths	43	76	85	21	83

Do these data provide sufficient evidence to indicate that the proportions of people dying of diseases A, B, C, and D at this hospital differ from the proportions accumulated for the population at large?

14.11 Schizophrenia Research has suggested a link between the prevalence of schizophrenia and birth during particular months of the year in which viral infections are prevalent. Suppose you are working on a similar problem and you suspect a linkage between a disease observed in later life and month of birth. You have records of 400 cases of the disease, and you classify them according to month of birth. The data appear in the table. Do the data present sufficient evidence to indicate that the proportion of cases of the disease per month varies from month to month? Test with $\alpha = .05$.

Month	Jan	Feb	Mar	Apr	May	June
Births	38	31	42	46	28	31

Month	July	Aug	Sept	Oct	Nov	Dec
Births	24	29	33	36	27	35

14.12 Snap Peas Suppose you are interested in following two independent traits in snap peas—seed texture (S = smooth, s = wrinkled) and seed color (Y = yellow, y = green)—in a second-generation cross of heterozygous parents. Mendelian theory states that the number of peas classified as smooth and yellow, wrinkled and yellow, smooth and green, and wrinkled and green should be in the ratio 9:3:3:1. Suppose that 100 randomly selected snap peas have 56, 19, 17, and 8 in these respective categories. Do these data indicate that the 9:3:3:1 model is correct? Test using $\alpha = .01$.

14.13 M&M'S The Mars, Incorporated website reported the following percentages of the various colors of its M&M'S candies for the "milk chocolate" variety:[2]

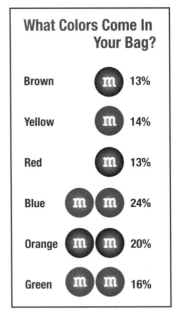

A 14-ounce bag of milk chocolate M&M'S is randomly selected and contains 70 brown, 72 yellow, 61 red, 118 blue, 108 orange, and 85 green candies. Do the data substantiate the percentages reported by Mars, Incorporated? Use the appropriate test and describe the nature of the differences, if there are any.

14.14 Peanut M&M'S The percentage of various colors are different for the "peanut" variety of M&M'S candies, as reported on the Mars, Incorporated website:[3]

What Colors Come In Your Bag?

Brown	ⓜ	12%
Yellow	ⓜ	15%
Red	ⓜ	12%
Blue	ⓜ ⓜ	23%
Orange	ⓜ ⓜ	23%
Green	ⓜ	15%

A 14-ounce bag of peanut M&M'S is randomly selected and contains 70 brown, 87 yellow, 64 red, 115 blue, 106 orange, and 85 green candies. Do the data substantiate the percentages reported by Mars, Incorporated? Use the appropriate test and describe the nature of the differences, if there are any.

14.15 Admission Standards Previous enrollment records at a large university indicate that of the total number of persons who apply for admission, 60% are admitted unconditionally, 5% are admitted on a trial basis, and the remainder are refused admission. Of 500 applications to date for the coming year, 329 applicants have been admitted unconditionally, 43 have been admitted on a trial basis, and the remainder have been refused admission. Do these data indicate a departure from previous admission rates? Test using $\alpha = .05$.

CONTINGENCY TABLES: A TWO-WAY CLASSIFICATION

14.4

In some situations, the researcher classifies an experimental unit according to *two qualitative variables* to generate *bivariate data,* which we discussed in Chapter 3.

- A defective piece of furniture is classified according to the type of defect and the production shift during which it was made.
- A professor is classified by professional rank and the type of university (public or private) at which she works.
- A patient is classified according to the type of preventive flu treatment he received and whether or not he contracted the flu during the winter.

When two *categorical variables* are recorded, you can summarize the data by counting the observed number of units that fall into each of the various intersections of category levels. The resulting counts are displayed in an array called a **contingency table.**

EXAMPLE 14.3

A total of $n = 309$ furniture defects were recorded and the defects were classified into four types: A, B, C, or D. At the same time, each piece of furniture was identified by the production shift in which it was manufactured. These counts are presented in a contingency table in Table 14.3.

TABLE 14.3 ● Contingency Table

Type of Defects	Shift 1	2	3	Total
A	15	26	33	74
B	21	31	17	69
C	45	34	49	128
D	13	5	20	38
Total	94	96	119	309

NEED A TIP?

With two-way classifications, we do not test hypotheses about specific probabilities.
We test whether the two methods of classification are independent.

When you study data that involves two variables, one important consideration is the *relationship between the two variables.* Does the proportion of measurements in the various categories for factor 1 depend on which category of factor 2 is being observed? For the furniture example, do the proportions of the various defects vary from shift to shift, or are these proportions the same, independently of which shift is observed? You may remember a similar phenomenon called *interaction* in the $a \times b$ factorial experiment from Chapter 11. In the analysis of a contingency table, the objective is to determine whether or not one method of classification is **contingent** or **dependent** on the other method of classification. If not, the two methods of classification are said to be **independent.**

The Chi-Square Test of Independence

The question of independence of the two methods of classification can be investigated using a test of hypothesis based on the chi-square statistic. These are the hypotheses:

$$H_0 : \text{The two methods of classification are independent}$$
$$H_a : \text{The two methods of classification are dependent}$$

Suppose we denote the observed cell count in row i and column j of the contingency table as O_{ij}. If you knew the expected cell counts ($E_{ij} = np_{ij}$) under the null hypothesis of independence, then you could use the chi-square statistic to compare the observed and expected counts. However, the expected values are not specified in H_0, as they were in previous examples.

To explain how to estimate these expected cell counts, we must revisit the concept of *independent events* from Chapter 4. Consider p_{ij}, the probability that an observation falls into row i and column j of the contingency table. If the rows and columns are independent, then

$p_{ij} = P(\text{observation falls in row } i \text{ and column } j)$

$\quad = P(\text{observation falls in row } i) \times P(\text{observation falls in column } j)$

$\quad = p_i p_j$

where p_i and p_j are the **unconditional** or **marginal probabilities** of falling into row i or column j, respectively. If you could obtain proper estimates of these marginal probabilities, you could use them in place of p_{ij} in the formula for the expected cell count.

Fortunately, these estimates do exist. In fact, they are exactly what you would intuitively choose:

NEED A TIP?

Degrees of freedom for an $r \times c$ contingency table: $df = (r - 1)(c - 1)$.

- To estimate a row probability, use

$$\hat{p}_i = \frac{\text{Total observations in row } i}{\text{Total number of observations}} = \frac{r_i}{n}$$

- To estimate a column probability, use

$$\hat{p}_j = \frac{\text{Total observations in column } j}{\text{Total number of observations}} = \frac{c_j}{n}$$

The estimate of the expected cell count for row i and column j follows from the independence assumption.

ESTIMATED EXPECTED CELL COUNT

$$\hat{E}_{ij} = n\left(\frac{r_i}{n}\right)\left(\frac{c_j}{n}\right) = \frac{r_i c_j}{n}$$

where r_i is the total for row i and c_j is the total for column j.

The chi-square test statistic for a contingency table with r rows and c columns is calculated as

$$X^2 = \Sigma\frac{(O_{ij} - \hat{E}_{ij})^2}{\hat{E}_{ij}}$$

and can be shown to have an approximate chi-square distribution with

$$df = (r - 1)(c - 1)$$

If the observed value of X^2 is too large, then the null hypothesis of independence is rejected.

EXAMPLE 14.4 Refer to Example 14.3. Do the data present sufficient evidence to indicate that the type of furniture defect varies with the shift during which the piece of furniture is produced?

Solution The estimated expected cell counts are shown in parentheses in Table 14.4. For example, the estimated expected count for a type C defect produced during the second shift is

$$\hat{E}_{32} = \frac{r_3 c_2}{n} = \frac{(128)(96)}{309} = 39.77$$

TABLE 14.4 ● **Observed and Estimated Expected Cell Counts**

	Shift			
Type of Defects	1	2	3	Total
A	15 (22.51)	26 (22.99)	33 (28.50)	74
B	21 (20.99)	31 (21.44)	17 (26.57)	69
C	45 (38.94)	34 (39.77)	49 (49.29)	128
D	13 (11.56)	5 (11.81)	20 (14.63)	38
Total	94	96	119	309

You can now use the values shown in Table 14.4 to calculate the test statistic as

$$X^2 = \Sigma\frac{(O_{ij} - \hat{E}_{ij})^2}{\hat{E}_{ij}}$$

$$= \frac{(15 - 22.51)^2}{22.51} + \frac{(26 - 22.99)^2}{22.99} + \cdots + \frac{(20 - 14.63)^2}{14.63}$$

$$= 19.18$$

When you index the chi-square distribution in Table 5 in Appendix I with

$$df = (r - 1)(c - 1) = (4 - 1)(3 - 1) = 6$$

the observed test statistic is greater than $\chi^2_{.005} = 18.5476$, which indicates that the p-value is less than .005. You can reject H_0 and declare the results to be highly significant ($P < .005$). There is sufficient evidence to indicate that the proportions of defect types vary from shift to shift.

Chi-Square Test of
Independence

The next obvious question you should ask involves the nature of the relationship between the two classifications. Which shift produces more of which type of defect? As with the factorial experiment in Chapter 11, once a dependence (or interaction) is found, you must look within the table at the relative or *conditional* proportions for each level of classification. For example, consider shift 1, which produced a total of 94 defects. These defects can be divided into types using the *conditional proportions* for this sample shown in the first column of Table 14.5. If you follow the same procedure for the other two shifts, you can then compare the distributions of defect types for the three shifts, as shown in Table 14.5.

Now compare the three sets of proportions (each sums to 1). It appears that shifts 1 and 2 produce defects in the same general order—types C, B, A, and D from most to least—though in differing proportions. Shift 3 shows a different pattern—the most type C defects again but followed by types A, D, and B, in that order. Depending on which type of defect is the most important to the manufacturer, each shift should be cautioned separately about the reasons for producing too many defects.

TABLE 14.5 ● **Conditional Probabilities for Types of Defect within Three Shifts**

	Shift		
Types of Defects	1	2	3
A	$\dfrac{15}{94} = .16$	$\dfrac{26}{96} = .27$	$\dfrac{33}{119} = .28$
B	$\dfrac{21}{94} = .22$	$\dfrac{31}{96} = .32$	$\dfrac{17}{119} = .14$
C	$\dfrac{45}{94} = .48$	$\dfrac{34}{96} = .35$	$\dfrac{49}{119} = .41$
D	$\dfrac{13}{94} = .14$	$\dfrac{5}{96} = .05$	$\dfrac{20}{119} = .17$
Total	1.00	1.00	1.00

 NEED TO KNOW...

How to Determine the Appropriate Number of Degrees of Freedom

Remember the general procedure for determining degrees of freedom:

1. Start with $k = rc$ categories or cells in the contingency table.

2. Subtract one degree of freedom because all of the rc cell probabilities must sum to 1.

3. You had to estimate $(r - 1)$ row probabilities and $(c - 1)$ column probabilities to calculate the estimated expected cell counts. (The last one of the row and column probabilities is determined because the *marginal* row and column probabilities must also sum to 1.) Subtract $(r - 1)$ and $(c - 1)$ *df*.

The total degrees of freedom for the $r \times c$ contingency table are

$$df = rc - 1 - (r - 1) - (c - 1) = rc - r - c + 1 = (r - 1)(c - 1)$$

EXAMPLE 14.5

A survey was conducted to evaluate the effectiveness of a new flu vaccine that had been administered in a small community. The vaccine was provided free of charge in a two-shot sequence over a period of 2 weeks. Some people received the two-shot sequence, some appeared for only the first shot, and others received neither. A survey of 1000 local residents the following spring provided the information shown in Table 14.6. Do the data present sufficient evidence to indicate that the vaccine was successful in reducing the number of flu cases in the community?

TABLE 14.6 ● **2 × 3 Contingency Table**

	No Vaccine	One Shot	Two Shots	Total
Flu	24	9	13	46
No Flu	289	100	565	954
Total	313	109	578	1000

Solution The success of the vaccine in reducing the number of flu cases can be assessed in two parts:

- If the vaccine is successful, the proportions of people who get the flu should vary, depending on which of the three treatments they received.

- Not only must this dependence exist, but the proportion of people who get the flu should decrease as the amount of flu prevention treatment increases—from zero to one to two shots.

The first part can be tested using the chi-square test with these hypotheses:

H_0 : No relationship between treatment and incidence of flu

H_a : Incidence of flu depends on amount of flu treatment

 NEED A TIP?

Use the value of X^2 and the p-value from the printout to test the hypothesis of independence.

As usual, computer software packages can eliminate all of the tedious calculations and, if the data are entered correctly, provide the correct output containing the observed value of the test statistic and its p-value. Such a printout, generated by *MINITAB*, is shown in Figure 14.1. You can find instructions for generating this printout in the section "Technology Today" at the end of this chapter. The observed value of the test statistic, $X^2 = 17.313$, has a p-value of .000 and the results are declared highly significant. That is, the null hypothesis is rejected. There is sufficient evidence to indicate a relationship between treatment and incidence of flu.

FIGURE 14.1 ●

MINITAB output for Example 14.5

Chi-Square Test: No Vaccine, One Shot, Two Shots

```
Expected counts are printed below observed counts
Chi-Square contributions are printed below expected counts

          No Vaccine One Shot Two Shots   Total
     1          24         9        13       46
             14.40      5.01     26.59
             6.404     3.169     6.944

     2         289       100       565      954
            298.60    103.99    551.41
             0.309     0.153     0.335

Total        313       109       578     1000

Chi-Sq =   17.313, DF = 2, P-Value = 0.000
```

What is the nature of this relationship? To answer this question, look at Table 14.7, which gives the *incidence* of flu in the sample for each of the three treatment groups. The answer is obvious. The group that received two shots was less susceptible to the flu; only one flu shot does not seem to decrease the susceptibility!

TABLE 14.7 ● **Incidence of Flu for Three Treatments**

No Vaccine	One Shot	Two Shots
$\frac{24}{313} = .08$	$\frac{9}{109} = .08$	$\frac{13}{578} = .02$

14.4 EXERCISES

BASIC TECHNIQUES

14.16 Calculate the value and give the number of degrees of freedom for X^2 for these contingency tables:

a.

	Columns			
Rows	1	2	3	4
1	120	70	55	16
2	79	108	95	43
3	31	49	81	140

b.

	Columns		
Rows	1	2	3
1	35	16	84
2	120	92	206

14.17 Suppose that a consumer survey summarizes the responses of $n = 307$ people in a contingency table that contains three rows and five columns. How many degrees of freedom are associated with the chi-square test statistic?

14.18 A survey of 400 respondents produced these cell counts in a 2×3 contingency table:

	Columns			
Rows	1	2	3	Total
1	37	34	93	164
2	66	57	113	236
Total	103	91	206	400

a. If you wish to test the null hypothesis of "independence"—that the probability that a response falls in any one row is independent of the column it falls in—and you plan to use a chi-square test, how many degrees of freedom will be associated with the χ^2 statistic?

b. Find the value of the test statistic.

c. Find the rejection region for $\alpha = .01$.

d. Conduct the test and state your conclusions.

e. Find the approximate p-value for the test and interpret its value.

14.19 Gender Differences Male and female respondents to a questionnaire on gender differences were categorized into three groups according to their answers on the first question:

	Group 1	Group 2	Group 3
Men	37	49	72
Women	7	50	31

Use the *MINITAB* printout to determine whether there is a difference in the responses according to gender. Explain the nature of the differences, if any exist.

MINITAB output for Exercise 14.19

Chi-Square Test: Group 1, Group 2, Group 3

```
Expected counts are printed below observed counts
Chi-Square contributions are printed below expected
counts

           Group 1    Group 2    Group 3      Total
    1           37         49         72        158
            28.26      63.59      66.15
            2.703      3.346      0.517

    2            7         50         31         88
            15.74      35.41      36.85
            4.853      6.007      0.927

Total           44         99        103        246

Chi-Sq =  18.352, DF = 2, P-Value = 0.000
```

APPLICATIONS

14.20 Health Care Reform Congress passed the Health Care Reform Act in 2010 that will require all small businesses to have a health plan in place by 2014 and included the provision that no one could be denied coverage due to pre-existing conditions. A *Newsweek* Poll based on $n = 848$ registered voters nationwide asked respondents to "Please tell me if you think the health care reform law passed earlier this year was good for the country or bad for the country in general."[4] The data that follow are based on the results of this poll.

Affiliation	Good	Bad	Unsure	Total
Republicans	32	301	17	350
Democrats	277	60	37	374
Independents	51	58	19	128

a. Are there significant differences in the proportions of those surveyed who think the health care reform law is good, bad, or unsure among the Republicans, Democrats, and Independents? Use $\alpha = .05$.

b. If significant differences exist, describe the nature of the differences by finding the proportions of those who think the law is good, bad, or unsure for each of the given affiliations.

14.21 Anxious Infants A study was conducted by Joseph Jacobson and Diane Wille to determine the effect of early child care on infant-mother attachment patterns.[5] In the study, 93 infants were classified as either "secure" or "anxious" using the Ainsworth strange situation paradigm. In addition, the infants were classified according to the average number of hours per week that they spent in child care. The data are presented in the table.

	Low (0–3 hours)	Moderate (4–19 hours)	High (20–54 hours)
Secure	24	35	5
Anxious	11	10	8

a. Do the data provide sufficient evidence to indicate that there is a difference in attachment pattern for the infants depending on the amount of time spent in child care? Test using $\alpha = .05$.

b. What is the approximate *p*-value for the test in part a?

14.22 Spending Patterns Is there a difference in the spending patterns of high school seniors depending on their gender? A study to investigate this question focused on 196 employed high school seniors. Students were asked to classify the amount of their earnings that they spent on their car during a given month:

	None or Only a Little	Some	About Half	Most	All or Almost All
Male	73	12	6	4	3
Female	57	15	11	9	6

A portion of the *MINITAB* printout is given here. Use the printout to analyze the relationship between spending patterns and gender. Write a short paragraph explaining your statistical conclusions and their practical implications.

Partial *MINITAB* output for Exercise 14.22

Chi-Square Test: None, Some, Half, Most, All

```
Chi-Sq = 6.696, DF = 4, P-Value = 0.153
2 cells with expected counts less than 5.
```

14.23 Hair Color The hair and eye color that follows was self-reported by a sample of Caucasian Americans born between 1957 and 1965 (currently 45–53 years old).[6] The following data was adapted from that study.

	Light Blond	Blond	Light Brown	Brown	Black	Red	Total
Males	4	46	45	176	23	10	304
Females	5	58	69	164	12	12	320

a. Is there sufficient evidence to conclude that the proportion of individuals with these hair colors differ for males and females? Use $\alpha = .05$.

b. Are there any cells with an expected number less than five? If so, combine those cells with those next to it and reanalyze the data. Do the end results differ?

14.24 The JFK Assassination Almost 50 years after the assassination of John F. Kennedy, a *FOX News* poll shows most Americans disagree with the government's conclusions about the killing. The *Warren Commission* found that Lee Harvey Oswald acted alone when he shot Kennedy, but many Americans are not so sure. Do you think that we know all the facts about the assassination of President John F. Kennedy or do you think there was a cover-up? Here are the results from a poll of 900 registered voters nationwide:[7]

	We Know All the Facts	There Was a Cover-Up	Not Sure
Democrats	42	309	31
Republicans	64	246	46
Independents	20	115	27

a. Do these data provide sufficient evidence to conclude that there is a difference in voters' opinions about a possible cover-up depending on the political affiliation of the voter? Test using $\alpha = .05$.

b. If there is a significant difference in part a, describe the nature of these differences.

Data set **14.25 Telecommuting** As an alternative to
EX1425 flextime, many companies allow employees to
do some of their work at home. Individuals in a random
sample of 300 workers were classified according to
salary and number of workdays per week spent at home.

	Workdays at Home per Week		
Salary	Less Than One	At Least One, but Not All	All at Home
Under $25,000	38	16	14
$25,000–$49,999	54	26	12
$50,000–$74,999	35	22	9
Above $75,000	33	29	12

a. Do the data present sufficient evidence to indicate
that salary is dependent on the number of workdays
spent at home? Test using $\alpha = .05$.

b. Use Table 5 in Appendix I to approximate the
p-value for this test of hypothesis. Does the p-value
confirm your conclusions from part a?

Data set **14.26 Telecommuting II** An article in
EX1426A *American Demographics* addressed the same
EX1426B telecommuting issue (Exercise 14.25) in a

slightly different way. They concluded that "people
who work exclusively at home tend to be older and
better educated than those who have to leave home to
report to work."[8] Use the data below based on random
samples of 300 workers each to either support or refute
their conclusions. Use the appropriate test of hypothesis, and explain why you either agree or disagree with
the *American Demographics* conclusions. Note that
"Mixed" workers are those who reported working at
home at least one full day in a typical week.

	Workers		
Age	Non-home	Mixed	Home
15–34	73	23	12
35–54	85	40	23
55 and Over	22	12	10

	Workers		
Education	Non-home	Mixed	Home
Less than H.S. Diploma	23	3	5
H.S. Graduate	54	12	11
Some College/Assoc. Degree	53	24	14
B.A. or More	41	42	18

COMPARING SEVERAL MULTINOMIAL POPULATIONS: A TWO-WAY CLASSIFICATION WITH FIXED ROW OR COLUMN TOTALS

14.5

An $r \times c$ contingency table results when each of n experimental units is counted as
falling into one of the rc cells of a multinomial experiment. Each cell represents a pair
of category levels—row level i and column level j. Sometimes, however, it is not
advisable to use this type of experimental design—that is, to let the n observations
fall where they may. For example, suppose you want to study the opinions of American families about their income levels—say, low, medium, and high. If you randomly
select $n = 1200$ families for your survey, you may not find any who classify themselves as low-income families! It might be better to decide ahead of time to survey
400 families in each income level. The resulting data will still appear as a two-way
classification, but the column totals are fixed in advance.

EXAMPLE **14.6**

In another flu prevention experiment like Example 14.5, the experimenter decides to
search the clinic records for 300 patients in each of the three treatment categories: no
vaccine, one shot, and two shots. The $n = 900$ patients will then be surveyed regarding
their winter flu history. The experiment results in a 2×3 table with the column totals fixed at 300, shown in Table 14.8. By fixing the column totals, the experimenter
no longer has a multinomial experiment with $2 \times 3 = 6$ cells. Instead, there are three
separate binomial experiments—call them 1, 2, and 3—each with a given probability
p_j of contracting the flu and q_j of not contracting the flu. (Remember that for a binomial population, $p_j + q_j = 1.$)

TABLE 14.8 • **Cases of Flu for Three Treatments**

	No Vaccine	One Shot	Two Shots	Total
Flu				r_1
No Flu				r_2
Total	300	300	300	n

Suppose you used the chi-square test to test for the independence of row and column classifications. If a particular treatment (column level) does not affect the incidence of flu, then each of the three binomial populations should have the same incidence of flu so that $p_1 = p_2 = p_3$ and $q_1 = q_2 = q_3$.

The 2×3 classification in Example 14.6 describes a situation in which the chi-square test of independence is equivalent to a test of the equality of $c = 3$ binomial proportions. Tests of this type are called **tests of homogeneity** and are used to compare several binomial populations. If there are *more than two* row categories with fixed column totals, then the test of independence is equivalent to a test of the equality of c sets of multinomial proportions.

You do not need to be concerned about the theoretical equivalence of the chi-square tests for these two experimental designs. Whether the columns (or rows) are fixed or not, the test statistic is calculated as

$$X^2 = \Sigma \frac{(O_{ij} - \hat{E}_{ij})^2}{\hat{E}_{ij}} \qquad \text{where } \hat{E}_{ij} = \frac{r_i c_j}{n}$$

which has an approximate chi-square distribution in repeated sampling with $df = (r - 1)(c - 1)$.

 NEED TO KNOW...

How to Determine the Appropriate Number of Degrees of Freedom

Remember the general procedure for determining degrees of freedom:

1. Start with the rc cells in the two-way table.
2. Subtract one degree of freedom for each of the c multinomial populations, whose column probabilities must add to one—a total of c df.
3. You had to estimate $(r - 1)$ row probabilities, but the column probabilities are fixed in advance and did not need to be estimated. Subtract $(r - 1)$ df.

The total degrees of freedom for the $r \times c$ (fixed-column) table are

$$rc - c - (r - 1) = rc - c - r + 1 = (r - 1)(c - 1)$$

EXAMPLE 14.7 A survey of voter sentiment was conducted in four midcity political wards to compare the fractions of voters who favor candidate A. Random samples of 200 voters were polled in each of the four wards with the results shown in Table 14.9. The values in parentheses in the table are the expected cell counts. Do the data present sufficient evidence to indicate that the fractions of voters who favor candidate A differ in the four wards?

TABLE 14.9 • **Voter Opinions in Four Wards**

	Ward 1	Ward 2	Ward 3	Ward 4	Total
Favor A	76 (59)	53 (59)	59 (59)	48 (59)	236
Do Not Favor A	124 (141)	147 (141)	141 (141)	152 (141)	564
Total	200	200	200	200	800

Solution   Since the column totals are fixed at 200, the design involves four binomial experiments, each containing the responses of 200 voters from each of the four wards. To test the equality of the proportions who favor candidate A in all four wards, the null hypothesis

$$H_0 : p_1 = p_2 = p_3 = p_4$$

is equivalent to the null hypothesis

H_0 : Proportion favoring candidate A is independent of ward

and will be rejected if the test statistic X^2 is too large. The observed value of the test statistic, $X^2 = 10.722$, and its associated p-value, .013, are shown in Figure 14.2. The results are significant ($P < .025$); that is, H_0 is rejected and you can conclude that there is a difference in the proportions of voters who favor candidate A among the four wards.

FIGURE 14.2

MINITAB output for
Example 14.7

•

Chi-Square Test: Ward 1, Ward 2, Ward 3, Ward 4,

```
Expected counts are printed below observed counts
Chi-Square contributions are printed below expected counts

          Ward 1   Ward 2   Ward 3   Ward 4    Total
    1         76       53       59       48      236
           59.00    59.00    59.00    59.00
           4.898    0.610    0.000    2.051

    2        124      147      141      152      564
          141.00   141.00   141.00   141.00
           2.050    0.255    0.000    0.858

Total      200      200      200      200      800

Chi-Sq = 10.722 DF = 3, P-Value = 0.013
```

What is the nature of the differences discovered by the chi-square test? To answer this question, look at Table 14.10, which shows the sample proportions who favor candidate A in each of the four wards. It appears that candidate A is doing best in the first ward and worst in the fourth ward. Is this of any *practical significance* to the candidate? Possibly a more important observation is that the candidate does not have a plurality of voters in any of the four wards. If this is a two-candidate race, candidate A needs to increase his campaigning!

TABLE 14.10 • **Proportions in Favor of Candidate A in Four Wards**

Ward 1	Ward 2	Ward 3	Ward 4
76/200 = .38	53/200 = .27	59/200 = .30	48/200 = .24

14.5 **EXERCISES**

BASIC TECHNIQUES

14.27 Random samples of 200 observations were selected from each of three populations, and each observation was classified according to whether it fell into one of three mutually exclusive categories:

	Category			
Population	1	2	3	Total
1	108	52	40	200
2	87	51	62	200
3	112	39	49	200

You want to know whether the data provide sufficient evidence to indicate that the proportions of observations in the three categories depend on the population from which they were drawn.

a. Give the value of X^2 for the test.

b. Give the rejection region for the test for $\alpha = .01$.

c. State your conclusions.

d. Find the approximate p-value for the test and interpret its value.

14.28 Suppose you wish to test the null hypothesis that three binomial parameters p_A, p_B, and p_C are equal versus the alternative hypothesis that at least two of the parameters differ. Independent random samples of 100 observations were selected from each of the populations. The data are shown in the table.

	Population			
	A	B	C	Total
Successes	24	19	33	76
Failures	76	81	67	224
Total	100	100	100	300

a. Write the null and alternative hypotheses for testing the equality of the three binomial proportions.

b. Calculate the test statistic and find the approximate p-value for the test in part a.

c. Use the approximate p-value to determine the statistical significance of your results. If the results are statistically significant, explore the nature of the differences in the three binomial proportions.

APPLICATIONS

14.29 The Sandwich Generation How do Americans in the "sandwich generation" balance the demands of caring for older and younger relatives?

In a telephone poll of Americans aged 45–55 years conducted by the *New York Times*,[9] the number providing financial support for their parents is listed in the next display.

Provide Financial Support	Yes	No
White Americans	40	160
African Americans	56	144
Hispanic Americans	68	132
Asian Americans	84	116

Is there a significant difference in the proportion of individuals providing financial support for their parents for these subpopulations of Americans? Use $\alpha = .01$.

14.30 Diseased Chickens A particular poultry disease is thought to be noncommunicable. To test this theory, 30,000 chickens were randomly partitioned into three groups of 10,000. One group had no contact with diseased chickens, one had moderate contact, and the third had heavy contact. After a 6-month period, data were collected on the number of diseased chickens in each group of 10,000. Do the data provide sufficient evidence to indicate a dependence between the amount of contact between diseased and nondiseased fowl and the incidence of the disease? Use $\alpha = .05$.

	No Contact	Moderate Contact	Heavy Contact
Disease	87	89	124
No Disease	9,913	9,911	9,876
Total	10,000	10,000	10,000

14.31 Long-Term Care A study conducted in northwest England made an assessment of long-term care facilities that have residents with dementia.[10] The homes included those that provided specialized services for elderly people with mental illness/health problems, known as "EMI homes," as well as others classified as "non-EMI homes." It was expected that the EMI homes would score higher on several measures of service quality for people with dementia. One measure included the structure of the home and the services provided, as given in the next table.

	Home Type		
Care Type	EMI	Non-EMI	Total
Nursing Care	54	22	76
Residential Care	59	77	136
Dual-Registered	49	26	75
Total	162	125	287

a. Describe the binomial experiments whose proportions are being compared in this experiment.

b. Do these data indicate that the type of care provided varies by the three types of home? Test at the $\alpha = .01$ level.

c. Based on the results of part b, explain the practical nature of the relationship between home type and care type.

14.32 Deep-Sea Research W.W. Menard has conducted research involving manganese nodules, a mineral-rich concoction found abundantly on the deep-sea floor.[11] In one portion of his report, Menard provides data relating the magnetic age of the earth's crust to the "probability of finding manganese nodules." The table gives the number of samples of the earth's core and the percentage of those that contain manganese nodules for each of a set of magnetic-crust ages. Do the data provide sufficient evidence to indicate that the probability of finding manganese nodules in the deep-sea earth's crust is dependent on the magnetic-age classification?

Age	Number of Samples	Percentage with Nodules
Miocene—recent	389	5.9
Oligocene	140	17.9
Eocene	214	16.4
Paleocene	84	21.4
Late Cretaceous	247	21.1
Early and Middle Cretaceous	1120	14.2
Jurassic	99	11.0

Data set **14.33 How Big Is the Household?** A local EX1433 chamber of commerce surveyed 120 households in their city—40 in each of three types of residence (apartment, duplex, or single residence)—and recorded the number of family members in each of the households. The data are shown in the table.

Family Members	Type of Residence		
	Apartment	Duplex	Single Residence
1	8	20	1
2	16	8	9
3	10	10	14
4 or More	6	2	16

Is there a significant difference in the family size distributions for the three types of residence? Test using $\alpha = .01$. If there are significant differences, describe the nature of these differences.

Data set **14.34 Evolution: Pro or Con?** According to EX1434 a poll by the Pew Research Center, 55% of young adults (ages 18–29) believe that evolution is the best explanation for the development of human life.[12] When the data are further categorized by whether or not the responders had a religious affiliation, this proportion changed for those not having a religious affiliation. The data that follow reflect the results of this poll.

	Religious Affiliation	Not Affiliated	Total
Yes	47	152	199
No	53	98	151
Total	100	250	350

a. Do the data indicate that the proportion of young adults who believe that evolution provides the best answer to the development of human life differ for those with a religious affiliation versus those without a religious affiliation? Use $\alpha = .05$.

b. If significant differences exist, explain what changes appear to have taken place when religious affiliation is included in the categorization.

14.6 THE EQUIVALENCE OF STATISTICAL TESTS

Remember that when there are only $k = 2$ categories in a multinomial experiment, the experiment reduces to a *binomial experiment* where you record the number of successes x (or O_1) in n (or $O_1 + O_2$) trials. Similarly, the data that result from *two binomial experiments* can be displayed as a two-way classification with $r = 2$ and $c = 2$, so that the chi-square test of *homogeneity* can be used to compare the two binomial proportions, p_1 and p_2. For these two situations, we have presented statistical tests for the binomial proportions based on the z-statistic of Chapter 9:

- **One sample:** $z = \dfrac{\hat{p} - p_0}{\sqrt{\dfrac{p_0 q_0}{n}}}$

$k = 2$	
Successes	Failures

- **Two samples:** $z = \dfrac{\hat{p}_1 - \hat{p}_2}{\sqrt{\hat{p}\hat{q}\left(\dfrac{1}{n_1} + \dfrac{1}{n_2}\right)}}$

$r = c = 2$	
Sample 1	**Sample 2**
Successes	Successes
Failures	Failures

NEED A TIP?

The one- and two-sample binomial tests from Chapter 9 are equivalent to chi-square tests—$z^2 = \chi^2$.

Why are there two different tests for the same statistical hypothesis? Which one should you use? For these two situations, you can use *either* the z-test *or* the chi-square test, and you will obtain identical results. For either the one- or two-sample test, we can prove algebraically that

$$z^2 = \chi^2$$

so that the test statistic z will be the square root (either positive or negative, depending on the data) of the chi-square statistic. Furthermore, we can show theoretically that the same relationship holds for the critical values in the z and χ^2 tables in Appendix I, which produces *identical p-values* for the two equivalent tests. To test a one-tailed alternative hypothesis such as $H_0: p_1 > p_2$, first determine whether $\hat{p}_1 - \hat{p}_2 > 0$, that is, if the difference in sample proportions has the appropriate sign. If so, the appropriate critical value of χ^2 from Table 5 in Appendix I will have one degree of freedom a right-tail area of 2α. For example, the critical χ^2 value with 1 *df* and $\alpha = .05$ will be $\chi^2_{.10} = 2.70554 = 1.645^2$.

In summary, you are free to choose the test (z or X^2) that is most convenient. Since most computer packages include the chi-square test, and most do not include the large-sample z-tests, the chi-square test may be preferable to you!

OTHER APPLICATIONS OF THE CHI-SQUARE TEST

The application of the chi-square test for analyzing count data is only one of many classification problems that result in multinomial data. Some of these applications are quite complex, requiring complicated or calculationally difficult procedures for estimating the expected cell counts. However, several applications are used often enough to make them worth mentioning.

- **Goodness-of-fit tests:** You can design a goodness-of-fit test to determine whether data are consistent with data drawn from a particular probability distribution—possibly the normal, binomial, Poisson, or other distributions. The cells of a sample frequency histogram correspond to the k cells of a multinomial experiment. Expected cell counts are calculated using the probabilities associated with the hypothesized probability distribution.

- **Time-dependent multinomials:** You can use the chi-square statistic to investigate the rate of change of multinomial (or binomial) proportions over time. For example, suppose that the proportion of correct answers on a 100-question exam is recorded for a student, who then repeats the exam in each of the next 4 weeks. Does the proportion of correct responses increase over time? Is learning taking place? In a process monitored by a quality control plan, is there a positive trend in the proportion of defective items as a function of time?

- **Multidimensional contingency tables:** Instead of only two methods of classification, you can investigate a dependence among three or more classifications. The two-way contingency table is extended to a table in more than two dimensions. The methodology is similar to that used for the $r \times c$ contingency table, but the analysis is a bit more complex.

- **Log-linear models:** Complex models can be created in which the logarithm of the cell probability ($\ln p_{ij}$) is some linear function of the row and column probabilities.

Most of these applications are rather complex and might require that you consult a professional statistician for advice before you conduct your experiment.

In all statistical applications that use *Pearson's chi-square statistic,* assumptions must be satisfied in order that the test statistic have an approximate chi-square probability distribution.

ASSUMPTIONS

- The cell counts $O_1, O_2, \ldots, O_k$ must satisfy the conditions of a multinomial experiment, or a set of multinomial experiments created by fixing either the row or column totals.
- The expected cell counts $E_1, E_2, \ldots, E_k$ should equal or exceed 5.

You can usually be fairly certain that you have satisfied the first assumption by carefully preparing and designing your experiment or sample survey. When you calculate the expected cell counts, if you find that one or more is less than 5, these options are available to you:

- Choose a larger sample size n. The larger the sample size, the closer the chi-square distribution will approximate the distribution of your test statistic X^2.
- It may be possible to combine one or more of the cells with small expected cell counts, thereby satisfying the assumption.

Finally, make sure that you are calculating the *degrees of freedom* correctly and that you carefully evaluate the statistical and practical conclusions that can be drawn from your test.

CHAPTER REVIEW

Key Concepts and Formulas

I. The Multinomial Experiment

1. There are n identical trials, and each outcome falls into one of k categories.

2. The probability of falling into category i is p_i and remains constant from trial to trial.

3. The trials are independent, $\Sigma p_i = 1$, and we measure O_i, the number of observations that fall into each of the k categories.

II. Pearson's Chi-Square Statistic

$$X^2 = \Sigma \frac{(O_i - E_i)^2}{E_i} \qquad \text{where } E_i = np_i$$

which has an approximate chi-square distribution with *degrees of freedom* determined by the application.

III. The Goodness-of-Fit Test

1. This is a one-way classification with cell probabilities specified in H_0.
2. Use the chi-square statistic with $E_i = np_i$ calculated with the hypothesized probabilities.
3. $df = k - 1 -$ (Number of parameters estimated in order to find E_i)
4. If H_0 is rejected, investigate the nature of the differences using the sample proportions.

IV. Contingency Tables

1. A two-way classification with n observations categorized into $r \times c$ cells of a two-way table using two different methods of classification is called a *contingency table*.
2. The test for independence of classification methods uses the chi-square statistic

$$X^2 = \Sigma \frac{(O_{ij} - \hat{E}_{ij})^2}{\hat{E}_{ij}}$$

with $\hat{E}_{ij} = \dfrac{r_i c_j}{n}$ and $df = (r - 1)(c - 1)$

3. If the null hypothesis of independence of classifications is rejected, investigate the nature of the dependence using conditional proportions within either the rows or columns of the contingency table.

V. Fixing Row or Column Totals

1. When either the row or column totals are fixed, the test of independence of classifications becomes a test of the homogeneity of cell probabilities for several multinomial experiments.
2. Use the same chi-square statistic as for contingency tables.
3. The large-sample z-tests for one and two binomial proportions are special cases of the chi-square statistic.

VI. Assumptions

1. The cell counts satisfy the conditions of a multinomial experiment, or a set of multinomial experiments with fixed sample sizes.
2. All expected cell counts must equal or exceed five in order that the chi-square approximation is valid.

 TECHNOLOGY TODAY

The Chi-Square Test—*Microsoft Excel*

The procedure for performing a chi-square test of independence in *MS Excel* requires that you enter both the observed and the expected cell counts into an *Excel* spreadsheet. If the *raw categorical data* have been stored in the spreadsheet rather than the *observed cell counts*, you may need to tally the data to obtain the cell counts before continuing.

 Suppose you have recorded the gender (M or F) and the college status (Fr, So, Jr, Sr, Grad) for 100 statistics students, as shown in the table below.

Gender	Fr	So	Jr	Sr	Grad
			Status		
F	16	8	8	8	4
M	9	11	12	12	12

1. Enter the observed values into the first five columns of an *Excel* spreadsheet.
2. Calculate (by hand) the 10 estimated expected cell counts and enter them into another range in the spreadsheet.

3. Place your cursor in an empty cell, and use **Formulas** ▶ **More Functions** ▶ **Statistical** ▶ **CHISQ.TEST** (**CHITEST** in earlier versions of *Excel*) to generate the Dialog box in Figure 14.3. Highlight or type in the cell ranges for the observed and expected cell counts.

FIGURE 14.3

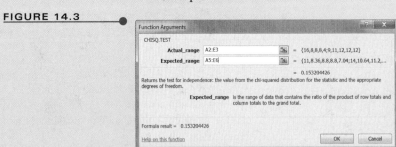

4. When you click **OK,** *MS Excel* will calculate the *p*-value associated with the chi-square test of independence. For this data, the large *p*-value (.153) indicates a nonsignificant result. There is insufficient evidence to indicate that a student's gender is dependent on class status.

NOTE: *MS Excel* does not provide a single command to allow you to perform the chi-square goodness-of-fit test; however, you could manually create formulas in *MS Excel* to perform this test and obtain the appropriate *p*-value.

The Chi-Square Test—*MINITAB*

Several procedures are available in the *MINITAB* package for analyzing categorical data. The appropriate procedure depends on whether the data represent a one-way classification (a single multinomial experiment) or a two-way classification or contingency table. If the *raw categorical data* have been stored in the *MINITAB* worksheet rather than the *observed cell counts,* you may need to tally or cross-classify the data to obtain the cell counts before continuing.

EXAMPLE  Suppose you have recorded the gender (M or F) and the college status (Fr, So, Jr, Sr, G) for 100 statistics students. The *MINITAB* worksheet would contain two columns of 100 observations each. Each row would contain an individual's gender in column 1 and college status in column 2.

1. To obtain the observed cell counts (O_{ij}) for the 2 × 5 contingency table, use **Stat** ▶ **Tables** ▶ **Cross Tabulation and Chi-Square** to generate the Dialog box shown in Figure 14.4(a).

FIGURE 14.4

(a)   (b)

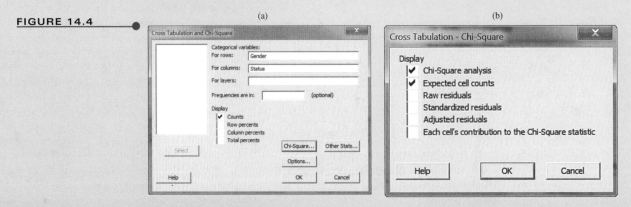

2. Under "Categorical Variables," select "Gender" for the row variable and "Status" for the column variable. Leave the boxes marked "For Layers" and "Frequencies are in:" blank. Make sure that the square labeled "Display Counts" is checked.

3. Click the **Chi-Square…** button to display the dialog box in Figure 14.4(b). Check the boxes for "Chi-Square Analysis" and "Expected Cell Counts." Click **OK** twice. This sequence of commands not only tabulates the contingency table but also performs the chi-square test of independence and displays the results in the Session window shown in Figure 14.5. For the gender/college status data, the large p-value ($P = .153$) indicates a nonsignificant result. There is insufficient evidence to indicate that a student's gender is dependent on class status.

FIGURE 14.5

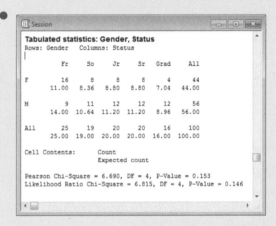

4. If the observed cell counts in the contingency table have already been tabulated, simply enter the counts into c columns of the *MINITAB* worksheet, use **Stat ▶ Tables ▶ Chi-Square Test (Two-Way Table in Worksheet),** and select the appropriate columns before clicking **OK.** For the gender/college status data, you can enter the counts into columns C3–C7 as shown in Figure 14.6. The resulting output will be labeled differently but will look exactly like the output in Figure 14.5.

FIGURE 14.6

EXAMPLE (14.10) — A simple test of a single multinomial experiment can be set up by considering whether the proportions of male and female statistics students are the same—that is, $p_1 = .5$ and $p_2 = .5$.

1. In *MINITAB 15* or *16*, use **Stat ▶ Tables ▶ Chi-Square Goodness-of-Fit Test (One Variable)** to display the dialog box in Figure 14.7. If you have raw categorical data in a column, click the "Categorical data:" button and enter the "Gender" column in the cell. If you have summary values of observed counts for each category, choose "Observed counts." Then enter the column containing the observed counts or type the observed counts for each category.

FIGURE 14.7

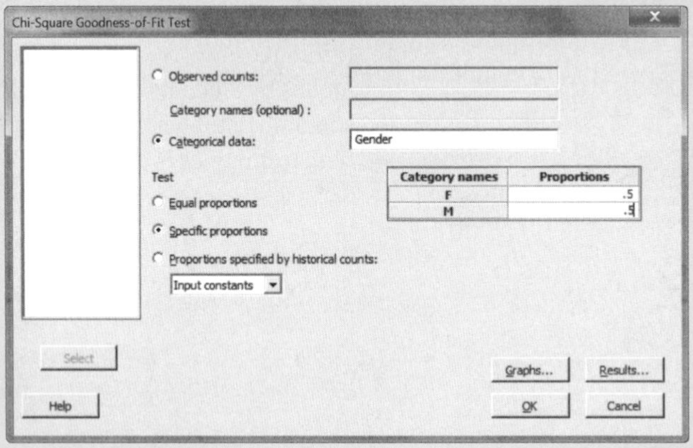

2. For this test, we can choose "Equal proportions" to test $H_0: p_1 = p_2 = .5$. When you have different proportions for each category, use "Specific proportions." You can store the proportions for each category in a column, choose "Input column" and enter the column. If you want to type the proportion for each category, choose "Input constants" and type the proportions for the corresponding categories. Click **OK.**

3. The resulting output will include several graphs along with the values for O_i and E_i for each category, the observed value of the test statistic, $X^2 = 1.44$, and its *p*-value $= 0.230$, which is not significant. There is insufficient evidence to indicate a difference in the proportion of male and female statistics students.

NOTE: If you are using a previous version of *MINITAB*, you will have to determine the observed and expected cell counts, and enter them into separate columns in the worksheet. Then use **Calc ▶ Calculator** and the expression **SUM(('O'− 'E')**2/'E')** to calculate the observed value of the test statistic.

Supplementary Exercises

Starred (*) exercises are optional.

14.35 Floor Polish A manufacturer of floor polish conducted a consumer preference experiment to see whether a new floor polish A was superior to those produced by four competitors, B, C, D, and E. A sample of 100 housekeepers viewed five patches of flooring that had received the five polishes, and each indicated the patch that he or she considered superior in appearance. The lighting, background, and so on were approximately the same for all five patches. The results of the survey are listed here:

Polish	A	B	C	D	E
Frequency	27	17	15	22	19

Do these data present sufficient evidence to indicate a preference for one or more of the polished patches of floor over the others? If one were to reject the hypothesis of no preference for this experiment, would this imply that polish A is superior to the others? Can you suggest a better way of conducting the experiment?

14.36 Physical Fitness in the United States

A survey was conducted to determine whether adult participation in physical fitness programs varies from one region of the United States to another. A random sample of people were interviewed in each of four states and these data were recorded:

	Rhode Island	Colorado	California	Florida
Participate	46	63	108	121
Do Not Participate	149	178	192	179

Do the data indicate a difference in adult participation in physical fitness programs from one state to another? If so, describe the nature of the differences.

14.37 Fatal Accidents

Accident data were analyzed to determine the numbers of fatal accidents for automobiles of three sizes. The data for 346 accidents are as follows:

	Small	Medium	Large
Fatal	67	26	16
Not Fatal	128	63	46

Do the data indicate that the frequency of fatal accidents is dependent on the size of automobiles? Write a short paragraph describing your statistical results and their practical implications.

14.38 Physicians and Medicare Patients

An experiment was conducted to investigate the effect of general hospital experience on the attitudes of physicians toward Medicare patients. A random sample of 50 physicians who had just completed 4 weeks of service in a general hospital were categorized according to their concern for Medicare patients before and after their general hospital experience. The data are shown in the table. Do the data provide sufficient evidence to indicate a change in "concern" after the general hospital experience? If so, describe the nature of the change.

	Concern After		
Concern Before	High	Low	Total
Low	27	5	32
High	9	9	18

Partial *MINITAB* output for Exercise 14.38

Chi-Square Test: High, Low

```
Chi-Sq = 6.752, DF = 1, P-Value = 0.009
```

14.39 Discovery-Based Teaching Two biology instructors set out to evaluate the effects of discovery-based teaching compared to the standard lecture-based teaching approach in the laboratory.[13] The standard lecture-based approach provided a list of instructions to follow at each step of the laboratory exercise, whereas the discovery-based approach asked questions rather than providing directions, and used small group reports to decide the best way to proceed in reaching the laboratory objective. One evaluation of the techniques involved written appraisals of both techniques by students at the end of the course. The comparison of the number of positive and negative responses for both techniques is given in the following table.

Group	Positive Evaluations	Negative Evaluations	Total
Discovery	37	11	48
Control	31	17	48

a. Is there a significant difference in the proportion of positive responses for each of the teaching methods? Use $\alpha = .05$. If so, how would you describe this difference?

b. What is the approximate *p*-value for the test in part a?

14.40 Baby's Sleeping Position Does a baby's sleeping position affect the development of motor skills? In one study, 343 full-term infants were examined at their 4-month checkup for various developmental milestones, such as rolling over, grasping a rattle, and reaching for an object.[14] The baby's predominant sleep position—either prone (on the stomach) or supine (on the back) or side—was determined by a telephone interview with the parent. The sample results for 320 of the 343 infants for whom information was received are shown in the table. The researcher reported that infants who slept in the side or supine position were less likely to roll over at the 4-month checkup than infants who slept primarily in the prone position ($P < .001$).

	Prone	Supine or Side
Number of Infants	121	199
Number Who Roll Over	93	119

a. Use a large-sample *z*-test to confirm or refute the researcher's conclusion.

b. Rewrite the sample data as a 2×2 contingency table. Use the chi-square test for homogeneity to confirm or refute the researcher's conclusion.

c. Compare the results of parts a and b. Confirm that the two test statistics are related as $z^2 = X^2$ and that the critical values for rejecting H_0 have the same relationship.

14.41 Refer to Exercise 14.40. Find the p-value for the large-sample z-test in part a. Compare this p-value with the p-value for the chi-square test, shown in the partial *MINITAB* printout.

Partial MINITAB output for Exercise 14.41

Chi-Square Test: Prone, Side

```
Chi-Sq = 9.795, DF = 1, P-Value = 0.002
```

14.42 Baby's Sleeping Position II The researchers in Exercise 14.40 also measured several other developmental milestones and their relationship to the infant's predominant sleep position.[14] The results of their research are presented in the table for the 320 infants at their 4-month checkup.

Milestone	Score	Prone	Supine or Side	P
Pulls to sit with no head lag	Pass	79	144	
	Fail	6	20	<.21
Grasps Rattle	Pass	102	167	
	Fail	3	1	<.13
Reaches for Object	Pass	107	183	
	Fail	3	5	<.97

Use your knowledge of the analysis of categorical data to explain the experimental design(s) used by the researchers. What hypotheses were of interest to the researchers, and what statistical test would the researchers have used? Explain the conclusions that can be drawn from the three p-values in the last column of the table and the practical implications that can be drawn from the statistical results. Have any statistical assumptions been violated?

14.43 Flower Color and Shape A botanist performs a secondary cross of petunias involving independent factors that control leaf shape and flower color, where the factor A represents red color, a represents white color, B represents round leaves, and b represents long leaves. According to the Mendelian model, the plants should exhibit the characteristics AB, Ab, aB, and ab in the ratio 9:3:3:1. Of 160 experimental plants, the following numbers were observed:

AB	Ab	aB	ab
95	30	28	7

Is there sufficient evidence to refute the Mendelian model at the $\alpha = .01$ level?

 14.44 Opportunities for Success A *CBS News Poll*[15] asked the question "Compared to
EX1423

your parents' generation, do you think in general your opportunities to succeed in life are better than theirs, about the same as theirs, or worse than theirs?" on three separate dates over a 10-year period. All surveys involved $n = 1048$ individuals.

Poll Date	Better	Same	Worse	Unsure	Totals
December 2009	493	252	283	20	1048
June 2007	650	189	189	20	1048
February 2000	755	231	52	10	1048

a. Is there a significant difference among the responses to this question over time? Use $\alpha = .05$.

b. If significant differences are found in part a, describe the nature of these differences.

14.45 An Arthritis Drug A study to determine the effectiveness of a drug (serum) for arthritis resulted in the comparison of two groups, each consisting of 200 arthritic patients. One group was inoculated with the serum; the other received a placebo (an inoculation that appears to contain serum but actually is nonactive). After a period of time, each person in the study was asked to state whether his or her arthritic condition had improved. These are the results:

	Treated	Untreated
Improved	117	74
Not Improved	83	126

You want to know whether these data present sufficient evidence to indicate that the serum was effective in improving the condition of arthritic patients.

a. Use the chi-square test of homogeneity to compare the proportions improved in the populations of treated and untreated subjects. Test at the 5% level of significance.

b. Test the equality of the two binomial proportions using the two-sample z-test of Section 9.6. Verify that the squared value of the test statistic $z^2 = X^2$ from part a. Are your conclusions the same as in part a?

14.46 Parking at the University A survey was conducted to determine student, faculty, and administration attitudes about a new university parking policy. The distribution of those favoring or opposing the policy is shown in the table. Do the data provide sufficient evidence to indicate that attitudes about the parking policy are independent of student, faculty, or administration status?

	Student	Faculty	Administration
Favor	252	107	43
Oppose	139	81	40

14.47* The chi-square test used in Exercise 14.45 is equivalent to the two-tailed z-test of Section 9.6 provided α is the same for the two tests. Show algebraically that the chi-square test statistic X^2 is the square of the test statistic z for the equivalent test.

14.48 Fitting a Binomial Distribution You can use a goodness-of-fit test to determine whether all of the criteria for a binomial experiment have actually been met in a given application. Suppose that an experiment consisting of four trials was repeated 100 times. The number of repetitions on which a given number of successes was obtained is recorded in the table:

Possible Results (number of successes)	Number of Times Obtained
0	11
1	17
2	42
3	21
4	9

Estimate p (assuming that the experiment was binomial), obtain estimates of the expected cell frequencies, and test for goodness-of-fit. To determine the appropriate number of degrees of freedom for X^2, note that p was estimated by a linear combination of the observed frequencies.

14.49 Antibiotics and Infection Infections sometimes occur when blood transfusions are given during surgical operations. An experiment was conducted to determine whether the injection of antibodies reduced the probability of infection. An examination of the records of 138 patients produced the data shown in the table. Do the data provide sufficient evidence to indicate that injections of antibodies affect the likelihood of transfusional infections? Test by using $\alpha = .05$.

	Infection	No Infection
Antibody	4	78
No Antibody	11	45

14.50 German Manufacturing U.S. labor unions have traditionally been content to leave the management of the company to managers and corporate executives. But in Europe, worker participation in management decision making is an accepted idea that is continually spreading. To study the effect of worker participation in managerial decision making, 100 workers were interviewed in each of two separate German manufacturing plants. One plant had active worker participation in managerial decision making; the other did not. Each selected worker was asked whether he or she generally approved of the manage-

rial decisions made within the firm. The results of the interviews are shown in the table:

	Participation	No Participation
Generally Approve	73	51
Do Not Approve	27	49

a. Do the data provide sufficient evidence to indicate that approval or disapproval of management's decisions depends on whether workers participate in decision making? Test by using the X^2 test statistic. Use $\alpha = .05$.

b. Do these data support the hypothesis that workers in a firm with participative decision making more generally approve of the firm's managerial decisions than those employed by firms without participative decision making? Test by using the z-test presented in Section 9.6. This problem requires a one-tailed test. Why?

14.51 Three Entrances An occupant-traffic study was conducted to aid in the remodeling of an office building that contains three entrances. The choice of entrance was recorded for a sample of 200 persons who entered the building. Do the data in the table indicate that there is a difference in preference for the three entrances? Find a 95% confidence interval for the proportion of persons favoring entrance 1.

Entrance	1	2	3
Number Entering	83	61	56

Data set
EX1452
14.52 Graduate Teaching Assistants Graduate students' responsibilities are often related to their roles as teaching assistants or research assistants. As part of a larger study, K.M. McGoldrick and her colleagues investigated the level of preparation of economics graduate students for their teaching-related duties for students at "top-tier" and those at "second-tier" schools.[16] The responses to the question "Are you satisfied with the level of preparation you have had for year teaching related duties?" follow.

	Top-Tier	Second-Tier
I am very satisfied	85	197
I am somewhat satisfied	102	171
I am unsatisfied	22	29
Total	209	397

a. Is there a significant difference in the responses to the question between students at "top-tier" schools compared to those at "second-tier" schools?

b. If significant, describe the nature of the differences in response for graduate students at "top-tier" versus "second-tier" schools.

14.53 Is Your Food Safe? How confident are you that the food you purchase is safe to eat? This question was asked in a *CBS News Poll:*[17] the data that follow reflect the results of the responses to this poll.

	Very Confident	Somewhat Confident	Not Too Confident	Not at All Confident	Total
Men	210	241	68	5	524
Women	129	306	73	16	524
Total	329	547	141	21	1048

a. Is there sufficient evidence to conclude that there are significant differences in responses between men and women at the $\alpha = .05$ level of significance?

b. Find the approximate *p*-value for the test.

14.54 Vehicle Colors Each model year seems to introduce new colors and different hues for a wide array of vehicles, from luxury cars, to full-size or intermediate models, to compacts and sports cars, to light trucks. However, white and silver/gray continue to make the top five or six colors across all of these categories of vehicles. The top six colors and their percentage of the market share for compact/sports cars are shown in the following table.[18]

Color	Silver	Black	Gray	Blue	Red	White
Percent	19	17	17	15	12	12

To verify the figures, a random sample consisting of 250 compact/sports cars was taken and the color of the vehicles recorded. The sample provided the following counts for the categories given above: 52, 43, 48, 41, 32, 19, respectively.

a. Is any category missing in the classification? How many vehicles belong to that category?

b. Is there sufficient evidence to indicate that our percentages of the colors for compact/sports cars differ from those given? Find the approximate *p*-value for the test.

14.55 Vehicle Colors, again Refer to Exercise 14.54. The researcher wants to see if there is a difference in the color distributions for compact/sports cars versus full/intermediate cars.[18]

Another random sample of 250 full/intermediate cars was taken and the color of the vehicles was recorded. The table below shows the results for both compact/sports and full/intermediate cars.

Color	Silver	Black	Gray	Blue	Red	White
Compact/Sports	52	43	48	41	32	19
Full/Intermediate	50	33	37	32	27	38

Do the data indicate that there is a difference in the color distributions depending on the type of vehicle? Use $\alpha = .05$. (HINT: Remember to include a column called "Other" for cars that do not fall into one of the six categories shown in the table.)

14.56 Good Tasting Medicine Pfizer Canada Inc. is a pharmaceutical company that makes azithromycin, an antibiotic in a cherry-flavored suspension used to treat bacterial infections in children. To compare the taste of their product with three competing medications, Pfizer tested 50 healthy children and 20 healthy adults. Among other taste-testing measures, they recorded the number of tasters who rated each of the four antibiotic suspensions as the best tasting.[19] The results are shown in the table. Is there a difference in the perception of the best taste between adults and children? If so, what is the nature of the difference, and why is it of practical importance to the pharmaceutical company?

	Flavor of Antibiotic			
	Banana	Cherry*	Wild Fruit	Strawberry-Banana
Children	14	20	7	9
Adults	4	14	0	2

*Azithromycin produced by Pfizer Canada Inc.

14.57 Rugby Injuries The prevalence and patterns of knee injuries among women collegiate rugby players were investigated using a sample questionnaire, to which 42 rugby clubs responded.[20] A total of 76 knee injuries were classified by type as well as the position (forward or back) of the player.

	Type of Knee Injury				
Position	Meniscal Tear	MCL Tear	ACL Tear	Patella Dislocation	PCL Tear
Forward	13	14	7	3	1
Back	12	9	14	2	1

MINITAB output for Exercise 14.57

Chi-Square Test: Men Tear, MCL Tear, ACL Tear, Patella, PCL Tear

```
Expected counts are printed below observed counts
Chi-Square contributions are printed below expected
counts

     Men Tear MCL Tear ACL Tear  Patella PCL Tear  Total
  1        13       14        7        3        1     38
       12.50    11.50    10.50     2.50     1.00
       0.020    0.543    1.167    0.100    0.000

  2        12        9       14        2        1     38
       12.50    11.50    10.50     2.50     1.00
       0.020    0.543    1.167    0.100    0.000

Total     25       23       21        5        2     76

Chi-Sq = 3.660, DF = 4, P-Value = 0.454
4 cells with expected counts less than 5.0
```

a. Use the *MINITAB* printout to determine whether there is a difference in the distribution of injury types for rugby backs and forwards. Have any of the assumptions necessary for the chi-square test been violated? What effect will this have on the magnitude of the test statistic?

b. The investigators report a significant difference in the proportion of MCL tears for the two positions ($P < .05$) and a significant difference in the proportion of ACL tears ($P < .05$), but indicate that all other injuries occur with equal frequency for the two positions. Do you agree with those conclusions? Explain.

Data set **14.58 Favorite Fast Foods** Is a customer's
EX1458 preference for a fast-food chain affected by the age of the customer? If so, advertising might need to target a particular age group. Suppose a random sample of 500 fast-food customers aged 16 and older was selected, and their favorite fast-food restaurants along with their age groups were recorded, as shown in the table:

Age Group	McDonald's	Burger King	Wendy's	Other
16–21	75	34	10	6
21–30	89	42	19	10
30–49	54	52	28	18
50+	21	25	7	10

Use an appropriate method to determine whether or not a customer's fast-food preference is dependent on age. Write a short paragraph presenting your statistical conclusions and their practical implications for marketing experts.

14.59 Catching a Cold Is your chance of getting a cold influenced by the number of social contacts you have? A study by Sheldon Cohen, a psychology

professor at Carnegie Mellon University, seems to show that the more social relationships you have, the *less susceptible* you are to colds.[21] A group of 276 healthy men and women were grouped according to their number of relationships (such as parent, friend, church member, neighbor). They were then exposed to a virus that causes colds. An adaptation of the results is shown in the table.

	Number of Relationships		
	Three or Fewer	Four or Five	Six or More
Cold	49	43	34
No Cold	31	57	62
Total	80	100	96

a. Do the data provide sufficient evidence to indicate that susceptibility to colds is affected by the number of relationships you have? Test at the 5% significance level.

b. Based on the results of part a, describe the nature of the relationship between the two categorical variables: cold incidence and number of social relationships. Do your observations agree with the author's conclusions?

Data set **14.60 Crime and Educational Achievement**
EX1460 A criminologist studying criminal offenders who have a record of one or more arrests is interested in knowing whether the educational achievement level of the offender influences the frequency of arrests. He has classified his data using four educational level classifications:

A: completed 6th grade or less

B: completed 7th, 8th, or 9th grade

C: completed 10th, 11th, or 12th grade

D: education beyond 12th grade

The contingency table shows the number of offenders in each educational category, along with the number of times they have been arrested.

	Educational Achievement			
Number of Arrests	A	B	C	D
1	55	40	43	30
2	15	25	18	22
3 or More	7	8	12	10

Do the data present sufficient evidence to indicate that the number of arrests is dependent on the educational achievement of a criminal offender? Test using $\alpha = .05$.

14.61 More Business on the Weekends A department store manager claims that her store has twice as many customers on Fridays and Saturdays than on any other day of the week (the store is closed on Sundays). That is, the probability that a customer visits the store Friday is 2/8, the probability that a customer visits the store Saturday is 2/8, while the probability that a customer visits the store on each of the remaining weekdays is 1/8. During an average week, the following numbers of customers visited the store:

Day	Number of Customers
Monday	95
Tuesday	110
Wednesday	125
Thursday	75
Friday	181
Saturday	214

Can the manager's claim be refuted at the $\alpha = .05$ level of significance?

CASE STUDY

 Working Women

Who is the Primary Breadwinner in Your Family?

How have the roles of working women changed in America? How many of the jobs in America are held by women? How has advertising refocused their ads to influence the 31% of women who are the primary breadwinners in their family? The latest numbers put women's share of the 130.2 million jobs in America at 49.8%. Mya Frazier has examined the role of working women in her white paper article "The Reality of the Working Woman: Her Impact on the Female Target Beyond Consumption."[22] The information that follows is adapted from a quantitative study of 1136 men and 795 women conducted April 7–14, 2010 by JWT and *Advertising Age* and discussed in her paper.

When asked "Who is the household breadwinner?" 100 men and 100 women responded as follows.

	You	Spouse or Significant Other	About Equal	Total
Men	64	16	20	100
Women	31	45	24	100

During the recent recession, 82% of pink slips went to men, reflecting men's dominance in sectors like construction and manufacturing. Anxieties during this time are listed in the next table for 100 men and 100 women.

	Most Anxiety					
	Finances	Out of Work	Family	Relationships	Health	Total
Men	42	24	12	12	10	100
Women	55	18	11	8	8	100

When asked if they had trouble "separating my work life from my personal life, and vice versa," $n = 300$ women respondents had a disparity by generations, as given in the next table.

	Millennials	Gen Xers	Boomers
Yes	47	30	24
No	53	70	76
Totals	100	100	100

The image of a working woman is nothing new in entertainment, but for the most part, many see the workplace as a man's world. Is a gender-balanced workforce a myth? The survey revealed that men and women appear to be in agreement on this issue. See the following table.

	Is a Gender-Balanced Workplace a Myth?		
	Yes	No	Total
Men	59	41	100
Women	63	37	100

1. Is there a significant difference in the proportions of men and women who identify themselves as the primary breadwinner in the family? Use $\alpha = .05$.

2. Is there a significant difference between men and women with respect to which facet of their lives produced the most anxiety during the recent economic downturn? Use $\alpha = .05$.

3. Does the proportion of women from each of the Millennial, Gen X or Boomer generations differ in their ability to separate their work lives from their personal lives? Use $\alpha = .05$.

4. Are men and women in agreement about a gender-balanced workplace? Use $\alpha = .05$.

5. Summarize the results of parts 1–4 as a written report of your findings.

Nonparametric Statistics

© Don Carstens/Brand X/CORBIS

How's Your Cholesterol Level?

What is your cholesterol level? Many of us have become more health conscious in the last few years as we read the nutritional labels on the food products we buy and choose foods that are low in fat and cholesterol and high in fiber. The case study at the end of this chapter involves a taste-testing experiment to compare three types of egg substitutes, using nonparametric techniques.

INTRODUCTION

Some experiments generate responses that can be ordered or ranked, but the actual value of the response cannot be measured numerically except with an arbitrary scale that you might create. It may be that you are able to tell only whether one observation is larger than another. Perhaps you can rank a whole set of observations without actually knowing the exact numerical values of the measurements. Here are a few examples:

- The sales abilities of four sales representatives are ranked from best to worst.
- The edibility and taste characteristics of five brands of raisin bran are rated on an arbitrary scale of 1 to 5.
- Five automobile designs are ranked from most appealing to least appealing.

NEED A TIP?

When sample sizes are small and the original populations are not normal, use nonparametric techniques.

How can you analyze these types of data? The small-sample statistical methods presented in Chapters 10–13 are valid only when the sampled population(s) are normal or approximately so. Data that consist of ranks or arbitrary scales from 1 to 5 *do not satisfy the normality assumption,* even to a reasonable degree. In some applications, the techniques are valid if the samples are randomly drawn from populations whose variances are equal.

When data do not appear to satisfy these and similar assumptions, an alternative method of analysis can be used—**nonparametric statistical methods.** Nonparametric methods generally specify hypotheses in terms of population distributions rather than parameters such as means and standard deviations. Parametric assumptions are often replaced by more general assumptions about the population distributions, and the ranks of the observations are often used in place of the actual measurements.

Research has shown that nonparametric statistical tests are almost as capable of detecting differences among populations as the parametric methods of preceding chapters when normality and other assumptions are satisfied. They may be, and often are, *more* powerful in detecting population differences when these assumptions are not satisfied. For this reason, some statisticians advocate the use of nonparametric procedures in preference to their parametric counterparts.

We will present nonparametric methods appropriate for comparing two or more populations using either independent or paired samples. We will also present a measure of association that is useful in determining whether one variable increases as the other increases or whether one variable decreases as the other increases.

THE WILCOXON RANK SUM TEST: INDEPENDENT RANDOM SAMPLES

In comparing the means of two populations based on independent samples, the pivotal statistic was the difference in the sample means. If you are not certain that the assumptions required for a two-sample *t*-test are satisfied, one alternative is to replace the values of the observations by their ranks and proceed as though the ranks were the actual observations. Two different nonparametric tests use a test statistic based on these sample ranks:

- Wilcoxon rank sum test
- Mann–Whitney *U*-test

They are *equivalent* in that they use the same sample information. The procedure that we will present is the Wilcoxon rank sum test, proposed by Frank Wilcoxon, which is based on the sum of the ranks of the sample that has the smaller sample size.

Assume that you have n_1 observations from population 1 and n_2 observations from population 2. The null hypothesis to be tested is that the two population distributions are identical versus the alternative hypothesis that the population distributions are different. These are the possibilities for the two populations:

- If H_0 is true and the observations have come from the same or identical populations, then the observations from both samples should be randomly mixed when jointly ranked from small to large. The sum of the ranks of the observations from sample 1 should be similar to the sum of the ranks from sample 2.
- If, on the other hand, the observations from population 1 tend to be smaller than those from population 2, then these observations would have the smaller ranks because most of these observations would be smaller than those from population 2. The sum of the ranks of these observations would be "small."
- If the observations from population 1 tend to be larger than those in population 2, these observations would be assigned larger ranks. The sum of the ranks of these observations would tend to be "large."

For example, suppose you have $n_1 = 3$ observations from population 1—2, 4, and 6—and $n_2 = 4$ observations from population 2—3, 5, 8, and 9. Table 15.1 shows seven observations ordered from small to large.

TABLE 15.1 ●	**Seven Observations in Order**						
Observation	x_1	y_1	x_2	y_2	x_3	y_3	y_4
Data	2	3	4	5	6	8	9
Rank	1	2	3	4	5	6	7

The smallest observation, $x_1 = 2$, is assigned rank 1; the next smallest observation, $y_1 = 3$, is assigned rank 2; and so on. The *sum of the ranks* of the observations from sample 1 is $1 + 3 + 5 = 9$, and the **rank sum** from sample 2 is $2 + 4 + 6 + 7 = 19$. How do you determine whether the rank sum of the observations from sample 1 is significantly small or significantly large? This depends on the probability distribution of the sum of the ranks of one of the samples. Since the ranks for $n_1 + n_2 = N$ observations are the first N integers, the sum of these ranks can be shown to be $N(N + 1)/2$. In this simple example, the sum of the $N = 7$ ranks is $1 + 2 + 3 + 4 + 5 + 6 + 7 = 7(8)/2$ or 28. Hence, if you know the rank sum for one of the samples, you can find the other by subtraction. In our example, notice that the rank sum for sample 1 is 9, whereas the second rank sum is $(28 - 9) = 19$. This means that only one of the two rank sums is needed for the test. To simplify the tabulation of critical values for this test, you should use the rank sum from the smaller sample as the test statistic. What happens if two or more observations are equal? Tied observations are assigned the average of the ranks that the observations would have had if they had been slightly different in value.

To implement the Wilcoxon rank sum test, suppose that independent random samples of size n_1 and n_2 are selected from populations 1 and 2, respectively. Let n_1 represent the *smaller* of the two sample sizes, and let T_1 represent the sum of the ranks of the observations in sample 1. If population 1 lies to the left of population 2, T_1 will be "small." T_1 will be "large" if population 1 lies to the right of population 2.

FORMULAS FOR THE WILCOXON RANK SUM STATISTIC (FOR INDEPENDENT SAMPLES)

Let

T_1 = Sum of the ranks for the first sample

$T_1^* = n_1(n_1 + n_2 + 1) - T_1$

T_1^* is the value of the rank sum for n_1 if the observations had been ranked from *large to small.* (It is *not* the rank sum for the second sample.) Depending on the nature of the alternative hypothesis, one of these two values will be chosen as the test statistic, T.

Table 7 in Appendix I can be used to locate *critical values* for the test statistic for four different values of one-tailed tests with $\alpha = .05$, .025, .01, and .005. To use Table 7 in Appendix I for a two-tailed test, the values of α are doubled—that is, $\alpha =$ .10, .05, .02, and .01. The tabled entry gives the value of a such that $P(T \le a) \le \alpha$. To see how to locate a critical value for the Wilcoxon rank sum test, suppose that $n_1 = 8$ and $n_2 = 10$ for a one-tailed test with $\alpha = .05$. You can use Table 7(a) in Appendix I, a portion of which is reproduced in Table 15.2. Notice that the table is constructed assuming that $n_1 \le n_2$. It is for this reason that we designate the population with the smaller sample size as population 1. Values of n_1 are shown across the top of the table, and values of n_2 are shown down the left side. The entry—$a = 56$, shaded—is the critical value for rejection of H_0. The null hypothesis of equality of the two distributions should be rejected if the observed value of the test statistic T is less than or equal to 56.

TABLE 15.2 **A Portion of the 5% Left-Tailed Critical Values, Table 7 in Appendix I**

	n_1						
n_2	2	3	4	5	6	7	8
3	—	6					
4	—	6	11				
5	3	7	12	19			
6	3	8	13	20	28		
7	3	8	14	21	29	39	
8	4	9	15	23	31	41	51
9	4	10	16	24	33	43	54
10	4	10	17	26	35	45	56

THE WILCOXON RANK SUM TEST

Let n_1 denote the smaller of the two sample sizes. This sample comes from population 1. The hypotheses to be tested are

H_0 : The distributions for populations 1 and 2 are identical

versus one of three alternative hypotheses:

H_a : The distributions for populations 1 and 2 are different (a two-tailed test)

H_a : The distribution for population 1 lies to the left of that for population 2
(a left-tailed test)

H_a : The distribution for population 1 lies to the right of that for population 2
(a right-tailed test)

Procedure

1. Rank all $n_1 + n_2$ observations from small to large.

2. Find T_1, the rank sum for the observations in sample 1. This is the test statistic for a left-tailed test.

3. Find $T_1^* = n_1(n_1 + n_2 + 1) - T_1$, the sum of the ranks of the observations from population 1 if the assigned ranks had been reversed from large to small. (The value of T_1^* is not the sum of the ranks of the observations in sample 2.) This is the test statistic for a right-tailed test.

4. The test statistic for a two-tailed test is T, the *minimum* of T_1 and T_1^*.

5. H_0 is rejected if the observed test statistic is less than or equal to the critical value found using Table 7 in Appendix I.

We illustrate the use of Table 7 with the next example.

EXAMPLE 15.1

The wing stroke frequencies of two species of Euglossine bees were recorded for a sample of $n_1 = 4$ *Euglossa mandibularis* Friese (species 1) and $n_2 = 6$ *Euglossa imperialis* Cockerell (species 2).[1] The frequencies are listed in Table 15.3. Can you conclude that the distributions of wing strokes differ for these two species? Test using $\alpha = .05$.

TABLE 15.3 ● **Wing Stroke Frequencies for Two Species of Bees**

Species 1	Species 2
235	180
225	169
190	180
188	185
	178
	182

Solution   You first need to rank the observations from small to large, as shown in Table 15.4.

TABLE 15.4 ● **Wing Stroke Frequencies Ranked from Small to Large**

Data	Species	Rank
169	2	1
178	2	2
180	2	3
180	2	4
182	2	5
185	2	6
188	1	7
190	1	8
225	1	9
235	1	10

The hypotheses to be tested are

H_0 : The distributions of the wing stroke frequencies are the same for the two
  species

versus

H_a : The distributions of the wing stroke frequencies differ for the two species

Since the sample size for individuals from species 1, $n_1 = 4$, is the smaller of the two
sample sizes, you have

$$T_1 = 7 + 8 + 9 + 10 = 34$$

and

$$T_1^* = n_1(n_1 + n_2 + 1) - T_1 = 4(4 + 6 + 1) - 34 = 10$$

For a two-tailed test, the test statistic is $T = 10$, the smaller of $T_1 = 34$ and $T_1^* = 10$.

For this two-tailed test with $\alpha = .05$, you can use Table 7(b) in Appendix I with
$n_1 = 4$ and $n_2 = 6$. The critical value of T such that $P(T \leq a) \leq \alpha/2 = .025$ is 12,
and you should reject the null hypothesis if the observed value of T is 12 or less. Since
the observed value of the test statistic—$T = 10$—is less than 12, you can reject the
hypothesis of equal distributions of wing stroke frequencies at the 5% level of significance.

A *MINITAB* printout of the Wilcoxon rank sum test (called Mann–Whitney by
MINITAB) for these data is given in Figure 15.1. You will find instructions for generating this output in the "Technology Today" section at the end of this chapter. Notice
that the rank sum of the first sample is given as W = 34.0, which agrees with our calculations. With a reported p-value of .0142 calculated by *MINITAB*, you can reject the
null hypothesis at the 5% level.

FIGURE 15.1

Printout for Example 15.1

```
Mann-Whitney Test and CI: Species 1, Species 2

               N      Median
Species 1      4      207.50
Species 2      6      180.00

Point estimate for ETA1-ETA2 is 30.50
95.7 Percent CI for ETA1-ETA2 is (5.99,56.01)
W = 34.0
Test of ETA1 = ETA2 vs ETA1 not = ETA2 is significant at 0.0142
The test is significant at 0.0139 (adjusted for ties)
```

Normal Approximation for the Wilcoxon Rank Sum Test

Table 7 in Appendix I contains critical values for sample sizes of $n_1 \leq n_2 = 3$,
4, . . . , 15. Provided n_1 is not too small,[†] approximations to the probabilities for the
Wilcoxon rank sum statistic T can be found using a normal approximation to the distribution of T. It can be shown that the mean and variance of T are

$$\mu_T = \frac{n_1(n_1 + n_2 + 1)}{2} \quad \text{and} \quad \sigma_T^2 = \frac{n_1 n_2(n_1 + n_2 + 1)}{12}$$

[†]Some researchers indicate that the normal approximation is adequate for samples as small as $n_1 = n_2 = 4$.

The distribution of

$$z = \frac{T - \mu_T}{\sigma_T}$$

is approximately normal with mean 0 and standard deviation 1 for values of n_1 and n_2 as small as 10.

If you try this approximation for Example 15.1, you get

$$\mu_T = \frac{n_1(n_1 + n_2 + 1)}{2} = \frac{4(4 + 6 + 1)}{2} = 22$$

and

$$\sigma_T^2 = \frac{n_1 n_2(n_1 + n_2 + 1)}{12} = \frac{4(6)(4 + 6 + 1)}{12} = 22$$

The p-value for this test is $2P(T \geq 34)$. If you use a .5 correction for continuity in calculating the value of z because n_1 and n_2 are both small,[†] you have

$$z = \frac{T - \mu_T}{\sigma_T} = \frac{(34 - .5) - 22}{\sqrt{22}} = 2.45$$

The p-value for this test is

$$2P(T \geq 34) \approx 2P(z \geq 2.45) = 2(.0071) = .0142$$

the value reported on the *MINITAB* printout in Figure 15.1.

THE WILCOXON RANK SUM TEST FOR LARGE SAMPLES: $n_1 \geq 10$ and $n_2 \geq 10$

1. Null hypothesis: H_0 : The population distributions are identical
2. Alternative hypothesis: H_a : The two population distributions are not identical (a two-tailed test). Or H_a : The distribution of population 1 is shifted to the right (or left) of the distribution of population 2 (a one-tailed test).
3. Test statistic: $z = \dfrac{T - n_1(n_1 + n_2 + 1)/2}{\sqrt{n_1 n_2(n_1 + n_2 + 1)/12}}$
4. Rejection region:
   a. For a two-tailed test, reject H_0 if $z > z_{\alpha/2}$ or $z < -z_{\alpha/2}$.
   b. For a one-tailed test in the right tail, reject H_0 if $z > z_\alpha$.
   c. For a one-tailed test in the left tail, reject H_0 if $z < -z_\alpha$.

   Or reject H_0 if p-value $< \alpha$.

   Tabulated values of z are found in Table 3 of Appendix I.

EXAMPLE 15.2

An experiment was conducted to compare the strengths of two types of kraft papers: one a standard kraft paper of a specified weight and the other the same standard kraft paper treated with a chemical substance. Ten pieces of each type of paper, randomly selected from production, produced the strength measurements shown in Table 15.5. Test the null hypothesis of no difference in the distributions of strengths for the two

---

[†]Since the value of $T = 34$ lies to the right of the mean 22, the subtraction of .5 in using the normal approximation takes into account the lower limit of the bar above the value 34 in the probability distribution of T.

types of paper versus the alternative hypothesis that the treated paper tends to be stronger (i.e., its distribution of strength measurements is shifted to the right of the corresponding distribution for the untreated paper).

TABLE 15.5 ●

Strength Measurements (and Their Ranks) for Two Types of Paper

Standard 1	Treated 2
1.21 (2)	1.49 (15)
1.43 (12)	1.37 (7.5)
1.35 (6)	1.67 (20)
1.51 (17)	1.50 (16)
1.39 (9)	1.31 (5)
1.17 (1)	1.29 (3.5)
1.48 (14)	1.52 (18)
1.42 (11)	1.37 (7.5)
1.29 (3.5)	1.44 (13)
1.40 (10)	1.53 (19)
Rank sum	$T_1 = 85.5$
	$T_1^* = n_1(n_1 + n_2 + 1) - T_1 = 210 - 85.5 = 124.5$

Solution Since the sample sizes are equal, you are at liberty to decide which of the two samples should be sample 1. Choosing the standard treatment as the first sample, you can rank the 20 strength measurements, and the values of T_1 and T_1^* are shown at the bottom of the table. Since you want to detect a shift in the standard (1) measurements to the left of the treated (2) measurements, you conduct a left-tailed test:

H_0 : No difference in the strength distributions

H_a : Standard distribution lies to the left of the treated distribution

and use $T = T_1$ as the test statistic, looking for an unusually small value of T.

To find the critical value for a one-tailed test with $\alpha = .05$, index Table 7(a) in Appendix I with $n_1 = n_2 = 10$. Using the tabled entry, you can reject H_0 when $T \le 82$. Since the observed value of the test statistic is $T = 85.5$, you are not able to reject H_0. There is insufficient evidence to conclude that the treated kraft paper is stronger than the standard paper.

To use the normal approximation to the distribution of T, you can calculate

$$\mu_T = \frac{n_1(n_1 + n_2 + 1)}{2} = \frac{10(21)}{2} = 105$$

and

$$\sigma_T^2 = \frac{n_1 n_2(n_1 + n_2 + 1)}{12} = \frac{10(10)(21)}{12} = 175$$

with $\sigma_T = \sqrt{175} = 13.23$. Then

$$z = \frac{T - \mu_T}{\sigma_T} = \frac{85.5 - 105}{13.23} = -1.47$$

The one-tailed p-value corresponding to $z = -1.47$ is

$$p\text{-value} = P(z \le -1.47) = .5 - .4292 = .0708$$

which is larger than $\alpha = .05$. The conclusion is the same. You cannot conclude that the treated kraft paper is stronger than the standard paper.

When should the Wilcoxon rank sum test be used in preference to the two-sample unpaired *t*-test? The two-sample *t*-test performs well if the data are normally distributed with equal variances. If there is doubt concerning these assumptions, a normal probability plot could be used to assess the degree of nonnormality, and a two-sample *F*-test of sample variances could be used to check the equality of variances. If these procedures indicate either nonnormality or inequality of variance, then the Wilcoxon rank sum test is appropriate.

15.2 EXERCISES

BASIC TECHNIQUES

15.1 Suppose you want to use the Wilcoxon rank sum test to detect a shift in distribution 1 to the right of distribution 2 based on samples of size $n_1 = 6$ and $n_2 = 8$.

a. Should you use T_1 or T_1^* as the test statistic?

b. What is the rejection region for the test if $\alpha = .05$?

c. What is the rejection region for the test if $\alpha = .01$?

15.2 Refer to Exercise 15.1. Suppose the alternative hypothesis is that distribution 1 is shifted either to the left or to the right of distribution 2.

a. Should you use T_1 or T_1^* as the test statistic?

b. What is the rejection region for the test if $\alpha = .05$?

c. What is the rejection region for the test if $\alpha = .01$?

15.3 Observations from two random and independent samples, drawn from populations 1 and 2, are given here. Use the Wilcoxon rank sum test to determine whether population 1 is shifted to the left of population 2.

Sample 1	1	3	2	3	5
Sample 2	4	7	6	8	6

a. State the null and alternative hypotheses to be tested.

b. Rank the combined sample from smallest to largest. Calculate T_1 and T_1^*.

c. What is the rejection region for $\alpha = .05$?

d. Do the data provide sufficient evidence to indicate that population 1 is shifted to the left of population 2?

15.4 Independent random samples of size $n_1 = 20$ and $n_2 = 25$ are drawn from nonnormal populations 1 and 2. The combined sample is ranked and $T_1 = 252$. Use the large-sample approximation to the Wilcoxon rank sum test to determine whether there is a difference in the two population distributions. Calculate the *p*-value for the test.

15.5 Suppose you wish to detect a shift in distribution 1 to the right of distribution 2 based on sample sizes $n_1 = 12$ and $n_2 = 14$. If $T_1 = 193$, what do you conclude? Use $\alpha = .05$.

APPLICATIONS

15.6 Alzheimer's Disease In some tests of healthy, elderly men, a new drug has restored their memory almost to that of young people. It will soon be tested on patients with Alzheimer's disease, the fatal brain disorder that destroys the mind. According to Dr. Gary Lynch of the University of California, Irvine, the drug, called ampakine CX-516, accelerates signals between brain cells and appears to significantly sharpen memory.[2] In a preliminary test on students in their early 20s and on men aged 65–70, the results were particularly striking. After being given mild doses of this drug, the 65–70-year-old men scored nearly as high as the young people. The accompanying data are the numbers of nonsense syllables recalled after 5 minutes for 10 men in their 20s and 10 men aged 65–70. Use the Wilcoxon rank sum test to determine whether the distributions for the number of nonsense syllables recalled are the same for these two groups.

20s	3	6	4	8	7	1	1	2	7	8
65–70s	1	0	4	1	2	5	0	2	2	3

15.7 Alzheimer's, continued Refer to Exercise 15.6. Suppose that two more groups of 10 men each are tested on the number of nonsense syllables they can remember after 5 minutes. However, this time the 65–70-year-olds are given a mild dose of ampakine CX-516. Do the data provide sufficient evidence to conclude that this drug improves memory in men aged 65–70 compared with that of 20-year-olds? Use an appropriate level of α.

20s	11	7	6	8	6	9	2	10	3	6
65–70s	1	9	6	8	7	8	5	7	10	3

15.8 Dissolved Oxygen Content The observations in the table are dissolved oxygen contents in water. The higher the dissolved oxygen content, the greater the ability of a river, lake, or stream to support aquatic life. In this experiment, a pollution-control inspector suspected that a river community was releasing semitreated sewage into a river. To check this theory, five randomly selected specimens of river water were selected at a location above the town and another five below. These are the dissolved oxygen readings (in parts per million):

Above Town	4.8	5.2	5.0	4.9	5.1
Below Town	5.0	4.7	4.9	4.8	4.9

a. Use a one-tailed Wilcoxon rank sum test with $\alpha = .05$ to confirm or refute the theory.

b. Use a Student's t-test (with $\alpha = .05$) to analyze the data. Compare the conclusion reached in part a.

15.9 Eye Movement In an investigation EX1509 of the visual scanning behavior of deaf children, measurements of eye movement were taken on nine deaf and nine hearing children. The table gives the eye-movement rates and their ranks (in parentheses). Does it appear that the distributions of eye-movement rates for deaf children and hearing children differ?

Deaf Children	Hearing Children
2.75 (15)	.89 (1)
2.14 (11)	1.43 (7)
3.23 (18)	1.06 (4)
2.07 (10)	1.01 (3)
2.49 (14)	.94 (2)
2.18 (12)	1.79 (8)
3.16 (17)	1.12 (5.5)
2.93 (16)	2.01 (9)
2.20 (13)	1.12 (5.5)
Rank Sum 126	45

15.10 Comparing NFL Quarterbacks How EX1510 does Aaron Rodgers, quarterback for the 2011 Super Bowl winners, the Minnesota Vikings, compare to Drew Brees, quarterback for the 2010 Super Bowl winners, the New Orleans Saints? The table below shows the number of completed passes for each athlete during the 2010 NFL football season:[3]

Aaron Rodgers			Drew Brees		
19	21	7	27	37	25
19	15	25	28	34	29
34	27	19	30	27	35
12	22		33	29	22
27	26		24	23	
18	21		21	24	

Use the Wilcoxon rank sum test to analyze the data and test to see whether the population distributions for the number of completed passes differ for the two quarterbacks. Use $\alpha = .05$.

15.11 Weights of Turtles The weights of EX1511 turtles caught in two different lakes were measured to compare the effects of the two lake environments on turtle growth. All the turtles were the same age and were tagged before being released into the lakes. The weights for $n_1 = 10$ tagged turtles caught in lake 1 and $n_2 = 8$ caught in lake 2 are listed here:

Lake	Weight (oz)									
1	14.1	15.2	13.9	14.5	14.7	13.8	14.0	16.1	12.7	15.3
2	12.2	13.0	14.1	13.6	12.4	11.9	12.5	13.8		

Do the data provide sufficient evidence to indicate a difference in the distributions of weights for the tagged turtles exposed to the two lake environments? Use the Wilcoxon rank sum test with $\alpha = .05$ to answer the question.

15.12 Chemotherapy Cancer treatment by EX1512 means of chemicals—chemotherapy—kills both cancerous and normal cells. In some instances, the toxicity of the cancer drug—that is, its effect on normal cells—can be reduced by the simultaneous injection of a second drug. A study was conducted to determine whether a particular drug injection reduced the harmful effects of a chemotherapy treatment on the survival time for rats. Two randomly selected groups of 12 rats were used in an experiment in which both groups, call them A and B, received the toxic drug in a dose large enough to cause death, but in addition, group B received the antitoxin, which was to reduce the toxic effect of the chemotherapy on normal cells. The test was terminated at the end of 20 days, or 480 hours. The survival times for the two groups of rats, to the nearest 4 hours, are shown in the table. Do the data provide sufficient evidence to indicate that rats receiving the antitoxin tend to survive longer after chemotherapy than those not receiving the antitoxin? Use the Wilcoxon rank sum test with $\alpha = .05$.

Chemotherapy Only A	Chemotherapy Plus Drug B
84	140
128	184
168	368
92	96
184	480
92	188
76	480
104	244
72	440
180	380
144	480
120	196

THE SIGN TEST FOR A PAIRED EXPERIMENT

The sign test is a fairly simple procedure that can be used to compare two populations when the samples consist of paired observations. This type of experimental design is called the **paired-difference** or **matched pairs** design, which you used to compare the average wear for two types of tires in Section 10.5. In general, for each pair, you measure whether the first response—say, A—exceeds the second response—say, B. The test statistic is x, the number of times that A exceeds B in the n pairs of observations.

When the two population distributions are identical, the probability that A exceeds B equals $p = .5$, and x, the number of times that A exceeds B, has a *binomial* distribution. Only pairs without ties are included in the test. Hence, you can test the hypothesis of identical population distributions by testing $H_0 : p = .5$ versus either a one- or two-tailed alternative. Critical values for the rejection region or exact p-values can be found using the cumulative binomial tables in Appendix I.

THE SIGN TEST FOR COMPARING TWO POPULATIONS

1. Null hypothesis: H_0 : The two population distributions are identical and $P(\text{A exceeds B}) = p = .5$

2. Alternative hypothesis:
   a. H_a : The population distributions are not identical and $p \neq .5$
   b. H_a : The population of A measurements is shifted to the right of the population of B measurements and $p > .5$
   c. H_a : The population of A measurements is shifted to the left of the population of B measurements and $p < .5$

3. Test statistic: For n, the number of pairs with no ties, use x, the number of times that $(\text{A} - \text{B})$ is positive.

4. Rejection region:
   a. For the two-tailed test $H_a : p \neq .5$, reject H_0 if $x \leq x_L$ or $x \geq x_U$, where $P(x \leq x_L) \leq \alpha/2$ and $P(x \geq x_U) \leq \alpha/2$ for x having a binomial distribution with $p = .5$.
   b. For $H_a : p > .5$, reject H_0 if $x \geq x_U$ with $P(x \geq x_U) \leq \alpha$.
   c. For $H_a : p < .5$, reject H_0 if $x \leq x_L$ with $P(x \leq x_L) \leq \alpha$.

   Or calculate the p-value and reject H_0 if the p-value $< \alpha$.

One problem that may occur when you are conducting a sign test is that the measurements associated with one or more pairs may be equal and therefore result in **tied observations.** When this happens, delete the tied pairs and reduce n, the total number of pairs. The following example will help you understand how the sign test is constructed and used.

EXAMPLE 15.3

The numbers of defective electrical fuses produced by two production lines, A and B, were recorded daily for a period of 10 days, with the results shown in Table 15.6. The response variable, the number of defective fuses, has an exact binomial distribution with a large number of fuses produced per day. Although this variable will have an approximately normal distribution, the plant supervisor would prefer a quick and easy statistical test to determine whether one production line tends to produce more defectives than the other. Use the sign test to test the appropriate hypothesis.

TABLE 15.6 ● **Defective Fuses from Two Production Lines**

Day	Line A	Line B	Sign of Difference
1	170	201	−
2	164	179	−
3	140	159	−
4	184	195	−
5	174	177	−
6	142	170	−
7	191	183	+
8	169	179	−
9	161	170	−
10	200	212	−

Solution   For this *paired-difference* experiment, x is the number of times that the observation for line A exceeds that for line B in a given day. If there is no difference in the distributions of defectives for the two production lines, then p, the proportion of days on which A exceeds B, is .5, which is the hypothesized value in a test of the binomial parameter p. Very small or very large values of x, the number of times that A exceeds B, are contrary to the null hypothesis.

Since $n = 10$ and the hypothesized value of p is .5, Table 1 of Appendix I can be used to find the exact p-value for the test of

$$H_0 : p = .5 \text{ versus } H_a : p \neq .5$$

The observed value of the test statistic—which is the number of "plus" signs in the table—is $x = 1$, and the p-value is calculated as

$$p\text{-value} = 2P(x \leq 1) = 2(.011) = .022$$

The fairly small p-value $= .022$ allows you to reject H_0 at the 5% level. There is significant evidence to indicate that the number of defective fuses is not the same for the two production lines; in fact, line B produces more defectives than line A. In this example, the sign test is an easy-to-calculate rough tool for detecting faulty production lines and works perfectly well to detect a significant difference using only a minimum amount of information.

Normal Approximation for the Sign Test

When the number of pairs n is large, the critical values for rejection of H_0 and the approximate p-values can be found using a normal approximation to the distribution of x, which was discussed in Section 6.4. Because the binomial distribution is perfectly symmetric when $p = .5$, this approximation works very well, even for n as small as 10.

For $n \geq 25$, you can conduct the sign test by using the z statistic,

$$z = \frac{x - np}{\sqrt{npq}} = \frac{x - .5n}{.5\sqrt{n}}$$

as the test statistic. In using z, you are testing the null hypothesis $p = .5$ versus the alternative $p \neq .5$ for a two-tailed test or versus the alternative $p > .5$ (or $p < .5$) for a one-tailed test. The tests use the familiar rejection regions of Chapter 9.

SIGN TEST FOR LARGE SAMPLES: $n \geq 25$

1. Null hypothesis: $H_0 : p = .5$ (one treatment is not preferred to a second treatment)

2. Alternative hypothesis: $H_a : p \neq .5$, for a two-tailed test (NOTE: We use the two-tailed test as an example. Many analyses might require a one-tailed test.)

3. Test statistic: $z = \dfrac{x - .5n}{.5\sqrt{n}}$

4. Rejection region: Reject H_0 if $z \geq z_{\alpha/2}$ or $z \leq -z_{\alpha/2}$, where $z_{\alpha/2}$ is the z-value from Table 3 in Appendix I corresponding to an area of $\alpha/2$ in the upper tail of the normal distribution.

EXAMPLE 15.4

A production superintendent claims that there is no difference between the employee accident rates for the day versus the evening shifts in a large manufacturing plant. The number of accidents per day is recorded for both the day and evening shifts for $n = 100$ days. It is found that the number of accidents per day for the evening shift x_E exceeded the corresponding number of accidents on the day shift x_D on 63 of the 100 days. Do these results provide sufficient evidence to indicate that more accidents tend to occur on one shift than on the other or, equivalently, that $P(x_E > x_D) \neq 1/2$?

Solution   This study is a paired-difference experiment, with $n = 100$ pairs of observations corresponding to the 100 days. To test the null hypothesis that the two distributions of accidents are identical, you can use the test statistic

$$z = \frac{x - .5n}{.5\sqrt{n}}$$

where x is the number of days in which the number of accidents on the evening shift exceeded the number of accidents on the day shift. Then for $\alpha = .05$, you can reject the null hypothesis if $z \geq 1.96$ or $z \leq -1.96$. Substituting into the formula for z, you get

$$z = \frac{x - .5n}{.5\sqrt{n}} = \frac{63 - (.5)(100)}{.5\sqrt{100}} = \frac{13}{5} = 2.60$$

Since the calculated value of z exceeds $z_{\alpha/2} = 1.96$, you can reject the null hypothesis. The data provide sufficient evidence to indicate a difference in the accident rate distributions for the day versus evening shifts.

When should the sign test be used in preference to the paired t-test? When only the *direction* of the difference in the measurement is given, *only* the sign test can be used. On the other hand, when the data are quantitative and satisfy the normality and constant variance assumptions, the paired t-test should be used. A normal probability plot can be used to assess normality, while a plot of the residuals $(d_i - \bar{d})$ can reveal large deviations that might indicate a variance that varies from pair to pair. When there are doubts about the validity of the assumptions, statisticians often recommend that both tests be performed. If both tests reach the same conclusions, then the parametric test results can be considered to be valid.

EXERCISES

BASIC TECHNIQUES

15.13 Suppose you wish to use the sign test to test $H_a : p > .5$ for a paired-difference experiment with $n = 25$ pairs.

a. State the practical situation that dictates the alternative hypothesis given.

b. Use Table 1 in Appendix I to find values of α ($\alpha < .15$) available for the test.

15.14 Repeat the instructions of Exercise 15.13 for $H_a : p \neq .5$.

15.15 Repeat the instructions of Exercises 15.13 and 15.14 for $n = 10, 15,$ and 20.

15.16 A paired-difference experiment was conducted to compare two populations. The data are shown in the table. Use a sign test to determine whether the population distributions are different.

Population	Pairs						
	1	2	3	4	5	6	7
1	8.9	8.1	9.3	7.7	10.4	8.3	7.4
2	8.8	7.4	9.0	7.8	9.9	8.1	6.9

a. State the null and alternative hypotheses for the test.

b. Determine an appropriate rejection region with $\alpha \approx .01$.

c. Calculate the observed value of the test statistic.

d. Do the data present sufficient evidence to indicate that populations 1 and 2 are different?

APPLICATIONS

15.17 Property Values In Exercise 10.46, you compared the property evaluations of two tax assessors, A and B. Their assessments for eight properties are shown in the table:

Property	Assessor A	Assessor B
1	276.3	275.1
2	288.4	286.8
3	280.2	277.3
4	294.7	290.6
5	268.7	269.1
6	282.8	281.0
7	276.1	275.3
8	279.0	279.1

a. Use the sign test to determine whether the data present sufficient evidence to indicate that one of the assessors tends to be consistently more conservative than the other; that is, $P(x_A > x_B) \neq 1/2$. Test by using a value of α near .05. Find the p-value for the test and interpret its value.

b. Exercise 10.46 uses the t statistic to test the null hypothesis that there is no difference in the mean property assessments between assessors A and B. Check the answer for Exercise 10.46 and compare it with your answer to part a. Do the test results agree? Explain why the answers are (or are not) consistent.

15.18 Gourmet Cooking Two gourmets, A and B, rated 22 meals on a scale of 1–10. The data are shown in the table. Do the data provide sufficient evidence to indicate that one of the gourmets tends to give higher ratings than the other? Test by using the sign test with a value of α near .05.

Meal	A	B	Meal	A	B
1	6	8	12	8	5
2	4	5	13	4	2
3	7	4	14	3	3
4	8	7	15	6	8
5	2	3	16	9	10
6	7	4	17	9	8
7	9	9	18	4	6
8	7	8	19	4	3
9	2	5	20	5	4
10	4	3	21	3	2
11	6	9	22	5	3

a. Use the binomial tables in Appendix I to find the exact rejection region for the test.

b. Use the large-sample z statistic. (NOTE: Although the large-sample approximation is suggested for $n \geq 25$, it works fairly well for values of n as small as 15.)

c. Compare the results of parts a and b.

15.19 Lead Levels in Blood A study reported in the *American Journal of Public Health (Science News)*—the first to follow blood lead levels in law-abiding handgun hobbyists using indoor firing ranges—documents a significant risk of lead poisoning.[4] Lead exposure measurements were made on 17 members of a law enforcement trainee class before, during, and after a 3-month period of firearm instruction at a state-owned indoor firing range. No trainee had elevated blood lead levels before the training, but 15 of the 17 ended their training with blood lead levels deemed "elevated" by the Occupational Safety and Health Administration (OSHA). If the use of an indoor firing range causes no increase in blood lead levels, then p, the probability that a person's blood lead level increases, is less than or equal to .5. If, however, use of the indoor firing range causes an increase in a person's blood lead levels, then $p > .5$. Use the sign test to determine whether using an indoor firing range has the effect of increasing a person's blood lead level with $\alpha = .05$. (HINT: The normal approximation to binomial probabilities is fairly accurate for $n = 17$.)

 15.20 Recovery Rates Clinical data concerning the effectiveness of two drugs in treating a particular disease were collected from 10 hospitals. The numbers of patients treated with the drugs varied from one hospital to another. You want to know whether the data present sufficient evidence to indicate a higher recovery rate for one of the two drugs.

a. Test using the sign test. Choose your rejection region so that α is near .05.

b. Why might it be inappropriate to use the Student's t-test in analyzing the data?

	Drug A				Drug B		
Hospital	Number in Group	Number Recovered	Percentage Recovered	Hospital	Number in Group	Number Recovered	Percentage Recovered
1	84	63	75.0	1	96	82	85.4
2	63	44	69.8	2	83	69	83.1
3	56	48	85.7	3	91	73	80.2
4	77	57	74.0	4	47	35	74.5
5	29	20	69.0	5	60	42	70.0
6	48	40	83.3	6	27	22	81.5
7	61	42	68.9	7	69	52	75.4
8	45	35	77.8	8	72	57	79.2
9	79	57	72.2	9	89	76	85.4
10	62	48	77.4	10	46	37	80.4

A COMPARISON OF STATISTICAL TESTS

The experiment in Example 15.3 is designed as a paired-difference experiment. If the assumptions of normality and constant variance, σ_d^2, for the differences were met, would the sign test detect a shift in location for the two populations as efficiently as the paired t-test? Probably not, since the t-test uses much more information than the sign test. It uses not only the sign of the difference but also the actual values of the differences. In this case, we would say that the sign test is not as *efficient* as the paired t-test. However, the sign test might be more efficient if the usual assumptions were not met.

When two different statistical tests can *both* be used to test a hypothesis based on the same data, it is natural to ask, Which is better? One way to answer this question would be to hold the sample size n and α constant for both procedures and compare β, the probability of a Type II error. Statisticians, however, prefer to examine the **power** of a test.

Definition Power $= 1 - \beta = P(\text{reject } H_0 \text{ when } H_a \text{ is true})$

Since β is the probability of failing to reject the null hypothesis when it is false, the **power** of the test is the probability of rejecting the null hypothesis when it is false and some specified alternative is true. It is the probability that the test will do what it was designed to do—that is, detect a departure from the null hypothesis when a departure exists.

Probably the most common method of comparing two test procedures is in terms of the relative efficiency of a pair of tests. **Relative efficiency** is the ratio of the sample sizes for the two test procedures required to achieve the same α and β for a given alternative to the null hypothesis.

In some situations, you may not be too concerned whether you are using the most powerful test. For example, you might choose to use the sign test over a more powerful competitor because of its ease of application. Thus, you might view tests as microscopes that are used to detect departures from an hypothesized theory. One need not know the exact power of a microscope to use it in a biological investigation, and the same applies to statistical tests. If the test procedure detects a departure from the null hypothesis, you are delighted. If not, you can reanalyze the data by using a more powerful microscope (test), or you can increase the power of the microscope (test) by increasing the sample size.

THE WILCOXON SIGNED-RANK TEST FOR A PAIRED EXPERIMENT

A signed-rank test proposed by Frank Wilcoxon can be used to analyze the paired-difference experiment of Section 10.5 by considering the paired differences of two treatments, 1 and 2. Under the null hypothesis of no differences in the distributions for 1 and 2, you would expect (on the average) half of the differences in pairs to be negative and half to be positive; that is, the expected number of negative differences between pairs would be $n/2$ (where n is the number of pairs). Furthermore, it follows that positive and negative differences of equal absolute magnitude should occur with equal probability. If one were to order the differences according to their absolute values and rank them from smallest to largest, the expected rank sums for the negative and positive differences would be equal. Sizable differences in the sums of the ranks assigned to the positive and negative differences would provide evidence to indicate a shift in location between the distributions of responses for the two treatments, 1 and 2.

If distribution 1 is shifted to the right of distribution 2, then more of the differences are expected to be positive, and this results in a small number of negative differences. Therefore, to detect this one-sided alternative, use the rank sum T^-—the sum of the ranks of the negative differences—and reject the null hypothesis for significantly small values of T^-. Along these same lines, if distribution 1 is shifted to the left of distribution 2, then more of the differences are expected to be negative, and the number of positive differences is small. Hence, to detect this one-sided alternative, use T^+—the sum of the ranks of the positive differences—and reject the null hypothesis if T^+ is significantly small.

CALCULATING THE TEST STATISTIC FOR THE WILCOXON SIGNED-RANK TEST

1. Calculate the differences $(x_1 - x_2)$ for each of the n pairs. Differences equal to 0 are eliminated, and the number of pairs, n, is reduced accordingly.

2. Rank the **absolute values** of the differences by assigning 1 to the smallest, 2 to the second smallest, and so on. Tied observations are assigned the average of the ranks that would have been assigned with no ties.

3. Calculate the **rank sum** for the **negative** differences and label this value T^-. Similarly, calculate T^+, the **rank sum** for the **positive** differences.

For a **two-tailed test,** use the **smaller of these two quantities T as a test statistic** to test the null hypothesis that the two population relative frequency histograms are identical. The smaller the value of T, the greater is the weight of evidence favoring rejection of the null hypothesis. **Therefore, you will reject the null hypothesis if T is less than or equal to some value—say, T_0.**

To detect the **one-sided alternative,** that **distribution 1 is shifted to the right of distribution 2, use the rank sum T^-** of the negative differences and reject the null hypothesis for small values of T^-—say, $T^- \le T_0$. If you wish to detect a **shift of distribution 2 to the right of distribution 1, use the rank sum T^+** of the positive differences as a test statistic and reject the null hypothesis for small values of T^+—say, $T^+ \le T_0$.

The probability that T is less than or equal to some value T_0 has been calculated for a combination of sample sizes and values of T_0. These probabilities, given in Table 8 in Appendix I, can be used to find the rejection region for the T test.

An abbreviated version of Table 8 is shown in Table 15.7. Across the top of the table you see the number of differences (the number of pairs) n. Values of α for a one-tailed test appear in the first column of the table. The second column gives values of α for a two-tailed test. Table entries are the critical values of T. You will recall that the critical value of a test statistic is the value that locates the boundary of the rejection region.

For example, suppose you have $n = 7$ pairs and you are conducting a two-tailed test of the null hypothesis that the two population relative frequency distributions are identical. Checking the $n = 7$ column of Table 15.7 and using the second row (corresponding to $\alpha = .05$ for a two-tailed test), you see the entry 2 (shaded). This value is T_0, the critical value of T. As noted earlier, the smaller the value of T, the greater is the evidence to reject the null hypothesis. Therefore, you will reject the null hypothesis for all values of T less than or equal to 2. The rejection region for the Wilcoxon signed-rank test for a paired experiment is always of the form: Reject H_0 if $T \le T_0$, where T_0 is the critical value of T. The rejection region is shown symbolically in Figure 15.2.

TABLE 15.7 ●

An Abbreviated Version of Table 8 in Appendix I; Critical Values of T

One-Sided	Two-Sided	$n = 5$	$n = 6$	$n = 7$	$n = 8$	$n = 9$	$n = 10$	$n = 11$
$\alpha = .050$	$\alpha = .10$	1	2	4	6	8	11	14
$\alpha = .025$	$\alpha = .05$		1	2	4	6	8	11
$\alpha = .010$	$\alpha = .02$			0	2	3	5	7
$\alpha = .005$	$\alpha = .01$				0	2	3	5

One-Sided	Two-Sided	$n = 12$	$n = 13$	$n = 14$	$n = 15$	$n = 16$	$n = 17$
$\alpha = .050$	$\alpha = .10$	17	21	26	30	36	41
$\alpha = .025$	$\alpha = .05$	14	17	21	25	30	35
$\alpha = .010$	$\alpha = .02$	10	13	16	20	24	28
$\alpha = .005$	$\alpha = .01$	7	10	13	16	19	23

FIGURE 15.2 ●

Rejection region for the Wilcoxon signed-rank test for a paired experiment (reject H_0 if $T \le T_0$)

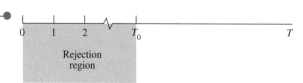

WILCOXON SIGNED-RANK TEST FOR A PAIRED EXPERIMENT

1. Null hypothesis: H_0 : The two population relative frequency distributions are identical

2. Alternative hypothesis: H_a : The two population relative frequency distributions differ in location (a two-tailed test). Or H_a : The population 1 relative frequency distribution is shifted to the right of the relative frequency distribution for population 2 (a one-tailed test).

3. Test statistic

    a. For a two-tailed test, use T, the smaller of the rank sum for positive and the rank sum for negative differences.

    b. For a one-tailed test (to detect the alternative hypothesis described above), use the rank sum T^- of the negative differences.

4. Rejection region

    a. For a two-tailed test, reject H_0 if $T \leq T_0$, where T_0 is the critical value given in Table 8 in Appendix I.

    b. For a one-tailed test (to detect the alternative hypothesis described above), use the rank sum T^- of the negative differences. Reject H_0 if $T^- \leq T_0$.[†]

    $$\left[\text{NOTE: It can be shown that } T^+ + T^- = \frac{n(n + 1)}{2}. \right]$$

EXAMPLE 15.5

An experiment was conducted to compare the densities of cakes prepared from two different cake mixes, A and B. Six cake pans received batter A, and six received batter B. Expecting a variation in oven temperature, the experimenter placed an A and a B cake side by side at six different locations in the oven. Test the hypothesis of no difference in the population distributions of cake densities for two different cake batters.

Solution   The data (density in ounces per cubic inch) and differences in density for six pairs of cakes are given in Table 15.8. The box plot of the differences in Figure 15.3 shows fairly strong skewing and a very large difference in the right tail, which indicates that the data may not satisfy the normality assumption. The sample of differences is too small to make valid decisions about normality and constant variance. In this situation, Wilcoxon's signed-rank test may be the prudent test to use.

As with other nonparametric tests, the null hypothesis to be tested is that the two population frequency distributions of cake densities are identical. The alternative hypothesis, which implies a two-tailed test, is that the distributions are different. Because the amount of data is small, you can conduct the test using $\alpha = .10$. From Table 8 in Appendix I, the critical value of T for a two-tailed test, $\alpha = .10$, is $T_0 = 2$. Hence, you can reject H_0 if $T \leq 2$.

[†]To detect a shift of distribution 2 to the right of distribution 1, use the rank sum T^+ of the positive differences as the test statistic and reject H_0 if $T^+ \leq T_0$.

TABLE 15.8 ● **Densities of Six Pairs of Cakes**

x_A	x_B	Difference $(x_A - x_B)$	Rank
.135	.129	.006	2
.102	.120	−.018	5
.098	.112	−.014	4
.141	.152	−.011	3
.131	.135	−.004	1
.144	.163	−.019	6

FIGURE 15.3

Box plot of differences for Example 15.5

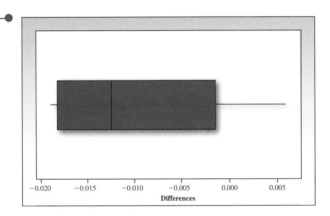

The differences $(x_1 - x_2)$ are calculated and ranked according to their absolute values in Table 15.8. The sum of the positive ranks is $T^+ = 2$, and the sum of the negative ranks is $T^- = 19$. The test statistic is the smaller of these two rank sums, or $T = 2$. Since $T = 2$ falls in the rejection region, you can reject H_0 and conclude that the two population frequency distributions of cake densities differ.

A *MINITAB* printout of the Wilcoxon signed-rank test for these data is given in Figure 15.4. You will find instructions for generating this output in the "Technology Today" section at the end of this chapter. You can see that the value of the test statistic agrees with the other calculations, and the *p*-value indicates that you can reject H_0 at the 10% level of significance.

FIGURE 15.4

MINITAB printout for Example 15.5

```
Wilcoxon Signed Rank Test: Difference

Test of median = 0.000000 versus median not = 0.000000

                    N for  Wilcoxon            Estimated
              N     Test   Statistic     P       Median
Difference    6      6         2.0    0.093    -0.01100
```

Normal Approximation for the Wilcoxon Signed-Rank Test

Although Table 8 in Appendix I has critical values for *n* as large as 50, T^+, like the Wilcoxon signed-rank test, will be approximately normally distributed when the null hypothesis is true and n is large—say, 25 or more. This enables you to construct a large-sample z-test, where

$$E(T) = \frac{n(n + 1)}{4}$$

$$\sigma_T^2 = \frac{n(n + 1)(2n + 1)}{24}$$

Then the z statistic

$$z = \frac{T^+ - E(T)}{\sigma_T} = \frac{T^+ - \dfrac{n(n + 1)}{4}}{\sqrt{\dfrac{n(n + 1)(2n + 1)}{24}}}$$

can be used as a test statistic. Thus, for a two-tailed test and $\alpha = .05$, you can reject the hypothesis of identical population distributions when $|z| \geq 1.96$.

A LARGE-SAMPLE WILCOXON SIGNED-RANK TEST FOR A PAIRED EXPERIMENT: $n \geq 25$

1. Null hypothesis: H_0 : The population relative frequency distributions 1 and 2 are identical.

2. Alternative hypothesis: H_a : The two population relative frequency distributions differ in location (a two-tailed test). Or H_a : The population 1 relative frequency distribution is shifted to the right (or left) of the relative frequency distribution for population 2 (a one-tailed test).

3. Test statistic: $z = \dfrac{T^+ - [n(n + 1)/4]}{\sqrt{[n(n + 1)(2n + 1)]/24}}$

4. Rejection region: Reject H_0 if $z > z_{\alpha/2}$ or $z < -z_{\alpha/2}$ for a two-tailed test. For a one-tailed test, place all of α in one tail of the z distribution. To detect a shift in distribution 1 to the right of distribution 2, reject H_0 when $z > z_\alpha$. To detect a shift in the opposite direction, reject H_0 if $z < -z_\alpha$.

Tabulated values of z are given in Table 3 in Appendix I.

15.5 EXERCISES

BASIC TECHNIQUES

15.21 Suppose you wish to detect a difference in the locations of two population distributions based on a paired-difference experiment consisting of $n = 30$ pairs.

a. Give the null and alternative hypotheses for the Wilcoxon signed-rank test.

b. Give the test statistic.

c. Give the rejection region for the test for $\alpha = .05$.

d. If $T^+ = 249$, what are your conclusions? [NOTE: $T^+ + T^- = n(n + 1)/2$.]

15.22 Refer to Exercise 15.21. Suppose you wish to detect only a shift in distribution 1 to the right of distribution 2.

a. Give the null and alternative hypotheses for the Wilcoxon signed-rank test.

b. Give the test statistic.

c. Give the rejection region for the test for $\alpha = .05$.

d. If $T^+ = 249$, what are your conclusions? [NOTE: $T^+ + T^- = n(n + 1)/2$.]

15.23 Refer to Exercise 15.21. Conduct the test using the large-sample z-test. Compare your results with the nonparametric test results in Exercise 15.22, part d.

15.24 Refer to Exercise 15.22. Conduct the test using the large-sample z-test. Compare your results with the nonparametric test results in Exercise 15.21, part d.

15.25 Refer to Exercise 15.16 and data set EX1516. The data in this table are from a paired-difference experiment with $n = 7$ pairs of observations.

Population	Pairs						
	1	2	3	4	5	6	7
1	8.9	8.1	9.3	7.7	10.4	8.3	7.4
2	8.8	7.4	9.0	7.8	9.9	8.1	6.9

a. Use Wilcoxon's signed-rank test to determine whether there is a significant difference between the two populations.

b. Compare the results of part a with the result you got in Exercise 15.16. Are they the same? Explain.

APPLICATIONS

15.26 Property Values II In Exercise 15.17, you used the sign test to determine whether the data provided sufficient evidence to indicate a difference in the distributions of property assessments for assessors A and B.

a. Use the Wilcoxon signed-rank test for a paired experiment to test the null hypothesis that there is no difference in the distributions of property assessments between assessors A and B. Test by using a value of α near .05.

b. Compare the conclusion of the test in part a with the conclusions derived from the t-test in Exercise 10.46 and the sign test in Exercise 15.17. Explain why these test conclusions are (or are not) consistent.

Data set **15.27 Machine Breakdowns** The number of
EX1527 machine breakdowns per month was recorded for 9 months on two identical machines, A and B, used to make wire rope:

Month	A	B
1	3	7
2	14	12
3	7	9
4	10	15
5	9	12
6	6	6
7	13	12
8	6	5
9	7	13

a. Do the data provide sufficient evidence to indicate a difference in the monthly breakdown rates for the two machines? Test by using a value of α near .05.

b. Can you think of a reason the breakdown rates for the two machines might vary from month to month?

15.28 Gourmet Cooking II Refer to the comparison of gourmet meal ratings in Exercise 15.18, and use the Wilcoxon signed-rank test to determine whether the data provide sufficient evidence to indicate a difference in the ratings of the two gourmets. Test by using a value of α near .05. Compare the results of this test with the results of the sign test in Exercise 15.18. Are the test conclusions consistent?

Data set **15.29 Traffic Control** Two methods for con-
EX1529 trolling traffic, A and B, were used at each of n $= 12$ intersections for a period of 1 week, and the numbers of accidents that occurred during this time period were recorded. The order of use (which method would be employed for the first week) was selected in a random manner. You want to know whether the data provide sufficient evidence to indicate a difference in the distributions of accident rates for traffic control methods A and B.

Intersection	Method A	Method B	Intersection	Method A	Method B
1	5	4	7	2	3
2	6	4	8	4	1
3	8	9	9	7	9
4	3	2	10	5	2
5	6	3	11	6	5
6	1	0	12	1	1

a. Analyze using a sign test.

b. Analyze using the Wilcoxon signed-rank test for a paired experiment.

Data set **15.30 Jigsaw Puzzles** Eight people were
EX1530 asked to perform a simple puzzle-assembly task under normal conditions and under stressful conditions. During the stressful time, a stimulus was delivered to subjects 3 minutes after the start of the experiment and every 30 seconds thereafter until the task was completed. Blood pressure readings were taken under both conditions. The data in the table are the highest readings during the experiment. Do the data present sufficient evidence to indicate higher blood pressure readings under stressful conditions? Analyze the data using the Wilcoxon signed-rank test for a paired experiment.

Subject	Normal	Stressful
1	126	130
2	117	118
3	115	125
4	118	120
5	118	121
6	128	125
7	125	130
8	120	120

Data set **15.31 Images and Word Recall** A psychol-
EX1531 ogy class performed an experiment to determine
whether a recall score in which instructions to form
images of 25 words were given differs from an initial
recall score for which no imagery instructions were
given. Twenty students participated in the experiment
with the results listed in the table.

Student	With Imagery	Without Imagery	Student	With Imagery	Without Imagery
1	20	5	11	17	8
2	24	9	12	20	16
3	20	5	13	20	10
4	18	9	14	16	12
5	22	6	15	24	7
6	19	11	16	22	9
7	20	8	17	25	21
8	19	11	18	21	14
9	17	7	19	19	12
10	21	9	20	23	13

a. What three testing procedures can be used to test
for differences in the distribution of recall scores
with and without imagery? What assumptions are
required for the parametric procedure? Do these
data satisfy these assumptions?

b. Use both the sign test and the Wilcoxon signed-
rank test to test for differences in the distributions
of recall scores under these two conditions.

c. Compare the results of the tests in part b. Are the
conclusions the same? If not, why not?

THE KRUSKAL–WALLIS *H*-TEST FOR COMPLETELY RANDOMIZED DESIGNS

15.6

Just as the Wilcoxon rank sum test is the nonparametric alternative to Student's *t*-test
for a comparison of population means, the Kruskal–Wallis *H*-test is the nonparametric alternative to the analysis of variance *F*-test for a completely randomized design.
It is used to detect differences in locations among more than two population distributions based on independent random sampling.

The procedure for conducting the Kruskal–Wallis *H*-test is similar to that used for
the Wilcoxon rank sum test. Suppose you are comparing *k* populations based on independent random samples n_1 from population 1, n_2 from population 2, . . . , n_k from
population *k*, where

$$n_1 + n_2 + \cdots + n_k = n$$

The first step is to rank all *n* observations from the smallest (rank 1) to the largest
(rank *n*). Tied observations are assigned a rank equal to the average of the ranks they
would have received if they had been nearly equal but not tied. You then calculate
the rank sums $T_1, T_2, \ldots, T_k$ for the *k* samples and calculate the test statistic

$$H = \frac{12}{n(n + 1)} \Sigma \frac{T_i^2}{n_i} - 3(n + 1)$$

which is proportional to $\Sigma\, n_i(\overline{T}_i - \overline{T})^2$, the sum of squared deviations of the rank means about the grand mean $\overline{T} = n(n + 1)/2n = (n + 1)/2$. The greater the differences in locations among the k population distributions, the larger is the value of the H statistic. Thus, you can reject the null hypothesis that the k population distributions are identical for large values of H.

How large is large? It can be shown (proof omitted) that when the sample sizes are moderate to large—say, each sample size is equal to five or larger—and when H_0 is true, the H statistic will have approximately a chi-square distribution with $(k - 1)$ degrees of freedom. Therefore, for a given value of α, you can reject H_0 when the H statistic exceeds χ_α^2 (see Figure 15.5).

FIGURE 15.5

Approximate distribution of the H statistic when H_0 is true

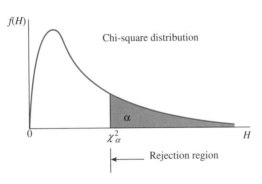

EXAMPLE 15.6

The data in Table 15.9 were collected using a completely randomized design. They are the achievement test scores for four different groups of students, each group taught by a different teaching technique. The objective of the experiment is to test the hypothesis of no difference in the population distributions of achievement test scores versus the alternative that they differ in location; that is, at least one of the distributions is shifted above the others. Conduct the test using the Kruskal–Wallis H-test with $\alpha = .05$.

TABLE 15.9

Test Scores (and Ranks) from Four Teaching Techniques

	1	2	3	4
	65 (3)	75 (9)	59 (1)	94 (23)
	87 (19)	69 (5.5)	78 (11)	89 (21)
	73 (8)	83 (17.5)	67 (4)	80 (14)
	79 (12.5)	81 (15.5)	62 (2)	88 (20)
	81 (15.5)	72 (7)	83 (17.5)	
	69 (5.5)	79 (12.5)	76 (10)	
		90 (22)		
Rank Sum	$T_1 = 63.5$	$T_2 = 89$	$T_3 = 45.5$	$T_4 = 78$

Solution Before you perform a nonparametric analysis on these data, you can use a one-way analysis of variance to provide the two plots in Figure 15.6. It appears that technique 4 has a smaller variance than the other three and that there is a marked deviation in the right tail of the normal probability plot. These deviations could be considered minor and either a parametric or nonparametric analysis could be used.

FIGURE 15.6

A normal probability plot
and a residual plot
following a one-way
analysis of variance for
Example 15.6

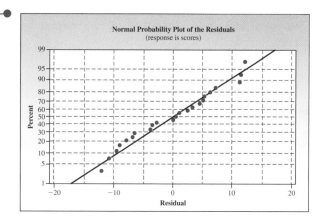

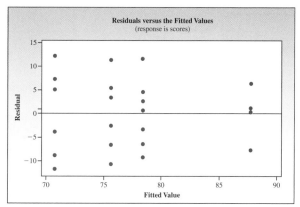

In the Kruskal–Wallis *H*-test procedure, the first step is to rank the $n = 23$ observations from the smallest (rank 1) to the largest (rank 23). These ranks are shown in parentheses in Table 15.9. Notice how the ties are handled. For example, two observations at 69 are tied for rank 5. Therefore, they are assigned the average 5.5 of the two ranks (5 and 6) that they would have occupied if they had been slightly different. The rank sums T_1, T_2, T_3, and T_4 for the four samples are shown in the bottom row of the table. Substituting rank sums and sample sizes into the formula for the *H* statistic, you get

$$H = \frac{12}{n(n + 1)} \Sigma \frac{T_i^2}{n_i} - 3(n + 1)$$

$$= \frac{12}{23(24)} \left[\frac{(63.5)^2}{6} + \frac{(89)^2}{7} + \frac{(45.5)^2}{6} + \frac{(78)^2}{4} \right] - 3(24)$$

$$= 79.775102 - 72 = 7.775102$$

The rejection region for the *H* statistic for $\alpha = .05$ includes values of $H \geq \chi^2_{.05}$, where $\chi^2_{.05}$ is based on $(k - 1) = (4 - 1) = 3$ *df*. The value of χ^2 given in Table 5 in Appendix I is $\chi^2_{.05} = 7.81473$. The observed value of the *H* statistic, $H = 7.775102$, does not fall into the rejection region for the test. Therefore, there is insufficient evidence to indicate differences in the distributions of achievement test scores for the four teaching techniques.

A *MINITAB* printout of the Kruskal–Wallis *H*-test for these data is given in Figure 15.7. Notice that the *p*-value, .051, is only slightly greater than the 5% level necessary to declare statistical significance.

FIGURE 15.7

MINITAB printout for the Kruskal–Wallis test for Example 15.6

Kruskal-Wallis Test: Scores versus Techniques

```
Kruskal-Wallis Test on Scores

Techniques    N     Median    Ave Rank         Z
1             6      76.00        10.6      -0.60
2             7      79.00        12.7       0.33
3             6      71.50         7.6      -1.86
4             4      88.50        19.5       2.43
Overall      23                  12.0

H = 7.78  DF = 3  P = 0.051
H = 7.79  DF = 3  P = 0.051 (adjusted for ties)

* NOTE * One or more small samples
```

EXAMPLE 15.7

Compare the results of the analysis of variance *F*-test and the Kruskal–Wallis *H*-test for testing for differences in the distributions of achievement test scores for the four teaching techniques in Example 15.6.

Solution   The *MINITAB* printout for a one-way analysis of variance for the data in Table 15.9 is given in Figure 15.8. The analysis of variance shows that the *F*-test for testing for differences among the means for the four techniques is significant at the .028 level. The Kruskal–Wallis *H*-test did not detect a shift in population distributions at the .05 level of significance. Although these conclusions seem to be far apart, the test results do not differ strongly. The *p*-value = .028 corresponding to $F = 3.77$, with $df_1 = 3$ and $df_2 = 19$, is slightly less than .05, in contrast to the *p*-value = .051 for $H = 7.78$, $df = 3$, which is slightly greater than .05. Someone viewing the *p*-values for the two tests would see little difference in the results of the *F*- and *H*-tests. However, if you adhere to the choice of $\alpha = .05$, you cannot reject H_0 using the *H*-test.

FIGURE 15.8

MINITAB printout for Example 15.7

One-way ANOVA: Scores versus Techniques

```
Source       DF       SS        MS       F       P
Techniques    3     712.6     237.5    3.77   0.028
Error        19    1196.6      63.0
Total        22    1909.2
```

THE KRUSKAL–WALLIS *H*-TEST FOR COMPARING MORE THAN TWO POPULATIONS: COMPLETELY RANDOMIZED DESIGN (INDEPENDENT RANDOM SAMPLES)

1. Null hypothesis: H_0 : The k population distributions are identical.
2. Alternative hypothesis: H_a : At least two of the k population distributions differ in location.
3. Test statistic: $H = \dfrac{12}{n(n+1)} \Sigma \dfrac{T_i^2}{n_i} - 3(n+1)$

where

n_i = Sample size for population i
T_i = Rank sum for population i
n = Total number of observations = $n_1 + n_2 + \cdots + n_k$

4. Rejection region for a given α: $H > \chi_\alpha^2$ with $(k - 1)$ df

Assumptions

- All sample sizes are greater than or equal to five.
- Ties take on the average of the ranks that they would have occupied if they had not been tied.

The Kruskal–Wallis H-test is a valuable alternative to a one-way analysis of variance when the normality and equality of variance assumptions are violated. Again, normal probability plots of residuals and plots of residuals per treatment group are helpful in determining whether these assumptions have been violated. Remember that a normal probability plot should appear as a straight line with a positive slope; residual plots per treatment groups should exhibit the same spread above and below the 0 line.

15.6 EXERCISES

BASIC TECHNIQUES

Data set EX1532 **15.32** Three treatments were compared using a completely randomized design. The data are shown in the table.

Treatment		
1	2	3
26	27	25
29	31	24
23	30	27
24	28	22
28	29	24
26	32	20
	30	21
	33	

Do the data provide sufficient evidence to indicate a difference in location for at least two of the population distributions? Test using the Kruskal–Wallis H statistic with $\alpha = .05$.

Data set EX1533 **15.33** Four treatments were compared using a completely randomized design. The data are shown here:

Treatment			
1	2	3	4
124	147	141	117
167	121	144	128
135	136	139	102
160	114	162	119
159	129	155	128
144	117	150	123
133	109		

Do the data provide sufficient evidence to indicate a difference in location for at least two of the population distributions? Test using the Kruskal–Wallis H statistic with $\alpha = .05$.

APPLICATIONS

15.34 Swampy Sites II Exercise 11.13 presents data (see data set EX1113) on the rates of growth of vegetation at four swampy underdeveloped sites. Six plants were randomly selected at each of the four sites to be used in the comparison. The data are the mean leaf length per plant (in centimeters) for a random sample of 10 leaves per plant.

Location	Mean Leaf Length (cm)					
1	5.7	6.3	6.1	6.0	5.8	6.2
2	6.2	5.3	5.7	6.0	5.2	5.5
3	5.4	5.0	6.0	5.6	4.9	5.2
4	3.7	3.2	3.9	4.0	3.5	3.6

a. Do the data present sufficient evidence to indicate differences in location for at least two of the distributions of mean leaf length corresponding to the four locations? Test using the Kruskal–Wallis H-test with $\alpha = .05$.

b. Find the approximate p-value for the test.

c. You analyzed this same set of data in Exercise 11.13 using an analysis of variance. Find the p-value for the F-test used to compare the four location means in Exercise 11.13.

d. Compare the p-values in parts b and c and explain the implications of the comparison.

15.35 Heart Rate and Exercise Exercise 11.60 presented data (data set EX1160) on the heart rates for samples of 10 men randomly selected from each of four age groups. Each man walked a treadmill at a fixed grade for a period of 12 minutes, and the increase in heart rate (the difference before and after exercise) was recorded (in beats per minute). The data are shown in the table.

	10–19	20–39	40–59	60–69
	29	24	37	28
	33	27	25	29
	26	33	22	34
	27	31	33	36
	39	21	28	21
	35	28	26	20
	33	24	30	25
	29	34	34	24
	36	21	27	33
	22	32	33	32
Total	309	275	295	282

a. Do the data present sufficient evidence to indicate differences in location for at least two of the four age groups? Test using the Kruskal–Wallis H-test with $\alpha = .01$.

b. Find the approximate p-value for the test in part a.

c. Since the F-test in Exercise 11.60 and the H-test in part a are both tests to detect differences in location of the four heart-rate populations, how do the test results compare? Compare the p-values for the two tests and explain the implications of the comparison.

15.36 pH Levels in Water A sampling of the acidity of rain for 10 randomly selected rainfalls was recorded at three different locations in the United States: the Northeast, the Middle Atlantic region, and the Southeast. The pH readings for these 30 rainfalls are shown in the table. (NOTE: pH readings range from 0 to 14; 0 is acid, 14 is alkaline. Pure water falling through clean air has a pH reading of 5.7.)

Northeast	Middle Atlantic	Southeast
4.45	4.60	4.55
4.02	4.27	4.31
4.13	4.31	4.84
3.51	3.88	4.67
4.42	4.49	4.28
3.89	4.22	4.95
4.18	4.54	4.72
3.95	4.76	4.63
4.07	4.36	4.36
4.29	4.21	4.47

a. Do the data present sufficient evidence to indicate differences in the levels of acidity in rainfalls in the three different locations? Test using the Kruskal–Wallis H-test.

b. Find the approximate p-value for the test in part a and interpret it.

15.37 Advertising Campaigns The results of an experiment to investigate product recognition for three advertising campaigns were reported in Example 11.14. The responses were the percentage of 400 adults who were familiar with the newly advertised product. The normal probability plot indicated that the data were not approximately normal and another method of analysis should be used. Is there a significant difference among the three population distributions from which these samples came? Use an appropriate nonparametric method to answer this question.

Campaign		
1	2	3
.33	.28	.21
.29	.41	.30
.21	.34	.26
.32	.39	.33
.25	.27	.31

THE FRIEDMAN F_r-TEST FOR RANDOMIZED BLOCK DESIGNS

15.7

The Friedman F_r-test, proposed by Nobel Prize–winning economist Milton Friedman, is a nonparametric test for comparing the distributions of measurements for k treatments laid out in b blocks using a randomized block design. The procedure for conducting the test is very similar to that used for the Kruskal–Wallis H-test. The first step in the procedure is to rank the k treatment observations within each block. Ties are treated in the usual way; that is, they receive an average of the ranks occupied by the tied observations. The rank sums $T_1, T_2, \ldots, T_k$ are then obtained and the test statistic

$$F_r = \frac{12}{bk(k + 1)} \Sigma T_i^2 - 3b(k + 1)$$

is calculated. The value of the F_r statistic is at a minimum when the rank sums are equal—that is, $T_1 = T_2 = \cdots = T_k$—and increases in value as the differences among the rank sums increase. When either the number k of treatments or the number b of blocks is larger than five, the sampling distribution of F_r can be approximated by a chi-square distribution with $(k - 1)$ df. Therefore, as for the Kruskal–Wallis H-test, the rejection region for the F_r-test consists of values of F_r for which

$$F_r > \chi_\alpha^2$$

EXAMPLE 15.8

Suppose you wish to compare the reaction times of people exposed to six different stimuli. A reaction time measurement is obtained by subjecting a person to a stimulus and then measuring the time until the person presents some specified reaction. The objective of the experiment is to determine whether differences exist in the reaction times for the stimuli used in the experiment. To eliminate the person-to-person variation in reaction time, four persons participated in the experiment and each person's reaction time (in seconds) was measured for each of the six stimuli. The data are given in Table 15.10 (ranks of the observations are shown in parentheses). Use the Friedman F_r-test to determine whether the data present sufficient evidence to indicate differences in the distributions of reaction times for the six stimuli. Test using $\alpha = .05$.

TABLE 15.10 ● **Reaction Times to Six Stimuli**

Subject	Stimulus					
	1	2	3	4	5	6
1	.6 (2.5)	.9 (6)	.8 (5)	.7 (4)	.5 (1)	.6 (2.5)
2	.7 (3.5)	1.1 (6)	.7 (3.5)	.8 (5)	.5 (1.5)	.5 (1.5)
3	.9 (3)	1.3 (6)	1.0 (4.5)	1.0 (4.5)	.7 (1)	.8 (2)
4	.5 (2)	.7 (5)	.8 (6)	.6 (3.5)	.4 (1)	.6 (3.5)
Rank Sum	$T_1 = 11$	$T_2 = 23$	$T_3 = 19$	$T_4 = 17$	$T_5 = 4.5$	$T_6 = 9.5$

Solution In Figure 15.9, the plot of the residuals for each of the six stimuli reveals that stimuli 1, 4, and 5 have variances somewhat smaller than the other stimuli. Furthermore, the normal probability plot of the residuals reveals a change in the slope of the line following the first three residuals, as well as curvature in the upper portion of the plot. It appears that a nonparametric analysis is appropriate for these data.

FIGURE 15.9

A plot of treatments versus residuals and a normal probability plot of residuals for Example 15.8

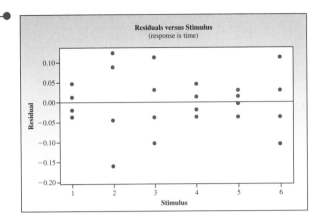

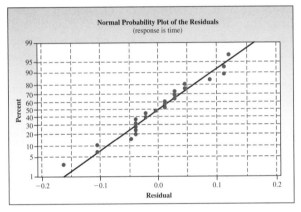

You wish to test

H_0 : The distributions of reaction times for the six stimuli are identical

versus the alternative hypothesis

H_a : At least two of the distributions of reaction times for the six stimuli differ in location

Table 15.10 shows the ranks (in parentheses) of the observations within each block and the rank sums for each of the six stimuli (the treatments). The value of the F_r statistic for these data is

$$F_r = \frac{12}{bk(k+1)} \Sigma T_i^2 - 3b(k+1)$$

$$= \frac{12}{(4)(6)(7)} [(11)^2 + (23)^2 + (19)^2 + \cdots + (9.5)^2] - 3(4)(7)$$

$$= 100.75 - 84 = 16.75$$

Since the number $k = 6$ of treatments exceeds five, the sampling distribution of F_r can be approximated by a chi-square distribution with $(k - 1) = (6 - 1) = 5$ df. Therefore, for $\alpha = .05$, you can reject H_0 if

$$F_r > \chi_{.05}^2 \qquad \text{where} \qquad \chi_{.05}^2 = 11.0705$$

This rejection region is shown in Figure 15.10. Since the observed value $F_r = 16.75$ exceeds $\chi^2_{.05} = 11.0705$, it falls in the rejection region. You can therefore reject H_0 and conclude that the distributions of reaction times differ in location for at least two stimuli. The *MINITAB* printout of the Friedman F_r-test for the data is given in Figure 15.11.

FIGURE 15.10

Rejection region for Example 15.8

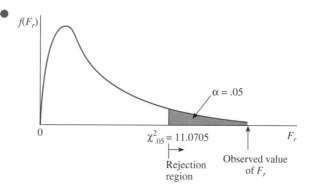

FIGURE 15.11

MINITAB printout for Example 15.8

Friedman Test: Time versus Stimulus blocked by Subject

```
S = 16.75   DF = 5   P = 0.005
S = 17.37   DF = 5   P = 0.004 (adjusted for ties)

                     Est      Sum of
Stimulus    N     Median      Ranks
1           4     0.6500       11.0
2           4     1.0000       23.0
3           4     0.8000       19.0
4           4     0.7500       17.0
5           4     0.5000        4.5
6           4     0.6000        9.5

Grand median  =   0.7167
```

 EXAMPLE 15.9

Find the approximate *p*-value for the test in Example 15.8.

Solution   Consulting Table 5 in Appendix I with 5 *df,* you find that the observed value of $F_r = 16.75$ exceeds the table value $\chi^2_{.005} = 16.7496$. Hence, the *p*-value is very close to, but slightly less than, .005.

THE FRIEDMAN F_r-TEST FOR A RANDOMIZED BLOCK DESIGN

1. Null hypothesis: H_0 : The k population distributions are identical

2. Alternative hypothesis: H_a : At least two of the k population distributions differ in location

3. Test statistic: $F_r = \dfrac{12}{bk(k+1)} \Sigma T_i^2 - 3b(k+1)$

where

$$b = \text{Number of blocks}$$

$$k = \text{Number of treatments}$$

$$T_i = \text{Rank sum for treatment } i, \; i = 1, 2, \ldots, k$$

4. Rejection region: $F_r > \chi_\alpha^2$, where χ_α^2 is based on $(k - 1)$ df

Assumption: Either the number k of treatments or the number b of blocks is greater than five.

15.7 EXERCISES

BASIC TECHNIQUES

15.38 A randomized block design is used to compare three treatments in six blocks.
EX1538

	Treatment		
Block	1	2	3
1	3.2	3.1	2.4
2	2.8	3.0	1.7
3	4.5	5.0	3.9
4	2.5	2.7	2.6
5	3.7	4.1	3.5
6	2.4	2.4	2.0

a. Use the Friedman F_r-test to detect differences in location among the three treatment distributions. Test using $\alpha = .05$.

b. Find the approximate p-value for the test in part a.

c. Perform an analysis of variance and give the ANOVA table for the analysis.

d. Give the value of the F statistic for testing the equality of the three treatment means.

e. Give the approximate p-value for the F statistic in part d.

f. Compare the p-values for the tests in parts a and d, and explain the practical implications of the comparison.

15.39 A randomized block design is used to compare four treatments in eight blocks.
EX1539

	Treatment			
Block	1	2	3	4
1	89	81	84	85
2	93	86	86	88
3	91	85	87	86
4	85	79	80	82
5	90	84	85	85
6	86	78	83	84
7	87	80	83	82
8	93	86	88	90

a. Use the Friedman F_r-test to detect differences in location among the four treatment distributions. Test using $\alpha = .05$.

b. Find the approximate p-value for the test in part a.

c. Perform an analysis of variance and give the ANOVA table for the analysis.

d. Give the value of the F statistic for testing the equality of the four treatment means.

e. Give the approximate p-value for the F statistic in part d.

f. Compare the p-values for the tests in parts a and d, and explain the practical implications of the comparison.

APPLICATIONS

15.40 Supermarket Prices In Exercise 11.43 (and data set EX1143), we compared the regular prices at four different grocery stores for eight items purchased on the same day. The prices are listed in the table.
EX1540

	Store			
Items	Vons	Ralphs	Stater Bros	WinCo
Salad mix, 12 oz. bag	3.99	2.79	1.99	1.78
Hillshire Farm® Beef Smoked Sausage, 14 oz.	4.29	4.29	3.99	2.50
Kellogg's Raisin Bran®, 25.5 oz.	4.49	5.49	4.49	3.15
Kraft® Philadelphia® Cream Cheese, 8 oz.	2.99	3.19	2.79	1.48
Kraft® Ranch Dressing, 16 oz.	3.19	3.49	3.49	1.48
TreeTop® Apple Juice, 64 oz.	2.99	3.49	3.49	1.58
Dial® Bar Soap, Gold, 8-4 oz.	5.99	6.49	5.79	5.14
Jif® Peanut Butter, Creamy, 28 oz.	5.15	5.49	4.79	4.34

a. Does the distribution of the prices differ from one supermarket to another? Test using the Friedman F_r-test with $\alpha = .05$.

b. Find the approximate p-value for the test and interpret it.

 15.41 Toxic Chemicals An experiment was
EX1541 conducted to compare the effects of three toxic chemicals, A, B, and C, on the skin of rats. One-inch squares of skin were treated with the chemicals and then scored from 0 to 10 depending on the degree of irritation. Three adjacent 1-inch squares were marked on the backs of eight rats, and each of the three chemicals was applied to each rat. Thus, the experiment was blocked on rats to eliminate the variation in skin sensitivity from rat to rat.

Rats

1	2	3	4	5	6	7	8
B	A	A	C	B	C	C	B
5	9	6	6	8	5	5	7
A	C	B	B	C	A	B	A
6	4	9	8	8	5	7	6
C	B	C	A	A	B	A	C
3	9	3	5	7	7	6	7

a. Do the data provide sufficient evidence to indicate a difference in the toxic effects of the three chemicals? Test using the Friedman F_r-test with $\alpha = .05$.

b. Find the approximate p-value for the test and interpret it.

15.42 Good Tasting Medicine In a study of
EX1542 the palatability of antibiotics in children, Dr. Doreen Matsui and colleagues used a voluntary sample

of healthy children to assess their reactions to the taste of four antibiotics.[5] The children's response was measured on a 10-centimeter (cm) visual analog scale incorporating the use of faces, from sad (low score) to happy (high score). The minimum score was 0 and the maximum was 10. For the accompanying data (simulated from the results of Matsui's report), each of five children was asked to taste each of four antibiotics and rate them using the visual (faces) analog scale from 0 to 10 cm.

	Antibiotic			
Child	1	2	3	4
1	4.8	2.2	6.8	6.2
2	8.1	9.2	6.6	9.6
3	5.0	2.6	3.6	6.5
4	7.9	9.4	5.3	8.5
5	3.9	7.4	2.1	2.0

a. What design is used in collecting these data?

b. Using an appropriate statistical package for a two-way classification, produce a normal probability plot of the residuals as well as a plot of residuals versus antibiotics. Do the usual analysis of variance assumptions appear to be satisfied?

c. Use the appropriate nonparametric test to test for differences in the distributions of responses to the tastes of the four antibiotics.

d. Comment on the results of the analysis of variance in part b compared with the nonparametric test in part c.

RANK CORRELATION COEFFICIENT

15.8

In the preceding sections, we used ranks to indicate the relative magnitude of observations in nonparametric tests for the comparison of treatments. We will now use the same technique in testing for a relationship between two ranked variables. Two common rank correlation coefficients are the **Spearman r_s** and the **Kendall τ**. We will present the Spearman r_s because its computation is identical to that for the sample correlation coefficient r of Chapters 3 and 12.

Suppose eight elementary school science teachers have been ranked by a judge according to their teaching ability and all have taken a "national teachers' examination." The data are listed in Table 15.11. Do the data suggest an agreement between the judge's ranking and the examination score? That is, is there a correlation between ranks and test scores?

TABLE 15.11 ● **Ranks and Test Scores for Eight Teachers**

Teacher	Judge's Rank	Examination Score
1	7	44
2	4	72
3	2	69
4	6	70
5	1	93
6	3	82
7	8	67
8	5	80

The two variables of interest are rank and test score. The former is already in rank form, and the test scores can be ranked similarly, as shown in Table 15.12. The ranks for tied observations are obtained by averaging the ranks that the tied observations would have had if no ties had been observed. The Spearman rank correlation coefficient r_s is calculated by using the ranks of the paired measurements on the two variables x and y in the formula for r (see Chapter 12).

TABLE 15.12 ● **Ranks of Data in Table 15.11**

Teacher	Judge's Rank, x_i	Test Rank, y_i
1	7	1
2	4	5
3	2	3
4	6	4
5	1	8
6	3	7
7	8	2
8	5	6

SPEARMAN'S RANK CORRELATION COEFFICIENT

$$r_s = \frac{S_{xy}}{\sqrt{S_{xx}S_{yy}}}$$

where x_i and y_i represent the ranks of the ith pair of observations and

$$S_{xy} = \Sigma\,(x_i - \bar{x})(y_i - \bar{y}) = \Sigma\,x_i y_i - \frac{(\Sigma\,x_i)(\Sigma\,y_i)}{n}$$

$$S_{xx} = \Sigma\,(x_i - \bar{x})^2 = \Sigma\,x_i^2 - \frac{(\Sigma\,x_i)^2}{n}$$

$$S_{yy} = \Sigma\,(y_i - \bar{y})^2 = \Sigma\,y_i^2 - \frac{(\Sigma\,y_i)^2}{n}$$

When there are no ties in either the x observations or the y observations, the expression for r_s algebraically reduces to the simpler expression

$$r_s = 1 - \frac{6\,\Sigma\,d_i^2}{n(n^2 - 1)} \qquad \text{where } d_i = (x_i - y_i)$$

If the number of ties is small in comparison with the number of data pairs, little error results in using this shortcut formula.

EXAMPLE 15.10 — Calculate r_s for the data in Table 15.12.

Solution The differences and squares of differences between the two rankings are provided in Table 15.13. Substituting values into the formula for r_s, you have

$$r_s = 1 - \frac{6 \Sigma d_i^2}{n(n^2 - 1)}$$

$$= 1 - \frac{6(144)}{8(64 - 1)} = -.714$$

TABLE 15.13 ● **Differences and Squares of Differences for the Teacher Ranks**

Teacher	x_i	y_i	d_i	d_i^2
1	7	1	6	36
2	4	5	−1	1
3	2	3	−1	1
4	6	4	2	4
5	1	8	−7	49
6	3	7	−4	16
7	8	2	6	36
8	5	6	−1	1
Total				144

The Spearman rank correlation coefficient can be used as a test statistic to test the hypothesis of no association between two populations. You can assume that the n pairs of observations (x_i, y_i) have been randomly selected and, therefore, no association between the populations implies a random assignment of the n ranks within each sample. Each random assignment (for the two samples) represents a simple event associated with the experiment, and a value of r_s can be calculated for each. Thus, it is possible to calculate the probability that r_s assumes a large absolute value due solely to chance and thereby suggests an association between populations when none exists.

The rejection region for a two-tailed test is shown in Figure 15.12. If the alternative hypothesis is that the correlation between x and y is negative, you would reject H_0 for negative values of r_s that are close to -1 (in the lower tail of Figure 15.12). Similarly, if the alternative hypothesis is that the correlation between x and y is positive, you would reject H_0 for large positive values of r_s (in the upper tail of Figure 15.12).

FIGURE 15.12 ●

Rejection region for a two-tailed test of the null hypothesis of no association, using Spearman's rank correlation test

r_s = Spearman's rank correlation coefficient

The critical values of r_s are given in Table 9 in Appendix I. An abbreviated version is shown in Table 15.14. Across the top of Table 15.14 (and Table 9 in Appendix I) are the recorded values of α that you might wish to use for a one-tailed test of the null hypothesis of no association between x and y. The number of rank pairs n appears at the left side of the table. The table entries give the critical value r_0 for a one-tailed test. Thus, $P(r_s \geq r_0) = \alpha$.

For example, suppose you have $n = 8$ rank pairs and the alternative hypothesis is that the correlation between the ranks is positive. You would want to reject the null hypothesis of no association for only large positive values of r_s, and you would use a one-tailed test. Referring to Table 15.14 and using the row corresponding to $n = 8$ and the column for $\alpha = .05$, you read $r_0 = .643$. Therefore, you can reject H_0 for all values of r_s greater than or equal to .643.

The test is conducted in exactly the same manner if you wish to test only the alternative hypothesis that the ranks are negatively correlated. The only difference is that you would reject the null hypothesis if $r_s \leq -.643$. That is, you use the negative of the tabulated value of r_0 to get the lower-tail critical value.

TABLE 15.14 ●

An Abbreviated Version of Table 9 in Appendix I; for Spearman's Rank Correlation Test

n	$\alpha = .05$	$\alpha = .025$	$\alpha = .01$	$\alpha = .005$
5	.900	—	—	—
6	.829	.886	.943	—
7	.714	.786	.893	—
8	.643	.738	.833	.881
9	.600	.683	.783	.833
10	.564	.648	.745	.794
11	.523	.623	.736	.818
12	.497	.591	.703	.780
13	.475	.566	.673	.745
14	.457	.545		
15	.441	.525		
16	.425			
17	.412			
18	.399			
19	.388			
20	.377			

To conduct a two-tailed test, you reject the null hypothesis if $r_s \geq r_0$ or $r_s \leq -r_0$. The value of α for the test is double the value shown at the top of the table. For example, if $n = 8$ and you choose the .025 column, you will reject H_0 if $r_s \geq .738$ or $r_s \leq -.738$. The α-value for the test is $2(.025) = .05$.

SPEARMAN'S RANK CORRELATION TEST

1. Null hypothesis: H_0 : There is no association between the rank pairs.
2. Alternative hypothesis: H_a : There is an association between the rank pairs (a two-tailed test). Or H_a : The correlation between the rank pairs is positive or negative (a one-tailed test).
3. Test statistic: $r_s = \dfrac{S_{xy}}{\sqrt{S_{xx}S_{yy}}}$

   where x_i and y_i represent the ranks of the ith pair of observations.
4. Rejection region: For a two-tailed test, reject H_0 if $r_s \geq r_0$ or $r_s \leq -r_0$, where r_0 is given in Table 9 in Appendix I. Double the tabulated probability to obtain the value of α for the two-tailed test. For a one-tailed test, reject H_0 if $r_s \geq r_0$ (for an upper-tailed test) or $r_s \leq -r_0$ (for a lower-tailed test). The α-value for a one-tailed test is the value shown in Table 9 in Appendix I.

EXAMPLE 15.11 — Test the hypothesis of no association between the populations for Example 15.10.

Solution The critical value of r_s for a one-tailed test with $\alpha = .05$ and $n = 8$ is .643. You may assume that a correlation between the judge's rank and the teachers' test scores could not possibly be positive. (A low rank means good teaching and should be associated with a high test score if the judge and the test measure teaching ability.) The alternative hypothesis is that the **population rank correlation coefficient** ρ_s is less than 0, and you are concerned with a one-tailed statistical test. Thus, α for the test is the tabulated value for .05, and you can reject the null hypothesis if $r_s \leq -.643$.

The calculated value of the test statistic, $r_s = -.714$, is less than the critical value for $\alpha = .05$. Hence, the null hypothesis is rejected at the $\alpha = .05$ level of significance. It appears that some agreement does exist between the judge's rankings and the test scores. However, it should be noted that this agreement could exist when *neither* provides an adequate yardstick for measuring teaching ability. For example, the association could exist if both the judge and those who constructed the teachers' examination had a completely erroneous, but similar, concept of the characteristics of good teaching.

What exactly does r_s measure? Spearman's correlation coefficient detects not only a linear relationship between two variables but also any other monotonic relationship (either y increases as x increases or y decreases as x increases). For example, if you calculated r_s for the two data sets in Table 15.15, both would produce a value of $r_s = 1$ because the assigned ranks for x and y in both cases agree for all pairs (x, y). It is important to remember that a significant value of r_s indicates a relationship between x and y that is either increasing or decreasing, but is not necessarily linear.

TABLE 15.15 ● **Twin Data Sets With $r_s = 1$**

x	$y = x^2$	x	$y = \log_{10}(x)$
1	1	10	1
2	4	100	2
3	9	1000	3
4	16	10,000	4
5	25	100,000	5
6	36	1,000,000	6

15.8 **EXERCISES**

BASIC TECHNIQUES

15.43 Give the rejection region for a test to detect positive rank correlation if the number of pairs of ranks is 16 and you have these α-values:

a. $\alpha = .05$ **b.** $\alpha = .01$

15.44 Give the rejection region for a test to detect negative rank correlation if the number of pairs of ranks is 12 and you have these α-values:

a. $\alpha = .05$ **b.** $\alpha = .01$

15.45 Give the rejection region for a test to detect rank correlation if the number of pairs of ranks is 25 and you have these α-values:

a. $\alpha = .05$ **b.** $\alpha = .01$

15.46 The following paired observations were obtained on two variables x and y:

x	1.2	.8	2.1	3.5	2.7	1.5
y	1.0	1.3	.1	-.8	-.2	.6

a. Calculate Spearman's rank correlation coefficient r_s.

b. Do the data present sufficient evidence to indicate a correlation between x and y? Test using $\alpha = .05$.

APPLICATIONS

15.47 Rating Political Candidates A political scientist wished to examine the relationship between the voter image of a conservative political candidate and the distance (in miles) between the residences of the voter and the candidate. Each of 12 voters rated the candidate on a scale of 1–20.

Voter	Rating	Distance
1	12	75
2	7	165
3	5	300
4	19	15
5	17	180
6	12	240
7	9	120
8	18	60
9	3	230
10	8	200
11	15	130
12	4	130

a. Calculate Spearman's rank correlation coefficient r_s.

b. Do these data provide sufficient evidence to indicate a negative correlation between rating and distance?

15.48 Competitive Running Is the number of years of competitive running experience related to a runner's distance running performance? The data on nine runners, obtained from the study by Scott Powers and colleagues, are shown in the table:[6]

Runner	Years of Competitive Running	10-Kilometer Finish Time (minutes)
1	9	33.15
2	13	33.33
3	5	33.50
4	7	33.55
5	12	33.73
6	6	33.86
7	4	33.90
8	5	34.15
9	3	34.90

a. Calculate the rank correlation coefficient between years of competitive running x and a runner's finish time y in the 10-kilometer race.

b. Do the data provide evidence to indicate a significant rank correlation between y and x? Test using $\alpha = .05$.

15.49 Tennis Racquets The data shown in the accompanying table give measures of bending stiffness and twisting stiffness as determined by engineering tests on 12 tennis racquets.

Racquet	Bending Stiffness	Twisting Stiffness
1	419	227
2	407	231
3	363	200
4	360	211
5	257	182
6	622	304
7	424	384
8	359	194
9	346	158
10	556	225
11	474	305
12	441	235

a. Calculate the rank correlation coefficient r_s between bending stiffness and twisting stiffness.

b. If a racquet has bending stiffness, is it also likely to have twisting stiffness? Use the rank correlation coefficient to determine whether there is a significant positive relationship between bending stiffness and twisting stiffness. Use $\alpha = .05$.

15.50 Student Ratings A school principal suspected that a teacher's attitude toward a first-grader depended on his original judgment of the child's ability. The principal also suspected that much of that judgment was based on the first-grader's IQ score, which was usually known to the teacher. After three weeks of teaching, a teacher was asked to rank the nine children in his class from 1 (highest) to 9 (lowest) as to his opinion of their ability. Calculate r_s for these teacher–IQ ranks:

Rank	1	2	3	4	5	6	7	8	9
IQ	3	1	2	4	5	7	9	6	8

15.51 Student Ratings, continued Refer to Exercise 15.50. Do the data provide sufficient evidence to indicate a positive correlation between the teacher's ranks and the ranks of the IQs? Use $\alpha = .05$.

15.52 Art Critics Two art critics each ranked 10 paintings by contemporary (but anonymous) artists in accordance with their appeal to the respective critics. The ratings are shown in the table. Do the critics seem to agree on their ratings of contemporary art? That is, do the data provide sufficient evidence to indicate a positive correlation between critics A and B? Test by using an α value near .05.

Painting	Critic A	Critic B
1	6	5
2	4	6
3	9	10
4	1	2
5	2	3
6	7	8
7	3	1
8	8	7
9	5	4
10	10	9

 15.53 Rating Tobacco Leaves An experi-
EX1553   ment was conducted to study the relationship
between the ratings of a tobacco leaf grader and the
moisture content of the tobacco leaves. Twelve leaves
were rated by the grader on a scale of 1–10, and corresponding readings of moisture content were made.

Leaf	Grader's Rating	Moisture Content
1	9	.22
2	6	.16
3	7	.17
4	7	.14
5	5	.12
6	8	.19
7	2	.10
8	6	.12
9	1	.05
10	10	.20
11	9	.16
12	3	.09

Calculate r_s. Do the data provide sufficient evidence to
indicate an association between the grader's ratings
and the moisture contents of the leaves?

15.54 Social Skills Training A social skills
EX1554   training program was implemented with seven
mildly challenged students in a study to determine
whether the program caused improvements in pre/post
measures and behavior ratings. For one such test, the
pre- and posttest scores for the seven students are
given in the table:

Student	Pretest	Posttest
Earl	101	113
Ned	89	89
Jasper	112	121
Charlie	105	99
Tom	90	104
Susie	91	94
Lori	89	99

a. Use a nonparametric test to determine whether there
is a significant positive relationship between the
pre- and posttest scores.

b. Do these results agree with the results of the parametric test in Exercise 12.54?

15.9 SUMMARY

The nonparametric tests presented in this chapter are only a few of the many nonparametric tests available to experimenters. The tests presented here are those for
which tables of critical values are readily available.

Nonparametric statistical methods are especially useful when the observations can
be rank ordered but cannot be located exactly on a measurement scale. Also, nonparametric methods are the only methods that can be used when the sampling designs
have been correctly adhered to, but the data are not or cannot be assumed to follow
the prescribed one or more distributional assumptions.

We have presented a wide array of nonparametric techniques that can be used when
either the data are not normally distributed or the other required assumptions are not
met. One-sample procedures are available in the literature; however, we have concentrated on analyzing two or more samples that have been properly selected using
random and independent sampling as required by the design involved. The nonparametric analogues of the parametric procedures presented in Chapters 10–14 are
straightforward and fairly simple to implement:

- The Wilcoxon rank sum test is the nonparametric analogue of the two-sample
  t-test.

- The sign test and the Wilcoxon signed-rank tests are the nonparametric
  analogues of the paired-sample *t*-test.

- The Kruskal–Wallis H-test is the rank equivalent of the one-way analysis of variance F-test.
- The Friedman F_r-test is the rank equivalent of the randomized block design two-way analysis of variance F-test.
- Spearman's rank correlation r_s is the rank equivalent of Pearson's correlation coefficient.

These and many more nonparametric procedures are available as alternatives to the parametric tests presented earlier. It is important to keep in mind that when the assumptions required of the sampled populations are relaxed, our ability to detect significant differences in one or more population characteristics is decreased.

CHAPTER REVIEW

Key Concepts and Formulas

I. Nonparametric Methods

1. These methods can be used when the data cannot be measured on a quantitative scale, or when

2. The numerical scale of measurement is arbitrarily set by the researcher, or when

3. The parametric assumptions such as normality or constant variance are seriously violated.

II. Wilcoxon Rank Sum Test: Independent Random Samples

1. Jointly rank the two samples. Designate the smaller sample as sample 1. Then

$$T_1 = \text{Rank sum of sample 1}$$
$$T_1^* = n_1(n_1 + n_2 + 1) - T_1$$

2. Use T_1 to test for population 1 to the left of population 2. Use T_1^* to test for population 1 to the right of population 2. Use the smaller of T_1 and T_1^* to test for a difference in the locations of the two populations.

3. Table 7 of Appendix I has critical values for the rejection of H_0.

4. When the sample sizes are large, use the normal approximation:

$$\mu_T = \frac{n_1(n_1 + n_2 + 1)}{2}$$
$$\sigma_T^2 = \frac{n_1 n_2(n_1 + n_2 + 1)}{12}$$
$$z = \frac{T - \mu_T}{\sigma_T}$$

III. Sign Test for a Paired Experiment

1. Find x, the number of times that observation A exceeds observation B for a given pair.

2. To test for a difference in two populations, test $H_0 : p = .5$ versus a one- or two-tailed alternative.

3. Use Table 1 of Appendix I to calculate the p-value for the test.

4. When the sample sizes are large, use the normal approximation:

$$z = \frac{x - .5n}{.5\sqrt{n}}$$

IV. Wilcoxon Signed-Rank Test: Paired Experiment

1. Calculate the differences in the paired observations. Rank the *absolute values* of the differences. Calculate the rank sums T^+ and T^- for the positive and negative differences, respectively. The test statistic T is the smaller of the two rank sums.

2. Table 8 in Appendix I has critical values for the rejection of H_0 for both one- and two-tailed tests.

3. When the sample sizes are large, use the normal approximation:

$$z = \frac{T^+ - [n(n + 1)/4]}{\sqrt{[n(n + 1)(2n + 1)]/24}}$$

V. Kruskal–Wallis *H*-Test: Completely Randomized Design

1. Jointly rank the n observations in the k samples. Calculate the rank sums, T_i = rank sum of sample i, and the test statistic

$$H = \frac{12}{n(n+1)} \Sigma \frac{T_i^2}{n_i} - 3(n+1)$$

2. If the null hypothesis of equality of distributions is false, H will be unusually large, resulting in a one-tailed test.

3. For sample sizes of five or greater, the rejection region for H is based on the chi-square distribution with $(k-1)$ degrees of freedom.

VI. The Friedman F_r-Test: Randomized Block Design

1. Rank the responses within each block from 1 to k. Calculate the rank sums, $T_1, T_2, \ldots, T_k$, and the test statistic

$$F_r = \frac{12}{bk(k+1)} \Sigma T_i^2 - 3b(k+1)$$

2. If the null hypothesis of equality of treatment distributions is false, F_r will be unusually large, resulting in a one-tailed test.

3. For block sizes of five or greater, the rejection region for F_r is based on the chi-square distribution with $(k-1)$ degrees of freedom.

VII. Spearman's Rank Correlation Coefficient

1. Rank the responses for the two variables from smallest to largest.

2. Calculate the correlation coefficient for the ranked observations:

$$r_s = \frac{S_{xy}}{\sqrt{S_{xx}S_{yy}}} \quad \text{or} \quad r_s = 1 - \frac{6 \Sigma d_i^2}{n(n^2-1)}$$

if there are no ties

3. Table 9 in Appendix I gives critical values for rank correlations significantly different from 0.

4. The rank correlation coefficient detects not only significant linear correlation but also any other monotonic relationship between the two variables.

 TECHNOLOGY TODAY

Nonparametric Procedures–*MINITAB*

Although there are no options for nonparametric procedures in *MS Excel,* many nonparametric procedures are available in the *MINITAB* package, including most of the tests discussed in this chapter. The Dialog boxes are all familiar to you by now, and we will discuss the tests in the order presented in the chapter.

To implement the Wilcoxon rank sum test for two independent random samples, enter the two sets of sample data into two columns (say, C1 and C2) of the *MINITAB* worksheet. The Dialog box in Figure 15.13 is generated using **Stat ▶ Nonparametrics ▶ Mann–Whitney.** Select C1 and C2 for the **First** and **Second Samples,** and indicate the appropriate confidence coefficient (for a confidence interval) and alternative hypothesis. Clicking **OK** will generate the output in Figure 15.1.

The sign test *and* the Wilcoxon signed-rank test for paired samples are performed in exactly the same way, with a change only in the last command of the sequence. Even the Dialog boxes are identical! Enter the data into two columns of the *MINITAB* worksheet (we used the cake mix data in Section 15.5). Before you can implement either test, you must generate a column of differences using **Calc ▶ Calculator,** as shown in Figure 15.14. Use **Stat ▶ Nonparametrics ▶ 1-Sample Sign** or **Stat ▶ Nonparametrics ▶ 1-Sample Wilcoxon** to generate the appropriate Dialog box shown in Figure 15.15.

FIGURE 15.13

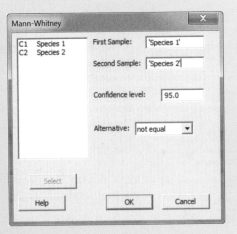

FIGURE 15.14

Remember that the median is the value of a variable such that 50% of the values are smaller and 50% are larger. Hence, if the two population distributions are the same, the median of the differences will be 0. This is equivalent to the null hypothesis

$$H_0 : P(\text{positive difference}) = P(\text{negative difference}) = .5$$

FIGURE 15.15

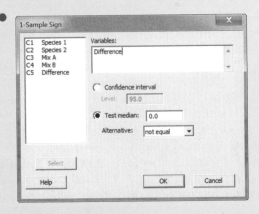

used for the sign test. Select the column of differences for the Variables box, and select the test of the median equals 0 with the appropriate alternative. Click **OK** to obtain the printout for either of the two tests. The Session window printout for the sign test, shown in Figure 15.16, indicates a nonsignificant difference in the distributions of densities for the two cake mixes. Notice that the *p*-value (.2188) is not the same as the *p*-value for the Wilcoxon signed-rank test (.093 from Figure 15.4). However, if you are testing at the 5% level, both tests produce nonsignificant differences.

FIGURE 15.16

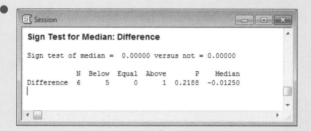

The procedures for implementing the Kruskal–Wallis *H*-test for *k* independent samples and Friedman's F_r-test for a randomized block design are identical to the procedures used for their parametric equivalents. Review the methods described in the "Technology Today" section in Chapter 11. Once you have entered the data as explained in that section, the commands **Stat ▶ Nonparametrics ▶ Kruskal–Wallis** or **Stat ▶ Nonparametrics ▶ Friedman** will generate a Dialog box in which you specify the Response column and the Factor column, or the response column, the treatment column and the block column, respectively. Click **OK** to obtain the outputs for these nonparametric tests.

Finally, you can generate the nonparametric rank correlation coefficient r_s if you enter the data into two columns and rank the data using **Data ▶ Rank.** For example, the data on judge's rank and test scores were entered into columns C6 and C7 of our *MINITAB* worksheet. Since the judge's ranks are already in rank order, we need only to rank C7 by selecting "Exam Score" and storing the ranks in C8 [named "Rank (y)" in Figure 15.17]. The commands **Stat ▶ Basic Statistics ▶ Correlation** will now produce the rank correlation coefficient when C6 and C8 are selected. However, the *p*-value that you see in the *output does not* produce exactly the same test as the critical values in Table 15.14. You should compare your value of r_s with the tabled value to check for a significant association between the two variables.

FIGURE 15.17

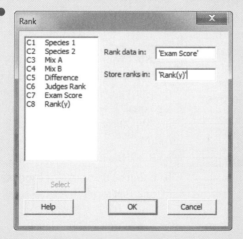

Supplementary Exercises

Data set EX1555 **15.55 Response Times** An experiment was conducted to compare the response times for two different stimuli. To remove natural person-to-person variability in the responses, both stimuli were presented to each of nine subjects, thus permitting an analysis of the differences between stimuli *within* each person. The table lists the response times (in seconds).

Subject	Stimulus 1	Stimulus 2
1	9.4	10.3
2	7.8	8.9
3	5.6	4.1
4	12.1	14.7
5	6.9	8.7
6	4.2	7.1
7	8.8	11.3
8	7.7	5.2
9	6.4	7.8

a. Use the sign test to determine whether sufficient evidence exists to indicate a difference in the mean response times for the two stimuli. Use a rejection region for which $\alpha \leq .05$.

b. Test the hypothesis of no difference in mean response times using Student's *t*-test.

15.56 Response Times, continued Refer to Exercise 15.55. Test the hypothesis that no difference exists in the distributions of response times for the two stimuli, using the Wilcoxon signed-rank test. Use a rejection region for which α is as near as possible to the α achieved in Exercise 15.55, part a.

Data set EX1557 **15.57 Identical Twins** To compare two junior high schools, A and B, in academic effectiveness, an experiment was designed requiring the use of 10 sets of identical twins, each twin having just completed the sixth grade. In each case, the twins in the same set had obtained their schooling in the same classrooms at each grade level. One child was selected at random from each pair of twins and assigned to school A. The remaining children were sent to school B. Near the end of the ninth grade, a certain achievement test was given to each child in the experiment. The test scores are shown in the table.

Twin Pair	School A	School B
1	67	39
2	80	75
3	65	69
4	70	55
5	86	74
6	50	52
7	63	56
8	81	72
9	86	89
10	60	47

a. Test (using the sign test) the hypothesis that the two schools are the same in academic effectiveness, as measured by scores on the achievement test, versus the alternative that the schools are not equally effective.

b. Suppose it was known that junior high school A had a superior faculty and better learning facilities. Test the hypothesis of equal academic effectiveness versus the alternative that school A is superior.

15.58 Identical Twins II Refer to Exercise 15.57. What answers are obtained if Wilcoxon's signed-rank test is used in analyzing the data? Compare with your earlier answers.

Data set EX1559 **15.59 Paper Brightness** The coded values for a measure of brightness in paper (light reflectivity), prepared by two different processes, are given in the table for samples of nine observations drawn randomly from each of the two processes. Do the data present sufficient evidence to indicate a difference in the brightness measurements for the two processes? Use both a parametric and a nonparametric test and compare your results.

Process	Brightness								
A	6.1	9.2	8.7	8.9	7.6	7.1	9.5	8.3	9.0
B	9.1	8.2	8.6	6.9	7.5	7.9	8.3	7.8	8.9

15.60 Precision Instruments Assume (as in the case of measurements produced by two well-calibrated measuring instruments) the means of two populations are equal. Use the Wilcoxon rank sum statistic for testing hypotheses concerning the population variances as follows:

a. Rank the combined sample.

b. Number the ranked observations "from the outside in"; that is, number the smallest observation 1, the largest 2, the next-to-smallest 3, the next-to-largest 4, and so on. This sequence of numbers induces an ordering on the symbols A (population A items) and B (population B items). If $\sigma_A^2 > \sigma_B^2$, one would expect to find a preponderance of A's near the beginning of the sequence, and thus a relatively small "sum of ranks" for the A observations.

c. Given the measurements in the table produced by well-calibrated precision instruments A and B, test at near the $\alpha = .05$ level to determine whether the more expensive instrument B is more precise than A. (Note that this implies a one-tailed test.) Use the Wilcoxon rank sum test statistic.

Instrument A	Instrument B
1060.21	1060.24
1060.34	1060.28
1060.27	1060.32
1060.36	1060.30
1060.40	

d. Test using the equality of variance F-test.

15.61 Meat Tenderizers An experiment was conducted to compare the tenderness of meat cuts treated with two different meat tenderizers, A and B. To reduce the effect of extraneous variables, the data were paired by the specific meat cut, by applying the tenderizers to two cuts taken from the same steer, by cooking paired cuts together, and by using a single judge for each pair. After cooking, each cut was rated by a judge on a scale of 1–10, with 10 corresponding to the most tender meat. The data are shown for a single judge. Do the data provide sufficient evidence to indicate that one of the two tenderizers tends to receive higher ratings than the other? Would a Student's t-test be appropriate for analyzing these data? Explain.

	Tenderizer	
Cut	A	B
Shoulder roast	5	7
Chuck roast	6	5
Rib steak	8	9
Brisket	4	5
Club steak	9	9
Round steak	3	5
Rump roast	7	6
Sirloin steak	8	8
Sirloin tip steak	8	9
T-bone steak	9	10

15.62 Interviewing Job Prospects A large corporation selects college graduates for employment using both interviews and a psychological achievement test. Interviews conducted at the home office of the company are far more expensive than the tests that can be conducted on campus. Consequently, the personnel office was interested in determining whether the test scores were correlated with interview ratings and whether tests could be substituted for interviews. The idea was not to eliminate interviews but to reduce their number. To determine whether the measures were correlated, 10 prospects were ranked during interviews and tested. The paired scores are as listed here:

Subject	Interview Rank	Test Score
1	8	74
2	5	81
3	10	66
4	3	83
5	6	66
6	1	94
7	4	96
8	7	70
9	9	61
10	2	86

Calculate the Spearman rank correlation coefficient r_s. Rank 1 is assigned to the candidate judged to be the best.

15.63 Interviews, continued Refer to Exercise 15.62. Do the data present sufficient evidence to indicate that the correlation between interview rankings and test scores is less than zero? If this evidence does exist, can you say that tests can be used to reduce the number of interviews?

15.64 Word Association Experiments A comparison of reaction times for two different stimuli in a psychological word-association experiment produced the accompanying results when applied to a random sample of 16 people:

Stimulus	Reaction Time (seconds)							
1	1	3	2	1	2	1	3	2
2	4	2	3	3	1	2	3	3

Do the data present sufficient evidence to indicate a difference in mean reaction times for the two stimuli? Use an appropriate nonparametric test and explain your conclusions.

15.65 Math and Art The table gives the scores of a group of 15 students in mathematics and art. Use Wilcoxon's signed-rank test to determine whether the median scores for these students differ significantly for the two subjects.

Student	Math	Art	Student	Math	Art
1	22	53	9	62	55
2	37	68	10	65	74
3	36	42	11	66	68
4	38	49	12	56	64
5	42	51	13	66	67
6	58	65	14	67	73
7	58	51	15	62	65
8	60	71			

15.66 Math and Art, continued Refer to Exercise 15.65. Compute Spearman's rank correlation coefficient for these data and test H_0 : no association between the rank pairs at the 10% level of significance.

15.67 Yield of Wheat Exercise 11.68 presented an analysis of variance of the yields of five different varieties of wheat, observed on one plot each at each of six different locations (see data set EX1168). The data from this randomized block design are listed here:

	Location					
Varieties	1	2	3	4	5	6
A	35.3	31.0	32.7	36.8	37.2	33.1
B	30.7	32.2	31.4	31.7	35.0	32.7
C	38.2	33.4	33.6	37.1	37.3	38.2
D	34.9	36.1	35.2	38.3	40.2	36.0
E	32.4	28.9	29.2	30.7	33.9	32.1

a. Use the appropriate nonparametric test to determine whether the data provide sufficient evidence to indicate a difference in the yields for the five different varieties of wheat. Test using $\alpha = .05$.

b. Exercise 11.68 presented a computer printout of the analysis of variance for comparing the mean yields for the five varieties of wheat. How do the results of the analysis of variance F test compare with the test in part a? Explain.

15.68 Learning to Sell In Exercise 11.61, you compared the numbers of sales per trainee after completion of one of four different sales training programs (see data set EX1161). Six trainees completed training program 1, eight completed program 2, and so on. The numbers of sales per trainee are shown in the table.

	Training Program			
	1	2	3	4
	78	99	74	81
	84	86	87	63
	86	90	80	71
	92	93	83	65
	69	94	78	86
	73	85		79
		97		73
		91		70
Total	482	735	402	588

a. Do the data present sufficient evidence to indicate that the distribution of number of sales per trainee differs from one training program to another? Test using the appropriate nonparametric test.

b. How do the test results in part a compare with the results of the analysis of variance F-test in Exercise 11.61?

15.69 Pollution from Chemical Plants In Exercise 11.66, you performed an analysis of variance to compare the mean levels of effluents in water at four different industrial plants (see data set EX1166). Five samples of liquid waste were taken at the output of each of four industrial plants. The data are shown in the table.

Plant	Polluting Effluents (lb/gal of waste)				
A	1.65	1.72	1.50	1.37	1.60
B	1.70	1.85	1.46	2.05	1.80
C	1.40	1.75	1.38	1.65	1.55
D	2.10	1.95	1.65	1.88	2.00

a. Do the data present sufficient evidence to indicate a difference in the levels of pollutants for the four different industrial plants? Test using the appropriate nonparametric test.

b. Find the approximate p-value for the test and interpret its value.

c. Compare the test results in part a with the analysis of variance test in Exercise 11.66. Do the results agree? Explain.

15.70 AIDS Research Scientists have shown that a newly developed vaccine can shield rhesus monkeys from infection by a virus closely related to the AIDS-causing human immunodeficiency virus (HIV). In their work, Ronald C. Resrosiers and his colleagues at the New England Regional Primate Research Center gave each of $n = 6$ rhesus monkeys five inoculations with the simian immunodeficiency virus (SIV) vaccine. One week after the last vaccination, each monkey received an injection of live SIV. Two of the six vaccinated monkeys showed no evidence of SIV infection for as long as a year and a half after the SIV injection.[7] Scientists were able to isolate the SIV virus from the other four vaccinated monkeys, although these animals showed no sign of the disease. Does this information contain sufficient evidence to indicate that the vaccine is effective in protecting monkeys from SIV? Use $\alpha = .10$.

15.71 Heavy Metal An experiment was performed to determine whether there is an accumulation of heavy metals in plants that were grown in soils amended with sludge and whether there is an accumulation of heavy metals in insects feeding on those plants.[8] The data in the table are cadmium concentrations (in μg/kg) in plants grown under six different rates of application of sludge for three different harvests. The rates of application are the treatments. The three harvests represent time blocks in the two-way design.

Rate	Harvest		
	1	2	3
Control	162.1	153.7	200.4
1	199.8	199.6	278.2
2	220.0	210.7	294.8
3	194.4	179.0	341.1
4	204.3	203.7	330.2
5	218.9	236.1	344.2

a. Based on the diagnostic plots, are you willing to assume that the normality and constant variance assumptions are satisfied?

Diagnostic plots for Exercise 15.71

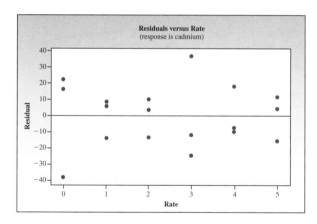

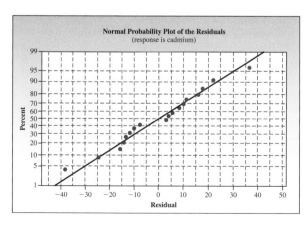

b. Using an appropriate method of analysis, analyze the data to determine whether there are significant differences among the responses due to rates of application.

15.72 Refer to Exercise 15.71. The data in this table are the cadmium concentrations found in aphids that fed on the plants grown in soil amended with sludge.

EX1572

Rate	Harvest		
	1	2	3
Control	16.2	55.8	65.8
1	16.9	119.4	181.1
2	12.7	171.9	184.6
3	31.3	128.4	196.4
4	38.5	182.0	163.7
5	20.6	191.3	242.8

a. Use the diagnostic plots to assess whether the assumptions of normality and constant variance are reasonable in this case.

b. Based on your conclusions in part a, use an appropriate statistical method to test for significant differences in cadmium concentrations for the six rates of application.

Diagnostic plots for Exercise 15.72

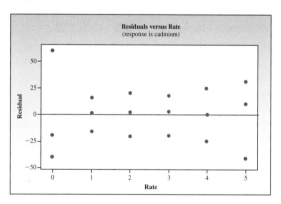

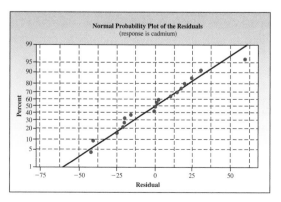

15.73 Rating Teaching Applicants Before filling several new teaching positions at the high school, the principal formed a review board consisting of five teachers who were asked to interview the 12 applicants and rank them in order of merit. Seven of the 12 applicants held college degrees but had limited teaching experience. Of the remaining five applicants, all had college degrees and substantial experience. The review board's rankings are given in the table.

Limited Experience	Substantial Experience
4	1
6	2
7	3
9	5
10	8
11	
12	

Do these rankings indicate that the review board considers experience a prime factor in the selection of the best candidates? Test using $\alpha = .05$.

15.74 Contaminants in Chemicals A
EX1574 manufacturer uses a large amount of a certain chemical. Since there are just two suppliers of this chemical, the manufacturer wishes to test whether the percentage of contaminants is the same for the two sources against the alternative that there is a difference in the percentages of contaminants for the two suppliers. Data from independent random samples are shown below:

Supplier			
1		**2**	
.86	.65	.55	.58
.69	1.13	.40	.16
.72	.65	.22	.07
1.18	.50	.09	.36
.45	1.04	.16	.20
1.41	.41	.26	.15

a. Use the Wilcoxon rank sum test to determine whether there is a difference in the contaminant percentages for the two suppliers. Use $\alpha = .05$.

b. Use the large-sample approximation to the Wilcoxon rank sum test to determine whether there is a difference in the contaminant percentages for the two suppliers. Use $\alpha = .05$. Compare your conclusions to the conclusions from part a.

15.75 Lighting in the Classroom The productivity of 35 students was observed and measured both before and after the installation of new lighting in their classroom. The productivity of 21 of the 35 students was observed to have improved, whereas the productivity of the others appeared to show no perceptible gain as a result of the new lighting. Use the normal approximation to the sign test to determine whether or not the new lighting was effective in increasing student productivity at the 5% level of significance.

15.76 Store Brands Save You $ Ron
EX1576 Marks[9] visited five supermarket chains in New York and New Jersey and compared store- and name-brand prices for 30 items listed below.

Product	Name Brand ($)	Store Brand ($)	Difference ($)
Aluminum foil	8.47	6.54	1.93
Baked beans	1.71	1.06	0.65
Bread crumbs	2.03	1.22	0.81
Butter quarters	4.03	3.03	1.00
Canned orange segments	2.07	1.22	0.85
Canola oil	4.28	3.34	0.94
Chocolate-flavored syrup	4.13	3.22	0.91
Cotton swabs	3.53	1.98	1.55
Cranberry juice cocktail	2.82	2.42	0.40
Cream cheese	2.05	1.35	0.70
Creamy peanut butter	2.85	2.05	0.80
Crescent rolls	2.71	1.98	0.73
Dry pasta	1.24	.87	0.37
Dry-roasted peanuts	3.53	2.79	0.74
Granulated sugar	4.14	2.65	1.49
Grape jelly	2.47	1.65	0.82
Ground black pepper	2.04	1.74	0.30
Half & Half (quart)	3.24	2.57	0.67
Pancake mix	2.84	1.88	0.96
Pancake syrup	3.45	2.25	1.20
Pretzel twists	2.91	1.41	1.50
Quick rice	2.55	1.97	0.58
Raisin bran cereal	3.65	2.63	1.02
Salad dressing	2.55	2.02	0.53
Shredded mozzarella	3.29	2.33	0.96
Sour cream	2.20	1.33	0.87
Spicy brown mustard	2.77	1.86	0.91
Steak sauce	4.05	2.21	1.84
Sugar substitute	2.70	2.11	0.59
Zippered sandwich bags	2.55	1.99	0.56

a. Use the sign test to determine whether or not store-brand items **cost less** than their name-brand counterparts using $\alpha = .01$.

b. Use the Wilcoxon signed-rank test to determine if store-brand items **cost less** than their name-brand counterparts using $\alpha = .01$.

c. Use the paired-t test to determine if the average cost of store-brand items **is less** than the average cost of their name-brand counterparts using $\alpha = .01$.

d. Are the conclusions the same for all three tests? Would you expect them to be? Why or why not?

15.77 Legos® The time required for kinder-
EX1577 garten children to assemble a specific Lego creation was measured for children who had been instructed for four different lengths of time. Four children were randomly assigned to each instructional group, but two were

eliminated during the experiment because of sickness. The length of time (in minutes) to assemble the Lego creation was recorded for each child in the experiment.

Training Period (hours)			
.5	1.0	1.5	2.0
8	9	4	4
14	7	6	7
9	5	7	5
12	8		

Use the Kruskal–Wallis H-Test to determine whether there is a differnce in the distribution of times for the four different lengths of instructional time. Use $\alpha = .01$.

 15.78 Worker Fatigue To investigate methEX1578 ods of reducing fatigue among employees whose jobs involve a monotonous assembly procedure, 12 randomly selected employees were asked to perform their usual job under each of three trial conditions. As a measure of fatigue, the experimenter used the number of assembly line stoppages during a 4-hour period for each trial condition.

Employee	Conditions		
	1	2	3
1	31	22	26
2	20	15	23
3	26	21	18
4	31	22	32
5	12	16	18
6	22	29	34
7	28	17	26
8	15	9	12
9	41	31	46
10	19	19	25
11	31	34	41
12	18	11	21

a. What type of experimental design has been used in this experiment?

b. Use the appropriate nonparametric test to determine whether the distribution of assembly line stoppages (and consequently worker fatigue) differs for these three conditions. Test at the 5% level of significance.

15.79 Ranking Quarterbacks A ranking of the quarterbacks in the top eight teams of the National Football League was made by polling a number of professional football coaches and sportswriters. This "true ranking" is shown below, together with "my ranking."

	Quarterback							
	A	B	C	D	E	F	G	H
True Ranking	1	2	3	4	5	6	7	8
My Ranking	3	1	4	5	2	8	6	7

a. Calculate r_s.

b. Do the data indicate a positive correlation between my ranking and that of the experts? Test at the 5% level of significance.

CASE STUDY How's Your Cholesterol Level?

Eggs

As consumers become more and more interested in eating healthy foods, many "light," "fat-free," and "cholesterol-free" products are appearing in the marketplace. One such product is the frozen egg substitute, a cholesterol-free product that can be used in cooking and baking in many of the same ways that regular eggs can—though not all. Some consumers even use egg substitutes for Caesar salad dressings and other recipes calling for raw eggs because these products are pasteurized and thus eliminate worries about bacterial contamination.

Unfortunately, the products currently on the market exhibit strong differences in both flavor and texture when tasted in their primary preparation as scrambled eggs. Five panelists, all experts in nutrition and food preparation, were asked to rate each of three egg substitutes on the basis of taste, appearance, texture, and whether they would buy the product.[10] The judges tasted the three egg substitutes and rated them on a scale of 0–20. The results, shown in the table, indicate that the highest rating, by 23 points, went to ConAgra's Healthy Choice Egg Product, which the tasters unanimously agreed most closely resembled eggs as they come from the hen. The second-place product, Morningstar Farms' Scramblers, struck several tasters as having an "oddly sweet flavor . . . similar to carrots." Finally, none of the tasters indicated that they would be willing to buy Fleishmann's Egg Beaters, which was described by the testers as "watery," "slippery," and "unpleasant." Oddly enough, these results are contrary to a similar taste test done 4 years earlier, in which Egg Beaters were considered better than competing egg substitutes.

Taster	Healthy Choice	Scramblers	Egg Beaters
Dan Bowe	16	9	7
John Carroll	16	7	8
Donna Katzl	14	8	4
Rick O'Connell	15	16	9
Roland Passot	13	11	2
Total	74	51	30

Source: Data from "Eggs Substitutes Range in Quality," by K. Sakekel, *The San Francisco Chronicle,* February 10, 1993, p. 8. Copyright © 1993 San Francisco Chronicle.

1. What type of design has been used in this taste-testing experiment?
2. Do the data satisfy the assumptions required for a parametric analysis of variance? Explain.
3. Use the appropriate nonparametric technique to determine whether there is a significant difference between the average scores for the three brands of egg substitutes.

Appendix I
Tables

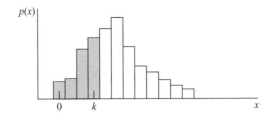

TABLE 1 Cumulative Binomial Probabilities

Tabulated values are $P(x \le k) = p(0) + p(1) + \cdots + p(k)$.
(Computations are rounded at the third decimal place.)

$n = 2$

k							p							k
	.01	.05	.10	.20	.30	.40	.50	.60	.70	.80	.90	.95	.99	
0	.980	.902	.810	.640	.490	.360	.250	.160	.090	.040	.010	.002	.000	0
1	1.000	.998	.990	.960	.910	.840	.750	.640	.510	.360	.190	.098	.020	1
2	1.000	1.000	1.000	1.000	1.000	1.000	1.000	1.000	1.000	1.000	1.000	1.000	1.000	2

$n = 3$

k							p							k
	.01	.05	.10	.20	.30	.40	.50	.60	.70	.80	.90	.95	.99	
0	.970	.857	.729	.512	.343	.216	.125	.064	.027	.008	.001	.000	.000	0
1	1.000	.993	.972	.896	.784	.648	.500	.352	.216	.104	.028	.007	.000	1
2	1.000	1.000	.999	.992	.973	.936	.875	.784	.657	.488	.271	.143	.030	2
3	1.000	1.000	1.000	1.000	1.000	1.000	1.000	1.000	1.000	1.000	1.000	1.000	1.000	3

$n = 4$

k							p							k
	.01	.05	.10	.20	.30	.40	.50	.60	.70	.80	.90	.95	.99	
0	.961	.815	.656	.410	.240	.130	.062	.026	.008	.002	.000	.000	.000	0
1	.999	.986	.948	.819	.652	.475	.312	.179	.084	.027	.004	.000	.000	1
2	1.000	1.000	.996	.973	.916	.821	.688	.525	.348	.181	.052	.014	.001	2
3	1.000	1.000	1.000	.998	.992	.974	.938	.870	.760	.590	.344	.185	.039	3
4	1.000	1.000	1.000	1.000	1.000	1.000	1.000	1.000	1.000	1.000	1.000	1.000	1.000	4

TABLE 1 *(continued)*

$n = 5$

							p								
k	.01	.05	.10	.20	.30	.40	.50	.60	.70	.80	.90	.95	.99	k
0	.951	.774	.590	.328	.168	.078	.031	.010	.002	.000	.000	.000	.000	0
1	.999	.977	.919	.737	.528	.337	.188	.087	.031	.007	.000	.000	.000	1
2	1.000	.999	.991	.942	.837	.683	.500	.317	.163	.058	.009	.001	.000	2
3	1.000	1.000	1.000	.993	.969	.913	.812	.663	.472	.263	.081	.023	.001	3
4	1.000	1.000	1.000	1.000	.998	.990	.969	.922	.832	.672	.410	.226	.049	4
5	1.000	1.000	1.000	1.000	1.000	1.000	1.000	1.000	1.000	1.000	1.000	1.000	1.000	5

$n = 6$

							p								
k	.01	.05	.10	.20	.30	.40	.50	.60	.70	.80	.90	.95	.99	k
0	.941	.735	.531	.262	.118	.047	.016	.004	.001	.000	.000	.000	.000	0
1	.999	.967	.886	.655	.420	.233	.109	.041	.011	.002	.000	.000	.000	1
2	1.000	.998	.984	.901	.744	.544	.344	.179	.070	.017	.001	.000	.000	2
3	1.000	1.000	.999	.983	.930	.821	.656	.456	.256	.099	.016	.002	.000	3
4	1.000	1.000	1.000	.998	.989	.959	.891	.767	.580	.345	.114	.033	.001	4
5	1.000	1.000	1.000	1.000	.999	.996	.984	.953	.882	.738	.469	.265	.059	5
6	1.000	1.000	1.000	1.000	1.000	1.000	1.000	1.000	1.000	1.000	1.000	1.000	1.000	6

$n = 7$

							p								
k	.01	.05	.10	.20	.30	.40	.50	.60	.70	.80	.90	.95	.99	k
0	.932	.698	.478	.210	.082	.028	.008	.002	.000	.000	.000	.000	.000	0
1	.998	.956	.850	.577	.329	.159	.062	.019	.004	.000	.000	.000	.000	1
2	1.000	.996	.974	.852	.647	.420	.227	.096	.029	.005	.000	.000	.000	2
3	1.000	1.000	.997	.967	.874	.710	.500	.290	.126	.033	.003	.000	.000	3
4	1.000	1.000	1.000	.995	.971	.904	.773	.580	.353	.148	.026	.004	.000	4
5	1.000	1.000	1.000	1.000	.996	.981	.938	.841	.671	.423	.150	.044	.002	5
6	1.000	1.000	1.000	1.000	1.000	.998	.992	.972	.918	.790	.522	.302	.068	6
7	1.000	1.000	1.000	1.000	1.000	1.000	1.000	1.000	1.000	1.000	1.000	1.000	1.000	7

$n = 8$

							p								
k	.01	.05	.10	.20	.30	.40	.50	.60	.70	.80	.90	.95	.99	k
0	.923	.663	.430	.168	.058	.017	.004	.001	.000	.000	.000	.000	.000	0
1	.997	.943	.813	.503	.255	.106	.035	.009	.001	.000	.000	.000	.000	1
2	1.000	.994	.962	.797	.552	.315	.145	.050	.011	.001	.000	.000	.000	2
3	1.000	1.000	.995	.944	.806	.594	.363	.174	.058	.010	.000	.000	.000	3
4	1.000	1.000	1.000	.990	.942	.826	.637	.406	.194	.056	.005	.000	.000	4
5	1.000	1.000	1.000	.999	.989	.950	.855	.685	.448	.203	.038	.006	.000	5
6	1.000	1.000	1.000	1.000	.999	.991	.965	.894	.745	.497	.187	.057	.003	6
7	1.000	1.000	1.000	1.000	1.000	.999	.996	.983	.942	.832	.570	.337	.077	7
8	1.000	1.000	1.000	1.000	1.000	1.000	1.000	1.000	1.000	1.000	1.000	1.000	1.000	8

TABLE 1 *(continued)*

n = 9

k	.01	.05	.10	.20	.30	.40	.50	.60	.70	.80	.90	.95	.99	*k*
0	.914	.630	.387	.134	.040	.010	.002	.000	.000	.000	.000	.000	.000	0
1	.997	.929	.775	.436	.196	.071	.020	.004	.000	.000	.000	.000	.000	1
2	1.000	.992	.947	.738	.463	.232	.090	.025	.004	.000	.000	.000	.000	2
3	1.000	.999	.992	.914	.730	.483	.254	.099	.025	.003	.000	.000	.000	3
4	1.000	1.000	.999	.980	.901	.733	.500	.267	.099	.020	.001	.000	.000	4
5	1.000	1.000	1.000	.997	.975	.901	.746	.517	.270	.086	.008	.001	.000	5
6	1.000	1.000	1.000	1.000	.996	.975	.910	.768	.537	.262	.053	.008	.000	6
7	1.000	1.000	1.000	1.000	1.000	.996	.980	.929	.804	.564	.225	.071	.003	7
8	1.000	1.000	1.000	1.000	1.000	1.000	.998	.990	.960	.866	.613	.370	.086	8
9	1.000	1.000	1.000	1.000	1.000	1.000	1.000	1.000	1.000	1.000	1.000	1.000	1.000	9

n = 10

k	.01	.05	.10	.20	.30	.40	.50	.60	.70	.80	.90	.95	.99	*k*
0	.904	.599	.349	.107	.028	.006	.001	.000	.000	.000	.000	.000	.000	0
1	.996	.914	.736	.376	.149	.046	.011	.002	.000	.000	.000	.000	.000	1
2	1.000	.988	.930	.678	.383	.167	.055	.012	.002	.000	.000	.000	.000	2
3	1.000	.999	.987	.879	.650	.382	.172	.055	.011	.001	.000	.000	.000	3
4	1.000	1.000	.998	.967	.850	.633	.377	.166	.047	.006	.000	.000	.000	4
5	1.000	1.000	1.000	.994	.953	.834	.623	.367	.150	.033	.002	.000	.000	5
6	1.000	1.000	1.000	.999	.989	.945	.828	.618	.350	.121	.013	.001	.000	6
7	1.000	1.000	1.000	1.000	.998	.988	.945	.833	.617	.322	.070	.012	.000	7
8	1.000	1.000	1.000	1.000	1.000	.998	.989	.954	.851	.624	.264	.086	.004	8
9	1.000	1.000	1.000	1.000	1.000	1.000	.999	.994	.972	.893	.651	.401	.096	9
10	1.000	1.000	1.000	1.000	1.000	1.000	1.000	1.000	1.000	1.000	1.000	1.000	1.000	10

n = 11

k	.01	.05	.10	.20	.30	.40	.50	.60	.70	.80	.90	.95	.99	*k*
0	.895	.569	.314	.086	.020	.004	.000	.000	.000	.000	.000	.000	.000	0
1	.995	.898	.697	.322	.113	.030	.006	.001	.000	.000	.000	.000	.000	1
2	1.000	.985	.910	.617	.313	.119	.033	.006	.001	.000	.000	.000	.000	2
3	1.000	.998	.981	.839	.570	.296	.113	.029	.004	.000	.000	.000	.000	3
4	1.000	1.000	.997	.950	.790	.533	.274	.099	.022	.002	.000	.000	.000	4
5	1.000	1.000	1.000	.988	.922	.754	.500	.246	.078	.012	.000	.000	.000	5
6	1.000	1.000	1.000	.998	.978	.901	.726	.467	.210	.050	.003	.000	.000	6
7	1.000	1.000	1.000	1.000	.996	.971	.887	.704	.430	.161	.019	.002	.000	7
8	1.000	1.000	1.000	1.000	.999	.994	.967	.881	.687	.383	.090	.015	.000	8
9	1.000	1.000	1.000	1.000	1.000	.999	.994	.970	.887	.678	.303	.102	.005	9
10	1.000	1.000	1.000	1.000	1.000	1.000	1.000	.996	.980	.914	.686	.431	.105	10
11	1.000	1.000	1.000	1.000	1.000	1.000	1.000	1.000	1.000	1.000	1.000	1.000	1.000	11

TABLE 1 *(continued)*

n = 12

							p							
k	.01	.05	.10	.20	.30	.40	.50	.60	.70	.80	.90	.95	.99	*k*
0	.886	.540	.282	.069	.014	.002	.000	.000	.000	.000	.000	.000	.000	0
1	.994	.882	.659	.275	.085	.020	.003	.000	.000	.000	.000	.000	.000	1
2	1.000	.980	.889	.558	.253	.083	.019	.003	.000	.000	.000	.000	.000	2
3	1.000	.998	.974	.795	.493	.225	.073	.015	.002	.000	.000	.000	.000	3
4	1.000	1.000	.996	.927	.724	.438	.194	.057	.009	.001	.000	.000	.000	4
5	1.000	1.000	.999	.981	.882	.665	.387	.158	.039	.004	.000	.000	.000	5
6	1.000	1.000	1.000	.996	.961	.842	.613	.335	.118	.019	.001	.000	.000	6
7	1.000	1.000	1.000	.999	.991	.943	.806	.562	.276	.073	.004	.000	.000	7
8	1.000	1.000	1.000	1.000	.998	.985	.927	.775	.507	.205	.026	.002	.000	8
9	1.000	1.000	1.000	1.000	1.000	.997	.981	.917	.747	.442	.111	.020	.000	9
10	1.000	1.000	1.000	1.000	1.000	1.000	.997	.980	.915	.725	.341	.118	.006	10
11	1.000	1.000	1.000	1.000	1.000	1.000	1.000	.998	.986	.931	.718	.460	.114	11
12	1.000	1.000	1.000	1.000	1.000	1.000	1.000	1.000	1.000	1.000	1.000	1.000	1.000	12

n = 15

							p							
k	.01	.05	.10	.20	.30	.40	.50	.60	.70	.80	.90	.95	.99	*k*
0	.860	.463	.206	.035	.005	.000	.000	.000	.000	.000	.000	.000	.000	0
1	.990	.829	.549	.167	.035	.005	.000	.000	.000	.000	.000	.000	.000	1
2	1.000	.964	.816	.398	.127	.027	.004	.000	.000	.000	.000	.000	.000	2
3	1.000	.995	.944	.648	.297	.091	.018	.002	.000	.000	.000	.000	.000	3
4	1.000	.999	.987	.836	.515	.217	.059	.009	.001	.000	.000	.000	.000	4
5	1.000	1.000	.998	.939	.722	.403	.151	.034	.004	.000	.000	.000	.000	5
6	1.000	1.000	1.000	.982	.869	.610	.304	.095	.015	.001	.000	.000	.000	6
7	1.000	1.000	1.000	.996	.950	.787	.500	.213	.050	.004	.000	.000	.000	7
8	1.000	1.000	1.000	.999	.985	.905	.696	.390	.131	.018	.000	.000	.000	8
9	1.000	1.000	1.000	1.000	.996	.966	.849	.597	.278	.061	.002	.000	.000	9
10	1.000	1.000	1.000	1.000	.999	.991	.941	.783	.485	.164	.013	.001	.000	10
11	1.000	1.000	1.000	1.000	1.000	.998	.982	.909	.703	.352	.056	.005	.000	11
12	1.000	1.000	1.000	1.000	1.000	1.000	.996	.973	.873	.602	.184	.036	.000	12
13	1.000	1.000	1.000	1.000	1.000	1.000	1.000	.995	.965	.833	.451	.171	.010	13
14	1.000	1.000	1.000	1.000	1.000	1.000	1.000	1.000	.995	.965	.794	.537	.140	14
15	1.000	1.000	1.000	1.000	1.000	1.000	1.000	1.000	1.000	1.000	1.000	1.000	1.000	15

TABLE 1 *(continued)*

$n = 20$

							p							
k	.01	.05	.10	.20	.30	.40	.50	.60	.70	.80	.90	.95	.99	k
0	.818	.358	.122	.012	.001	.000	.000	.000	.000	.000	.000	.000	.000	0
1	.983	.736	.392	.069	.008	.001	.000	.000	.000	.000	.000	.000	.000	1
2	.999	.925	.677	.206	.035	.004	.000	.000	.000	.000	.000	.000	.000	2
3	1.000	.984	.867	.411	.107	.016	.001	.000	.000	.000	.000	.000	.000	3
4	1.000	.997	.957	.630	.238	.051	.006	.000	.000	.000	.000	.000	.000	4
5	1.000	1.000	.989	.804	.416	.126	.021	.002	.000	.000	.000	.000	.000	5
6	1.000	1.000	.998	.913	.608	.250	.058	.006	.000	.000	.000	.000	.000	6
7	1.000	1.000	1.000	.968	.772	.416	.132	.021	.001	.000	.000	.000	.000	7
8	1.000	1.000	1.000	.990	.887	.596	.252	.057	.005	.000	.000	.000	.000	8
9	1.000	1.000	1.000	.997	.952	.755	.412	.128	.017	.001	.000	.000	.000	9
10	1.000	1.000	1.000	.999	.983	.872	.588	.245	.048	.003	.000	.000	.000	10
11	1.000	1.000	1.000	1.000	.995	.943	.748	.404	.113	.010	.000	.000	.000	11
12	1.000	1.000	1.000	1.000	.999	.979	.868	.584	.228	.032	.000	.000	.000	12
13	1.000	1.000	1.000	1.000	1.000	.994	.942	.750	.392	.087	.002	.000	.000	13
14	1.000	1.000	1.000	1.000	1.000	.998	.979	.874	.584	.196	.011	.000	.000	14
15	1.000	1.000	1.000	1.000	1.000	1.000	.994	.949	.762	.370	.043	.003	.000	15
16	1.000	1.000	1.000	1.000	1.000	1.000	.999	.984	.893	.589	.133	.016	.000	16
17	1.000	1.000	1.000	1.000	1.000	1.000	1.000	.996	.965	.794	.323	.075	.001	17
18	1.000	1.000	1.000	1.000	1.000	1.000	1.000	.999	.992	.931	.608	.264	.017	18
19	1.000	1.000	1.000	1.000	1.000	1.000	1.000	1.000	.999	.988	.878	.642	.182	19
20	1.000	1.000	1.000	1.000	1.000	1.000	1.000	1.000	1.000	1.000	1.000	1.000	1.000	20

TABLE 1 *(continued)*

$n = 25$

k	.01	.05	.10	.20	.30	.40	.50	.60	.70	.80	.90	.95	.99	k
0	.778	.277	.072	.004	.000	.000	.000	.000	.000	.000	.000	.000	.000	0
1	.974	.642	.271	.027	.002	.000	.000	.000	.000	.000	.000	.000	.000	1
2	.998	.873	.537	.098	.009	.000	.000	.000	.000	.000	.000	.000	.000	2
3	1.000	.966	.764	.234	.033	.002	.000	.000	.000	.000	.000	.000	.000	3
4	1.000	.993	.902	.421	.090	.009	.000	.000	.000	.000	.000	.000	.000	4
5	1.000	.999	.967	.617	.193	.029	.002	.000	.000	.000	.000	.000	.000	5
6	1.000	1.000	.991	.780	.341	.074	.007	.000	.000	.000	.000	.000	.000	6
7	1.000	1.000	.998	.891	.512	.154	.022	.001	.000	.000	.000	.000	.000	7
8	1.000	1.000	1.000	.953	.677	.274	.054	.004	.000	.000	.000	.000	.000	8
9	1.000	1.000	1.000	.983	.811	.425	.115	.013	.000	.000	.000	.000	.000	9
10	1.000	1.000	1.000	.994	.902	.586	.212	.034	.002	.000	.000	.000	.000	10
11	1.000	1.000	1.000	.998	.956	.732	.345	.078	.006	.000	.000	.000	.000	11
12	1.000	1.000	1.000	1.000	.983	.846	.500	.154	.017	.000	.000	.000	.000	12
13	1.000	1.000	1.000	1.000	.994	.922	.655	.268	.044	.002	.000	.000	.000	13
14	1.000	1.000	1.000	1.000	.998	.966	.788	.414	.098	.006	.000	.000	.000	14
15	1.000	1.000	1.000	1.000	1.000	.987	.885	.575	.189	.017	.000	.000	.000	15
16	1.000	1.000	1.000	1.000	1.000	.996	.946	.726	.323	.047	.000	.000	.000	16
17	1.000	1.000	1.000	1.000	1.000	.999	.978	.846	.488	.109	.002	.000	.000	17
18	1.000	1.000	1.000	1.000	1.000	1.000	.993	.926	.659	.220	.009	.000	.000	18
19	1.000	1.000	1.000	1.000	1.000	1.000	.998	.971	.807	.383	.033	.001	.000	19
20	1.000	1.000	1.000	1.000	1.000	1.000	1.000	.991	.910	.579	.098	.007	.000	20
21	1.000	1.000	1.000	1.000	1.000	1.000	1.000	.998	.967	.766	.236	.034	.000	21
22	1.000	1.000	1.000	1.000	1.000	1.000	1.000	1.000	.991	.902	.463	.127	.002	22
23	1.000	1.000	1.000	1.000	1.000	1.000	1.000	1.000	.998	.973	.729	.358	.026	23
24	1.000	1.000	1.000	1.000	1.000	1.000	1.000	1.000	1.000	.996	.928	.723	.222	24
25	1.000	1.000	1.000	1.000	1.000	1.000	1.000	1.000	1.000	1.000	1.000	1.000	1.000	25

TABLE 2 Cumulative Poisson Probabilities

Tabulated values are $P(x \leq k) = p(0) + p(1) + \cdots + p(k)$.
(Computations are rounded at the third decimal place.)

						μ						
k	.1	.2	.3	.4	.5	.6	.7	.8	.9	1.0	1.5
0	.905	.819	.741	.670	.607	.549	.497	.449	.407	.368	.223
1	.995	.982	.963	.938	.910	.878	.844	.809	.772	.736	.558
2	1.000	.999	.996	.992	.986	.977	.966	.953	.937	.920	.809
3		1.000	1.000	.999	.998	.997	.994	.991	.987	.981	.934
4				1.000	1.000	1.000	.999	.999	.998	.996	.981
5							1.000	1.000	1.000	.999	.996
6										1.000	.999
7											1.000

						μ					
k	2.0	2.5	3.0	3.5	4.0	4.5	5.0	5.5	6.0	6.5	7.0
0	.135	.082	.050	.033	.018	.011	.007	.004	.003	.002	.001
1	.406	.287	.199	.136	.092	.061	.040	.027	.017	.011	.007
2	.677	.544	.423	.321	.238	.174	.125	.088	.062	.043	.030
3	.857	.758	.647	.537	.433	.342	.265	.202	.151	.112	.082
4	.947	.891	.815	.725	.629	.532	.440	.358	.285	.224	.173
5	.983	.958	.916	.858	.785	.703	.616	.529	.446	.369	.301
6	.995	.986	.966	.935	.889	.831	.762	.686	.606	.563	.450
7	.999	.996	.988	.973	.949	.913	.867	.809	.744	.673	.599
8	1.000	.999	.996	.990	.979	.960	.932	.894	.847	.792	.729
9		1.000	.999	.997	.992	.983	.968	.946	.916	.877	.830
10			1.000	.999	.997	.993	.986	.975	.957	.933	.901
11				1.000	.999	.998	.995	.989	.980	.966	.947
12					1.000	.999	.998	.996	.991	.984	.973
13						1.000	.999	.998	.996	.993	.987
14							1.000	.999	.999	.997	.994
15								1.000	.999	.999	.998
16									1.000	1.000	.999
17											1.000

TABLE 2 *(continued)*

					μ				
k	7.5	8.0	8.5	9.0	9.5	10.0	12.0	15.0	20.0
0	.001	.000	.000	.000	.000	.000	.000	.000	.000
1	.005	.003	.002	.001	.001	.000	.000	.000	.000
2	.020	.014	.009	.006	.004	.003	.001	.000	.000
3	.059	.042	.030	.021	.015	.010	.002	.000	.000
4	.132	.100	.074	.055	.040	.029	.008	.001	.000
5	.241	.191	.150	.116	.089	.067	.020	.003	.000
6	.378	.313	.256	.207	.165	.130	.046	.008	.000
7	.525	.453	.386	.324	.269	.220	.090	.018	.001
8	.662	.593	.523	.456	.392	.333	.155	.037	.002
9	.776	.717	.653	.587	.522	.458	.242	.070	.005
10	.862	.816	.763	.706	.645	.583	.347	.118	.011
11	.921	.888	.849	.803	.752	.697	.462	.185	.021
12	.957	.936	.909	.876	.836	.792	.576	.268	.039
13	.978	.966	.949	.926	.898	.864	.682	.363	.066
14	.990	.983	.973	.959	.940	.917	.772	.466	.105
15	.995	.992	.986	.978	.967	.951	.844	.568	.157
16	.998	.996	.993	.989	.982	.973	.899	.664	.221
17	.999	.998	.997	.995	.991	.986	.937	.749	.297
18	1.000	.999	.999	.998	.996	.993	.963	.819	.381
19		1.000	.999	.999	.998	.997	.979	.875	.470
20			1.000	1.000	.999	.998	.988	.917	.559
21					1.000	.999	.994	.947	.644
22						1.000	.997	.967	.721
23							.999	.981	.787
24							.999	.989	.843
25							1.000	.994	.888
26								.997	.922
27								.998	.948
28								.999	.966
29								1.000	.978
30									.987
31									.992
32									.995
33									.997
34									.999
35									.999
36									1.000

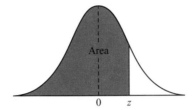

TABLE 3   Areas under the Normal Curve

z	.00	.01	.02	.03	.04	.05	.06	.07	.08	.09
−3.4	.0003	.0003	.0003	.0003	.0003	.0003	.0003	.0003	.0003	.0002
−3.3	.0005	.0005	.0005	.0004	.0004	.0004	.0004	.0004	.0004	.0003
−3.2	.0007	.0007	.0006	.0006	.0006	.0006	.0006	.0005	.0005	.0005
−3.1	.0010	.0009	.0009	.0009	.0008	.0008	.0008	.0008	.0007	.0007
−3.0	.0013	.0013	.0013	.0012	.0012	.0011	.0011	.0011	.0010	.0010
−2.9	.0019	.0018	.0017	.0017	.0016	.0016	.0015	.0015	.0014	.0014
−2.8	.0026	.0025	.0024	.0023	.0023	.0022	.0021	.0021	.0020	.0019
−2.7	.0035	.0034	.0033	.0032	.0031	.0030	.0029	.0028	.0027	.0026
−2.6	.0047	.0045	.0044	.0043	.0041	.0040	.0039	.0038	.0037	.0036
−2.5	.0062	.0060	.0059	.0057	.0055	.0054	.0052	.0051	.0049	.0048
−2.4	.0082	.0080	.0078	.0075	.0073	.0071	.0069	.0068	.0066	.0064
−2.3	.0107	.0104	.0102	.0099	.0096	.0094	.0091	.0089	.0087	.0084
−2.2	.0139	.0136	.0132	.0129	.0125	.0122	.0119	.0116	.0113	.0110
−2.1	.0179	.0174	.0170	.0166	.0162	.0158	.0154	.0150	.0146	.0143
−2.0	.0228	.0222	.0217	.0212	.0207	.0202	.0197	.0192	.0188	.0183
−1.9	.0287	.0281	.0274	.0268	.0262	.0256	.0250	.0244	.0239	.0233
−1.8	.0359	.0351	.0344	.0336	.0329	.0322	.0314	.0307	.0301	.0294
−1.7	.0446	.0436	.0427	.0418	.0409	.0401	.0392	.0384	.0375	.0367
−1.6	.0548	.0537	.0526	.0516	.0505	.0495	.0485	.0475	.0465	.0455
−1.5	.0668	.0655	.0643	.0630	.0618	.0606	.0594	.0582	.0571	.0559
−1.4	.0808	.0793	.0778	.0764	.0749	.0735	.0722	.0708	.0694	.0681
−1.3	.0968	.0951	.0934	.0918	.0901	.0885	.0869	.0853	.0838	.0823
−1.2	.1151	.1131	.1112	.1093	.1075	.1056	.1038	.1020	.1003	.0985
−1.1	.1357	.1335	.1314	.1292	.1271	.1251	.1230	.1210	.1190	.1170
−1.0	.1587	.1562	.1539	.1515	.1492	.1469	.1446	.1423	.1401	.1379
−0.9	.1841	.1814	.1788	.1762	.1736	.1711	.1685	.1660	.1635	.1611
−0.8	.2119	.2090	.2061	.2033	.2005	.1977	.1949	.1922	.1894	.1867
−0.7	.2420	.2389	.2358	.2327	.2296	.2266	.2236	.2206	.2177	.2148
−0.6	.2743	.2709	.2676	.2643	.2611	.2578	.2546	.2514	.2483	.2451
−0.5	.3085	.3050	.3015	.2981	.2946	.2912	.2877	.2843	.2810	.2776
−0.4	.3446	.3409	.3372	.3336	.3300	.3264	.3228	.3192	.3156	.3121
−0.3	.3821	.3783	.3745	.3707	.3669	.3632	.3594	.3557	.3520	.3483
−0.2	.4207	.4168	.4129	.4090	.4052	.4013	.3974	.3936	.3897	.3859
−0.1	.4602	.4562	.4522	.4483	.4443	.4404	.4364	.4325	.4286	.4247
−0.0	.5000	.4960	.4920	.4880	.4840	.4801	.4761	.4721	.4681	.4641

TABLE 3 *(continued)*

z	.00	.01	.02	.03	.04	.05	.06	.07	.08	.09
0.0	.5000	.5040	.5080	.5120	.5160	.5199	.5239	.5279	.5319	.5359
0.1	.5398	.5438	.5478	.5517	.5557	.5596	.5636	.5675	.5714	.5753
0.2	.5793	.5832	.5871	.5910	.5948	.5987	.6026	.6064	.6103	.6141
0.3	.6179	.6217	.6255	.6293	.6331	.6368	.6406	.6443	.6480	.6517
0.4	.6554	.6591	.6628	.6664	.6700	.6736	.6772	.6808	.6844	.6879
0.5	.6915	.6950	.6985	.7019	.7054	.7088	.7123	.7157	.7190	.7224
0.6	.7257	.7291	.7324	.7357	.7389	.7422	.7454	.7486	.7517	.7549
0.7	.7580	.7611	.7642	.7673	.7704	.7734	.7764	.7794	.7823	.7852
0.8	.7881	.7910	.7939	.7967	.7995	.8023	.8051	.8078	.8106	.8133
0.9	.8159	.8186	.8212	.8238	.8264	.8289	.8315	.8340	.8365	.8389
1.0	.8413	.8438	.8461	.8485	.8508	.8531	.8554	.8577	.8599	.8621
1.1	.8643	.8665	.8686	.8708	.8729	.8749	.8770	.8790	.8810	.8830
1.2	.8849	.8869	.8888	.8907	.8925	.8944	.8962	.8980	.8997	.9015
1.3	.9032	.9049	.9066	.9082	.9099	.9115	.9131	.9147	.9162	.9177
1.4	.9192	.9207	.9222	.9236	.9251	.9265	.9279	.9292	.9306	.9319
1.5	.9332	.9345	.9357	.9370	.9382	.9394	.9406	.9418	.9429	.9441
1.6	.9452	.9463	.9474	.9484	.9495	.9505	.9515	.9525	.9535	.9545
1.7	.9554	.9564	.9573	.9582	.9591	.9599	.9608	.9616	.9625	.9633
1.8	.9641	.9649	.9656	.9664	.9671	.9678	.9686	.9693	.9699	.9706
1.9	.9713	.9719	.9726	.9732	.9738	.9744	.9750	.9756	.9761	.9767
2.0	.9772	.9778	.9783	.9788	.9793	.9798	.9803	.9808	.9812	.9817
2.1	.9821	.9826	.9830	.9834	.9838	.9842	.9846	.9850	.9854	.9857
2.2	.9861	.9864	.9868	.9871	.9875	.9878	.9881	.9884	.9887	.9890
2.3	.9893	.9896	.9898	.9901	.9904	.9906	.9909	.9911	.9913	.9916
2.4	.9918	.9920	.9922	.9925	.9927	.9929	.9931	.9932	.9934	.9936
2.5	.9938	.9940	.9941	.9943	.9945	.9946	.9948	.9949	.9951	.9952
2.6	.9953	.9955	.9956	.9957	.9959	.9960	.9961	.9962	.9963	.9964
2.7	.9965	.9966	.9967	.9968	.9969	.9970	.9971	.9972	.9973	.9974
2.8	.9974	.9975	.9976	.9977	.9977	.9978	.9979	.9979	.9980	.9981
2.9	.9981	.9982	.9982	.9983	.9984	.9984	.9985	.9985	.9986	.9986
3.0	.9987	.9987	.9987	.9988	.9988	.9989	.9989	.9989	.9990	.9990
3.1	.9990	.9991	.9991	.9991	.9992	.9992	.9992	.9992	.9993	.9993
3.2	.9993	.9993	.9994	.9994	.9994	.9994	.9994	.9995	.9995	.9995
3.3	.9995	.9995	.9995	.9996	.9996	.9996	.9996	.9996	.9996	.9997
3.4	.9997	.9997	.9997	.9997	.9997	.9997	.9997	.9997	.9997	.9998

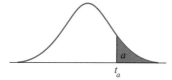

TABLE 4
Critical Values
of *t*

df	$t_{.100}$	$t_{.050}$	$t_{.025}$	$t_{.010}$	$t_{.005}$	df
1	3.078	6.314	12.706	31.821	63.657	1
2	1.886	2.920	4.303	6.965	9.925	2
3	1.638	2.353	3.182	4.541	5.841	3
4	1.533	2.132	2.776	3.747	4.604	4
5	1.476	2.015	2.571	3.365	4.032	5
6	1.440	1.943	2.447	3.143	3.707	6
7	1.415	1.895	2.365	2.998	3.499	7
8	1.397	1.860	2.306	2.896	3.355	8
9	1.383	1.833	2.262	2.821	3.250	9
10	1.372	1.812	2.228	2.764	3.169	10
11	1.363	1.796	2.201	2.718	3.106	11
12	1.356	1.782	2.179	2.681	3.055	12
13	1.350	1.771	2.160	2.650	3.012	13
14	1.345	1.761	2.145	2.624	2.977	14
15	1.341	1.753	2.131	2.602	2.947	15
16	1.337	1.746	2.120	2.583	2.921	16
17	1.333	1.740	2.110	2.567	2.898	17
18	1.330	1.734	2.101	2.552	2.878	18
19	1.328	1.729	2.093	2.539	2.861	19
20	1.325	1.725	2.086	2.528	2.845	20
21	1.323	1.721	2.080	2.518	2.831	21
22	1.321	1.717	2.074	2.508	2.819	22
23	1.319	1.714	2.069	2.500	2.807	23
24	1.318	1.711	2.064	2.492	2.797	24
25	1.316	1.708	2.060	2.485	2.787	25
26	1.315	1.706	2.056	2.479	2.779	26
27	1.314	1.703	2.052	2.473	2.771	27
28	1.313	1.701	2.048	2.467	2.763	28
29	1.311	1.699	2.045	2.462	2.756	29
∞	1.282	1.645	1.960	2.326	2.576	∞

Source: From "Table of Percentage Points of the *t*-Distribution," *Biometrika* 32 (1941):300. Reproduced by permission of the *Biometrika* Trustees.

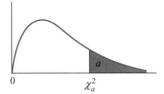

df	$\chi^2_{.995}$	$\chi^2_{.990}$	$\chi^2_{.975}$	$\chi^2_{.950}$	$\chi^2_{.900}$
1	.0000393	.0001571	.0009821	.0039321	.0157908
2	.0100251	.0201007	.0506356	.102587	.210720
3	.0717212	.114832	.215795	.351846	.584375
4	.206990	.297110	.484419	.710721	1.063623
5	.411740	.554300	.831211	1.145476	1.61031
6	.675727	.872085	1.237347	1.63539	2.20413
7	.989265	1.239043	1.68987	2.16735	2.83311
8	1.344419	1.646482	2.17973	2.73264	3.48954
9	1.734926	2.087912	2.70039	3.32511	4.16816
10	2.15585	2.55821	3.24697	3.94030	4.86518
11	2.60321	3.05347	3.81575	4.57481	5.57779
12	3.07382	3.57056	4.40379	5.22603	6.30380
13	3.56503	4.10691	5.00874	5.89186	7.04150
14	4.07468	4.66043	5.62872	6.57063	7.78953
15	4.60094	5.22935	6.26214	7.26094	8.54675
16	5.14224	5.81221	6.90766	7.96164	9.31223
17	5.69724	6.40776	7.56418	8.67176	10.0852
18	6.26481	7.01491	8.23075	9.39046	10.8649
19	6.84398	7.63273	8.90655	10.1170	11.6509
20	7.43386	8.26040	9.59083	10.8508	12.4426
21	8.03366	8.89720	10.28293	11.5913	13.2396
22	8.64272	9.54249	10.9823	12.3380	14.0415
23	9.26042	10.19567	11.6885	13.0905	14.8479
24	9.88623	10.8564	12.4011	13.8484	15.6587
25	10.5197	11.5240	13.1197	14.6114	16.4734
26	11.1603	12.1981	13.8439	15.3791	17.2919
27	11.8076	12.8786	14.5733	16.1513	18.1138
28	12.4613	13.5648	15.3079	16.9279	18.9392
29	13.1211	14.2565	16.0471	17.7083	19.7677
30	13.7867	14.9535	16.7908	18.4926	20.5992
40	20.7065	22.1643	24.4331	26.5093	29.0505
50	27.9907	29.7067	32.3574	34.7642	37.6886
60	35.5346	37.4848	40.4817	43.1879	46.4589
70	43.2752	45.4418	48.7576	51.7393	55.3290
80	51.1720	53.5400	57.1532	60.3915	64.2778
90	59.1963	61.7541	65.6466	69.1260	73.2912
100	67.3276	70.0648	74.2219	77.9295	82.3581

Source: From "Tables of the Percentage Points of the χ^2-Distribution," *Biometrika Tables for Statisticians,* Vol. 1, 3rd ed. (1966). Reproduced by permission of the *Biometrika* Trustees.

TABLE 5
(continued)

$\chi^2_{.100}$	$\chi^2_{.050}$	$\chi^2_{.025}$	$\chi^2_{.010}$	$\chi^2_{.005}$	df
2.70554	3.84146	5.02389	6.63490	7.87944	1
4.60517	5.99147	7.37776	9.21034	10.5966	2
6.25139	7.81473	9.34840	11.3449	12.8381	3
7.77944	9.48773	11.1433	13.2767	14.8602	4
9.23635	11.0705	12.8325	15.0863	16.7496	5
10.6446	12.5916	14.4494	16.8119	18.5476	6
12.0170	14.0671	16.0128	18.4753	20.2777	7
13.3616	15.5073	17.5346	20.0902	21.9550	8
14.6837	16.9190	19.0228	21.6660	23.5893	9
15.9871	18.3070	20.4831	23.2093	25.1882	10
17.2750	19.6751	21.9200	24.7250	26.7569	11
18.5494	21.0261	23.3367	26.2170	28.2995	12
19.8119	22.3621	24.7356	27.6883	29.8194	13
21.0642	23.6848	26.1190	29.1413	31.3193	14
22.3072	24.9958	27.4884	30.5779	32.8013	15
23.5418	26.2962	28.8485	31.9999	34.2672	16
24.7690	27.8571	30.1910	33.4087	35.7185	17
25.9894	28.8693	31.5264	34.8053	37.1564	18
27.2036	30.1435	32.8523	36.1908	38.5822	19
28.4120	31.4104	34.1696	37.5662	39.9968	20
29.6151	32.6705	35.4789	38.9321	41.4010	21
30.8133	33.9244	36.7807	40.2894	42.7956	22
32.0069	35.1725	38.0757	41.6384	44.1813	23
33.1963	36.4151	39.3641	42.9798	45.5585	24
34.3816	37.6525	40.6465	44.3141	46.9278	25
35.5631	38.8852	41.9232	45.6417	48.2899	26
36.7412	40.1133	43.1944	46.9630	49.6449	27
37.9159	41.3372	44.4607	48.2782	50.9933	28
39.0875	42.5569	45.7222	49.5879	52.3356	29
40.2560	43.7729	46.9792	50.8922	53.6720	30
51.8050	55.7585	59.3417	63.6907	66.7659	40
63.1671	67.5048	71.4202	76.1539	79.4900	50
74.3970	79.0819	83.2976	88.3794	91.9517	60
85.5271	90.5312	95.0231	100.425	104.215	70
96.5782	101.879	106.629	112.329	116.321	80
107.565	113.145	118.136	124.116	128.299	90
118.498	124.342	129.561	135.807	140.169	100

TABLE 6 Percentage Points of the *F* Distribution

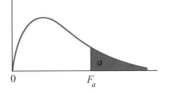

df_2	a					df_1				
		1	2	3	4	5	6	7	8	9
1	.100	39.86	49.50	53.59	55.83	57.24	58.20	58.91	59.44	59.86
	.050	161.4	199.5	215.7	224.6	230.2	234.0	236.8	238.9	240.5
	.025	647.8	799.5	864.2	899.6	921.8	937.1	948.2	956.7	963.3
	.010	4052	4999.5	5403	5625	5764	5859	5928	5982	6022
	.005	16211	20000	21615	22500	23056	23437	23715	23925	24091
2	.100	8.53	9.00	9.16	9.24	9.29	9.33	9.35	9.37	9.38
	.050	18.51	19.00	19.16	19.25	19.30	19.33	19.35	19.37	19.38
	.025	38.51	39.00	39.17	39.25	39.30	39.33	39.36	39.37	39.39
	.010	98.50	99.00	99.17	99.25	99.30	99.33	99.36	99.37	99.39
	.005	198.5	199.0	199.2	199.2	199.3	199.3	199.4	199.4	199.4
3	.100	5.54	5.46	5.39	5.34	5.31	5.28	5.27	5.25	5.24
	.050	10.13	9.55	9.28	9.12	9.01	8.94	8.89	8.85	8.81
	.025	17.44	16.04	15.44	15.10	14.88	14.73	14.62	14.54	14.47
	.010	34.12	30.82	29.46	28.71	28.24	27.91	27.64	27.49	27.35
	.005	55.55	49.80	47.47	46.19	45.39	44.84	44.43	44.13	43.88
4	.100	4.54	4.32	4.19	4.11	4.05	4.01	3.98	3.95	3.94
	.050	7.71	6.94	6.59	6.39	6.26	6.16	6.09	6.04	6.00
	.025	12.22	10.65	9.98	9.60	9.36	9.20	9.07	8.98	8.90
	.010	21.20	18.00	16.69	15.98	15.52	15.21	14.98	14.80	14.66
	.005	31.33	26.28	24.26	23.15	22.46	21.97	21.62	21.35	21.14
5	.100	4.06	3.78	3.62	3.52	3.45	3.40	3.37	3.34	3.32
	.050	6.61	5.79	5.41	5.19	5.05	4.95	4.88	4.82	4.77
	.025	10.01	8.43	7.76	7.39	7.15	6.98	6.85	6.76	6.68
	.010	16.26	13.27	12.06	11.39	10.97	10.67	10.46	10.29	10.16
	.005	22.78	18.31	16.53	15.56	14.94	14.51	14.20	13.96	13.77
6	.100	3.78	3.46	3.29	3.18	3.11	3.05	3.01	2.98	2.96
	.050	5.99	5.14	4.76	4.53	4.39	4.28	4.21	4.15	4.10
	.025	8.81	7.26	6.60	6.23	5.99	5.82	5.70	5.60	5.52
	.010	13.75	10.92	9.78	9.15	8.75	8.47	8.26	8.10	7.98
	.005	18.63	14.54	12.92	12.03	11.46	11.07	10.79	10.57	10.39
7	.100	3.59	3.26	3.07	2.96	2.88	2.83	2.78	2.75	2.72
	.050	5.59	4.74	4.35	4.12	3.97	3.87	3.79	3.73	3.68
	.025	8.07	6.54	5.89	5.52	5.29	5.12	4.99	4.90	4.82
	.010	12.25	9.55	8.45	7.85	7.46	7.19	6.99	6.84	6.72
	.005	16.24	12.40	10.88	10.05	9.52	9.16	8.89	8.68	8.51
8	.100	3.46	3.11	2.92	2.81	2.73	2.67	2.62	2.59	2.56
	.050	5.32	4.46	4.07	3.84	3.69	3.58	3.50	3.44	3.39
	.025	7.57	6.06	5.42	5.05	4.82	4.65	4.53	4.43	4.36
	.010	11.26	8.65	7.59	7.01	6.63	6.37	6.18	6.03	5.91
	.005	14.69	11.04	9.60	8.81	8.30	7.95	7.69	7.50	7.34
9	.100	3.36	3.01	2.81	2.69	2.61	2.55	2.51	2.47	2.44
	.050	5.12	4.26	3.86	3.63	3.48	3.37	3.29	3.23	3.18
	.025	7.21	5.71	5.08	4.72	4.48	4.32	4.20	4.10	4.03
	.010	10.56	8.02	6.99	6.42	6.06	5.80	5.61	5.47	5.35
	.005	13.61	10.11	8.72	7.96	7.47	7.13	6.88	6.69	6.54

Source: A portion of "Tables of Percentage Points of the Inverted Beta (*F*) Distribution," *Biometrika,* Vol. 33 (1943) by M. Merrington and C.M. Thompson and from Table 18 of *Biometrika Tables for Statisticians,* Vol. 1, Cambridge University Press, 1954, edited by E.S. Pearson and H.O. Hartley. Reproduced with permission of the authors, editors, and *Biometrika* Trustees.

TABLE 6 *(continued)*

	10	12	15	20	24	30	40	60	120	∞	a	df₂
df_1												
	60.19	60.71	61.22	61.74	62.00	62.26	62.53	62.79	63.06	63.33	.100	1
	241.9	243.9	245.9	248.0	249.1	250.1	251.2	252.2	253.3	254.3	.050	
	968.6	976.7	984.9	993.1	997.2	1001	1006	1010	1014	1018	.025	
	6056	6106	6157	6209	6235	6261	6287	6313	6339	6366	.010	
	24224	24426	24630	24836	24940	25044	25148	25253	25359	25465	.005	
	9.39	9.41	9.42	9.44	9.45	9.46	9.47	9.47	9.48	9.49	.100	2
	19.40	19.41	19.43	19.45	19.45	19.46	19.47	19.48	19.49	19.50	.050	
	39.40	39.41	39.43	39.45	39.46	39.46	39.47	39.48	39.49	39.50	.025	
	99.40	99.42	99.43	99.45	99.46	99.47	99.47	99.48	99.49	99.50	.010	
	199.4	199.4	199.4	199.4	199.5	199.5	199.5	199.5	199.5	199.5	.005	
	5.23	5.22	5.20	5.18	5.18	5.17	5.16	5.15	5.14	5.13	.100	3
	8.79	8.74	8.70	8.66	8.64	8.62	8.59	8.57	8.55	8.53	.050	
	14.42	14.34	14.25	14.17	14.12	14.08	14.04	13.99	13.95	13.90	.025	
	27.23	27.05	26.87	26.69	26.60	26.50	26.41	26.32	26.22	26.13	.010	
	43.69	43.39	43.08	42.78	42.62	42.47	42.31	42.15	41.99	41.83	.005	
	3.92	3.90	3.87	3.84	3.83	3.82	3.80	3.79	3.78	3.76	.100	4
	5.96	5.91	5.86	5.80	5.77	5.75	5.72	5.69	5.66	5.63	.050	
	8.84	8.75	8.66	8.56	8.51	8.46	8.41	8.36	8.31	8.26	.025	
	14.55	14.37	14.20	14.02	13.93	13.84	13.75	13.65	13.56	13.46	.010	
	20.97	20.70	20.44	20.17	20.03	19.89	19.75	19.61	19.47	19.32	.005	
	3.30	3.27	3.24	3.21	3.19	3.17	3.16	3.14	3.12	3.10	.100	5
	4.74	4.68	4.62	4.56	4.53	4.50	4.46	4.43	4.40	4.36	.050	
	6.62	6.52	6.43	6.33	6.28	6.23	6.18	6.12	6.07	6.02	.025	
	10.05	9.89	9.72	9.55	9.47	9.38	9.29	9.20	9.11	9.02	.010	
	13.62	13.38	13.15	12.90	12.78	12.66	12.53	12.40	12.27	12.14	.005	
	2.94	2.90	2.87	2.84	2.82	2.80	2.78	2.76	2.74	2.72	.100	6
	4.06	4.00	3.94	3.87	3.84	3.81	3.77	3.74	3.70	3.67	.050	
	5.46	5.37	5.27	5.17	5.12	5.07	5.01	4.96	4.90	4.85	.025	
	7.87	7.72	7.56	7.40	7.31	7.23	7.14	7.06	6.97	6.88	.010	
	10.25	10.03	9.81	9.59	9.47	9.36	9.24	9.12	9.00	8.88	.005	
	2.70	2.67	2.63	2.59	2.58	2.56	2.54	2.51	2.49	2.47	.100	7
	3.64	3.57	3.51	3.44	3.41	3.38	3.34	3.30	3.27	3.23	.050	
	4.76	4.67	4.57	4.47	4.42	4.36	4.31	4.25	4.20	4.14	.025	
	6.62	6.47	6.31	6.16	6.07	5.99	5.91	5.82	5.74	5.65	.010	
	8.38	8.18	7.97	7.75	7.65	7.53	7.42	7.31	7.19	7.08	.005	
	2.54	2.50	2.46	2.42	2.40	2.38	2.36	2.34	2.32	2.29	.100	8
	3.35	3.28	3.22	3.15	3.12	3.08	3.04	3.01	2.97	2.93	.050	
	4.30	4.20	4.10	4.00	3.95	3.89	3.84	3.78	3.73	3.67	.025	
	5.81	5.67	5.52	5.36	5.28	5.20	5.12	5.03	4.95	4.86	.010	
	7.21	7.01	6.81	6.61	6.50	6.40	6.29	6.18	6.06	5.95	.005	
	2.42	2.38	2.34	2.30	2.28	2.25	2.23	2.21	2.18	2.16	.100	9
	3.14	3.07	3.01	2.94	2.90	2.86	2.83	2.79	2.75	2.71	.050	
	3.96	3.87	3.77	3.67	3.61	3.56	3.51	3.45	3.39	3.33	.025	
	5.26	5.11	4.96	4.81	4.73	4.65	4.57	4.48	4.40	4.31	.010	
	6.42	6.23	6.03	5.83	5.73	5.62	5.52	5.41	5.30	5.19	.005	

TABLE 6 *(continued)*

df_2	a	1	2	3	4	5	6	7	8	9
						df_1				
10	.100	3.29	2.92	2.73	2.61	2.52	2.46	2.41	2.38	2.35
	.050	4.96	4.10	3.71	3.48	3.33	3.22	3.14	3.07	3.02
	.025	6.94	5.46	4.83	4.47	4.24	4.07	3.95	3.85	3.78
	.010	10.04	7.56	6.55	5.99	5.64	5.39	5.20	5.06	4.94
	.005	12.83	9.43	8.08	7.34	6.87	6.54	6.30	6.12	5.97
11	.100	3.23	2.86	2.66	2.54	2.45	2.39	2.34	2.30	2.27
	.050	4.84	3.98	3.59	3.36	3.20	3.09	3.01	2.95	2.90
	.025	6.72	5.26	4.63	4.28	4.04	3.88	3.76	3.66	3.59
	.010	9.65	7.21	6.22	5.67	5.32	5.07	4.89	4.74	4.63
	.005	12.23	8.91	7.60	6.88	6.42	6.10	5.86	5.68	5.54
12	.100	3.18	2.81	2.61	2.48	2.39	2.33	2.28	2.24	2.21
	.050	4.75	3.89	3.49	3.26	3.11	3.00	2.91	2.85	2.80
	.025	6.55	5.10	4.47	4.12	3.89	3.73	3.61	3.51	3.44
	.010	9.33	6.93	5.95	5.41	5.06	4.82	4.64	4.50	4.39
	.005	11.75	8.51	7.23	6.52	6.07	5.76	5.52	5.35	5.20
13	.100	3.14	2.76	2.56	2.43	2.35	2.28	2.23	2.20	2.16
	.050	4.67	3.81	3.41	3.18	3.03	2.92	2.83	2.77	2.71
	.025	6.41	4.97	4.35	4.00	3.77	3.60	3.48	3.39	3.31
	.010	9.07	6.70	5.74	5.21	4.86	4.62	4.44	4.30	4.19
	.005	11.37	8.19	6.93	6.23	5.79	5.48	5.25	5.08	4.94
14	.100	3.10	2.73	2.52	2.39	2.31	2.24	2.19	2.15	2.12
	.050	4.60	3.74	3.34	3.11	2.96	2.85	2.76	2.70	2.65
	.025	6.30	4.86	4.24	3.89	3.66	3.50	3.38	3.29	3.21
	.010	8.86	6.51	5.56	5.04	4.69	4.46	4.28	4.14	4.03
	.005	11.06	7.92	6.68	6.00	5.56	5.26	5.03	4.86	4.72
15	.100	3.07	2.70	2.49	2.36	2.27	2.21	2.16	2.12	2.09
	.050	4.54	3.68	3.29	3.06	2.90	2.79	2.71	2.64	2.59
	.025	6.20	4.77	4.15	3.80	3.58	3.41	3.29	3.20	3.12
	.010	8.68	6.36	5.42	4.89	4.56	4.32	4.14	4.00	3.89
	.005	10.80	7.70	6.48	5.80	5.37	5.07	4.85	4.67	4.54
16	.100	3.05	2.67	2.46	2.33	2.24	2.18	2.13	2.09	2.06
	.050	4.49	3.63	3.24	3.01	2.85	2.74	2.66	2.59	2.54
	.025	6.12	4.69	4.08	3.73	3.50	3.34	3.22	3.12	3.05
	.010	8.53	6.23	5.29	4.77	4.44	4.20	4.03	3.89	3.78
	.005	10.58	7.51	6.30	5.64	5.21	4.91	4.69	4.52	4.38
17	.100	3.03	2.64	2.44	2.31	2.22	2.15	2.10	2.06	2.03
	.050	4.45	3.59	3.20	2.96	2.81	2.70	2.61	2.55	2.49
	.025	6.04	4.62	4.01	3.66	3.44	3.28	3.16	3.06	2.98
	.010	8.40	6.11	5.18	4.67	4.34	4.10	3.93	3.79	3.68
	.005	10.38	7.35	6.16	5.50	5.07	4.78	4.56	4.39	4.25
18	.100	3.01	2.62	2.42	2.29	2.20	2.13	2.08	2.04	2.00
	.050	4.41	3.55	3.16	2.93	2.77	2.66	2.58	2.51	2.46
	.025	5.98	4.56	3.95	3.61	3.38	3.22	3.10	3.01	2.93
	.010	8.29	6.01	5.09	4.58	4.25	4.01	3.84	3.71	3.60
	.005	10.22	7.21	6.03	5.37	4.96	4.66	4.44	4.28	4.14
19	.100	2.99	2.61	2.40	2.27	2.18	2.11	2.06	2.02	1.98
	.050	4.38	3.52	3.13	2.90	2.74	2.63	2.54	2.48	2.42
	.025	5.92	4.51	3.90	3.56	3.33	3.17	3.05	2.96	2.88
	.010	8.18	5.93	5.01	4.50	4.17	3.94	3.77	3.63	3.52
	.005	10.07	7.09	5.92	5.27	4.85	4.56	4.34	4.18	4.04
20	.100	2.97	2.59	2.38	2.25	2.16	2.09	2.04	2.00	1.96
	.050	4.35	3.49	3.10	2.87	2.71	2.60	2.51	2.45	2.39
	.025	5.87	4.46	3.86	3.51	3.29	3.13	3.01	2.91	2.84
	.010	8.10	5.85	4.94	4.43	4.10	3.87	3.70	3.56	3.46
	.005	9.94	6.99	5.82	5.17	4.76	4.47	4.26	4.09	3.96

TABLE 6 *(continued)*

10	12	15	20	24	30	40	60	120	∞	a	df_2
						df_1					
2.32	2.28	2.24	2.20	2.18	2.16	2.13	2.11	2.08	2.06	.100	10
2.98	2.91	2.85	2.77	2.74	2.70	2.66	2.62	2.58	2.54	.050	
3.72	3.62	3.52	3.42	3.37	3.31	3.26	3.20	3.14	3.08	.025	
4.85	4.71	4.56	4.41	4.33	4.25	4.17	4.08	4.00	3.91	.010	
5.85	5.66	5.47	5.27	5.17	5.07	4.97	4.86	4.75	4.64	.005	
2.25	2.21	2.17	2.12	2.10	2.08	2.05	2.03	2.00	1.97	.100	11
2.85	2.79	2.72	2.65	2.61	2.57	2.53	2.49	2.45	2.40	.050	
3.53	3.43	3.33	3.23	3.17	3.12	3.06	3.00	2.94	2.88	.025	
4.54	4.40	4.25	4.10	4.02	3.94	3.86	3.78	3.69	3.60	.010	
5.42	5.24	5.05	4.86	4.76	4.65	4.55	4.44	4.34	4.23	.005	
2.19	2.15	2.10	2.06	2.04	2.01	1.99	1.96	1.93	1.90	.100	12
2.75	2.69	2.62	2.54	2.51	2.47	2.43	2.38	2.34	2.30	.050	
3.37	3.28	3.18	3.07	3.02	2.96	2.91	2.85	2.79	2.72	.025	
4.30	4.16	4.01	3.86	3.78	3.70	3.62	3.54	3.45	3.36	.010	
5.09	4.91	4.72	4.53	4.43	4.33	4.23	4.12	4.01	3.90	.005	
2.14	2.10	2.05	2.01	1.98	1.96	1.93	1.90	1.88	1.85	.100	13
2.67	2.60	2.53	2.46	2.42	2.38	2.34	2.30	2.25	2.21	.050	
3.25	3.15	3.05	2.95	2.89	2.84	2.78	2.72	2.66	2.60	.025	
4.10	3.96	3.82	3.66	3.59	3.51	3.43	3.34	3.25	3.17	.010	
4.82	4.64	4.46	4.27	4.17	4.07	3.97	3.87	3.76	3.65	.005	
2.10	2.05	2.01	1.96	1.94	1.91	1.89	1.86	1.83	1.80	.100	14
2.60	2.53	2.46	2.39	2.35	2.31	2.27	2.22	2.18	2.13	.050	
3.15	3.05	2.95	2.84	2.79	2.73	2.67	2.61	2.55	2.49	.025	
3.94	3.80	3.66	3.51	3.43	3.35	3.27	3.18	3.09	3.00	.010	
4.60	4.43	4.25	4.06	3.96	3.86	3.76	3.66	3.55	3.44	.005	
2.06	2.02	1.97	1.92	1.90	1.87	1.85	1.82	1.79	1.76	.100	15
2.54	2.48	2.40	2.33	2.29	2.25	2.20	2.16	2.11	2.07	.050	
3.06	2.96	2.86	2.76	2.70	2.64	2.59	2.52	2.46	2.40	.025	
3.80	3.67	3.52	3.37	3.29	3.21	3.13	3.05	2.96	2.87	.010	
4.42	4.25	4.07	3.88	3.79	3.69	3.58	3.48	3.37	3.26	.005	
2.03	1.99	1.94	1.89	1.87	1.84	1.81	1.78	1.75	1.72	.100	16
2.49	2.42	2.35	2.28	2.24	2.19	2.15	2.11	2.06	2.01	.050	
2.99	2.89	2.79	2.68	2.63	2.57	2.51	2.45	2.38	2.32	.025	
3.69	3.55	3.41	3.26	3.18	3.10	3.02	2.93	2.84	2.75	.010	
4.27	4.10	3.92	3.73	3.64	3.54	3.44	3.33	3.22	3.11	.005	
2.00	1.96	1.91	1.86	1.84	1.81	1.78	1.75	1.72	1.69	.100	17
2.45	2.38	2.31	2.23	2.19	2.15	2.10	2.06	2.01	1.96	.050	
2.92	2.82	2.72	2.62	2.56	2.50	2.44	2.38	2.32	2.25	.025	
3.59	3.46	3.31	3.16	3.08	3.00	2.92	2.83	2.75	2.65	.010	
4.14	3.97	3.79	3.61	3.51	3.41	3.31	3.21	3.10	2.98	.005	
1.98	1.93	1.89	1.84	1.81	1.78	1.75	1.72	1.69	1.66	.100	18
2.41	2.34	2.27	2.19	2.15	2.11	2.06	2.02	1.97	1.92	.050	
2.87	2.77	2.67	2.56	2.50	2.44	2.38	2.32	2.26	2.19	.025	
3.51	3.37	3.23	3.08	3.00	2.92	2.84	2.75	2.66	2.57	.010	
4.03	3.86	3.68	3.50	3.40	3.30	3.20	3.10	2.99	2.87	.005	
1.96	1.91	1.86	1.81	1.79	1.76	1.73	1.70	1.67	1.63	.100	19
2.38	2.31	2.23	2.16	2.11	2.07	2.03	1.98	1.93	1.88	.050	
2.82	2.72	2.62	2.51	2.45	2.39	2.33	2.27	2.20	2.13	.025	
3.43	3.30	3.15	3.00	2.92	2.84	2.76	2.67	2.58	2.49	.010	
3.93	3.76	3.59	3.40	3.31	3.21	3.11	3.00	2.89	2.78	.005	
1.94	1.89	1.84	1.79	1.77	1.74	1.71	1.68	1.64	1.61	.100	20
2.35	2.28	2.20	2.12	2.08	2.04	1.99	1.95	1.90	1.84	.050	
2.77	2.68	2.57	2.46	2.41	2.35	2.29	2.22	2.16	2.09	.025	
3.37	3.23	3.09	2.94	2.86	2.78	2.69	2.61	2.52	2.42	.010	
3.85	3.68	3.50	3.32	3.22	3.12	3.02	2.92	2.81	2.69	.005	

TABLE 6 *(continued)*

df_2	a					df_1				
		1	2	3	4	5	6	7	8	9
21	.100	2.96	2.57	2.36	2.23	2.14	2.08	2.02	1.98	1.95
	.050	4.32	3.47	3.07	2.84	2.68	2.57	2.49	2.42	2.37
	.025	5.83	4.42	3.82	3.48	3.25	3.09	2.97	2.87	2.80
	.010	8.02	5.78	4.87	4.37	4.04	3.81	3.64	3.51	3.40
	.005	9.83	6.89	5.73	5.09	4.68	4.39	4.18	4.01	3.88
22	.100	2.95	2.56	2.35	2.22	2.13	2.06	2.01	1.97	1.93
	.050	4.30	3.44	3.05	2.82	2.66	2.55	2.46	2.40	2.34
	.025	5.79	4.38	3.78	3.44	3.22	3.05	2.93	2.84	2.76
	.010	7.95	5.72	4.82	4.31	3.99	3.76	3.59	3.45	3.35
	.005	9.73	6.81	5.65	5.02	4.61	4.32	4.11	3.94	3.81
23	.100	2.94	2.55	2.34	2.21	2.11	2.05	1.99	1.95	1.92
	.050	4.28	3.42	3.03	2.80	2.64	2.53	2.44	2.37	2.32
	.025	5.75	4.35	3.75	3.41	3.18	3.02	2.90	2.81	2.73
	.010	7.88	5.66	4.76	4.26	3.94	3.71	3.54	3.41	3.30
	.005	9.63	6.73	5.58	4.95	4.54	4.26	4.05	3.88	3.75
24	.100	2.93	2.54	2.33	2.19	2.10	2.04	1.98	1.94	1.91
	.050	4.26	3.40	3.01	2.78	2.62	2.51	2.42	2.36	2.30
	.025	5.72	4.32	3.72	3.38	3.15	2.99	2.87	2.78	2.70
	.010	7.82	5.61	4.72	4.22	3.90	3.67	3.50	3.36	3.26
	.005	9.55	6.66	5.52	4.89	4.49	4.20	3.99	3.83	3.69
25	.100	2.92	2.53	2.32	2.18	2.09	2.02	1.97	1.93	1.89
	.050	4.24	3.39	2.99	2.76	2.60	2.49	2.40	2.34	2.28
	.025	5.69	4.29	3.69	3.35	3.13	2.97	2.85	2.75	2.68
	.010	7.77	5.57	4.68	4.18	3.85	3.63	3.46	3.32	3.22
	.005	9.48	6.60	5.46	4.84	4.43	4.15	3.94	3.78	3.64
26	.100	2.91	2.52	2.31	2.17	2.08	2.01	1.96	1.92	1.88
	.050	4.23	3.37	2.98	2.74	2.59	2.47	2.39	2.32	2.27
	.025	5.66	4.27	3.67	3.33	3.10	2.94	2.82	2.73	2.65
	.010	7.72	5.53	4.64	4.14	3.82	3.59	3.42	3.29	3.18
	.005	9.41	6.54	5.41	4.79	4.38	4.10	3.89	3.73	3.60
27	.100	2.90	2.51	2.30	2.17	2.07	2.00	1.95	1.91	1.87
	.050	4.21	3.35	2.96	2.73	2.57	2.46	2.37	2.31	2.25
	.025	5.63	4.24	3.65	3.31	3.08	2.92	2.80	2.71	2.63
	.010	7.68	5.49	4.60	4.11	3.78	3.56	3.39	3.26	3.15
	.005	9.34	6.49	5.36	4.74	4.34	4.06	3.85	3.69	3.56
28	.100	2.89	2.50	2.29	2.16	2.06	2.00	1.94	1.90	1.87
	.050	4.20	3.34	2.95	2.71	2.56	2.45	2.36	2.29	2.24
	.025	5.61	4.22	3.63	3.29	3.06	2.90	2.78	2.69	2.61
	.010	7.64	5.45	4.57	4.07	3.75	3.53	3.36	3.23	3.12
	.005	9.28	6.44	5.32	4.70	4.30	4.02	3.81	3.65	3.52
29	.100	2.89	2.50	2.28	2.15	2.06	1.99	1.93	1.89	1.86
	.050	4.18	3.33	2.93	2.70	2.55	2.43	2.35	2.28	2.22
	.025	5.59	4.20	3.61	3.27	3.04	2.88	2.76	2.67	2.59
	.010	7.60	5.42	4.54	4.04	3.73	3.50	3.33	3.20	3.09
	.005	9.23	6.40	5.28	4.66	4.26	3.98	3.77	3.61	3.48
30	.100	2.88	2.49	2.28	2.14	2.05	1.98	1.93	1.88	1.85
	.050	4.17	3.32	2.92	2.69	2.53	2.42	2.33	2.27	2.21
	.025	5.57	4.18	3.59	3.25	3.03	2.87	2.75	2.65	2.57
	.010	7.56	5.39	4.51	4.02	3.70	3.47	3.30	3.17	3.07
	.005	9.18	6.35	5.24	4.62	4.23	3.95	3.74	3.58	3.45

TABLE 6 *(continued)*

10	12	15	20	24	30	40	60	120	∞	a	df_2
					df_1						
1.92	1.87	1.83	1.78	1.75	1.72	1.69	1.66	1.62	1.59	.100	21
2.32	2.25	2.18	2.10	2.05	2.01	1.96	1.92	1.87	1.81	.050	
2.73	2.64	2.53	2.42	2.37	2.31	2.25	2.18	2.11	2.04	.025	
3.31	3.17	3.03	2.88	2.80	2.72	2.64	2.55	2.46	2.36	.010	
3.77	3.60	3.43	3.24	3.15	3.05	2.95	2.84	2.73	2.61	.005	
1.90	1.86	1.81	1.76	1.73	1.70	1.67	1.64	1.60	1.57	.100	22
2.30	2.23	2.15	2.07	2.03	1.98	1.94	1.89	1.84	1.78	.050	
2.70	2.60	2.50	2.39	2.33	2.27	2.21	2.14	2.08	2.00	.025	
3.26	3.12	2.98	2.83	2.75	2.67	2.58	2.50	2.40	2.31	.010	
3.70	3.54	3.36	3.18	3.08	2.98	2.88	2.77	2.66	2.55	.005	
1.89	1.84	1.80	1.74	1.72	1.69	1.66	1.62	1.59	1.55	.100	23
2.27	2.20	2.13	2.05	2.01	1.96	1.91	1.86	1.81	1.76	.050	
2.67	2.57	2.47	2.36	2.30	2.24	2.18	2.11	2.04	1.97	.025	
3.21	3.07	2.93	2.78	2.70	2.62	2.54	2.45	2.35	2.26	.010	
3.64	3.47	3.30	3.12	3.02	2.92	2.82	2.71	2.60	2.48	.005	
1.88	1.83	1.78	1.73	1.70	1.67	1.64	1.61	1.57	1.53	.100	24
2.25	2.18	2.11	2.03	1.98	1.94	1.89	1.84	1.79	1.73	.050	
2.64	2.54	2.44	2.33	2.27	2.21	2.15	2.08	2.01	1.94	.025	
3.17	3.03	2.89	2.74	2.66	2.58	2.49	2.40	2.31	2.21	.010	
3.59	3.42	3.25	3.06	2.97	2.87	2.77	2.66	2.55	2.43	.005	
1.87	1.82	1.77	1.72	1.69	1.66	1.63	1.59	1.56	1.52	.100	25
2.24	2.16	2.09	2.01	1.96	1.92	1.87	1.82	1.77	1.71	.050	
2.61	2.51	2.41	2.30	2.24	2.18	2.12	2.05	1.98	1.91	.025	
3.13	2.99	2.85	2.70	2.62	2.54	2.45	2.36	2.27	2.17	.010	
3.54	3.37	3.20	3.01	2.92	2.82	2.72	2.61	2.50	2.38	.005	
1.86	1.81	1.76	1.71	1.68	1.65	1.61	1.58	1.54	1.50	.100	26
2.22	2.15	2.07	1.99	1.95	1.90	1.85	1.80	1.75	1.69	.050	
2.59	2.49	2.39	2.28	2.22	2.16	2.09	2.03	1.95	1.88	.025	
3.09	2.96	2.81	2.66	2.58	2.50	2.42	2.33	2.23	2.13	.010	
3.49	3.33	3.15	2.97	2.87	2.77	2.67	2.56	2.45	2.33	.005	
1.85	1.80	1.75	1.70	1.67	1.64	1.60	1.57	1.53	1.49	.100	27
2.20	2.13	2.06	1.97	1.93	1.88	1.84	1.79	1.73	1.67	.050	
2.57	2.47	2.36	2.25	2.19	2.13	2.07	2.00	1.93	1.85	.025	
3.06	2.93	2.78	2.63	2.55	2.47	2.38	2.29	2.20	2.10	.010	
3.45	3.28	3.11	2.93	2.83	2.73	2.63	2.52	2.41	2.29	.005	
1.84	1.79	1.74	1.69	1.66	1.63	1.59	1.56	1.52	1.48	.100	28
2.19	2.12	2.04	1.96	1.91	1.87	1.82	1.77	1.71	1.65	.050	
2.55	2.45	2.34	2.23	2.17	2.11	2.05	1.98	1.91	1.83	.025	
3.03	2.90	2.75	2.60	2.52	2.44	2.35	2.26	2.17	2.06	.010	
3.41	3.25	3.07	2.89	2.79	2.69	2.59	2.48	2.37	2.25	.005	
1.83	1.78	1.73	1.68	1.65	1.62	1.58	1.55	1.51	1.47	.100	29
2.18	2.10	2.03	1.94	1.90	1.85	1.81	1.75	1.70	1.64	.050	
2.53	2.43	2.32	2.21	2.15	2.09	2.03	1.96	1.89	1.81	.025	
3.00	2.87	2.73	2.57	2.49	2.41	2.33	2.23	2.14	2.03	.010	
3.38	3.21	3.04	2.86	2.76	2.66	2.56	2.45	2.33	2.21	.005	
1.82	1.77	1.72	1.67	1.64	1.61	1.57	1.54	1.50	1.46	.100	30
2.16	2.09	2.01	1.93	1.89	1.84	1.79	1.74	1.68	1.62	.050	
2.51	2.41	2.31	2.20	2.14	2.07	2.01	1.94	1.87	1.79	.025	
2.98	2.84	2.70	2.55	2.47	2.39	2.30	2.21	2.11	2.01	.010	
3.34	3.18	3.01	2.82	2.73	2.63	2.52	2.42	2.30	2.18	.005	

TABLE 6 *(continued)*

df_2	a	1	2	3	4	5	6	7	8	9
						df_1				
40	.100	2.84	2.44	2.23	2.09	2.00	1.93	1.87	1.83	1.79
	.050	4.08	3.23	2.84	2.61	2.45	2.34	2.25	2.18	2.12
	.025	5.42	4.05	3.46	3.13	2.90	2.74	2.62	2.53	2.45
	.010	7.31	5.18	4.31	3.83	3.51	3.29	3.12	2.99	2.89
	.005	8.83	6.07	4.98	4.37	3.99	3.71	3.51	3.35	3.22
60	.100	2.79	2.39	2.18	2.04	1.95	1.87	1.82	1.77	1.74
	.050	4.00	3.15	2.76	2.53	2.37	2.25	2.17	2.10	2.04
	.025	5.29	3.93	3.34	3.01	2.79	2.63	2.51	2.41	2.33
	.010	7.08	4.98	4.13	3.65	3.34	3.12	2.95	2.82	2.72
	.005	8.49	5.79	4.73	4.14	3.76	3.49	3.29	3.13	3.01
120	.100	2.75	2.35	2.13	1.99	1.90	1.82	1.77	1.72	1.68
	.050	3.92	3.07	2.68	2.45	2.29	2.17	2.09	2.02	1.96
	.025	5.15	3.80	3.23	2.89	2.67	2.52	2.39	2.30	2.22
	.010	6.85	4.79	3.95	3.48	3.17	2.96	2.79	2.66	2.56
	.005	8.18	5.54	4.50	3.92	3.55	3.28	3.09	2.93	2.81
∞	.100	2.71	2.30	2.08	1.94	1.85	1.77	1.72	1.67	1.63
	.050	3.84	3.00	2.60	2.37	2.21	2.10	2.01	1.94	1.63
	.025	5.02	3.69	3.12	2.79	2.57	2.41	2.29	2.19	2.11
	.010	6.63	4.61	3.78	3.32	3.02	2.80	2.64	2.51	2.41
	.005	7.88	5.30	4.28	3.72	3.35	3.09	2.90	2.74	2.62

TABLE 6 *(continued)*

				df$_1$							
10	12	15	20	24	30	40	60	120	∞	a	df$_2$
1.76	1.71	1.66	1.61	1.57	1.54	1.51	1.47	1.42	1.38	.100	40
2.08	2.00	1.92	1.84	1.79	1.74	1.69	1.64	1.58	1.51	.050	
2.39	2.29	2.18	2.07	2.01	1.94	1.88	1.80	1.72	1.64	.025	
2.80	2.66	2.52	2.37	2.29	2.20	2.11	2.02	1.92	1.80	.010	
3.12	2.95	2.78	2.60	2.50	2.40	2.30	2.18	2.06	1.93	.005	
1.71	1.66	1.60	1.54	1.51	1.48	1.44	1.40	1.35	1.29	.100	60
1.99	1.92	1.84	1.75	1.70	1.65	1.59	1.53	1.47	1.39	.050	
2.27	2.17	2.06	1.94	1.88	1.82	1.74	1.67	1.58	1.48	.025	
2.63	2.50	2.35	2.20	2.12	2.03	1.94	1.84	1.73	1.60	.010	
2.90	2.74	2.57	2.39	2.29	2.19	2.08	1.96	1.83	1.69	.005	
1.65	1.60	1.55	1.48	1.45	1.41	1.37	1.32	1.26	1.19	.100	120
1.91	1.83	1.75	1.66	1.61	1.55	1.50	1.43	1.35	1.25	.050	
2.16	2.05	1.94	1.82	1.76	1.69	1.61	1.53	1.43	1.31	.025	
2.47	2.34	2.19	2.03	1.95	1.86	1.76	1.66	1.53	1.38	.010	
2.71	2.54	2.37	2.19	2.09	1.98	1.87	1.75	1.61	1.43	.005	
1.60	1.55	1.49	1.42	1.38	1.34	1.30	1.24	1.17	1.00	.100	∞
1.83	1.75	1.67	1.57	1.52	1.46	1.39	1.32	1.22	1.00	.050	
2.05	1.94	1.83	1.71	1.64	1.57	1.48	1.39	1.27	1.00	.025	
2.32	2.18	2.04	1.88	1.79	1.70	1.59	1.47	1.32	1.00	.010	
2.52	2.36	2.19	2.00	1.90	1.79	1.67	1.53	1.36	1.00	.005	

TABLE 7 Critical Values of T for the Wilcoxon Rank Sum Test, $n_1 \leq n_2$

TABLE 7(a)
5% Left-Tailed
Critical Values

n_2	\multicolumn{14}{c}{n_1}
	2	3	4	5	6	7	8	9	10	11	12	13	14	15
3	—	6												
4	—	6	11											
5	3	7	12	19										
6	3	8	13	20	28									
7	3	8	14	21	29	39								
8	4	9	15	23	31	41	51							
9	4	10	16	24	33	43	54	66						
10	4	10	17	26	35	45	56	69	82					
11	4	11	18	27	37	47	59	72	86	100				
12	5	11	19	28	38	49	62	75	89	104	120			
13	5	12	20	30	40	52	64	78	92	108	125	142		
14	6	13	21	31	42	54	67	81	96	112	129	147	166	
15	6	13	22	33	44	56	69	84	99	116	133	152	171	192

TABLE 7(b)
2.5% Left-Tailed
Critical Values

n_2	\multicolumn{14}{c}{n_1}
	2	3	4	5	6	7	8	9	10	11	12	13	14	15
4	—	—	10											
5	—	6	11	17										
6	—	7	12	18	26									
7	—	7	13	20	27	36								
8	3	8	14	21	29	38	49							
9	3	8	14	22	31	40	51	62						
10	3	9	15	23	32	42	53	65	78					
11	3	9	16	24	34	44	55	68	81	96				
12	4	10	17	26	35	46	58	71	84	99	115			
13	4	10	18	27	37	48	60	73	88	103	119	136		
14	4	11	19	28	38	50	62	76	91	106	123	141	160	
15	4	11	20	29	40	52	65	79	94	110	127	145	164	184

TABLE 7(c)
1% Left-Tailed
Critical Values

								n_1						
n_2	2	3	4	5	6	7	8	9	10	11	12	13	14	15
3	—	—												
4	—	—	—											
5	—	—	10	16										
6	—	—	11	17	24									
7	—	6	11	18	25	34								
8	—	6	12	19	27	35	45							
9	—	7	13	20	28	37	47	59						
10	—	7	13	21	29	39	49	61	74					
11	—	7	14	22	30	40	51	63	77	91				
12	—	8	15	23	32	42	53	66	79	94	109			
13	3	8	15	24	33	44	56	68	82	97	113	130		
14	3	8	16	25	34	45	58	71	85	100	116	134	152	
15	3	9	17	26	36	47	60	73	88	103	120	138	156	176

TABLE 7(d)
.5% Left-Tailed
Critical Values

								n_1						
n_2	3	4	5	6	7	8	9	10	11	12	13	14	15
3	—												
4	—	—											
5	—	—	15										
6	—	10	16	23									
7	—	10	16	24	32								
8	—	11	17	25	34	42							
9	6	11	18	26	35	45	56						
10	6	12	19	27	37	47	58	71					
11	6	12	20	28	38	49	61	73	87				
12	7	13	21	30	40	51	63	76	90	105			
13	7	13	22	31	41	53	65	79	93	109	125		
14	7	14	22	32	43	54	67	81	96	112	129	147	
15	8	15	23	33	44	56	69	84	99	115	133	151	171

TABLE 8
Critical Values
of *T* for the
Wilcoxon
Signed-Rank
Test, *n* = 5(1)50

One-Sided	Two-Sided	$n = 5$	$n = 6$	$n = 7$	$n = 8$	$n = 9$	$n = 10$
$\alpha = .050$	$\alpha = .10$	1	2	4	6	8	11
$\alpha = .025$	$\alpha = .05$		1	2	4	6	8
$\alpha = .010$	$\alpha = .02$			0	2	3	5
$\alpha = .005$	$\alpha = .01$				0	2	3

One-Sided	Two-Sided	$n = 11$	$n = 12$	$n = 13$	$n = 14$	$n = 15$	$n = 16$
$\alpha = .050$	$\alpha = .10$	14	17	21	26	30	36
$\alpha = .025$	$\alpha = .05$	11	14	17	21	25	30
$\alpha = .010$	$\alpha = .02$	7	10	13	16	20	24
$\alpha = .005$	$\alpha = .01$	5	7	10	13	16	19

One-Sided	Two-Sided	$n = 17$	$n = 18$	$n = 19$	$n = 20$	$n = 21$	$n = 22$
$\alpha = .050$	$\alpha = .10$	41	47	54	60	68	75
$\alpha = .025$	$\alpha = .05$	35	40	46	52	59	66
$\alpha = .010$	$\alpha = .02$	28	33	38	43	49	56
$\alpha = .005$	$\alpha = .01$	23	28	32	37	43	49

One-Sided	Two-Sided	$n = 23$	$n = 24$	$n = 25$	$n = 26$	$n = 27$	$n = 28$
$\alpha = .050$	$\alpha = .10$	83	92	101	110	120	130
$\alpha = .025$	$\alpha = .05$	73	81	90	98	107	117
$\alpha = .010$	$\alpha = .02$	62	69	77	85	93	102
$\alpha = .005$	$\alpha = .01$	55	68	68	76	84	92

One-Sided	Two-Sided	$n = 29$	$n = 30$	$n = 31$	$n = 32$	$n = 33$	$n = 34$
$\alpha = .050$	$\alpha = .10$	141	152	163	175	188	201
$\alpha = .025$	$\alpha = .05$	127	137	148	159	171	183
$\alpha = .010$	$\alpha = .02$	111	120	130	141	151	162
$\alpha = .005$	$\alpha = .01$	100	109	118	128	138	149

One-Sided	Two-Sided	$n = 35$	$n = 36$	$n = 37$	$n = 38$	$n = 39$	
$\alpha = .050$	$\alpha = .10$	214	228	242	256	271	
$\alpha = .025$	$\alpha = .05$	195	208	222	235	250	
$\alpha = .010$	$\alpha = .02$	174	186	198	211	224	
$\alpha = .005$	$\alpha = .01$	160	171	183	195	208	

One-Sided	Two-Sided	$n = 40$	$n = 41$	$n = 42$	$n = 43$	$n = 44$	$n = 45$
$\alpha = .050$	$\alpha = .10$	287	303	319	336	353	371
$\alpha = .025$	$\alpha = .05$	264	279	295	311	327	344
$\alpha = .010$	$\alpha = .02$	238	252	267	281	297	313
$\alpha = .005$	$\alpha = .01$	221	234	248	262	277	292

One-Sided	Two-Sided	$n = 46$	$n = 47$	$n = 48$	$n = 49$	$n = 50$	
$\alpha = .050$	$\alpha = .10$	389	408	427	446	466	
$\alpha = .025$	$\alpha = .05$	361	379	397	415	434	
$\alpha = .010$	$\alpha = .02$	329	345	362	380	398	
$\alpha = .005$	$\alpha = .01$	307	323	339	356	373	

SOURCE: From "Some Rapid Approximate Statistical Procedures" (1964) 28, by F. Wilcoxon and R.A. Wilcox. Reproduced with the kind permission of Lederle Laboratories, a division of American Cyanamid Company.

TABLE 9
Critical Values of Spearman's Rank Correlation Coefficient for a One-Tailed Test

n	$\alpha = .05$	$\alpha = .025$	$\alpha = .01$	$\alpha = .005$
5	.900	—	—	—
6	.829	.886	.943	—
7	.714	.786	.893	—
8	.643	.738	.833	.881
9	.600	.683	.783	.833
10	.564	.648	.745	.794
11	.523	.623	.736	.818
12	.497	.591	.703	.780
13	.475	.566	.673	.745
14	.457	.545	.646	.716
15	.441	.525	.623	.689
16	.425	.507	.601	.666
17	.412	.490	.582	.645
18	.399	.476	.564	.625
19	.388	.462	.549	.608
20	.377	.450	.534	.591
21	.368	.438	.521	.576
22	.359	.428	.508	.562
23	.351	.418	.496	.549
24	.343	.409	.485	.537
25	.336	.400	.475	.526
26	.329	.392	.465	.515
27	.323	.385	.456	.505
28	.317	.377	.448	.496
29	.311	.370	.440	.487
30	.305	.364	.432	.478

SOURCE: From "Distribution of Sums of Squares of Rank Differences for Small Samples" by E.G. Olds, *Annals of Mathematical Statistics* 9 (1938). Reproduced with the permission of the editor, *Annals of Mathematical Statistics*.

TABLE 10 **Random Numbers**

Line	1	2	3	4	5	6	7	8	9	10	11	12	13	14
							Column							
1	10480	15011	01536	02011	81647	91646	69179	14194	62590	36207	20969	99570	91291	90700
2	22368	46573	25595	85393	30995	89198	27982	53402	93965	34095	52666	19174	39615	99505
3	24130	48360	22527	97265	76393	64809	15179	24830	49340	32081	30680	19655	63348	58629
4	42167	93093	06243	61680	07856	16376	39440	53537	71341	57004	00849	74917	97758	16379
5	37570	39975	81837	16656	06121	91782	60468	81305	49684	60672	14110	06927	01263	54613
6	77921	06907	11008	42751	27756	53498	18602	70659	90655	15053	21916	81825	44394	42880
7	99562	72905	56420	69994	98872	31016	71194	18738	44013	48840	63213	21069	10634	12952
8	96301	91977	05463	07972	18876	20922	94595	56869	69014	60045	18425	84903	42508	32307
9	89579	14342	63661	10281	17453	18103	57740	84378	25331	12566	58678	44947	05585	56941
10	84575	36857	53342	53988	53060	59533	38867	62300	08158	17983	16439	11458	18593	64952
11	28918	69578	88231	33276	70997	79936	56865	05859	90106	31595	01547	85590	91610	78188
12	63553	40961	48235	03427	49626	69445	18663	72695	52180	20847	12234	90511	33703	90322
13	09429	93969	52636	92737	88974	33488	36320	17617	30015	08272	84115	27156	30613	74952
14	10365	61129	87529	85689	48237	52267	67689	93394	01511	26358	85104	20285	29975	89868
15	07119	97336	71048	08178	77233	13916	47564	81056	97735	85977	29372	74461	28551	90707
16	51085	12765	51821	51259	77452	16308	60756	92144	49442	53900	70960	63990	75601	40719
17	02368	21382	52404	60268	89368	19885	55322	44819	01188	65255	64835	44919	05944	55157
18	01011	54092	33362	94904	31273	04146	18594	29852	71585	85030	51132	01915	92747	64951
19	52162	53916	46369	58586	23216	14513	83149	98736	23495	64350	94738	17752	35156	35749
20	07056	97628	33787	09998	42698	06691	76988	13602	51851	46104	88916	19509	25625	58104
21	48663	91245	85828	14346	09172	30168	90229	04734	59193	22178	30421	61666	99904	32812
22	54164	58492	22421	74103	47070	25306	76468	26384	58151	06646	21524	15227	96909	44592
23	32639	32363	05597	24200	13363	38005	94342	28728	35806	06912	17012	64161	18296	22851
24	29334	27001	87637	87308	58731	00256	45834	15398	46557	41135	10367	07684	36188	18510
25	02488	33062	28834	07351	19731	92420	60952	61280	50001	67658	32586	86679	50720	94953
26	81525	72295	04839	96423	24878	82651	66566	14778	76797	14780	13300	87074	79666	95725
27	29676	20591	68086	26432	46901	20849	89768	81536	86645	12659	92259	57102	80428	25280
28	00742	57392	39064	66432	84673	40027	32832	61362	98947	96067	64760	64585	96096	98253
29	05366	04213	25669	26422	44407	44048	37937	63904	45766	66134	75470	66520	34693	90449
30	91921	26418	64117	94305	26766	25940	39972	22209	71500	64568	91402	42416	07844	69618
31	00582	04711	87917	77341	42206	35126	74087	99547	81817	42607	43808	76655	62028	76630
32	00725	69884	62797	56170	86324	88072	76222	36086	84637	93161	76038	65855	77919	88006
33	69011	65795	95876	55293	18988	27354	26575	08625	40801	59920	29841	80150	12777	48501
34	25976	57948	29888	88604	67917	48708	18912	82271	65424	69774	33611	54262	85963	03547
35	09763	83473	73577	12908	30883	18317	28290	35797	05998	41688	34952	37888	38917	88050
36	91567	42595	27958	30134	04024	86385	29880	99730	55536	84855	29080	09250	79656	73211
37	17955	56349	90999	49127	20044	59931	06115	20542	18059	02008	73708	83517	36103	42791
38	46503	18584	18845	49618	02304	51038	20655	58727	28168	15475	56942	53389	20562	87338
39	92157	89634	94824	78171	84610	82834	09922	25417	44137	48413	25555	21246	35509	20468
40	14577	62765	35605	81263	39667	47358	56873	56307	61607	49518	89656	20103	77490	18062
41	98427	07523	33362	64270	01638	92477	66969	98420	04880	45585	46565	04102	46880	45709
42	34914	63976	88720	82765	34476	17032	87589	40836	32427	70002	70663	88863	77775	69348
43	70060	28277	39475	46473	23219	53416	94970	25832	69975	94884	19661	72828	00102	66794
44	53976	54914	06990	67245	68350	82948	11398	42878	80287	88267	47363	46634	06541	97809
45	76072	29515	40980	07391	58745	25774	22987	80059	39911	96189	41151	14222	60697	59583
46	90725	52210	83974	29992	65831	38857	50490	83765	55657	14361	31720	57375	56228	41546
47	64364	67412	33339	31926	14883	24413	59744	92351	97473	89286	35931	04110	23726	51900
48	08962	00358	31662	25388	61642	34072	81249	35648	56891	69352	48373	45578	78547	81788
49	95012	68379	93526	70765	10592	04542	76463	54328	02349	17247	28865	14777	62730	92277
50	15664	10493	20492	38391	91132	21999	59516	81652	27195	48223	46751	22923	32261	85653

SOURCE: From *Handbook of Tables for Probability and Statistics,* 2nd ed., edited by William H. Beyer (CRC Press). Used by permission of William H. Beyer.

TABLE 10 *(continued)*

Line	Column													
	1	2	3	4	5	6	7	8	9	10	11	12	13	14
51	16408	81899	04153	53381	79401	21438	83035	92350	36693	31238	59649	91754	72772	02338
52	18629	81953	05520	91962	04739	13092	97662	24822	94730	06496	35090	04822	86774	98289
53	73115	35101	47498	87637	99016	71060	88824	71013	18735	20286	23153	72924	35165	43040
54	57491	16703	23167	49323	45021	33132	12544	41035	80780	45393	44812	12515	98931	91202
55	30405	83946	23792	14422	15059	45799	22716	19792	09983	74353	68668	30429	70735	25499
56	16631	35006	85900	98275	32388	52390	16815	69298	82732	38480	73817	32523	41961	44437
57	96773	20206	42559	78985	05300	22164	24369	54224	35033	19687	11052	91491	60383	19746
58	38935	64202	14349	82674	66523	44133	00697	35552	35970	19124	63318	29686	03387	59846
59	31624	76384	17403	53363	44167	64486	64758	75366	76554	31601	12614	33072	60332	92325
60	78919	19474	23632	27889	47914	02584	37680	20801	72152	39339	34806	08930	85001	87820
61	03931	33309	57047	74211	63445	17361	62825	39908	05607	91284	68833	25570	38818	46920
62	74426	33278	43972	10119	89917	15665	52872	73823	73144	88662	88970	74492	51805	99378
63	09066	00903	20795	95452	92648	45454	09552	88815	16553	51125	79375	97596	16296	66092
64	42238	12426	87025	14267	20979	04508	64535	31355	86064	29472	47689	05974	52468	16834
65	16153	08002	26504	41744	81959	65642	74240	56302	00033	67107	77510	70625	28725	34191
66	21457	40742	29820	96783	29400	21840	15035	34537	33310	06116	95240	15957	16572	06004
67	21581	57802	02050	89728	17937	37621	47075	42080	97403	48626	68995	43805	33386	21597
68	55612	78095	83197	33732	05810	24813	86902	60397	16489	03264	88525	42786	05269	92532
69	44657	66999	99324	51281	84463	60563	79312	93454	68876	25471	93911	25650	12682	73572
70	91340	84979	46949	81973	37949	61023	43997	15263	80644	43942	89203	71795	99533	50501
71	91227	21199	31935	27022	84067	05462	35216	14486	29891	68607	41867	14951	91696	85065
72	50001	38140	66321	19924	72163	09538	12151	06878	91903	18749	34405	56087	82790	70925
73	65390	05224	72958	28609	81406	39147	25549	48542	42627	45233	57202	94617	23772	07896
74	27504	96131	83944	41575	10573	08619	64482	73923	36152	05184	94142	25299	84387	34925
75	37169	94851	39117	89632	00959	16487	65536	49071	39782	17095	02330	74301	00275	48280
76	11508	70225	51111	38351	19444	66499	71945	05422	13442	78675	84081	66938	93654	59894
77	37449	30362	06694	54690	04052	53115	62757	95348	78662	11163	81651	50245	34971	52924
78	46515	70331	85922	38329	57015	15765	97161	17869	45349	61796	66345	81073	49106	79860
79	30986	81223	42416	58353	21532	30502	32305	86482	05174	07901	54339	58861	74818	46942
80	63798	64995	46583	09785	44160	78128	83991	42865	92520	83531	80377	35909	81250	54238
81	82486	84846	99254	67632	43218	50076	21361	64816	51202	88124	41870	52689	51275	83556
82	21885	32906	92431	09060	64297	51674	64126	62570	26123	05155	59194	52799	28225	85762
83	60336	98782	07408	53458	13564	59089	26445	29789	85205	41001	12535	12133	14645	23541
84	43937	46891	24010	25560	86355	33941	25786	54990	71899	15475	95434	98227	21824	19585
85	97656	63175	89303	16275	07100	92063	21942	18611	47348	20203	18534	03862	78095	50136
86	03299	01221	05418	38982	55758	92237	26759	86367	21216	98442	08303	56613	91511	75928
87	79626	06486	03574	17668	07785	76020	79924	25651	83325	88428	85076	72811	22717	50585
88	85636	68335	47539	03129	65651	11977	02510	26113	99447	68645	34327	15152	55230	93448
89	18039	14367	61337	06177	12143	46609	32989	74014	64708	00533	35398	58408	13261	47908
90	08362	15656	60627	36478	65648	16764	53412	09013	07832	41574	17639	82163	60859	75567
91	79556	29068	04142	16268	15387	12856	66227	38358	22478	73373	88732	09443	82558	05250
92	92608	82674	27072	32534	17075	27698	98204	63863	11951	34648	88022	56148	34925	57031
93	23982	25835	40055	67006	12293	02753	14827	23235	35071	99704	37543	11601	35503	85171
94	09915	96306	05908	97901	28395	14186	00821	80703	70426	75647	76310	88717	37890	40129
95	59037	33300	26695	62247	69927	76123	50842	43834	86654	70959	79725	93872	28117	19233
96	42488	78077	69882	61657	34136	79180	97526	43092	04098	73571	80799	76536	71255	64239
97	46764	86273	63003	93017	31204	36692	40202	35275	57306	55543	53203	18098	47625	88684
98	03237	45430	55417	63282	90816	17349	88298	90183	36600	78406	06216	95787	42579	90730
99	86591	81482	52667	61582	14972	90053	89534	76036	49199	43716	97548	04379	46370	28672
100	38534	01715	94964	87288	65680	43772	39560	12918	86737	62738	19636	51132	25739	56947

TABLE 11(a)
Percentage
Points of the
Studentized
Range,
$q_{.05}(k, df)$;
Upper 5%
Points

					k					
df	2	3	4	5	6	7	8	9	10	11
1	17.97	26.98	32.82	37.08	40.41	43.12	45.40	47.36	49.07	50.59
2	6.08	8.33	9.80	10.88	11.74	12.44	13.03	13.54	13.99	14.39
3	4.50	5.91	6.82	7.50	8.04	8.48	8.85	9.18	9.46	9.72
4	3.93	5.04	5.76	6.29	6.71	7.05	7.35	7.60	7.83	8.03
5	3.64	4.60	5.22	5.67	6.03	6.33	6.58	6.80	6.99	7.17
6	3.46	4.34	4.90	5.30	5.63	5.90	6.12	6.32	6.49	6.65
7	3.34	4.16	4.68	5.06	5.36	5.61	5.82	6.00	6.16	6.30
8	3.26	4.04	4.53	4.89	5.17	5.40	5.60	5.77	5.92	6.05
9	3.20	3.95	4.41	4.76	5.02	5.24	5.43	5.59	5.74	5.87
10	3.15	3.88	4.33	4.65	4.91	5.12	5.30	5.46	5.60	5.72
11	3.11	3.82	4.26	4.57	4.82	5.03	5.20	5.35	5.49	5.61
12	3.08	3.77	4.20	4.51	4.75	4.95	5.12	5.27	5.39	5.51
13	3.06	3.73	4.15	4.45	4.69	4.88	5.05	5.19	5.32	5.43
14	3.03	3.70	4.11	4.41	4.64	4.83	4.99	5.13	5.25	5.36
15	3.01	3.67	4.08	4.37	4.60	4.78	4.94	5.08	5.20	5.31
16	3.00	3.65	4.05	4.33	4.56	4.74	4.90	5.03	5.15	5.26
17	2.98	3.63	4.02	4.30	4.52	4.70	4.86	4.99	5.11	5.21
18	2.97	3.61	4.00	4.28	4.49	4.67	4.82	4.96	5.07	5.17
19	2.96	3.59	3.98	4.25	4.47	4.65	4.79	4.92	5.04	5.14
20	2.95	3.58	3.96	4.23	4.45	4.62	4.77	4.90	5.01	5.11
24	2.92	3.53	3.90	4.17	4.37	4.54	4.68	4.81	4.92	5.01
30	2.89	3.49	3.85	4.10	4.30	4.46	4.60	4.72	4.82	4.92
40	2.86	3.44	3.79	4.04	4.23	4.39	4.52	4.63	4.73	4.82
60	2.83	3.40	3.74	3.98	4.16	4.31	4.44	4.55	4.65	4.73
120	2.80	3.36	3.68	3.92	4.10	4.24	4.36	4.47	4.56	4.64
∞	2.77	3.31	3.63	3.86	4.03	4.17	4.29	4.39	4.47	4.55

TABLE 11(a) *(continued)*

k									
12	13	14	15	16	17	18	19	20	*df*
51.96	53.20	54.33	55.36	56.32	57.22	58.04	58.83	59.56	1
14.75	15.08	15.38	15.65	15.91	16.14	16.37	16.57	16.77	2
9.95	10.15	10.35	10.52	10.69	10.84	10.98	11.11	11.24	3
8.21	8.37	8.52	8.66	8.79	8.91	9.03	9.13	9.23	4
7.32	7.47	7.60	7.72	7.83	7.93	8.03	8.12	8.21	5
6.79	6.92	7.03	7.14	7.24	7.34	7.43	7.51	7.59	6
6.43	6.55	6.66	6.76	6.85	6.94	7.02	7.10	7.17	7
6.18	6.29	6.39	6.48	6.57	6.65	6.73	6.80	6.87	8
5.98	6.09	6.19	6.28	6.36	6.44	6.51	6.58	6.64	9
5.83	5.93	6.03	6.11	6.19	6.27	6.34	6.40	6.47	10
5.71	5.81	5.90	5.98	6.06	6.13	6.20	6.27	6.33	11
5.61	5.71	5.80	5.88	5.95	6.02	6.09	6.15	6.21	12
5.53	5.63	5.71	5.79	5.86	5.93	5.99	6.05	6.11	13
5.46	5.55	5.64	5.71	5.79	5.85	5.91	5.97	6.03	14
5.40	5.49	5.57	5.65	5.72	5.78	5.85	5.90	5.96	15
5.35	5.44	5.52	5.59	5.66	5.73	5.79	5.84	5.90	16
5.31	5.39	5.47	5.54	5.61	5.67	5.73	5.79	5.84	17
5.27	5.35	5.43	5.50	5.57	5.63	5.69	5.74	5.79	18
5.23	5.31	5.39	5.46	5.53	5.59	5.65	5.70	5.75	19
5.20	5.28	5.36	5.43	5.49	5.55	5.61	5.66	5.71	20
5.10	5.18	5.25	5.32	5.38	5.44	5.49	5.55	5.59	24
5.00	5.08	5.15	5.21	5.27	5.33	5.38	5.43	5.47	30
4.90	4.98	5.04	5.11	5.16	5.22	5.27	5.31	5.36	40
4.81	4.88	4.94	5.00	5.06	5.11	5.15	5.20	5.24	60
4.71	4.78	4.84	4.90	4.95	5.00	5.04	5.09	5.13	120
4.62	4.68	4.74	4.80	4.85	4.89	4.93	4.97	5.01	∞

SOURCE: From *Biometrika Tables for Statisticians,* Vol. 1, 3rd ed., edited by E.S. Pearson and H.O. Hartley (Cambridge University Press, 1966). Reproduced by permission of the *Biometrika* Trustees.

TABLE 11(b)
Percentage Points of the Studentized Range, $q_{.01}(k, df)$; Upper 1% Points

df						k				
	2	3	4	5	6	7	8	9	10	11
1	90.03	135.0	164.3	185.6	202.2	215.8	227.2	237.0	245.6	253.2
2	14.04	19.02	22.29	24.72	26.63	28.20	29.53	30.68	31.69	32.59
3	8.26	10.62	12.17	13.33	14.24	15.00	15.64	16.20	16.69	17.13
4	6.51	8.12	9.17	9.96	10.58	11.10	11.55	11.93	12.27	12.57
5	5.70	6.98	7.80	8.42	8.91	9.32	9.67	9.97	10.24	10.48
6	5.24	6.33	7.03	7.56	7.97	8.32	8.61	8.87	9.10	9.30
7	4.95	5.92	6.54	7.01	7.37	7.68	7.94	8.17	8.37	8.55
8	4.75	5.64	6.20	6.62	6.96	7.24	7.47	7.68	7.86	8.03
9	4.60	5.43	5.96	6.35	6.66	6.91	7.13	7.33	7.49	7.65
10	4.48	5.27	5.77	6.14	6.43	6.67	6.87	7.05	7.21	7.36
11	4.39	5.15	5.62	5.97	6.25	6.48	6.67	6.84	6.99	7.13
12	4.32	5.05	5.50	5.84	6.10	6.32	6.51	6.67	6.81	6.94
13	4.26	4.96	5.40	5.73	5.98	6.19	6.37	6.53	6.67	6.79
14	4.21	4.89	5.32	5.63	5.88	6.08	6.26	6.41	6.54	6.66
15	4.17	4.84	5.25	5.56	5.80	5.99	6.16	6.31	6.44	6.55
16	4.13	4.79	5.19	5.49	5.72	5.92	6.08	6.22	6.35	6.46
17	4.10	4.74	5.14	5.43	5.66	5.85	6.01	6.15	6.27	6.38
18	4.07	4.70	5.09	5.38	5.60	5.79	5.94	6.08	6.20	6.31
19	4.05	4.67	5.05	5.33	5.55	5.73	5.89	6.02	6.14	6.25
20	4.02	4.64	5.02	5.29	5.51	5.69	5.84	5.97	6.09	6.19
24	3.96	4.55	4.91	5.17	5.37	5.54	5.69	5.81	5.92	6.02
30	3.89	4.45	4.80	5.05	5.24	5.40	5.54	5.65	5.76	5.85
40	3.82	4.37	4.70	4.93	5.11	5.26	5.39	5.50	5.60	5.69
60	3.76	4.28	4.59	4.82	4.99	5.13	5.25	5.36	5.45	5.53
120	3.70	4.20	4.50	4.71	4.87	5.01	5.12	5.21	5.30	5.37
∞	3.64	4.12	4.40	4.60	4.76	4.88	4.99	5.08	5.16	5.23

TABLE 11(b)
(continued)

k: 12	13	14	15	16	17	18	19	20	*df*
260.0	266.2	271.8	277.0	281.8	286.3	290.0	294.3	298.0	1
33.40	34.13	34.81	35.43	36.00	36.53	37.03	37.50	37.95	2
17.53	17.89	18.22	18.52	18.81	19.07	19.32	19.55	19.77	3
12.84	13.09	13.32	13.53	13.73	13.91	14.08	14.24	14.40	4
10.70	10.89	11.08	11.24	11.40	11.55	11.68	11.81	11.93	5
9.48	9.65	9.81	9.95	10.08	10.21	10.32	10.43	10.54	6
8.71	8.86	9.00	9.12	9.24	9.35	9.46	9.55	9.65	7
8.18	8.31	8.44	8.55	8.66	8.76	8.85	8.94	9.03	8
7.78	7.91	8.03	8.13	8.23	8.33	8.41	8.49	8.57	9
7.49	7.60	7.71	7.81	7.91	7.99	8.08	8.15	8.23	10
7.25	7.36	7.46	7.56	7.65	7.73	7.81	7.88	7.95	11
7.06	7.17	7.26	7.36	7.44	7.52	7.59	7.66	7.73	12
6.90	7.01	7.10	7.19	7.27	7.35	7.42	7.48	7.55	13
6.77	6.87	6.96	7.05	7.13	7.20	7.27	7.33	7.39	14
6.66	6.76	6.84	6.93	7.00	7.07	7.14	7.20	7.26	15
6.56	6.66	6.74	6.82	6.90	6.97	7.03	7.09	7.15	16
6.48	6.57	6.66	6.73	6.81	6.87	6.94	7.00	7.05	17
6.41	6.50	6.58	6.65	6.72	6.79	6.85	6.91	6.97	18
6.34	6.43	6.51	6.58	6.65	6.72	6.78	6.84	6.89	19
6.28	6.37	6.45	6.52	6.59	6.65	6.71	6.77	6.82	20
6.11	6.19	6.26	6.33	6.39	6.45	6.51	6.56	6.61	24
5.93	6.01	6.08	6.14	6.20	6.26	6.31	6.36	6.41	30
5.76	5.83	5.90	5.96	6.02	6.07	6.12	6.16	6.21	40
5.60	5.67	5.73	5.78	5.84	5.89	5.93	5.97	6.01	60
5.44	5.50	5.56	5.61	5.66	5.71	5.75	5.79	5.83	120
5.29	5.35	5.40	5.45	5.49	5.54	5.57	5.61	5.65	∞

SOURCE: From *Biometrika Tables for Statisticians,* Vol. 1, 3rd ed., edited by E.S. Pearson and H.O. Hartley (Cambridge University Press, 1966). Reproduced by permission of the *Biometrika* Trustees.

Data Sources

Introduction

1. Thomas Watkins, "Rowdy crowd jeers Whitman," *The Press-Enterprise* (Riverside, CA), 30 October 2010, p. A14.

2. Trevor Hunnicutt, "Fiorina calls herself similar to Feinstein, who supports Boxer," *The Press-Enterprise* (Riverside, CA), 30 October 2010, p. A14.

3. Jim Miller, "Race for attorney general tight," *The Press-Enterprise* (Riverside, CA), 30 October 2010, p. A14

4. *Fox News,* http://www.foxnews.com/story/0,2933,102511,00.html, 10 February 2004.

5. "Hot News: 98.6 Not Normal," *The Press-Enterprise* (Riverside, CA), 23 September 1992.

Chapter 1

1. "Election 2012." CNN/Opinion Research Corporation Poll, http://www.pollingreport.com/2012.htm, 9–11 April 2010.

2. "Run the Country? Most Teens Would Pass," http://abcnews.go.com/images/pdf/943a1TeensandthePresidency.pdf, 22 January 2004.

3. "Facebook Demographics and Statistics Report 2010—145% Growth in 1 Year," http://www.istrategylabs.com/2010/01/facebook-demographics-and-statistics-report-2010-145-growth-in-1-year/, 7 July 2010.

4. "Getting Back to Work," http://www.usatoday.com/snapshot/news/2001-07-17-backtowork.htm, 2 October 2001.

5. Sarah Janssen, Ed., *The World Almanac and Book of Facts 2011* (New York, NY: Infobase Publishing), 2011.

6. Adapted from: Arlene Eisenberg, Heidi Murkoff, and Sandee Hathaway. *What to Expect the First Year* (New York: Workman Publishing), 2003.

7. D. G. Altman and J.M. Bland. "Time to Survival Data," *British Medical Journal BMJ* 1998; 317:468–469 (15 August) at http://bmj.bmjjournals.com/cgi/content/full/317/7156/4687.

8. "Employment Projections: Education Pays. . .," http://www.bls.gov/emp/ep_chart_001.htm, 7 July 2010.

9. "Major Religions of the World Ranked by Number of Adherents," http://www.adherents.com/Religions_By_Adherents.html, 6 July 2010.

10. "U.S. Box Office Actuals—Weekend of June 25, 2010," http://www.radiofree.com/mov-tops.shtml, 6 July 2010.

11. Robert P. Wilder, D. Brennan, and D.E. Schotte, "A Standard Measure for Exercise Prescription for Aqua Running," *American Journal of Sports Medicine* 21, no. 1 (1993):45.

12. http://www.kentuckyderby.info/kentuckyderby-history2006.php and www.kentuckyderby.com/news.

13. Bryan Walsh, "The Spreading Stain," *Time,* 21 June 2010, pp. 51–59.

14. "Election Center 2008," http://www.cnn.com/ELECTION/2008/results/president/, 6 July 2010.

15. "Dealing with Mugabe's Diamonds," *Times,* 5 July 2010, p. 14. (no author listed)

16. "Starbucks—Riverside, CA," www.insiderpages.com/s/CA/Riverside/Starbucks, 6 July 2010.

17. David Von Drehle, "The Other Financial Crisis," *Time,* 28 June 2010, p. 22.

18. A. Tubb, A.J. Parker, and G. Nickless, "The Analysis of Romano-British Pottery by Atomic Absorption Spectrophotometry," *Archaeometry* 22 (1980):153.

19. Mike Schwartz and Mark Kendall, "The Great Calorie Debate," *The Press-Enterprise* (Riverside, CA), 10 February 2004, p. El.

20. Lawrence E. Levine and Victorina Wasmuth, "Laptops, Technology, and Algebra 1: A Case Study of an Experiment," *Mathematics Teacher* 97, no. 2 (February 2004):136.

21. A. Azzalini and A.W. Bowman, "A Look at Some Data on the Old Faithful Geyser," *Applied Statistics* (1990):57.

22. P.A. Mackowiak, S.S. Wasserman, and M.M. Levine, "A Critical Appraisal of 98.6 Degrees F, the Upper Limit of the Normal Body Temperature, and Other Legacies of Carl Reinhold August Wunderlich," *Journal of the American Medical Association* (268):1578–1580.

23. Allen L. Shoemaker, "What's Normal? Temperature, Gender, and Heart Rate," *Journal of Statistics Education* (1996).

Chapter 2

1. "2010 Automobile Insurance," 8 July 2010, http://interactive.web.insurance.ca.gov/survey/survey?type=autoSurvey&event=autoStart.

2. "Fortune 500," http://money.cnn.com/magazines/fortune/fortune500/2010/full_list/, 8 July 2010.

3. "Birth Order and the Baby Boom," *American Demographics* (Trend Cop), March 1997, p. 10.

4. "Tuna Goes Upscale," *Consumer Reports,* June 2001, p. 19.

5. "Starbucks—Riverside, CA," www.insiderpages.com/s/CA/Riverside/Starbucks, 6 July 2010.

6. "Nintendo Wii 45496880019 Gaming Console," http://videogames.pricegrabber.com/wii-console-accessories/Nintendo-Console/m27737371.html/st=pop/sv=title, 8 July 2010.

7. A. Tubb, A.J. Parker, and G. Nickless, "The Analysis of Romano-British Pottery by Atomic Absorption Spectrophotometry," *Archaeometry* 22 (1980):153.

8. A. Azzalini and A.W. Bowman, "A Look at Some Data on the Old Faithful Geyser," *Applied Statistics* (1990):57.

9. Sarah Janssen, Ed., *The World Almanac and Book of Facts 2011* (New York, NY: Infobase Publishing), 2011.

10. "Aaron Rodgers #12 QB," http://sports.espn.go.com/nfl/players/gamelog?playerId=8439, 18 March 2011.

11. "Aaron Rodgers #12 QB," http://sports.espn.go.com/nfl/players/gamelog?playerId=8439 and "Drew Brees #9 QB," http://sports.espn.go.com/nfl/players/gamelog?playerId=2580, 18 March 2011.

12. D.G. Altman and J.M. Bland. "Time to Survival Data," *British Medical Journal BMJ* 1998; 3 17:468–469 (15 August) at http://bmj.bmjjournals.com/cgi/content/full/317/7156/4687.

13. Allen L. Shoemaker, "What's Normal? Temperature, Gender, and Heart Rate," *Journal of Statistics Education* (1996).

14. "New SUV Ratings & Reliability," www.consumerreports.org/cro/cars/new-cars/suvs/ratings-reliability/specs.htm, 20 August 2010.

15. "Favorite Camping Activity," http://www.usatoday.com/snapshot/news/2001-05-22-camping.htm, 26 September 2001 (*Source:* Wirthlin Worldwide for Coleman Company).

16. www.mlb.com, 25 August 2010.

17. "How Much We Volunteer," http://www.usatoday.com/news/snapshot.htm, 25 August 2010.

18. "How Old is Your Vehicle?" http://www.usatoday.com/news/snapshot.htm, 6 July 2010.

19. "What Fans are Willing to Pay for a Concert Ticket," http://www.usatoday.com/news/snapshot.htm, 25 August 2010.

Chapter 3

1. Adapted from http://nces.ed.gov/programs/digest/d09/tables/dt09_258.asp, *Digest of Educational Statistics,* 5 July 2010.
2. Adapted from Michael J. Weiss, "The New Summer Break," *American Demographics,* August 2001, p. 55.
3. "Consumer Price Index—All Urban Consumers," http://data.bls.gov/cgi-bin/surveymost, 13 July 2010.
4. "U.S. Facebook Audience," *EContent Magazine,* Vol. 33, No. 2, March 2010, p. 4.
5. Gregory K. Torrey, S.F. Vasa, J.W. Maag, and J.J. Kramer, "Social Skills Interventions Across School Settings: Case Study Reviews of Students with Mild Disabilities," *Psychology in the Schools* 29 (July 1992):248.
6. "LCD TV Ratings & Reliability," http://www.consumerreports.org/cro/electronics-computers/tvs-services/tvs/lcd-tv-ratings/ratings-overview.htm, 13 July 2010.
7. "New SUV Ratings & Reliability," www.consumerreports.org/cro/cars/new-cars/suvs/ratings-reliability/specs.htm, 20 August 2010.
8. Stellan Ohlsson, "The Learning Curve for Writing Books: Evidence from Professor Asimov," *Psychological Science* 3, no. 6 (1992):380–382.
9. "Facebook Demographics and Statistics Report 2010—145% Growth in 1 Year," http://www.istrategylabs.com/2010/01/facebook-demographics—and-statistics-report-2010-145-growth-in-1-year/, 7 July 2010.
10. "Weekend Box Office," http://boxofficemojo.com/weekend/chart/, 25 August 2010.
11. Sarah Janssen, Ed., *The World Almanac and Book of Facts 2011* (New York, NY: Infobase Publishing), 2011.
12. "Aaron Rodgers #12 QB," http://sports.espn.go.com/nfl/players/gamelog?playerId=8439, 18 March 2011.
13. A. Tubb, A.J. Parker, and G. Nickless, "The Analysis of Romano-British Pottery by Atomic Absorption Spectrophotometry," *Archaeometry* 22 (1980):153.
14. Celia Chen, "2010: Housing Recuperates," http://www.nabe.com/rt/real/documents/Chen_NABE_01212010.pdf, 23 August 2010, p. 12.
15. Borgna Brunner, Ed., *Time Almanac 2007* (Boston, MA: Pearson Education, Inc.), 2006.
16. "Cell Phones and Services: Smart Phone Ratings," http://www.consumerreports.org/cro/electronics-computers/phones-mobile-devices/cell-phones-services/smart-phone-ratings/ratings-overview.htm, 23 August 2010.
17. "Dishwashers," *Consumer Reports,* August 2010, p. 34.

Chapter 4

1. "Racial and Ethnic Distribution of ABO Bloodtypes: Bloodbook.com," http://www.bloodbook.com/world-abo.html, 25 August 2010.
2. Table adapted from http://www.pollingreport.com/science.htm#Stem, 30 October 2006.
3. "WNBA Teams," http://espn.go.com/wnba/teams, 26 August 2010.
4. Bruce E. Morgan and Michael A. Oberlander, "An Examination of Injuries in Major League Soccer," *The American Journal of Sports Medicine,* 29(4), 2001, pp. 426–429.
5. http://sports.espn.go.com/nba/teams/stats, 24 August 2010.
6. Adapted from "Demo Memo," *American Demographics,* May 1997, and http://factfinder.census.gov/servlet/STTable?_bm=y&-geo_id=01000US&-qr_name=ACS_2008_1YR_G00_S0101&-ds_name=ACS_2008_1YR_G00_&-_lang=en&-redoLog=false&-state=st&-CONTEXT=st, 26 August 2010.
7. "Health: Latest Findings," *Time,* 5 July 2010, p. 18.
8. "Coffee Breaks Daily," http://www.usatoday.com/news/snapshot.htm?section=L&label=2006-10-26-cell, 26 October 2006.

9. Andrew S. Levy, M.J. Wetzler, M. Lewars, and W. Laughlin, "Knee Injuries in Women Collegiate Rugby Players," *The American Journal of Sports Medicine* 25, no. 3 (1997):360.

10. P.D. Franklin, R.A. Lemon, and H.S. Barden, "Accuracy of Imaging the Menisci on an In-Office, Dedicated, Magnetic Resonance Imaging Extremity System," *The American Journal of Sports Medicine* 25, no. 3 (1997):382.

11. Mya Frazier, "The Reality of the Working Woman," *Ad Age Insights*—White Paper, Spring, 2010, http://adage.com/images/bin/pdf/aa_working_women_whitepaper_web.pdf, p. 4.

12. Michael Crichton, *Congo* (New York: Knopf, 1980).

Chapter 5

1. http://testprep.about.com/od/sat/f/SATFAQ_GoodSAT.htm, 27 August 2010.

2. "Percentage of Major League Sports Players Born Outside the USA," http://www.usatoday.com/news/snapshot.htm, 6 July 2010.

3. "How back pain limits sports" http://www.usatoday.com/news/snapshot.htm?section=L&label=2006-10-27-wash.

4. "U.S. Pet Ownership Statistics," http://www.humanesociety.org/issues/pet_overpopulation/facts/pet_ownership_statistics.html, 27 August 2010.

5. http://www.foodsafetynews.com/2010/07/three-daycares-closed-due-to-e-coli-outbreak/, 26 August 2010.

6. http://www.cnn.com/2010/HEALTH/04/15/foodborne.illness.cdc/index.html?iref=allsearch, 26 August 2010.

7. "Checking In On Vacation," http://www.usatoday.com/news/snapshot.htm?section=L&label=2006-10-27-wash.

8. "Call It in the Air," *The Press-Enterprise* (Riverside, CA), 19 October 1992.

9. Mark A. Atkinson, "Diet, Genetics, and Diabetes," *Food Technology* 51, no. 3 (March 1997), p. 77.

10. "SquareTrade's Report on Simple Vs. Smartphone Reliability," http://blog.squaretrade.com/2008/05/smartphone-reli.html, 26 August 2010.

11. "Most popular chocolate," http://usatoday.com/news/snapshot.htm?section=L&label=2006-10-27-wash.

12. "What Would You Do First if You Won $1 Million Dollars Tomorrow?" http://www.usatoday.com/news/snapshot.htm, 25 August 2010.

13. "How women eat on the run," http://www.usatoday.com, 1 January 2004.

14. "Upgrades Drivers Say Would Make Their Communities More Drivable," http://www.usatoday.com/news/snapshot.htm, 6 July 2010.

15. "Consumers Who are Committed to Living with Fewer Credit Cards," http://www.usatoday.com/news/snapshot.htm, 25 August 2010.

16. "Top destinations for vacationers," http://www.usatoday.com/news/snapshots, 2 November 2006.

17. Matthew L. Wald, "Cancers Near a Reactor: A Mystery and a Debate," *New York Times*, 21 May 1987, p. A-22.

Chapter 6

1. http://en.wikipedia.org/wiki/Human_height and http://en.wikipedia.org/wiki/Heights_of_Presidents_of_the_United_States_and_presidential_candidates

2. Adapted from A.M. Goodman and A.R. Ennos, "The Response of Field-Grown Sunflower and Maize to Mechanical Support," *Annals of Botany* 79 (1997):703.

3. John Fetto, "Shop Around the Clock," *American Demographics,* September 2003, p. 18.

4. "Medical Encyclopedia: Pulse," *Medline Plus: Trusted Health Information for You,* http://www.nlm.nih.gov/medlineplus/ency/article/003399.htm#Normal%20Values, 2 April 2004.

5. "Top Eco-actions Taken to Help the Environment," http://www.usatoday.com/news/ snapshot.htm, 25 August 2010.

6. "Pepsico (PEP)," http://www.wikinvest.com/stock/Pepsico_(PEP)#_note-PEP7, 27 August 2010.

7. Sonja Steptoe, "Ready, Set, Relax!" *Time,* 27 October 2003, p. 38.

8. Philip A. Altman and D.S. Dittmer, *The Biology Data Book,* 2nd ed., Vol I. (Bethesda, MD: Federation of American Societies for Experimental Biology, 1964), p. 137.

9. http://lib.stat.cmu.edu/DASL/Datafiles/MilitiamenChests.html, 27 September 2010.

10. Allen L. Shoemaker, "What's Normal? Temperature, Gender, and Heart Rate," *Journal of Statistics Education* (1996).

11. "How polite are cellphone users?" http://usatoday/com/news/snapshot.htm?section= L&label=2006-10-27.

12. "Average Salary of Full Time Instructional Faculty on 9-Month Contracts in Degree-granting Institutions," *Digest of Educational Statistics,* http://nces.ed.gov/programs/digest/d10/tables/ dt10_268.asp, 5 May 2011.

Chapter 7

1. Alice Park, "Omega-3 May Reduce Heart Risks Less Than Thought." http://www.time.com/ time/health/article/0,8599,2014603,00.html, 30 August 2010.

2. Chris Gilberg, J.L. Cos, H. Kashima, and K. Eberle, "Survey Biases: When Does the Interviewer's Race Matter?" *Chance,* Fall 1996, p. 23.

3. Chery Smith and Stefanie Fila, "Comparison of the Kid's Block Food Frequency Questionnaire to the 24-Hour Recall in Urban Native American Youth," *American Journal of Human Biology,* 18:706–709 (2006).

4. Liz Szabo,"Study: Tai Chi Could Ease Fibromyalgia Pain," *The Press-Enterprise* (Riverside, CA), 19 August 2010, p. 4D.

5. "Space Exploration," CNN/*USA Today*/Gallup Poll, http://www.pollingreport.com/ science.htm#Space, 5 April 2004.

6. "ASK AMERICA: 2003 Nationwide Policy Survey," Congressional District #44, 23 June 2003.

7. "Average Salary of Full Time Instructional Faculty on 9-Month Contracts in Degree-granting Institutions," *Digest of Educational Statistics,* http://nces.ed.gov/programs/digest/d09/tables/ dt09_258.asp, 5 July 2010.

8. "USDA National Nutrient Database: Bananas, raw," http://www.nal.usda.gov/fnic/foodcom/ cgi-bin/list_nut_edit.pl, 1 September 2010.

9. Allen L. Shoemaker, "What's Normal? Temperature, Gender, and Heart Rate," *Journal of Statistics Education* (1996).

10. Nicola Maffulli, V. Testa, G. Capasso, and A. Sullo, "Calcific Insertional Achilles Tendinopathy," *The American Journal of Sports Medicine* 32, no. 1 (January/February 2004):174.

11. "Top Eco-actions Taken to Help the Environment," http://www.usatoday.com/news/ snapshot.htm, 25 August 2010.

12. "Must Have Accessories on Family Road Trips," http://www.usatoday.com/news/snapshot.htm, 25 August 2010.

13. Meghan Baker, "We're Driven to Distraction When Fido is Co-Pilot, Study Finds," www.foxnews.com/us/2010/08/19/driven-distraction-pets-car-just-dangerour-texting/, 20 August 2010.

14. Adam Fernandez, "Nuts About You," *American Demographics* 26, no. 1 (February 2004):14.

15. P.C. Karalekas, Jr., C.R. Ryan, and F.B. Taylor, "Control of Lead, Copper, and Iron Pipe Corrosion in Boston," *American Water Works Journal,* February 1983.

16. *Science News* 136 (19 August 1989):124.

17. "Same-Sex Marriage, Gay Rights," *CBS News Poll,* http://www.pollingreport.com/civil.htm, 20–24 August 2010.

18. Catherine M. Santaniello and R.E. Koning, "Are Radishes Really Allelopathic to Lettuce?" *The American Biology Teacher* 58, no. 2 (February 1996):102.

19. http://www.gallup.com/poll/indicators/indairlines.asp#RelatedAnalyses. Gallup Poll News Service, 16 October 2001.

20. J. Hackl, *Journal of Quality Technology,* April 1991.

21. Daniel Seligman, "The Road to Monte Carlo," *Fortune,* 15 April 1985.

Chapter 8

1. Adapted from "Polar Bear," http://en.wikipedia.org/wiki/Polar_Bear#Size_and_weight, 9 November 2006.

2. *Science News* 136 (19 August 1989):124.

3. "Immigration," *ABC News/Washington Post Poll,* 3–6 June 2010, http://www.pollingreport.com/immigration.htm, 2 September 2010.

4. "Hotels for any Budget," *Consumer Reports,* June 2010 and "Hotel Ratings," http://consumerreports.org/cro/magazine-archive/2010/june/shopping/hotels/ratings/index.htm, 2 September 2010.

5. "Space Exploration," Associated Press Poll, http://www.pollingreport.com/science.htm#Space, 5 April 2004.

6. Mya Frazier, "The Reality of the Working Woman," *Ad Age Insights: White Paper,* http://adage.com/images/bin/pdf/aa_working_women_whitepaper_web.pdf, 7 June 2010.

7. "Same-Sex Marriage, Gay Rights," *CBS News Poll,* http://www.pollingreport.com/civil.htm, 20–24 August 2010.

8. Brian Dumaine, "Can the Volt Charge GM?" *Time*, 2 July 2010, p. 40.

9. Allen L. Shoemaker, "What's Normal? Temperature, Gender, and Heart Rate," *Journal of Statistics Education* (1996).

10. "Voting Intentions Even, Turnout Indicators Favor GOP," http://people-press.org/report/630/, 1 July 2010.

11. William Leonard, Barbara Speziale, and John Pernick, "Performance Assessment of a Standards-Based High School Biology Curriculum," *The American Biology Teacher* 63, no. 5 (2001):310–316.

12. "Engineering, Computer Science Students Have Highest Salary Expectations," http://www.naceweb.org/so/08022010/computer_science_engineering_salary/, 2 September 2010.

13. "Problems and Priorities," *FOX News/Opinion Dynamics Poll*, 4–5 May 2010, http://www.pollingreport.com/prioriti3.htm, 4.

14. Mark Gillespie, "Baseball Fans Overwhelmingly Want Mandatory Steroid Testing," Gallup News Service, http://gallup.com/content/print.aspx?ci=11245, 14 February 2004.

15. "When Bargaining Pays Off," *Consumer Reports*, August 2009, http://www.consumerreports.org/cro/magazine-archive/august-2009/money/bargaining/overview/bargaining-ov.htm, 2 September 2010.

16. David L. Wheeler, "More Social Roles Means Fewer Colds," *Chronicle of Higher Education* XLIII, no. 44 (11 July 1997):A13.

17. "Generation Next: A Snapshot," www.pewtrusts.org/ideas, 9 November 2006 and "The American Freshman: National Norms for Fall 2005," http://www.gseis.ucla.edu/heri/PDFs/ResearchBrief05.pdf, 9 November 2006.

18. "Per Capita Consumption of Major Food Commodities," and "Per Capita Consumption of Selected Beverages by Type," *2010 Statistical Abstract of the United States*, U.S. Census Bureau, http://www.census.gov/compendia/statab/cats/prices/food_cost_and_prices.html.

19 "Eating More: Enjoying Less," 19 April 2006, *Social and Demographic Trends, Pew Research Center*, http://pewsocialtrends.org/pubs/309/eating-more-enjoying-less, 3 September 2010.

20. Adapted from A.M. Goodman and A.R. Ennos, "The Responses of Field-Grown Sunflower and Maize to Mechanical Support," *Annals of Botany* 79 (1997):703.

21. http://wiki.answers.com/Q/What_is_the_average_price_of_an_NBA_ticket, 3 September 2010.

22. G. Wayne Marino, "Selected Mechanical Factors Associated with Acceleration in Ice Skating," *Research Quarterly for Exercise and Sport* 54, no. 3 (1983).

23. "Collective Bargaining Agreement Between Mount Baker School District and Mount Baker: September 1, 2008 to August 31, 2012," http://www.pseclassified.org, 4 September 2010.

24. http://cbsnews.com/stories/2005/11/20/opinion/polls/printaale1060315.shtml, 10 November 2005.

Chapter 9

1. Jan Pergl, Irena Perglova, Per Pysek, and Hansjorg Dietz, "Population Age Structure and Reproductive Behavior of the Monocarpic Perennial *Heraculaneum Mantegazzianum* (Apiaceae) in its Native and Invaded Distribution Ranges," *American Journal of Botany,* 93(7):1018–1028, 2006.

2. "America by the Numbers," *Time,* 30 October 2006, pp. 43–55.

3. Allen L. Shoemaker, "What's Normal? Temperature, Gender, and Heart Rate," *Journal of Statistics Education* (1996).

4. "Hot News: 98.6 Not Normal," *The Press-Enterprise* (Riverside, CA), 23 September 1992.

5. Nicola Maffulli, V. Testa, G. Capasso, and A. Sullo, "Calcific Insertional Achilles Tendinopathy," *The American Journal of Sports Medicine* 32, no. 1 (January/February 2004):174.

6. "Engineering, Computer Science Students Have Highest Salary Expectations," http://www.naceweb.org/so/08022010/computer_science_engineering_salary/, 2 September 2010.

7. "Hotels for any Budget," *Consumer Reports*, June 2010 and "Hotel Ratings," http://consumerreports.org/cro/magazine-archive/2010/june/shopping/hotels/ratings/index.htm, 2 September 2010.

8. "8 Ways to Land a Great Airfare," *Consumer Reports,* June 2010 and www.consumerreports.org/cro/magazine-archive/20l0/june/money/airfares/overview/index.htm, 2 September 2010.

9. Dianne Hales, "We're Changing the Way We Eat," *PARADE,* November 12, 2006, pp. 4–5.

10. "It Pays to Buy Store Brands," *Consumer Reports*, October 2009 and www.consumerreports.org/cro/magazine-archive/october-2009/shopping/buying-storebrands/overview/buying-store-brands-ov.htm, 4 September 2010.

11. Paul Taylor and Wendy Wang, "The Fading Glory of the Television and Telephone," http://pewsocialtrends.org/pubs/762/fading-glory-television-telephone-luxury-necessity, 19 August 2010.

12. "U.S. Pet Ownership Statistics," http://www.humanesociety.org/issues/pet_overpopulation/facts/pet_ownership_statistics.html, 3 September 2010.

13. Liz Szabo, "Study: Tai Chi Could Ease Fibromyalgia Pain," *The Press-Enterprise* (Riverside, CA), 19 August 2010, p. 4D.

14. Denise Grady, "Study Finds Alzheimer's Danger in Hormone Therapy," *The Press-Enterprise* (Riverside, CA), 28 May 2003.

15. *Heart Healthy Women* website, http://www.hearthealthywomen.org/patients/medications/blood_thinners__aspirin_6.html.

16. Jonathan W. Jantz, C.D. Blosser, and L.A. Fruechting, "A Motor Milestone Change Noted with a Change in Sleep Position," *Archives of Pediatric Adolescent Medicine* 151 (June 1997):565.

17. "Generation Next: A Snapshot," www.pewtrusts.org/ideas, 9 November 2006; and "The American Freshman: National Norms for Fall 2005," http://www.gseis.ucla.edu/heri/PDFs/ResearchBrief05.PDF, 9 November 2006.

18. Loren Hill, *Bassmaster,* September/October 1980.

19. Mya Frazier, "The Reality of the Working Woman," *Ad Age Insights: White Paper,* http://adage.com/images/bin/pdf/aa_working_women_whitepaper_web.pdf, 7 June 2010.

20. Charles Dickey, "A Strategy for Big Bucks," *Field and Stream,* October 1990.

21. *Science News* 136 (19 August 1989):124.

22. Jeeseung Choi, Janet Meininger, and Robert E. Roberts, "Ethnic Differences in Adolescents' Mental Distress, Social Stress, and Resources," *Adolescence,* Vol. 41, no. 162, Summer 2006, pp. 263–278.

23. "2010 College-Bound Seniors: Total Group Profile Report," http://professionals.collegeboard .com/profdownload/2010-total-group-profile-report-cbs.pdf, 14 September 2010.

24. "Brides-to-be Pick Their Wedding Sites," http://www.usatoday.com/news/snapshot.htm, 6 July 2010.

25. "California English Language Development Test," California Department of Education, http://dq.cde.ca.gov/dataquest/CELDT/results.aspx?year=2009-2010&level=district &assessment=2&subgroup=1&entity=33-67215-0000, 4 September 2010.

26. Kurt Grote, T.L. Lincoln, and J.G. Gamble, "Hip Adductor Injury in Competitive Swimmers," *The American Journal of Sports Medicine* 32, no. 1 (January/February 2004):104.

27. Joel B. Greenhouse and Samuel W. Greenhouse, "An Aspirin a Day . . . ?" *Chance: New Directions for Statistics and Computing* 1, no. 4 (1988):24–31.

Chapter 10

1. "Pricing of Tuna," *Consumer Reports,* June 2001.

2. W.B. Jeffries, H.K. Voris, and C.M. Yang, "Diversity and Distribution of the Pedunculate Barnacles *Octolasmis* Gray, 1825 Epizoic on the Scyllarid Lobster *Thenus orientalis* (Lund, 1793)," *Crustaceana* 46, no. 3 (1984).

3. "Ben Roethlisberger #7 QB," http://sports.espn.go.com/nfl/players/gamelog?playerId=5536, 19 March 2011.

4. Wendy K. Baell and E.H. Wertheim, "Predictors of Outcome in the Treatment of Bulimia Nervosa," *British Journal of Clinical Psychology* 31 (1992):330–332.

5. "L.A. Heart Data." Adapted from data found at http://www-unix.oit.umass.edu/~statdata/ statdata/data/laheart.dat.

6. Jan D. Lindhe, "Clinical Assessment of Antiplaque Agents," *Compendium of Continuing Education in Dentistry,* Supplement 5 (1984).

7. Susan J. Beckham, W.A. Grana, P. Buckley, J.E. Breasile, and P.L. Claypool, "A Comparison of Anterior Compartment Pressures in Competitive Runners and Cyclists," *American Journal of Sports Medicine* 21, no. 1 (1992):36.

8. Michael A. Brehm, J.S. Buguliskis, D.K. Hawkins, E.S. Lee, D. Sabapathi, and R.A. Smith, "Determining Differences in Efficacy of Two Disinfectants Using *t*-tests," *The American Biology Teacher* 58, no. 2 (February 1996):111.

9. "Aaron Rodgers #12 QB," http://sports.espn.go.com/nfl/players/gamelog?playerId=8439 and "Drew Brees #9 QB," http://sports.espn.go.com/nfl/players/gamelog?playerId=2580, 18 March 2011.

10. A. Tubb, A.J. Parker, and G. Nickless, "The Analysis of Romano-British Pottery by Atomic Absorption Spectrophotometry," *Archaeometry* 22 (1980):153.

11. "2010 Automobile Insurance," http://interactive.web.insurance.ca.gov/survey/ survey?type=autoSurvey&event=autoStart, 8 July 2010.

12. "2010 College-Bound Seniors: Total Group Profile Report," http://professionals.collegeboard .com/profdownload/20l0-total-group-profile-report-cbs.pdf, 14 September 2010.

13. "Aaron Rodgers #12 QB," http://sports.espn.go.com/nfl/players/gamelog?playerId=8439 and "Ben Roethlisberger #7 QB," http://sports.espn.go.com/nfl/players/gamelog?playerId=5536, 18 March 2011.

14. Carlos E. Macellari, "Revision of Serpulids of the Genus *Rotularia* (*Annelida*) at Seymour Island (Antarctic Peninsula) and Their Value in Stratigraphy," *Journal of Paleontology* 58, no. 4 (1984).

15. T.M. Casey, M.L. May, and K.R. Morgan, "Flight Energetics of Euglossine Bees in Relation to Morphology and Wing Stroke Frequency," *Journal of Experimental Biology* 116 (1985).

16. Karl J. Niklas and T.G. Owens, "Physiological and Morphological Modifications of *Plantago Major* (*Plantaginaceae*) in Response to Light Conditions," *American Journal of Botany 76*, no. 3 (1989):370–382.

17. "Pricing and Ordering Comparisons," http://www.pharmacychecker.com/Pricing.asp?DrugName=Buspar&DrugId=29296&DrugStrengthId=49856, 16 September 2010.

18. "TicketNetwork Advisory—Lakers vs. Celtics NBA Finals Ticket Prices Lower Than 2008," http://www.marketwire.com/press-release/TicketNetwork-Advisory-Lakers-vs-Celtics-NBA-Finals-Ticket-Prices-Lower-Than-2008-1269109.htm, 1 June 2010.

19. John Fetto, "Shop Around the Clock," *American Demographics* 25, no. 7 (September 2003):18.

20. Dayna Straehley, "No Algebra Book: Two Classes at Amerlia Earhart Middle School are in a Pilot Study," *The Press-Enterprise* (Riverside, CA), 9 September 2010, p. A3.

21. "2009–2010 Accountability Progress Report," *The Press-Enterprise* (Riverside, CA), 14 September 2010, p. A8. (Original source: California Department of Education.)

Chapter 11

1. "Pricing of Tuna," *Consumer Reports,* June 2001.

2. "2010–2011: Interpreting Your GRE Scores" http://www.ets.org/s/gre/pdf/gre_interpreting_scores.pdf, 16 September 2010.

3. "2011 Automobile Insurance," http://interactive.web.insurance.ca.gov/survey/survey?type=auto Survey&event=autoStart, 22 March 2011.

4. H.F. Barsam and Z.M. Simutis, "Computer-Based Graphics for Terrain Visualization Training," *Human Factors,* no. 26, 1984. Copyright 1984 by the Human Factors Society, Inc. Reproduced by permission.

5. Ciril Rebetez, Mireille Betrancourt, Mirweis Sangin, and Pierre Dillenbourg, "Learning from Animation Enabled by Collaboration," *Instructional Science*, V 35, No. 5, pp. 471–485, September 2010.

6. Russell R. Pate, Chia-Yih Wang, Marsha Dowda, Stephen W. Farrell, and Jennifer R. O'Neill, "Cardiorespiratory Fitness Levels Among U.S. Youth 12 to 19 Years of Age," *Archives of Pediatric Adolescent Medicine,* Vol. 160, October 2006, pp. 1005–1011.

7. Based on data from "Average Salary of Full Time Instructional Faculty on 9-Month Contracts in Degree-granting Institutions," *Digest of Educational Statistics,* http://nces.ed.gov/programs/digest/d09/tables/dt09_257.asp, 5 July 2010.

8. A. Tubb, A.J. Parker, and G. Nickless, "The Analysis of Romano-British Pottery by Atomic Absorption Spectrophotometry," *Archaeometry* 22 (1980):153.

9. "Smart Phone Ratings," http://www.consumerreports.org/cro/electronics-computers/phones-mobile-devices/cell-phones-services/smart-phone-ratings/ratings-overview.htm, 23 August 2010.

10. Roberta R. Bailey and Craig Idlebrook, "Save Money on Groceries," www.MotherEarthNews.com, August/September 2010.

Chapter 12

1. Stellan Ohlsson, "The Learning Curve for Writing Books: Evidence from Professor Asimov," *Psychological Science* 3, no. 6 (1992):380–382.

2. Daniel C. Harris, *Quantitative Chemical Analysis,* 3rd ed. (New York: Freeman, 1991).

3. "2009–2010 Accountability Progress Reporting (APR)," http://api.cde.ca.gov/AcntRpt2010/2010GrthAPICo.aspx?cYear=2009-10&cSelect=33, RIVERSIDE, 23 September 2010.

4. Sarah Janssen, Ed., *The World Almanac and Book of Facts 2011* (New York, NY: Infobase Publishing), 2011.

5. Lawrence E. Levine and Victorina Wasmuth, "Laptops, Technology, and Algebra 1: A Case Study of an Experiment," *Mathematics Teacher* 97, no. 2 (February 2004):136.

6. "LCD TV Ratings & Reliability," http://www.consumerreports.org/cro/electronics-computers/tvs-services/tvs/lcd-tv-ratings/ratings-overview.htm, 13 July 2010.

7. "Drew Brees #9 QB," http://sports.espn.go.com/nfl/players/gamelog?playerId=2580, 22 March 2011.

8. W.B. Jeffries, H.K. Voris, and C.M. Yang, "Diversity and Distribution of the Pedunculate Barnacles *Octolasmis* Gray, 1825 Epizoic on the Scyllarid Lobster, *Thenus orientalis* (Lund, 1793)," *Crustaceana* 46, no. 3 (1984).

9. Gregory K. Torrey, S.F. Vasa, J.W. Maag, and J.J. Kramer, "Social Skills Interventions Across School Settings: Cast Study Reviews of Students with Mild Disabilities," *Psychology in the Schools* 29 (July 1992):248.

10. G. Wayne Marino, "Selected Mechanical Factors Associated with Acceleration in Ice Skating," *Research Quarterly for Exercise and Sport* 54, no. 3 (1983).

11. A.J. Ellis, "Geothermal Systems," *American Scientist,* September/October 1975.

12. "Ice Cream Nutritional Statement," *Coldstone Creamery,* http://www.coldstonecreamery.com/nutritional/nutrition_information.html, 27 September 2010.

13. Allen L. Shoemaker, "What's Normal? Temperature, Gender, and Heart Rate," *Journal of Statistics Education* (1996).

14. http://espn.go.com/mlb/stats/team/_/stat/batting/year/2010/seasontype/2, 5 November 2010.

15. David R. McAllister et al., "A Comparison of Preoperative Imaging Techniques for Predicting Patellar Tendon Graft Length before Cruciate Ligament Reconstruction," *The American Journal of Sports Medicine,* 20(4):461–465.

16. Henry Gleitman, *Basic Psychology,* 4th ed. (New York: Norton, 1996).

17. http://www.the-numbers.com/charts/daily/201/20100923.php, 27 September 2010.

18. "Lexus GX," http://en.wikipedia.org/wiki/Lexus_GX, 27 September 2010.

19. "Explore Our Menu," http://www.starbucks.com/menu/catalog/nutrition?paging=false&drink=all&page=2#view_control=nutrition, 27 September 2010.

20. *Automotive News: 1997 Market Data Book,* 28 May 1997, and Sarah Janssen, Ed., *The World Almanac and Book of Facts 2011* (New York, NY: Infobase Publishing), 2011.

Chapter 13

1. W.S. Good, "Productivity in the Retail Grocery Trade," *Journal of Retailing* 60, no. 3 (1984).

2. "Digital Cameras," *Consumer Reports,* August 2010, p. 50.

3. "Lexus GX," http://en.wikipedia.org/wiki/Lexus_GX, 27 September 2010.

4. Kimberly Pierceall, "S.B. Airport Plans for Big Future," *The Press-Enterprise* (Riverside, CA), 24 September 2010, p. D1.

5. "2009–2010 Accountability Progress Reporting (APR)," http://api.cde.ca.gov/AcntRpt2010/2010GrthAPICo.aspx?cYear=2009-10&cSelect=33, RIVERSIDE, 23 September 2010.

6. R. Blair and R. Miser, "Biotin Bioavailability from Protein Supplements and Cereal Grains for Growing Broiler Chickens," *International Journal of Vitamin and Nutrition Research* 59 (1989):55–58.

7. "2010–2011: Interpreting Your GRE Scores," http://www.ets.org/s/gre/pdf/gre_interpreting_scores.pdf, 16 September 2010.

8. "All Season Tire Ratings." *Consumer Reports,* www.consumerreports.org/cro/cars/tires-auto-parts/tires/all-season-tire-ratings/ratings-overview.htm, 28 September 2010.

9. "Tuna Goes Upscale," *Consumer Reports,* June 2001, p. 19.

10. *Automotive News: 1997 Market Data Book,* 28 May 1997, and Sarah Janssen, Ed., *The World Almanac and Book of Facts 2011* (New York, NY: Infobase Publishing), 2011.

Chapter 14

1. Daniel Q. Haney, "Mondays May Be Hazardous," *The Press-Enterprise* (Riverside, CA), 17 November 1992, p. A16.

2. "What Colors Come in Your Bag?" http://us.mms.com/us/about/products/milkchocolate/, 4 January 2007.

3. "What Colors Come in Your Bag?" http://us.mms.com/us/about/products/peanut/, 4 January 2007.

4. http://www.pollingreport.com/health.htm, *Newsweek Poll* conducted by Princeton Survey Research Associates International. 20–21 October, 2010.

5. Adapted from Linda Schmittroth, Ed., *Statistical Record of Women Worldwide* (Detroit and London: Gale Research, 1991).

6. Jason Malloy, "Gene Expression: NLSY Blogging: Eye and Hair Color of Americans," http://www.gnxp.com/blog/2008/12/nlsy-blogging-eye-and-hair-color-of.php, 31 December 2008.

7. Adapted from Dana Blanton, "Poll: Most Believe 'Cover-Up' of JFK Assassination Facts," http://www.foxnews.com/story/0,2933,102511,00.html, 10 February 2004.

8. "No Shows," *American Demographics*, 25, no. 9 (November 2003):11.

9. Adapted from Tamar Lewin, "Report Looks at a Generation, and Caring for Young and Old," *The New York Times on the Web*, 11 July 2001.

10. Siobhan Reilly, Michele Abendstern, Jane Hughes, David Challis, Dan Venables, and Irene Pedersen, "Quality in Long-Term Care Home for People with Dementia: An Assessment of Specialist Provision," *Aging and Society*, 26(2006):649–668.

11. W.W. Menard, "Time, Chance and the Origin of Manganese Nodules," *American Scientist*, September/October, 1976.

12. "Religion Among the Millennials: Less Religiously Active Than Older Americans, But Fairly Traditional In Other Ways," http://pewforum.org/Age/Religion-Among-the-Millennials.aspx, 17 February 2010.

13. Thomas Lord and Terri Orkwiszewski, "Moving from Didactic to Inquiry-Based Instruction in a Science Laboratory," *American Journal of Primatolgy*, 68 (October 2006).

14. Jonathan W. Jantz, C.D. Blosser, and L.A. Fruechting, "A Motor Milestone Change Noted with a Change in Sleep Position," *Archives of Pediatric Adolescent Medicine* 151 (June 1997):565.

15. "Next Generation of Americans," *CBS News Poll*, www.pollingreport.com, 17–22 December 2009.

16. Kim Marie McGoldrick, Gail Hoyt, and David Colander, "The Professional Development of Graduate Students for Teaching Activities: The Students' Perspective," *Journal of Economic Education*, 41(2): 194–201, 2010.

17. "Food Safety," *CBS News Poll*, www.pollingreport.com, 17–22 December 2009.

18. Sarah Janssen, Ed., *The World Almanac and Book of Facts 2011* (New York, NY: Infobase Publishing), 2011.

19. Doreen Matsui, R. Lim, T. Tschen, and M.J. Rieder, "Assessment of the Palatability of b-Lactamase-Resistant Antibiotics in Children," *Archives of Pediatric Adolescent Medicine* 151 (June 1997):599.

20. Andrew S. Levy, M.J. Wetzler, M. Lewars, and W. Laughlin, "Knee Injuries in Women Collegiate Rugby Players," *The American Journal of Sports Medicine* 25, no. 3 (1997):360.

21. Adapted from David L. Wheeler, "More Social Roles Means Fewer Colds," *Chronicle of Higher Education* XLIII, no. 44 (July 11, 1997):A13.

22. Mya Frazier, "The Reality of the Working Woman," Ad Age Insights: White Paper, http://adage.com/images/bin/pdf/aa_working_women_whitepaper_web.pdf, 7 June 2010.

Chapter 15

1. T.M. Casey, M.L. May, and K.R. Morgan, "Flight Energetics of Euglossine Bees in Relation to Morphology and Wing Stroke Frequency," *Journal of Experimental Biology* 116 (1985).

2. "Alzheimer's Test Set for New Memory Drug," *The Press-Enterprise* (Riverside, CA), 18 November 1997, p. A-4.

3. "Aaron Rodgers #12 QB," http://sports.espn.go.com/nfl/players/gamelog?playerId=8439 and "Drew Brees #9 QB," http://sports.espn.go.com/nfl/players/gamelog?playerId=2580, 18 March 2011.

4. *Science News* 136 (August 1989):126.

5. D. Matsui et al., "Assessment of the Palatability of b-Lactamase-Resistant Antibiotics in Children," *Archives of Pediatric Adolescent Medicine* 151 (1997):559–601.

6. Scott K. Powers and M.B. Walker, "Physiological and Anatomical Characteristics of Outstanding Female Junior Tennis Players," *Research Quarterly for Exercise and Sport* 53, no. 2 (1983).

7. *Science News,* 1989, p. 116.

8. G. Merrington, L. Winder, and I. Green, "The Uptake of Cadmium and Zinc by the Birdcherry Oat Aphid *Rhopalosiphum Padi* (*Homoptera:Aphididae*) Feeding on Wheat Grown on Sewage Sludge Amended Agricultural Soil," *Environmental Pollution* 96, no. 1 (1997):111–114.

9. Ron Marks, "Store brands save up to 52%," http://www.consumerreports.org/cro/magazine-archive/2010/october/shopping/store-brands-vs-name-brands/store-brands-saved/index.htm, October 2010.

10. Karola Sakekel, "Egg Substitutes Range in Quality," *San Francisco Chronicle,* 10 February 1993, p. 8.

Answers to Selected Exercises

Chapter 1

1.1 a. the student **b.** the exam **c.** the patient
d. the plant **e.** the car

1.3 a. discrete **b.** continuous **c.** continuous
d. discrete

1.5 a. vehicle **b.** type (qualitative); make
(qualitative); carpool (qualitative); distance
(quantitative continuous); age (quantitative
continuous) **c.** multivariate

1.7 The population is the set of voter preferences for
all voters in the state. Voter preferences may
change over time.

1.9 a. score on the reading test; quantitative
b. the student **c.** the set of scores for all
deaf students who hypothetically might take
the test

1.11 a. a pair of jeans **b.** the state in which the
jeans are produced; qualitative **e.** 8/25
f. California **g.** The three states produce
roughly the same numbers of jeans.

1.13 a. no; add a category called "Other"

1.15 a. no **b.** not quite **c.** the bar chart

1.17 a. eight to ten class intervals
c. 43/50 **d.** 33/50 **e.** yes

1.19 b. .30 **c.** .70 **d.** .30
e. relatively symmetric; no

1.23 a. roughly mound-shaped **b.** .20

1.25 a–b. skewed left **c.** 8 and 11

1.27 a. bar chart

1.29 c. the Pareto chart

1.31 a. skewed right; several outliers

1.33 b.
```
Stem-and-leaf of Ages N = 38
Leaf Unit = 1.0
   2   4   69
   3   5   3
   7   5   6678
  13   6   003344
  19   6   567778
  19   7   011234
  13   7   7889
   9   8   013
   6   8   58
   4   9   0033
```
relatively mound-shaped, with a slight peak
in the right tail **c.** Kennedy, Garfield, and
Lincoln were assassinated.

1.35 b. 0.05

1.37 a. number of hazardous waste sites
(discrete) **b.** skewed right; MI, NY, CA,
PA, NJ **c.** size of the state; amount of
industrial activity

1.39 a. skewed **b.** symmetric **c.** symmetric
d. symmetric **e.** skewed **f.** skewed

1.41 a. continuous **b.** continuous **c.** discrete
d. discrete **e.** discrete

1.43
```
 7 | 8  9
 8 | 0  1  7
 9 | 0  1  2  4  4  5  6  6  6  8  8
10 | 1  7  9
11 | 2
```

1.45 b. skewed right

1.49 a. no **b.** roughly mound-shaped

1.51 a. skewed right **c.** yes; large states

1.53 a. Popular vote is skewed right; percent vote is
relatively mound-shaped. **b.** yes
c. Once the size of the state is removed, each
state will be measured on an equal basis.

1.55 d. pie chart or Pareto chart.

1.57 a. skewed left; three stores within one mile of UCR   **b.** as the distance from UCR increases, each successive area becomes larger.

1.59 b. bimodal distribution, outliers; different kiln sites   **c.** yes

1.63 a.
```
Stem-and-leaf of Tax N = 51
   Leaf Unit = 1.0
    1    2   6
    3    3   22
   16    3   5557778888999
  (15)   4   000011111223333
   20    4   566689
   14    5   00111234
    6    5   58
    4    6   133
    1    6   7
```
b. roughly mound-shaped   **c.** no

1.65 Use a pie chart or a bar chart.

1.67 a. approximately mound-shaped
b. bar centered at 100.8
c. slightly above the center

Chapter 2

2.1 b. $\bar{x} = 2$; $m = 1$; mode $= 1$ **c.** skewed

2.3 a. 5.8   **b.** 5.5   **c.** 5 and 6

2.5 a. slightly skewed right   **c.** $\bar{x} = 1.08$; $m = 1$; mode $= 1$

2.7 2.5 is an average number calculated (or estimated) for all families in a particular category.

2.9 The median, because the distribution is highly skewed to the right.

2.11 a. $\bar{x} = 5.476$; $m = 4$; 2 modes (1 and 2)
b. skewed right   **c.** yes

2.13 a. 2.4   **b.** 2.8   **c.** 1.673

2.15 a. 3   **b.** 2.125   **c.** $s^2 = 1.2679$; $s = 1.126$

2.17 a. 1.11   **b.** $s^2 = .19007$, $s = .436$
c. $R \approx 2.5s$

2.19 a. $s \approx 1.67$ **b.** $s = 1.75$ **c.** no
d. yes   **e.** no

2.21 a. approximately .68   **b.** approximately .95
c. approximately .815   **d.** approximately .16

2.23 a. $s \approx .20$ **b.** $\bar{x} = .76$; $s = .165$

2.25 a. approximately .68   **b.** approximately .95
c. approximately .003

2.27 a. relatively flat; $\bar{x} \approx 4.5$ **b.** ≈ 2.25
c. $\bar{x} = 4.586$; $s = 2.892$

2.29 a. skewed right   **b.** 0 to 104 days

2.31 b. $\bar{x} = 7.729$ **c.** $s = 1.985$

k	$\bar{x} \pm ks$	Actual	Tchebysheff	Empirical Rule
1	(5.744, 9.714)	.71	At least 0	Approx. .68
2	(3.759, 11.699)	.96	At least 3/4	Approx. .95
3	(1.774, 13.684)	1.00	At least 8/9	Approx. .997

2.33 a. 42   **b.** $s \approx 10.5$ **c.** $s = 13.10$
d. 1.00; 1.00; yes

2.35 a. $s \approx .444$ **b.** $s = .436$

2.37 a–b. $\bar{x} = 1.4$; $s^2 = 1.4$

2.39 a. $\bar{x} = 2.04$; $s = 2.806$
b–c.

k	$\bar{x} \pm ks$	Actual	Tchebysheff	Empirical Rule
1	$(-.766,\ 4.846)$	.84	At least 0	Approx. .68
2	$(-3.572,\ 7.652)$	.92	At least 3/4	Approx. .95
3	$(-6.378, 10.458)$	1.00	At least 8/9	Approx. .997

2.41 min $= 0$, $Q_1 = 6$, $m = 10$, $Q_3 = 14$, max $= 19$; IQR $= 8$

2.43 a. $Q_1 = .3125$; $Q_3 = .7150$ **b.** .4025
c. lower and upper fences: $-.29125$ and 1.31875; no

2.45 lower and upper fences: -2.25 and 15.75; $x = 22$ is an outlier

2.47 a. min $= 1.70$, $Q_1 = 130.5$, $m = 246.5$, $Q_3 = 317.5$, max $= 485$
b. lower and upper fences: -150 and 598
c–d. No, but there are four extremely small observations, not identified by the box plot as outliers.

2.49 a.
```
Variable  Minimum   Q1   Median    Q3   Maximum
Rodgers     7.00  18.00  21.00  26.00   34.00
Brees      21.00  24.00  27.50  32.25   37.00
```
b. *Rodgers:* lower and upper fences: 6 and 38; no outliers; relatively symmetric. *Brees:* upper and lower fences: 11.625 and 44.625; no outliers; relatively symmetric.

2.51 a. skewed left   **b.** $\bar{x} = 108.15$; $m = 123.5$; mean < median implies skewed left
c. lower and upper fences: -43.125 and 259.875; skewed left, no outliers.

2.53 Female temperatures have a higher center (median) and are more variable; three outliers in the female group.

2.55 a. *Generic:* $m = 26$, $Q_1 = 25$, $Q_3 = 27.25$, IQR $= 2.25$; *Sunmaid:* $m = 26$, $Q_1 = 24$, $Q_3 = 28$, IQR $= 4$ **b.** *Generic:* lower and upper fences: 21.625 and 30.625; *Sunmaid:* lower and upper fences: 18 and 34   **c.** yes **d.** The average size is nearly the same; individual raisin sizes are more variable for Sunmaid raisins.

2.57 a. $R = 32.1$ **b.** $s \approx 8.025$ **c.** $s = 7.671$

2.59 $m = 6.35$, $Q_1 = 2.325$, $Q_3 = 12.825$; lower and upper fences: -13.425 and 28.575; one outlier ($x = 32.3$).

2.61 a–b.

k	$\bar{x} \pm ks$	Tchebysheff	Empirical Rule
1	(.16, .18)	At least 0	Approx. .68
2	(.15, .19)	At least 3/4	Approx. .95
3	(.14, .20)	At least 8/9	Approx. .997

c. No, distribution of $n = 4$ measurements cannot be mound-shaped.

2.63 68%; 95%

2.65 a. 27; 20.2; 6.8   **b.** slightly skewed left **c.** 23.96; 1.641   **d.** largest $x = 27$, z-score $= 1.85$; smallest $x = 20.2$, z-score $= -2.29$; no **e.** 24.3   **f.** 22.95 and 24.85

2.67 a. $s \approx 7.75$ **b.** $\bar{x} = 59.2$; $s = 10.369$ **c.** $m = 60$, $Q_1 = 51.25$, $Q_3 = 69.75$; lower and upper fences: 23.5 and 97.5; no outliers.

2.69 $\sigma \approx 100$

2.71 a. 16%   **b.** 81.5%

2.73 a. .9735   **b.** .16

2.75 a. .025   **b.** .84

2.77 a. At least 3/4 have between 145 and 205 teachers.   **b.** .16

2.81 a. 8.36   **b.** 4   **c.** skewed right   **d.** lower and upper fences: -24.375 and 42.625; no; yes

2.83 b. yes   **c.** more than 2 or 3 standard deviations from the mean

2.85 a. 2.5, 3.75, 4.2, 4.75, 5.7   **b.** lower and upper fences: 2.25 and 6.25   **c.** no **d.** mound-shaped; yes

Chapter 3

3.3 a. comparative pie charts; side-by-side or stacked bar charts   **c.** Proportions spent in all four categories are substantially different for men and women.

3.5 a. *Population:* responses to free time question for all parents and children in the United States. *Sample:* responses for the 398 people in the survey.   **b.** bivariate data, measuring relationship (qualitative) and response (qualitative)   **c.** the number of people who fall into that relationship-opinion category **e.** stacked or side-by-side bar charts

3.9 a. .5   **b.** increases   **c.** $y = 2.0$; y-intercept **d.** 3.25; 4.0

3.11 b. As x increases, y increases.   **c.** .903 **d.** $y = 3.58 + .815x$; yes

3.13 b. As x increases, y decreases.   **c.** $-.987$

3.15 a. $y = 56.11 + 23.83x$ **c.** $199.06 (using full accuracy); no

3.17 b. slight positive trend   **c.** $r = .760$

3.19 a. price = dependent variable; size = independent variable   **b.** no

3.21 b. The professor's productivity appears to increase, with less time required to write later books; no.

3.23 a. number of users (quantitative), year (qualitative, since used as a category), education level (qualitative), city (qualitative)   **b.** the population of responses for all Facebook users in 2009 and 2010; population at a fixed point in time   **c.** side-by-side bar charts; stacked bar charts or comparative pie charts

3.27 a. .944 (.941 using printout)   **b.** $x =$ week, $y =$ total gross   **c.** .777; .436

3.29 a. no   **b.** $r = -.028$; yes   **c.** Large cluster in lower left corner shows no apparent relationship; 7 to 10 states form a cluster with a negative linear trend   **d.** local environmental regulations; population per square mile; geographic region

3.31 a. aluminum oxide (quantitative), site (qualitative)   **b.** higher levels of aluminum oxide at Ashley Rails and Island Thorns

3.33 b. relatively strong-linear or possibly curvilinear **c.** hippo, African elephant, Asian elephant **d.** no, scatterplot appears more curvilinear.

3.35 a. strong positive linear relationship **b.** .946   **c.** $b \approx 1$ **d.** $y = 12.221 + .815x$

3.37 a. .635; relatively strong   **b.** $y = 8.730 + .849x$ **c.** 80.895

3.39 a. $r = .971$ **b.** yes

3.41 a. relatively strong positive linear   **b.** .668 **c.** $y = 67.955 + .028x$

Chapter 4

4.1 a. {1, 2, 3, 4, 5, 6}   **c.** 1/6
d. $P(A) = 1/6$; $P(B) = 1/2$; $P(C) = 2/3$;
$P(D) = 1/6$; $P(E) = 1/2$; $P(F) = 0$

4.3 $P(E_1) = .45$; $P(E_2) = .15$; $P(E_i) = .05$ for
$i = 3, 4, \ldots, 10$

4.5 a. {NDQ, NDH, NQH, DQH}   **b.** 3/4
c. 3/4

4.9 a. .58   **b.** .14   **c.** .46

4.11 a. randomly selecting three people and
recording their gender   **b.** {FFF, FMM,
MFM, MMF, MFF, FMF, FFM, MMM}
c. 1/8   **d.** 3/8   **e.** 1/8

4.13 a. rank A, B, C
b. {ABC, ACB, BAC, BCA, CAB, CBA}
d. 1/3, 1/3

4.15 a. .467   **b.** .513   **c.** .533

4.17 80

4.19 a. 60   **b.** 3,628,800
c. 720   **d.** 20

4.21 6720

4.23 216

4.25 120

4.27 720

4.29 a. 140,608   **b.** 132,600   **c.** .00037
d. .943

4.31 a. 2,598,960   **b.** 4   **c.** .000001539

4.33 $5.720645 \times (10^{12})$

4.35 a. 36   **b.** 1/36   **c.** 5/6

4.37 1/56

4.39 $\dfrac{4!(3!)^4}{12!}$

4.41 a. 3/5   **b.** 4/5

4.43 a. 1   **b.** 1/5   **c.** 1/5

4.45 a. .05   **b.** yes   **c.** no   **d.** no

4.47 a. no; no   **b.** no; yes

4.49 a. .08   **b.** .52

4.51 a. .3   **b.** no   **c.** yes

4.53 a. no, since $P(A \cap B) \neq 0$
b. no, since $P(A) \neq P(A|B)$

4.55 a. .14   **b.** .56   **c.** .30

4.59 a. $P(A) = .9918$; $P(B) = .0082$
b. $P(A) = .9836$; $P(B) = .0164$

4.61 .05

4.63 a. .99   **b.** .01

4.65 a. 154/256   **b.** 155/256   **c.** 88/256
d. 88/154   **e.** 44/67   **f.** 23/35
g. 12/101   **h.** 189/256

4.67 a. .7225   **b.** .4712   **c.** .1043

4.69 a. .23   **b.** .6087; .3913

4.71 .38

4.73 .012

4.75 a. .6585   **b.** .3415   **c.** left

4.77 .3130

4.79 a. $P(D) = .10$; $P(D^C) = .90$; $P(N | D^C) = .94$;
$P(N | D) = .20$ **b.** .023   **c.** .023
d. .056   **e.** .20   **f.** false negative

4.81 a. continuous   **b.** continuous
c. discrete   **d.** discrete   **e.** continuous

4.83 a. .2   **c.** $\mu = 1.9$; $\sigma^2 = 1.29$; $\sigma = 1.136$
d. .3   **e.** .9

4.85 1.5

4.87 a. $p(x) = C_x^3 (.47)^x (.53)^{3-x}$
c. .396   **d.** $\mu = 1.41$; $\sigma = .864$

4.89 a. $p(0) = 3/10$; $p(1) = 6/10$; $p(2) = 1/10$

4.91 a. .1; .09; .081   **b.** $p(x) = (.9)^{x-1}(.1)$

4.93 a. 4.0656   **b.** 4.125   **c.** 3.3186
d. $E(x)$ decreases as $P(A)$ increases

4.95 $1500

4.97 a. .28   **b.** .18   **c.** $\mu = 1.32$; $\sigma = 1.199$
d. .94

4.99 $20,500

4.101 .0713

4.103 $P(A) = 1/2$; $P(B) = 2/3$; $P(A \cap B) = 1/3$;
$P(A \cup B) = 5/6$; $P(C) = 1/6$; $P(A \cap C) = 0$;
$P(A \cup C) = 2/3$

4.105 2/7

4.107 $p(0) = .0256$; $p(1) = .1536$; $p(2) = .3456$;
$p(3) = .3456$; $p(4) = .1296$; .4752

4.109 −$0.26

4.111 3/10; 6/10

4.113 a. .73   **b.** .27

4.115 .999999

4.117 8

4.119 a. .3582   **b.** .4883   **c.** .4467

4.121 a. 1/8   **b.** 1/64   **c.** Not necessarily;
they could have studied together, and so on.

4.123 a. 5/6 **b.** 25/36 **c.** 11/36

4.125 a. .8 **b.** .64 **c.** .36

4.127 .0256; .1296

4.129 .2; .1

4.131 a. .5182 **b.** .1136 **c.** .7091
d. .3906

4.133 a. .0625 **b.** .25

4.135 a.

x	0	1	2
$p(x)$	6/15	8/15	1/15

b. 1/15 **c.** $\mu = 2/3$; $\sigma^2 = 16/45$

4.137 a. .48 **b.** .10 **c.** .262

Chapter 5

5.1 a. .058 **b.** .989 **c.** .011 **d.** .047
e. .437

5.3 a. .2965 **b.** .8145 **c.** .1172
d. .3670

5.5 a. .097 **b.** .329 **c.** .671 **d.** 2.1
e. 1.212

5.7 $p(0) = .000$; $p(1) = .002$; $p(2) = .015$;
$p(3) = .082$; $p(4) = .246$; $p(5) = .393$;
$p(6) = .262$

5.9 a. .251 **b.** .618 **c.** .367 **d.** .633
e. 4 **f.** 1.549

5.11 a. .901 **b.** .015 **c.** .002 **d.** .998

5.13 a. .748 **b.** .610 **c.** .367 **d.** .966
e. .656

5.15 a. 1; .99 **b.** 90; 3 **c.** 30; 4.58
d. 70; 4.58 **e.** 50; 5

5.17 a. .9568 **b.** .957 **c.** .9569
d. $\mu = 2$; $\sigma = 1.342$ **e.** .7455; .9569; .9977
f. yes; yes

5.19 binomial; $n = 2$; $p = .6$

5.21 no; the variable is not the number of successes in
n trials. Instead, the number of trials n is variable.

5.23 a. 1.000 **b.** .997 **c.** .086

5.25 a. .098 **b.** .991 **c.** .098 **d.** .138
e. .430 **f.** .902

5.27 a. .0081 **b.** .4116 **c.** .2401

5.29 a. $\mu = 10$ **b.** 4 to 16 **c.** If this unlikely
value were actually observed, it might be possible
that the trials (fields) are not independent.

5.31 a. .016796 **c.** .98320

5.33 a. .107 **b.** .762

5.35 a. .082085 **b.** .205212 **c.** .256516
d. .543813

5.37 a. .647 **b.** .353 **c.** .224 **d.** .493

5.39 a. .135335 **b.** .27067 **c.** .593994
d. .036089

5.41 a. .677 **b.** .6767 **c.** yes

5.43 a. .0067 **b.** .1755 **c.** .560

5.45 a. .271 **b.** .594 **c.** .406

5.47 $P(x > 5) = .017$; unlikely.

5.49 a. 2/3 **b.** 1/15 **c.** 1/2

5.51 a. .6 **b.** .5143 **c.** .0714

5.53 a. $p(0) = .36$; $p(1) = .48$; $p(2) = .15$; $p(3) = .01$
c. $\mu = .8$, $\sigma^2 = .50286$ **d.** .99; .99; yes

5.55 $p(0) = .2$; $p(1) = .6$; $p(2) = .2$

5.57 a. hypergeometric **b.** .1786 **c.** .01786
d. .2857

5.63 a. $p(0) = .729$; $p(1) = .243$; $p(2) = .027$;
$p(3) = .001$ **c.** .3; .520 **d.** .729; .972

5.65 a. .234 **b.** .136 **c.** Claim is not unlikely.

5.67 a. .228 **b.** no indication that people are
more likely to choose middle numbers

5.69 a. 20 **b.** 4 **c.** .006 **d.** Psychiatrist
is incorrect.

5.71 a. $\mu = 50$; $\sigma = 6.124$
b. The value $x = 35$ lies 2.45 standard deviations
below the mean. It is somewhat unlikely that the
25% figure is representative of this campus.

5.73 a. .5 **b.** $\mu = 12.5$; $\sigma = 2.5$
c. There is a preference for the second design.

5.75 a. yes; $n = 10$; $p = .75$ **b.** .2440
c. .0000296 **d.** Yes; genetic model is not
behaving as expected.

5.77 a. yes **b.** $1/8192 = .00012$

5.79 a. $p(x) = \dfrac{C_x^M C_{10-x}^{50-M}}{C_{10}^{50}}$ where M = # of defectives
in the carton and x = # of defectives in the sample.
b. .6367 **c.** .3968; .2415

5.81 a. .015625 **b.** .421875 **c.** .25

5.83 a. $p = 1/3$ **b.** .3292 **c.** .8683

5.85 a. 14 **b.** 2.049 **c.** no; $x = 10$ is only
1.95 standard deviations below the mean

5.87 a. .135335 **b.** .676676

5.89 .655

5.91 a. .794　　**b.** .056　　**c.** −0.82 to 3.82 or 0 to 3

5.93 a. 36　　**b.** 4.8
c. Yes, since $x = 49$ lies 2.71 standard deviations above the mean.

5.95 a. 240　　**b.** 9.798　　**c.** 221 to 259
d. $x = 200$ lies more than 4 standard deviations below the mean; perhaps the 60% figure is too high.

5.99 a. .172　　**b.** .656　　**c.** .656

Chapter 6

6.1 a. .9772　　**b.** .1230　　**c.** .9802　　**d.** .9699
6.3 a. .9452　　**b.** .9664　　**c.** .8159
d. ≈ 1.0000

6.5 a. .6753　　**b.** .2401　　**c.** .2694
d. .0901　　**e.** ≈ 0

6.7 a. 1.96　　**b.** 1.44

6.9 a. 1.65　　**b.** −1.645

6.11 a. 1.28　　**b.** 1.645　　**c.** 2.05　　**d.** 2.33

6.13 a. .1596　　**b.** .1151　　**c.** .1359

6.15 58.3

6.17 $\mu = 8$; $\sigma = 2$

6.19 a. .2389　　**b.** .5077　　**c.** no
d. somewhat; $y = 18$ lies 2.76 standard deviations above the mean.

6.21 a. .4586　　**b.** .0526　　**c.** .0170

6.23 .1562; .0012

6.25 a. .0475　　**b.** .00226　　**c.** 29.12 to 40.88
d. 38.84

6.27 a. .9938　　**b.** .0301

6.29 .0475

6.31 63,550

6.33 a. .3085　　**b.** .2417　　**c.** .0045

6.35 a. yes　　**b.** 15; 2.449　　**c.** .9878

6.37 a. yes　　**b.** $\mu = 7.5$; $\sigma = 2.291$　　**c.** .6156
d. .618

6.39 a. .2676　　**b.** .3520　　**c.** .3208 (use even-odd rule for rounding)　　**d.** .9162

6.41 a. .178　　**b.** .392

6.43 a. .245　　**b.** .2483

6.45 a. .0006　　**b.** .3050　　**c.** .5675

6.47 .9441

6.49 a. .3859　　**b.** They do not consider height when casting their ballot.

6.51 a. .0014　　**b.** .7114　　**c.** .9943
d. yes; Pepsi's market share is higher than claimed.

6.53 a. 31　　**b.** 3.432　　**c.** no; $x = 25$ is only 1.75 standard deviations below the mean.

6.55 a. .9544　　**b.** .0561

6.57 a. $z_0 = -1.96$　　**b.** $z_0 = .36$

6.59 a. .9651　　**b.** .1056　　**c.** .0062

6.61 a. .8849　　**b.** .1841　　**c.** .9279
d. .3372

6.63 a. .7734　　**b.** .9115

6.65 .8612

6.67 a. .1056　　**b.** .8944　　**c.** .1056

6.69 .16

6.71 a. .0778　　**b.** .0274

6.73 .0344

6.75 .3859

6.77 a. no　　**b.** .0179

6.79 383.5 hours

6.81 .8980

6.83 a. binomial, $n = 100$, $p = .75$　　**b.** yes
c. .7960　　**d.** ≈ 0

6.85 no; for $x = 19$, $z = 1.461$

6.87 a. .1587　　**b.** 7.935　　**c.** no; $z = 2.73$

6.89 87.48

6.91 a. .9107　　**b.** 35.812 and 43.848
c. .921; .953

6.93 .0446

6.95 a. 1.27　　**b.** .1020

6.97 .1251

Chapter 7

7.1 1/500

7.11 a. convenience sample　　**c.** Yes, but only if the students behave like a random sample from the general population of Native American youth.

7.13 a. first question　　**b.** Percent favoring the program decreased, perhaps due to the "spending billions of dollars" wording in the question.

7.15 a. $\mu = 10$; $\sigma/\sqrt{n} = .5$
b. $\mu = 5$; $\sigma/\sqrt{n} = .2$
c. $\mu = 120$; $\sigma/\sqrt{n} = .3536$

7.17 c. roughly mound-shaped

7.19 a. 1   **b.** .707   **c.** .500   **d.** .333
     e. .250   **f.** .200   **g.** .100

7.21 a. approximately normal   **b.** 53; 3

7.23 a. approximately normal   **b.** 100; 3.16

7.25 a. 106; 2.4   **b.** .0475   **c.** .9050

7.27 b. a large number of replications

7.31 a. 1266; 22.517   **b.** .0655

7.33 a. ≈ 0 **b.** yes; the value $\bar{x} = 98.25$ is almost 5 standard deviations below the assumed mean, $\mu = 98.6$.

7.35 a. $p = .3$; SE = .0458   **b.** $p = .1$; SE = .015   **c.** $p = .6$; SE = .0310

7.37 a. .7019   **b.** .5125

7.39 a. .0099   **b.** .03   **c.** .0458   **d.** .05
     e. .0458   **f.** .03   **g.** .0099

7.41 a. approximately normal   **b.** .25; .0484
     c. .9265

7.43 a. yes; $\mu = .78$ and $\sigma = .0414$ **b.** .2358
     c. .2090   **d.** $z = -3.14$; perhaps p is less than .78.

7.45 a. approximately normal with mean .13 and standard deviation .0453   **b.** .9382
     c. ≈ 0 **d.** .04 to .22

7.47 a. approximately normal with mean .75 and standard deviation .0306   **b.** .0516
     c. .69 to .81

7.49 a. LCL = 150.13; UCL = 161.67

7.51 a. LCL = 0; UCL = .090

7.53 a. LCL = 8598.7; UCL = 12,905.3

7.55 LCL = .078; UCL = .316

7.57 LCL = .0155; UCL = .0357

7.59 mean too large at hours 2, 3, and 4

7.63 a. .4938   **b.** .0062   **c.** .0000

7.65 a. ≈ 12.5 **b.** .9986   **c.** They are probably correct.

7.67 c. no

7.73 a. cluster sample   **b.** 1-in-10 systematic sample   **c.** stratified sample
     d. 1-in-10 systematic sample   **e.** simple random sample

7.75 a. 131.2; 3.677   **b.** yes   **c.** .1515

7.77 a. LCL = 0; UCL = .0848   **b.** $\hat{p} > .0848$

7.79 yes

7.83 a. approximately normal with mean 288 and standard deviation .9798   **b.** .0207   **c.** .0071

7.85 UCL = .2273; LCL = 0

Chapter 8

8.3 a. .160   **b.** .339   **c.** .438

8.5 a. .554   **b.** .175   **c.** .055

8.7 a. .179   **b.** .098   **c.** .049   **d.** .031

8.9 a. .0588   **b.** .0898   **c.** .098   **d.** .0898
     e. .0588   **f.** $p = .5$

8.11 $\hat{p} = .728$; margin of error (MOE) = .029

8.13 $\hat{p} = .90$; MOE = .0263

8.15 $\bar{x} = 39.8$; MOE = 4.768

8.17 $\bar{x} = 7.2\%$; MOE = .776

8.19 a. $\hat{p} = .75$; MOE = .0268

     b. $1.96 \sqrt{\dfrac{.5(.5)}{1004}} = .031$; no, unless the poll has

rounded up to the next half-percentage point.

8.21 a. no   **b.** nothing; no

8.23 Point estimate is $\bar{x} = 19.3$ with margin of error = 1.86.

8.25 a. (.797, .883)   **b.** (21.469, 22.331)
     c. Intervals constructed in this way enclose the true value of μ 90% of the time in repeated sampling.

8.27 (.846, .908)

8.29 a. 3.92   **b.** 2.772   **c.** 1.96

8.31 a. 3.29   **b.** 5.16   **c.** The width increases.

8.33 (3.496, 3.904); random sample

8.35 a. (.932, 1.088)   **c.** no; $\mu = 1$ is a possible value for the population mean

8.37 a. (.106, .166)   **b.** Increase the sample size and/or decrease the confidence level.

8.39 a. $98.085 < \mu < 98.415$ **b.** no; perhaps the value 98.6 is not the true average body temperature for healthy people.

8.41 a. (4.61, 5.99)   **b.** yes

8.43 a. $(-.77, 3.77)$ **b.** no

8.45 (15.463, 36.937)

8.47 a. (17.676, 19.324)   **b.** (15.710, 17.290)
     c. (.858, 3.142)   **d.** yes

8.49 a. $\bar{x}_1 - \bar{x}_2 = 5545$; MOE = 902.08   **b.** yes

8.51 a. $(-22.85, -7.15)$ **b.** (30.56, 49.44)
     c. no   **d.** yes; yes

8.53 a. $(-.528, -.032)$; yes, since $\mu_1 - \mu_2 = 0$ is not in the interval.

8.55 a. $(-.203, -.117)$ **b.** random and independent samples from binomial distributions

8.57 a. $(-.221, .149)$ **b.** no

8.59 a. $(-.118, -.002)$ **b.** Yes, since $p_1 - p_2 = 0$ is not in the interval.

8.61 a. $(.095, .445)$ **b.** yes

8.63 $(.061, .259)$

8.65 a. $(-.082, .022)$ **b.** No, since $p_1 - p_2 = 0$ is in the interval.

8.67 a. $\mu < 76.63$ **b.** $\mu < 1.89$

8.69 $\mu_1 - \mu_2 < 4$

8.71 505

8.73 $n_1 = n_2 = 1086$

8.75 b. 9604

8.77 $n_1 = n_2 = 360$

8.79 97

8.81 $n_1 = n_2 = 136$

8.83 $n_1 = n_2 = 98$

8.85 a. $\bar{x} = 29.1$; MOE $= .9555$ **b.** $(28.298, 29.902)$ **c.** $\mu > 28.48$ **d.** 234

8.87 $n_1 = n_2 = 224$

8.89 1083

8.91 $n_1 = n_2 = 925$

8.93 a. $\hat{p}_1 = .37$; $\hat{p}_2 = 13$ **b.** $(.097, .383)$ **c.** there is sufficient evidence to indicate a difference in the proportions for the two age groups.

8.95 $(8.087, 11.313)$

8.97 97

8.99 $(33.41, 34.59)$

8.101 b. MOE $= .021$ **c.** $n = 9604$

8.103 a. skewed right   **b.** Central Limit Theorem   **c.** $(61.64, 103.36)$; support the claim

8.105 at least 1825

8.107 .3874; .651

8.109 a. $(2.837, 3.087)$ **b.** 276

8.111 ($12.52, $14.38); no; average hourly wage is higher in Auburn, WA

8.113 $(2.694, 2.716)$

8.115 $(.161, .239)$

8.117 at least 97

Chapter 9

9.1 a. $z > 2.33$ **b.** $|z| > 1.96$

9.3 a. $z < -2.33$ **b.** $|z| > 2.58$ **c.** Reject H_0 at the 1% level; do not reject H_0.

9.5 a. Do not reject H_0; results are not statistically significant.   **b.** Reject H_0; results are highly significant.   **c.** Reject H_0; results are statistically significant.

9.7 a. .0207   **b.** Reject H_0; results are statistically significant.   **c.** yes

9.9 p-value $= .0644$; do not reject H_0; results are not statistically significant.

9.11 a. H_0: $\mu = 1$; H_a: $\mu \neq 1$ **b.** p-value $= .7414$; do not reject H_0 **c.** There is no evidence to indicate that the average weight is different from 1 pound.

9.13 a. H_0: $\mu = 80$ **b.** H_a: $\mu \neq 80$ **c.** $z = -3.75$; reject H_0

9.15 a. $z = 2.63$; p-value $= .0043$; reject H_0 at the 1% and 5% levels of significance

9.17 yes; $z = 10.94$

9.19 no; $z = -1.334$ with p-value $= .0918$; do not reject H_0

9.21 a. H_0: $\mu_1 - \mu_2 = 0$; H_a: $\mu_1 - \mu_2 > 0$; one-tailed   **b.** $z = 2.074$; reject H_0

9.23 a. $z = -2.26$; p-value $= .0238$; reject H_0 **b.** $(-3.55, -.25)$ **c.** no

9.25 a. H_0: $\mu_1 - \mu_2 = 0$; H_a: $\mu_1 - \mu_2 \neq 0$ **b.** yes; $z = -3.75$ **c.** p-value ≈ 0

9.27 a. yes; $z = 6.93$; p-value ≈ 0 **b.** $(44.47, 79.53)$; yes

9.29 a. $z = -2.22$ with p-value $= .0264$ **b.** significant at the 5% but not the 1% level.

9.31 H_0: $p = .4$; H_a: $p \neq .4$ **b.** p-value $= .093$; not statistically significant   **c.** no

9.33 a. H_0: $p = .15$; H_a: $p < .15$ **b.** Reject H_0; $z = -4.53$.   **c.** ≈ 0

9.35 a. H_a: $p > 2/3$ **b.** H_0: $p = 2/3$ **c.** yes; $z = 4.6$ **d.** p-value $< .0002$

9.37 a. H_0: $p = 5$; H_a: $p > .5$ **b.** Do not reject H_0; $z = -3.21$ (wrong tail)

9.39 no; $z = -1.06$

9.41 no; $z = -.71$

9.43 a. H_0: $p_1 - p_2 = 0$; H_a: $p_1 - p_2 < 0$ **b.** one-tailed   **c.** Do not reject H_0; $z = -.84$

9.45 a. yes; $z = -2.40$ **b.** $(-.43, -.05)$

9.47 Do not reject H_0; $z = -.39$; there is insufficient evidence to indicate a difference in the two population proportions.

9.49 $p_1 - p_2 > .001$(sample proportions calculated using three-decimal place accuracy); the risk is at least 1/1000 higher when taking *Prempro*.

9.51 Reject H_0; $z = 3.14$ with p-value $= .0008$; researcher's conclusions are confirmed.

9.55 The power increases.

9.57 a. p-value $< .0002$ (or p-value approximately 0)
b. Reject H_0; $z = 4.47$

9.59 a. H_0: $\mu = 7.5$; H_a: $\mu < 7.5$
b. one-tailed   **d.** $z = -5.477$; reject H_0

9.61 a. H_0: $\mu_1 - \mu_2 = 0$; H_a: $\mu_1 - \mu_2 \neq 0$
b. two-tailed   **c.** no; $z = -.954$

9.63 no; do not reject H_0; $z = 1.684$

9.65 a. no, $z = .16$ **b.** .4364
c. Do not reject H_0.

9.67 yes; $z = 4$; reject H_0

9.69 yes; $z = 4.00$

9.71 a. yes; $z = 4.33$ **b.** (7.12, 18.88)

9.73 a. no; $z = -.20$ **b.** no; $z = -.21$ **c.** no

9.75 no; $z = 2.19$

9.77 no; $z = .61$

9.79 yes; $z = 7.48$

9.81 a. (1447.49, 4880.51)   **b.** between 1500 and 5000 more meters per week; they only have one stroke to practice.

Chapter 10

10.1 a. 2.015   **b.** 2.306   **c.** 1.330
d. 1.96

10.3 a. $.02 < p$-value $< .05$
b. p-value $< .005$ **c.** p-value $> .20$
d. p-value $< .005$

10.5 a. $\bar{x} = 7.05$; $s = .4994$ **b.** 7.496
c. Reject H_0; $t = -2.849$ **d.** Yes

10.7 no; $t = -1.195$

10.9 a. yes; $t = -3.044$ **b.** 98.316; yes

10.11 (3.652, 3.912)

10.13 a. Reject H_0; $t = -4.31$.   **b.** (23.23, 29.97)
c. The pretreatment mean looks smaller than the other two means.

10.17 (233.98, 259.94)

10.19 a. 3.775   **b.** 21.2258

10.21 a. H_0: $\mu_1 - \mu_2 = 0$; H_a: $\mu_1 - \mu_2 \neq 0$
b. $|t| > 2.771$ **c.** $t = 2.795$
d. p-value $< .01$ **e.** Reject H_0

10.23 a. yes; larger s^2/smaller $s^2 = 1.36$
b. $t = .06$ with p-value $= .95$ **c.** 19.1844
d. Do not reject H_0.   **e.** $(-5.223, 5.503)$; yes

10.25 a. H_0: $\mu_1 - \mu_2 = 0$ versus H_a: $\mu_1 - \mu_2 > 0$
b. yes; $t = 2.806$ **c.** $.005 < p$-value $< .01$

10.27 a. Do not reject H_0; $t = 1.92$
b. $(-3.32, 4.72)$ **c.** unpooled

10.29 a. Do not reject H_0; $t = -1.68$
b. $(-.0029, .0003)$; yes

10.31 a. no; do not reject H_0; $t = 1.606$
b. $(-.061, .341)$

10.33 $\mu_1 - \mu_2 > -.118$; yes

10.35 no; $t = .79$

10.37 a. H_0: $\mu_1 - \mu_2 = 0$ versus H_a: $\mu_1 - \mu_2 > 0$
b. Do not reject H_0; $t = 1.511$

10.39 b. no; $t = .168$ **c.** p-value $> .20$
d. $(-818.84, 867.34)$ **e.** no

10.41 a. no; $t = -3.00$ **b.** $.05 < p$-value $< .10$

10.43 paired analysis

10.45 no; $t = 2.29$

10.47 yes; $t = 11.32$

10.49 no; $\chi^2 = 34.24$

10.51 a. $s^2 = .6990476$ **b.** (.291, 3.390)
c. do not reject H_0; $\chi^2 = 5.24$
d. p-value $> .20$

10.53 a. no; $t = -.232$ **b.** yes; $\chi^2 = 20.18$

10.55 a. no   **b.** yes; $z = 3.262$

10.57 no; $\chi^2 = 29.433$

10.59 (.667, 4.896)

10.61 $F = 1.22$ with p-value $> .20$; do not reject H_0: $\sigma_1^2 = \sigma_2^2$.

10.63 a. no; $F = 2.66$ **b.** yes

10.65 Rest: $F = 1.03$ with p-value $> .20$; 80% maximal O_2: $F = 2.01$ with p-value $> .20$; maximal O_2: $F = 14.29$ with p-value $< .01$; use the unpooled t-test for maximal O_2.

10.71 a. p-value $> .10$ **b.** $.05 < p$-value $< .10$
c. $.005 < p$-value $< .01$ **d.** p-value $> .10$

10.73 yes; $t = 2.108$; $.025 < p$-value $< .05$

10.75 (56.223, 99.303)

10.77 (3.545, 4.975)

10.79 a. (169.1, 199.9)   **b.** (69.43, 76.57)
c. (2.28, 2.80)   **d.** no

10.81 no; $t = -1.49$

10.83 a. yes; $t = -3.354$ **b.** p-value $< .01$
c. $(-10.246, -2.354)$; yes

10.85 a. Reject H_0: $\sigma_1^2 = \sigma_2^2$; $F = 3.88$
    b. Reject H_0: $\mu_1 - \mu_2 = 300$; $t = 2.13$; there is sufficient evidence to indicate that $(\mu_1 - \mu_2) > 300$.

10.87 a. (.02698, .02808)

10.89 a. no; $F = 2.21$ **b.** (.975, 5.03); yes

10.91 (35.845, 48.405)

10.93 a. normality assumption is valid
    b. (5.12, 5.67)

10.95 yes; $t = -2.39$

10.97 a. $t = 9.5641$ with p-value = .0000; there is sufficient evidence to indicate a difference in the average strengths.

10.101 yes; $F = 2.407$

10.103 no; $t = -1.8$

10.105 a. $(-11.414, -8.958)$ **b.** yes

10.107 a. no; $F = 1.922$ with p-value $>.20$
    b. (.452, 8.1685)

10.109 no; $t = 3.038$

10.111 (24.582, 73.243)

10.113 a. yes; $\chi^2 = 24.73$ **b.** (.0284, .1318)

10.115 (3.873, 4.519)

10.117 a. yes; $F = 1.21$ **b.** $t = 1.65$; there is insufficient evidence to indicate a difference in the two population means.

Chapter 11

11.1

Source	df
Treatments	5
Error	54
Total	59

11.3 a. (2.731, 3.409)   **b.** (.07, 1.03)

11.5 a.

Source	df	SS	MS	F
Treatments	3	339.8	113.267	16.98
Error	20	133.4	6.67	
Total	23			

    b. $df_1 = 3$ and $df_2 = 20$ **c.** $F > 3.10$
    d. yes, $F = 16.98$ **e.** p-value $< .005$; yes

11.7 a. CM = 103.142857; Total SS = 26.8571
    b. SST = 14.5071; MST = 7.2536
    c. SSE = 12.3500; MSE = 1.1227
    d. Analysis of Variance

Source	DF	SS	MS	F	P
Trts	2	14.51	7.25	6.46	0.014
Error	11	12.35	1.12		
Total	13	26.86			

    f. $F = 6.46$; reject H_0 with .01 $< p$-value $< .025$.

11.9 a. (1.95, 3.65)   **b.** (.27, 2.83)

11.11 a. (67.86, 84.14)   **b.** (55.82, 76.84)
    c. $(-3.629, 22.963)$ **d.** No, they are not independent.

11.13 a. Each observation is the mean length of 10 leaves.   **b.** yes, $F = 57.38$ with p-value = .000   **c.** Reject H_0; $t = 12.09$.
    d. (1.810, 2.924)

11.15

Analysis of Variance for Percent

Source	DF	SS	MS	F	P
Method	2	0.0000041	0.0000021	16.38	0.000
Error	12	0.0000015	0.0000001		
Total	14	0.0000056			

11.17 a. completely randomized design

    b.

Source	DF	SS	MS	F	P
State	3	3272.2	1090.73	26.44	0.000
Error	16	660.0	41.25		
Total	19	3932.2			

    c. $F = 26.44$; reject H_0 with p-value $< .005$.

11.19 Sample means must be independent; equal sample sizes.

11.21 a. $1.878s$ **b.** $2.1567s$

11.23 $\bar{x}_1$ $\bar{x}_2$ $\bar{x}_3$ $\bar{x}_4$

11.25 a. no; $F = .60$ with p-value = .562
    b. no differences

11.27 a. yes; $F = 4.47$, p-value $< .05$
    b. $(-128.946, 38.946)$ **c.** $\bar{x}_{SS}$ $\bar{x}_{LS}$ $\bar{x}_{PS}$

11.29

Source	df	SS	MS	F
Treatments	2	11.4	5.70	4.01
Blocks	5	17.1	3.42	2.41
Error	10	14.2	1.42	
Total	17	42.7		

11.31 $(-3.833, -.767)$

11.33 a. yes; $F = 19.19$ **b.** yes; $F = 135.75$
    c. $\bar{x}_1$ $\bar{x}_3$ $\bar{x}_4$ $\bar{x}_2$ **d.** $(-5.332, -2.668)$
    e. yes

11.35 a. 7   **b.** 7   **c.** 5   **e.** yes; $F = 9.68$
    f. yes; $F = 8.59$

11.37

Two-way ANOVA: y versus Blocks, Chemicals
Analysis of Variance for y

Source	DF	SS	MS	F	P
Blocks	2	7.1717	3.5858	40.21	0.000
Chemical	3	5.2000	1.7333	19.44	0.002
Error	6	0.5350	0.0892		
Total	11	12.9067			

11.39 a. yes; $F = 10.06$ **b.** yes; $F = 10.88$
    c. $\omega = 2.98$; preparations 1 and 3 are not significantly different.   **d.** (1.12, 5.88)

11.41

Two-way ANOVA: Cost versus Estimator, Job

Analysis of Variance for Cost

Source	DF	SS	MS	F	P
Estimator	2	10.862	5.431	7.20	0.025
Job	3	37.607	12.536	16.61	0.003
Error	6	4.528	0.755		
Total	11	52.997			

11.43 a. Blocks are items; treatments are stores.
b. yes, $F = 25.53$; p-value $= .000$
c. yes, $F = 29.99$; p-value $= .000$

11.45 a. 20 **b.** 60 **c.**

Source	df
A	3
B	4
AB	12
Error	40
Total	59

11.47 $(-1.11, 5.11)$

11.49 a. strong interaction present
b. $F = 37.85$ with p-value $= .000$; yes **d.** no

11.51 b. yes **c.** Since the interaction is significant, attention should be focused on means for the individual factor level combinations.
d. Training: $.05 < p$-value $< .10$; ability: p-value $< .005$; interaction: $.01 < p$-value $< .025$

11.53 a. 2×4 factorial; students; gender at two levels, schools at four levels **c.** no; $F = 1.19$ **e.** Main effect for schools is significant; $F = 27.75$; Tukey's $\omega = 82.63$.

11.55 a.

Source	DF	SS	MS	F	P
Training	1	4489.00	4489.00	117.49	0.000
Situation	1	132.25	132.25	3.46	0.087
Interaction	1	56.25	56.25	1.47	0.248
Error	12	458.50	38.21		
Total	15	5136.00			

b. No. $F = 1.47$; p-value $= .248$. **c.** no
$F = 3.46$; p-value $= .087$. **d.** yes
$F = 117.49$; p-value $= .000$.

11.57 significant differences between treatments A and C, B and C, C and E, and D and E

11.59 a. significant difference in treatment means; $F = 27.776$ **b.** Tukey's $\omega = .190$
c. yes; $F = 6.588$

11.61

One-way ANOVA: Sales versus Program

Analysis of Variance for Sales

Source	DF	SS	MS	F	P
Program	3	1385.8	461.9	9.84	0.000
Error	23	1079.4	46.9		
Total	26	2465.2			

11.63 a. no; $F = 1.40$ **b.** p-value $> .10$
c. yes; $F = 6.51$ **d.** yes; $F = 7.37$

11.65 a. 2×3 factorial experiment **b.** no;
$F = .452$ with p-value $= .642$
d. $(-22.56, -5.84)$

11.67 a. randomized block design (or paired difference)
b.

Two-way ANOVA: Total versus Week, Store

Source	DF	SS	MS	F	P
Week	3	139.708	46.569	0.74	0.593
Store	1	562.298	562.298	8.99	0.058
Error	3	187.654	62.551		
Total	7	889.660			

c. no, $F = 8.99$ with p-value $= .058$

11.69 a. factorial experiment **b.** yes; $F = 7.61$
c. $\omega = 2.67$

11.71 a. completely randomized design
b. Yes, there is a significant difference.
$F = 126.85$, p-value $= .000$

Source	DF	SS	MS	F	P
Site	2	132.277	66.139	126.85	0.000
Error	21	10.950	0.521		
Total	23	143.227			

11.73 There is no evidence of nonnormality. There appears to be a difference in the variability within some of the factor level combinations.

Chapter 12

12.1 y-intercept $= 1$, slope $= 2$

12.3 $y = 3 - x$

12.7 a. $\hat{y} = 6.00 - .557x$ **c.** 4.05
d. Analysis of Variance

Source	DF	SS	MS
Regression	1	5.4321	5.4321
Residual Error	4	0.1429	0.0357
Total	5	5.5750	

12.9 a. $\hat{y} = 195.90 + 67x$
c.

Source	df	SS	MS
Regression	1	43146.9296	43146.9296
Error	3	1860.2704	620.0901
Total	4	45007.2000	

12.11 a. 10 **b.** 9
c. **Regression Analysis: y versus x**
The regression equation is
y = 3.00 + 0.475 x

Predictor	Coef	SE Coef	T	P
Constant	3.000	2.127	1.41	0.196
x	0.4750	0.1253	3.79	0.005

S = 2.24165 R-Sq = 64.2% R-Sq(adj) = 59.8%

Analysis of Variance

Source	DF	SS	MS	F	P
Regression	1	72.200	72.200	14.37	0.005
Residual Error	8	40.200	5.025		
Total	9	112.400			

d. $\hat{y} = 3.00 + .475x$ **e.** 7.75

12.13 a. $y = $ API; $x = $ EL   **b.** yes
c. $\hat{y} = 847.468 - 1.7045x$ **d.** yes

12.15 a. yes   **b.** $\hat{y} = -11.665 + .755x$
c. $\hat{y} = 52.51$

12.17 a. strong positive linear relationship
b. approximately 1   **c.** $\hat{y} = 12.221 + .815x$
d. $\hat{y} = 62.75$

12.19 a. yes, $t = 5.20$ **b.** $F = 27.00$
c. $t_{.025} = 3.182$; $F_{.05} = 10.13$

12.21 a. yes, $F = 152.10$ with p-value $= .000$
b. $r^2 = .974$

12.23 a. $\hat{y} = -.257 + .215x$ **b.** yes; $t = 5.90$
c. $r^2 = .813$

12.25 a. No. Reject H_0; $t = 8.34$ with p-value $< .005$
b. $r^2 = .959$ **c.** pattern indicates that the
relationship may be curvilinear

12.27 a. $MSE = .08333$ **b.** yes; $t = -12.124$
c. $r^2 = .98$ **d.** the total variation has been
reduced by 98%

12.29 a. $r^2 = .757$ **b.** 75.7%

12.31 normal probability plot of residuals; points
should approximate a straight line sloping
upward

12.33 plot of residuals versus fits; random scatter of
points, free of patterns.

12.35 no extreme violations of regression
assumptions

12.37 b. $MSE = 58.1$ **c.** slight deviation from
normality; possibly one unusual observation,
but no extreme violations of the regression
assumptions

12.39 a. (3.259, 5.141)   **b.** (2.24, 6.16)

12.41 a. $\hat{y} = 4.3 + 1.5x$ **c.** $s^2 = 1.53$
d. yes, $t = 3.83$ **e.** p-value $< .01$
f. no violations of regression assumptions
g. (8.11, 9.49)

12.43 a. (21.613, 33.199)   **b.** (304,676, 307,360)
c. 167.739   **d.** (295,826, 304,151)

12.45 a. $\hat{y} = 156.13 + 4.844x$ **b.** $r^2 = .163$
c. no obvious violations

12.49 a. $r = 1$ **b.** $r = -1$

12.51 a. $r = -.982$ **c.** 96.47%

12.53 a. negatively correlated   **b.** $H_a: \rho < 0$
c. Reject H_0; $t = -1.872$

12.55 a. yes, $t = -3.260$ **b.** p-value $< .01$

12.57 no, $t = .92$ with p-value $> .10$

12.59 a. $r = .1741$ **b.** no, $t = .559$

12.61 a. $\hat{y} = 46 - .317x$
c.

Source	df	SS	MS
Regression	1	601.6667	601.6667
Error	10	190.3333	19.0333
Total	11	792.0000	

d. slight irregularities   **e.** $(-.442, -.192)$
f. (27.09, 33.24)   **g.** (19.97, 40.36)

12.63 Answers will vary.

12.65 a. all three are significant
b. radiographs, 3-D MRIs, standard MRIs
(using coefficient of determination)
c. consistent conclusions (cannot differentiate
between radiographs and 3-D MRIs)

12.67 Answers will vary.

12.71 Yes; $r = .562$ with p-value $= .036$

12.75 a. $\hat{y} = 7 + 15.4x$
b.

Source	df	SS	MS
Regression	1	2371.6	2371.6
Error	6	50.4	8.4
Total	7	2422.0	

c. yes; $t = 16.80$ **d.** (42.99, 48.01)
e. (13.16, 17.64)   **f.** $r^2 = .979$

12.77 a. curvilinear
b. $\hat{y} = 2,309,189.75 - 1140.595x$
c. no, $t = -.56$ **d.** wrong model has
been fit

Chapter 13

13.1 b. parallel lines

13.3 a. yes, $F = 57.44$ with p-value $< .005$
b. $R^2 = .94$

13.5 a. quadratic   **b.** $R^2 = .815$; relatively good fit
c. yes, $F = 37.37$ with p-value $= .000$

13.7 a. $b_0 = 10.5638$ **b.** yes, $t = 15.20$ with
p-value $= .000$

13.9 b. $t = -8.11$ with p-value $= .000$; reject H_0:
$\beta_2 = 0$ in favor of $H_a: \beta_2 < 0$.

13.11 a. $R^2 = .9955$ **b.** $R^2(\text{adj}) = 99.25\%$
c. The quadratic model fits slightly better.

13.13 a. Use variables x_1, x_3, and x_5.   **b.** no

13.15 a. $\hat{y} = -8.177 + .292x_1 + 4.434x_2$
b. Reject H_0, $F = 16.28$ with p-value $= .002$.
The model contributes significant information
for the prediction of y.   **c.** yes, $t = 5.54$
with p-value $= .001$ **d.** $R^2 = .823$; 82.3%

13.17 a. quantitative **b.** quantitative
c. qualitative; $x_1 = 1$ if plant B, 0 otherwise;
$x_2 = 1$ if plant C, 0 otherwise **d.** quantitative
e. qualitative; $x_1 = 1$ if day shift, 0 if night shift

13.19 a. x_2 **b.** $\hat{y} = 12.6 + 3.9x_2^2$ or
$\hat{y} = 13.14 - 1.2x_2 + 3.9x_2^2$

13.21 a. $y = \beta_0 + \beta_1 x_1 + \beta_2 x_2 + \beta_3 x_1 x_2 + \epsilon$ with
$x_2 = 1$ if cucumber, 0 if cotton **c.** No,
the test for interaction yields $t = .63$ with
p-value = .533. **d.** yes

13.23 $y = \beta_0 + \beta_1 x_1 + \beta_2 x_1^2 + \beta_3 x_2 + \beta_4 x_1 x_2$
$+ \beta_5 x_1^2 x_2 + \varepsilon$

13.25 a. $\hat{y} = 8.585 + 3.8208x - 0.21663x^2$
b. $R^2 = .944$ **c.** yes; $F = 33.44$
d. yes; $t = -4.93$ with p-value = .008 **e.** no

13.27 b. $\hat{y} = 4.10 + 1.04x_1 + 3.53x_2 + 4.76x_3 -$
$0.43x_1 x_2 - 0.08x_1 x_3$ **c.** yes; $t = -2.61$
with p-value = .028 **d.** no; $F = 3.86$;
consider eliminating the interaction terms.

13.29 a. $y = \beta_0 + \beta_1 x_1 + \beta_2 x_2 + \beta_3 x_1^2 + \beta_4 x_1 x_2$
$+ \beta_5 x_1^2 x_2 + \epsilon$
b. $F = 25.85$; $R^2 = .768$
c. $\hat{y} = 4.51 + 6.394x_1 + .1318x_1^2$
d. $\hat{y} = -46.34 + 23.458x_1 - .3707x_1^2$
e. no; $t = .78$ with p-value = .439

13.31 a. yes, price and overall score are correlated; y
is correlated with x_1 and x_3
b. $y = \beta_0 + \beta_1 x_1 + \beta_2 x_2 + \beta_3 x_3 + \beta_4 x_4 + \beta_5 x_5$
$+ \epsilon$ **c.** $R^2 = .471$ and R^2(adj) = .339; no
d. x_1 and x_3; $y = \beta_0 + \beta_1 x_1 + \beta_2 x_3 + \epsilon$; R^2
= .468 and R^2(adj) = .421; slightly better than
full model.

13.35 a. 99.85%
b. yes; $F = 1676.61$ with p-value = .000
c. yes; $t = -2.652$ with p-value = .045
d. yes; $t = 15.138$ with p-value = .000
e. Linear: R^2(adj) = 91.87%, quadratic:
R^2(adj) = 99.79%, quadratic term is significant.
f. Quadratic term is missing.

Chapter 14

14.3 a. $X^2 > 12.59$ **b.** $X^2 > 21.666$

14.5 a. $H_0: p_1 = p_2 = p_3 = p_4 = p_5 = 1/5$
b. 4 **c.** 9.4877 **d.** $X^2 = 8.00$
e. Do not reject H_0.

14.7 Yes, $X^2 = 24.48$; drivers tend to prefer the
inside lanes.

14.9 no, $X^2 = 3.63$

14.11 no, $X^2 = 13.58$

14.13 yes; do not reject H_0; $X^2 = 1.247$

14.15 yes; reject H_0; $X^2 = 28.386$

14.17 8

14.19 Reject H_0; $X^2 = 18.352$ with p-value = .000.

14.21 a. yes; $X^2 = 7.267$
b. $.025 < p$-value $< .05$

14.23 a. no, $X^2 = 10.207$ with p-value = .07
b. yes, $X^2 = 10.207$ with p-value = .037

14.25 a. no; do not reject H_0; $X^2 = 6.447$
b. p-value $> .10$; yes

14.27 a. $X^2 = 10.597$ **b.** $X^2 > 13.2767$
c. Do not reject H_0. **d.** $.025 < p$-value $< .05$

14.29 yes; $X^2 = 24.31$

14.31 a. Each care type represents a binomial
population in which we measure the presence
or absence of EMI services.
b. yes; $X^2 = 18.446$

14.33 Yes, reject H_0; $X^2 = 36.499$.

14.35 no, $X^2 = 4.4$ with p-value $> .10$

14.37 no, $X^2 = 1.89$ with p-value $> .10$

14.39 a. no, $X^2 = 1.815$ **b.** p-value $> .10$

14.43 no; do not reject H_0; $X^2 = 1.311$.

14.45 a. Reject H_0; $X^2 = 18.527$. **b.** Reject H_0;
$z = 4.304$; yes.

14.49 yes, $X^2 = 7.488$ with $.005 < p$-value $< .01$

14.51 yes, $X^2 = 6.190$ with $.025 < p$-value $< .05$;
(.347, .483)

14.53 a. yes, $X^2 = 33.017$ **b.** p-value $< .005$

14.55 yes, $X^2 = 17.395$ with p-value = .008

14.57 a. Do not reject H_0; $X^2 = 3.660$ with
p-value = .454; yes.

14.59 a. yes; $X^2 = 12.182$ with p-value = .002
b. The susceptibility to a cold seems to
decrease as the number of relationships
increases.

14.61 Yes, reject H_0; $X^2 = 16.535$.

Chapter 15

15.1 a. T_1^* **b.** $T \le 31$ **c.** $T \le 27$

15.3 a. H_0: population distributions are identical;
H_a: population 1 shifted to the left of
population 2. **b.** $T_1 = 16$; $T_1^* = 39$
c. $T \le 19$ **d.** yes; reject H_0

15.5 Do not reject H_0; $z = -1.59$.

15.7 Do not reject H_0; $T = 102$.

15.9 yes; reject H_0; $T = 45$

15.11 yes; reject H_0; $T = 44$

15.13 **b.** $\alpha = .002, .007, .022, .054, .115$

15.15 one-tailed: **n = 10:** $\alpha = .001, .011, .055$; **n = 15:** $\alpha = .004, .018, .059$; **n = 20:** $\alpha = .001, .006, .021, .058, .132$; two-tailed: **n = 10:** $\alpha = .002, .022, .110$; **n = 15:** $\alpha = .008, .036, .118$; **n = 20:** $\alpha = .002, .012, .042, .116$

15.17 **a.** H_0: $p = \frac{1}{2}$; H_a: $p \neq \frac{1}{2}$; rejection region: $\{0, 1, 7, 8\}$; $x = 6$; do not reject H_0 at $\alpha = .07$; p-value $= .290$.

15.19 $z = 3.15$; reject H_0.

15.21 **b.** $T = \min\{T^+, T^-\}$ **c.** $T \leq 137$ **d.** Do not reject H_0.

15.23 Do not reject H_0; $z = -.34$.

15.25 **a.** Reject H_0; $T = 1.5$ **b.** Results do not agree.

15.27 **a.** no; $T = 6.5$

15.29 **a.** Do not reject H_0; $x = 8$. **b.** Do not reject H_0; $T = 14.5$.

15.31 **a.** paired difference test, sign test, Wilcoxon signed-rank test **b.** Reject H_0 with both tests; $x = 0$ and $T = 0$.

15.33 yes, $H = 13.90$

15.35 **a.** no; $H = 2.63$ **b.** p-value $> .10$ **c.** p-value $> .10$

15.37 no; $H = 2.54$ with p-value $> .10$

15.39 **a.** Reject H_0; $F_r = 21.19$. **b.** p-value $< .005$ **d.** $F = 75.43$ **e.** p-value $< .005$ **f.** Results are identical.

15.41 **a.** Do not reject H_0; $F_r = 5.81$. **b.** $.05 < p$-value $< .10$

15.43 **a.** $r_s \geq .425$ **b.** $r_s \geq .601$

15.45 **a.** $|r_s| \geq .400$ **b.** $|r_s| \geq .526$

15.47 **a.** $-.593$ **b.** yes

15.49 **a.** $r_s = .811$ **b.** yes

15.51 yes

15.53 yes, $r_s = .9118$

15.55 **a.** Do not reject H_0; $x = 2$. **b.** Do not reject H_0; $t = -1.646$.

15.57 **a.** Do not reject H_0; $x = 7$. **b.** Do not reject H_0; $x = 7$.

15.59 Do not reject H_0 with the Wilcoxon rank sum test ($T = 77$) or the paired difference test ($t = .30$).

15.61 Do not reject H_0 using the sign test ($x = 2$); no.

15.63 yes; $r_s = -.845$

15.65 Reject H_0; $T = 14$.

15.67 **a.** Reject H_0; $F_r = 20.13$. **b.** The results are the same.

15.69 **a.** Reject H_0; $H = 9.08$. **b.** $.025 < p$-value $< .05$ **c.** The results are the same.

15.71 **a.** no **b.** significant differences among the responses to the three rates of application; $F_r = 10.33$ with p-value $= .006$

15.73 $T = 19$. $T_{.05} = 21$ ($T_{.01} = 18$.) Reject H_0.

15.75 $z = 1.18 < z_{.05} = 1.645$; lighting not effective

15.77 $H = 7.43$ $df = 3$; $.05 < p$-value $< .10$; no significant difference

15.79 **a.** $r_s = .738$. **b.** p-value $= .025 < .05$; yes, positive correlation

Index

TABLE 3 *(continued)*

z	.00	.01	.02	.03	.04	.05	.06	.07	.08	.09
0.0	.5000	.5040	.5080	.5120	.5160	.5199	.5239	.5279	.5319	.5359
0.1	.5398	.5438	.5478	.5517	.5557	.5596	.5636	.5675	.5714	.5753
0.2	.5793	.5832	.5871	.5910	.5948	.5987	.6026	.6064	.6103	.6141
0.3	.6179	.6217	.6255	.6293	.6331	.6368	.6406	.6443	.6480	.6517
0.4	.6554	.6591	.6628	.6664	.6700	.6736	.6772	.6808	.6844	.6879
0.5	.6915	.6950	.6985	.7019	.7054	.7088	.7123	.7157	.7190	.7224
0.6	.7257	.7291	.7324	.7357	.7389	.7422	.7454	.7486	.7517	.7549
0.7	.7580	.7611	.7642	.7673	.7704	.7734	.7764	.7794	.7823	.7852
0.8	.7881	.7910	.7939	.7967	.7995	.8023	.8051	.8078	.8106	.8133
0.9	.8159	.8186	.8212	.8238	.8264	.8289	.8315	.8340	.8365	.8389
1.0	.8413	.8438	.8461	.8485	.8508	.8531	.8554	.8577	.8599	.8621
1.1	.8643	.8665	.8686	.8708	.8729	.8749	.8770	.8790	.8810	.8830
1.2	.8849	.8869	.8888	.8907	.8925	.8944	.8962	.8980	.8997	.9015
1.3	.9032	.9049	.9066	.9082	.9099	.9115	.9131	.9147	.9162	.9177
1.4	.9192	.9207	.9222	.9236	.9251	.9265	.9279	.9292	.9306	.9319
1.5	.9332	.9345	.9357	.9370	.9382	.9394	.9406	.9418	.9429	.9441
1.6	.9452	.9463	.9474	.9484	.9495	.9505	.9515	.9525	.9535	.9545
1.7	.9554	.9564	.9573	.9582	.9591	.9599	.9608	.9616	.9625	.9633
1.8	.9641	.9649	.9656	.9664	.9671	.9678	.9686	.9693	.9699	.9706
1.9	.9713	.9719	.9726	.9732	.9738	.9744	.9750	.9756	.9761	.9767
2.0	.9772	.9778	.9783	.9788	.9793	.9798	.9803	.9808	.9812	.9817
2.1	.9821	.9826	.9830	.9834	.9838	.9842	.9846	.9850	.9854	.9857
2.2	.9861	.9864	.9868	.9871	.9875	.9878	.9881	.9884	.9887	.9890
2.3	.9893	.9896	.9898	.9901	.9904	.9906	.9909	.9911	.9913	.9916
2.4	.9918	.9920	.9922	.9925	.9927	.9929	.9931	.9932	.9934	.9936
2.5	.9938	.9940	.9941	.9943	.9945	.9946	.9948	.9949	.9951	.9952
2.6	.9953	.9955	.9956	.9957	.9959	.9960	.9961	.9962	.9963	.9964
2.7	.9965	.9966	.9967	.9968	.9969	.9970	.9971	.9972	.9973	.9974
2.8	.9974	.9975	.9976	.9977	.9977	.9978	.9979	.9979	.9980	.9981
2.9	.9981	.9982	.9982	.9983	.9984	.9984	.9985	.9985	.9986	.9986
3.0	.9987	.9987	.9987	.9988	.9988	.9989	.9989	.9989	.9990	.9990
3.1	.9990	.9991	.9991	.9991	.9992	.9992	.9992	.9992	.9993	.9993
3.2	.9993	.9993	.9994	.9994	.9994	.9994	.9994	.9995	.9995	.9995
3.3	.9995	.9995	.9995	.9996	.9996	.9996	.9996	.9996	.9996	.9997
3.4	.9997	.9997	.9997	.9997	.9997	.9997	.9997	.9997	.9997	.9998